ROUTLEDGE HANDBOOK OF GENDER AND FEMINIST GEOGRAPHIES

This handbook provides a comprehensive analysis of contemporary gender and feminist geographies in an international and multi-disciplinary context. It features 48 new contributions from both experienced and emerging scholars, artists and activists who critically review and appraise current spatial politics. Each chapter advances the future development of feminist geography and gender studies, as well as empirical evidence of changing relationships between gender, power, place and space. Following an introduction by the Editors, the handbook presents original work organized into four parts which engage with relevant issues including violence, resistance, agency and desire:

- Establishing feminist geographies
- Placing feminist geographies
- Engaging feminist geographies
- Doing feminist geographies

The *Routledge Handbook of Gender and Feminist Geographies* will be an essential reference work for scholars interested in feminist geography, gender studies and geographical thought.

Anindita Datta is an associate professor at the Department of Geography, Delhi School of Economics, University of Delhi, India.

Peter Hopkins is a professor of Social Geography in the School of Geography, Politics and Sociology at Newcastle University, UK.

Lynda Johnston is a professor of Geography at the University of Waikato in Tauranga, Aotearoa New Zealand.

Elizabeth Olson is a professor of Geography and Global Studies at the University of North Carolina – Chapel Hill, USA.

Joseli Maria Silva is a professor of Geography at the State University of Ponta Grossa, Brazil.

ROUTLEDGE HANDBOOK OF GENDER AND FEMINIST GEOGRAPHIES

Edited by Anindita Datta, Peter Hopkins, Lynda Johnston, Elizabeth Olson and Joseli Maria Silva

LONDON AND NEW YORK

First published 2020
by Routledge
4 Park Square, Milton Park, Abingdon, Oxon OX14 4RN
605 Third Avenue, New York, NY 10017

First issued in paperback 2023

Routledge is an imprint of the Taylor & Francis Group, an informa business

British Library Cataloguing-in-Publication Data
A catalogue record for this book is available from the British Library

Library of Congress Cataloging-in-Publication Data
Names: Datta, Anindita, 1968- editor.
Title: Routledge handbook of gender and feminist geographies/
edited by Anindita Datta, [and four others].
Description: Milton Park, Abingdon, Oxon; New York, NY: Routledge, [2020] |
Includes bibliographical references and index.
Identifiers: LCCN 2019055881
Subjects: LCSH: Feminist geography. | Feminist theory. | Spatial behavior.
Classification: LCC HQ1233 .R68 2020 | DDC 305.42–dc23
LC record available at https://lccn.loc.gov/2019055881

ISBN: 978-1-03-257002-0 (pbk)
ISBN: 978-1-138-05768-5 (hbk)
ISBN: 978-1-315-16474-8 (ebk)

DOI: 10.4324/9781315164748

Typeset in Bembo
by Newgen Publishing UK

Cover image *Hyde Street Student Party* by artist Judy Wilson, 2019

Publisher's Note
The publisher has gone to great lengths to ensure the quality of this reprint but points out that some imperfections in the original copies may be apparent.

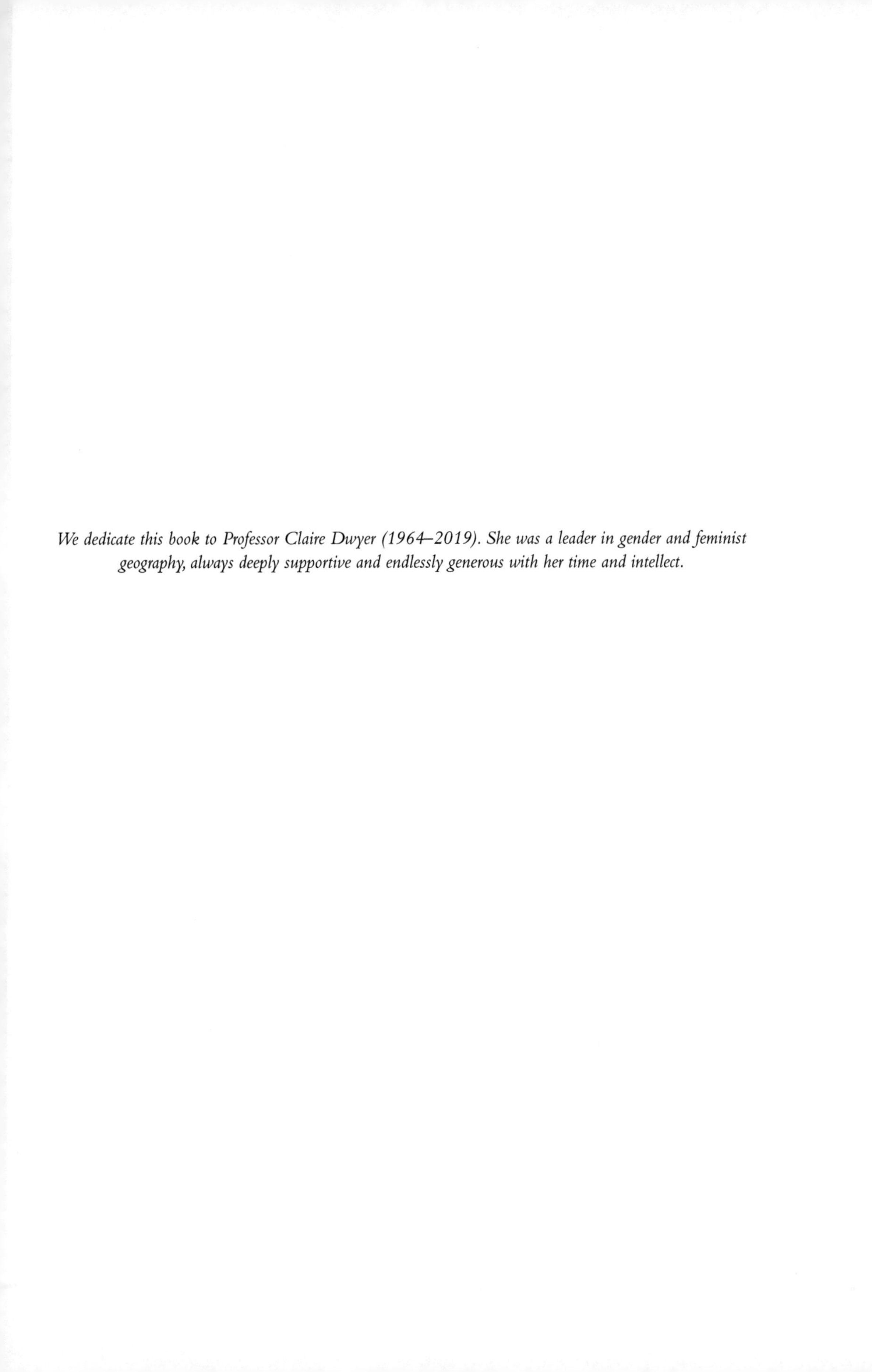

We dedicate this book to Professor Claire Dwyer (1964–2019). She was a leader in gender and feminist geography, always deeply supportive and endlessly generous with her time and intellect.

CONTENTS

ILLUSTRATIONS

Figures

Map

CONTRIBUTORS

About the editors

Anindita Datta is an associate professor in the Department of Geography, Delhi School of Economics at the University of Delhi. Her research interests are broadly in the area of feminist geographies, in particular the issues of gendered and epistemic violence, indigenous feminisms, spaces of resistance and agency and geographies of care. Anindita is a member of the Steering Committee of the International Geographical Union's Commission on Gender and Geography (2012–2020), has been on the international editorial board of Gender, *Place and Culture* and is currently on the international editorial board of *Social and Cultural Geography*.

Peter Hopkins is a professor of Social Geography in the School of Geography, Politics and Sociology, Newcastle University, England, UK. His research interests focus upon: masculinities, ethnicities and place; young people, place and identity; intersectionality, equality and diversity; and racism, Islamophobia and Muslim identities. He previously served as the managing editor of *Gender, Place and Culture.*

Lynda Johnston is a professor of Geography at the University of Waikato, Tauranga, with research interests in feminist, embodied and queer geographies. For almost two decades Lynda has pursued research and conducted activism on the challenges and spatial complexities of inequality. Lynda is the Chair of the Gender and Geography Commission for the International Geographical Union (2016–2020) and the President of the New Zealand Geographical Society (2017–2019).

Elizabeth Olson is a professor of Geography and Global Studies at the University of North Carolina – Chapel Hill. Her research engages with themes of gender in relation to structural inequality, ethics, religion and young people, with a more recent focus on the ethics of care in the context of contemporary caregiving economies. She has served on the editorial board of *Gender, Place and Culture,* and is a member of the Scientific Committee for the International Young Carers Conference and the AAG Climate Action Task Force.

Joseli Maria Silva is a professor of Geography in the postgraduate programme at the State University of Ponta Grossa, Brazil. She is coordinator of the Group of Territorial Studies at the same university and chief editor of the *Revista Latino-americana de Geografia e Gênero (Latin American Journal of Geography and Gender)*. She is one of the founders of the Ibero- Latin American Network for the Study of Geography, Gender and Sexualities. Joseli is a member of the Steering Committee of the International Geographical Union Gender and Geography Commission. Her research is focused on the relationship between space, gender and sexualities, with special attention on trans-sexualities.

About the authors

Gail Adams-Hutcheson is a teaching fellow at the University of Waikato, Aotearoa New Zealand. Her research/teaching focuses on feminist framings of emotion and affect. Gail's most recent articles, 'Farming in the Troposphere' and 'Challenging the Masculinist Framing of Disaster Research', critically analyse spaces of emotion and affect.

Tamir Arviv is a postdoctoral fellow in the Faculty of Architecture and Town Planning at the Technion, Israel Institute of Technology. His current research explores the relationship between the planning and design of high-rise complexes in Israeli cities and the everyday socio-spatial practices and interactions among Arab and Jewish residents.

Nazgol Bagheri is an associate professor of geography at University of Texas at San Antonio. She is interested in the unique relationship between the aesthetics of modern planning, the gendering of spatial boundaries and the contingent nature of public space in Middle Eastern contexts. Recent publications include 'Tehran's Subway: Gender, Mobility, and the Adaptation of the "Proper" Muslim Woman' and 'Avoiding the "F" Word: Feminist Geography in Iran'.

Ann E. Bartos is a senior lecturer in the School of Environment at the University of Auckland, Aotearoa New Zealand. She has an interest in questions around political agency, politics of embodiment, political ecology and geographies of care. Her research has focused on sexual and gender-based violence, agriculture and food politics and children's environmental agency.

Roxane Bettinger first trained as a social worker and now teaches research methodologies. Her political engagements are radical materialist feminist struggles and advocacy for LGBT rights, both understood in intersectional perspectives. She writes for collective organizations and in November 2018 authored an autobiographical fiction about sex work and feminism.

Carl Bonner-Thompson is a departmental lecturer in Human Geography at the University of Oxford having previously been a research assistant affiliated with the School of Geography, Politics and Sociology at Newcastle University, UK. Carl's ESRC-funded project explores the gendered, sexualized and digital geographies of men who use Grindr.

Kai Bosworth is an assistant professor at the School of World Studies, Virginia Commonwealth University, USA. He completed a PhD in the Department of Geography, Environment and Society at the University of Minnesota, where he studies the politics of environmentalism, especially concerning natural resource extraction and transportation as an aspect of settler colonialism. His publications have appeared in *Environment and Planning D: Society & Space*; *Antipode* and *The Annals of the American Association of Geographers*.

Alessandro Boussalem is a PhD student affiliated with the School of Geography, Politics and Sociology at Newcastle University, UK. Alessandro's research explores LGBTQ subjectivities among LGBTQ Muslims who currently live in Brussels.

Kate Boyer is a senior lecturer in the School of Geography and Planning at Cardiff University, UK. She has been researching and publishing on motherhood and breastfeeding over the last ten years, and in 2018 published *Spaces and Politics of Motherhood*.

Katherine Brickell is a professor of Human Geography at Royal Holloway, University of London. She is editor of the journal *Gender, Place and Culture* and is former chair of the Gender and Feminist Geographies Research Group (GFGRG) of the Royal Geographical Society. Katherine's research focuses on geographies of violence and domestic life.

Cathrine Brun is a human geographer and director of the Centre for Development and Emergency Practice (CENDEP), the School of Architecture at Oxford Brookes University, UK. Her research interests concern forced migration and conflict, housing and home; and the theory, ethics and practice of humanitarianism.

Caitlin Cahill is an associate professor of Urban Geography and Politics at Pratt Institute, Brooklyn, New York. Caitlin engages in critical participatory action research with communities, focused on the everyday intimate experiences of neoliberal racial capitalism, specifically as it concerns gentrification, immigration, education and zero tolerance/policing policies.

Rebecca Campbell is a PhD candidate in the School of Geography and Sustainable Communities at the University of Wollongong, Australia. Her current research interests are in embodiment, emotion, gender, ethnic diversity and sustainability.

Martina Angela Caretta is a feminist geographer researching the human dimensions of water. She is a coordinating lead author of the forthcoming 6th UN Intergovernmental Panel on Climate Change Assessment report. She also explores how the neoliberal turn of academia affects early-career female geographers.

David Alberto Quijada Cerecer is an associate professor of Ethnic Studies at Saint Mary's College of California. His research interests include inter-cultural/ethnic youth alliances, cultural citizenship, feminist ethnographic methods and youth participatory action research.

Yvette Sonia González Coronado is an assistant professor/lecturer of Field Education at the University of Utah College of Social Work. Yvette's practice centres on the belief that individuals and communities are holders of knowledge and light. Yvette exists and resists at the tension between academia and community.

Laura Crawford is a lecturer in Human Geography at the University of Northampton having recently completed a PhD at Loughborough University. Her research explores the cultural historical geographies of Leonard Cheshire Disability, a UK charity historically associated with residential care homes for disabled people. Her research interests include geographies of disability, the home, care and belonging.

Dana Cuomo is an assistant professor at Lafayette College. Her current project investigates the role of technology in facilitating abuse, and assesses the needs of domestic violence survivors whose safety and security are compromised as a result. Dana's research has been published in *Gender, Place & Culture* and in *Progress in Human Geography*.

Julie Cupples is a professor of Human Geography and Cultural Studies at the University of Edinburgh. Her research interests include cultural politics, especially in Central America, Indigenous and Afro-descendant media and geopolitical television. She is the author or editor of six books, the most recent being the *Routledge Handbook of Latin American Development*.

Sarah de Leeuw is an award-winning researcher and author (poetry and literary non-fiction), focusing on unsettling power and the role of humanities in making biomedical and health sciences socially accountable. A Canada Research Chair in Humanities and Health Inequities, she is an associate professor with the University of Northern British Columbia's Northern Medical Program, the University of British Columbia's Faculty of Medicine.

Lydia Delicado-Moratalla has a PhD in gender studies. She is currently a researcher–collaborator at the Gender Studies Research Institute at the University of Alicante, Spain. Her work focuses on sex trafficking and forced prostitution of Nigerian women in Spain, and is published in *Gender, Place and Culture.*

Danielle Drozdzewski is a senior lecturer in human geography at Stockholm University, Sweden. Her main research areas are cultural geography, cultural memory and geographies of ethno-cultural and national identities, especially people's interactions with memorials in everyday locations and how a politics of memory influences memory selection. She edited *Memory, Place and Identity: Commemoration and Remembrance of War and Conflict* and *Doing Memory Research: New Methods and Approaches*.

Isabel Dyck is a professor emeritus, School of Geography, Queen Mary University of London. Her work focuses on a range of settlement issues for international migrants, both in Canada and the UK, including the reconstitution of home and family, health and care practices and household economic strategies.

Kim England is a professor in the Department of Geography, University of Washington. Her research addresses the geographies of social and economic restructuring, especially around the interfaces of social reproduction and production, neoliberalization and globalization, and the home and the body.

Anita H. Fábos is a professor of International Development and Social Change at Clark University's International Development, Community, and Environment department. She is an anthropologist who has worked and conducted research together with Muslim Arab Sudanese refugees in the Middle East, Europe, and the US. Formerly Director of Forced Migration and Refugee Studies at the American University in Cairo, and later Programme Coordinator for MA Refugee Studies at the University of East London, Fábos integrates teaching, research and participatory programs that have incorporated refugee and forced migrant perspectives. Fábos and her writing partner Cathrine Brun are working on a book on home and home-making for people in circumstances of long-term displacement, entitled *Constellations of Home.*

Bisola Falola is a feminist geographer and graduate of the Department of Geography and the Environment at the University of Texas at Austin. She is a programme officer at the Open Society Foundation, working on youth, urban precarity and marginalization, and specializing in narrative and cultural change strategies.

Maria Fannin is a reader in Human Geography at the University of Bristol, UK. Her research focuses on cultural aspects of human tissue use in medicine and the life sciences. She is the co-editor, with Marcia R. England and Helen Hazen, of *Reproductive Geographies: Bodies, Places and Politics* (Routledge, 2018).

Caroline Faria is a feminist geographer and an assistant professor in the Department of Geography and the Environment and the Center for Women's and Gender Studies at the University of Texas at Austin. She examines the intersectional work of nationalism and neoliberal globalization, with a current focus on the Gulf–East African hair and beauty industry.

May Farrales is a postdoctoral fellow with the Health Arts Research Centre and ECHO (Environment, Community, Health Observatory) project at the University of Northern British Columbia on the traditional territory of the Lheidli T'enneh people. Her research interests are nested in understanding how different forms of colonialism work over time and space.

Tovi Fenster is a professor and head of the Department of Geography and Human Environment at Tel Aviv University. She is the founder and head of PECLAB; former head of the Institute of Diplomacy and Regional Cooperation and the NCJW Women and Gender Studies programme; and former chair of IGU Gender and Geography Commission. In 1999, she initiated and was the first chair of *Bimkom-Planners for Planning Rights* in Israel (NGO) and now serves as its president.

Joos Droogleever Fortuijn retired as an associate professor in human geography and served as chair of the Department of Geography, Planning and International Development Studies of the University of Amsterdam. She is the first vice-president of the International Geographical Union. She has published on urban and rural geography, gender, ageing and feminist geography teaching.

Kevin Glynn is an associate professor in the Department of Geography and Environmental Sciences at Northumbria University in the UK. Much of his research examines the politics of popular media cultures. His books include *Tabloid Culture: Trash Taste, Popular Power, and the Transformation of American Television* and *Communications/Media/Geographies.*

Leticia Alvarez Gutiérrez is an associate professor in the Department of Education, Culture and Society at the University of Utah. She teaches and mentors linguistically and culturally diverse students, working extensively with societally marginalized young people, teacher educators and teachers who strive to transform education. She directs the Westside Pathways and is faculty advisor to Mestizo Arts and Activism (MAA) Collective.

Claire Hancock is a professor of Geography at the University Paris-Est Créteil, a member of the Lab'Urba and joint head of the JEDI (*Justice, Espace, Discriminations, Inégalités*) working group. Her recent work has focused on issues of the right to the city for gender minorities in Paris and other European cities.

Nancy Hansen is associate professor and director of the Interdisciplinary Master's Program in Disability Studies at the University of Manitoba. She has a PhD (Human Geography) from the University of Glasgow. She is a co-editor of two academic volumes, has written numerous book chapters and contributed to various academic journals.

Harriet Hawkins researches the geographies of artworks and art worlds. She is currently working on a project on the underground, including caves and grottos. She is the author of several monographs ('For Creative Geographies' and 'Creativity: Live, Work, Create'), and works collaboratively with artists and curators around the world. Harriet is currently a professor of GeoHumanities at Royal Holloway, University of London.

Allison Hayes-Conroy is an associate professor of Geography and Urban Studies at Temple University, Philadelphia. Her research has centred on visceral geography, an approach that has allowed her to explore interests in the role of the feeling/sensing body in political life, as well as different forms of knowledge production about the body.

Jessica Hayes-Conroy is an associate professor of Women's Studies at Hobart and William Smith Colleges, New York. Her scholarship has worked to develop new ways of theorizing and practising the food–body relationship through the lens of social inequality.

José Hernández Zamudio is an associate instructor in the Office of Undergraduate Studies at the University of Utah. His research areas of interest include early college education, race, class and gender.

Shirlena Huang is an associate professor at the Department of Geography, National University of Singapore. Her research and publications focus on issues at the intersection of transnational migration, gender and family, with particular interest in the themes of care labour migration and transnational families within the Asia-Pacific region.

Symon James-Wilson is a graduate of the Department of Geography and Planning at the University of Toronto. Her current research investigates historical and contemporary geographies of school segregation in the US. Symon has contributed to the *Wiley Blackwell Encyclopedia of Urban and Regional Studies* and several other forthcoming books.

Eden Kinkaid is a doctoral candidate in the departments of Geography and Women's, Gender, and Sexuality Studies at Pennsylvania State University. Her research concerns the intersections of queer theory, phenomenology and geography, with a particular interest in philosophies of space and the subject.

Eleonore Kofman is a professor of Gender, Migration and Citizenship at Middlesex University London. She has research interests in gender and family, skilled and intra-European migration. With Parvati Raghuram she has co-authored *Gendered Migration and Global Social Reproduction* and 'Women, Migration, and Care: Explorations of Diversity and Dynamism in the Global South' (*Social Politics*).

Keichi Kumagai is a professor in the Department of Geography and Environmental Studies at Ochanomizu University, Japan. His interests are in social and cultural geography, Oceanic area studies and shifting masculinities. Publications include 'Place, Body and Nature: Rethinking Japanese *fudo* (milieu) and Minamata Disease' and *Papua Nyu Ginia no 'Basho' no Monogatari: Doutai Chishi to Firudowaku (Stories of 'Places' in Papua New Guinea: Dynamic Regional Geography and Fieldwork).*

Carina Listerborn is a professor in Urban Planning, Department of Urban Studies, Malmö University. Research interests are gender and urban neoliberalism, housing inequality and urban transformation. Her latest publications include 'The Flagship Concept of "The 4th Urban Environment": Branding and Visioning in Malmö, Sweden'.

Robyn Longhurst is a deputy vice-chancellor academic and professor of Geography at the University of Waikato. Her research coalesces around the theme of bodies, space and place. Robyn is author of *Skype: Bodies, Screens, Space* (2017), *Maternities: Gender, Bodies and Space* (2008) and *Bodies: Exploring Fluid Boundaries* (2001); and co-author of *Space, Place and Sex* (2010) and *Pleasure Zones: Bodies Cities, Spaces* (2001).

Jessa M. Loomis received her doctorate from the University of Kentucky in 2018 and is a lecturer of Economic Geography at Newcastle University. Loomis uses qualitative methods and feminist theory to examine social and economic change. Her research interests include financialization, economic subjectivity, inequality and the social and scalar relations of debt.

Sophia Maalsen is a lecturer in urbanism at the School of Architecture, Design and Planning, University of Sydney. Prior to this post, Sophia was a postdoctoral researcher on the European Union-funded Programmable City Project, where she investigated the digital transformation of cities and urban governance. Her particular expertise is in understanding the intersection of gender, the material, digital and the human and how this affects lived experience.

Avril Maddrell is a professor of Social and Cultural Geography at the University of Reading. She is a feminist geographer, with research interests in spaces, landscapes and practices of death, mourning and remembrance; sacred mobilities; and gendered historiography. She is an Editor of *Social and Cultural Geography*, former Editor of *Gender, Place and Culture*, and author/co-author/co-editor of several books, including *Complex Locations. Women's Geographical Work in the UK 1850–1970* (Maddrell 2009) and *Contemporary Encounters in Gender and Religion* (Gemzoe, Keinanen and Maddrell (Eds).

Sofia Manseri is a long-time resident of the Paris suburbs. Besides her regular office job, she works on the policy and promotion of equality, as an elected city councillor, in Gennevilliers. She collaborates with the research collective Les Urbaines (https://urbaines.hypotheses.org) and writes about equality in critical perspectives on her blog, http://loubiaconnection.blogspot.com.

Sallie A. Marston is a human geographer whose work is located at the intersection of socio-spatial theory and politics. She investigates through everyday life and the seemingly mundane practices that constitute it. In addition to being a professor in the School of Geography and Development, she is the director of the University of Arizona's Community and School Garden Program.

Andréanne Martel is a programme officer at the Canadian Council for International Development (CCIC) in Ottawa on a joint initiative with the Canadian Association for the Study of International Development. Before, she coordinated a research centre on international development. Andréanne holds a Master's degree in Political Science from the University of Québec in Montreal.

Jarred Martinez is an education partnership manager with University Neighborhood Partners at the University of Utah. Jarred's praxis as a bridge-builder and youth worker is rooted in creating and supporting spaces that are humanizing, liberatory and transformational for young people.

Doreen Massey was born on 3 January 1944 and passed away on 11 March 2016. She was a Professor of Geography at the Open University for most of her career and published widely in human geography, including in debates about gender, capitalism and place.

Robyn Mayes is an academic in the School of Management at Queensland University of Technology. Her research interests encompass gender, labour migration, rurality and community. This work is grounded in long-standing empirical work critically examining the Australian mining sector.

Jessica McLean is a geographer in the Department of Geography and Planning at Macquarie University. Her research focuses on digital geographies and water cultures. Both call for critical, engaged and situated research praxis, building on partnership approaches with community groups and individuals who are working in Indigenous, feminist and digital rights contexts.

Nicole McNamara is a geographer and planner, working to integrate the two disciplines in her professional planning practice. She recently completed her PhD entitled 'Understanding Cycling: Practices and Experiences in Sydney' at Macquarie University. Nicole's interdisciplinary research focuses on emotional and affectual geographies, mobilities, social practice theory and feminist geographies.

Graeme William Mearns is a queer geographer who has spent several years working outside his home discipline as part of an interdisciplinary research, with colleagues in computing. His core interests are sexual citizenship, gender and emerging technologies.

Lisa Melville is a PhD student at the University of Waikato and is researching lesbians' experiences of conception, pregnancy and mothering. Lisa wrote '"Who's the Dad?" And Other Things Not to Say to Lesbian Mums', published in *The Spinoff* and *The New Zealand Herald e-Edition* (2018), and 'Lesbians Making Babies: Why Research on Sperm, Space and Decisions Matters' in *Te Kua Kete Aronui* (2016).

Paula Meth is a reader in Urban Studies and Planning at the University of Sheffield and an associate fellow of the School of Architecture and Planning at Wits University in South Africa. She focuses on social and everyday lives within cities of the Global South, exploring how changes to their material and infrastructural elements (housing, for example) shape, and in turn are shaped by, social and political processes. She co-authored *Geographies of Developing Areas* (2014, Routledge).

Sarah Mills is a reader in human geography at Loughborough University. Her research interests include the geographies of youth citizenship, informal education and volunteering in both historical and contemporary contexts. She is the co-editor of *Informal Education, Childhood and Youth: Geographies, Histories, Practices* (2014, Palgrave Macmillan) and *Politics, Citizenship and Rights* (2016, Springer).

Chen Misgav is currently a postdoctoral fellow in the Open University, Israel, and a research fellow in Minerva Humanities Center in Tel-Aviv University. He was awarded his PhD by the Department of Geography and Human Environment/Planning with Communities for the Environment at Tel-Aviv University, Israel. Chen has published on the intersections of space, politics, social and protest movements, gender, sexuality and planning.

Sharlene Mollett is an associate professor in the Department of Geography and Centre for Critical Development Studies at the University of Toronto. Her work is positioned at the intersection of postcolonial political ecologies and critical feminist racial studies and is published in such venues as *Cultural Geographies and Antipode.*

Janice Monk is a research professor at the University of Arizona, in the School of Geography and Development and Research Social Scientist Emerita of the Southwest Institute for Research on Women, and also a Fellow of the American Association of Geographers. She is Adjunct Professor at Macquarie University, working with colleagues in Indigenous Studies. Janice is active in the Commission on Gender and Geography of the IGU and the Gender Group in Geography at the Autonomous University of Barcelona.

Richa Nagar is a professor of the College in the College of Liberal Arts at University of Minnesota and holds an honorary professorship at Rhodes University in Grahamstown, South Africa. Working in the areas of critical development studies, feminist epistemologies and questions of solidarity work across borders, she writes in multiple genres and co-creates theatre in English and Hindi.

Lise Nelson is an associate professor in the Departments of Geography and Women's, Gender, and Sexuality Studies at Pennsylvania State University. She researches labour, identity, place, and citizenship in the context of neoliberal globalization. Nelson is co-editor of the *Companion to Feminist Geography* and has articles in *Gender, Place, and Culture* and the *Annals of the American Association of Geographers.*

Christopher Neubert is a PhD candidate in geography at the University of North Carolina (UNC) at Chapel Hill. He conducted research in Sri Lanka as a Fulbright fellow, and completed his MA thesis at UNC on the materiality of livestock waste in 2016. His PhD is on the intersections of Whiteness, masculinity and nationalism as they play out in the US through the intimate geopolitics of agricultural policy and practice.

Paula Ngaire Smith has a Master's degree in geography from the University of Waikato, Aotearoa New Zealand, exploring the changes when homes are made, unmade and remade following relationship challenges. Feminist, queer and poststructuralist geographical theories analyse the connection between the relationships, emotions and materialities of home spaces.

Ann M. Oberhauser is a professor of Sociology and Director of Women's and Gender Studies at Iowa State University. She conducts research in the areas of feminist geography, gender and development, qualitative methods and feminist pedagogy. She is co-editor of *Global Perspectives on Gender and Space* (2014) and co-author of *Feminist Spaces* (2017).

Devin Oliver is both a geographer and urban planner, receiving his BA in geography from the Ohio State University, and his MS in Community and Regional Planning and MA in Latin American Studies from the University of Texas at Austin. Devin is equally committed to human geography and Black studies, which influence his work in environmental and energy policy. He currently works for low-income solar and energy-efficiency programmes across the California.

Iliana Ortega-Alcázar is a research fellow in human geography at the University of Southampton. Iliana's research interests are in housing, home, migration, belonging and visual research methods. Her book, *Self-help Housing: Space and Family Life in Mexico City*, was published in Spanish by Flacso-UNAM.

Rachel Pain is a professor of human geography at Newcastle University in the UK. Informed by feminist and participatory theory, practice and activism, her research focuses on international and intimate violence, fear and trauma. She is author of *Fear: Critical Geopolitics and Everyday Life* and the *Sage Handbook of Social Geographies*.

Naimah Petigny received her BA in Women's Studies and Sociology from Vassar College and is currently a doctoral candidate in Gender, Women and Sexuality Studies at the University of Minnesota. Working at the intersections of Black feminist theory and performance studies, her dissertation centres on the peripheralized queer Black femme and the unruly dimensions of Black subjectivity.

Beaudelaine Pierre is a published writer and doctoral candidate in Gender Women and Sexuality Studies at the University of Minnesota; she is also co-editor of *How to Write an Earthquake*, a trilingual volume in response to the catastrophe that Haiti suffered on 12 January 2010.

Barbara Pini is a professor in the School of Humanities, Languages and Social Science at Griffith University, Brisbane, Australia.

Geraldine Pratt is a professor of Geography and Canada Research Chair in Transnationalism and Precarious Labour at the University of British Columbia. She has collaborated with Filipino migrant organizations in Vancouver and Manila to research migrant labour and to create a site-specific testimonial play on this issue.

Ximena Quintero Saavedra is a planning and social development professional and a legal representative for and director of Corporación Casa Mía in Medellin, Colombia. She is a perpetual learner and lover of life.

Parvati Raghuram is a professor in Geography and Migration at the Open University. Her most recent ESRC-funded projects are 'Gender, Skilled Migration and the IT Sector: A Comparative Study of India and the UK' and 'Facilitating Equitable Access and Quality Education for Development: South African International Distance Education'. She co-edits the Palgrave Pivot series Mobility and Politics.

Menah Raven-Ellison is a practising NHS cognitive behavioural psychotherapist and registered occupational therapist, with a PhD in human geography from Queen Mary, University of London. Her qualitative research was concerned with the experiences and well-being of forced migrant women who had been detained in UK immigration removal centres.

Alonso R. Reyna Rivarola is an assistant director for the Office of Diversity & Multicultural Affairs at Salt Lake Community College. Growing up undocumented in Utah, his academic, research and professional interests are found in the raced-gendered intersections of education, labor, and immigration within the US context.

Nahid Rezwana is an associate professor at the Department of Geography and Environment of Dhaka University in Bangladesh. Her field of interest is gender, disasters and climate change. She is author of *Disasters, Gender and Healthcare Access: Women in Coastal Bangladesh* (2018) and *Social Formation in Dhaka 1985–2005* (2016).

Ged Ridley is affiliated with the School of Geography, Politics and Sociology at Newcastle University. Ged's ESRC-funded PhD focuses on trans experiences of public bathrooms in the north of England, a collaborative project with the *Yorkshire Trans* support and advocacy network.

Maria Rodó-Zárate is a *Juan de la Cierva* postdoctoral researcher at the Gender and ICT research group, Universitat Oberta de Catalunya. She holds a PhD in geography (Autonomous University of Barcelona), and her topics of interest are feminist geographies, geographies of sexualities and youth and intersectionality.

Laura Rodriguez Castro is a PhD candidate for the Doctor of Philosophy at Griffith University, Australia. Her research interests are in the areas rural studies and decolonialism. Her participatory research involves visual ethnography with *campesina* women in Colombia in the context of globalization.

Zuriatunfadzliah Sahdan is a lecturer in human geography at Universiti Pendidikan Sultan Idris, Malaysia. Dr. Sahdan received her PhD from Durham University. Her research focuses on postcolonialism, space and cultures of domestic violence in Malaysia, using participatory methods and storytelling.

Vanessa Sloan Morgan researches how dispossession is normalized interpersonally and through structures of power, and how creative methods can work to dismantle these dynamics. Vanessa is currently a postdoctoral fellow at the University of Northern British Columbia.

Sara H. Smith is an associate professor of geography at the University of North Carolina at Chapel Hill. She is interested in the relationship between territory, bodies and the everyday and how politics and geopolitics are constituted or disrupted through love, friendship and birth. Her project on the intimate politics of Himalayan youth and the wider context is developing work on race, biopolitics and the future.

Elizabeth R. Straughan is currently working in teaching support at the University of Melbourne. Her research unravels the volatile nature of the body through attendance to the skin, as well as the material and metaphorical dynamics of touch. She has published in international journals such as *Cultural Geography*; *Emotion, Space and Society*; and *Geography Compass*.

Corrinne Sullivan is an Aboriginal scholar from the Wiradjuri nation, a senior lecturer at Western Sydney University and a PhD student in geography at Macquarie University. Her research interests focus broadly on experiences and effects of body and identity in relation to Indigenous peoples, and on how these people are affected by their experiences of space and place.

Sophie Tamas is an assistant professor at Carleton University in Ottawa, Canada, in the Department of Geography and Environmental Studies and the School of Indigenous and Canadian Studies. She teaches emotional geographies and qualitative research methods. Her publications in various genres mostly focus on autoethnographic ethics, trauma and over-shares.

Qian Hui Tan is a PhD student in the Department of Geography, National University of Singapore. She is a social-cultural geographer with a keen interest in feminist and queer theories.

Anna Tarrant is an associate professor in Sociology at the University of Lincoln. She has taken published research about a range of themes from an interdisciplinary, feminist perspective, including men and masculinities; caregiving; and family life. She is also interested in qualitative longitudinal methodologies, including the potential of qualitative secondary analysis.

Pavithra Vasudevan is assistant professor of Women's and Gender Studies and African & African Diaspora Studies at the University of Texas at Austin. She is a scholar-activist whose research focuses on the stuff of environmental justice: race, water and power. Her current project explores radicalized toxicity as a mode of twentieth-century racial capitalism, incorporating Black feminism, materialist science studies, feminist geopolitics and cultural studies.

Laura Vaz-Jones is a PhD student in human geography at the University of Toronto. A feminist and urban geographer, her research argues for an intersectional perspective in 'right to the city struggles in Salvador da Bahia, Brazil'. She has published in *Signs: Journal of Women in Culture and Society*.

Gordon Waitt is a professor in the School of Geography and Sustainable Communities at the University of Wollongong, Australia. His current research is underpinned by a corporeal feminist geographical approach to address questions of fuel poverty, everyday mobilities and household sustainability.

Margaret Walton-Roberts is a professor in the Geography and Environmental Studies Department at Wilfrid Laurier University, and is affiliated to the Balsillie School of International Affairs, Waterloo, Ontario. Current research focuses on the international migration of healthcare professionals. She has published widely on immigration and citizenship policy, settlement and integration and transnational community formation.

Yoshiko Yamasaki has worked on women's health and early childhood education, and was previously the chef and owner of a restaurant. She is passionate about Farm to Early Childhood Education, knows the importance of setting early foundations of wellness and has worked with families experiencing the complex effects that poverty has on well-being.

ACKNOWLEDGEMENTS

This Handbook has been a long time in the making, and we would like to acknowledge all of those who were involved in enabling it to reach publication. We thank all of the contributors for taking the time to write for this Handbook, often bringing in new and emerging scholars and writing about contested and challenging issues. We thank our co-editors for their solidarity, political insight and commitments to gender equality and feminist geographies during challenging times; our different locations across the globe and the struggles found within each of these contexts shaped our conversations and sparked new ideas and new ways of thinking.

We thank Matthew Shahin Richardson, Mark Ortiz and Sertanya Reddy, who helpfully assisted with the compiling of the Handbook chapters. Massive thanks also to Alison Williamson, for her outstanding proofreading skills.

1
INTRODUCTION
Establishing, placing, engaging and doing feminist geographies

Lynda Johnston, Anindita Datta, Peter Hopkins, Joseli Maria Silva and Elizabeth Olson

Introduction

This Handbook reflects the immense depth of interest, developments, directions and tensions in gender and feminist geographies. It brings together 48 chapters by new, emerging and established scholars, activists and artists in order to highlight original international work in gender and feminist geographies. For nearly five decades – from the 1970s onwards – geographers have employed feminist and other critical social theories to understand gender, power, place and space. There is now a vast and considerable literature in gender and feminist geographies, and also a growing number of scholars who bring feminist theory and praxis to diverse topics and locations. There are continuities between earlier compendiums, such as *Geographies of New Femininities* (Laurie et al. 2014), *Feminist Geography in Practice* (Moss 2002) and *A Companion to Feminist Geography* (Nelson and Seager 2005) and this Handbook. The differences between earlier volumes and this one suggest that the field of scholarship continues to mature in theory and practice, in part by diversifying the voices and perspectives that refocus its scholarship towards new questions, new approaches and new critical theories.

This Handbook aims to provide a window into established gender and feminist geographies while pointing readers towards new directions. We, as editors, are deeply honoured to be caretakers of the chapters. From the outset, our main goal has been to represent the diversity of research, different theoretical frameworks, the variety of methodological tools and the multiplicity of practices within gender and feminist geographies. The range and merit of different approaches to gender and feminist geographies are based on different interpretations of what is regarded as salient and significant for scholars and activists. Where we have succeeded in achieving our goal, we are indebted to the contributors to this compendium. Where we have failed, we look with hope towards the scholars who will generate their own inspired points of departure.

Early in our collaboration as editors, we recognized that producing this Handbook provides the opportunity to encourage contributors to bring their commitments into their authorship. Gender and feminist geographers were invited to contribute a chapter and, wherever possible, to co-author the chapter with others, such as established or new and emerging scholars, students, activists, community groups and artists. We are delighted to have 100 authors from 18 countries. This Handbook, then, provides a comprehensive statement, thematic overview and reference point for contemporary feminist geographies and gender studies in an international and multi-disciplinary context. Specifically, it provides critical reviews and appraisals of the current state of the art and future development of conceptual and theoretical approaches, as well as empirical knowledges and understandings of feminist geographies and gender studies.

As editors and feminist scholars from different parts of the world – India, Britain, Aotearoa New Zealand, the USA and Brazil – our experiences of feminist theory reflect the ways that our own biographies in place and space intersect with dominant forms of knowledge. Our conversations have raised questions about what it means to publish a feminist handbook in English with a Western press or what to recommend by way of authors' writings, which use a specialist, theoretical language that can be inaccessible even to seasoned scholars. We have worked to try to decentre Anglo-American and Eurocentric Western knowledge. The politics of knowledge production is crucial to feminism, and it is a central concern within this Handbook. We hope this collection continues to transform the discipline of geography through building capacity with new practitioners and early career researchers. We also hope this collection enables geographers to critique and question static concepts and paradigms (Johnson 2009, 2012; Peake 2015).

Our diverse positionalities in the global networks of scientific production allow us to reflectively engage with the advantages and disadvantages of a coloniality of knowledge (language skills, access to economic and technological resources and networks of personal and professional relationships). The solidary conciliation of our simultaneous strengths and weaknesses, not always easy, was the energy that created cracks in power structures so that we could advance towards *pluri-* and *multi-*located geographies.

Gender and feminist geography scholarship is, of course, dependent on time and place. Knowledge, therefore, is diverse in its aims, articulations and practices. We embrace this diversity and the many links that the authors make to other critical geographies, such as queer, social, cultural, anti-racist and post-colonial geographies. What unites this diverse scholarship is the disruption of inequalities and an articulation of difference. This collection, then, enables us to showcase the different places and concepts identified and detailed by contributors as vital to the establishment and development of gender and feminist geographies. Living a feminist life (Ahmed 2017) happens somewhere, as this Handbook shows.

The Handbook focuses on contemporary social and political places and gives an ongoing feminist commentary on gender, places and spaces at a variety of scales. Our aims for the Handbook were ambitious. They were to: establish thematic overviews of gender and feminist geographies; provide critical reviews and appraisals of the current state of the art and future development of conceptual, theoretical and methodological approaches, as well as empirical knowledge and understanding of gender and feminist geographies; engage simultaneously with different geographical scales and societal issues, such as violence, resistance, agency and desire; reflect the politics, methods, theories and practices involved in feminist geographies; and showcase the ongoing transformative research that arises from feminist geographical knowledges. As we write this Introduction and review the chapters one more time, we are truly delighted with the exciting, engaging and challenging content that authors have brought to the Handbook.

In what follows, we briefly outline some existing research and literature upon which this Handbook rests and extends, noting also the current social and political context in which these chapters were written. We have called this 'Changing places, politics and gender' to recognize the powerful gendered geopolitical geographies across the globe. Second, we offer an overview of each part: Establishing feminist geographies; Placing feminist geographies; Engaging feminist geographies; and Doing feminist geographies. By introducing each part, we hope to clarify connections and divergences between chapters. Finally, we offer personal and political reflections about the editorial work that we did in and around our lives in five different countries. The 'personal is political' is a well-established feminist theoretical tool, and we highlight events that impacted on the making of this Handbook.

Changing places, politics and gender

It is generally agreed that feminist geography took root in the 1970s, along with the rise of social movements and when the concept of gender began to take hold (Mackenzie 1984; Nelson and Seager 2005). The 'strange case of the missing female geographer' (Zelinsky 1973) was not an isolated incident (see, for example: McDowell's 1979 *Women in British Geography*; Momsen's 1980 *Women in Canadian Geography*; the Women and Geography Study Group of the Institute of British Geographers' *Geography and Gender*, IBG 1984; and García Ramón et al.'s 1988 *Women and Geography in Spanish Universities*). Yet, while this groundbreaking work is important, feminist geographer Janice Monk's (2004) research shows us that since the late-nineteenth century there have been hundreds of women professional geographers (see also Maddrell 2009). Monk (2004, 1) points out that 'histories of American geography have tended to concentrate on geographic thought and on the men who have been seen as major figures in research'. Monk (2004; see also Peake 2015) highlighted the women geographers working in and beyond the margins of the academy, but at the centre of social movements, many of whom, such as African–American Thelma Glass (1916–2012), were civil rights activists.

These early reflections – perhaps better framed as an insistent intervention – serve as a reminder that feminist geographers have long been rethinking key geographical concepts, such as: class (Gibson-Graham 2006); work (McDowell 1997); development (Momsen and Kincaid 1993); migration (Pratt 2012); mobility (Hanson 2010); methodologies (Kindon and Cupples 2014); and, 'space, place and knowledge' (Moss and Al Hindi 2007). Feminist geographers have developed new subject areas and foci of inquiry for the broader discipline, for example: bodies (Longhurst 2001); home (Blunt and Dowling 2006); emotion and affect (Davidson and Bondi 2004; Thien 2005; Tolia-Kelly 2006); political ecology (Mollett and Faria 2013); sexuality and space (Bell and Valentine, 1995); Indigenous (Radcliffe 2014; Simmonds 2011); and, transgender and gender-variant geographies (Browne et al. 2010; Doan 2010; Johnston 2018; Sullivan 2019).

During the half-century that gender and feminist geography has been recognized as a field of study, scholars have cultivated three recognizable strands of inquiry. In-depth detail of the history of feminist geography can be found elsewhere (see Nelson and Seager 2005), so we provide only a brief outline here. It is generally agreed that the first strand – the geography of women – grew from the women's liberation movement of the 1960s and 1970s. This prompted geographers to address gender inequalities, particularly the absence of women from geographical research and teaching (Bowlby et al. 1989; Johnston et al. 2000; Tivers 1978). Linda McDowell (1992) identifies four reasons for the absence of women from geography, and related issues, in the 1970s. She notes:

> What tend, somewhat dismissively, to be termed 'women's issues' were excluded from consideration for many years on one or several of four grounds – that they are trivial; that they are set at the wrong spatial scale, for example the domestic; that the methods used to examine these issues are not respectable (not science, inappropriate to geography); that the work is biased, subjective or, worse, political.
>
> *McDowell 1992, 404*

The second strand – associated with scholarship emerging first in the 1980s and 1990s – focused on feminist geographies of gender, work, place and space. This 'socialist feminist geography' (Johnston et al. 2000) was produced at the intersection of gender and class relations. The relationship between patriarchy and capitalism received a great deal of interest, particularly in the UK (see debates in the journal *Antipode*). A landmark publication – Doreen Massey's 1994 book *Space, Place and Gender* – unravels the accepted distinctions and categories of the time, such as gender versus class, economic versus cultural, feminine versus masculine, local versus global, space versus time, partial versus universal and political versus academic. Such was the influence of Massey's early work that we reprint from this book a chapter on politics and space/time that was also published in *New Left Review* in 1992.

Postmodernism shaped the third stand of feminist geography – a feminist geography of difference. From the mid-1990s, feminist geographers have been producing a substantial amount of research focused on the intersection of bodies, identities, place and space. Early and important publications – such as: *BodySpace* (Duncan 1996); *Places through the Body* (Nast and Pile 1998); *Mind and Body Spaces: Geographies of Illness, Impairment and Disability* (Butler and Parr 1999); and *Bodies: Exploring Fluid Boundaries* (Longhurst 2000) – reshaped the geographical discipline, and some were the forerunners of queer geographies. Early landmark publications on sexuality, place and space are *Mapping Desire* (Bell and Valentine 1995) and *Closet Space: Geographies of Metaphor from the Body to the Globe* (Brown 2000).

Before we outline each part of the Introduction, we first consider the ways that place matters to the construction of knowledge (Monk 1994). There is, however, a risk in presenting gender and feminist geography as a single, universal field. The history of feminist geography, itself, is beset with an historical unevenness that reinforces Monk's (1994) emphasis on the significance of place to the construction of knowledge. The uptake of feminist geography and the concept of gender are deployed unevenly by geographers, and this is not surprising. Power relations in specific places influence knowledge construction. What it means to do research on gender issues and to be a feminist and/or queer geographer varies across time and place. Attention to local, regional, national and international contexts clearly illustrates the heterogeneity of feminist and gender geography.

Maria Dolors García Ramón and Janice Monk (2007, 247) note: 'It is now widely acknowledged that knowledge is "situated", reflecting its cultural, political and intellectual contexts as well as the personal values of those engaged in its creation. This recognition presents an especially interesting perspective for geographers.' Often, as noted by García Ramón and Monk (2007), those feminist or gender geographies not produced in the 'Anglo-American centre' are marginalized and/or deemed less important. This happens in several ways. For example, research publications that do not rely on Anglo-American and European texts or theories are constructed as lacking theoretical sophistication, and empirical evidence that is not situated in Anglo-American or European places is often deemed 'irrelevant' (Longhurst and Johnston 2015).

Even within Anglo-American or European scholarly communities, Black geographies of gender, sex and sexuality have only recently been imperfectly integrated into the canons of feminist theory. Where they have, feminist geography has led the way in opening conversations for discussions of race, racism and the historical (Pulido 2002), theoretical and topical innovations in intersectionality and space by scholars, including Katherine McKittrick (2006), Ruth Wilson Gilmore (2007) and Claire Dwyer and Caroline Bressey (2008), among others. Today, a new generation of Black geographies of gender and feminist thought is evident in the work of LaToya Eaves (2019), Caroline Faria and Sharlene Mollett (2014), Priscilla McCutcheon (2019), and Camilla Hawthorne and Brittany Meché (2016) who, along with other contemporaries offer new topics, places, and theories that are moving feminist geographies towards emerging intersections (Gökarıksel et al. forthcoming). Indigenous geographies of gender and sexuality are also providing new intersections with post-colonial and decolonizing research in new models of feminist praxis (Hunt and Holmes 2015; de Leeuw 2016; Radcliffe 2017; Simmonds 2011; Sullivan 2019 and in this Handbook). Many of these influences are evident in the chapters of this Handbook.

Early in 2017, when we started contacting hundreds of potential Handbook contributors, we asked them to write a short chapter that 'will engage scholars who might be new to our field, while also signalling new directions and excitements. In the spirit of the publication, we encourage you to invite a junior scholar, non-academic collaborator, or scholar from another global region to join you as co-author.' This invitation was an attempt to 'widen the net' of gender and feminist geography. As a result, in terms of *where* contributors are based, the Handbook represents scholars and collaborators who live in the following countries: Australia; Bangladesh; Brazil; Canada; Colombia; France; India, Israel; Japan; Malaysia; Netherlands; New Zealand; Norway; Singapore; Spain; Sweden; the UK; and the USA. Countries such as the UK and the USA are still over-represented and, while we have not necessarily succeeded in destabilizing Anglo-American and Eurocentric feminist geographies, we hope that we have heightened the awareness of the value of considering the importance of place, gender and geography knowledge construction.

While contributors were writing their chapters in 2017 and 2018, a number of significant global and national events – historic moments – with heightened traditionalism and fear circulated in and through bodies and places. The legal rights of women, people of colour, Muslims, LGBTIQ and other marginalized people across the globe came under attack. For example, in January of 2017 US President Trump rescinded funding for reproductive healthcare when he signed a decree – known as the 'global gag rule' – barring US federal funding for foreign NGOs that support abortion (McGinley and Goldstein 2017). The implications are significant for women in countries that depend on development assistance for planned parenthood and reproductive health services.

In India, the return of a right-wing government with a thumping majority in the 17th Lok Sabha elections in 2019 signalled a new normal that was unmistakably 'saffron', or militant Hindu. An unprecedented social polarization plays out against the backdrop of crony capitalism, massive agrarian distress, farmer suicides, rising unemployment and a weak economy (Banerjee et al. 2019; Chacko 2018).

In Brazil, like most of Latin America, the recent electoral processes evidenced the fast growth of conservative and extreme rightist forces, deepening an openly anti-feminist, racist, anti-LGBTIQ and neoliberal political culture. The disbelief in progressive ideas such as freedom, equality and responsibility of the State for promoting social well-being and preserving lives, no matter whose lives, has devastating consequences for a colonized continent that is already marked by the precarious tools of environmental protection, huge income inequality and high indices of violence in poor areas, femicide and LGBTIQ murders.

In Aotearoa New Zealand, early in 2019, a White supremist man filled with hatred and violence killed 51 Muslims and injured another 49 at the Masjid Al Noor and Linwood mosques in Christchurch. As religious- and racist-motivated violence and hate crimes continue to occur around the world – Negombo in Sri Lanka, San Diego in the US – it is clear that gender and place matter. These events are not just about religious and/or ethnic violence, rather, they are also about gendered violence. Nearly all debates circulating about this violence in the media and by politicians cast it as a problem existing within, and/or because of, ethnic or religious groups. As feminist geographers, we have a vital role to play in analysing and interpreting the gendered nature, identities and stereotypes of hate crimes. As this Handbook shows, many of us are directly engaged in research and teaching to understand the diversity and intersections of race, ethnicity, faith, gender, power and the different spatial activity, behaviour and experiences of place.

Meanwhile, the UK electorate chose to vote to leave the European Union (EU) at the Referendum in June 2016, against a backdrop of increasing racism, Islamophobia and xenophobia (see Burrell et al. 2019). Recorded hate crime soared after the Brexit vote, and Tell MAMA (a national organization that records and analyses incidents of anti-Muslim hatred) reported an increase in 475% of reports of anti-Muslim hatred and Islamophobia in the week following the vote (Tell MAMA 2017). We know that Islamophobia is gendered in nature and that Muslim women are the main victims (e.g. Najib and Hopkins 2019). Furthermore, shortly following the Brexit result the UK Prime Minister, David Cameron, resigned. In the contest to replace him, Andrea Leadsom claimed that she would be a better prime minister than Theresa May (who was ultimately elected) because, unlike May, Leadsom was a mother and so was deemed to have a very real stake in the future of the country. Gender norms, inequalities and everyday sexism permeate these actions. As the chapters in this Handbook demonstrate, gender inequalities and everyday sexism are part of politics and place, embedded within people's lives, and need to be challenged and overcome.

These events are not some kind of anomaly, rather, they are part of particular trends. The rise of the far right has prompted some – deeply concerned about and impacted by xenophobia, homophobia, transphobia and sexism – to mobilize against far-right local and global movements (Gökarıksel and Smith 2016; Page and Dittmer 2016). The most visible articulation of resistance to the far right was the 2017 Women's March on Washington, DC. This was not an isolated event. People of all genders, generations, sexualities, classes and ethnicities gathered in rural villages and city centres on every continent under the banner of 'women'. The sense of urgency to demonstrate expanded to include, for example, Black Lives Matter, transgender rights and asylum for refugees. Gender-based inequalities – which intersect with sex, sexuality, race and so on – remain one of the greatest global injustices. We hope that this Handbook will make readers more impatient for change, yet also that readers will turn to the Handbook when it feels that change is sometimes impossible.

Overview

The Handbook is structured into four themed parts that profile the distinct contributions that geographers make to the study of gender and feminism: Establishing feminist geographies; Placing feminist geographies; Engaging feminist geographies; and Doing feminist geographies. Each part brings together different ways of thinking that extend existing geographical analysis, as well as feminism and gender studies. This overview reviews the core concepts and debates in the four parts.

Part 1: Establishing feminist geographies

The first part of this Handbook contains 11 chapters based on feminist geographies of difference. Authors draw on a wide range of social and cultural theories in order to be attentive to intersectional understandings of bodies, identities, places and spaces. The first three chapters address diverse genders and sexualities from different perspectives. Beginning with 'Indigenous Australian sexualities explored through the lens of sex work', Corrinne Sullivan provides an affirmation of Indigenous rights to self-determination and a celebration of Indigenous sexualities and gender diversity. This chapter acknowledges the importance of centring Indigenous standpoint theories of sexuality and gender, as these are vital if we are to have a multifaceted gender and feminist geography. Also concerned with destablizing the field of sexuality and geography are Carl Bonner-Thompson, Graeme William Mearns, Ged Ridley and Alessandro Boussalem. Their chapter, 'From order to chaos: geographies of sexualities', urges us to consider the diversity of trans, intersex, drag, cross-dressing and other gender and sexuality subjectivities in order to challenge spaces and places of heteronormativity.

Black feminist interventions in LGBT/queer studies are considered by Devin Oliver and Caroline Faria in their chapter, 'Hip-hop urbanism, placemaking, and community-building among Black LGBT youth in Rio de Janeiro, Brazil'. They complicate power in queer politics by examining how Rio's Black LGBT youth negotiate violence in everyday life and use creative expression, physical and virtual spaces to produce inclusive urban space.

Rethinking masculinities in Japan and Papua New Guinea – places where there has been little or no sustained attention to gender and geography – is the topic of Keichi Kumagai's chapter, 'Shifting multiple masculinities'. Kumagai highlights the ways in which it is possible to resist hegemonic masculinities. Moving into the spaces of Western universities, Nancy Hansen discusses the presence of disabled women in her chapter 'Disabled women academics reshaping the landscape of the academy'. Drawing attention to the physical, social and psychological needs of her body, and of other disabled people at work, the chapter shows the ways in which institutions with disability policies and programmes are, indeed, rife with embedded ableism. Also examining the academy, Martina Angela Caretta and Avril Maddrell in Chapter 7 map relational networks within Western feminist geography. Networks play a significant role in monitoring and publicizing inequalities based on gender and intersecting identities.

Gail Adams-Hutcheson's and Paula Smith's chapter, 'Skin, sweat and materiality: feminist geographies of emotion and affect', urges us to examine emotion and affect across a range of feminist geographies and scales. Emotional and affectual geographies of intimacy, shame, love, sweat, abjection and bodies are fleshed out in Chapter 8.

Performative, Politics and space/time, Economy and Globalization themed chapters bring Part 1 to an end. Revisiting Judith Butler's influence on feminist geographies, Eden Kinkaid and Lise Nelson critically assess the various ways in which geographers have employed performativity. They also sketch out new and innovative connections between performativity, space, place and subjectivities. In Chapter 10, we reprint Massey's influential essay about politics and space/time. In Chapter 11, Jessa M. Loomis and Ann M. Oberhauser consider the multiple and contested meanings of 'the economy', showing how feminist geographers have challenged conventional and masculinist approaches to its study. They highlight the importance of race, class, gender and sexuality, and show how the economy is redefined through alternative systems and/or practices of exchange that place human relationships at their centre. The final chapter of Part 1, 'Disentangling globalization', applies a feminist commodity-chain analysis of the Gulf-East Africa hair and beauty trade. Caroline Faria and Bisola Falola focus on: post-colonial disruption narratives; connections between gender, race, class and sexual power; and global-intimate relational understandings of spatial scale. The multi-billion dollar hair and beauty industry reflects and drives the rising economies of Africa and the Middle East.

Part 2: Placing feminist geographies

The 13 chapters in Part 2 provide a wealth of place-specific feminist and gender geographies, and together illustrate the ways that feminist geographers link the personal to the political across distance and in the constitution of scale. Robyn Longhurst and Lisa Melville illustrate the ways in which embodied geographies have developed over a 10-year period. Such scholarship frames Melville's doctoral project on lesbians' experiences of becoming mothers. Staying with the 'intimate' and connecting to the 'geopolitical', Chris Neubert, Sara Smith and Pavithra Vasudevan focus on gender and race in rural Iowa and North Carolina in the US. They trace White agricultural masculinities alongside the disproportionate toxic burden borne by bodies and communities of colour. The intimate concept of home is the topic of Cathrine Brun and Anita Fábos' chapter, 'Home-keeping in long-term displacement'. They critically analyse narrow conceptions of home in refugee policies and show ways to reconfigure home as constellations of political and feminist spaces. Thinking environmentally, Gordon Waitt and Rebecca Campbell consider how different feminist concepts – ecofeminist, post-Marxist feminist and visceral – are used to make everyday lives more environmentally sustainable. Empirical examples of tasting jam, eating kangaroo and showering are offered to show the importance of body and environmental relationality.

A feminist critique of neoliberal planning practices – in Carina Listerborn's chapter – shows the need for more research to understand how gender manifests in planning and urban development. In cities, issues such as safety are being addressed yet gendered and racialized poverty are not. Planning is the topic of Tovi Fenster's and Chen Misgav's chapter, entitled 'Gender and sexuality in participatory planning in Israel: a journey between discourses'. They chart three decades of changes in relation to identity (from women to gender to LBGTQ identities) and in geographical focus (from the peripheries to the inner city). Moving from the city to the rural, Barbara Pini, Robyn Mayes and Laura Rodriguez-Castro consider the troubled relationships between rurality, geography and feminism. Butler's (1990) concept of 'gender trouble' is extended to rural geographies, and the authors trouble universalist constructions of feminism and rurality.

Turning to nationhood, Maria Rodó-Zárate considers feminist approaches, emancipatory processes and intersecting identities. The sexual and gendered dimension of nations are highlighted alongside feminist movements. Pro-independence feminist groups in Catalan, for example, may open new ways of thinking about nations and nationalism. Thinking across nations, May Farrales and Geraldine Pratt provide a range of examples from Filipino diaspora and assert that transnationalism holds both promises and problems for gender and sexual subjectivities. Feminist, anti-racist, decolonial and queer approaches to transnationalism show both new and congealed patterns and (un)certainties. An Indigenous critique of nation reveals that transnational is already embedded within settler colonial states. Expanding the notion of citizenship, Tamir Arviv and Symon James-Wilson present case studies from Toronto, Canada, to highlight the myriad techniques that diverse groups of people use to create spaces of political and social belonging. Feminist geographies of citizenship – particularly centred on migrants – show the polyvariance of political identities, allegiances and practices. Gendered migration is the topic of the following chapter, by Eleonore Kofman and Parvati Raghuram. Feminist analyses of migration have meant that, over the past two decades, our understanding of gendered migrations has increased enormously. There are still gaps in this research, however, particularly when thinking about immobilities, materialities and beyond South-North migration.

The last two chapters of Part 2 address the research associated with landscape and political ecology. Memory landscapes are the topic of the chapter by Danielle Drozdzewski and Janice Monk. Weaving in and out of monuments and memorials, the chapter highlights the gendered landscapes of public memory. Finally, Sharlene Mollett, Laura Vaz-Jones and Lydia Delicado-Moratalla bring together feminist political ecology, decolonial and post-colonial thought and Black feminist thinking about bodies and slavery. They show how young, poor, ethnic minority women in Nigeria are vulnerable to sex slavery and that this violence is an extension of environment degradation of their homelands.

Part 3: Engaging feminist geographies

The 11 chapters in Part 3 consider the engaging relationships, experiences and understandings between bodies, places and feminist geographies. Authors are attentive to the interrelationships between gendered subjectivities and conceptual issues related to place and space. The first three chapters address feminist engagement with violence, trauma and survival. Chapter 26, 'Trauma, gender and space', considers the ways in which gender-based violence (GBV) in Bangladesh, Malaysia and the UK takes on different forms. Rachel Pain, Nahid Rezwana and Zuriatunfadzliah Sahdan use case studies to illustrate the influence of physical, social and political spaces and places on deeply felt trauma following GBV. The second chapter of Part 3 engages with geographies of violence to give visibility to the intersection of violences, injustices, space and place. Katherine Brickell and Dana Cuomo use feminist geopolitical approaches to shed light on masculinized 'hot' spaces (for example, warzones) and feminized 'banal', emotional and intimate violence (for example, sexual assault in the military and on college campuses). 'Scaling a survivor-centric approach for survivors of sexual violence' – Chapter 28 – is based on a collaboration between a Canadian funder and an Indian NGO. Andréanne Martel and Margaret Walton-Roberts illustrate the transformational potential of scaling in disseminating and implementing action-based feminist research.

Engagement can often require different kinds of interface with the subject of study, and traditional subjects of feminist theory, such as reproduction, motherhood, care and labour, have themselves undergone notable transitions. The 'Motherhood in feminist geography' chapter offers an overview of trends on mothering, place and space, together with a selection of key Anglophone theoretical influences. Kate Boyer shows the ways in which the field is both socially and politically engaged, while at the same time is engaging with (and defining) critical conceptual scholarship. The work is being pushed by the structuring of the economy and its relationship to the science of life itself. Embodied labour and the bioeconomy is the topic of Maria Fannin's chapter. Bioeconomy is an emerging field in economic geography and geographies of science and technology. Fannin argues that it is crucial that feminists engage with and develop theories of labour and gather empirical evidence to study emergent forms of embodied labour.

Care and caregiving are addressed in the chapter 'Care, health and migration' by Kim England, Isabel Dyck, Iliana Ortega-Alcázar and Menah Raven-Ellison. Themes of care, citizenship and belonging in healthcare practices and policy are addressed using three groups of migrants to the UK: well-established migrant communities who came from India and West Indies; international nurses recruited to address a 'nurse shortage'; and, asylum seekers who had been released from detention centres. Staying with concepts of care, Anna Tarrant's chapter 'Contexts of "caring masculinities"' draws on two UK studies – the everyday geographies of grandfatherhood and the care responsibilities of men living in low-income families and localities in a norther English city – to shed light on the spatial and temporal dynamics of men's invisible care practices across the lifecourse. The chapter shows how caring and masculinity are constructed, viewed and maintained. 'Looking in, out and back' at the fields of feminist geography and geographies of children, youth and family, Annie Bartos charts the successful engagement with, and cross-pollination of, ideas, people, discourses and arguments. Chapter 33, 'Giving birth to geographies of young people', shows the potential to reframe debates when two subfields are explicitly interconnected.

Paula Meth's chapter, 'Gendered geographies of development', explores the definitions of gender used in development contexts and charts the areas of development most commonly concerned with gender, for example the policy response of mainstreaming. The chapter argues that more engagement with transformational and structural inequalities, for example global economic practices that undermine women's experiences, is needed if gender equality is to be achieved.

Allison Hayes-Conroy, Jessica Hayes-Conroy, Yoshiko Yamasaki and Ximena Quintero Saavedra's chapter, 'Feminist visceral politics: from taste to territory', focuses on the messy engagements that arose while doing research about food and bodies, tastes of community and rebuilding territory through attentiveness to feeling. Ephemeral happenings, affective traces, building partnerships and imaging futures matter when research is framed with a feminist geography of visceral politics.

The last chapter in Part 3, 'Engaging feminist geographies', is by Shirlena Huang and Qian Hui Tan. 'Feminist perspectives on neoliberal globalization, (post)feminisms and (homo) normativities' considers key issues, such as development and women as neoliberal subjects, the proliferation of (post)feminist identities and queer politics of neoliberalism. The authors review current debates on development, (post)feminism and homonormativity to show how feminist and queer culture has been 'colonized' by neoliberal discourses of gender and sexualities. Critical scholarship, they argue, must animate collective resistance to neoliberal individualism.

Part 4: Doing feminist geographies

Reflecting on the doing of feminist geography, Part 4 contributes to discussions about knowledge production and positionalities in the context of doing fieldwork on, with and as feminist, queer and critical scholars. As producers of knowledge, we are situated within specific sets of power relations, and this prompts us to consider knowledge as always relational, that is, made and shared with others.

In Chapter 37, Beaudelaine Pierre, Naimah Petigny and Richa Nagar co-author, co-travel and co-make knowledges. Under the title 'Embodied translations: decolonizing methodologies of knowing and being', they consider the site of knowledge and the struggle of bodies at the intersections of axes of race, religion, caste, gender, sexuality, place and citizenship.

Participatory action research (PAR) is the topic of the next chapter, by Caitlin Cahill, David Alberto Quijada Cerecer, Leticia Alvarez Gutiérrez, Yvette Sonia González Coronado, José Hernández Zamudio, Jarred Martinez and Alonso R. Reyna Rivarola. '"Still we rise": critical participatory action research towards justice' draws on the authors' work with the Mestizo Arts & Activism Collective, an intergenerational social justice think-tank based in Salt Lake City, Utah, in the US. They discuss the relationships between critical participatory action research (PAR) and activism, drawing on specific projects.

The next chapter in the Handbook bring insights into feminist geographies of activism. Claire Hancock, Roxane Bettinger and Sofia Manseri discuss the 'Spaces and scales of feminist activism'. Their chapter maps the spaces and places of feminist activism, from the specific microcosm of French, particularly Parisian, feminist activism. They draw on scholarly work and also work by activists, journalists and bloggers, all of whom contribute to the advancement of feminist knowledge and social justice.

Imagining a feminist geopolitics of climate change through the study of artworks is the focus of the chapter by Sallie A. Marston, Harriet Hawkins and Elizabeth Straughan. The authors offer critical insights into art–science collaborators who are producing a cultural response to climate change.

Turning to the Anthropocene, Kai Bosworth's chapter shows how feminist geographers are crucial to the contestation of gendered and racialized ecologies. Interdisciplinary feminist scholarship on the Anthropocene is reviewed, as well as feminist geophilosophy and new materialism approaches. The chapter brings to the fore the importance of Black, Indigenous and subaltern feminisms as a way to find innovative and creative mutuality and interdependence of the physical and human.

Qualitative Geographic Information Systems (QGIS) and their applications can enrich feminist geographers' research, as Nazgol Bagheri's Chapter 42 tells us. She outlines the opportunities and limits of QGIS, noting that it can enable visualization and analysis yet it is not a complete picture of the complexities of subjectivities, spaces and places. Also in digital space, Jessica McLean, Sophia Maalsen and Nicole McNamara consider, in Chapter 43, the way in which feminist geographic digital methods are redefining the research. Additionally, feminist geographers are attuned to the complexities of gendered, emotional and affective relationships within digital geographies. Turning to another media source, the gendered geopolitics of television drama is the theme of Julie Cupples and Kevin Glynn's chapter, 'Drone queen of the Homeland'. They problematize the geopolitics of gender and show how television is a key site of popular cultural citizenship and feminist geographical investigation.

Providing an overview of doing historical research, the next chapter, by Laura Crawford and Sarah Mills, outlines important feminist debates on the practice and politics of archival fieldwork. Themes of activism, curation and advocacy are also discussed. Inside learning spaces, Joos Droogleever Fortuijn reviews changes to feminist geography teaching and highlights diverse practices and perspectives in Chapter 46, 'Teaching feminist geography'. What and how we teach depends on where we teach feminist geography.

'Autogeography: placing research in the first-person singular' by Sophie Tamas speaks directly to the reader, as she writes about herself, her place and her intimate maternal geographies. Sketching the risks and the possibilities of autogeography, this chapter illustrates the power of performative and personal writing. Finally, Chapter 48, 'Narrating new spaces: theories and practices of storytelling in feminist geographies', by Sarah de Leeuw and Vanessa Sloan Morgan, discusses storytelling. Feminist, anti-racist, queer, Indigenous and critical geographers are transforming geography through the powerful art of storytelling.

To us, as editors, it seems highly appropriate to end the Handbook with chapters about writing ourselves/places and storytelling. Through the process of being critical scholars, feminist and queer geographers, this Handbook is a way to own our stories and actions. We bring together different ways of knowing and transforming the understanding of ourselves and relationships with each other, places and spaces.

Conclusion

We hope that you will dive into this Handbook, as excited as we were, to read new contributions from experienced, creative and promising scholars in the field who reflect the diversity and variety of conceptual, theoretical and practical approaches to feminist geographies and gender studies. Our multidisciplinary ethos – with contributors from feminist geographers – should appeal to scholars in a range of fields such as anthropology, development studies, global studies, human geography, planning, sociology and urban studies, gender and sexuality studies, media studies, and law and legal studies, as well as others.

With a Handbook of this size, we have had a number of challenges. One challenge that all large, edited collections face is the grouping of contributions together in a coherent and thematic way. From the outset, we proposed our four parts – establishing, placing, engaging, doing – of feminist geographies. These are broad enough not to foreclose possibilities. Yet, this broadness means that some chapters could fit into several parts. Indeed, our own research, as editors, spans these various articulations of feminist geographies. From start to end, we have been strong in our desire for a broad and inclusive approach, one which seeks to embrace authors and their multi-scalar work across a number of disciplinary boundaries.

We invite you to dip into the Handbook to a part or chapter that excites you. The collection is unique in that it includes work on bodies, ethnicities and Indigeneity from scholars who reside outside of Anglophone centres. Some chapters provide contextual and gendered understandings of, for example, women from the Global South. Throughout the Handbook, we have sought to highlight the ways in which gendered and sexed subjectivities, places and spaces intersect and intertwine. There are also suggestions for further reading that accompany each chapter, should you be intrigued to read more about a specific topic or issue. This comprehensive overview of current debates and issues in gender and feminist geographies will be a key reference point for researchers and students for many years to come. Feminist books, such as this one, 'have a special agency, all of their own' (Ahmed 2017, 240) and are crucial when naming and calling attention to injustices and inequities. Take it with you wherever you go.

References

Ahmed, S. 2017. *Living a Feminist Life*. Durham, NC: Duke University Press.

Banerjee, A., A. Gethin, and T. Piketty. 2019. "Growing Cleavages in India?" *Economic and Political Weekly* 54 (11): 34–44.

Bell, D., and G. Valentine, eds. 1995. *Mapping Desire: Geographies of Sexualities*. New York: Routledge.

Blunt, A., and R. Dowling. 2006. *Home*. New York: Routledge.

Bowlby, S., J. Lewis, L. McDowell, and J. Foord. 1989. "The Geography of Gender." In: *New Models in Geography*, volume 1, edited by R. Peet and N. Thrift, 157–175. Boston: Unwin & Hyman.

Brown, M. 2000. *Closet Space: Geographies of Metaphor from the Body to the Globe*. New York: Routledge.

Browne, K., C.J. Nash, and S. Hines. 2010. "Introduction: Towards Trans Geographies." *Gender, Place and Culture* 17 (5): 573–577.

Burrell, K., P. Hopkins, A. Isakjee, C. Lorne, C. Nagel, R. Finlay, A. Nayak, M.C. Benwell, R. Pande, M. Richardson, K. Botterill, and B. Rogaly. 2019. "Brexit, Race and Migration." *Environment and Planning C: Politics and Space* 37 (1): 3–40.

Butler, J. 1990. *Gender Trouble*. London: Routledge.

Butler, R., and H. Parr, eds. 1999. *Mind and Body Spaces: Geographies of Illness, Impairment and Disabilities*. London: Routledge.

Chacko, P. 2018. "The Right Turn in India: Authoritarianism, Populism and Neoliberalisation." *Journal of Contemporary Asia* 48 (4): 541–565.

Datta, A. 2016. "The Genderscapes of Hate: On Violence Against Women in India." *Dialogues in Human Geography* 6 (20): 178–181.

Davidson, J., and L. Bondi. 2004. "Spatializing Affect, Affecting Space: An Introduction." *Gender, Place and Culture* 11: 373–374.

de Leeuw, S. 2016. "Tender Grounds: Intimate Visceral Violence and British Columbia's Colonial Geographies." *Political Geography* 52: 14–23.

Doan, P. 2010. "The Tyranny of Gendered Spaces – Reflections from Beyond the Gender Dichotomy." *Gender, Place and Culture* 17 (5): 1314–1321.

Duncan, N., ed. 1996. *BodySpace*. Abingdon: Routledge.

Dwyer, C. 2016. "Why Does Religion Matter for Cultural Geographers?" *Social and Cultural Geography* 17 (6): 758–762.

Dwyer, C., and C. Bressey, eds. 2008. *New Geographies of Race and Racism*. Aldershot: Ashgate.

Eaves, L.E. 2019. "The Imperative of Struggle: Feminist and Gender Geographies in the United States." *Gender, Place & Culture*. Available at: https://doi.org/10.1080/0966369X.2018.1552564.

Faria, C., and S. Mollett. 2014. "Critical Feminist Reflexivity and the Politics of Whiteness in the 'Field'." *Gender, Place and Culture* 23 (1): 1–15.

García Ramón, M.D., M. Castaner, and N. Centelles. 1988. "Women and Geography in Spanish Universities." *Professional Geographer* 40: 307–315.
García Ramon, M.D., and J. Monk. 2007. "Gender and Geography: World Views and Practices." *Belgeo* 3: 247–259.
Gibson-Graham, J.K. 2006. *A Postcapitalist Politics*. Minneapolis: University of Minnesota Press.
Gilmore, R.W. 2007. *Golden Gulag: Prisons, Surplus, Crisis, and Opposition in Globalizing California*. Berkeley: University of California Press.
Gökarıksel, B., and S. Smith. 2016. "'Making America Great Again'?: The Fascist Body Politics of Donald Trump." *Political Geography* 54: 79–81.
Gökarıksel, B., and S. Smith. 2017. "Intersectional Feminism Beyond U.S. Flag: Hijab and Pussy Hats in Trump's America." *Gender, Place and Culture* 24: 628–644.
Gökarıksel, B., Hawkins, M., Neubert, C. and S. Smith. Forthcoming. *Feminist Geography Unbound: Discomfort, Bodies, and Prefigured Futures*. Morgantown: West Virginia University Press.
Hanson, S. 2010. "Gender and Mobility: New Approaches for Informing Sustainability." *Gender, Place and Culture* 17 (1): 5–23.
Hawthorne, C., and B. Meché. 2016. "Making Room for Black Feminist Praxis in Geography." *Society and Space* open site. Available at: http://societyandspace.org/2016/09/30/making-room-for-black-feminist-praxis-in-geography/.
Hopkins, P. 2018. "Feminist Geographies and Intersectionality." *Gender, Place and Culture* 25(4): 585–590.
Hopkins, P. 2019. "Social Geography II: Islamophobia, Transphobia, and Sizism." *Progress in Human Geography*. Available at: https://doi-org.ezproxy.waikato.ac.nz/10.1177/0309132519833472.
Hopkins, P., and R. Gale, eds. 2009. *Muslims in Britain: Race, Place and Identities*. Edinburgh: Edinburgh University Press.
Hunt, S., and C. Holmes. 2015. "Everyday Decolonization: Living a Decolonizing Queer Politics." *Journal of Lesbian Studies* 19: 154–172.
Jamieson, S. 2016. "People Scared to Leave the House after Rise in Hate Crime Post-Brexit." Available at: www.telegraph.co.uk/news/2016/06/30/people-scared-to-leave-the-house-after-rise-in-hate-crime-post-b/.
Johnson, L. 2009. "Feminism/Feminist Geography." In: *International Encyclopedia of Human Geography*, edited by R. Kitchin and N. Thrift, 44–58. London: Elsevier.
Johnson, L. 2012. "Feminist Geography 30 Years On – They Came, They Saw But Did They Conquer?" *Geographical Research* 50 (4): 345–355. doi:10.1111/j.1745-5871.2012.00785.x.
Johnston, L. 2018. *Transforming Gender, Sex and Place: Gender Variant Geographies*. London: Routledge.
Johnston, R.J., D. Gregory, D.G. Pratt, and M. Watts. 2000. *The Dictionary of Human Geography*, 4th edition. Oxford: Blackwell.
Kindon, S., and J. Cupples. 2014. "Anything to Declare? The Politics of Leaving the Field." In: *Development Fieldwork: A Practical Guide*, 2nd edition, edited by R. Scheyvans, 217–235. London: Sage.
Kingsolver, B. 2016. "End this Misogynistic Horror Show. Put Hillary Clinton in the White House." *The Guardian* 6 November. Available at: www.theguardian.com/commentisfree/2016/nov/06/hillary-clinton-white-house-donald-trump-bullying-barbara-kingsolver.
LaFrance, A. 2016. "Donald Trump's Warning for Women." *The Atlantic* 14 October. Available at: www.theatlantic.com/politics/archive/2016/10/donald-trumps-warning-for-women/504251/.
Laurie, N., C. Dwyer, S.L. Holloway, and F. Smith. 2014. *Geographies of New Femininities*. New York: Routledge.
Longhurst, R. 2001. *Bodies: Exploring Fluid Boundaries*. London: Routledge.
Longhurst, R. and Johnston, L. 2015. "Recollecting and Reflecting on Feminist Geography in Aotearoa New Zealand and Beyond." *Women's Studies Journal* 29 (1): 21–33.
Mackenzie, S. 1984. "Editorial Introduction: Women and the Environment." *Antipode* 16 (3): 3–10.
Maddrell, A. 2009. *Complex Locations: Women's Geographical Work in the UK 1850–1970*. Oxford: Wiley Blackwell.
Massey, D. 1994. *Space, Place and Gender*. Minneapolis: University of Minneapolis Press.
McCutcheon, P. 2019. "Fannie Lou Hamer's Freedom Farms and Black Agrarian Geographies." *Antipode* 51 (1): 207–224.
McDowell, L. 1979. "Women in British Geography." *Area* 11 (2): 151–154.
McDowell, L. 1992. "Doing Gender: Feminism, Feminists and Research Methods in Human Geography." *Transactions of the Institute of British Geographers* 17: 399–416.

McGinley, L., and A. Goldstein. 2017. "Trump Reverses Abortion-related US Policy, Bans Funding to International Health Groups." *The Washington Post,* 23 January 2019. Available at: www.washingtonpost.com/news/to-your-health/wp/2017/01/23/trump-reverses-abortion-related-policy-to-ban-funding-to-international-health-groups/?noredirect=on&utm_term=.0954cfb0dbcb.
McKittrick, K. 2006. *Demonic Grounds: Black Women and the Cartographies of Struggle.* Minneapolis: Minnesota Press.
Mollett, S., and C. Faria. 2013. "Messing with Gender in Feminist Political Ecology. *Geoforum* 45: 116–125.
Momsen, J. 1980. "Women in Canadian Geography." *Canadian Geographer* 24: 177–183.
Momsen, J., and V. Kincaid. 1993. *Different Places, Different Voices: Gender and Development in Africa, Asia and Latin America.* London: Routledge.
Monk, J. 1994. "Place Matters: Comparative International Perspectives on Feminist Geography." *Professional Geographer* 46 (3): 277–288.
Monk, J. 2004. "Women, Gender, and the Histories of American Geography." *Annals of the Association of American Geographers* 94 (1): 1–22.
Moss, P., and K. Al-Hindi, eds. 2007. *Feminisms in Geography: Rethinking Space, Place and Knowledges.* Lanham, MD: Rowman & Littlefield.
Moss, P., K.F. Al-Hindi, and H. Kawabata. 2002. *Feminist Geography in Practice: Research and Methods.* Hoboken, NJ: Wiley-Blackwell.
Najib, K., and P. Hopkins. 2019. "Veiled Muslim Women's Strategies in Response to Islamophobia in Paris." *Political Geography* **73**: 103–111.
Nast, H., and S. Pile, eds. 1998. *Places through the Body.* New York: Routledge.
Nelson, L., and J. Seager. 2005. *A Companion to Feminist Geography.* Chichester: Blackwell.
Olson, E. 2016. "Gender and Geopolitics in 'Secular Time'." *Area* 45 (2): 148–154.
Page, S., and Dittmer, J. 2016. "Donald Trump and the White-Male Dissonance Machine." *Political Geography* 54: 76–78.
Peake, L. 2015. "The Suzanne Mackenzie Memorial Lecture: Rethinking the Politics of Feminist Knowledge Production in Anglo-American Geography." *The Canadian Geographer* 59 (3): 257–266. doi:10.1111/cag.12174.
Pratt, G. 2012. *Families Apart: Migrant Mothers and the Conflicts of Labor and Love.* Minneapolis: University of Minnesota Press.
Pulido, L. 2002. Reflections on a White Discipline. *Professional Geographer* 54 (1): 42–49.
Radcliffe, A. 2014. "Gendered Frontiers of Land Control: Indigenous Territory, Women and Contests over Land in Ecuador." *Gender, Place and Culture* 21 (7): 854–871.
Radcliffe, S. 2017. "Geography and Indigeneity: Critical Geographies of Indigenous Body Politics." *Progress in Human Geography* 42 (3): 436–445.
Silva, J.M., M.J. Ornat, and A.B. Chimin Jr, eds. 2017. *Geografias Feministas e das Sexualidades: Encontros e Diferenças.* Todapalavre: Ponta Grossa/PR.
Simmonds, N. 2011. "*Mana wahine*: Decolonising Politics." *Women's Studies Journal* 25 (2): 11–25.
Sullivan, C. 2019. "Majesty in the City: Experiences of an Aboriginal Transgender Sex Worker in Sydney, Australia. *Gender, Place and Culture.* doi:org.ezproxy.waikato.ac.nz/10.1080/0966369X.2018.1553853.
Tell MAMA. 2017. *Beyond the Incident: Outcomes for Victims of Anti-Muslim Prejudice.* London: Tell MAMA.
Thien, D. 2005. "After or Beyond Feeling? A Consideration of Affect and Emotion in Geography." *Area* 37: 450–456.
Tivers, J. 1978. "How the Other Half Lives: The Geographical Study of Women." *Area* 10 302–306.
Tolia-Kelly, D. 2006. "Affect – An Ethnocentric Encounter? Exploring the 'Universalist' Imperative of Emotional/Affectual Geographies." *Area* 38: 213–217.
Women and Geography Study Group of the IBG. 1984. *Geography and Gender: An Introduction to Feminist Geography.* London: Hutchinson.
Zelinsky, W. 1973. "The Strange Case of the Missing Female Geographer." *Professional Geographer* 25(2): 101–105.

PART 1

Establishing feminist geographies

2

INDIGENOUS AUSTRALIAN SEXUALITIES EXPLORED THROUGH THE LENS OF SEX WORK

Corrinne Sullivan

Introduction

The concept of Indigenous sexuality often produces an abundance of recirculated ideas and (mis)conceptions which, on interrogation, tend to show more about the minds that wrote them than they do about Indigenous Australians or Indigenous sexuality. Further, there may well be a connection between the type of narratives and the silences that occur in Indigenous, geographical and feminist scholarship, as intimated in this chapter. Representations of Indigenous sexuality, particularly Indigenous women's sexuality, are usually linked to violence and exploitation. There is limited writing on pleasure, desire or enjoyment. My research with Indigenous Australian sex workers brings to light the limited perspectives presented in the academic literature that discusses Indigenous sexuality. Through my research, Indigenous male, female and transgender sex workers provide counter-narratives that are not solely reacting to colonial and/or cultural constructions of sexuality. Rather, these counter-narratives centre Indigenous Australian experiences, rendering colonial and cultural constructions as a process to be understood, not as the defining factor in the way in which Indigenous sexuality, gender diversity and identity is represented. As an Indigenous person from the Wiradjuri nation in central-western New South Wales, I draw on my experiences as both an Indigenous woman and as a sexual being to view, analyse and interpret the narratives and the literature.

Engaging with the lives of Aboriginal Australian sex workers propelled me back to the imperialist nineteenth and twentieth centuries, when hierarchies of race dominated public discourse. The conceptual terrain of historical works provides a familiar narrative of Indigenous people who suffered sexual exploitation and victimhood, and many such stories are produced and reproduced in academic literature (Langton 2008; Moreton-Robinson 1998; Ryan 2016) and, although sexual exploitation did and does occur in Indigenous communities, such positions are not taken from an Indigenous point of view, nor do they include counter-narratives to these assertions. Barker observes:

> These representational practices suppress Indigenous epistemologies, histories, and cultural practices regarding gender and sexuality while also concealing the historical and social reality of patriarchy, sexism, and homophobia within Indigenous communities.
>
> *2017a, 13*

Scholars may well be continuing these tracks of thought without question or interrogation. I argue that, in large measure, this blindness stems from the fact that, again and again, scholars of Indigenous histories and contemporalities sidestep an essential truth: as human beings, our sexuality as Indigenous people is central to our sense of self and our desires. Indigenous sexuality and gender diversity are often ignored, silenced or misunderstood in Indigenous, geographic and feminist scholarship (Brown 2012; Moreton-Robinson 2000, 2013; Sullivan 2018) and are rarely written from the point of view of Indigenous people. It is striking that questions of Indigenous sex, sexuality and gender are rarely evident in geographical, feminist and Indigenous literature, as such matters should be intertwined with the most central topics related to Indigenous being, such as identity, body and emotion, as well as the very nature of Indigenous people as social and cultural beings. Though scholars have had much to say about Indigenous ways of life, for the most part they have had little to say in regard to Indigenous sexuality.

One way to map Indigenous sexuality is to chronicle Indigenous peoples' lives in sexual spaces. In this chapter, I engage with Indigenous Australian people in the sphere of sex work. The space of the body and its geographies of sexuality and gender are highlighted within the bounds of sex-work labour, economics and sexual autonomy. Such research on Indigenous Australians' sexual relations aims to encourage new understandings of sex, sexuality and gender and to stimulate different ways of (re)imagining Indigenous bodies. This chapter is offered as an affirmation of Indigenous rights to self-determination, as well as a form of resistance against the misrepresentation of Indigenous sexualities and gender diversity. Indigenous sexuality is not just about having sex; it is about identity and self-determination (Barker 2017b). It is about gender, body and the expression of those two things.

The colonial hangover

This section provides an account of the dominant Australian historical discourses, finding that Indigenous people, in particular Indigenous women, were viewed as exotic, erotic and something to be desired, and yet simultaneously caused anxiety and were objects to be feared. Indigenous Australian people were described as savage, promiscuous and primitive (Moreton-Robinson 1998). In order for the colonizing male to maintain control, Indigenous Australian people's disturbing and disruptive sexual energies had to be contained, and they became increasingly targeted in violent interventions and racialized legislation and policies. Assimilationist federal and state policies and legislation were central to a regime of sexual surveillance, and this control of supposedly degenerate sexuality became pivotal to the portrayal of Indigenous people. These anxieties and fears emerged from the moment of the invasion of Australia. A key feature of colonial anxiety was the fabrication of the sexualities of Indigenous people – a subject well documented by Indigenous feminist scholars (Barker 2017b; Langton 1993, 2008; Moreton-Robinson 1998, 2000; Sullivan 2018). Colonial discourses on Indigenous people's sexuality led to the objectification and ensuing dehumanization of Indigenous people globally (Barker 2017b; Smith 2012). In Australia, Indigenous people were, and often continue to be, situated on the lowest rung of the class ladder, our social standing a critical factor underpinning social inequality, a social position closely linked to long-held beliefs regarding people of colour globally (McKittrick 2006; Moreton-Robinson 2000; Sullivan 2018).

Representations of Indigenous sexualities are shaped almost entirely by historically and colonially constituted narratives that determine/explain/describe the deepest intimacies of our lives at any given time and place. These narratives are socially constructed and are bounded legally, socially, culturally and sexually, trapping Indigenous sexuality within the imagination of predominantly White people, a position that is shaped and reinforced by history, racism and discrimination and that renders the nuances of particular people's experiences invisible – actually, not invisible; rather silenced and unacknowledged. Indeed, racialized constructions of sexuality and bodies were essential to the rationale for invasion and colonization, fashioning Indigenous bodies as savage and primitive in order to justify and reinforce the imposition of Western superiority and civilization (Levine 2008). Moreover, the bodies and physical appearances of Indigenous peoples were unfairly forged through a colonizing Western gaze, based on Eurocentric aesthetic standards (Conor 2016). Indigenous people have been portrayed as 'animalistic, not quite human' and the 'most docile creature lacking agency' (Bond 2015, para 4). Indigenous Australian artist Troy-Anthony Baylis surmises:

> it is as if history has constructed Aboriginality as being so pure and so savage ... that if tainted by the complexity of sexuality and gender, mixed ethnographies, mixed geographies and mixed appearances, the whole look would be ruined.
>
> *2015, 1*

What better way to colonize a people than to make them ashamed of their bodies and the expressions of those bodies? Perhaps the 'original violence' of colonialism was 'to cover our being with its rules and regulations' (Watson 1998, 2). The covering of Indigenous bodies in clothing, in legislation, in practice and in policies transpired in ways that were pathologized, exoticized and fetishized, obscuring Indigenous sexuality (Franklin 2014). Obscuring Indigenous sexuality, in this context, is described by Behrendt as 'neo-colonial power relationships [which] carry the baggage and the legacy of frontier and colonial power relations' (2000, 365). These analyses place Indigenous people in a context of heterosexuality, a binary that does not afford the voices of those who are both and neither, or, one without the other. These are racially and gendered snapshots of sexuality, stagnant images of femininity and masculinity that stifle Indigenous peoples' knowledges and power over their own culture and identity. These representations, enacted through racialized, gendered and sexualized images of Indigenous people, heterosexual and pan-Aboriginal (meaning all the same), are not a novelty or coincidence. They serve a purpose. Colonialism demands that Indigenous people fit within the heterosexual 'norm'.

The sexual and gender diversity of Indigenous peoples remains mostly absent from the records and interpretations of Australian histories, and these absences reinforce a hetero-centric reading of Indigenous Australian cultures. These representations often frame sexuality in terms of gender, and the performance of gender. The favouring of Western, social and cultural constructive discourses of gender, in particular of Indigenous Australian gender roles, tends to inscribe Indigenous people as oppressed, subordinated by patriarchal structures or performances (Clark 2017; Huggins 1991; Moreton-Robinson 1998, 2000). Rarely were Indigenous bodies and sexualities granted an agentic, diverse or self-determined presence in White Australian imaginaries.

The reluctance of scholars to discuss Indigenous sexuality is apparent by its lack of inclusion in the literature. While some studies are explicitly concerned with geographies of race and sexuality (McKittrick and Peake 2005; Oliver, Flicker et al. 2015; Peake 2010), geographers studying 'race' or 'Indigeneity' are culpable of often neglecting sexuality in their discussions or viewing it through a White Anglophone lens. To shift away from Western discourses and engage with Indigenous ways of knowing and seeing the world would assist in dismantling 'a priori categories but could also help undo dominant constructions of race, sexuality and gender that hide from view more humane and just ways of organizing the world' (Peake 2010, 70). Unfortunately, we have yet to witness much work on the geography of sexuality conducted in this way: the ethnocentricity of the literature on sexuality and space remains largely unchallenged, despite the growth of material on the intersections of race and sexuality elsewhere (Hopkins 2018).

Although a relatively large body of work by Indigenous scholars aims to reposition representations of Indigenous Australian people as agentic and resistant (see, for example, Behrendt 2000; Langton 2008; Moffatt 1987; Moreton-Robinson 2000; Sullivan 2018), only a small number of ground-breaking seminal texts from Indigenous Australian scholars, artists and filmmakers provide Indigenous representations that pay specific attention to sexuality, bodies and gender and disrupt previously held beliefs. In her essay, 'Well I Heard it on the Television' (1993), Langton's anti-colonial critique of colonial narratives, highlights 'distorted' and narrow representations of Indigenous Australian sexuality and gender. In the film *Nice Coloured Girls,* world-renowned artist and filmmaker Tracey Moffatt generates a powerful commentary of the way in which Indigenous women are represented through stereotypes, clichés and colonial moralism (1987). This body of work provides formidable and challenging observations in relation to representational politics. However, there is a glaring absence from broader research that attends to the voices of Indigenous Australian people, as well as those of Indigenous people globally, and their views on sex and sexuality.

Engaging with sexuality with Indigenous Australian sex workers

The narratives of Indigenous sex workers invigorate understandings of bodies and sexualities; their perspectives attend to concepts of Indigenous sexuality and, in doing so, transcend colonial limitations and enable a rethinking of sex, sexuality, gender and race. The inclusion of Indigenous female, male, trans and queer sex-worker narratives unsettles heteronormative conceptualizations of sexually based services and complicates depictions of sex workers as victims. These narratives are derived from a doctoral study investigating the everyday lives and experiences of seven Aboriginal sex workers. The participants in this study have self-selected a pseudonym to protect their identities. I employed a qualitative narrative approach to explore the stories that sex workers shared about their entry into sex work, their experiences within the industry and the implications of these experiences. The accounts were analysed thematically from an Indigenous Standpoint (Moreton-Robinson 2013; Nakata 2007). Instead of articulating Indigenous sexual and gender identities as a categorically imposed colonial demarcation, Indigenous Standpoint centralizes and positions Indigenous ways of seeing, doing and knowing as situated knowledges. Indigenous sexuality and gender diversity are hence positioned as a constitutive method of seeing, doing and knowing. Knowledges from the body are a form of understanding, a way of perceiving the world that occurs between and across bodies, cultures and geographies (Louis 2007; Martin and Mirraboopa 2003; Moreton-Robinson 2013).

For the purpose of this chapter, sex work is defined as an occupation where a sex worker is hired to provide sexual services for monetary considerations (Sanders 2013). Terms such as prostitute, sex work and sex worker generally invoke an image of a female and of sexual interaction with a client, who is presumed to be male, and is often seen as an entirely heterosexual affair. This fixation on female sex workers excludes other involvement in sex work, reflecting a particular void in sexual and scholarly interest. The inclusion of Indigenous, female, male, trans and queer sex-worker narratives unsettles the heteronormative thinking of sexually based services and complicates depictions of sex workers as victims. Although my research intention was to explore the experiences of Aboriginal people in the sex industry and to fill gaps in this area of knowledge, it soon became clear that there were further gaps in scholarly thought, particularly in regard to Indigenous Australians' contemporary sexual relations.

From an academic perspective, Indigenous Australian sex workers are an under-researched group and therefore an unknown group of people. Recommendation number 5 of the 2012 New South Wales Sex Industry Report states that 'Data on the sexual health of regional and rural, Aboriginal, street-based, male, and gender diverse sex workers should be sought and collated' (Donovan et al. 2012, 8). Despite this recommendation, there remains a paucity of research in this area. The exclusion of diverse identities in scholarship on the sex industry restricts the political agency of Indigenous sex workers yet also reinforces the very gender dualisms that many feminist and queer scholars wish to challenge. The lives and voices of Indigenous Australian sex workers are concealed by discourses of exploitation and victimization that draw attention to marginalization and colonio-centric views and fail to include diverse stories and perspectives. Despite recent work that highlights the intrinsic nature of sex work as not all oppressive and as involving different kinds of worker and client experiences and varying degrees of victimization, exploitation, agency and choice, there remains a distinct silence in the literature (Sullivan 2018).

Indigenous sex-worker views on sex and sexuality

Like many previous studies of sex workers (see Vanwesenbeeck 2001; Weitzer 2005, 2010), the sex workers who participated in this study stated that money was their primary motivation for entering sex work. Many felt that the money earned in sex work was better than the earnings they could command in other jobs:

> At the time I wasn't very skilled. I was still a uni student. I mean I could have got skilled in something else and got working but … I had no idea … no one was going to let me work the hours I wanted, when I wanted, and they didn't pay me enough, I like the satisfaction of having money. It's not easy being poor.
>
> *Majesty, trans*

> I could do other work, I could be a cleaner. But most jobs aren't very flexible, I choose my hours, the clients I see. And if I need time off like for school holidays, or one of my kids get sick, I can. I don't want to do anything else, at least for now, this works for me.
>
> *Moira, female, heterosexual*

> I work when I need the money, it's straight like that. I only work part time so when I need extra cash or when my family is hustlin' [for money] well I get my ass on the corner. It's easier, it's straightforward. I work till I get the money I need, sometimes a bit more and that's it.
>
> *Bianca, female, heterosexual*

> I like having extra money because I have a few jobs. But if I need extra money or I want extra money that would be a reason now that I work. Whereas maybe in the past it also might have been for my own entertainment, depending on the client I guess.
>
> *JJ, queer*

> I'm a good lookin' guy. I know what I have and I know what others want. This is my time to make as much money as I can. I'm not gonna be pretty forever you know … my body and face would be wasted in an office, what's the point! Later when I'm old I will go back to office work. I've been doing some consultancy work to keep my 'normal' career going for later, but for now I just want to make a tonne of cash and have fun.
>
> *Jack, male, gay*

These narratives show that sex work was not chosen due to a lack of other choices. This observation is supported by the findings of Sanders (2005), who observed that, while most sex workers were likely from low-economic backgrounds, their entry may have been more related to a desire for economic and social independence than a means of survival.

Although economic benefit was the primary reason for sex work, there are indications that money was not the only motivator. JJ's and Jack's examples above highlight that interest, desire, intimacy and fantasy are also involved. Majesty highlights the difference between working in her male identity and in her trans identity: when she worked as a male, it was 'about having money so that we could go to breakfast in the morning', whereas as a trans woman Majesty's sex work is far more complex and nuanced than it is about money, opportunity, emotion and validation. The work became:

> less cash based, yeah. Although when the cash would present itself, why not? Is there something else sometimes? … there might be. There's a possibility. This is what I need, you possibly need something, swapsies?
>
> *Majesty*

In regard to sexuality, body and desire, the participants disclosed that providing sexually based services resulted in an improved sense of sexuality and bodily awareness. Although none of the participants were asked direct questions in relation to sex or sexuality, it emerged as a key theme within the research. Expression of sexuality was one of the foremost rewards of their line of work, underscoring the notion that, for some people, sex work has significant bodily implications that impart meaning and self-determination to their lives. For example, Jeremy, who identifies as a brotherboy (trans) and is bisexual, reveals:

> it's actually been a positive thing for me, I reckon. It's just given me more positivity about sex and given me more confidence in myself as a trans guy to – well, just to be more open about myself sexually … I do it because I do like sex. I'm pretty horny right now. I want to fuck someone. So I'll do it for that reason. [Or] it's just about the money kind of thing, you know. You take the opportunity if it comes along.
>
> *Jeremy, brotherboy (trans), bisexual*

Another example comes from Moira:

> I was raised to not talk about sex, or even think about it really. Sex for women was something that we endured. Never enjoyed. Sex work has been eye-opening for me, I have learnt what I like and what I don't like. I feel like I have more control over my body … I used to be real shamed of my body, but now I love it. Men love it. It makes me feel good. Sometimes I get lucky and the client is hot, like yes I would be

> interested if I met you at a bar kinda hot. When that happens I just enjoy it. It's even better if he's good at it.
>
> *Moira, female, heterosexual*

There were other accounts that highlighted not only expressions of sexuality but also the fluidity of sexuality. For example:

> sometimes do doubles, you know, with another girl, I consider myself to be straight, totally 100 per cent straight, but you know in those moments it can feel good. Maybe I am not so 100 per cent sure about sex anymore.
>
> *Bianca, female, heterosexual*

Another participant intimated the sense of power and control that she derived from sex work. Isabelle says:

> I thought sex was dirty, or that there was something wrong with you if you wanted, or liked, to have sex. I thought there was something wrong with me. I loved [sex work], I felt powerful, sexual and in control. There aren't too many places where you can feel that way, as a Black woman, and be paid.
>
> *Isabelle, female, heterosexual*

Isabelle expressed feeling affirmed by the positive attention from clients who valued her appearance, although she also offered reflections about the fetishized nature of her Black body. She said,

> I would often get chosen for being 'the Black one'; the girls get lined up for the client to choose or we meet him one after the other until he chooses one, you know, and they won't remember my name, just my skin colour. But I'm OK with that; in the brothel, that makes me special, a bit exotic.

For Isabelle, the financial aspects of sex work led to her fiscal self-sufficiency and her social interaction, assisted by her exploration of her identities as both a sexual being and a Black woman. Sex work, for Isabelle, was a means by which to validate and explore aspects of those identities. Her satisfaction and enjoyment of sex work were linked to the financial benefits of the work, but also the feelings of power and control she felt.

For the most part, participants articulated feelings of satisfaction about their work, the way in which they used their bodies and their ability to utilize sex work to help to stabilize their lives financially. Although Indigenous sex workers fitting the popular negative stereotypes do exist and their experiences need to be recognized and addressed, those descriptions do not recognize the diversity of lifestyles and experiences that constitute Indigenous sex workers' lives. Furthermore, such representations ignore broader structural understandings of the sex industry, which tend to portray Indigenous sex workers as oppressed victims who are incapable of rational choice. Rather, what can be ascertained from the voices of the participants in this study is that their involvement in the sex industry is predominantly about economic freedom, and it is also in some cases about sexual desire and sexual expression. The evidence provided by participants in my research is not of victimhood, exploitation or abuse. Rather, it is overwhelmingly a narrative of agency, bodily autonomy and self-determination. For the participants in this study, entering into sex work is the outcome of a dignified rational choice for financial gain that has also had positive impacts on their sexual and gender identities.

Indigenous sexuality, as demonstrated by the narratives highlighted here, is not synonymous with biological sex (being female or male) or gender or defined by sexual practice (heterosexual, homosexual, bisexual). It is matter of shifting relationships between bodies, desires, emotions, selves and others. It is something that should be self-determined and realized, rather than socially and culturally predetermined. Indigenous identities are not fixed; rather, they are evolving entities that are transformative, fluid, collective, personal and ambiguous. The narratives provided in this chapter refute the colonial claims of deviancy, victimization and social control that are used to explain and discipline and gender. Rather, the narratives illustrate that sex work is chosen by some Indigenous Australian people as a means of expressing agency and empowerment, and communicates a political statement through challenging the cultural, sexual and gendered identities that Indigenous people are normatively ascribed in society.

Conclusion

The ways that Indigenous people experience sexuality and gender in sex work are one path of inquiry into how gender and other identities are mapped onto bodies. While academic scholarship is increasingly addressing issues of sexual and gender identity at a macro level, its particular attention remains largely on a White or Western focus, therefore a conceptual deconstruction is required. The challenge of unpacking colonial lines of thinking on sex and sexuality in Indigenous Australia, it would seem, halts discussion or imagination of Indigenous sexuality. As Shino Konishi points out, ethnographies of Indigenous sexuality rely exclusively on European records to construct their images of Indigenous society and practices, however unstable (2008). While there is a considerable amount of material on Western, Eurocentric perspectives of Indigenous people and their sexualities and genders, it is difficult to find Indigenous people's accounts of their views and perspectives of sexual relations, sexualities and gender diversity within the literature.

This chapter presents some of the first accounts of Indigenous Australian sex workers shared in the academic literature. Such voices are powerful, representing a discussion of sexual freedom and expression that is under-narrated throughout history and across written texts. Acknowledging, including and centring Indigenous perspectives of sexuality and gender are necessary in Indigenous, feminist and geographical scholarship. The narratives of Indigenous people contribute important understandings and perspectives, and such inclusion promotes exciting theoretical innovation with highly valuable insights. The narratives of Indigenous sex workers open the door for conversations around Indigenous sexuality being heterosexual, homosexual, bisexual, cisgender, transgender, queer and non-binary. Such conversations challenge notions of sexuality, gender, identity and race. The future of geographical research must do more to engage with and include Indigenous people and perspectives. Indigenous people and their knowledges need to be centred, agentic and free to express themselves in their own voices. There is a growing insurgence of Indigenous academics actively discussing matters of relevance to Indigenous people globally. They have created a space for their voices where there was none, and the strength of this scholarship needs to be readily and actively represented and taken into account. The inclusion of Indigenous people's academic literature, their voices, knowledges and perspectives in geographic and feminist literature serve to strengthen the discipline/s – if only they would listen.

A provocation …

What if 'we', as scholars, refuse to accept the equation of Indigenous people and sex with shame, disease, victimization, exploitation or moral degeneracy? What if, let's say, scholars stop reproducing and recirculating colonio-centric ideologies of who and what Indigenous Australia is? What if 'we' listen and include the voices of Indigenous Australians, centring them as the privileged holders of their own views and perspectives of the world? Contemplation of an alternative reality such as this might help to foster 'our' imaginations as 'we' try to envision a world in which Indigenous sex, sexuality and gender diversity are not shameful, taboo or to be hidden or silenced; where Indigenous sexualities and expression of those sexualities are not just considered, but where multifaceted, complex and nuanced views and perspectives are accepted – and, dare I say, even highly respected and valued. As scholars interested in sex, sexuality and gender research, 'we' need to open 'our' imaginations and 'our' ears, then perhaps 'we' will no longer be challenged by Indigenous Australians' right to fuck and to be fucked, without being fucked over.

Key readings

McKittrick, K., and L. Peake. 2005. "What Difference Does Difference Make to Geography." In: *Questioning Geography,* edited by N. Castree, A. Rogers and D. Sherman, 39–54. Oxford: Blackwell.

Moreton-Robinson, A. 2013. "Towards an Australian Indigenous Women's Standpoint Theory: A Methodological Tool." *Australian Feminist Studies* 28 (78): 331–347.

Sullivan, C.T. 2018. "Indigenous Australian Women's Colonial Sexual Intimacies: Positioning Indigenous Women's Agency." *Culture, Health & Sexuality* 20 (4): 397–410.

References

Barker, J., ed. 2017a. *Critically Sovereign: Indigenous Gender, Sexuality, and Feminist Studies.* Durham, NC: Duke University Press.

Barker, J. 2017b. Introduction. In *Critically Sovereign: Indigenous Gender, Sexuality, and Feminist Studies.* Durham, NC: Duke University Press.

Baylis, T.-A. 2015. "Introduction: Looking into the Mirror." In: *Colouring the Rainbow: Blak Queer and Trans Perspectives: Life Stories and Essays by First Nations People of Australia,* edited by D. Hodge. South Australia: Wakefield Press.

Behrendt, L. 2000. "Consent in a (Neo) Colonial Society: Aboriginal Women as Sexual and Legal 'Other'." *Australian Feminist Studies* 15: 353–367.

Bond, C. 2015. "The 'Rise' of Aboriginal Women: From Domesticated Cow to Cash Cow" [online]. 14 July, New Matilda. Available at: https://newmatilda.com/2015/07/14/rise-aboriginal-women-domesticated-cow-cash-cow/ (accessed 1 January 2018).

Brown, M. 2012. "Gender and Sexuality I: Intersectional Anxieties." *Progress in Human Geography* 36: 541–550.

Clark, M. 2017. "Becoming-with and Together: Indigenous Transgender and Transcultural Practices." *Artlink* 37: 76–81.

Conor, L. 2016. "Skin Deep: Settler Impressions of Aboriginal Women." Crawley: UWA.

Donovan, B., C. Harcourt, S. Egger, L. Watchirs Smith, K. Schneider, J. Kaldor, M. Chen, C. Fairley, and S. Tabrizi. 2012. *The Sex Industry in New South Wales: A Report to the NSW Ministry of Health.* Sydney: Kirby Institute, University of New South Wales.

Franklin, C. 2014. "Belonging to Bad: Ambiguity, Parramatta Girls and the Parramatta Girls Home." *Geographical Research* 52: 157–167.

Hopkins, P. 2018. "Feminist Geographies and Intersectionality." *Gender, Place & Culture* 25: 585–590.

Huggins, J., ed. 1991. *Black Women and Women's Liberation.* London: Routledge.

Konishi, S. 2008. "'Wanton with Plenty': Questioning Ethno-historical Constructions of Sexual Savagery in Aboriginal Societies, 1788–1803." *Australian Historical Studies* 39: 356–372.

Langton, M. 1993. "'Well, I Heard It on the Radio and I Saw It on the Television...': An Essay for the Australian Film Commission on the Politics and Aesthetics of Filmmaking by and about Aboriginal People and Things." Sydney: Australian Film Commission.
Langton, M. 2008. "Trapped in the Aboriginal Reality Show." *Griffith Review* 19: 143–159.
Levine, P. 2008. "States of Undress: Nakedness and the Colonial Imagination." *Victorian Studies* 50: 189–219.
Louis, R.P. 2007. "'Can You Hear Us Now?'Voices from the Margin: Using Indigenous Methodologies in Geographic Research." *Geographical Research* 45: 130–139.
Martin, K., and B. Mirraboopa. 2003. "Ways of Knowing, Being and Doing: A Theoretical Framework and Methods for Indigenous and Indigenist Re-search." *Journal of Australian Studies* 27: 203–214.
McKittrick, K. 2006. *Demonic Grounds: Black Woman and the Cartographies of Struggle.* Minneapolis: University of Minnesota Press.
McKittrick, K., and L. Peake. 2005. "What Difference Does Difference Make to Geography." In: *Questioning Geography*, edited by N. Castree, A. Rogers, and D. Sherman, 39–54. Oxford: Blackwell.
Nice Coloured Girls. 1987. Film directed by T. Moffatt. Kanopy Films Australia.
Moreton-Robinson, A. 1998. "When the Object Speaks: A Postcolonial Encounter: Anthropological Representations and Aboriginal's Women's Self-Presentations." *Discourse: Studies in the Cultural Politics of Education* 19: 275–298.
Moreton-Robinson, A. 2000. *Talkin' Up to the White Woman: Indigenous Women and Feminism.* Brisbane: University of Queensland Press.
Moreton-Robinson, A. 2013. "Towards an Australian Indigenous Women's Standpoint Theory: A Methodological Tool." *Australian Feminist Studies* 28: 331–347.
Nakata, M. 2007. "An Indigenous Standpoint Theory." In: *Disciplining the Savages Savaging the Disciplines*, edited by M. Nakata, 213–217. Sydney: Aboriginal Studies Press.
Oliver, V., S. Flicker, J. Danforth, E. Konsmo, C. Wilson, R. Jackson, J.-P. Restoule, T. Prentice, J. Larkin, and C. Mitchell. 2015. "'Women are Supposed to be the Leaders': Intersections of Gender, Race and Colonisation in HIV Prevention with Indigenous Young People." *Culture, Health & Sexuality* 17: 906–919.
Peake, L. 2010. "Gender, Race, Sexuality." In: *The SAGE Handbook of Social Geographies,* edited by S. Smith, R. Pain, S. Marston, and J.P. Jones, 55–77. London: Sage.
Ryan, T. 2016. "Seen But Unseen: Missing Visible Indigenous Women in the Media and what it Means for Leadership in Indigenous Australia." *PLATFORM: Journal of Media & Communication* 7.
Sanders, T. 2005. "'It's Just Acting': Sex Workers' Strategies for Capitalizing on Sexuality." *Gender, Work & Organization* 12: 319–342.
Sanders, T. 2013. *Sex Work.* Portland, OR: Routledge.
Smith, L.T. 2012. *Decolonizing Methodologies: Research and Indigenous Peoples.* London: Zen Books.
Sullivan, C.T. 2018. "Indigenous Australian Women's Colonial Sexual Intimacies: Positioning Indigenous Women's Agency." *Culture, Health & Sexuality* 20: 397–410.
Vanwesenbeeck, I. 2001. "Another Decade of Social Scientific Work on Sex Work: A Review of Research 1990–2000." *Annual Review of Sex Research* 12: 242.
Watson, I. 1998. "Naked Peoples: Rules and Regulations." *Law Text Culture* 4: 1.
Weitzer, R. 2005. "New Directions in Research on Prostitution." *Crime, Law and Social Change* 43: 211–235.
Weitzer, R. 2010. *Sex for Sale: Prostitution, Pornography, and the Sex Industry.* New York: Routledge.

3

FROM ORDER TO CHAOS

Geographies of sexualities

Carl Bonner-Thompson, Graeme William Mearns, Alessandro Boussalem and Ged Ridley

Section 1: mapping geographies of sexualities

Several years ago, one of us had our attention directed to a new digital mapping service, 'I Just Made Love' (Gray, 2011). Now reworked into a dating service that resembles countless others that are often now also normalised into daily routines, the 'unique' feature of this service remains a function that allows the inscription of the places where users have sex onto the Mercator projection. People use 'pins' to indicate the gender(s) of the person(s) they have sex with, accompanied by descriptions of the encounter. In thinking about geographies of sexualities in 2018, such a service still spurs manifold questions for the field. What motivates people to indicate the location of sex on a map using a smartphone? What visceral and/or bodily pleasures are gained from mapping sex in this manner? What spatial patterns are observable? How does the prevalence of same-sex sexual activity in and by this 'global' map nuance thinking about sexual identities, practices and politics in this semi-public community? Do people consider safety, especially in countries where same-sex sexual activity and/or non-marital heterosexual sex remains criminalized? In short, 'I Just Made Love' indicates just how much intimate worlds have transformed since the 1990s, when the field was cemented. This chapter covers key transformations, while delimiting some possible future directions.

Burgess (1929) is often cited as *the* ethnographer who first linked sexuality to urban processes in a study of 'vice' in the 1920s Chicago School. Later, Loyd and Rowntree (1978), Levine (1979) and Weightman (1981) mapped spatial aggregations of gay men (and some lesbians) in a handful of US neighbourhoods. Nevertheless, Castells' (1983) *City and the Grassroots* garners the most attention in terms of the field's origins, perhaps as this was less positivist in efforts to explain the lesbian and gay clusters identified in San Francisco to the othering of homosexuality and overt discrimination (Adler and Brenner 1992).

By the mid-1990s, 'sexuality and space' was an important subfield of feminist geographies. David Bell and Gill Valentine published *Mapping Desire* in 1995, an edited volume credited as a milestone in its utility of queer theory to move beyond lesbian and gay geographies of the city towards a radical project of destabilizing sexual and spatial norms (Peake 2016). Motivated by post-structuralism, its contributors foreground a range of non-heterosexualities, utilizing

Butler's (1990) concept of performativity to disrupt the assumed 'naturalness' of male/female, gay/straight and mind/body dichotomies (Johnston 2015). Debates around performativity featured heavily in early geographies of sexualities commentary (Bell et al. 1994). Exclusions prevalent in consumer spaces, such as on the bar and club scene (Skeggs et al. 2004) and Pride (Browne 2007), began to be unpacked. Backed by critiques of the neoliberal processes prevalent in urban and economic geography, such work challenged the myth of lesbian and gay affluence while adopting what might now be deemed an 'intersectional' framework to highlight fluidity and diversity within queer lives according to gender, race, ethnicity, class and other differences (Duggan 2003).

Performativity also motivated debate about the risk of 'disembodying' desire, vis-à-vis the importance of representation and discourse in and to these processes (Johnston 2018a). Misgav and Johnston (2014) spoke of the dilemma of reproducing heteronormativity by subjugating bodies beneath identities, contending that the ways in which the participants in their research attempted to stabilize gender and sexuality by regulating bodily fluids, such as sweat, nuanced the readings of discursive identity as fluid. Brown's (2008) study of gay men's cruising culture does similar, engaging with non-representational and affective theories (an emphasis on 'happenings' or 'practices') to discuss overlooked potentials of desire; positing that smells, objects and flesh (as well as identity) highlight how desire can unfold in ways that undo assumed identity categories (see also Binnie 1997). Concurrently, Bell (2007) courageously stressed the importance of talking more about sexual acts within the discipline to counter from within the prevailing squeamishness. This, he argues, often works to underpin 'heteronormativity' (see Section 2, 'Queer geographies').

Queer perspectives now traverse multiple paths, but typically share a concern with undoing binaries of gender and sexuality rather than 'just' representing 'abject' and 'abnormal' lives (Browne 2006). More scholars are doing this with a focus on heterosexuality, especially in terms of marriage, sex work and normative readings of family. The last has involved queering spaces of domesticity largely assumed to be private, like the home (Beasley et al. 2015; Pilkey 2014). Binnie and Valentine's (1999) review of the field through to the late-1990s remains vital reading, especially in introducing sexual citizenship. Johnston (2015, 2017, 2018a) provides a contemporary take on the field's standing now.

For over three decades, geographers have investigated how certain sexualities (re)produce social and spatial orders, imbued with uneven power relations. During what is now a lively period of debate, far removed from simple correlations of sexuality-space, the visualizations of 'chaos' and 'complexity' apparent across 'I Just Made Love' could arguably not have been foreseen by those working in the face of considerable stigma to push at scholarly boundaries by mapping sexuality in early research. This once-academic quest to deduce spatial order from what we now realize as multifaceted assemblages of bodies imbued with discursive codings of genders and sexualities has largely been democratized through digital technologies – wherein services such as OKCupid, Tinder or Grindr overlay space with a digital grid, rupturing how sexualized spaces are experienced and theorized.

While significant legal challenges have been 'won', the geographies of sexualities remain uneven and discrimination continues against lesbian, gay, bisexual, trans and other queer (LGBTQ) lives in all places, but to a varying intensity, ranging from death and violence through to subtler prejudices. The changing face of marriage, family and relationships, growth in trans activism, the normalization of online sexual networks, queer lives in the countryside and LGBTQ religiosity are a tiny selection of the topics now studied (Binnie 2016; Gray et al. 2016; Smart and Whittemore 2016; Taylor and Snowdon 2014; Valentine et al. 2016). We can provide only a snapshot, but the remainder of this chapter highlights a raft of things still waiting to be understood by those within the field. Mindful of initial efforts to correlate sexuality and space, the thrust of this chapter is not to offer a corrective to the chaos we now see as unfolding but to assist those wishing to navigate and contribute to this diverse field of scholarship for the first time.

We start by highlighting some of the limits to performativity, with the aim of bringing readers up to date with complementary approaches that may take the geographies of sexualities forward, alongside queer theory. We then explore the rise of trans geographies before mapping some futures we see for the field. The emergence of queer theorizations and their incorporation into geographical analyses represents an important turn. Over the last couple of decades, studies of spaces and sexualities that adopt queer perspectives and concepts have multiplied. In the next section, we present the 'sparks' of queer inquiry.

Section 2: queer geographies

The concept of heteronormativity has been central to the subfield of queer geographies, and can be summarized as 'the set of norms that make heterosexuality seem natural or right and organize homosexuality as its binary opposite' (Corber and Valocchi 2003, 4). Butler's (1993) formulation of queer theory seeks to disrupt binary configurations of sex, gender and sexualities, showing their social construction through regulatory heteronormative laws, institutions and systems. Considering this deconstructive theory exposes 'the limits and instabilities of a binary identity figure', but Seidman (1995, 131) also suggests that queer theory risks failing to observe specific axes of domination that differentially influence the lives of individuals and groups. Consequently, a total dismissal of identity categories and their associated politics can be counterproductive in cases where one wishes to attend to how the lives of oppressed people (and their praxes towards justice) are to be understood by social actors themselves (Collins and Bilge 2016).

With focus on how different modes of power – ableism, classism, homophobia, racism and sexism – interlock to produce a *specific* kind of oppression for those who occupy one or more difference(s), intersectionality has been noted as being a potentially useful approach that might help to overcome some of the limits of queer theory. Intersectionality helps geographers in exploring the variety of ways in which processes of marginalization are lived out and felt according to particular subjective positions – say, those pertaining to being a British lesbian *but also* being Black in predominantly White spaces, while also maintaining a complex affiliation to one's working-class background; a subjective position that would likely reveal a layered oppression that differs from those faced by a White working-class lesbian. A number of geographers have outlined how intersectionality might be deployed to understand how such variances unfold and/or are reproduced spatially so that they can be tackled (Hopkins 2017; Valentine 2007). For Oswin (2008, 100), it remains necessary to move past framings of queer theory as focused solely on non-heterosexual lives to 'examine sexuality's deployment in concert with racialized, classed and gendered processes'. Topics that could benefit from an intersectional queer perspective include 'transnational labour flows, diaspora, immigration, public health, globalization, domesticity, geopolitics and poverty' (Oswin 2008, 100). These topics are still comparatively marginal to those illuminated in Section 1 of this chapter, which dominated the geographies of sexualities until the century's turn. To challenge the normalization of queer identities and knowledge, some geographers turned to a politics of homonormativity.

Section 3: homonormativity and homonationalism

Duggan (2003, 179) defined homonormativity as a 'politics that does not contest dominant heteronormative assumptions and institutions but upholds and sustains them whilst promising the possibility of a demobilized gay constituency and a privatized, depoliticized gay culture anchored in domesticity and consumption'. Deployment of the concept promotes consideration of how some LGBTQ people are included as citizens who are active in upholding neoliberal modes of consumption, allowing the observation of queer complicities with patriarchal, classist and racial systems of oppression (Nast 2002; Oswin 2004). In her analysis of such complicities, Puar (2006) coins the term 'homonationalism' to stress how particular queer discourses in the West can produce and sustain nationalist ideology that works to oppress the racialized other. Deployed within an assemblage framework, homonationalism can aid critical geographical analyses of discourses and practices that are revealing of the prevailing – and often subtle – prejudices that mark diverse queer lives: for example, how well-intended 'hate crime' policies around sexual orientation and gender identity can work to oppress racialized communities (Haritaworn 2010); how political demonstrations and parades can use limited LGBTQ rights discourses that legitimize Islamophobic stances (Kehl 2018); and how 'gay-friendly' city-branding initiatives can bolster regulatory and normative discourses that are detrimental to others, particularly those of lower economic status who may have limited means to participate in the consumer practices that such initiatives tend to encompass (Hubbard and Wilkinson 2015). Homonationalism prompts manifold questions of sexual citizenship that need answering to diversify the multi-scalar processes that produce 'acceptable' citizens, identities and bodies.

Section 4: expanding queer – bodies, affect and emotion

During the past two decades, the scale of the body (including emotional responses) has gained prominence in feminist geographies in efforts to understand the complexity of the social world (Davidson and Milligan 2004). This focus has allowed geographers to emphasize aspects of human experience that are not (yet) cognitive. Affective theories are part of this shift. Lim (2007, 68) productively combines queer and affective theories, understanding affect as 'the capability of a body to impact other bodies and to be affected by them'. In turn, he suggests that 'body' can refer to both human and nonhuman composites. The workings of emotions, embodiments and affects as critical aspects of human experience have enabled queer geographers to examine relationships between sexualities and space in new ways.

Gorman-Murray (2009) has applied an embodied framework to his study of queer migration, while Brown (2008) combines queer and affective theories to observe 'cruising' practices among men seeking sex. Both of these geographers also reveal possibilities to engage corporeal feminist perspectives with haptic geographies (those of 'touch'), as Waitt and Stains (2015) have done in their study of sweat management among young people in Australia. Both Johnston (2012) and Morrison (2012) have brought the geographies of sexualities into conversation with haptic geographies to help to understand how touch contributes to the making and remaking of bodily boundaries, which impacts on sexual and gendered expressions. Greater attention to the body is arguably important in furthering development of the field, while also being central to the emergence of trans geographies, which are summarized in the subsequent section of the chapter.

Section 5: trans geographies

Trans geographies have emerged as an important and lively subfield as the geographies of sexualities have become more closely aligned with the concepts, theories and epistemological standpoints of gender studies, particularly research at the folds of queer theory and other post-structural standpoints (Stryker and Whittle 2006). The latter often used trans subjectivities to illuminate gender as an unstable and fluid social construction, as explained in Section 2 of this chapter (West and Zimmerman 1987). Ideas of gender as a malleable construct underpin considerable scholarship, an idea dependent on a range of factors, presentations, embodiments and contexts rather than a fixed binary. Butler (1993) talks about trans bodies in aforementioned work on performativity, using trans women's bodies to explain the constructed nature of gender. Arguably, such views are now seen as problematic as they can deny varied trans subjective, lived and felt experiences (Hines 2010).

Poststructuralist thought reintroduced the body as a focus of research in terms of (gender) transitional care, perceptions of trans bodies and the violence experienced by particular gendered embodiments (Namaste 2000). Current work tends to utilize trans bodies to demonstrate gender 'fluidity', but it is crucial to remember that this fluidity applies to all bodies and that trans identities can be just as fixed or fluid as those of cis-gender people (Cornwell 2014). Geographers must therefore work not just to examine what trans bodies can teach us about gender but also to acknowledge diversity within the community itself, rather than using trans merely as a point in teaching to a predominantly cis-gender audience (Johnston 2015). Furthermore, trans studies are, currently, US Anglocentric, although there is growing work on trans women of colour and a body of work engaging post-colonial approaches to understand gendered cultures and histories outside the Western conceptualizations that have dominated scholarship within and outside the geographies of sexualities (Ahmed 2000).

A tension remains between discursive and embodied theories of trans geographies. Trans people should not be reduced to bodily differences, despite these differences playing a significant role in trans lives. Within the field, there is considerable criticism of the reductive way that some research exclusively focuses on bodily difference rather than the effect(s) that these differences come to have on how bodies inhabit and move through spaces (Ellis et al. 2014). Accordingly, approaches that centre only on the body can be voyeuristic and risk othering the actualities of trans lives further (Stryker 1994). Trans geographies are beginning to encompass a mix of theoretical and methodological perspectives; from Doan's (2010) feminist autoethnographic accounts through to Browne, Nash and Hine's (2010) larger studies on the diverse views and identities of trans people in Brighton, UK. A commonality in the emerging scholarship appears to be the foregrounding of subjective experiences to understand how trans people live with and respond to gender binaries in the range of spaces through which everyday lives play out. For instance, Crawford (2014) addresses the multiple ways that trans people encounter the space of the bathroom, while Doan (2010) explores how her own transition impacted on the ways in which she moves through spaces, which have become more difficult to use following her transition, illuminating further the ways in which a two-pronged sex/gender logic structures the spaces that most (often, cis people) take for granted. Doan (2010) calls this 'the tyranny of gender'. While Crawford's (2014) framework is a more conceptual, Foucauldian spatial analysis, both she and Doan (2010) use the spatial exclusion of trans bodies to highlight the marginalization of certain identities and bodies through the reinforcement of rigid gender binaries (Hines 2018; Johnston 2018b).

As the subfield develops, further attention to who is writing about trans lives may be useful. Many have critiqued a view of 'nothing about us without us' in disabilities studies (Shakespeare 1993). Arguably, part of the reason why so little has been done by trans people is the hostile response that this type of research garners. Hines (2018) is among the feminist scholars who have been met with significant personal criticism for engaging with this work – focusing not just on the content of her research itself but the very existence of trans people as viable identities. Looking forward, this is a broader issue than just who is writing. Institutionally, more support is needed to allow these marginalized voices to emerge, the voices of those who disproportionately face all kinds of educational and employment barriers, precarious living conditions, poverty, a lack of parental support and a lack of, or limited access to, healthcare and housing (Serano 2007).

Section 6: future geographies of sexualities

By means of a conclusion, we wish to highlight some of the ways in which the geographies of sexualities might continue to trouble heteronormative knowledge production. Longhurst and Johnston (2014) have argued that, in spite of a proliferation of scholarship on bodies, masculinist, heteronormative and sexist power structures remain intact. We agree with this sentiment and suggest that a starting point for those considering engagement with the field is the need for a careful appreciation of the messiness of sexual and gender identities and bodies vis-à-vis the relationships that these have with place. Binnie (in Skeggs et al. 2004) suggested that geographers work to challenge 'squeamish' queer and feminist epistemologies and continually push the boundaries of 'acceptable' knowledge. A way in which to do this might be to focus more on 'touch', as this is still largely absent from work within the field and 'sex itself is a series of touches, feelings and embodied sensations' (Morrison 2012, 11). Therefore, geographers working within the field could push to understand how places impact on how sex is understood, mediated and felt. The latter means understanding how people can break or reform spatial and bodily power relations to advance queer perspectives.

Assemblage can also be deemed to be important to the future of the field. Nash and Gorman-Murray (2017, 6–7) recently called for closer attention to assemblage in terms of urban sexualities, contending that it:

> Might help us to explain what places and why, what subjects and why, and to consider these assemblages as events, non-binding and ephemeral. Identity and subjectivity are not pre-given but are a 'sexuality/gender' coming into being through the viscosity of bodies, non-human actants, objects, ideas, capital and constituting, we can hope, a proliferation of sexualities and genders that are nevertheless unbounded, while tentatively (and recursively) formulated in and through place.

Assemblage is potentially useful in enabling understanding of the multiple bodies, objects and affects that make sexualities meaningful through everyday lived experiences of places. By understanding the viscosity of bodies, we can understand how subjectivities emerge through places, rather than being pre-given and bounded. In this sense, the fleshy physiology of people and bodies can be appreciated alongside their social and cultural meaning – they do not exist in opposition but work in combination to shape everyday lives. Bonner-Thompson (2017) is informed by assemblage to understand how masculinity and sexuality emerge for men who use Grindr, through their engagement with digital spaces. He explores how locations, skin, flesh and desire are assembled by gender and sexuality in Grindr profiles. Subjectivities are not formed through singular bodies but multiple ones. The presence of technologies in our lives is always increasing, becoming further entangled in our sexualities. Therefore, the importance of nonhuman 'things' in constituting sexual subjectivities requires further investigation. This also provides an interesting way to understand the rising interest in digital technologies. A *Gender, Place and Culture* special issue entitled 'Queer Code/Space' raises important questions in exploring sex, sexuality and desire through emerging technologies (Cockayne and Richardson 2017). Many of the topics covered in this chapter could be revisited through greater consideration of the role of technology in assemblages that are generative of genders and sexualities.

To further destabilize the field, we must consider the proliferation of trans, intersex, drag, cross-dressing and other subjectivities, which tells us more about how places are brought into being and the ways in which these can conform or resist gender binaries (Johnston 2015). Diversifying work in this manner will enable a wider populous of voices to challenge heteronormativity. However, these developments should be focused not only on categories and epistemologies of the West. Efforts to decolonize the geographies of sexualities are continuing, recognizing that the field has largely been a White and Anglophone project until recent years, as with other parts of the discipline. The word 'queer', for example, is itself understood in many places as a Western concept. Not acknowledging and unpacking how Western categories are used to understand sexualities in different times and places renders a risk of complicity in neo-colonialism (Puar 2007). Queer, by some, is understood as a colonial tool, reproducing the epistemologies of the Global North (Silva and Vieira 2014). Adopting a non-White and non-Western view may mean radically changing the ways in which those of us in the field *see* genders and sexualities; a difficult, yet necessary, process if geography is to be refined in its abilities to analyse the social world beyond the West. The geographies of sexualities are at an important moment. To continue disruptive agendas, multiple dimensions must be considered. The field must continually build on and push at the boundaries of feminist and queer epistemologies to combat normative processes and challenge intersecting power relations of ableism, classism, homophobia, racism and sexism, while not forgetting the neoliberal processes that such geographical knowledges are all too often entrapped within.

Key readings

Bell, D., and G. Valentine, eds. 1995. *Mapping Desire: Geographies of Sexualities.* London: Routledge.

Binnie, J., and G. Valentine. 1999. "Geographies of Sexuality – A Review of Progress." *Progress in Human Geography* 23 (2): 175–187. doi:10.1177/030913259902300202.

Johnston, L. 2015. "Gender and Sexuality I: Genderqueer Geographies?" *Progress in Human Geography* 40 (5): 668–678. doi:10.1177/0309132515592109.

References

Adler, S., and J. Brenner. 1992. "Gender and Space: Lesbians and Gay Men in the City." *International Journal of Urban and Regional Research* 16 (1): 24–34. doi:10.1111/j.1468-2427.1992.tb00463.x.

Ahmed, S. 2000. *Strange Encounters: Embodied Others in Post-coloniality*. London: Routledge.

Beasley, C., M. Holmes, and H. Brook. 2015. "Heterodoxy: Challenging Orthodoxies about Heterosexuality." *Sexualities* 18 (5–6): 681–697. doi:10.1177/1363460714561714.

Bell, D. 2007. "Fucking Geography, Again." In: *Geographies of Sexualities: Theory, Practice and Politics,* edited by K. Browne, J. Lim and G. Brown, 81–88. Surrey: Ashgate.

Bell, D., J. Binnie, J. Cream, and G. Valentine. 1994. "All Hyped Up and No Place to Go." *Gender, Place and Culture* 1 (1): 31–47. doi:10.1080/09663699408721199.

Bell, D., and G. Valentine. 1995. *Mapping Desire: Geographies of Sexualities.* London: Routledge.

Binnie, J. 1997. "Coming Out of Geography: Towards a Queer Epistemology." *Environment and Planning D: Society and Space* 15 (2): 223–237. doi:10.1068/d150223.

Binnie, J. 2016. "Critical Queer Regionality and LGBTQ Politics in Europe." *Gender, Place and Culture* 23 (11): 1631–1642. doi:10.1080/0966369X.2015.1136812.

Binnie, J., and G. Valentine. 1999. "Geographies of Sexuality – A Review of Progress." *Progress in Human Geography* 23 (2): 175–187. doi:10.1177/030913259902300202.

Bonner-Thompson, C. 2017. "'The Meat Market': Production and Regulation of Masculinities on the Grindr Grid in Newcastle-upon-Tyne, UK." *Gender, Place and Culture* 24 (11): 1611–1625. doi:10.1080/0966369X.

Brown, G. 2008. "Ceramics, Clothing and Other Bodies: Affective Geographies of Homoerotic Cruising Encounters." *Social & Cultural Geography* 9 (8): 915–932.

Browne, K., and J. Lim. 2010. "Trans Lives in the 'Gay Capital' of the UK." *Gender, Place and Culture* 17 (5): 615–633. doi:10.1080/0966369X.2010.503118.
Browne, K. 2006. "Challenging Queer Geographies." *Antipode* 38 (5): 885–893. doi:10.1111/j.1467-8330.2006.00483.x.
Browne, K. 2007. "A Party with Politics? (Re)making LGBTQ Pride Spaces in Dublin and Brighton." *Social and Cultural Geography* 8 (1): 63–87. doi:10.1080/1469360701251817.
Browne, K., C.J. Nash, and S. Hines. 2010. "Introduction: Towards Trans Geographies." *Gender Place and Culture* 17 (5): 573–577. doi:10.1080/0966369X.2010.503104.
Burgess, E.W. 1929. "Urban Areas." In: *Chicago: An Experiment in Social Science Research*, edited by T.V. Smith and L. White, 46–69. Chicago, IL: University of Chicago Press.
Butler, J. 1990. *Gender Trouble: Feminism and the Subversion of Identity*. New York: Routledge.
Butler, J. 1993. *Bodies That Matter: On the Discursive Limits of Sex*. New York: Routledge.
Castells, M. 1983. *The City and the Grassroots*. London: Edward Arnold.
Cockayne, D.G., and L. Richardson. 2017. "Queering Code/Space: The Co-production of Socio-Sexual Codes and Digital Technologies." *Gender, Place and Culture* 24 (1): 1642–1658. doi:10.1080/0966369X.2017.1339672.
Collins, P.H., and S. Bilge. 2016. *Intersectionality*. Cambridge: Polity Press.
Corber, R.J., and S. Valocchi. 2003. *Queer Studies: An Interdisciplinary Reader*. London: Blackwell.
Cornwell, S. 2014. "Recognising the Full Spectrum of Gender? Transgender, Intersex and the Futures of Feminist Theology." *Feminist Theology* 20 (3): 236–241. doi:10.1177/0966735012436895.
Crawford, L.C. 2014. "Derivative Plumbing: Redesigning Washrooms, Bodies, and Trans Affects in Ds+r's Brasserie." *Journal of Homosexuality* 61 (5): 621–635. doi:10.1080/00918369.2014.865475.
Davidson, J., and C. Milligan. 2004. "Embodying Emotion Sensing Space: Introducing Emotional Geographies." *Social and Cultural Geography* 3 (4): 523–532. doi:10.1080/1464936042000317677.
Doan, P.L. 2010. "The Tyranny of Gendered Spaces – Reflections from Beyond the Gender Dichotomy." *Gender, Place and Culture* 17 (5): 635–654. doi:10.1080/0966369X.2010.503121.
Duggan, L. 2003. *The New Homonormativity: The Sexual Politics of Neoliberalism*. Durham, NC: Duke University Press.
Ellis, S.J., J. McNeil, and L. Bailey. 2014. "Gender, Stage of Transition and Situational Avoidance: A UK Study of Trans People's Experiences." *Sexual and Relationship Therapy* 29 (3): 351–364. doi:10.1080/14681994.2014.902925.
Gorman-Murray, A. 2009. "Intimate Mobilities: Emotional Embodiment and Queer Migration." *Social and Cultural Geography* 10 (4): 441–460. doi:10.1080/14649360902853262.
Gray, E. 2011. "'I Just Made Love' App Lets Users Track Their Sexual Encounters." *Huffpost* online. Available at: www.huffingtonpost.co.uk/entry/i-just-made-love-app_n_1158266?guccounter=1&guce_referrer_us=aHR0cHM6Ly93d3cuZ29vZ2xlLmNvLnVrLw&guce_referrer_cs=K-YuROa-qwP7fV2YEaCQ8Q.
Gray, M.L., C.R. Johnson, and B.J. Gilley. 2016. *Queering the Countryside*. New York: New York University Press.
Haritaworn, J. 2010. "Queer Injuries: The Racial Politics of 'Homophobic Hate Crime' in Germany." *Social Justice* 37 (1): 69–89.
Hines, S. 2010. "Queerly Situated? Exploring Negotiations of Trans Queer Subjectivities at Work and Within Community Spaces in the UK." *Gender, Place and Culture* 17 (5): 597–613. doi:10.1080/0966369X.2010.503116.
Hines, S. 2018. "Trans and Feminist Rights Have Been Falsely Cast in Opposition." *Economist*, 13 July. Available at: www.economist.com/open-future/2018/07/13/trans-and-feminist-rights-have-been-falsely-cast-in-opposition.
Hopkins, P. 2017. "Social Geography I: Intersectionality." *Progress in Human Geography*. doi:10.1177/0309132517743677.
Hubbard, P., and E. Wilkinson. 2015. "Welcoming the World? Hospitality, Homonationalism, and the London 2012 Olympics." *Antipode* 47 (3): 598–615. doi:10.1111/anti.12082.
Johnston, L. 2012. "Sites of Excess: The Spatial Politics of Touch for Drag Queens in Aotearoa New Zealand." *Emotion, Space and Society* 5 (1): 1–9. doi:10.1016/j.emospa.2010.02.003.
Johnston, L. 2015. "Gender and Sexuality I: Genderqueer Geographies?." *Progress in Human Geography* 40 (5): 668–678. doi:10.1177/0309132515592109.
Johnston, L. 2017. "Gender and Sexuality II: Activism." *Progress in Human Geography* 41 (5): 648–656. doi:10.1177/0309132516659569.

Johnston, L. 2018a. *Transforming Gender, Sex, and Place: Gender Variant Geographies:* (Gender, Space and Society series). London: Routledge.
Johnston, L. 2018b. "Gender and Sexuality III: Precarious Places." *Progress in Human Geography.* doi:10.1177/0309132517731256.
Kehl, K. 2018. "'In Sweden, Girls are Allowed to Kiss Girls, and Boys are Allowed to Kiss Boys': Pride Järva and the Inclusion of 'LGBT Other' in Swedish Nationalist Discourses." *Sexualities* 21 (4): 674–691. doi:10.1177/1363460717748621.
Levine, M.P. 1979. *Gay Men: The Sociology of Male Homosexuality.* New York: Harper and Row.
Lim, J. 2007. "Queer Critique and the Politics of Affect." In: *Geographies of Sexualities: Theory, Practices and Politics,* edited by K. Browne, J. Lim and G. Brown, 267–278. London: Routledge.
Longhurst, R., and L. Johnston. 2014. "Bodies, Gender, Place and Culture: 21 Years On." *Gender, Place and Culture* 21 (3): 267–278. doi:10.1080/0966369X.2014.897220.
Loyd, B., and L. Rowntree. 1978. "Radical Feminists and Gay Men in San Francisco: Social Space in Dispersed Communities." In: *Invitation to Geography,* edited by D. Lanegran and R. Palm, 78–88. New York: McGraw Hill.
Misgav, C., and L. Johnston. 2014. "Dirty Dancing: The (Non)Fluid Embodied Geographies of a Queer Nightclub in Tel Aviv." *Social and Cultural Geography* 15 (7): 730–746. doi:10.1080/14649365.2014.916744.
Morrison, C.-A. 2012. "Heterosexuality and Home: Intimacies of Space and Spaces of Touch." *Emotion, Space and Society* 5 (1): 10–18. doi:10.1016/j/emospa.2010.09.001.
Namaste, V. 2000. *Invisible Lives: The Erasure of Transsexual and Transgendered People.* Chicago, IL: University of Chicago Press.
Nast, H.J. 2002. "Queer Patriarchies, Queer Racisms, International." *Antipode* 34 (5): 874–909.
Nash, H.J, and A. Gorman-Murray. 2017. "Sexualities, Subjectivities and Urban Spaces: A Case for Assemblage Thinking." *Gender, Place and Culture* 34 (5): 874–909. doi:10.1111/1467-8330.00281.
Oswin, N. 2004. "Towards Radical Geographies of Complicit Queer Futures." *ACME: An International E-Journal for Critical Geographies* 3 (2): 79–86.
Oswin, N. 2008. "Critical Geographies and the Uses of Sexuality: Deconstructing Queer Space." *Progress in Human Geography* 32 (1): 89–103. doi:10.1177/0309132507085213.
Peake, L. 2016. "Classics in Human Geography: David Bell and Gill Valentine (eds) (1995) Mapping Desire: Geographies of Sexualities." *Progress in Human Geography* 40 (4): 574–578. doi:10.1177/0309132515585060.
Pilkey, B. 2014. "Queering Heteronormativity at Home: Older Gay Londoners and the Negotiation of Domestic Materiality." *Gender, Place and Culture* 21 (9): 1142–1157. doi:10.1080/0966369X.2013.832659.
Puar, J. 2006. "Mapping US Homonormativities." *Gender, Place and Culture* 13 (1): 76–88. doi:10.1080/09663690500531014.
Puar, J. 2007. *Terrorist Assemblages: Homonationalism in Queer Times.* Durham, NC: Duke University Press.
Seidman, S. 1995. "Deconstructing Queer Theory or the Under-theorization of the Social and the Ethical." In: *Social Postmodernism: Beyond Identity Politics,* edited by L. Nicholson, 116–141. Cambridge: Cambridge University Press.
Serano, J. 2007. *Whipping Girl: A Transsexual Woman on the Sexism and Scapegoating of Femininity.* Berkeley, CA: Seal Press.
Shakespeare, T. 1993. "Disabled People's Self-organisation: A New Social Movement?" *Disability, Handicap and Society* 8 (3): 249–264. doi:10.1080/02674649366780261.
Silva, M. J., and P.J. Vieira. 2014. "Geographies of Sexualities in Brazil: Between National Invisibility and Subordinate Inclusion in Postcolonial Networks of Knowledge Production." *Geography Compass* 8 (10): 767–777. doi:10.1111/gec3.12165.
Skeggs, B., L. Moran, P. Tyer, and J. Binnie. 2004. "Queer as Folk: Producing the Real of Urban Space." *Urban Studies* 41 (9): 1839–1856. doi:10.1080/0042098042000243183.
Smart, M.J., and A.H. Whittemore. 2016. "There Goes the Gayborhood? Dispersion and Clustering in a Gay and Lesbian Real Estate Market in Dallas, TX, 1986–2012." *Urban Studies* 54 (3): 600–615. doi:10.1177/0042098016650154.
Stryker, S., and S. Whittle. 2006. *The Transgender Studies Reader.* New York: Routledge.
Stryker, S. 1994. "My Words to Victor Frankenstein Above the Village of Chamounix: Performing Transgender Rage." *GLQ: A Journal of Lesbian and Gay Studies* 1 (3): 237–254.
Taylor, Y., and R. Snowdon. 2014. *Queering Religion, Religious Queers.* New York: Routledge.

Valentine, G. 2007. "Theorizing and Researching Intersectionality: A Challenge for Feminist Geography." *Professional Geographer* 59 (1): 10–21. doi:10.1111/j.1467-9272.2007.00587.x.
Valentine, G, R.M. Vanderbeck, J. Sadgrove, J. Andersson, and K. Ward. 2016. "Transnational Religious Networks: Sexuality and the Changing Power Geometries of the Anglican Community." *Transactions of the Institute of British Geographers* 38: 50–64. doi:10.1111/j.1475-5661.2012.00507.x.
Waitt, G., and E.R. Stanes . 2015. "Sweating Bodies: Men, Masculinities, Affect, Emotion." *Geoforum* 59: 30–38. doi:10.1016/j.geoforum.2014.12.001.
Weightman, B.A. 1981. "Commentary: Towards a Geography of the Gay Community." *Journal of Cultural Geography* 1 (2): 106–112. doi:10.1080/08873638109478645.
West, C., and D.H. Zimmerman. 1987. "Doing Gender." *Gender and Society* 1 (2): 125–151.

4

HIP-HOP URBANISM, PLACEMAKING AND COMMUNITY-BUILDING AMONG BLACK LGBT YOUTH IN RIO DE JANEIRO, BRAZIL

Devin Oliver and Caroline Faria

Introduction

Rio de Janeiro brands itself as an inclusive haven for racial-sexual diversity, yet anti-Black racism, homo- and trans-phobia and misogyny are central in ordering urban space there. In fact, societal institutions, practices and discourses operate in tandem to transform Rio de Janeiro into a hostile racial-sexual terrain that directs violence toward Black, young and non-heteronormative people. They face devastating rates of lethal and non-lethal violence, with some of the highest recorded statistics emerging in recent years as the country has moved politically to the right.

Such violence has a long history in Brazil. However, it is at its height at the contemporary moment. Nothing exemplifies the racist and misogynistic, trans- and homo-phobic violence facing Black LGBT communities in Brazil as much as the assassination of Rio City Councilwoman Marielle Franco and her driver, Anderson Pedro Gomes, on the night of 14 March 2018. Elected in 2016, Franco was a radical Black lesbian and feminist activist-politician, mother and advocate for the poor. In addition to her activism, Franco remained a fearless critic of police corruption and brutality against the city's most marginalized residents; that is, until 13 bullets pierced her and her driver. Nearly one year after these slayings, state officials have failed to make progress in any investigations, despite nationwide and global protests demanding justice for Franco and honouring her legacy under the hashtag #MariellePresente. Activists who, like Franco, navigate the intersections of Black, LGBT and womanhood understand Franco's assassination as a symptom of global systems of oppression. From the perspective of Black feminist activists across and beyond Brazil, the gendered, anti-Black violence evidenced in Franco's killing are:

> part of a global phenomenon of anti-Blackness, manifested through routine violence against Black peoples, suppression of Black political voices, displacement from Black lands, exploitation of Black labor, erasure of Black cultures and histories, and gender and sexual violence against Black women, queer and transgender people.
>
> *AAG 2018*

Franco's murder forms part of a wider rise in violence against Black LGBT lives following the 2016 impeachment of leftist President Dilma Rousseff and the subsequent right-wing incumbency of President Michel Temer. Following impeachment, Temer's administration made swift, severe cuts to education, social services and human rights institutions[1] that all support long-marginalized Black LGBT youth. These cuts will surely continue under the new far-right President Jair Bolsonaro. The rhetoric of several politicians – including Bolsonaro – reflect and reinforce state violence. For example, these politicians have stated that Black women are thug-producing 'factories' and that they would 'rather have a dead child than a gay child'.[2]

In the face of these daily assaults, *carioca* Black LGBT youth face a recurring dilemma: how can they build political power when they are virtually locked out of state institutions and formal politics? This chapter details both the daily violences and spatial strategies of resistance among Black LGBT youth in Rio. Through Oliver's ethnographic research, we show that these youth are consistent placemakers across spatial realms and scales, creating sites of self-making and political intervention through social media, art, popular education and celebrations. These spaces serve as political sites from which youth exchange experiential knowledge, share resources and position themselves both within and against the state in order to address the institutional forces that create precarious livelihoods. The chapter builds on and connects scholarship in feminist political geography, Black feminist theory and radical planning. We make space there for intersectional queer critique while, in turn, pushing queer scholarship towards an anti-racist and spatial sensitivity.

Black feminist interventions in LGBT/queer studies

Brazilianist scholars have attempted to decentre queer and sexuality studies from the urban centres of the Global North through ethnographic accounts of sexualities. Specifically in the context of Brazil, Parker (1999) and Kulick (1998) contextualize in historical and cultural terms their analyses of non-normative genders and sexualities across several Brazilian cities. Both offer readings of such lived experiences largely without assuming a queer politics as it takes place in North America. However, works like these are often problematic through failing to consider the authors' own positionalities as White, cis-gender gay men, initially drawn to Brazil as titillated tourists, as well as the effects of anti-Black racism on the lives of the majority-Black LGBT people whom they study. These shortcomings result in a typical White masculinist ethnographic gaze that seems 'colour blind', leaving the authors unmarked and unaccountable and rendering LGBT people as 'incomplete' subjects, identifiable only by sexual and gender identities (see Stout 2014).

One main reason for the centring of White queer experiences – or elision of those of Black queer folks – in this literature can be explained by what Cohen (1997, 438) calls an 'uncomplicated understanding of power' and identity categories in mainstream queer politics. What is created as a result of 'activating only one characteristic of their identity' (ibid.) is a simplistic dichotomy between everything that is queer being assumed to be radical and just, and everything that is heterosexual being inherently oppressive and despised (see Oswin 2008). Left to grapple with this divide, while caught in the middle of it, are Black and other non-White, non-normative people.

Black feminist and queer scholars have long problematized simplistic understandings of power and identity categories and the tendency of queer theorists and activists to isolate and prioritize sexuality as the 'primary frame through which they pursue their politics' (Cohen 1997, 440). They have advanced forceful critiques that build a broadened, intersectional understanding of queerness, violence, power and identities via 'standpoint epistemologies', which call for critical knowledge production that is situated and grounded in everyday, lived experience. This knowledge must be centred in the experiential knowledge of non-normative folk in an 'attempt to locate authority or expertise with those who experience a circumstance' (Miller-Young 2014; Richie 2012, 129–130).

Black feminist geographies: interventions on space, race, gender and violence

Critical analyses of space and place-making are vital here. Black feminist geographers contend that Black people's sense of place is informed by racism, sexism and homophobia: present-day violences through which we can 'locate and speak back to the geographies of modernity … slavery, and colonialism' (McKittrick 2006, xiii). As McKittrick states, 'Black matters are spatial matters', in that space and place factor centrally in Black lives and, in turn, render us geographic actors whose negotiations can result in reformulations of their own subjectivities and space itself (McKittrick 2006, xii). In other words, Black geographies *necessarily* centre Black agency in geographies of both domination and resistance.

For example, Black feminist activist-scholars like Keisha-Khan Perry (2013), Erica Williams (2014) and Christen Smith (2016) offer important insights into how Black Brazilian subjects – Black women in particular – must make strategic choices to fashion themselves, cope and intervene to secure daily survival and livability. Working with Black sex workers, neighbourhood women and activist performers, all three see Blackness and Black spatial consciousness not as a mere effect of exclusion but rather as central to the constructions of the spaces, places and society that one takes for granted daily – like tourist landscapes, plazas and even celebratory 'gay spaces'. Smith (2016) sees this as a dangerous paradox, which she calls 'afro-paradise'. This is a 'gendered, sexualized and racialized imaginary', through which Salvador becomes a place of Black fantasy and consumption, as well as Black death (Smith 2016, 3). Black people in highly exoticized sites like Rio de Janeiro simultaneously become hyper-sexualized, romanticized, consumed and exterminated objects. However, this paradox is intentional; it is precisely this Black fantasy that serves as a smokescreen to facilitate the state's extermination of Black bodies. Brazilian society, media and the state actively choreograph these celebratory performances and routine killings. It is in the midst of this erotic, genocidal landscape that Black people – especially Black women and youth – must carve out spaces of everyday survival, consciousness and political intervention.

Similarly, Katherine McKittrick (2013) understands this contradiction as indicative of a 'plantation future'. Through this lens, the plantation becomes a material and symbolic site of violence and resistance, where the built environment, urban social processes and Blackness become inseparable. Because of and despite the violence that the plantation symbolizes, 'painful racial histories hold in them the possibility to organize our collective futures' (McKittrick 2013, 3).

Anti-racist feminist scholars of space also offer us pathways of resistance. Miraftab (2009) provides a conceptual framework through which we can begin to decolonize planners' imaginations, which idolize the spatially 'ordered' and 'legible' Western city as the ultimate object of desire. With more decolonized constructions of the city, we can begin envisioning and enacting a collective 'liberation' that decolonizes the mind and our surrounding environments. Whether through rent parties, shelter provision, parades or outright occupation, what youth is doing in the city is often transgressive, counter-hegemonic and imaginative in ways that disrupt common assumptions about the roles of planning in shaping urban space. Black queer insurgent planning and placemaking constitute spatial critiques through everyday subaltern life, which illustrate how gendered anti-Black racism and homophobia are central in ordering urban space – in effect, acting to decolonize the mind, the built environment and urbanism itself. In connection, in a more intimate sense, Sweet's analysis of 'kitchen-table' resistance is useful. Here, the 'kitchen table' is an intentional metaphor, representing the literal and political site from which women of colour – Black young transgender women, in this case – exchange experiential knowledge in order to collectively address the institutional structures and forces that impact on everyday life, at both individual and communal levels. As such, 'kitchen-table planning' offers an alternative to planning as a White, Eurocentric, masculinist field (ibid.)

Intersectional feminist and queer scholarship across the Americas has thus made productive interventions that view the conditions of Black people's lives as always already ambiguous, embedded in epistemological tensions. In such ambiguity, where trauma, pleasure, exploitation and survival coexist in daily life, Black people exercise their agency by appropriating and refashioning negative representations in hopes of building socio-spatial alternatives on their own terms – a 'remix', in a way, of the violence that centrally defines our existence (Bailey and Shabazz 2014; Walcott 2007). As spatial subjects, then, Black people engage in building sites of consciousness, survival and resistance in the face of trauma and violence. The geographical imaginations and practices of Black LGBT youth should, then, be understood as 'social processes that make [space and place] a racial-sexual terrain' (McKittrick 2006, xiv).

In what follows, we first highlight the form and nature of anti-Black and homophobic violence in Rio and its devastating impacts. We show how this violence is enabled by, and reproduced through, the particular racialized, gendered and sexualized imaginaries of the city. Recognizing Black LGBT youth as geographical actors who also have a significant stake in the everyday production of space, we turn to their work in creating spaces of tolerance, focusing on the creation of Casa Nem. This project brings to the fore the spatial agency and imaginations of Black LGBT youth, which in turn provide insurgent, reimagined geographies of violence, struggle and pleasure.

Gendered racism as homophobic and transphobic violence

The urban space of Rio de Janeiro, both material and imagined, is a hostile landscape through which Black people must constantly navigate and struggle to make life livable. Here, racism intersects and affects Black people differentially in relation to gender, class and sexual identities, as well as age. Black women and Black LGBT people remain the key target of gendered, anti-Black racism, which Gilmore (2007) defines as state-sanctioned or extralegal subjection of Black people to conditions that lead to their premature death. In 2016, 347 LGBT people – an unprecedented recorded number – were murdered in Brazil. The annual total has more than doubled since 2000 and is steadily increasing (Mott and Paulinho 2017). Of the 347 reported homicides, 50 per cent were of gay men and 36 per cent of transgender people and *travestis*. Higher proportions of transgender people and *travestis* were murdered, and are thus the most victimized. Many of those murdered were both poor and Black. The data provided by Grupo Gay da Bahia suggest that in Brazil a transgender woman is 14 times more likely to be killed than a gay man. Compared to the US, Brazilian trans women face a nine-times greater chance of being murdered (Mott and Paulinho 2017). Indeed, according to the international agency Transgender Europe (2016), between 2008 and 2016 Brazil witnessed 802 transphobic murders: more than any other country (in terms of relative numbers). Mexico and the US follow Brazil, with 229 and 132 trans murders respectively. Of the homicides of LGBT Brazilians last year, 32 per cent were between 19 and 30 at the time of their death, and 21 per cent were minors. In other words, young LGBT people constitute the majority of reported LGBT homicides. Similar to cases of racist hate crime, many of these murders go unpunished due to the police and judges' dismissal of these cases as potential hate crime.

Homicide rates aside, according to the Global Rights Partners for Justice and the Black LGBT Network for Brazil, 90 per cent of Brazilian transgender women are illiterate, due to discrimination and social isolation, and 68 per cent of transgender Brazilians are living with HIV (Inter-American Commission on Human Rights 2015). These groups denounce the lack of federal legal protections for Black transgender women's rights in Brazil, due to 'discrimination and acts of violence based on homophobia and transphobia [i.e. sexuality and gender identity], which affects people's rights to gender expression as well as effective access to jobs, education, and health services' (Pomykata 2013; ibid.). Thus, particular gendered, sexualized, classed and age-associated racial formations transform the city into a battleground for survival for young Black LGBT people.

Rio de Janeiro as a 'gay-friendly' afro-paradise

Rio de Janeiro is not only a place of Black death, but is associated with sexual freedom and deviance. In fact, these spaces are co-produced, with carnivalesque Black fantasy being part of the process of the devaluation of Black life. These spatial representations and performances are not coincidental, but rather processes that re-present Rio de Janeiro as a sort of 'gay-friendly' afro-paradise. The gay tourism industry and the mainstream LGBT movement appropriate Black bodies and culture to demonstrate sexual diversity and tolerance. Yet, in doing so, they rearticulate racial-sexual hierarchies rooted in colonial violence and are enabling violence in the present day.

The Brazilian state, city government and tourism industry all strategically appropriate and resignify Rio's racialized landscape as a 'site of desire' (Manderson and Jolly 1997) and of 'erotic possibilities' (Binnie 1997). Here, Blackness is central, being simultaneously celebrated, commodified and concealed from touristic sites and narratives (see Williams 2014). We see this simultaneous desire and disgust for Blackness in Rio's city-sponsored gay tourism campaigns, with an all-White cast, set in Ipanema, with rainbow flags waving below palm trees and Afro-Brazilian samba[3] music playing in the background to entice the prospective tourist. However, what is concealed is the gratuitous violence with which the police 'pacify' the neighbouring *favelas,* as well as the communities from which many low-wage employees commute daily. This performative production of touristic and gay leisure spaces relies upon the concealment of Blackness, on the one hand, but also on an eroticized hypervisibility of Black *cariocas'* bodies and cultural labour (see Kelley 1994).

Carioca LGBT activists have increasingly appropriated key tourist landscapes for political action and visibility, such as Posto 8.5 – 'the gay beach' – in Ipanema, as well as Avenida Atlântica, which runs parallel to Copacabana Beach, developing key partnerships with city agencies to spearhead gay tourism and anti-homophobia campaigns. Rainbow flags, bronzed, muscular bodies, the boardwalk and palm trees are prominent in the city-sponsored gay tourism brochures. The spatial imaginary of gay tourism materials, gay nightlife sites and maps and word of mouth locate most 'gay spaces' within the affluent neighbourhoods of the Southern Zone: Ipanema, Leblon, Botafogo and Copacabana, with 'the gayest street in Rio' being Farme de Amoedo Street in Ipanema.

Yet these allegedly 'queer' spaces of tolerance and racial-sexual diversity are embedded in an urban terrain starkly segregated by race, class and gender. Simplistic binary notions of 'queer' space and 'straight' space, which are present in most of this literature, fail to recognize that queer spaces are also contested along lines of race and class (see Oswin 2008). Often, gayness in the city becomes mapped or demarcated in the Southern Zone, far from where most Black *cariocas* – including Black LGBT youth – live. As a result, even when Black LGBT people are present in these gay spaces, they are 'out of place' unless they are working as mobile vendors, service employees in bars and restaurants, domestic workers, sex workers or criminals from nearby *favelas* – the racial-sexual 'spectacle' that Beatriz Nascimento (1982) witnessed in Copacabana.

Reterritorializing anti-Black landscapes: Casa Nem

As documented among many queer, trans and non-heteronormative groups throughout the Americas, Rio's Black LGBT youth have gone to great lengths to affirm themselves by reclaiming and occupying urban spaces. Black LGBT youth in both Brazil and the US have repeatedly demonstrated that 'occupying space has [been] an important queer tactic' (Valentine 2003, 417) – a process known 'reterritorialization of heterosexual space' (Oswin 2008, 90) – and, as we argue, a reterritorialization of anti-Black landscapes. The social nature of these places can vary widely, including liminal spaces like highway underpasses, highly public spaces like central plazas and neighbourhood streets, or conventionally private spaces such as residences and art studios. A small house, a highway underpass or even an alleyway can become a site of celebration, performance, entrepreneurship and political activism, often spilling over into adjacent streets and sidewalks. As Wesley Miranda, a São Paulo-based DJ and artist explained in a personal interview in 2017, 'We can make something out of nothing, anywhere – what matters is that we find each other and create something together for ourselves'.

The potential of Rio's Black LGBT youth to carve out and reclaim urban spaces becomes apparent through the case of Casa Nem. Under a name that itself presents a critique of the gender binary – *nem* means 'neither/nor' – gender-non-conforming people have occupied and transformed a small house in Lapa, a neighbourhood known for nightlife and sex work, into a multipurpose site. While the house's name alludes to diversity and inclusion, its slogan states it clearly, declaring that 'There are faggots, dykes, trans men, and there will also be trans women and *travestis*, too!' (Casa Nem, nd). According to Leonardo, a frequent guest of the house and Black transgender activist, the house had originally been rented and managed by middle-class White gay men who used the house as a residence and arts space. After experiencing a series of transphobic attacks by other house guests (and the complicity of the gay men running the space), the survivors – all transgender women and *travestis* – occupied the house and refused to leave until the tenants agreed to apologize and do right by them in future. Instead of reconciling and creating a more inclusive space, the former occupants vacated the property, leaving the house to the new occupants.

Since then, transgender women (including *travestis*) have become the primary occupants, akin to house 'mothers' and elders, and continue to run the house as a repurposed shelter for transient youth needing shelter from the violence of discriminatory police, clients and bystanders. The house has now become a base for community-building, artistic expression and political organizing. It has also become a refuge for transgender women and *travestis*, many of whom identify as Black and are (in)voluntarily engaged in sex work, largely due to extreme barriers to formal employment and education opportunities. The house's location in Lapa gives many sex workers, whether trans-feminine or as young cis-gender men, direct access to shelter, food and communal support in times of need. The house has been increasingly useful to sex workers as the demand for sex work, and the consequent violent policing of said sex work, have both drastically increased in the midst of recent mega-events and growing tourism.

While Casa Nem began primarily as a reclaimed refuge for homeless trans youth, the house has quickly evolved to serve the multiple collective needs of Rio's marginalized LGBT

Figure 4.1 Front entrance of Casa Nem, stating, 'Respect the girls, queens, transgender women, *travestis* and whores!'.
Source: Devin Oliver.

community – all under the leadership of trans and gender-nonconforming people of various ages, races and walks of life. House programming has used the arts, education and celebration to equitably fundraise and train its LGBT (and largely Black) constituency. A salient example of such programming is PreparaNem. With all-transgender pupils, PreparaNem is a free, intensive preparatory course for the ENEM,[4] the universal entrance exam for all of Brazil's competitive federal universities. Many of the students on the course have either lived in or frequented Casa Nem at some point. The course's LGBT instructors, all volunteers, have attended elite public universities and/or have multiple degrees, equipping the students with the knowledge and strategies needed to overcome the barrier of the ENEM – and all free of charge. Despite its main objective, PreparaNem has better-than-superior test outcomes and has achieved more transgender youth entering university. In 2016, the course organizers and the first class of students collaborated with local photographers and graphic designers to create and sell calendars featuring the students, dressed in attire attributed to their dream profession. Using their aspirations, networks and creativity, the house youths have actively participated and have helped to 'pay it forward' by fundraising for the coming classes.

Another way that the house collective has reclaimed spaces in the service of community needs is through celebrations akin to the 'rent parties' in the Black, working-class neighbourhoods of the US. At these monthly parties, house residents and members fully staff the function, deploying their craft as bartenders, event planners, entertainers, security, DJs or visual artists in order to create a fun, discrimination-free atmosphere for attendees. This intersectional feminist ethos is literally etched into the physical structure of the house, with both interior and exterior walls marked heavily with art and phrases against racism, homo- and trans-phobia, ableism, the criminalization of sex work and HIV, and other forms of violence and discrimination. At the same time, the house fulfils a second objective: to create a mechanism that provides equitable monetary support for house beneficiaries and programming. Rent parties pay the bulk of the R$5000 monthly rent, food and living supplies for house residents; the parties are fully staffed by house members. Non-transgender attendees pay a cover of R$10, while transgender and *travesti* attendees can enjoy the parties free of charge. However, as research partner Leonardo explained, many of the trans and gender-non-comforming attendees help to run the party in some way, taking shifts to sell beer, *catuaba* and *caipirinhas* as a means to raise additional house funds.

Casa Nem has evolved into a haven for political organizing around issues of racism, heteropatriarchy and human rights, as well as an electoral campaign base for socialist PSOL-endorsed candidates Marielle Franco, Marcelo Freixo (runner-up in Rio's mayoral election) and Indianara Siqueira (house mother at Casa Nem and Rio city councilmember candidate). As a political organizing base, Casa Nem has been behind several protests, celebrations and broader organizing networks. Through Casa Nem's community organizing work, Rodrigo Luther King, a 28-year-old trans man from the nearby *favela* of São Carlos, has connected with other Black LGBT activists, in particular with assassinated councilwoman Marielle Franco. Rodrigo describes Casa Nem as:

> a space where I felt like I belonged. It is a shelter for people in vulnerable situations and living on the street. I lived there for a month, I learned, I grew, because inside the *favela* we do not get as much information. There I learned to stand up for myself and to fight for my rights, and now I … can pass this on in a clear way to my brothers, in a way that means people from my surroundings in São Carlos can understand me, embark forward from these ideas, and learn to stand up for themselves as well, seeing what their rights are and fighting for them.
>
> *Maxx 2018*

In times of increased state violence and rollbacks of civil and human rights, youth's experiences at Casa Nem have reaffirmed their existence and have equipped them with the conceptual tools needed to articulate their reasons for resistance. The April 2018 assassination of Marielle Franco demonstrated the benefits of such a support system. As Rodrigo further explains,

> What happened to my sister Marielle can happen to any of us, even more because we are Black … So it is very important for us to have a representative like Marielle was. She was not only having an impact by being a city councilmember, but by representing exactly this excluded minority.
>
> *Maxx 2018*

Conclusion

Violence and resistance are both spatial acts through which Black people must assert their own place in the world. Amid the contradictory landscapes of Rio de Janeiro, Black people must carve out spaces of everyday survival and political intervention. In doing so they create, and remake, place.

What then are the broader implications of Black LGBT youth's placemaking tactics, for geographers and other social scientists and theorists as well as for urban policy practitioners? How do youth's spatial practices challenge us to reconsider what counts as legitimate 'politics', 'community planning' and policy best practices? As researchers and practitioners, we must recognize our own implicit biases when defining politics, stakeholders and policy insights. What we may write off as mere child's play or feminized social work may be the key to restorative, transformative community development.

The placemaking of Black LGBT youth, in the wake of sustained and lethal violent attacks, teaches us to question seriously the conventional power relations between local governments and their constituencies. In Black feminist tradition, the source of critical knowledge needed to design socially equitable places lies not with state officials but rather with the disenfranchised constituents most directly affected by the socioeconomic issue at hand. What this implies for practitioners is that constituents are not mere tokens to provide passive 'buy-in' for the policy programmes predesigned for them; the problem, solutions and budgets must be directly informed by the everyday realities of the most disenfranchised. In essence, the task is the operationalization of what Black feminist scholars call 'standpoint epistemology': critical knowledge production, situated and grounded in everyday lived experience. This knowledge must be centred around the experiential knowledge of non-normative folk.

The infrapolitics of Black LGBT youth's placemaking also constitute what Miraftab (2009) and other critical planning scholars call 'insurgent planning': a broader political project, rooted in the everyday practices of subaltern subjects, that responds to the crises of neoliberal capitalism – crises that set up Rio de Janeiro as a touristic 'afro-paradise'. Everyday acts of resistance and survival through pleasure, play, dance, fashion, performance or popular education create 'hidden transcripts', outside of the realm of formal politics, which challenge us to rethink what counts as strategies of resistance, creative responses to spatial inequalities and social movements overall. The redemptive spaces that youth produce serve as political sites from which youth can exchange knowledge, imagine and enact alternative worlds and position themselves, both within and against the state, in order to address the institutional forces that create precarious livelihoods.

The communal work of Black transgender women at Casa Nem, and related subjects and spaces, could easily be feminized and thus considered more akin to social work than community planning. Such a perspective is masculinist and misogynistic, missing the value of such cultural and economic labour as an intersectional critique of the politics of urban space and Brazilian society. As an intersectional spatial critique, the members of Casa Nem engage in insurgent 'kitchen-table planning' (Sweet 2015), providing a specifically intersectional critique of planning history, theory and practice. Sites like Casa Nem represent variants of a Black queer 'kitchen table', so to speak, from which Black LGBT youth can plan for themselves and their futures. Time and time again, youth use everything within their reach to create redemptive places amid a hostile landscape in ways that acknowledge diverse lived experiences and seek to enact more livable futures for LGBT people of all races, genders and backgrounds.

Notes

1 From this moment, the Ministries of Culture and of Women, Racial Equality, and Youth, which had long championed progressive social policy programmes, were terminated.
2 Right-wing Congressman – and current President-elect – Jair Bolsonaro from São Paulo stated publicly to journalists (in Portuguese), '*Prefiro um filho morto do que homossexual!*'. In a similar anti-Black and misogynistic vein, former Rio de Janeiro State Governor Sérgio Cabral stated that poor (assumedly Black) communities are 'factories that produce marginal people', and that '*Do ventre da mulher negra nasce bandido*' ('thugs are born from Black women's bellies').
3 *Samba* is a musical and cultural tradition with roots deeply grounded in the cultures and religious practices of working-class Black Brazilians. Once banned as a moral hazard, it was appropriated in 1935 as central to the national identity of Brazilians, catalyzing its transformation into what we witness annually as the world-renowned *samba* school parades in the Samba Dome, a key tourist attraction for Rio de Janeiro during Carnival.
4 The *Exame Nacional do Ensino Médio* (ENEM), or the National High School Exam, is the principal barrier to a solid university education in Brazil. Due to disinvestment in public education and the lack of school resources, the vast majority of public high-school students do not achieve a high enough score to gain admission to Brazil's prestigious public universities. The ENEM has impeded the vast majority of poor, Black and transgender youth from entering public universities and thus their socioeconomic mobility.

Key readings

McKittrick, K. 2006. *Demonic Grounds: Black Women and the Cartographies of Struggle.* Minneapolis: University of Minnesota Press.
Oswin, N. 2008. "Critical Geographies and the Uses of Sexuality: Deconstructing Queer Space." *Progress in Human Geography* 32 (1): 89–103.
Smith, C. 2016. *Afro-Paradise: Blackness, Violence and Performance in Brazil.* Chicago: University of Illinois Press.

References

Association of American Geographers (AAG). 2018. #MariellePresente. "Statement on Franco's Assassination and Gendered Racism across the Black Diaspora" presented by Black Geographies, Geographic Perspectives on Women, Latin America, Latinx Geographies, and Sexuality and Space Speciality Groups, in May.
Bailey, M.M., and R. Shabazz. 2014. "Gender and Sexual Geographies of Blackness: New Black Cartographies of Resistance and Survival (part 2)." *Gender, Place & Culture* 21 (4): 449–452.
Binnie, J. 1997. "Coming Out of Geography: Towards a Queer Epistemology?" *Environment and Planning D* 15 (2): 223–237.
Casa Nem. (n.d.). "Sobre/About". Facebook fan page. Available at: www.facebook.com/casanemcasaviva/.
Cohen, C. 1997. "Punks, Bulldaggers and Welfare Queens: The Radical Potential of Queer Politics?" *GLQ: A Journal of Lesbian and Gay Studies* 3 (4): 437–465.
Gilmore, R.W. 2007. *Golden Gulag: Prisons, Surplus, Crisis and Oppression in Globalizing California.* Berkeley: University of California Press.
Inter-American Commission on Human Rights – CIDH. 2015. *Violência contra pessoas Lésbicas, Gays, Bissexuais, Trans e Intersexo nas Américas.* Available at: www.oas.org/pt/cidh/docs/pdf/ViolenciaPessoasLGBTI.pdf (accessed 13 November 2018).
Kelley, R.D.G. 1994. *Race Rebels: Culture, Politics, and the Black Working Class.* New York: Free Press.
Kulick, D. 1998. *Travesti: Sex, Gender, and Culture Among Brazilian Transgendered Prostitutes.* Chicago, IL: University of Chicago Press.
Manderson, L., and M. Jolly. 1997. *Sites of Desire/Economies of Pleasure: Sexualities in Asia and the Pacific.* Chicago, IL: University of Chicago Press.
Maxx, M. 2018. "Jovens lideranças periféricas falam sobre Marielle Franco." *Vice,* 22 March. Available at: www.vice.com/pt_br/article/j5a8mp/jovens-liderancas-perifericas-marielle-franco (accessed 13 November 2018).

McKittrick, K. 2006. *Demonic Grounds: Black Women and the Cartographies of Struggle*. Minneapolis: Minnesota Press.
McKittrick, K. 2013. "Plantation Futures." *Small Axe* 17 (3): 1–15.
Miller-Young, M. 2014. *A Taste for Brown Sugar: Black Women in Pornography*. Durham, NC: Duke University Press.
Miraftab, F. 2009. "Insurgent Planning: Situating Radical Planning in the Global South." *Planning Theory* 8 (1): 32–50.
Mott, L., and E.M. Paulinho. 2017. *Mortes Violentas de LGBT no Brasil: Relatório 2017*. Grupo Gay da Bahia. https://homofobiamata.files.wordpress.com/2017/12/relatorio-2081.pdf.
Nascimento, B. 1982. "Kilombo e memória comunitária – um estudo de caso." *Estudos Afro-Asiáticos* 6 (7): 259–265.
Oswin, N. 2008. "Critical Geographies and the Uses of Sexuality: Deconstructing Queer Space. *Progress in Human Geography* 32 (1): 89–103.
Parker, R. 1999. *Beneath the Equator: Cultures of Desire, Male Homosexuality and Emerging Gay Communities in Brazil*. New York: Routledge.
Perry, K.K.Y. 2013. *Black Women Against the Land Grab: The Fight for Racial Justice in Brazil*. Minneapolis: Minnesota Press.
Pomykata, S. 2013. "The Human Rights Situation of Trans Persons of African Descent in Brazil. Human Rights Brief." *Human Rights Brief*, 31 October. Available at: http://hrbrief.org/hearings/the-human-rights-situation-of-trans-persons-of-african-descent-in-brazil (accessed 13 November 2018).
Richie, B.E. 2012. *Arrested Justice: Black Women, Violence, and America's Prison Nation*. New York: New York University Press.
Smith, C. 2016. *Afro-Paradise: Blackness, Violence and Performance in Brazil*. Chicago, IL: University of Illinois Press.
Stout, N.M. 2014. *After Love: Queer Intimacy and Erotic Economies in Post-Soviet Cuba*. Durham, NC: Duke University Press.
Sweet, E.L. 2015. "Latina Kitchen Table Planning Saving Communities: Intersectionality and Insurgencies in an Anti-immigrant City." *Local Environment* 20 (6): 728–743.
Transgender Europe. 2016. "Transgender Day ofVisibility 2016 – Trans Murder Monitoring Update." *Trans Respect Versus Transphobia Worldwide*, 29 March. Available at: https://transrespect.org/en/tdov-2016-tmm-update (accessed 13 November 2018).
Valentine, G. 2003. "In Pursuit of Social Justice: Ethics and Emotions in Geographies of Health and Disability." *Progress in Human Geography* 27 (3): 375–380.
Walcott, R. 2007. "Black Queer Diaspora." In: *Black Geographies and the Politics of Place*, edited by C. Woods and K. McKittrick, 233–245. New York: South End Press.
Williams, E.L. 2014. "Sex Work and Exclusion in the Tourist Districts of Salvador, Brazil." *Gender, Place and Culture* 21 (4): 453–470.

5

SHIFTING MULTIPLE MASCULINITIES

Alternative views from Japan and Papua New Guinea

Keichi Kumagai

Introduction: globalization and the 'crisis' among men

Today's globalizing world has brought hegemonic masculinity to the fore in diplomacy, as seen in exclusionary nationalist fervour in the US and Europe, as well as Japan and China. Such developments call for an earnest consideration of masculinity. As globalization continues, men and masculinities are undergoing major changes in the Global North and South (Chant and Gutmann 2000; Kimmel 2003, 2013). Global capitalism in the post-Fordist era has given rise to the feminization of labour, thereby diminishing men's identities as breadwinners. The proliferation of non-regular employment has led to declining upward social mobility, loss of status and declining social power (Kumagai 2012). The recent literature explores how such conditions have been behind a crisis in masculinity and men, as well as swelling radical movements built on masculinist politics of xenophobia, White supremacy and terrorism (Kimmel 2003, 605). Such movements deploy masculinity as a form of symbolic capital and restore a safe space for men in crisis, where they may perceive themselves as thriving and in power (Kumagai 2012). Current changes, along with renewed attention to masculinity, beg the question of whether a crisis is truly occurring. In this chapter, I argue that while recent shifts have, indeed, led to increased tension in models of hegemonic masculinity, they have also opened spaces for alternative masculinities to emerge.

Changes in men and masculinities vary by place, culture and society. Masculinity is temporally and geographically contingent (Berg and Longhurst 2003). As such, geography is uniquely equipped to explore not only how masculinities play out in different spaces but how those spaces shape the very experience of masculinity and how it relates to other key dimensions of social relations (Hopkins and Noble 2009). Amid these changes, 'hard' masculinities rooted in strength, violence and the domination of women and others continue to prevail, while 'soft' masculinities (Louie and Low 2003) or alternative masculinities are nevertheless emergent. Here, I present spatially diverse and temporally variable masculinities as I review how men and masculinities are constructed differently, according to the national and local context.

Geography has been criticized as a masculine discipline (Rose 1993), as well as one where the perspective of White men implicitly prevails (Blunt and Dowling 2006; Blunt and Rose

1994; Longhurst 2001). Berg and Longhurst (2003) also note that, while knowledge produced from work in metropolitan Anglo-American sites is regarded as universal, or theory building, the knowledge from work in non-metropolitan 'peripheries' is typically regarded as local, or a mere case study. Instead, I highlight masculinities in Papua New Guinea, where I have been doing fieldwork since 1979, as well as Japan and its neighbouring East Asian countries. They should be understood not as local case studies but as dynamic models for situating masculinities in globalizing societies.

Rethinking concepts of masculinity

What is masculinity? The debate surrounding this very point is far-reaching, as masculinity is hard to define, while the binary construct of male and female is itself undergoing transformation.

There are differences in the lived realities and geographies of masculinity, manliness and maleness. Nevertheless, it is clear that masculinity is constructed relationally to femininity (Connell 1995). What typifies the negation, or opposite, of manliness? In Japan, if a man is deemed to be 'too feminine', he may be constructed as uncertain, inconsistent, unprincipled and effeminate and/or ruled by emotion. In other words, masculinity subsumes positively viewed traits such as decisiveness, rationality and authenticity, and necessarily stands opposite to negative traits or qualities associated with feminine (or degraded) persons. Masculinity is produced from homosociality, which is part of a matrix of homophobia and misogyny (Sedgwick 1985; Ueno 2010). For my purposes here, masculinity – when rooted in the binary construction of gender – may be understood as qualities believed to be inherent to 'men's' bodies, differentiated through society as being opposite to 'women'.

Much has been written on the relationship between men and masculinities on the one hand, and bodies and nature on the other. Feminist geographers and others have exposed the ways in which men, unlike women, are viewed as possessing a rational consciousness that transcends the body and nature and are therefore upheld as creators and proprietors of culture (Longhurst 2001; Ortner 1974; Ortner and Whitehead 1981 ; Rose 1993). Western culture rests on the notion that White men transcend physical conditions and existence through the use of this consciousness, while the same is not presumed for women, ethnic and racial minorities, people with disabilities and/or non-heteronormative identities, the elderly and children, among others (Longhurst 2001). However, masculinity entails the demonstration of physical strength. Sports – that is, competitions involving rivalry and body contact – as well as military and/or physical training and discipline are typically viewed as testimony to one's masculinity (Connell 1995). Such a relationship between masculinities and physical strength is a transcultural and also an historically evolving construct.

In Japan, for example, masculinity has shifted over time. Prior to the modern era, *Bushido* was the individual relationship of loyalty between a vassal and his lord. A chain of such lord/vassal relationships structured the warrior (*bushi*) class. This was quite different in nature from the modern military unit, whose lines of authority and responsibility make it a bureaucratic type of organization (Watanabe 2012, 31–32). Moreover, historically, *Bushido* was only relevant to the *bushi* class of Japanese society. However, the *Meiji* state restructured and appropriated this idea under the modern imperial system to form a new, ideological universe embracing the Japanese state as a household and the emperor as its patriarch, thereby requiring the people (as children of the emperor) to pledge their loyalty to the emperor. In pre-war Japan, there was a cultural loathing of the physically weak ('weakness-phobia'), and the Japanese state cultivated it by tying it into the conscription system to nurture strong, imperialist soldiers, while the media, especially boys' magazines, further amplified it (Uchida 2010).

The literature on sexuality reveals a fluidity between the categories of male and female (Butler 1990), which amplifies queer geographies (Horschelmann and van Hoven 2005; Johnston 2005). At the same time, masculinities and femininities exist as qualities separate from the physical body. For example, the literature covers a spectrum including female masculinity (Halberstam 1998), butch identity (Butler 1990; Sedgwick 1995) and male femininity, embodied in man-breasts (Longhurst 2005), among other things. In addition, additional genders beyond the male and female dichotomy have been recognized in numerous societies around the world, such as with the *hijra* in India and the *fa'fafine* in Samoa (Johnston 2019; Shore 1981).

Rethinking men's violence and bodies from Papua New Guinea: Papua New Guinea's masculine societies

Academic exploration of masculinity aside from its physical body/embodiment remains underdeveloped, at present. More must be understood about how masculinity is constructed, apart from its essential physical qualities. The following case from Papua New Guinea stands as an example of such an inquiry.

While Papua New Guinea has a multicultural society, with over 700 distinct ethno-linguistic communities and pronounced geographic diversity, it is regarded as an extremely patriarchal society dominated by men. As of 2012, only three women were among the country's 109 parliament members. Violence, including sexual assault, is widespread in urban areas and unmarried men constitute the majority of the offenders. Papua New Guinea is also home to the practice of strict gender segregation and antagonism toward women, especially in the Highland regions (Gelber 1986), owing to the belief that masculinity is violated and damaged through excessive contact with women.

However, one should not interpret men's violence in urban space as an expression of the inherent violence of Papua New Guinean men. Rather, they reflect the real impacts of a gradual othering of ethnic groups, or *wantoks,* in urban space. *Wantoks* (a pidgin word derived from 'one talk') are associations formed of people of the same linguistic or geographic roots, and they serve as social safety nets in urban areas. *Wantoks* are crucial providers of housing and jobs in a state that offers insufficient public assistance. Many migrants to Port Moresby, Papua New Guinea's capital, are living in spontaneous (squatter) settlements comprised mostly of *wantoks.* Moreover, against a background of competition and conflict among *wantoks*, violence has become a means for making visible the cohesion within groups (Kumagai 2016).

The spatial structure of the city reinforces these conditions. The urban spatial layout of Port Moresby carries the vestiges of Australian colonial rule, seen in the diffusion of urban zones and the reliance on private vehicles for travel, which is not only inconvenient for ordinary people but hinders the formation of autonomous local communities. Public transport stops running in the evening and the city streets never attract crowds at night. Street vending, mostly undertaken by women, and other informal income opportunities have been suppressed in the name of urban beautification, and the city lacks even competitive sports or movies at evening events. Young men are frustrated by their circumstances and are practically channelled into crime by the lack of public space for interaction and avenues for earning an honest living to support families (Kumagai 2016).

In recent decades, urban life has been imbued with two forms of hegemonic masculinities: the *wantok* system governing grassroots social organizations, and Western masculinist systems (encapsulated in the postcolonial city) governing urban design and management in ways that exclude people's survival strategies. The development of a new 'commons', one that connects people beyond the present *wantok* system, would carry great potential for overcoming both masculinist systems and, in particular, the physical and economic violence that they normalize. Frazer (1992, 126) states that the idea of an egalitarian, multicultural society only makes sense if we suppose a plurality of public arenas in which diverse values and rhetoric participate. However, in urban Papua New Guinea, such socially interactive and inclusive public spaces have been hampered by necessarily exclusive hegemonic violent masculinities in parochialism that perpetuate antagonism and isolation. As a result, alternative public spaces – as well as alternative masculinities – that allow for the inclusion of more women and younger men in social arrangements are thus far absent, yet are as crucial as ever (Kumagai 2016).

Masculinity and the male body in rural Papua New Guinea

Rural Papua New Guinea also shows signs of continuation and change in men and masculinities, not unlike the cities. In coming of age ceremonies, for example, boys endure physically taxing rites of passage to manhood, as a parallel to the strength that women show during childbirth. These rites of passage in the Sepik River Basin of northern Papua New Guinea include skin cutting, one of the world's most grueling rituals, which uses incisions and scarification to produce patterns similar to crocodile scales along the young men's stomachs and backs (see Figure 5.1). While these rites were eliminated from many areas in and around the Sepik River Basin following the arrival of Christian missionaries, a few villages in the Black Water Basin that I have visited almost every year since 1986 continue these rites as the Catholic missionaries

Figure 5.1 Four young men on the morning after the skin-cutting ritual. Kraimbit village, Papua New Guinea, August 2018.
Photograph: K. Kumagai.

had been lenient over their practice. This rite has three cultural meanings: first, one is seen as becoming a man through a process of expelling the blood inherited from one's mother; second, strength of manhood is achieved by enduring the ritual physical pain; third, one is transformed into a brave man and a warrior by carving into one's body the scales of the crocodile, a symbol for the region.

When I observed this ritual in 2018, it was more than simply the performance of a traditional cultural ceremony exclusively for men. One neighbouring community has in recent years increasingly commercialized its traditions to earn income through tourism. Therefore, the re-enactment of the coming-of-age ceremony serves as affirmation of a collective identity and a source of pride for the community, as well as its authenticity. Not only men but the whole community gathers for the ceremony, to act as witnesses and to participate. On the day after the skin cutting, the expressions on the faces of the newly initiated men were most extraordinary: radiant and beaming with a quiet pride. The male elders who led the ritual, the relatives of the initiated and the community as a whole, including women and children, carried warm expressions of gratification. Here, in a region infiltrated by globalization and monetization of its economy, the ritual of masculinity serves as an opportunity for the local society to affirm and re-assert its own cultural fortitude and strength.

Cases of masculinities in urban and rural Papua New Guinea reveal that issues involving men and masculinities are necessarily produced at the intersections of culture, ethnicity, class, economy, society, (post-)coloniality and space and place, among other things, thereby underscoring the importance of intersectionality in their consideration.

Shifting masculinities in Japan and East Asia: corporate masculinity as hegemonic masculinity and its transformation

Japan's post-war economy experienced high rates of growth through the 1980s. During this era, Japanese men were expected to uphold both nation and economy through dedicated labour, with the paradigmatic example being the *kigyo-senshi*, the corporate warriors employed by large corporations. These employees spent most of their time away from families, working long hours and commuting long distances to and from suburban homes. The totality of men's work was complemented by women's sole responsibility as housewives, providing reproductive labour and unpaid domestic work. Men's long hours of hard work were compensated for by a Japanese management system that promised stable lifetime employment and was promoted by a social order that advocated the ideological view of corporation as a family. Consequently, workplaces and *tsukiai* (socializing after work) were distinctively masculine places that have always been hostile to women and made their career advancement difficult. These types of 'corporate masculinity', 'salaryman masculinity' (Dasgupta 2003) and 'salaryman doxa' (Roberson and Suzuki 2003) shaped the everyday experiences of Japanese men and served as the foundation for Japan's hegemonic masculinity at the time.

From the late-1990s, however, non-regular employment rose dramatically, predominantly affecting Japan's youngest generations as they entered the labour force. Further, while women make up a greater proportion of non-regular employees across the whole of the Japanese workforce, the gap between women and men is narrowest in the youngest cohort. These young men in non-regular employment are, by definition, less financially secure than their counterparts in regular employment, and research shows that they are less likely to marry, as well (Fujimori 2010).

In 1999, the Basic Law for a Gender-Equal Society and the revised Equal Employment Opportunity Law were passed. These laws covered the scope of employment and increased women's opportunities for promotion while reducing gender segregation in the labour market, which was a compromise between the femocrats who sought gender equality and neo-liberal labour policy (Ueno 2013). Ultimately, some frustrated young men who felt deprived of job and status entitlements responded by blaming the gender-equality movement for their hardships, such that the 'feminist-phobic' sentiment escalated (Kaizuma 2005). At the same time, gender norms are unchanged; men are still expected to maintain the overarching role of breadwinning to establish their masculinity. These changes have provoked conflict and frustration among the many young men who fail to satisfy the strict codes inherent to the hegemonic image of Japanese masculinity (Kumagai 2012). Hence, new masculinities are appearing, such as *ikumen*, *soshokukei-danshi*, *otaku* and *neto-uyo*, which I discuss further below.

Ikumen

Ikumen[1] refers to men who enjoy childrearing, as well as seeking their own, independent growth. Such a recognition of fatherhood was captured in a project launched by the Ministry of Health, Labor and Welfare (June 2010) to promote men's participation in childrearing and the use of childcare leave. In addition, it came about because of a critical turn against traditional masculinity during Japan's period of high economic growth. In 1980, a movement advocating for paternity leave was established by *Otoko mo Onna mo Ikuji Jikan o Renrakukai* (Child Care Hours for Men and Women Network), known as *Ikujiren*. *Ikujiren* worked tirelessly to make its agenda known in the mass media and to influence government, thereby contributing to the creation of the 1992 Parental Leave Law, which was inclusive of paternity leave (Ishii-Kuntz 2003, 203). While *Ikujiren* members tend to be white-collar workers of high status, and thus not marginal men, they are not associated with hegemonic masculinity (Ishii-Kuntz 2003, 212). Resistance to their platform came from masculinist work environments, which value conformity and uniformity above all else, making it difficult for men to receive paternity leave. In addition, many young men have found that their precarious employment situation has complicated their ability to build partnerships.

Soshokukei danshi

The expression *soshokukei danshi* (literally 'herbivorous boys') first appeared in print in 2006 when columnist Maki Fukazawa coined it to refer young men who are 'unconcerned with sex' in an article on trends in youth culture. Since 2008, this concept has moved on from the pages of women's magazines and gained popularity in a wide range of print and television media. The magazine *Hanako for Men* (published intermittently), launched in November 2009 for *soshokukei danshi* readers, offers key clues to how the media shape the images of these young men. The magazine pages contain numerous articles and photographs depicting life indoors and in Japan. Articles about and images of outdoors or exotic overseas destinations rarely make an appearance; instead, the feature articles are on how to decorate with greenery, such as bonsai, and how to best clean and polish a kitchen, subjects that had never appeared in men's magazines before. In my interview with the magazine's chief editor, he remarked that the word *soshokukei* typically carries negative connotations but, to him, it simply encapsulates the gentleness and grace of Japanese culture. *Soshokukei* are, in fact, a majority among young men at present, and the number of men who consider cleaning, cooking and bonsai to be their hobbies is on the rise. Men are incorporating activities that are considered feminine into their interests.

To illustrate the experiences of young men known as *soshokukei danshi*, I offer the story of 'A'. 'A' is 29 years old (in 2010), single and working as a hairdresser. He said that, should he ever marry, he would like to be with someone who has a good relationship with their family. He explains that, by doing this, he wishes to give back to his own family, which has provided him with support and understanding over the years. He describes himself as footloose and fancy-free:

> We (*soshokukei danshi*) are unable to be commanding. We might actually be closer to women's ideal men. The kinds of men that declare to women, 'I'll support you' are already a minority.

Otaku

The term *otaku*[2] is used to describe youth (commonly men) and the world that they create through their obsession with virtual images from manga, animation and computer games. Although the word *otaku* is generally associated with men, due to the prominence of young female characters in animation and computer games, its cultural universe has spread in recent years to include women and international fans interested in Japanese subcultures.

An *otaku*'s place or home is their own private room, where they keep collections of their favourite products or items. One of their regular gathering places is Akihabara, which has recently been transformed from a shopping district for electronic goods to the holy land for *otaku,* with a heavy concentration of computer goods and game software. Akihabara's very first 'maid café' opened in 2001. One defining characteristic of these cafes is how customers are greeted upon entering: *Okaerinasai goshujin-sama*! ('Welcome home, master/husband').[3] For the *otaku*, they represent a home substitute in which traditional (colonialist) gender relations are played out with flourish in the exaggerated exchanges. However, these maids, who elicit feelings of *moe* (affection) in the *otaku,* appear to be nothing more than real-life portrayals of the popular two-dimensional images of women.

I interviewed a self-professed maid-cafe *otaku*, 'B', in Akihabara. 'B' was in his thirties (as of 2010) and unmarried. He said that he had worked as an analyst, commuting to work in Marunouchi, Tokyo's central business district, while living with his parents in Saitama prefecture. He entered his current job after unsuccessfully applying to work at a games-software company. According to 'B', there is a difference between *otaku* of the 1980s and *otaku* today:

> Men in their fifties are the first generation of *otaku*. Those of us in our late twenties and early thirties are the third generation. We are interested in animation, PC games, music and anything computer related. The games we play are computer-based and tend to feature pretty young girls. College students today are the newest generation. The greatest difference is that they have no reservations about engaging in the *otaku* culture. Their interests revolve around animation, computer games, figurines and dolls, among other things. These dolls are made to look like women, with movable parts, and can be carried around. Whether these dolls are replacements (for women) or not, I think maybe fifty-fifty. Men clearly have sexual feelings. And they crave things with more realism, which is why their desires shift from two-dimensional to three-dimensional objects. Yet, I assume *otaku* women would cynically counter that 'fiction is fiction' and 'reality is reality'.

'B' also participates in gatherings where he can practise speaking in a woman's voice.

> When I like a girl it is not so much because I want to have her, but because I want to be like her. Rather than possess her, I wish I could become her. Actual girls aren't perfect. They have pores and split ends. And their feet aren't so slender. I can't become

> an actual, realistic girl, but when it comes to voice, I can become more like them. This is not real; I'm not saying that I want to become a girl in reality. And this is not a fetish where I want to have their things. This is (creating) a woman of myth. I'm often asked my type, but if I were to pursue my ideal, she would not exist in the real world. In the future, I would like to marry and have children. I'd like to make a family. However, I don't feel like women excite me. As dull as it is, the reality is that I don't have a choice.

Neto-uyo/Petit (neo)nationalists

Neto-uyo, formed from the words *netto* (internet) and *uyoku* (right-wing), refers to people who use the internet to circulate discriminatory, far-right discourse. Owing to the anonymity provided by online spaces, it is difficult to identify who the users really are but, judging from the language they use and the subjects they project, it is conceivable that the majority are relatively young men.

In recent years, and especially since 2000, support for neo-nationalist discourse has spread, owing in part to a string of neoliberal policies and political shows unrolled under the Liberal Democratic Party government[4] of Prime Ministers Koizumi (2001–2006) and Abe (2006–2007; 2012–present). The *neto-uyo* movement has links to numerous outside groups and ideological orientations. These include media that stir up animosity against China and Korea; historical revisionists such as the Japanese Society for History Textbook Reform, who criticize 'self-hating' views of history; neo-fundamentalist politicians who enthusiastically bash 'gender-free' and gender-equality movements; and racists such as the *Zainichi Tokken wo Yurusanai Simin no Kai* (Association of Citizens against the Special Privileges of the Korean in Japan), popularly known as *Zaitokukai,* who take to the streets with hate speech. It is impossible to cover the full range of trends and groups proliferating via the internet, and the word *neto-uyo* is used here as a general term.

At the foundation of the *neto-uyo* ideology is ethno-supremacism and a desire to bring it to bear on reality. This typically links to revisions of history, as well as the romanticization and embrace of wartime Japan, as fanatically expressed in Yoshinori Kobayashi's manga *Gomanism Sengen* (*Declaration of Arrogance-ism*). In his book, the young first-time author, Akagi (2007), yearns for war and violence as he anticipates that these will bring better opportunities to lower-class young men through class and generational reshuffling.

Oguma and Ueno (2003) use interviews and participant observation to glimpse into the lives of youth (mostly male) who congregate at the Kanagawa Office of the Japanese Society for History Textbook Reform. At the Society, words such as 'sensible', 'common sense', 'healthy nationalism', 'common people', 'beacon of the Japanese' and 'tradition' are often heard. These youths see themselves not as adherents to a right-wing ideology but as simply embracing a 'healthy common sense' or realism (Oguma and Ueno 2003, 89–91). What is interesting is the gap between the generation of those who experienced World War II and the youths in their teens or twenties. While the older generation seeks a place to talk of their experiences of war, the younger generation gathers to seek a place to relieve their anxiety over not having a core identity (204).

According to Koichi Yasuda, participants in the *Zaitokukai* see themselves as marginalized and embrace anti-elitist sentiments (Yasuda 2012). As in Oguma and Ueno's observations, Yasuda points out that youths involved in *Zaitokukai* wish to find a place of their own or a home. Yasuda elaborates that, as youths struggle with isolation amid rising labour exploitation, inequality and social division, some find that they can claim identity and belonging through the category of being 'Japanese' (Yasuda 2012, 353). While *neto-uyo* emphasize Japanese strength and the central importance of preserving national integrity, what is required for inclusion in this category of 'Japanese' is not obvious, neither now nor ever. The discourse claiming that Japan is a homogenous country is a fiction created after Japan lost World War II and its colonial territories such as Korea and Taiwan (Oguma 1995). Moreover, the boundaries defining who is Japanese are shifting as international marriages increase.[5]

One may say that the reality of this growing diversity, along with the shifting boundaries of 'being Japanese', has given rise to a reactionary movement of people seeking to reconstruct the Japanese people and reserve the space of Japan as home to, and a place for, exclusively 'pure' Japanese. Because the *neto-uyo* see Japan as their home, they seek and strike out at people whom they believe should be expelled, such as the political left, feminists, foreigners, Koreans and Chinese. Here, they demonstrate fear and anxiety related to mixing and fluidity.

East Asian masculinities and their changes

At present, political and international relations between the East Asian countries (namely China, North Korea, South Korea and Japan) are imbued with hegemonic masculinity. People inherently develop a sense of belonging to certain places, but actively tying such belonging to a nationalist territoriality triggers violence and exclusion. As seen with Japan's *neto-uyo*, feelings of frustration among middle- and lower-class youth in Korea and China, who are sidelined by advancing globalization, have convened with nationalist sentiment and spurred on 'anti-Japanese' movements (Takahara 2006).

Nevertheless, in reality, Japanese, Chinese and Korean societies have been continually gaining more in the way of shared resources linking the countries, as seen in international marriages or popular culture such as Japanese manga, *kanryu* (Korean-style) culture, shown in television dramas, and K-pop (Korean pop music). Interestingly, in each of Japan, Korea and China, today's young generation has a greater interest in 'beautiful' and 'gentle' men. In Korea, where there is universal male conscription and high value is placed on macho men, nevertheless a Korean equivalent of *soshokukei danshi* exists, known as *choi-sik-nam*. Moreover, the handsome young men of K-pop are gaining popularity with Chinese and Japanese women. In China, men known as *nuǎ nán* (暖男) are admired for their gentle nature in relation to not just women but all people.[6]

There is a hope that macro-level politics in East Asia, infused as it is with a 'hard' and hegemonic masculinity and the pursuit of dominance over land and soldiers, is being eroded and redirected by the micro-level emergence of such 'soft' and alternative masculinities.

Conclusion

Masculinities, as situated in society, are multiple and shifting, in line with socio-economic and cultural transformation. In Papua New Guinea, masculinities have served as crucial cultural resources for group cohesion and personal power among men competing for leadership. This is visible in many aspects of urban multicultural society, where the lack of national integration is pronounced. For example, 'hard' masculinity, expressed in part through violence based on *wantok* solidarity, tends to heighten hostilities and acts of 'othering' among ethnic groups, while women are often rendered victims of violence and urban crime. The ideas propping up a hegemonic masculinity that defies the construction of an urban commons or space for mutual interaction have been passed down for generations through (post-)colonial urban planning and have produced today's discrete urban communities.

This hegemonic masculinity shapes a space suited to crime, together with social frustration, especially among the unmarried men who form the majority of criminal offenders and who should have been the key agents in the development of a new urban way of life. On the other hand, in Papua New Guinea's marginalized rural villages, physical bodies and strength have acquired alternative meanings. Masculinity, as recreated through traditional initiation rituals, has enhanced community solidarity not by using violence or hostility but rather by fulfilling self-confidence and resisting marginalization to amplify the community's collective sense of autonomy under globalism.

In Japan, drastic economic transformations under globalization have made it difficult for young men to attain the ideals of hegemonic masculinity through becoming breadwinners. New types of masculinities are emerging, such as *ikumen*, *soshokukei-danshi*, *otaku* and *neto-uyo*, of which the first three reveal a reconsideration of – or distancing from – the hegemonic masculinity of older generations. However, with the exception of *ikumen*, who seek equal partnership with women regarding childcare, these new masculinities lack avenues for personal growth and positive social interaction. Instead, they are characterized by tendencies to remain confined in limited places or home spaces, such as private rooms, houses and nations like Japan, actively excluding others (Kumagai 2015). Interestingly, soft masculinities such as *soshokukei-danshi* are appearing and gaining popularity in both Korea and China. While foreign relations between Japan, Korea and China are confrontational, the popular cultures of these East Asian countries are infiltrating each other and increasing sympathies.

Masculinities, as well as their meanings and contexts, vary and change over time in both Papua New Guinea and Japan. The (re)construction of new masculinities for overcoming hegemonic masculinities is closely related to space and place, as well as time. Competition, domination and emotional suppression (hooks 2004) were once characteristic qualities of masculinity and, as such, must be addressed.

The ideas of economic rationalism underlying global market capitalism are a masculine fiction that presumes that all individuals are self-reliant, thereby disregarding the critical importance of care. In the neoliberal era, 'care' is once again being increasingly transferred to families. If gender norms do not change, care work will continue to be thrust onto women and a minority of men, who may be compelled to leave the waged labour force to fulfill these roles (Fineman 2004), or female migrant workers, who sacrifice caring for their own families to serve wealthier families in the Global North. Care, and the socialization of care, are seen as extensions of 'women's work' and are, therefore, undervalued and disproportionately pushed onto feminized labour forces. Remedying this involves improving the status of care workers and ensuring that men actively participate in care work. This is not only for the purpose of achieving gender equality in the cost of care but also to enable us – by making child and elderly care universal – to collectively reclaim the 'right to care' (Cornell 2002; Muta 2006), which has been simultaneously diminished and made more onerous through its unequal distribution. Doing so would bolster the public's awareness surrounding the importance of our affective and, thus, ethical attunement (Cornell 2002, 67) to one another. By looking within ourselves, we can collectively reckon with the limitations imposed by binary gender and discover possibilities apart from a social order founded on male domination and female care.

It is incumbent on all people to resist masculine constructions that perpetuate exclusion and domination and, instead, to enhance mutual interaction and collaboration. Modern Western logocentrism, which is also masculine, tends to obscure diverse realities through the use of monolithic categories (Ito 1993). As Ito, a pioneer scholar of Japanese men's studies, has suggested, we ought to resist classifying others or allowing ourselves to be classified as such (Ito

1993, 197). Instead, men and masculinities should be reconstructed as agentic actors in a diverse, fluid movement to 'multifly' (multiply and diversify) gender – and masculinity and femininity.

Notes

1 *Ikumen* is a compound formed from *ikuji* (childcare) and men, and is a play on the expression *ikemen* (good-looking men).
2 The word *otaku* is a formal term meaning 'you', and is used regularly within the community. The term was first used by Akio Nakamori in 1983 when referring to the mania surrounding the comic market.
3 The Japanese word *goshujin-sama*, or *shujin*, which means master, is also traditionally used to refer to one's husband with respect.
4 While Koizumi advanced privatization and labour deregulation, the younger generation enthusiastically supported his government because of his overt assertions that he would destroy vested rights, including those of public servants. Koizumi also promoted Japanese nationalism by visiting the Yasukuni shrine (a religious symbol of World War II), despite Korean and Chinese disapproval.
5 The percentage of international marriages of all marriages rose to 6.1% in 2006. Many of those couples were of Japanese men and Asian (particularly Chinese) women.
6 Masculinity in China once held as its ideal men who possessed both literary (文*wen*) and martial (武 *wu*) accomplishments. Also, as found in traditions of Chinese imperial exams, the ideals of masculinity encompassed both the spirituality of literary endeavours and the physicality of martial endeavours, unlike Japanese masculinity formed from a heavy focus on the martial. Greater examination would be required to determine whether or not these historical masculinities have been carried into the present.

Key readings

Kam, L., and M. Low. 2005. *Asian Masculinities: The Meaning and Practice of Manhood in China and Japan.* London: Routledge Curzon.
Kumagai, K. 2012. "Floating Young Men: Globalization and the Crisis of Masculinity." *HAGAR (Studies in Culture, Policy and Identities)* 10 (2): 157–165.
Roberson, J.E., and N. Suzuki, eds. 2003. *Men and Masculinities in Contemporary Japan: Dislocating the Salaryman Doxa.* London: Routledge Curzon.

References

Akagi, T. 2007. *Wakamono o Migoroshi ni Suru Kuni: Watashi o Senso ni Mukawaserumono wa Nani ka? (The Country that Leaves its Youth Behind: What Turned me toward War?).* Tokyo: Sofusha.
Berg, L.D., and R. Longhurst. 2003. "Placing Masculinities and Geography." *Gender, Place and Culture* 10 (4): 351–360.
Blunt, A., and G. Rose. 1994. *Writing Women and Space: Colonial and Postcolonial Geographies.* London: Guildford Press.
Blunt, A., and R. Dowling. 2006. *Home.* Oxford: Routledge.
Butler, J. 1990. *Gender Trouble: Feminism and Subversion of Identity.* New York: Routledge.
Chant, S., and M. Gutmann. 2000. *Mainstreaming Men into Gender and Development: Debates, Reflections and, Experiences.* Oxford: Oxfam Working Papers.
Connell, R.W. 1995. *Masculinities.* NSW, Australia: Allen & Unwin.
Cornell, D. 2002. *Between Women and Generations: Legacies of Dignity.* New York: Palgrave.
Dasgupta, R. 2003. "Creating Corporate Warriors: The 'Salaryman' and Masculinity in Japan." In: *Asian Masculinities: The Meaning and Practice of Manhood in China and Japan,* edited by K. Louie and M. Low, 118–134. London: Routledge Curzon.
Fineman, M.A. 2004. *The Autonomy Myth: A Theory of Dependency.* New York: New Press.
Frazer, N. 1992. "Rethinking the Public Sphere: A Contribution to the Critique of Actually Existing Democracy." In: *Habermas and the Public Sphere,* edited by C. Calhoun, 109–142. Cambridge, MA: MIT Press.
Fujimori, K. 2010. *Tanshin Kyuzo Shakai no Shogeki (A Shock by the Change of Society with Rapidly Increasing Single Persons).* Tokyo: Nippon Keizai Shimbun Shuppansha.

Gelber, M.G. 1986. *Gender and Society in the New Guinea Highlands: An Anthropological Perspective on Antagonism toward Women*. Boulder, CO: WestView Press.

Halberstam, J. 1998. *Female Masculinities*. Durham, NC: Duke University Press.

hooks, b. 2004. *The Will to Change: Men, Masculinity, and Love*. New York: Washington Square Press.

Hopkins, P., and G. Noble. 2009. "Masculinities in Place: Situated Identities, Relations and Intersectionality." *Social & Cultural Geography* 10 (8): 811–819.

Horschelmann, K., and B. van Hoven, eds. 2005. *Spaces of Masculinities*. Abingdon: Routledge.

Ishii-Kuntz, M. 2003. "Balancing of Fatherhood and Work: Emergence of Diverse Masculinities in Contemporary Japan." In: *Men and Masculinities in Contemporary Japan: Dislocating the Salaryman Doxa*, edited by J.E. Roberson and N. Suzuki, 198–216. London: Routledge.

Ito, K. 1993. *Otokorashisa-no-Yukue* (*Directions for Masculinities*). Tokyo: Shinyosha.

Johnston, L. 2005. *Queering Tourism: Paradoxical Performances at Gay Pride Parades*. London: Routledge.

Johnston, L. 2019. *Gender, Sex and Place: Gender Variant Geographies*. London: Routledge.

Kaizuma, K. 2005. "Taikō Bunka toshiteno Han-Femi-Nachi: Nihon ni okeru Dansei no Shūenka to Bakkurasshu" ("Anti-'femi-Nazis' as a Culture of Resistance: The Marginalization of Japanese Males and Backlash"). In: *Jenda-Furii Toraburu: Basshingu Genshō o Kenshō Suru* (*Gender-Free Trouble: Verifying the Phenomenon of Bashing*), edited by Ryōko Kimura, 35–53. Tokyo: Gendai Shokan.

Kam, L., and M. Low, eds. 2003. *Asian Masculinities: The Meaning and Practice of Manhood in China and Japan*. London: Routledge Curzon.

Kimmel, M. 2003. "Globalization and its Mal(e) Contents: The Gendered Moral and Political Economy of Terrorism." *International Sociology* 18 (3): 603–620.

Kimmel, M. 2013. *Angry White Men: American Masculinity at the End of an Era*. New York: Nation Books.

Kumagai, K. 2012. "Floating Young Men: Globalization and the Crisis of Masculinity." *HAGAR (Studies in Culture, Policy and Identities)* 10 (2): 157–165.

Kumagai, K. 2015. "Gendai Nihon no shakaikeizai henka to dannsei-sei no henyo o meguru shiron: basho to homu no siten kara (Changing Masculinities in Japanese Youth: Viewing from Pace, Home and Domesticity)." *Journal of Gender Studies* 17: 87–98.

Kumagai, K. 2016. "Can *Wantok* Networks be Counter-publics? Development and Public Space in Urban Papua New Guinea." *Ochanomizu Chiri* (*Annals of Ochanomizu Geographical Society*) 55: 49–58.

Longhurst, R. 2001. *Bodies: Exploring Fluid Boundaries*. Abingdon: Routledge.

Longhurst, R. 2005. 'Man-breasts': Spaces of Sexual Difference, Fluidity and Abjection." In: *Spaces of Masculinities*, edited by K. Horschelmann and B. van Hoven, 165–178. Abingdon: Routledge.

Louie, K., and M. Low. 2003. *Asian Masculinities: The Meaning and Practice of Manhood in China and Japan*. London: Routledge Curzon.

Muta, K. 2006. *Jenda Kazoku o Koete: Kingendai no Sei/Sei no Seiji to Feminizum*. (*Beyond the Gendered Family: Modern Bio/Sexual Politics and Feminism*). Tokyo: Shinyosha.

Oguma, E. 1995. *Tan'itsu Minzoku Shinwa no Kigen: "Nipponjin" no Jigazo no Keihu (The Origin of the Myth of a Homogenous Nation: A Genealogy of Japanese Self Images)*. Tokyo: Shinyosha.

Oguma, E., and Y. Ueno 2003. *Iyashi no Nashonarizumu: Kusa no Ne Hoshu Undo no Jissho Kenkyu (A Healing Nationalism: A Substantive Study into Grassroots Conservative Movements)*. Tokyo: Keio University Press.

Ortner, S.B. 1974. "Is Female to Male as Nature is to Culture?" In: *Women, Culture and Society*, edited by M.Z. Rosald and L. Lamphere, 68–87. Stanford, CA: Stanford University Press.

Ortner, S.B., and H. Whitehead, eds. 1981. *Sexual Meanings: The Cultural Construction of Gender and Sexuality*. New York: Cambridge University Press.

Roberson, J.E., and N. Suzuki, eds. 2003. *Men and Masculinities in Contemporary Japan: Dislocating the Salaryman Doxa*. London: Routledge Curzon.

Rose, G. 1993. *Feminism and Geography: The Limits of Geographical Knowledge*. Cambridge: Polity Press.

Sedgwick, E.K. 1985. *Between Men: English Literature and Male Homosocial Desire*. New York: Columbia University Press.

Sedgwick, E.K. 1995. "Gosh, Boy George, You Must Be Awfully Secure in Your Masculinity!" In: *Constructing Masculinity*, edited by M. Berger, B. Wallis, and S. Watson.

Shore, B. 1981. "Sexuality and Gender in Samoa." In: *Sexual Meaning*, edited by S. Ortner and H. Whitehead, 192–215. Cambridge: Cambridge University Press.

Takahara, M. 2006. *Fuangata Nashonarizumu no Jidai: Nikkanchu no Netto Sedai ga Nikushimiau Honto no Riyu (The Age of Insecure Nationalism: The Real Reason Why Japan's, Korea's and China's Net Generations Hate Each Other)*. Tokyo: Yosensha.

Uchida, M. 2010. *Dainippon Teikoku no "Shonen" to "Danseisei": Shonen Shojo Zashi ni Miru Wikunes Fobia (Youth and Masculinity in Imperial Japan: Weakness-phobia According to Boys and Girls Magazines)*. Tokyo: Akashi Shoten.
Ueno, C. 2010. *Onna Girai: Nippon no Misojini (Hating Women: Misogyny in Japan)*. Tokyo: Kinokuniya shoten.
Ueno, C. 2013. *Onnatachi no Sabaibaru Sakusen (Survival Strategies of Women)*. Tokyo: Bungeishunju sha.
Watanabe, H. 2012. *A History of Japanese Political Thought, 1600–1901*, translated by D. Noble. Tokyo: International House of Japan.
Yasuda, K. 2012. *Netto to Aikoku: Zaitokukai no Yami wo Oikakete (Internet and Patriotism: Searching for the Zaitokukai in the Dark)*. Tokyo: Kodansha.

6

DISABLED WOMEN ACADEMICS RESHAPING THE LANDSCAPE OF THE ACADEMY

Nancy Hansen

Introduction

As a human geographer with cerebral palsy who uses crutches to aid my mobility, I am highly accustomed to and comfortable with the physical uniqueness of my non-conformist body (Hansen and Philo 2007). I have always been keenly aware of the peaks and valleys of my workplace landscape. I routinely work my way around, through and over the workspace environment, all the time trying to make my activities appear seamless in the academy, where my presence is, in many ways, unexpected (Titchkosky 2011). Although some progress has been made in addressing the needs of disabled students in post-secondary education, the same cannot be said for those of disabled academics (Butler and Parr 1999; Chouinard 1996, 2010). In many ways, the academy has yet to understand that disability and impairment can be found on both sides of the desk (Hansen 2010). To date, comparatively little is known or has been written about disabled academics in the academy (Chouinard 1996, 2018; Kosanic et al. 2018). Far less has been written about disabled academics who are women (Chouinard 1996; Foster 2017). Indeed, while writing this chapter I have spoken with numerous disabled women academics in various educational institutions who said they would not write about life in the academy for fear of jeopardizing their career prospects.

Established academy practices often fail to account for the possibility that members of that elite group may *themselves* have disabilities (Chouinard 1996; Grigely 2017; Stannett 2006). With the increasing corporate focus within universities, the nature of the academy is changing (Berg and Seeber 2016). The *business* of education is coming to the forefront (Foster 2017). The presence of 'otherness' in any form is often regarded as unsettling to the 'regular', 'natural' ebb and flow of established patterns and practices of daily living (Jones and Calafell 2012). The academy is an institution built on long-established traditions and, in the main, a reflection of majority beliefs and values of usually those who are White, male, hetero-normative and non-disabled (Eased 2004). There is a certain level of comfort in dealing with the known (Hansen and Philo 2007), what Eased (2004) refers to as 'cultural cloning', replicating what has gone before. Indeed, the academy, as an institution with institutionalized practices, has functioned in such a way for decades. The majority of academic posts still reflect this (Berg and Seeber 2016).

Although some incremental gains have been made in recent years, most full professorships still belong to White, straight men (Berg and Seeber 2016). To talk simply in terms of singular identities negates the complexity of what is actually happening here (Foster 2017). We are not simply one block identity; there is a constant intersection of two or more identities at any given time. I am not simply a disabled person, although disability does impact my life: I am also a White, straight, upper-middle-class woman and, at one time or another, two or more of these identities intersect (Butler 1990; Butler and Bowlby 1997; Peters 1996; Vernon 1996, 1998). Policies and practices – although modified with regard to language – remain largely unchanged, imbued with privileged majority understandings of the academy and the world around it (Vellani 2013). For those of us who have been marginalized or 'othered', for whatever reason, by that ever-present but invisible binary against which we are measured (in ability, gender, sexuality and race), the academy can be an elusive and isolating place (Chouinard 1996, 2010). While some degree of understanding has been achieved in relation to gender, sexuality and race (Kobayashi 1994), as yet the concept of ableism –or disableism (disability discrimination), as it is known in some regions – is not readily recognized or well understood (Foster 2017; Vellani 2013). Indeed, established academy policies and practices are often imbued with ableism (Foster 2017).

While a social citizenship rights-based approach to disabled students in the academy has gradually moved forward, with improved physical access and policy accommodation in recent years – fuelled to some degree by advancements in human rights legislation – this is not the case for disabled academics (Grigely 2017). We have as yet to realize appreciable gains in this arena and, in many ways, we remain exotic, disruptive strangers, constantly adapting to an alien landscape. What follows is an ongoing chronicle of personal adaptation.

Navigating the neo-liberal landscape

The neo-liberal academy is a strange place: rarefied, elite, individualized and isolating, all at the same time (Berg and Seeber 2016). The academic community is fast becoming an industrialized complex, meaning that corporate business practices are being adopted (Berg and Seeber 2016). As the academy is *modernized,* underscoring the necessity for resourcefulness, office support personnel are being removed and academics are required to undertake increasing amounts of time-consuming administration (Berg and Seeber 2016). As part of this reformation process, various types of bulk software have been adopted for everything from travel to office supplies and room bookings. The software is rarely accessible and, as a result, we are forced to develop (if possible), individualized work-around methods in order to use it, taking up yet more time and energy.

Academic ability is regularly judged in tandem with an arbitrary to exceptional level of physicality and energy, which is another manifestation of an exceptional status provision (Berg and Seeber 2016). With the presence of disabled faculty members often treated as exceptional or problematic (Dolmage 2017), I am faced with having to demonstrate everyday management skills regularly, to legitimize myself. I am always factoring in extra time (on my own private timeline) in order to manage in a workspace illustrative of hyper non-disability (Dolmage 2017). Whereas non-disabled academics are provided with access to the academy without this added hurdle, it is as if disabled academics must provide this insurance as added proof of the right to be in this rarified space (Hansen 2008).

A more troubling element is the arrogance often brought about by the unquestioned power and privilege within the academic institution itself, which effectively silences those individuals located on the margins (Berg and Seeber 2016; Sang 2017). Scholars maintain that the invisibility experienced by disabled women and other socially disadvantaged groups reflects a lack of 'space' within the academy (Vellani 2013). They are not 'permitted' to become academics (Grigely 2017). As a woman academic with a disability, I am placed in an interesting situation, diminished within the academy and having a fragmented visibility in a social context (Hansen 2008; Sang 2017). On the academic's side of the desk, disability workplace accommodation is, in large measure, non-existent. At present, very few colleges and universities provide support to their disabled faculty or staff (Chouinard 1996; Grigely 2017), unlike the rights-based options presented to faculty that are reframed as health, well-being and resilience (Berg and Seeber 2016; Foster 2017). This perspective is individualized and focused on rehabilitation (Grigely 2017). The need for disability accommodation (physical and/or attitudinal adaptations) may be confused with some form of personal weakness (Foster 2017; Sang 2017). This version of academy appears to place a premium on speed, quantification, chronic adaptability and a constant readiness to work, coupled with an unrelenting sense of urgency (Berg and Seeber 2016). This places disabled academics at a distinct disadvantage (Sang 2017). Furthermore, this widespread negative association may account for a genuine reluctance to self-identify as having a disability (Grigely 2017; Sang 2017).

The academic community is fast becoming an industrialized complex in which corporate business practices are adopted (Berg and Seeber 2016).

The work of 'passing' in the academy

An acceptable shape, size, colour, height, sexuality and physicality is culturally mediated (Butler 1990; Young 1997). The perception of the body as a 'working' machine is a fixture buried deep in the collective social consciousness (Hansen 2002). Western society's acceptance and understanding of disability are driven by history, economics and productivity, based on dogma rather than wisdom or facts (Davis 1995). The discourse is of the 'average' capacity of the body as machine, in an industrial society (Hahn 1989). Our understandings of the machine would appear to be somewhat superficial, as is Western society's understanding of disability and impairment (Foster 2017). Based on this current logic, disabled women represent extreme bodily non-conformity and non-male imperfection (Hansen 2002).

Consequently, similar to Chouinard and Grant (1997), at times I feel 'nowhere near the project', not because I feel out of place or that I do not belong but rather because I have to contend with an institutional environment whose policies and programmes that are aimed at disability 'management' are rife with embedded ableism.

Preconceptions of physical incapacity and access may be used as a convenient means to mask the deeper underlying social insecurities or objections about sharing non-disabled spaces with disabled people. It is almost as if the presence of disabled people compromises or contaminates these higher education elite 'public' non-disabled spaces (Hansen 2002, 2008). This unfounded belief would appear to be well entrenched, given British Chancellor Philip Hammond's recent assertion that disabled people are responsible for Britain's stagnant economy (Slawson 2017). The Chancellor appeared in front of the Treasury select committee to answer questions on the November budget and said that 'high levels of engagement in the workforce, for example of disabled people', may be one of the factors keeping down UK productivity levels (Slawson 2017, np).

Traditionally, Western society concerns itself with visible, physical bodily mechanics. Agility, mobility, dexterity, speed and spatial relations are usually considered solely in the context of the individual's condition. Individual physicality and work activity are looked at solely within the structural confines of the public physicality of the work environment. Factors that are usually considered include the availability of: wheelchair-accessible toilets; ramped building access; adapted workstations; and technological devices. While these elements do play a significant role in day-to-day working life, they should not be viewed in isolation from other equally important but less visible considerations, such as anxiety, fatigue or pain (Chouinard 2010; Foster 2017).

Fear of job loss or misperceptions of inability or a lack of professionalism stemming from negative social attitudes and stereotypes often lead disabled academics to minimize or conceal the reality of their disability or chronic illness in the academic arena. 'Passing' and resisting can be both physically and psychologically demanding:

> 'I spend a lot of time – hours and hours – advocating for myself' (William Peace). This may be one reason so few graduate students with disabilities pursue professional careers in academe: The task of having to advocate for yourself is a thankless professional obligation.
>
> *Grigely 2017*

Time–space continuum

Time is a genuine concern for me, as the nature of my disability is such that any physical activity takes significantly more time and effort than it would for a non-disabled academic. As a disabled woman and an academic, I make visible what is commonly not visible, because disabled women are 'not visible'. The planning, organizing and spacing of my daily activity are of paramount importance if I am to achieve the appearance of moving through a largely inflexible (at least, commonly perceived as such) environment with apparent ease. Frequently, it is so much a part of the process of life that it takes place at an unconscious level. Those unrecognized and/or unquestioned concepts that are accepted as intrinsic are the ones that provide the greatest difficulty for disabled people, being presented as the 'natural' order of things (Hansen and Philo 2007).

Timing, organization and spatial awareness gain an enhanced significance when my movements are measured and my energy levels are distinct. Curbs, steps, doors, signs and toilets assume a crucial importance. These seemingly mundane factors have a profound impact on the daily lives of disabled academics. Taken-for-granted facilities of non-disabled life experience are often unexpected premiums, for disabled people.

According to Aimee Louw, a Canadian access advocate speaking on radio (CBC Radio 2017), ableism (privileging non-disability) is pervasive in that it is present in more places than is governed by legislation. The use of legislation does not guarantee a successful outcome. Most disability within the academy, just as it is outside the academy, is invisible in that it is not readily apparent to the observer; that is, most impairments do not require assistive devices such as wheelchairs, canes or crutches.

I would argue that the current *business* focus of education makes it less likely that the academy will readily recruit or accommodate disabled academics and, with the steady encroachment of neo-liberal ideology on the academy, it is now increasingly unlikely to do so. Production, throughput and publication are now the markers we live by. Louw (CBC Radio 2017) argues that the speed–time focus that is now paramount in all sectors means that it is much more difficult to negotiate times and timelines for the work to be done. Laptop computers may be bolted into place on trolleys, making it imperative to adopt a standing, right-handed position, hence there are times when I need a right-handed *driver* (student or assistant) to use the computer during seminars.

The centralized travel-booking software at the university does not have a disability access option to book business travel, consequently I must book through a travel agent. Similarly, there is no access option for the centrally booked classrooms, although my name is now the default for the faculty-booked classroom option. Accessible transportation on campus is through student accessibility services. This is yet another example of the lack of preparation for, or expectation of, disabled faculty and staff.

Accessing campus services is often not a straightforward process, and proximity may not solve the associated access difficulties, whether structural or bureaucratic (Grigely 2017). Hence, my non-conformist physicality (mobility disability) may be perceived as disruptive simply by being present. Access is often treated as an add-on or afterthought. The growth of metrication has meant that these creative efforts remain largely unknown in striving to 'pass', minimize and make it look seamless (Diaz Merced 2016).

My situation is not unique. Sally French (1994, 157–158), a university lecturer with a vision impairment, has expressed similar concerns:

> Another vital issue for me is time. Equal opportunities policies never tackle this issue, even though it is so crucial to visually impaired and other disabled people. … My job involves a lot of reading and my reading speed is slow. I have never calculated the exact hours I work. Perhaps if I did I would have to acknowledge how little time there is left for me beyond my employment.

We are expected to adjust and to produce to a non-disabled timeline (not taking into account the extra time required to shoulder daily activities or to risk being labelled as somehow lacking in professional competence). Despite my best efforts, I am sometimes perceived as being 'severely disabled', although I do not see myself in that way.

Often it is the disabled academic who is expected to possess the necessary disability information, on the job, and this situation can create additional pressures in the workplace. Many disabled academics must contend with the embedded social practices, cultural attitudes and prejudices of non-disabled co-workers and employers. Well-meaning colleagues have said to me that watching me 'struggle' makes them appreciate their life experience. I think this has been meant as a compliment.

There is a false sense of security in accepted knowledge that, in itself, imposes a form of social distancing. More disturbing, however, is a fixation on the personal mechanics of disability in the workplace, which is the focus on individualized 'difference'. There are few, if any, support systems for disabled people in the academic workplace, and often employees are alone and isolated (French 1994). Indeed, most universities remain poorly equipped to address the needs of disabled academics (Grigely 2017; Sang 2017). Consequently, disabled academics work very hard to adapt to the established academic workplace infrastructure (Kosanic et al. 2018). There is often a constant scrambling of 'private' and 'public' here that may cause unease, arising whenever there is a failure to conceal something 'private' in 'public' (Hansen and Philo 2007). The result leads some to set up boundaries against the offending person/body, while others start to treat that person/body as if they were in their intimate private circle (Hansen 2008).

My four-leggedness (use of crutches) is readily visible and almost always on public display. My colitis, however, remains for the most part outwardly invisible, yet it is always present for me. I'm reminded that 'this is your body, you must find some way to live within all of it' (Coates, 2015, 12). Ironically, I work in a geographic region recognized as having one of the highest incidences of ulcerative colitis in the world (Blanchard et al. 2001), yet this reality is largely unknown and rarely publicly discussed. In order to manage better the possibility of intestinal volatility, I have adapted my wardrobe and diet accordingly. Furthermore, the visible measured speed and spacing realities of cerebral palsy, which are more 'publically understood', can aid in maintaining the invisibility of colitis, which is a condition that is hardly ever understood. By chance, my classroom is located near a toilet, but my body is highly visible; the heart of the matter is obviously to do with appearances and how these can be managed. I understand the necessity of 'keeping up appearances' or of looking as 'normal' as possible in order to make my way through the world on a daily basis with the least amount of difficulty. There is irony in that this kind of action may at times create a greater personal obstacle and create further complications. In some ways, it may also reflect a larger internalized social pressure to conform to what is expected and thereby to keep what is often perceived as a personal 'problem' private. Chronic illness sometimes involves what is conventionally defined as a 'private' concern (Hansen 2008).

Although some progress has been made, there appears to be a reluctance to theorize disability and gender, as though they were a lesser element of feminist studies (Erevelles 2001). Disabled feminist scholars have assumed a leadership role in this process. The task has not been an easy one, as Chouinard (1999, 146) explains:

> Struggles to empower the disabled are waged largely by people who have committed themselves to the struggle for disability rights; and because of this their personal passages are out of step with those in power. Perhaps this is what struggles to make space for disabled women and men in academia and beyond are all about: *disrupting* spaces of power and privilege to the point where those who dominate such spaces are forced to recognise the difference that disabling differences make, and are confronted with the roles they play in sustaining such cultural oppression.

Disabled academics are restricted in the wider socialization process, and non-disabled values remain just below the surface (Sang 2017). While some disabled women academics may have found a 'voice', much of the disability experience remains shrouded in the community of disabled women and caught up in the public/private dichotomy of tentative social acceptance (Chouinard 1999; Wendell 1996, 1997). Certain aspects of disability are thought to be too disturbing for mainstream sensibilities. Distorted preconceptions frequently permeate discussions concerning autonomy, independence, choice and pain, often reflecting more the fears of the non-disabled than the reality experienced by disabled women (Garland-Thomson 1997; Wendell 1997).

Theoretical in-depth analysis and synthesis remain limited. To fully develop perspectives on disability requires that disablement and impairment are carefully integrated within the race, class and gender continuum in order to reflect a more inclusive intersectionality. A broader spectrum of analysis is necessary in order to achieve a complete analytical framework, incorporating the numerous factors that impact on gender and disability issues. Validation of the lived experience of disabled academics is pivotal to this process, coupled with a recognition that disability and impairment are important elements in the academy's reality (Sang 2017).

Conclusion

The amount of 'work' required in order for many academics to 'go to work' in the academy is daunting (Berg and Seeber 2016). To do so with a disability adds yet another layer of complexity (Sang 2017). Constantly working against physical, social and emotional barriers exacts a heavy personal toll (Diaz Merced 2016). Rarely do disabled academics speak of the realities of their daily work–life experience (Kosanic et al. 2018). Indeed, I have never addressed the invisible aspects of my chronic condition before in the context of my own personal space. I have great privilege, by virtue of community membership. Therefore, I have a responsibility to be forthright in my disclosure.

As yet, the basic tenets and structures of academia have not been seriously examined or modified in relation to the physical, social and psychological needs of disabled people (Foster 2017; Grigely 2017). For the most part, we find ourselves in an alien academic landscape. We navigate the spaces created by incremental shifts that focus on form over a substance that we have not created (Kosanic et al. 2018). The presence of disabled women is gradually reshaping the academic landscape. Although we adopt various survival mechanisms to function in mainstream society, for the most part we do not construct the broader economic, political, social and cultural contours of the world (Chouinard 1999). Our bodies are regularly marginalized, while others are privileged.

Our work environment can often present challenges that may not necessarily be accepting or accommodating of disabled academics' impairments and chronic conditions (Sang 2017). We spend a great deal of energy to create and maintain our physical 'space' in the academic workplace (Foster 2017). Small changes make a big difference. So, we take heart from small victories and we recognize that we are getting there by degrees.

Key readings

Chouinard, V. 2010. "'Like Alice Through the Looking Glass' II: The Struggle for Accommodation Continues." *Resources for Feminist Research* 33 (3–4): 161–178.

Grigely, J. 2017. "The Neglected Demographic: Faculty Members with Disabilities." *The Chronicle of Higher Education*, 27 June. Available at: www.chronicle.com/article/The-Neglected-Demographic-/240439.

Hansen, N., and C. Philo. 2007. "The Normality of Doing Things Differently: Bodies, Spaces and Disability Geography." *Tijdschrift voor Economische en Sociale Geografie* 98 (4): 493–506. doi:10.1111/j.1467-9663.2007.00417.x.

References

Berg, K., and B. Seeber. 2016. *The Slow Professor: Challenging the Culture of Speed in the Academy*. Toronto: University of Toronto Press.

Blanchard, J., C. Bernstein, A. Wajda, and P. Rawsthorne. 2001. "Small-area Variations and Sociodemographic Correlates for the Incidence of Crohn's Disease and Ulcerative Colitis." *American Journal of Epidemiology* 154 (4): 328–335.

Butler, J. 1990. *Gender Trouble: Feminism and the Subversion of Identity*. London: Routledge.

Butler, R., and S. Bowlby. 1997. "Bodies and Spaces: An Exploration of Disabled People's Experiences of Public Spaces." *Environment and Planning D: Society and Space* 15 (4): 411–433.

Butler, R., and H. Parr, eds. 1999. *Mind and Body Spaces: Geographies of Illness, Impairment and Disability*. London: Routledge.

CBC Radio. 2017. The Next 150, Episode 1: Aimee Louw: "To Me, the Future is Accessible... I Have a Very Good Imagination." *The Walrus Talks*, 7 June. Available at: www.cbc.ca/player/play/966292547984 (accessed 10 June 2017).

Chouinard, V. 1996. "Like Alice Through the Looking Glass: Accommodation in Academia." *Resources for Feminist Research* 24 (3–4): 3–11.

Chouinard, V. 1999. "Life at the Margins Disabled Women's Explorations of Ableist Spaces." In: *Embodied Geographies Spaces, Bodies and Rites of Passage*, edited by E. Kenworthy Teather, 142–156. London: Routledge.

Chouinard, V. 2010. "'Like Alice Through the Looking Glass' II: The Struggle for Accommodation Continues." *Resources for Feminist Research* 33 (3–4): 161–178.

Chouinard, V. 2018. "'Like Alice through the Looking Glass' II: The Struggle for Accommodation Continues." In: *Untold Stories, A Canadian Disability History Reader*, edited by N. Hansen, R. Hanes, and D. Driedger, 320–338. Toronto: Canadian Scholars' Press.

Chouinard, V., and A. Grant. 1997. "On Not Being Anywhere Near the Project: Revolutionary Ways of Putting Ourselves in the Picture." In: *Space, Gender and Knowledge: Feminist Readings,* edited by L. McDowell and J. Sharp, 147–170. London: Arnold.

Coates, T. 2015. *Between the World and Me.* New York: Spiegel and Grau.

Davis, L. J. 1995. *Enforcing Normalcy: Disability, Deafness, and the Body.* New York: Verso.

Diaz Merced, W. 2016. "Metrics of Productivity." *Chronically Academic.* Available at: https://chronicallyacademic.blogspot.com/2016/08/metrics-of-productivity.html.

Dolmage, J. 2017. *Academic Ableism: Disability and Higher Education.* Ann Arbor: University of Michigan Press.

Erevelles, N. 2001. "In Search of the Disabled Subject." In: *Embodied Rhetorics Disability in Language and Culture,* edited by J. Wilson and C. Lewiecki-Wilson, 92–111. Carbondale: University of Southern Illinois Press.

Essed, P. 2004. "Cloning Amongst Professors: Normativities and Imagined Homogeneities." *NORA: Nordic Journal of Women's Studies* 12 (2): 113–122. doi: 10.1080/08038740410004588.

Foster, D. 2017. "The Health and Well-Being at Work Agenda: Good News for (Disabled) Workers or Just a Capital Idea?" *Work, Employment and Society* 32 (1): 186–197. DOI: 10.1177/0950017016682458.

French, S. 1994. "Equal Opportunities... Yes, Please." In: *Mustn't Grumble,* edited by L. Keith, 154–160. London: Women's Press.

Garland-Thomson, R. 1997. *Extraordinary Bodies: Figuring Physical Disability in American Culture and Literature.* New York: Columbia University Press.

Grigely, J. 2017. "The Neglected Demographic: Faculty Members with Disabilities." *Chronicle of Higher Education* 27 June. Available at: www.chronicle.com/article/The-Neglected-Demographic-/240439 (accessed 1 November 2017).

Hahn, H. 1989. "Disability and the Reproduction of Bodily Images: The Dynamics of Human Appearances." In: *The Power of Geography: How Territory Shapes Social Life,* edited by J. Wolch and M. Dear, 370–388. London: Unwin Hyman.

Hansen, N. 2002. "Passing Through Other People's Spaces: Disabled Women, Geography, and Work." Unpublished PhD dissertation, University of Glasgow, Scotland. Available at: http://theses.gla.ac.uk/3245/.

Hansen, N. 2008. "A Delicate Balance. Chronic Conditions and Workspace." In: *Dissonant Disabilities. Women with Chronic Illness Explore Their Lives,* edited by D. Driedger and M. Owen, 131–148. Tornoto: Women's Press, Toronto Press.

Hansen, N. 2010. "Making it Work: Disabled Women Shaping Spaces in Education and Employment." *Equity Matters,* 30 November. Available at: www.ideas-idees.ca/blog/making-it-work-disabled-women-shaping-spaces-education-and-employment (accessed 1 October 2017).

Hansen, N., and C. Philo. 2007. "The Normality of Doing Things Differently: Further Thoughts on Bodies, Spaces and Disability Geography." *Tijdschrift voor Economische en Sociale Geografie* 98 (4): 493–506.

Jones Jr., R., and M. Calafell. 2012. "Contesting Neoliberalism Through Critical Pedagogy, Intersectional Reflexivity, and Personal Narrative: Queer Tales of Academia." *Journal of Homosexuality* 57 (7): 957–981.

Kosanic, A., N. Hansen, S. Zimmermann-Janschitz, and V. Chouinard. 2018. *Researchers with Disabilities in the Academic System.* American Association of Geographers Op-Ed, 1 September. Available at: http://news.aag.org/2018/09/researchers-with-disabilities-in-the-academic-system/. doi: 10.14433/2017.0042.

Kobayashi, A. 1994. "Colouring the Field: Gender, Race and the Politics of Field Work." *Professional Geographer* 46 (1): 73–80.

Peters, S. 1996. "The Politics of Disability Identity." In: *Disability and Society: Emerging Issues and Insights,* edited by L. Barton, 215–234. London: Longman.

Sang, K. 2017. *Disability and Academic Careers.* Edinburgh: Heriot Watt: University.

Slawson, N. 2017. "Philip Hammond Causes Storm with Remarks about Disabled Workers." *The Guardian,* 7 December 2017. Available at: www.theguardian.com/politics/2017/dec/07/ philip-hammond-causes-storm-with-remarks-about-disabled-workers.

Stannett, P. 2006. "Disabled and Graduated: Barriers and Dilemmas for the Disabled Psychology Graduate." In: *Disability and Psychology: Critical Introductions and Reflections,* edited by D. Goodley and R. Lawthom, 71–83. New York: Palgrave Macmillan.

Titchkosky, T. 2011. *The Question of Access: Disability, Space, Meaning.* Toronto: University of Toronto Press.

Vellani, F. 2013. *Understanding Disability Discrimination Law Through Geography.* London: Routledge.

Vernon, A. 1996. "A Stranger in Many Camps: The Experience of Disabled Black and Ethnic Minority Women." In: *Encounters with Strangers: Feminism and Disability,* edited by J. Morris, 47–68. London: Women's Press.

Vernon, A. 1998. "Multiple Oppression and the Disabled Peoples Movement." In: *The Disability Reader: Social Science Perspectives*, edited by T. Shakespeare, 201–210. London: Cassell.
Wendell, S. 1996. *The Rejected Body: Feminist Philosophical Reflections on Disability*. New York: Routledge.
Wendell, S. 1997. "Towards a Feminist Theory of Disability." In: *The Disability Studies Reader,* edited by L. Davies, 260–278. New York: Routledge.
Young, I.M. 1997. "The Scaling of Bodies and the Politics of Identity." In: *Space, Gender and Knowledge: Feminist Readings*, edited by L. McDowell and J. Sharp, 218–231. London: Arnold.

7

GENDER AND THE DISCIPLINE OF GEOGRAPHY

Case studies of relational networks of support in Western academia

Martina Angela Caretta and Avril Maddrell

Introduction

Networks can be thought of in a variety of ways, for example: as rail or road maps; made up of key nodes and links; the dendritic pattern of a family tree; or a spider's web, centred on a single focal point and entity. However, in reality these sets of connections and relations (in the widest sense) are often much more irregular than the diagrams suggest. The iconic topological map of the London Tube is an example par excellence of an irregular network represented in simplified form. The public map represents the 'need to know' information for users and gives the impression that only the public stations and their connecting tracks exist, and that these are all on the same level and broadly equidistant. In fact, the underground train lines are a complex system of multilayered tracks, with maintenance-only and disused tunnels and an irregular pattern of stations. Simply travelling on the Tube, moving between the different lines via multiple escalators and stairs, quickly belies the apparent simplicity of the public map.

Nonetheless, even if they are more irregular, dynamic and contingent than at first glance, human networks are ultimately about connections, relationality and – to continue the London Tube analogy – getting to your destination, wherever that may be. For the purposes of this chapter, networks are understood as associations of academics linked through common interests, goals and needs, within and beyond the discipline of geography and the institutions they are affiliated to. Spatialities, including these networks, can be place-based or virtual; they are embodied and experiential, with emotional-affective dimensions shaped in part by intersectional identity attributes, such as alma mater, subdiscipline, gender, race-ethnicity, socio-economic class and sexuality (Maddrell et al. 2016). Within the social sciences, networks have typically been studied as sets of interactions through the lens of actor network theory, but this approach has been criticized for its lack of attention to the power relations and politics inherent within networks. Therefore, understanding the mechanisms of gendered networks requires a feminist analysis of power relations (Quinlan 2012).

Here, we explore the nature and practice of networks, sensitive to their gendered power relations. Such networks can be inclusionary or exclusionary, both of which will be discussed below. The chapter moves from reflections on historic and contemporary networks and their gendered implications for the discipline of geography, to a discussion of how networks can be deployed to enhance (or hinder) gender equality within geography. The chapter is grounded in the specific geographies of our experience and research, principally through studying and working in the UK, Western Europe and the US, as well as engaging with wider personal and formal international networks, such as the International Geographical Union (IGU) and the international journal *Gender, Place and Culture* (*GPC*). Indeed, it was through *GPC* that we authors first encountered each other (as editor and author), followed by meeting in person at an IGU Gender Commission conference. The term 'Western' has been adopted here as a flag of convenience to represent these largely Anglophone confluences of contexts and experiences, but with a conscious awareness of its limitations.

Historic examples of support networks for women in British geography

Various histories of geography as a scholarly endeavour have highlighted disparities in gendered access to geographical knowledge, institutions, opportunities to publish, employment, and so on, in both the US and the UK (Bell and McEwan 1996; Domosh and Morin 2003; Maddrell 1997, 2009). Geographical societies are key examples of networks that have been central to bringing people together to learn about, engage in and promote geographical knowledge. In the case of UK geography's founding institution, the Royal Geographical Society, from its foundation in 1830 until 1913 women were barred from full membership, with the exception of a cohort of 22 who were admitted during a brief window from 1892 to 1893. The Society's council agreed to admit women in 1892, but this was challenged by those who thought the presence of women as anything other than occasional guests would undermine the status of the subject, still in its infancy in universities, and feminize the homosocial space that facilitated particular types of masculine networking seen as antithetical to women's interests and capabilities: political, commercial and exploratory (Bell and McEwan 1996; Blunt 1994; Maddrell 2007, 2009).

Despite this exclusionary tactic in the institutional hub of British geography, women were central to the establishment of the modern discipline of geography in British universities during the late nineteenth and early twentieth century. They were vital to boosting first the attendance of initial lecture courses and then the registered student numbers when the first accredited courses in geography were established (the diploma in geography at the University of Oxford in 1902, and first-degree courses at Liverpool University and University College Aberystwyth, both in 1917). These early courses catered primarily for schoolteachers, as did the associated summer schools, and both accredited courses and short summer schools were grounded in physical geography, regional and field studies. A number of leading early twentieth-century male geographers appointed women to posts in these new geography departments (e.g. Halford Mackinder, A.J. Herbertson, H.J. Fleure and L.W. Lyde) (Maddrell 2009). Successful women students were recruited as junior staff (typically teaching practical classes before being given responsibility for lectures and fieldwork). For instance, at the University of Oxford, Nora MacNunn was appointed as a demonstrator then lecturer (1906–1935) and Eva Taylor as a research assistant to A.J. Herbertson. Taylor went on to become the first UK female Professor of Geography (1930, University of London). Oxford, as a key hub for the establishment of geography as a university subject in the UK, was a central node in the national geographical community network through its open-access summer schools and field courses, as well as its formal taught programme and alumni.

Fieldwork is widely recognized as structuring social relations, as well as developing geographical knowledge and skills, and a number of important fieldwork networks arose, particularly through the summer schools and related activities which were organized for mixed groups of women and men students/teachers in the early 1900s. During the first half of the twentieth century, university summer-school activities were curtailed as more accredited courses were established and faculty staff time became more stretched. This prompted more geographers to turn to the sociological Le Play Society field group, fieldtrips organized by the Geographical Association (GA) (largely catering for schoolteachers) and, later, the Geographical Field Group (GFG). These trips demanded varying levels of physical fitness and geographical field skills, and were attended by women, men and married couples.

Vignette of a networker

Joan Fuller, who lectured in geography at the University of Nottingham from 1942 to 1967, was a networker par excellence. She built on her school-teaching experience prior to her appointment to the university by establishing links with the university's Institute of Education and by chairing the Nottingham branch of the GA. After completing four years of temporary appointment during the war years, she was formally appointed as a lecturer on a permanent contract in 1946. Later, she initiated the departmental newsletter for staff and alumni that she produced annually for 11 years, drawing on student news sent to her in Christmas cards. Her last was produced in 1969, shortly after retirement. The newsletter represented a record of staff's and alumni's professional and personal activities, circulating information about appointments, marriages, births, and so on. This newsletter can be seen as a form of emotional labour that maintained and perpetuated the network of connection between staff and past students – something which would be the envy of many alumni officers today.

Fuller also played a central role in mid-century field-study networks. She had participated in the Le Play Society field study student group led by her head of department, K. C. Edwards, then continued with the GFG from 1946 onwards. During the 1950s and 1960s, Fuller led two British and at least six foreign field courses under the auspices of the GFG. A full half of all GFG courses were run by women during this period (Maddrell 2009; Wheeler 1967). The field courses were numerically dominated by women participants (e.g. Fuller's 1957 trip to the Jura, Switzerland, included 15 women and three men). Travelling and working as a group for a shared purpose was both supportive and efficient, with combined skills facilitating a detailed and rigorous geographical study of a particular locality, at home or abroad. While there was an overall leader, these trips were very communitarian, with sub-team leaders, e.g. for geomorphological or economic geography data collection. Ultimately, a collaborative report would be co-produced and circulated throughout the whole group membership, informing many an evening's reading, geography class and public lecture. These international field courses were largely populated by graduates, likely to be teachers and junior lecturers, as well as those working in other occupations, and can be seen as a form of serious leisure (Rojek 2000) and continuous professional development. They were reportedly favoured by women because they offered opportunities for like-minded people to travel and undertake demanding international study as part of a supportive group. The example of the GFG also highlights the role and disciplinary legacy of women academics whose influence on their peers was significant yet often intangible – not something that could be measured by publications. The next section turns to another 'GFG' in British geography.

From the Women and Geography Study Group to the Gender and Feminist Geography Research Group debate in RGS–IBG 2012–2013

As geography became an established discipline within the British university system in the twentieth century, it was professionalized and masculinized, with women teachers and researchers in the minority, and only four women had been appointed to the promoted post of professor by 1970 (Maddrell 2009). It is in this context that the Women and Geography Study Group (WGSG) was established, and it was formally recognized by the Institute of British Geographers (IBG) in 1982. At a time when the academy was experienced by many women as a patriarchal hegemony (Leigh McDowell 1990), the group's stated aims welcomed women and men from the geographical community in order to provide 'a forum for geographers interested in gender issues', as well as 'a network for women in a male dominated discipline' (Wyse 2013). These aims were pursued via conference sessions, collaborative publications, reading weekends and social events, each element contributing to the development of feminist geography scholarship, networks and women's sense of belonging within the academic community of geographers. The group's activities continued after the merger of the IBG with the Royal Geographical Society in 1994.

The WGSG's work helped to counter and mitigate patriarchal networks within geography, for example offering a space that recognized and valued feminist scholarship through organizing academic retreats, dedicated workshops, conference sessions and related publications, as well as functioning as an alternative forum for social interaction and the sort of informal mentoring accessed by some male graduates through masculine academic networks (e.g. social sporting activities with senior male colleagues [Maddrell et al. 2016, 53]). Historically, credentials have played a vital role in validating women's presence, work and influence within the discipline (e.g. the attainment of degrees and publications in recognized journals). The establishment of the WGSG within the UK's professional body for academic geography, not uncontested, was vital to creating a professional, metaphorical and epistemological space for women, gender studies and feminist analysis. Publications were also significant in materializing a tangible network of validated scholarship on which other feminist geographers could build, including joint publications by founder members and pivotal collective publications published under the group's name (e.g. WGSG 1984, 1997). However, despite publications, conference sessions, effective advocacy, and so on, the WGSG's name, stated aims of gender-inclusive membership, focus on researching and teaching gender issues, and providing a network for women were, for some, conflictual in practice. For example, by 2012, a growing number of British geographers were researching masculinities, sexualities and wider gender issues, but new researchers were not necessarily engaging with the WGSG and only four men were affiliated to the group. This prompted a debate about the group's name and purpose, summarized in a set of short pieces published in *Area*, including those favouring a more inclusive name (Brickell et al. 2013), those in support of the existing name (Browne et al. 2013), those who could see useful points across both sides of the debate (Evans et al. 2013), and men who did/did not feel excluded by the WGSG's name (Hopkins and Jackson 2013).

Such debates are not unique, similar debates have occurred in other feminist networks, particularly around the dialectics and praxis of challenging exclusively male networks while maintaining explicit or implicit women-only networks. Ultimately, the WGSG became the GFG (Gender and Feminist Geography) Research Group in 2013, when the group also convened a host of '100+' activities at RGS–IBG's annual conference to mark the centenary of women's formal membership in the RGS. This is not the place to revisit those debates in detail, but it highlights the value of geography's feminist networks' ongoing commitment to being reflexive about what is taken as axiomatic and who is explicitly/implicitly included, excluded or privileged within their own activities – and the associated implications. Likewise, within the day-to-day arena of universities as workplaces, those who organize departmental committees, faculty social events, and national or international conferences have an obligation to reflect on the ethos, 'spirit' and implicit cultures of the shifting networks of inclusion-exclusion, power and opportunity thereby created, as explored below in relation to early-career support networks.

Current networks of support for early-career women in geography

Professional networks in geography remain fundamental, currently, to counteract, mitigate and guide early-career geographers, particularly women, through the increasing neoliberalization of the academic job market (Oberhauser and Caretta 2019). In fact, female PhDs in geography who have received their doctorate since the 2000s are faced with a job market very different from that of their predecessors (Pitt and Mewburn 2016; Thwaites and Pressland 2017). Numerous authors (Bosanquet 2017; Caretta et al. 2018; González Ramos and Vergés Bosch 2013; Strauss 2013) have spoken about the ongoing neoliberalization of academia and its gendered consequences. Most research-led universities in the Global North are increasingly assuming managerial leadership patterns, whereby productivity is held to be the highest consideration (Martínez Alemán 2014). In this context, tenure-required productivity is typically measured in terms of: 1) the number of peer-reviewed articles published in high impact-factor journals; 2) the value of external research grants acquired; and, in some cases, 3) the average grade received on students' evaluation of teaching. Faculty and researchers are held to account against this scale each year and are given a grade themselves, ranging from unsatisfactory/poor to excellent. Having a few years of poor student feedback has been shown to lead to academics being laid off (Moosa 2018; SIGJ2 Writing Collective 2012; Taylor and Lahad 2018).

Precarity of employment is a key issue. Contingent/adjunct/postdoctoral appointments have increased exponentially in recent decades. For instance, 'while the total number of college instructors in the United States has more than doubled since 1980 from 675,000 to 1.4 million, the total number of tenure track faculty only increased by 22 percent' (Griffey 2016, np). Thus, for those in temporary research or teaching positions, unable to ensure a constant inflow of external grants essentially to acquire the funds to self-employ themselves in the long term, secure employment is a chimera. In this context, geographical mobility and flexibility are key to enable them to take up positions (Archer 2008; González Ramos and Vergés Bosch 2013; Pitt and Mewburn 2016; Thwaites and Pressland 2017), possibly including successive annual or short-notice moves to short-term/soft-funded posts in geographically disparate locations (Caretta et al. 2018; Thwaites and Pressland 2017).

These changes to employment practices in the neoliberalized university sector have increased job precarity, notably – but not only – for early-career scholars, with gendered implications (Maddrell et al. 2016; Strauss 2013). Women are often ill-placed to compete in this highly precarious and temporary job market, for several reasons. First, there is evidence of an intrinsic bias against women in science hiring (Moss-Racusin et al. 2012), which impacts on geography posts. Secondly, women's reproductive age tends to be closely aligned with the time when they acquire their PhD and/or they enter the job market, resulting in family responsibilities that may not be easily reconcilable with a flexible and geographically mobile career (González Ramos and Vergés Bosch 2013). Finally, in keeping with gendered social roles, women tend to do more service and teaching than their male colleagues, neither of which are given the same recognition in the hiring process as publishing and securing grants, thereby placing them at a disadvantage

(Flaherty 2017; Maddrell 2009). Geography is no stranger to these ongoing challenges to gender equality, and women, as in other disciplines, are under-represented in appointments in all stages of career compared to their male counterparts (Caretta and Webster 2016; Maddrell et al. 2016; Mott and Cockayne 2017).

For these reasons, mentoring and networks of support by peers and senior colleagues are crucial in keeping women in academia and on a path to tenure (Darwin and Palmer 2009; Macoun and Miller 2014; Mountz et al. 2016; Mullings et al. 2016; Oberhauser and Caretta 2019). In recent decades, circles of feminist geographers have been strong and consistent advocates for women's retention in our discipline by challenging the normalization of productivity metrics as the main indicator of academic success (Domosh 2015; Kobayashi 2006; Maddrell et al. 2016; Mountz et al. 2016; Mullings et al. 2016; Winkler 2000).

Numerous initiatives and networks have emerged internationally and locally to support early-career women geographers. In fact, multiple sources of support are needed in current times to enter and stay in academia, remain motivated and to continue on a career trajectory that does not necessarily look any brighter or less precarious in the longer term (e.g. an analysis of male and female representation in senior UK geography posts showed an approximately four-to-one gender discrepancy in 2010 (Maddrell et al. 2016)). Multiple-source 'mosaic' mentoring, as it is defined in the literature (de Janasz and Sullivan 2004), can provide effective support not only for early-career scholars but potentially for those at *all* career stages who would benefit from guidance, peer support, peer-monitored targets, mentoring and more on a range of fronts: from research strategies, effective teaching and service roles, to balancing work and personal life.

Informal mentoring is a common phenomenon in feminist geography. Junior female geographers often encounter their first informal mentors when attending international conferences, such as the International Geographical Union Gender Commission conferences, the American Association of Geographers Geographic Perspectives on Women specialty group (GPOW) or the Royal Geographical Society–IBG Gender and Feminist Geography Research Group (GFGRG) meetings and workshops (Evans and Maddrell 2019; Hardwick 2005; Huang et al. 2017; Oberhauser and Caretta 2019). In these informal contexts, at conference sessions, meals and excursions, committee meetings and book receptions, the conversations with like-minded colleagues can include the personal and can break down junior/senior boundaries. These occasions are paramount for early-career female geographers to start building an international network of scholars who work on related topics and to form the basis for collaborations, which can enhance their visibility, productivity and, hence, career. Additionally, these settings have increasingly become avenues for the creation of solidarity and support circles in the context of the ongoing neoliberalization of academia across national and age boundaries (Kern et al. 2014; Mountz et al. 2016; Parizeau et al. 2016).

Formal mentoring, that is, when a junior faculty member is paired up with a mentor through an institutional scheme, is something that few women in geography can count upon. Often it is the purview of big research universities, where the tenure-track system is still considered an asset to the institution (de Janasz and Sullivan 2004). In the case of the UK, some departments and universities are now developing mentoring schemes when competing for national accreditation for gender equality through the Athena Swan award, which is a prerequisite for key research fund applications. However, mentoring schemes can represent little more than cynical institutional lip service to improving gender equality and, in fact, may reinforce the exploitation of women's 'service' roles if such schemes fail to resource and recognize mentors' and mentees' time within their workload and evaluations of service, continuous professional development and leadership. In some US schemes, junior faculty members receive funding to work with a senior mentor, for a fixed period, often at the beginning of their tenure-track period, to strategize about the best course of action to attain tenure by ticking all the boxes for research, service and teaching (Berk et al. 2005; Oberhauser and Caretta 2019).

Mentoring is not an essentialist feminine attribute: men can make good mentors (as shown in the discussion of the historical examples earlier in this chapter); and not all women make good mentors. Nonetheless, this type of mentoring relationship has been shown to be particularly effective in developing female faculty members' productivity and self-confidence, especially if the mentor is also a woman (Allen and Eby 2004). In this sense, there is an added value in having a mentor whom the mentee can relate to, and who can provide insights into how to balance a similar successful career trajectory with a personal life, including possibly having children (Bosanquet 2017).

Mentoring circles are another format of peer support. They include women in different phases of their career, ranging from PhD candidates to tenured faculty, who meet regularly in groups to discuss the personal and systemic challenges that they face in their daily work life (Darwin and Palmer 2009; Macoun and Miller 2014; Mullings et al. 2016). There are different ways in which these circles can be structured and function. They can be as informal as a weekly support lunch group or as formal as a series of seminars around gendered strategies to succeed through tenure. These groups are a way to hold their members accountable, with weekly check-ins on work-related goals and tasks to be accomplished before the next meeting. Additionally, they provide support as a forum to share strategies to facilitate a work–life balance, especially pre-tenure. These groups are deemed to be especially beneficial if members vary in career stage, as they create a sense of community among women within a discipline regardless of age, career stage, and so on. On the one hand, PhD candidates become aware early on of the long-term challenges they will face if they stay in academia (Berg 2015); on the other hand, senior tenured faculty members, often in positions of power, can remain in touch with the struggles of contingent and junior faculty members, and be their advocates and role models for rising to the higher ranks in the university. While formal and informal mentoring arrangements are very much dependent on the consistency and initiative of both mentor and mentee, the advantage of a mentoring circle is that it generally does not require much preparation by the participants, and its leadership, and logistical matters can be shared (Mountz et al. 2016).

E-mentoring is becoming increasingly popular (Headlam-Wells, Gosland and Craig 2005; Single and Single 2005). This mode of mentoring does not simply include long-distance informal relationships but incorporates a mentoring circle on an online platform and, at times, involves a weekly phone call. An example is the US National Center for Faculty Diversity and Success, which pays for a semester-long group of four junior faculty members, across disciplines, to be coached by a mentor. This programme, called the Faculty Success Program, is a mentoring circle whereby a mentor leads a group of mentees, but it is also an 'accountability' group among mentees, who register their writing time and their goals on an online platform (Facultydiversity.org). Although this programme might be perceived as quite impersonal and, given its time and logistical constraints, may not consolidate mentoring relationships, it still provides mentees with a valuable set of time-management tools and productivity strategies, which will facilitate meeting the targets required in neoliberal academia.

Conclusion

Historical and contemporary studies highlight the barriers faced by, and opportunities for, women in the academy. Barriers are particularly acute at present, in the context of the neoliberalization of the university sector in many Global North 'Western' countries, particularly those experiencing reduced external funding or increased staff-student ratios as a result of economic recession or other shifts in funding streams. While these pressures are not unique to women, they can have a disproportionate effect on women within the academy, especially those in precarious early-career positions, as well as for those of all genders with caring responsibilities for dependents (Maddrell et al. 2016). This chapter has demonstrated ways in which geographers are using existing formal and informal feminist and other professional networks in order to counter disadvantage and an uneven playing field in employment and career opportunities. However, not all women are equally disadvantaged, nor all men equally advantaged. The research on networks indicates the need for networks to be responsive to specific contingent needs and the dynamic shifting complexities of their constituents, including multifaceted discrimination that reflects intersectional identity, for instance women of colour, those requiring flexible working hours in order to meet caring responsibilities or religious obligations, and so on.

Feminist geographical networks are frequently grounded in and dependent upon generous leadership, yet activities such as mentoring or skills-sharing do not necessarily have to reinscribe the senior–junior hierarchies or narrow assumptions concerning gender issues or who is or is not a feminist. This chapter has highlighted professional enrichment through participation in networks, and how mentoring and networks of support by both peers and senior colleagues have been crucial in keeping women in geography and on a path to secure employment and mid-career progression. These networks have opened up spaces for mutual support and collaborations, often across countries, building solidarity among feminist geographers at different stages of their career, counteracting and mitigating the personal consequences of an entrenched neoliberal system of knowledge production. While not a panacea, at their best these genuinely collaborative networks can do much more. Participation in networks can provide opportunities to practise leadership as well as to gain support; they can act as a seedbed for innovation, as well as collaborative research projects and publications; they can act as a space to establish a collective voice and influence; they can reach beyond international boundaries and the academy to become sites of effective praxis, contributing to work to address wider inequalities.

Key readings

Faria, C., B. Falola, J. Henderson, and R.M. Torres. 2019. "A Long Way to Go: Collective Paths to Racial Justice in Geography." *Professional Geographer* 71 (2): 364–376. doi:10.1080/00330124.2018.1547977.

Fem-Mentee Collective: A.L. Bain, R. Baker, N. Laliberté, A. Milan, W.J. Payne, L. Ravensbergen, and D. Saad. 2017. "Emotional Masking and Spillouts in the Neoliberalized University: A Feminist Geographic Perspective on Mentorship." *Journal of Geography in Higher Education* 41 (4): 590–607. doi: 10.1080/03098265.2017.1331424.

Oberhauser, A.M., and M.A. Caretta. 2018. "Mentoring Early Career Women Geographers in the Neoliberal Academy: Dialogue, Reflexivity, and Ethics of Care." *Geografiska Annaler: Series B, Human Geography* 101 (1): 1–12. https://doi.org/10.1080/04353684.2018.1556566.

References

Allen, T.D., and L.T. Eby. 2004. "Factors Related to Mentor Reports of Mentoring Functions Provided: Gender and Relational Characteristics." *Sex Roles* 50 (1–2): 129–139.

Archer, L. 2008. "The New Neoliberal Subjects? Young/er Academics' Constructions of Professional Identity." *Journal of Education Policy* 23: 265–285. doi: 10.1080/02680930701754047.

Bell, M., and C. McEwan. 1996. "The Admission of Women Fellows to the Royal Geographical Society, 1892–1914: The Controversy and the Outcome." *Geographical Journal*, 162 (3): 295–312.

Berg, L.D. 2015. "Rethinking the PhD in the Age of Neoliberalization." *GeoJournal* 80 (2): 219–224. doi: 10.1007/s10708-014-9574-6.

Berk, R.A., J. Berg, R. Mortimer, B. Walton-Moss, and T. Yao. 2005. "Measuring the Effectiveness of Faculty Mentoring Relationships." *Academic Medicine* 80 (1): 66–71.
Blunt, A. 1994. "*Travel, Gender and Imperialism. Mary Kingsley and West Africa.*" New York: Guilford.
Bosanquet, A. 2017. "Academic, Woman, Mother: Negotiating Multiple Subjectivities During Early Career." In: *Being an Early Career Feminist Academic*, edited by R. Thwaites and A. Pressland, 73–91. London: Palgrave Macmillan. Available at: http://link.springer.com/chapter/10.1057/978-1-137-54325-7_4.
Brickell, K., A. Maddrell, A. Martin, and L. Price. 2013a. "By Any Other Name? The Women and Geography Study Group." *Area* 45 (1), 11–12.
Browne, K., J. Norcup, E. Robson, and J. Sharp. 2013b. "What's in a Name? Removing Women from the Women and Geography Study Group." *Area* 45 (1): 7–8. doi:10.1111/area.12007.
Caretta, M.A., D. Drozdzewski, J.C. Jokinen, and E. Falconer. 2018. "'Who Can Play This Game?' The Lived Experiences of Doctoral Candidates and Early Career Women in the Neoliberal University." *Journal of Higher Education in Geography* 42 (2): 261–275. https://doi.org/10.1080/03098265.2018.1434762.
Caretta, M.A., and N.A. Webster. 2016. "'What Kept Me Going Was Stubbornness'? Perspectives from Swedish Early Career Women Academics in Geography." *Investigaciones Feministas* 7 (2): 89–113. https://doi.org/10.5209/INFE.52911.
Darwin, A., and E. Palmer. 2009. "Mentoring Circles in Higher Education." *Higher Education Research & Development* 28 (2): 125–136. https://doi.org/10.1080/07294360902725017.
de Janasz, S.C., and S.E. Sullivan. 2004. "Multiple Mentoring in Academe: Developing the Professorial Network." *Journal of Vocational Behavior* 64 (2): 263–283. https://doi.org/10.1016/j.jvb.2002.07.001.
Domosh, M. 1991. "Towards A Feminist Historiography of Geography." *Transactions of the Institute of British Geographers* 16: 95–104.
Domosh, M., and K.M. Morin. 2003. "Travels with Feminist Historical Geography." *Gender, Place and Culture* 10 (3): 257–264.
Evans, S.L., K. Maclean, J. MacLeavy, M. Stepney, K. Strauss, A. Tarrant, I. Wallace, and K. Brickell. 2013. "Naming the Next Generation: Early Career Perspectives on the Future of the Women and Geography Study Group." *Area* 45: 13–15. doi:10.1111/area.12008.
Evans, S.L., and A. Maddrell. 2019. "Feminist Geography in the UK: The Dialectics of Women-Gender-Feminism-Intersectionality and Praxis." *Gender, Place & Culture*, 26 (7–9): 1304–1313. doi: 10.1080/0966369X.2019.1567475.
"Feminists and Feminism in the Academy." 1992. *Antipode* 24: 218–237. doi:10.1111/j.1467-8330.1992.tb00443.x.
Flaherty, C. 2017. "Study Finds Male PhD Candidates Submit and Publish Papers at Significantly Higher Rates than Female Peers on the Same Campus." Available at: www.insidehighered.com/news/2017/11/08/study-finds-male-phd-candidates-submit-and-publish-papers-significantly-higher-rates (accessed 5 December 2017).
González Ramos, A.M., and N.V. Bosch. 2013. "International Mobility of Women in Science and Technology Careers: Shaping Plans for Personal and Professional Purposes." *Gender, Place & Culture* 20: 613–629. doi: 10.1080/0966369X.2012.701198.
Griffey, T. 2016. "*Decline of Tenure for Higher Education Faculty: An Introduction.*" Available at: www.lawcha.org/2016/09/02/decline-tenure-higher-education-faculty-introduction/ (accessed 5 December 2017).
Hardwick, S.W. 2005. "Mentoring Early Career Faculty in Geography: Issues and Strategies." *Professional Geographer* 57 (1): 21–27.
Headlam-Wells, J., J. Gosland, and J. Craig. 2005. "There's Magic in the Web": E-Mentoring for Women's Career Development. *Career Development International* 10 (6/7): 444–459. https://doi.org/10.1108/13620430510620548.
Hopkins, P., and P. Jackson. 2013. "Researching Masculinities and the Future of the WGSG." *Area* 45 (1): 9–10.
Huang, S., J. Monk, J.D. Fortuijn, M.D. Garcia-Ramon, and J.H. Momsen. 2017. "A Continuing Agenda for Gender: The Role of the IGU Commission on Gender and Geography." *Gender, Place and Culture* 24 (7): 919–938. doi: 10.1080/0966369X.2017.1343283.
Kern, L., R. Hawkins, K.F. Al-Hindi, and P. Moss. 2014. "A Collective Biography of Joy in Academic Practice." *Social & Cultural Geography* 15 (7): 834–851. https://doi.org/10.1080/14649365.2014.929729.
Kobayashi, A. 2006. "Why Women of Colour in Geography?" *Gender, Place & Culture* 13 (1): 33–38. https://doi.org/10.1080/09663690500530941.
Leigh McDowell, L. 1990. "Sex and Power in Academia." *Area* 22 (4): 323–332.

Macoun, A., and D. Miller. 2014. "Surviving (Thriving) in Academia: Feminist Support Networks and Women ECRs." *Journal of Gender Studies* 23: 287–301. doi: 10.1080/09589236.2014.909718.
Maddrell, A. 1997. "Marion Newbigin and the Scientific Discourse." *Scottish Geographical Magazine* 113: 33–41.
Maddrell, A. 2007. "Teaching a Contextual History of Geography Through Role Play: Women's Membership of the Royal Geographical Society (1892–3)." *Journal of Geography in Higher Education* 31: 393–412.
Maddrell, A. 2009. *Complex Locations: Women's Geographical Work in the UK 1850–1970*. Royal Geographical Society–IBG. Chichester: Wiley-Blackwell.
Maddrell, A., K. Strauss, N. Thomas, and S. Wyse. 2016. "Mind the Gap: Gender Disparities Still to be Addressed in UK Higher Education Geography." *Area* 48 (1): 48–56.
Martínez Alemán, A.M. 2014. "Managerialism as the 'New' Discursive Masculinity in the University." *Feminist Formations* 26: 107–134. doi: 10.1353/ff.2014.0017.
Moosa, I.A. 2018. *Publish or Perish: Perceived Benefits Versus Unintended Consequences*. Cheltenham: Edward Elgar.
Moss-Racusin, C.A., J.F. Dovido, V.L. Brescoll, M.J. Graham, and J. Handelsman. 2012. "Science Faculty's Subtle Gender Biases Favor Male Students." *Proceedings of the National Academy of Sciences* 109 (41): 16474–16479. https://doi.org/10.1073/pnas.1211286109.
Mott, C., and D. Cockayne. 2017. "Citation Matters: Mobilizing the Politics of Citation Toward a Practice of 'Conscientious Engagement'." *Gender, Place & Culture* 24 (7): 954–973. https://doi.org/10.1080/0966369X.2017.1339022.
Mountz, A., A. Bonds, B. Mansfield, J. Loyd, J. Hyndman, M. Walton-Roberts, R. Basu, R. Whitson, R. Hawkins, T. Hamilton, and W. Curran. 2016. "For Slow Scholarship: A Feminist Politics of Resistance through Collaborative Action in the Neoliberal University." *ACME* 14: 1235–1259.
Mullings, B., L. Peake, and K. Parizeau. 2016. "Cultivating an Ethic of Wellness in Geography." *Canadian Geographer* 60 (2): 161–167. doi: 10.1111/cag.12275.
Oberhauser, A., and M.A. Caretta. 2019. "Mentoring for Early Career Women Geographers in the Neoliberal Academy: Dialogue, Reflexivity, and Ethics of Care." *Geografiska Annaler: Series B, Human Geography* 101 (1): 56–67. https://doi.org/10.1080/04353684.2018.1556566.
Parizeau, K., L. Shillington, R. Hawkins, F. Sultana, A. Mountz, B. Mullings, and L. Peake. 2016. "Breaking the Silence: A Feminist Call to Action." *Canadian Geographer/Le Géographe Canadien* 60 (2): 192–204.
Pitt, R., and I. Mewburn. 2016. "Academic Superheroes? A Critical Analysis of Academic Job Descriptions." *Journal of Higher Education Policy and Management* 38 (1): 88–101. doi: 10.1080/1360080X.2015.1126896.
Quinlan, A. 2012. "Imagining a Feminist Actor-Network Theory". *International Journal of Actor-Network Theory and Technological Innovation* 4 (2): 1–9.
Rojek, C. 2000. "Leisure and the Rich Today." In *Work, Leisure and Wellbeing*, edited by J.T. Haworth, 117–130. London: Routledge.
SIGJ2 Writing Collective. 2012. "What Can We Do? The Challenge of Being New Academics in Neoliberal Universities." *Antipode* 44 (4): 1055–1058. https://doi.org/10.1111/j.1467-8330.2012.01011.x.
Single, P.B., and R.M. Single. 2005. "E-mentoring for Social Equity: Review of Research to Inform Program Development." *Mentoring and Tutoring*, 13 (2): 301–320.
Strauss, K. 2013. "Internationalisation and the Neoliberal University." *Querelles: Almanac for Women and Gender Studies*, 2012 edition.
Taylor, Y., and K. Lahad. 2018. *Feeling Academic in the Neoliberal University: Feminist Flights, Fights and* Failures. Cham: Palgrave Macmillan.
Thwaites, R., and A. Pressland. eds. 2017. "*Being an Early Career Feminist Academic." Global Perspectives, Experiences and Challenges.* London: Palgrave Macmillan.
Wheeler, P.T. 1967. "The Development and Role of the Geographical Field Group." *East Midland Geographer* 4: 185–195.
Winkler, J.A. 2000. "Focus Section: Women in Geography in the 21st Century: Faculty Reappointment, Tenure, and Promotion: Barriers for Women." *Professional Geographer* 52 (4): 737–750. https://doi.org/10.1080/00330124.2000.9628423.
Women and Geography Study Group of the IBG. 1984. *Geography and Gender: An Introduction to Feminist Geography*. London: Hutchinson.
Women and Geography Study Group. 1997. *Feminist Geographies: Explorations in Diversity and Difference*. London: Longman.
Wyse, S. 2013. "The Founding of the Women and Geography Study Group." *Area* 45: 4–6. doi:10.1111/area.12010.

8
SKIN, SWEAT AND MATERIALITY
Feminist geographies of emotion and affect

Gail Adams-Hutcheson and Paula Smith

Introduction

The tapestry of geographical ideas on emotion and affect is rich. In writing this chapter on emotion and affect, we are acutely aware of the vast canon of literature built around each, as well as both, of the terms. Nevertheless, to make the most effective use of the concepts, it is worth being explicit about how one intends to use the expressions.

Affect theory's application is eclectic. It traverses psychoanalysis, psychology and psychiatry, and the literature has an interdisciplinary and malleable 'feel'. In the past three decades, geography has seen an abundance of theoretically astute research on affect.[1] These works sit beside interdisciplinary borrowings and geographical analyses of affect in philosophy (Ahmed 2004; Deleuze and Guattari 1988; Massumi 2002; Sedgwick and Frank 1995), in political theory (Bennett 2010; Williams 2007) and feminist studies (Berlant 2000; Probyn 2005; Riley 2005). Emotion and affect are themes that have attracted attention in human geography since the early 2000s (Pile 2010). Feminist geographers have sought to critically absorb and promote the strength of these frames into their work. However, scholars have considered emotion and/or affect individually and together; some cleave the terms, while others stake a claim to usefully connect the two.

From early on, feminist geographers argued for emotion to have a place in 'serious' scholarship (Davidson and Milligan 2004; Davidson et al. 2005; McDowell 1992; Smith et al. 2009). Drawing on poststructuralist ideas, feminist geographers destabilized the binaries (a set of interrelated and mutually reinforcing dualisms that explain the 'naturalization' of hegemonic gender oppressions) in which knowledge production takes place. Accordingly, men equate with traits such as rationality, superiority, science, independence, public space and culture, while women equate with traits such as irrationality, emotion, dependence and private space and nature (Longhurst 2001; McDowell 1999; Women, Geography Study Group (WGSG) 1997).

With a shared ontology of fluidity, emotion appears the more accessible subject. Affect, conversely, can be somewhat difficult to articulate clearly, given its ephemeral character. Thus, Joanne Sharp (2009) discusses a debate that kindled over the distinction between emotion and affect, such that emotion is often times considered as individual/personal and affect as transhuman/political. We consider that the separation of emotion and affect follows masculinist traditions, relegating and feminizing emotion to the personal level and favouring affect as a theoretically more advanced concept (Thien 2005). With terms such as *immediacy*, *immanence* and *the virtual*, the focus on the abstract centre of affect theory (McCormack 2012) is proposed here to be inattentive to power and geopolitics, and strangely disembodied.

Accordingly, we share the position that feminist scholars present, such that emotion and affect are mutually sustaining (see Ahmed 2004; Bondi and Davidson 2011; Thien 2005; Wright 2010). Indeed, Liz Bondi and Joyce Davidson (2011) surmise that emotion and affect do not quite map directly, one over the other, when they argue that 'emotions and affects might be considered disagreeable in many ways, but their rough edges, the very wildness that frustrates domestication is precisely what gives them such power' (Bondi and Davidson 2011, 595). We reflect that theories of emotion and affect are understandings of orientation, or starting points (Ahmed 2006), and draw on critical feminist scholarship to explain human and nonhuman complexities.

This chapter deliberates on feminist geographers' scholarship on emotion and affect, including our own. Paula Smith undertook a Masters in social science on the emotional and affective impacts of relationship challenges on the meanings of home. Her thesis examines practices of homemaking, unmaking and remaking of ten heterosexual individuals who have experienced relationship challenges and have homes in the Waikato region of Aotearoa New Zealand. Our connection is through supervision. Gail Adams-Hutcheson was one of Paula's supervisors (the other being Lynda Johnston). Gail's research is on emotion, affect and critical feminist analyses of transient communities. The communities included post-disaster survivors who relocated from Christchurch to Waikato after the devastating Christchurch earthquakes.[2] More recently, she has been working with sharemilkers (non-landowners who milk dairy cows), a transitory community in the Waikato dairy industry.

In each of the following sections, we draw on our own, plus others', research to reflect on the gendered geographies of feeling, emotion and affect. Throughout this chapter, we understand emotion and affect to be intimately connected and use the terms interchangeably. Therefore, we review how emotion and affect apply in feminist geographical scholarship at different spaces and scales.[3]

The first section pays critical attention to the body as a spatial scale, and we describe feelings, emotion and affect in relation to skin, trauma, sweat and breastfeeding. In the second section, Sara Ahmed's (2008) work is crucial to linking bodies to public space and collective feelings. Managing emotions and the (un)comfortability of sweating and breastfeeding draw attention to the politics of public space and how emotion and affect are managed. We then move to explore private space via a focus on home, love and relationship breakup. The scale of the home is important to feminist geographers (Blunt and Dowling 2006). In particular, Paula's work on emotion and affect when homes 'dissolve' and couple relationships end allows inspection of material items and their entanglement in intergenerational bonds. The final section overviews emotions in the field, their stickiness and 'doing' affect. Feminist geographers have been at the forefront of interrogating research spaces, ethically, emotionally and affectually for close to three decades. Here, we draw on Gail's contemporary methodological discussion on transference, suicide and rhythm to explore the transformative potential of research methods, including Skype and semi-structured interviews.

The body: affective politics of shame and disgust

The body is a central locus for critical feminist geographies in a number of crucial ways. Feminist poststructural analyses of bodies and space (Longhurst 1995, 2001, 2004; McDowell 1999; Rose 1993) incorporated performance, representation, surveillance and identity politics. Of further importance was to dismantle the dualistic structure of Western thinking that sustained an identity politics of man/woman, rationality/irrationality, production/reproduction and mind/body. Contemporarily, though, scholars seek to include the fleshy body, in all its messy materiality (Longhurst and Johnston 2014) and push further to see what bodies can do (Colls 2012; Slocum 2008) in a 'Deleuze-inspired corporeal feminism' (Waitt and Stanes 2015, 31). In all its complexity, affect is frequently used in geographical literature in the Spinozian sense as the bodily capacity to affect and be affected. Bharuch Spinoza (2000/1677) understands the body as a series of intensities that are constantly in relation and in connection to other bodies (as opposed to 'the [static] body'). Using the fleshy body, affected and affecting and struck through with emotion, feminist scholars draw on the transformative potential of Julia Kristeva's (1982), Elspeth Probyn's (2005) and Iris Marion Young's (1990) work to examine the affective politics of shame and abjection. It is Young's (1990, 145) early work that links body aesthetics to 'racism, sexism, homophobia, ageism and ableism which, are partially structured by abjection – an involuntary, unconscious judgement of ugliness and loathing'. Abjection is the affect or feeling of anxiety. Kristeva (1982) argues that the abject provokes fear and disgust because it exposes a border between self and Other; it also exposes the fragility of borders and how they can be threatened (by dirt, disorder, disease, pain and trauma).

Robyn Longhurst (2001) has examined the aversion to body fluids, as fluids rupture boundaries that are supposed to be shored up, solid and impermeable. More recently, these ideas have inspired Adams-Hutcheson (2016) to examine the affective disruption of trauma as something that defies boundaries, playing across the skin, unsettling the spatial connections between people and places. Ideas of skin are utilized in two key ways: first, as a metaphorical container for elucidating how respondents kept trauma at bay and their fear of trauma escaping beyond bounded skin; second, the skin (or crust) of the earth is deemed to be abject and traumatic when its boundaries are ruptured through the seeping of liquefaction in the post-disaster city of Christchurch. When elements (like liquefaction) flow across usually augmented boundaries, they pose a challenge to the dominant symbolic order; it then becomes the mark of the abject (Adams-Hutcheson 2016). Significantly, disgust and abjection are not necessarily voluntary responses, as bodily reactions to certain Others are sometimes unconscious. Often, we feel shame at our own feelings of disgust. It is imperative that feminist geographers recognize the affective responses of shame when confronting that which is Other (Longhurst et al. 2008) because, as Probyn (2000, 2005) states, this might pave the way to understanding and acceptance.

Shame has been a key lens through which to consider emotions and the affective capacities of bodies. Similar to abjection, shame is a visceral response, an affect that forces people to confront 'the proximity of ourselves to Others' and prompts reflection on subjectivities, individually and collectively (Probyn 2000, 132, 2005). The notion of shame is integral to feminist geographers' work on fat bodies, where 'fat shaming' and 'fat phobia' reconstitute bodies in public spaces (Evans 2006; Hopkins 2012; Longhurst 2005). Fat bodies are marginalized and stigmatized, and the distaste for fat continues. Hopkins (2012) links this to affective bodily responses when he interrogates the idea that, for many, fat people occupy the position of having 'ugly, fearful, or loathsome bodies' (Young 1990, 124). These ideas draw together in Gordon Waitt and Elyse Stanes' (2015) work on sweat and masculinities. Sweaty bodies can be sites of shame or pride, with the fit sweaty body often revered while the fat sweaty body is reviled. Sweat leaks from the body, it smells and exposes the fragility of bodily spatial boundaries. Sweaty bodies are central in continuing to pay attention to the messy, fleshy material body often omitted from mainstream geography. Crucially though, these works offer a stark glimpse into how emotion and affect structure spatial relationships that expose the dominant discourses of acceptability and Otherness and how place plays a central role. Signalling embodiment also serves to underscore the eminence of place. The sweaty body is 'in place' in the gym, for example, but 'out of place' in crowded transport. Whether public or private, spaces script emotion, affect and their collective transfer in different, ambivalent and complex ways.

In the next two sections, we describe public and private space and the connection to encountering emotion and affect.

Public space: breasts, sweat and (dis)connection

Emotions and affects are bound up with how we inhabit the world 'with' others. They are about the intimate relations between selves, objects and others (Ahmed 2004), and they are not divorced from their public and/or private situated context. Jennifer Harding and Deidre Pribram (2002) argue that affective geographies should draw our attention exactly because they dissolve public/private boundaries, working across individual bodies to affect and change collective ones. However, we wish to underscore that place matters and is crucial when managing emotions and the opaqueness of affect.

Public space often denotes a collective atmosphere. People, emotion and affect converge in interesting ways, as when Teresa Brennan (2004, 1) asks, 'Is there anyone who has not, at least once, walked into a room and "felt" the atmosphere?' Affect links to something like 'collective impressions', a 'felt' atmosphere or perhaps an overall mood. It may be understood as the experience of being 'sapped', 'tired' or 'depressed' in the company of someone, while conversely feeling 'energized', 'inspired' or 'invigorated' by others (Brennan 2004). In this way, Brennan (2004) sees affect as a process that is transmittable, transmutable, picked up, transformed, reshaped and reshared between and among a collective. However, we want to complicate the idea of affective translation from body to body.

We contemplate what happens when feelings and affects are discordant or are not returned nor shared, perhaps because they are masked, misinterpreted or require management. Collective impressions may alienate, confound or confuse a sense of social sharing among the collective. In Gail's PhD research, Christchurch participants acted out their responses to stress and anxiety in ambivalent ways. In some cases, emotion and affect were deployed strategically to mask fear, such as, for example, Liz's mother and brothers, who 'acted tough and nonchalant' during the earthquake's aftershocks (Adams-Hutcheson 2014, 159). Their nonchalant behaviour, however, alienated Liz, who felt her alarm and panic rising after each bout of shaking. Liz ultimately felt that the family was dismissive of her terror. Ahmed (2008) uses the idea of an affect 'alien' to examine how affective practice can alienate as well as connect; that is, sometimes emotion and affect simply do not line up with the collective impression from others. Sadly, as a result, Liz left Christchurch without saying a last goodbye at the airport, because she thought that the family did not support or understand her emotionally.

Managing emotions and affect is linked to the idea of 'masking', or the conscious projection of particular emotions by creating a façade. Being outwardly calm while seething, for example, challenges the argument that emotion is tied only to the body and that affect is non-conscious. Paula interviewed women who had experienced relationship break up and the division of home and contents. She found that, for some respondents, their public performances of post-relationship breakdown were complicated. Participants' public performances – healed, strong and capable – contrasted markedly to those in private moments that were understood as emotionally raw, shaky and containing emotional outbursts. Paula (Smith 2017, 41) notes that:

> While I stayed with Coco, she was candid about her experiences, but I felt she projected a mostly accepting and positive outlook about her situation as a single mum

> dealing with significant challenges. But when we got to her [empty at the weekend] work space – a space she helped design and adores – I noticed a change. Her vulnerability and deep sadness were more apparent, with tears, and wobbly voice as she articulated her feelings.

The most revealing aspect of these examples is the complexity of emotion and affect, which illustrates bodily projection as one state while simultaneously feeling the opposite. Public actions and reactions embed Byzantine-like webs of interactions choreographed by space. Therefore, people are able to knowingly direct their reactions, bodily compositions and demeanour to portray a particular outer countenance (rather than non-conscious affect), such as remaining outwardly calm while crumbling with terror inside. As Ahmed (2008, 11) argues, 'we might even become strangers or affect aliens, at such moments'. How these affective registers are drawn on and received collectively remains variable and is far from transparent.

Kate Boyer's (2012) work on breastfeeding in public spaces unpacks public attitudes and comfortability by building on feminist geographers' work that frames subjectivity and the politics of public space. High levels of commitment to breastfeeding and a 'willingness to engage in counter-normative behaviour' (Boyer 2012, 554) shaped women's experiences, not least breastmilk's 'suspect' (that is, abject) status as a potentially contaminating fluid, which may leak or transgress bodily boundaries. Breastfeeding can raise deep-seated anxieties about bodily fluids, and it highlights the ways in which difference is materially and affectively experienced. Breastfeeding women are marked and marginalized in the public sphere, and are 'responsible for the comfort of others through risk censure' (Boyer 2012, 553). Both Boyer's (2012) and Waitt and Stanes' (2015) work situates the importance of public space(s) as crucial to comfortability – or not. Ahmed (2010) observes that maintaining public comfort requires certain (other) bodies 'to go along with it' (2010, 584). The sweating or breastfeeding body is not problematic in itself but can become so in public places, and is managed in various ways. Disentangling space is not straightforward, and we do not wish to (re)inscribe a dichotomous separation of public from private space. Yet, in doing so, the strength of gendered analysis of space and place retains its political integrity. As outlined above, public space can be problematic for some bodies (frequently marginalized through different forms of oppression) and not others.

Next, we attend to private space, the quintessential feminine space of home and love, albeit one that has had to be fought to be included in the mainstream academy.

Private space: love and materiality in unmade homes

Home is a key geographical site for feminist geographers to examine gender roles and relations, sexuality, embodied differences and identity politics (Blunt and Dowling 2006). More recently, love (Morrison, Johnston and Longhurst 2012), domesticity (Gorman-Murray 2008) and material objects (Morrison 2010, 2012) have been key sites of framing for feminist geographers engaged with emotion and affect at the scale of home.

When thinking of broken hearts and broken homes (Smith 2017), love has been considered a contentious term academically, a feminized topic associated with private spaces, individualized feelings and perhaps 'gossip' (Morrison, Johnston and Longhurst 2012). Margaret Toye (2010, 41) argues that, rather than avoiding the concept of love, 'feminist scholars should have a special interest in the topic because of the ways the discourse of love has not just been associated with women, but has been used against them'. Feminist theorists, then, have illustrated that the concept – and its hegemonic bedfellow, marriage – is a means by which women are restricted and subordinated (Ahmed 2004; D'Emilio and Freedman 2012; Jackson 2014; Miller 1996).

This ideology, while illustrating the gendered binary of power relations present in so many heterosexual intimate relationships, stops short of describing love as 'spatial, relational and political' (Morrison, Johnston and Longhurst 2012, 506). Love is complex, contradictory and political; it is a queer feeling (Berlant 2001) that destabilizes binaries, such that love engulfs anger, shame and fighting into its sphere. Disruptions to our most intimate relationships create (emotional) movement between joy and sadness, in either and both directions (Ahmed 2004). With such turbulent emotions circulating, there is considerable transference of affect, not just between bodies but also between matter and humans (Bennett 2010).

Like Jane Bennett (2010), Paula and Gail's research uses Spinozian thinking on affect in two ways; that is, Spinoza's explanation of the ability or force within bodies to affect and are affected, and the ways in which he alludes to vibrant matter. In co-joining the relations between materiality and immateriality and subjective affective atmospheres (bodies impact bodies à la Spinoza), affective materiality is not reducible to individual subjective experience. Examining her work with sharemilkers, Adams-Hutcheson (2017) describes how fluctuating weather conditions work their way across human and nonhuman bodies (researcher, researched and cows) in a farming context. The troposphere leaves an impression, such that a sunny day is affectively different from wind and rain. The weather creates an *affective atmosphere* that explains how materiality and immateriality mesh, transfer and, importantly, are shared (Adams-Hutcheson 2017). These ideas gather in Paula's examination of the affective resonance of material items that travel among homes and relationships. Thus, both Spinoza and Bennett (2010) are drawn on to describe how bodies and household items have a particular vitality. Again, by linking materiality and immateriality, it is argued that people feel and are affected by material items.

By extrapolating the material geographies of home, Andrew Gorman-Murray (2008, 283) points to this call to '"re-materialize" research in social and cultural geography' as a way of investigating the role of domestic spaces in identity formation and management. Within a feminist framework, the concept of 'stickiness' of emotions (Ahmed 2004) is noted. This stickiness carries through and attaches itself to materialities, due to the emotions that summon and attach themselves to material items and contribute to identity (re)formation. Things such as household items are passed on. Items may be handed on generationally, between friends, or may be returned, sold, loved, pre-loved, mended and marked. In Paula's project, materials took a central role. Restored chairs, furniture, travel souvenirs, statues, paintings and a rug were weighted with emotional and affective significance. Smith (2017, 98) explains that, for respondent Jen, who let go of some material items but not others in divorce settlement, a rug holds a strong attachment. The emotional and affective transfer that the rug signifies allows Jen to re-engage with the past, a time of building her first home with her husband and the arrival of their first child. Loved items are able to 'matter' affect. Household items transferred in these ways are not necessarily of great financial value; rather, the worth of the items is determined by emotional attachment. The process of dividing the material objects and collections often adds affective layers of pain to an already distressing breakup. Material objects are frequently the focus of legal proceedings, as networks of feelings are disentangled, cleaving what was once imagined as one entity back into two. Accordingly, power struggles are observed to be enacted through 'rights' to such items (Goode 2007).

When intimate couple relationships and normative constructs of love are challenged, home is changed in a myriad of ways both materially and in the imaginary. These changes may be positive or negative, or fluctuate ferociously between the two. The increasingly 'fragmented and reconstituted families or intra-familial negotiations across multiple generations' (Valentine 2008, 2101) demonstrate that material items become emotional and affectual detritus floating between home(s). Passing items from generation to generation is common practice, and is viewed by many as a means of displaying love for and of home and, by association, family (Morrison 2013). Changes may take the form of spatial changes – in the ways in which the materials are removed, replaced, or rearranged – but, also, the home itself may be vacated, to be rebuilt and reimagined. Reconstruction of the home could be in an entirely new geographical location, or else remade in the current space, in both cases with new or reconstructed meanings and materialities, each step saturated with emotion and affect (Smith 2017).

In the final section, we discuss emotion and affect as a highly useful frame to investigate research spaces, human connection and the feminist politics of 'doing' research.

'Doing' feminist research

A core strength of feminist geography has been the critical attention paid to research moments, power relations and the ethical politics of conducting research with others. In utilizing emotion and affect in feminist scholarship there is a sense of attempting to, in Eve Sedgwick's (2003, 62) words, 'touch the textures of social life', which brings enchanting, funny, humbling and harsh emotion and affect into the realm of academic scrutiny. Emotion and affect are judged as being 'free radicals' (Sedgwick 2003), in that there can never be a carefully circumscribed emotional or affectual geography, neatly separate from other geographies. When scholars work with microscopic intensity, there is sensitivity to the researcher and the researched's 'volatile body' (Grosz 1994), whose experiences in place emerge 'though the sensation of spacing that is material and immaterial, human and animal, organic and inorganic' (Dewsbury 2010, 36). Such experiences allow insight into how bodies extend into and through the corporeality of the researcher, affecting and affective of each other. Emotion and affect draw in different intensities and ways of paying attention to space and place through: rhythm (Duffy et al. 2011); the body (Longurst, Ho and Johnston 2008); emotions (Davidson and Bondi 2004; Sharp 2009; Widdowfield 2000); listening and music (Duffy 2013; Wood and Smith 2004); and through technologies (Longhurst 2013, 2016), to name but a few.

Acknowledging Liz Bondi's (2004, 2005) work on psychoanalysis and geography, the interview space is shared as an oscillation of thoughts and feelings. The intersubjectivity that weaves in and out of the interview conversation takes on an almost felt atmosphere or 'presence'. In most human interactions there is either a conscious or subconscious desire to connect and share in some way. Then again, research is a purposeful interaction with various levels of power-laden differences that weave in and out of the exchange and colour the experience at every level. Feminist geographers are cognizant of the affective dimensions within research as a political motivation for conducting ethical analyses of space and place.

Bondi (2005) discusses the researcher and some of the unconscious impacts that another person's distress or happiness can have on a researcher. Using emotion and affect as a frame highlights moments of lightness, hope and laughter, sadness and trauma, and so on, which exchange through non-conscious impacts. Bondi (2005, 71) maintains that most of these personal interchanges within interview moments are done non-cognitively and non-verbally, and that the participant can then feel understood 'emotionally and experientially as well as cognitively'. Using empathy and psychoanalytical methods, feminist scholars have honed in on emotion and affect and how these transfer from the researcher to the researched (and back) in oscillation. For instance, Gail's PhD participants sweated, shifted in their seats, laughed, cried, fell silent, reddened, sighed and shared a myriad of other intensities, beyond language, which help to create momentary collectivities or shared spaces of learning (Hutcheson 2013). An investigation of research interactions and relationships also entails thinking about how interviews and focus group meetings are dynamic exchanges of intersubjectivity.

The spaces of research and how intersubjectivity may be achieved have been further interrogated by examination of research technologies. A feminist analysis of Skype challenges the separation of public and private space, as well as proximity and distance. Skype, it is argued, creates definitive affective atmospheres but may also be mistranslated when there is disembodiment and lack of 'touch' (Adams-Hutcheson and Longhurst 2016). Conversation, control of routines, speaking spaces and emotions are sometimes fractured and discordant via technologies such as Skype. Rhythms, when disrupted, are often difficult to re-establish without familiar, embodied social cues. This lack of the organizational processes of rhythm can lead to discomfort and anxiety, which are generally difficult to smooth from behind a screen or at a distance (Adams-Hutcheson and Longhurst 2016; Longhurst 2013, 2016).

These touching moments and staying in touch via Skype build on Lynda Johnston's (2012) work in considering the haptic geographies of drag queens in Aotearoa New Zealand. Lynda argues that touch reminds us of the fragility of femininities and masculinities and points to the affectual language of 'excess'. Drag queens utilize the performing body excessively to evoke strong emotions. Audiences, through their complex emotional responses, highlight how emotions and bodily sensations are used strategically to seek political and erotic justice (Johnston 2012): 'A focus on drag queens and touch creates a space in which sexualities can be explored in each affectual and emotional encounter. Body to body touch is an intimate sensual encounter which is always situated somewhere' (Johnston 2012, 8). Place is crucial to the ways in which bodies may, or may not touch, connect or disconnect. At the heart of conversations about emotion and affect are similarity and difference, affinity and distance.

Why was it that some experiences are intimately shared and yet, with others, the 'sharing' is on a relatively superficial level? Using the body as a research instrument (Longhurst et al. 2008) and taking on a sensorial knowledge framework mean admitting desire, disgust, angst, friendliness and a range of complex emotions and affects, which ebb and flow as forces of knowledge construction within the research environment. Emotion and affect are then, political, gendered and spatially articulated in both obvious and less-obvious ways. Gorton (2007) argues that our actions are guided not just by what we think but also by how we feel and our bodily response to those feelings.

Recently, feminist scholars have begun to examine intimate moments of research spaces and how these leave impressions emotionally and affectively on researchers (Moss and Donovan 2017). Attending to the circulation and stickiness of emotion and affect gives insight into how researchers are entangled in the field (Ahmed 2004), and also once one has left the field. Adams-Hutcheson and Longhurst (2017) describe a research encounter that happened during the data-gathering phase, four years prior to the dissemination of Gail's PhD project. It took years to want to pay close attention to a respondent's admission of her intention to commit suicide and the unknown outcome of her phone call to Gail. The authors tackle the consequences of suppressing data, as well as studying what it may mean to illuminate such a taboo topic as suicide (Adams-Hutcheson and Longhurst 2017). Feminist geographers (like many others) wield enormous power at the editing table, and questions are beginning to be asked about emotion and affect that are too raw or intimate to be included in analyses. Consequently, they argue that grittier, messy, unsettling and intimate work remains (Adams-Hutcheson 2017) to fully grapple with emotion and affect that are deemed to be too revealing yet that continue to challenge reason and masculinism in the academy.

Conclusions

There is important geographical work done on emotion and affect. In this chapter, we have framed the complex and vibrant field from the critical feminist geography perspective. We have deliberately encompassed critical works that pay attention to the power, ethics, intersubjectivity and spatial scales that are important to feminist geographers. Terms such as becoming, potentiality, immediacy, immanent and the virtual all hold together complex literatures that seem strangely disembodied and perhaps, at times, drift toward universalist rather than geopolitical framings. Space in these literatures is ontogenetically multiple, rhizomatic and fascinating (Deleuze and Guattari 1988) yet, often, the bodily, emotionally messy research practices never make it to print.

Feminist geographers have responded to the lack of geopolitical and embodied research that uses emotion and affect as a lens. Emotion and affect are examined within politically active and critical frames of analyses. Thus, we have chosen to illuminate power geometries, intimacy, shame, love, sweat, abjection and fleshy bodies, among other things, as empirically important to feminist geographers. Space and place, too, make a difference to how bodies can be touched, read, managed and/or perceived. Emotion and affect simply do not always line up and are purposefully directed at times, leading to opaqueness in public spaces. We employ the idea of a façade that may be presented outwardly, while inner turmoil roils. This discussion embeds Ahmed's (2008) ideas of the affect 'alien', by absorbing emotion as a wilful and directed manifestation of conscious affect. Furthermore, we direct our attention to nonhuman objects, materials, relations and technologies, which embrace and return emotion and affect in circular patterns. We extrapolate from all sorts of unlikely materials that are caught up in vital connections to and with individuals.

This chapter promotes an active, material, sensing (nonhuman) body with a progressive politics that desires to open up lines of engagement, connection and communication among feminist geographers and others.

Notes

1 See Patricia Clough and Jean Halley (2007), *The Affective Turn: Theorising the Social*; Melissa Gregg and Gregory Seigworth (2010), *The Affect Theory Reader*, Kathleen Stewart (2007), *Ordinary Affects;* and the journal *Body and Society* (2010) 16 (1): special issue on the theme of affect. Affect is also a chapter entry in a number of geographically focused books such as: Ben Anderson's (2013) 'Affect and emotion' in the *Wiley Blackwell Companion to Cultural Geography;* John David Dewsbury's (2010) 'Affect' in the *International Encyclopaedia of Human Geography*; Keith Woodward and Jennifer Lea's (2009) 'Geographies of Affect' in the *SAGE Handbook of Social Geographies.* See also Ben Anderson and Paul Harrison's (2006) 'Questioning Emotion and Affect'; Steve Pile's (2010) 'Emotion and Affect in Recent Human Geography' and his (2011) 'For a Geographical Understanding of Emotion and Affect'; and Deborah Thien's (2005) 'After or Beyond Feeling? A Consideration of Affect and Emotion in Geography'.

2 The Waikato area was chosen as a research location, as Paula and Gail live and attend university in the area. Further, Gail's respondents moved out of post-disaster Christchurch (South Island) to the Waikato (North Island) largely due to its geological stability. For more information and a map, see Hutcheson 2013.

3 In choosing the order of this chapter, we realize that scales are unable to be discrete and neatly contained (Marston, Jones and Woodward 2005). The material leaks across these conceptual boundaries with ease. We also acknowledge the selective use of scholarship, which is far from exhaustive and risks glossing over deeply rich work.

Key readings

Ahmed, S. 2004. *The Cultural Politics of Emotion.* Edinburgh: Edinburgh University Press.

Gorton, K. 2007. "Theorizing Emotion and Affect." *Feminist Engagements, Feminist Theory* 8 (3): 333–348.

Probyn, E. 2005. *Blush: Faces of Shame.* Minneapolis: Minnesota Press.

References

Adams-Hutcheson, G. 2014. "Stories of Relocation to the Waikato: Spaces of Emotion and Affect in the 2010/2011 Canterbury Earthquakes, Aotearoa New Zealand." PhD dissertation, University of Waikato.

Adams-Hutcheson, G. 2016. "Spatialising Skin: Pushing the Boundaries of Trauma Geographies." *Emotion, Space and Society* 24: 105–112.
Adams-Hutcheson, G. 2017. "Farming in the Troposphere: Drawing Together Affective Atmospheres and Elemental Geographies." *Social and Cultural Geography.* Advance online. doi:10.1080/14649365.2017.1406982.
Adams-Hutcheson, G., and R. Longhurst. 2016. "'At Least in Person There Would Have Been a Cup of Tea': Interviewing via Skype." *Area* 49 (2): 148–155.
Adams-Hutcheson, G., and R. Longhurst. 2017. "'I'm Here, I Hate It and I Just Can't Cope Anymore': Writing about Suicide." In: *Writing Intimacy into Feminist Geography,* edited by P. Moss and C. Donovan, 41–47. London: Routledge.
Ahmed, S. 2004. *The Cultural Politics of Emotion.* Edinburgh: Edinburgh University Press.
Ahmed, S. 2006. *Queer Phenomenology: Orientations, Objects, Others.* Durham, DC: Duke University Press.
Ahmed, S. 2008. "Sociable Happiness." *Emotion, Space and Society* 1: 10–13.
Ahmed, S. 2010. "Killing Joy: Feminism and the History of Happiness." *Signs* 35 (3): 571–594.
Anderson, B. 2013. "Affect and Emotion." In: *The Wiley-Blackwell Companion to Cultural Geography*, edited by N. Johnson, R. Schein, and J. Winders, 452–464. London: Wiley-Blackwell Press.
Anderson, B., and P. Harrison. 2006. "Questioning Affect and Emotion." *Area* 38: 333–335.
Bennett, J. 2010. *Vibrant Matter: A Political Ecology of Things.* Durham, DC: Duke University Press.
Berlant, L., ed. 2000. *Intimacy.* Chicago, IL: University of Chicago Press.
Berlant, L. 2001. "Love, a Queer Feeling." In: *Homosexuality and Psychoanalysis*, edited by T. Dean and C. Lane, 430–452. Chicago, IL: Chicago University Press.
Blunt, A., and R. Dowling. 2006. *Home.* Abington: Routledge.
Bondi, L. 2004. "10th Anniversary Address for a Feminist Geography of Ambivalence." *Gender, Place and Culture: A Journal of Feminist Geography* 11 (1): 3–15.
Bondi, L. 2005. "Making Connections and Thinking Through Emotions: Between Geography and Psychotherapy." *Transactions of the Institute of British Geographers* 30: 433–448.
Bondi, L., and J. Davidson. 2011. "Lost in Translation." *Transactions of the Institute of British Geographers* 36 (4): 595–598.
Boyer, K. 2012. "Affect, Corporeality and the Limits of Belonging: Breastfeeding in Public in the Contemporary UK." *Health and Place* 18: 552–560.
Brennan, T. 2004. *The Transmission of Affect.* Ithaca, NY: Cornell University Press.
Clough, P., and J. Halley, eds. 2007. *The Affective Turn Theorizing the Social.* Durham, DC: Duke University Press.
Colls, R. 2012. "Feminism, Bodily Difference and Non-Representational Geographies." *Transactions of the Institute of British Geographers* 37: 430–445.
Davidson, J., and L. Bondi. 2004. "Spatializing Affect; Affecting Space: An Introduction." *Gender, Place and Culture: A Journal of Feminist Geography* 11 (3): 373–374.
Davidson, J., L. Bondi, and M. Smith, eds. 2005. *Emotional Geographies.* Aldershot: Ashgate.
Davidson, J., and C. Milligan. 2004. "Embodying Emotion Sensing Space: Introducing Emotional Geographies." *Social and Cultural Geography* 5 (4): 523–532.
D'Emilio, J., and E. Freedman. 2012. *Intimate Matters: A History of Sexuality in America.* Chicago, IL: University of Chicago Press.
Deleuze, G., and F. Guattari. 1988. *A Thousand Plateaus: Capitalism and Schizophrenia*, translated by B. Massumi. London: Althone Press (originally published 1980).
Dewsbury, J.D. 2010. "Performative, Non-Representational, and Affect-Based Research: Seven Injunctions." In: *The SAGE Handbook of Qualitative Geography*, edited by D. De Lyser, S. Herbert, S. Aitken, M. Crang, and L. McDowell, 321–334. London: Sage.
Duffy, M. 2013. "The Requirement of Having a Body." *Geographical Research* 51 (2): 130–136.
Duffy, M., G. Waitt, A. Gorman-Murray, and C. Gibson. 2011. "Bodily Rhythms: Corporeal Capacities to Engage with Festival Spaces." *Emotion, Space and Society* 4: 17–24.
Evans, B. 2006. "'Gluttony or Sloth': Critical Geographies of Bodies and Morality in (Anti) Obesity Policy." *Area* 38 (3): 259–267.
Goode, J. 2007. "Whose Collection is it Anyway? An Autoethnographic Account of 'Dividing the Spoils' Upon Divorce." *Cultural Sociology* 1 (3): 365–382.
Gorman-Murray, A. 2008. "Masculinity and the Home: A Critical Review and Conceptual Framework." *Australian Geographer* 39 (3): 367–379.
Gorton, K. 2007. "Theorizing Emotion and Affect." *Feminist Engagements, Feminist Theory* 8 (3): 333–348.

Gregg, M., and G. Seigworth, eds. 2010. *The Affect Theory Reader*. Durham, DC: Duke University Press.
Grosz, E. 1994. *Volatile Bodies: Toward a Corporeal Feminism*. Bloomington: Indiana University Press.
Harding, J., and D. Pribram. 2002. "The Power of Feeling: Locating Emotions in Culture." *European Journal of Cultural Studies* 5 (4): 407–426.
Hochschild, A.R. 2003. *The Managed Heart*, twentieth anniversary ed. Oakland: University of California Press.
Hopkins, P. 2012. "Everyday Politics of Fat." *Antipode* 44 (4), 1227–1246.
Hutcheson, G. 2013. "Methodological Reflections on Transference and Countertransference in Geographical Research: Relocation Experiences from Post-Disaster Christchurch, Aotearoa New Zealand." *Area* 45 (4): 477–484.
Jackson, S. 2014. "Love, Social Change, and Everyday Heterosexuality." In: *Love: A Question for Feminism in the Twenty-first Century*, edited by A. Jónasdóttir and A. Ferguson, 33–47. London: Routledge.
Johnston, L. 2012. "Sites of Excess: The Spatial Politics of Touch for Drag Queens in Aotearoa New Zealand." *Emotion, Space and Society* 5 (1): 1–9.
Kristeva, J. 1982. *Powers of Horror*, translated by L. Roudiez. New York: Colombia University Press.
Longhurst, R. 1995. "VIEWPOINT: The Body and Geography." *Gender, Place & Culture* 2 (1): 97–106.
Longhurst, R. 2001. "Geography and Gender: Looking Back, Looking Forward." *Progress in Human Geography* 25 (1): 641–648.
Longhurst, R. 2004. *Bodies: Exploring Fluid Boundaries*. London: Routledge.
Longhurst, R. 2005. "Fat Bodies: Developing Geographical Research Agendas." *Progress in Human Geography* 29 (3): 247–259.
Longhurst, R. 2008. "The Geography Closest in – The Body… The Politics of Pregnability." *Australian Geographical Studies* 32 (2): 214–223.
Longhurst, R. 2013. "Using Skype to Mother: Bodies, Emotions, Visuality, and Screens." *Environment and Planning D: Society and Space* 31: 664–679.
Longhurst, R. 2016. *Skype: Bodies, Screens, Space*. London: Routledge.
Longhurst, R., E. Ho, and L. Johnston. 2008. "Using 'the Body' as an 'Instrument of Research': Kimch'i and Pavlova." *Area* 40 (2): 208–217.
Longhurst, R., and L. Johnston. 2014. "Bodies, Gender, Place and Culture: 21 Years On." *Gender, Place and Culture: A Journal of Feminist Geography* 21 (3): 267–278.
Marston, S.A., J.P. Jones III, and K. Woodward. 2005. "Human Geography without Scale." *Transactions of the Institute of British Geographers* 30 (4): 416–432.
Massumi, B. 2002. *Parables for the Virtual: Movement, Affect, Sensation*. Durham, DC: Duke University Press.
McCormack, D. 2012. "Geography and Abstraction: Towards an Affirmative Critique." *Progress in Human Geography* 36 (6): 715–734.
McDowell, L. 1992. "Doing Gender: Feminists and Research Methods in Human Geography." *Transactions of the Institute of British Geographers* 17: 399–416.
McDowell, L. 1999. *Gender, Identity and Place*. Cambridge: Polity.
Miller, M. 1996. *Intimate Terrorism: The Crisis of Love in an Age of Disillusion*. London: W.W. Norton.
Morrison, C.-A. 2010. "'Home is Where the Heart Is': Everyday Geographies of Young Heterosexual Couples' Love in and of Homes." PhD dissertation, University of Waikato.
Morrison, C.-A. 2012. "Heterosexuality and Home: Intimacies of Space and Spaces of Touch." *Emotion, Space and Society* 5 (1): 10–18.
Morrison, C.-A. 2013. "Homemaking in New Zealand: Thinking Through the Mutually Constitutive Relationship Between Domestic Material Objects, Heterosexuality and Home." *Gender, Place and Culture: A Journal of Feminist Geography* 20 (4): 413–431.
Morrison, C.-A., L. Johnston, and R. Longhurst (2012) "Critical Geographies of Love as Spatial, Relational and Political." *Progress in Human Geography* 37 (4): 505–521.
Moss, P., and C. Donovan, eds. 2017. *Writing Intimacy into Feminist Geography*. London: Routledge.
Pile, S. 2010. "Emotions and Affect in Recent Human Geography." *Transactions of the Institute of British Geographers* 35: 5–20.
Pile, S. 2011. "For a Geographical Understanding of Affect and Emotions." *Transactions of the Institute of British Geographers* 36 (4): 603–606.
Probyn, E. 2000. *Carnal Appetites: FoodSexIdentities*. London: Routledge.
Probyn, E. 2005. *Blush: Faces of Shame*. Minneapolis: Minnesota Press.
Riley, D. 2005. *Impersonal Passion: Language as Affect*. Durham, DC: Duke University Press.
Rose, G. 1993. *Feminism and Geography: The Limits of Geographical Knowledge*. Cambridge: Polity Press.
Sedgwick, E.K. 2003. *Touching Feeling: Affect, Pedagogy, Performativity*. Durham, DC: Duke University Press.

Sedgwick, E.K., and A. Frank, eds. 1995. *Shame and Its Sisters: A Silvan Tomkins Reader.* Durham, NC: Duke University Press.
Sharp, J. 2009. "Geography and Gender: What Belongs to Feminist Geography? Emotion, Power and Change." *Progress in Human Geography* 33: 1–7.
Slocum, R. 2008. "Thinking Race through Corporeal Feminist Theory: Divisions and Intimacies at the Minneapolis Farmers' Market." *Social and Cultural Geography* 9 (8): 849–869.
Smith, M., J. Davidson, L. Cameron, and L. Bondi, eds. 2009. *Emotion, Place and Culture.* Farnham: Ashgate.
Smith, P. 2017. "Home is Where the Heart is Broken? Examining the Impact of Intimate Relationship Challenges on Meanings of Home." Unpublished Masters dissertation, University of Waikato.
Spinoza, B. 1677/2000. *Ethics,* translated by G.H.R. Parkinson. Oxford: Oxford University Press.
Stewart, K. 2007. *Ordinary Affects.* Durham, DC: Duke University Press.
Thien, D. 2005. "After or Beyond Feeling? A Consideration of Affect and Emotion in Geography." *Area* 37: 450–456.
Toye, M.E. 2010. Towards a Poethics of Love: Poststructuralist Feminist Ethics and Literary Creation. *Feminist Theory* 11 (1): 39–55.
Valentine, G. 2008. "The Ties That Bind: Towards Geographies of Intimacy." *Geography Compass* 2 (6): 2097–2110.
Waitt, G., and E. Stanes. 2015. "Sweating Bodies: Men, Masculinities, Affect, Emotion." *Geoforum* 59: 30–38.
Widdowfield, R. 2000. "The Place of Emotions in Academic Research." *Area* 32 (2): 199–208.
Williams, C. 2007. "Thinking the Political in the Wake of Spinoza: Power, Affect and Imagination." *Ethics, Contemporary Political Theory* 6: 349–369.
Women's Geography Study Group (WGSG). 1997. *Feminist Geographies: Explorations in Diversity and Difference.* London: Longman.
Wood, N., and S. Smith. 2004. "Instrumental Routes to Emotional Geographies." *Social and Cultural Geography* 5: 533–548.
Woodward, K., and J. Lea. 2009. "*Geographies of Affect.*" In: *The Sage Handbook of Social Geographies*, edited by S. Smith, R. Pain, S. Marston, and J.P. Jones, 154–175. London: Sage.
Wright, M. 2010. "Geography and Gender: Feminism and a Feeling of Justice." *Progress in Human Geography* 34 (6): 818–827.
Young, I.M. 1990. *Throwing Like a Girl, and Other Essays in Feminist Philosophy and Social Theory.* Bloomington: Indiana University Press.

9
ON THE SUBJECT OF PERFORMATIVITY
Judith Butler's influence in geography

Eden Kinkaid and Lise Nelson

Introduction

Since its articulation in 1988, Judith Butler's concept of performativity has been employed, critiqued and transformed by scholars in countless fields. Feminist geographers continue to explore the term's potentials and limitations for theoretical, methodological and empirical applications in diverse modes of geographic scholarship. This chapter charts the development of performativity and its uptake in geographic scholarship, with particular attention to the theorization of the subject in these applications. We begin with a summary of performativity in Butler's early work, before turning to geographic critiques and applications and, finally, to future directions for geographic inquiry. Throughout this discussion, we endeavour to keep geographic debates in conversation with Butler's evolving work on performativity, as it continues to provide direction in ongoing debates.

Philosophical foundations of performativity

Butler first articulates a performative theory of gender in 'Performative Acts and Gender Constitution' (1988). Against then-common understandings of gender as an internal essence linked to biological sex, Butler argues that gender is a discursive production and 'must be understood as the mundane way in which bodily gestures, movements and enactments of various kinds constitute the illusion of an abiding and gendered self' (1988, 519) by way of repetition and sedimentation. Instead of a stable, sexed self 'expressing' gender as an internal truth, Butler considers how various acts, comportments and discourses of the body actually *produce* the subject that they supposedly reveal. Rather than naming a gendered subject that pre-exists its constitution in language, discourse performatively produces this subject.

Butler further elaborates this argument in *Gender Trouble* (1990), clarifying the implications of gender performativity. She describes how 'the gendered body ... has no ontological status apart from the various acts which constitute its reality' (1990, 185). Thus 'gender is always a doing, though not a doing by a subject who might be said to preexist the deed' (1990, 34). This doing relies on the citation, or repetition, of norms and previous conventions, but may also break with these norms, given the possibility for slippage or parody to reinscribe their historical meanings.

In her next major work, *Bodies that Matter* (1993), it is clear that Butler is responding to critiques of *Gender Trouble*, which was read as being overly reliant on the analysis of the operations of discourse, at the expense of attending to the embodied practices of concrete subjects (see 1993, viii–xxiv). In *Bodies that Matter*, Butler attempts to further clarify the role of discourse in her argument:

> [t]o claim that discourse is formative is not to claim that it originates, causes, or exhaustively composes that which it concedes; rather, it is to claim that there is no reference to a pure body which is not at the same time a further formation of that body.
>
> *1993, xix*

Bodies that Matter, then, is concerned with illuminating the ways in which the 'materiality' of the body (i.e. the category of sex) is nonetheless inscribed with cultural meanings that actually mobilize and (re)produce the distinction between the material and the discursive that is at stake in these discussions.

Butler's concept of gender performativity has wide-ranging implications for how we think about issues of subjectivity and agency in relation to gender and other discourses. In her critique of the sovereign subject and voluntaristic notions of agency, Butler argues that we cannot imagine a subject outside of discourses of gender, as gender is one of the terms in which we recognize ourselves and others and become intelligible subjects in the first place. Gender is an 'enabling cultural condition' that provides the terms through which we might articulate an intention or desire as 'our own' (1993, xvii). Quoting Derrida, Butler describes: 'in such a typology, the category of intention will not disappear; it will have its place, but from that place it will no longer be able to govern the entire scene and system of utterance' (qtd. in 1993, xxi). In *Bodies that Matter*, Butler thus gestures toward the idea that agency and intention must be understood as situated within the limits of recognition and cultural intelligibility, *as constrained and enabled by existing meanings and practices* (e.g. those of determining and investing 'sex' with meaning and material consequence) rather than as a property of an individual. At times, Butler's own slippages and incomplete distinctions between performativity and performance, reliant on an ascription of intention and self-consciousness to the latter, have made her interventions less than perfectly clear.

Butler continued to engage with these questions of power, agency and subjectivity throughout the 1990s. In her most theoretically explicit discussion of performativity, *Excitable Speech* (1997), Butler elaborates upon the relationship between the body, speech, identity and agency. She again emphasizes that an attention to the linguistic does not occlude the body: 'if we consider that the *habitus* operates according to a performativity, then it would appear that the theoretical distinction between the social and the linguistic is difficult, if not impossible, to sustain' (1997, 153). Here, she takes up the issue of hate speech to consider how speech can cause bodily harm to others, demonstrating how speech constitutes a bodily act and how the linguistic and social cannot be separated.

In *Excitable Speech*, Butler also elaborates upon the performative power of discourse to produce the subjects that it names. She explains how one's interpellation by language (even when it constitutes misrecognition or harm) constitutes the founding of oneself as a named and speaking subject, establishing the conditions of possibility to redeploy and re-appropriate discourses. Butler describes:

> [t]he paradox of subjectivation is precisely that the subject who would resist such norms is itself enabled, if not produced, by such norms. Although this constitutive constraint does not foreclose the possibility of agency, it does locate agency as a reiterative or re-articulatory practice, immanent to power, and not a relation of external opposition to power.
>
> *1997, xxiii*

Excitable Speech, then, clarifies the necessity of a significant focus on language in Butler's account, in that the intersubjective linguistic realm must be seen as the conditions of possibility and, simultaneously, the historical limitations, of one's agency as a (non-sovereign) speaking subject. 'Whereas some critics mistake the critique of sovereignty for the demolition of agency,' Butler explains, 'I propose that agency begins where sovereignty wanes' (1997, 16).

These provocative and challenging interventions into issues of gender, the subject and agency have excited and puzzled scholars attempting to employ Butler's work in geographic research. Many of the aforementioned tensions in Butler's articulation – particularly as they concern the subject, body and agency – are highlighted and negotiated in subsequent applications by feminist and other geographers.

The next section presents geographic critiques of performativity and surveys the ways in which performativity has been used in geography. The questions posed here – the problem of the subject and agency – do not disappear, but continue to shape and trouble inquiries into the intersections of subjectivity, space and meaning.

Performativity in geography: space and the subject of performativity

As Butler's concept of performativity became widely circulated and debated throughout the 1990s, scholars began to explore its potential uses and limitations within particular disciplines. For feminist geographers concerned with the production of identity, discourse and social power, performativity offered a powerful new way to think about these processes. The concept, however, also presented stumbling blocks, as Butler's articulation did not speak much to the importance of space, place or other matters of concern to geographers.

In an early critique of performativity and its engagement by geographers, one of us – Nelson (1999) – raises concerns that geographers at that time relied exclusively on *Gender Trouble* and *Bodies that Matter* to theorize performativity, leaving most with an 'abstracted' subject and few ways to account for how historically and geographically concrete subjects, individually or collectively, might consciously navigate, resist or re-signify hegemonic norms. In these two early works, Nelson and others (see Magnus 2006; McNay 1999) take issue with Butler's formulation of resistance as the re-signification of dominant discourses, which seems to occur solely through 'accidental slippages' by unwitting subjects within an otherwise 'compelled repetition' of norms (Nelson 1999). Thus, for Nelson, '[w]ithout a critical reworking, Butler's notion of performativity actually undermines attempts to imagine a historically and geographically concrete subject that is constituted by dominant discourses, but is potentially able to reflect upon and actively negotiate, appropriate or resist them' (1999, 332).

Nelson's critique also draws attention to the paradoxical manner in which geographers working with performativity were sidestepping the issue of the subject, often by assuming a voluntaristic subject at odds with Butler's formulation. An early attempt to apply performative thinking to gendered and sexed space by Bell et al. (1994) has been critiqued by both Nelson (1999, 2014) and Lloyd (1999) for overemphasizing a voluntaristic subject without attending to the concrete contexts in which identities become expressed and read. This points to larger issues at play in formulations and applications of performativity; for Lloyd, '[w]hat is occluded ... is the space within which performance occurs, the others involved in or implicated by the production, and how they receive and interpret what they see' (1999, 210). Here, both authors emphasize problems and ambiguities in Butler's formulation, how these issues are amplified in subsequent applications and the need to take seriously the geographic specificity in which meanings are performed and produced.

Despite these concerns, Butler's concept of performativity has been widely engaged in a range of geographical scholarship because it provides powerful ways to conceptualize the embodied and continual re-enactment of identity by a subject constituted by – not separate from– wider power relations and discourses. In bringing performativity into geographic debates, both the subject matter of performativity has shifted (from gender to a larger consideration of space and social relations) and the philosophical 'subject' of performativity has been revised with a finer attention to how space and the subject are performatively co-constituted. As Gregson and Rose argue, 'a notion of performance is indeed crucial for a critical human geography concerned to understand the construction of social identity, social difference, and social power relations, and the way space might articulate all of these' (2000, 38). Further, they suggest, '[s]pace too needs to be thought of as brought into being through performances and as a performative articulation of power' (2000, 38).

Keeping these theoretical debates and interventions in mind, we now turn to a survey of applications of performativity in geography. Given the diversity of geographic approaches to thinking performatively, we thematize this work into three broad domains: critical identity; political geography; and economic geography. This survey is certainly not exhaustive, but we hope it will give some indication of the ways in which geographers are utilizing the concept of performativity and suggest some directions for future work.

Geographers engaging issues of identity, social space and power (notably feminist geographers) have employed performativity closest to its original foundations (i.e. as a critique of identity formation and subjectivity), yet have elaborated upon the geographic meanings of the concept to connect processes of subject formation to specific sites and spaces. Geographers of space and sexuality have used performativity to think through the co-constitution of space, normativities, sexual identities and practices (see contributions to Bell and Valentine 1995). Other scholars have employed the term in reference to masculinities (Barber 2015); beauty, race and place (Tate 2013); labour and gender identity (McDowell 1995; Secor 2003); and racial identity and sexual identity in particular spaces (Thomas 2004, 2005). As pointed out by Nelson (2014), these critiques of identity formation and subjectivity struggle to varying extents with the tensions present in Butler's original account by either overemphasizing the power of discourse and leaving little room for agency and resistance or by mistakenly importing a voluntaristic subject into their arguments. A closer empirical and theoretical attention to subjectivity as relational and located within spatial and place-based relations and meanings may help resolve some of these issues in future work.

Political geographers have employed performativity to consider how political identities, spaces and ideas are performatively co-constituted; that is, how political identities, practices and institutions are continually (re)articulated within specific struggles for political, representational and social space. Political geographers have applied performative thinking to private/public space (Sullivan 2016); electoral politics (Schurr 2013); nationality (Benwell 2014); citizenship and borders (Kaiser 2014); property and cartography (Blomley 2014); regionalism and development (Glass 2014); environmental governance (Cohen and Harris 2014); scale (Cohen and Harris 2014; Kaiser and Nikiforova 2008); and urban (re)development (Rose-Redwood 2014), among many other topics. For political geographers, then, the 'subject of performativity' is rather expansive; rather than focus exclusively on the production of gendered and sexed subjects through norms, they are more concerned with how discourses, representations and other political processes across scale define and, in doing so, call into being particular spaces and conceptions of the political. Treatments of resistance and contestation in these accounts provide various opportunities for thinking through issues of agency and intention within larger socio-spatial and political realms. While these developments toward notions of political performativity present interesting opportunities to engage the concept, it is important to note that these reworkings have often elided the importance of thinking about sex and gender, reinscribing sexless and genderless bodies as the subjects of political performativity.

Similarly, economic geographers have expanded the referent of performativity to think about the performative elements of economic discourses, practices and institutions. In doing so, they have urged against a simplistic and taken-for-granted approach to economies, pointing to how specific practices of exchange, representation and institutionalization performatively enact economies at various scales. Feminist economic geographers J.K. Gibson-Graham have argued that totalizing discourses of global capitalism erase the diversity of economic forms and unevenness of capitalist 'penetration' and invisibilize already-existing alternatives to capitalist economies (2008). Similarly, Steinfort, Hendrikx and Pijpers (2017) consider 'communal performativity' as a means of resisting neoliberalism. Others have used innovative methodologies, like actor-network theory, to attend to the specific relations and processes that make economies what they are (Berndt and Boeckler 2009; Callon 2007). Here, we see the terms of performativity intersecting with concepts and methodologies of practice, leading to a redefinition of the subject, sites and scale of the performative.

From the preceding discussion, it is clear that performativity has deeply impacted on geographic thought in feminist geographies and beyond, and that the concept has transformed in some surprising ways. After decades of scholarship and debate, many of the original tensions presented in Butler's early work are still visible, though they have become refracted, redefined and relocated in various geographical and epistemological sites. In a reflection of the 'ongoing limits' of performative thinking in geography, Nelson (2014) partly attributes these remaining tensions to an overreliance on Butler's early work and a general lack of engagement with the manner in which Butler herself has revised her thinking. Engaging later works of Butler's – ones more concerned with relations, ethics and precarity (2001 onward) – Nelson argues, might change the way we imagine the subject of performativity, leading to more productive engagements with performativity in geography.

We argue here that yet another shift is visible in Butler's most recent scholarship, which more squarely attempts to bring together earlier investigations into gender performativity with the aforementioned shift toward ethics, human community and political struggles. These later works offer some interesting and unanticipated responses to the issues that have organized and polarized debates about performativity in geography. A renewed attention to Butler's work and the ways in which she attempts to rework the terms of the discussion will certainly be instructive for ongoing applications of performativity in geography, especially given the 'relational turn' evident in both Butler's thought and emerging subfields of geography.

Future directions: taking the 'relational turn' with performativity

Having summarized the development of performativity and how it has been employed in geography, in this final section we consider some potential sites of innovation in performative thinking and future directions for geographic work. We briefly engage Butler's recent work to consider how she has reworked performativity, before turning to forward-looking experiments with performativity that, like Butler's later work, signal a 'relational turn': a theoretical reorientation that has illuminated new conceptual terrain in the social sciences and humanities (for an overview of this 'relational turn' in geography, see Jones 2009). While many of these promising geographic applications of performativity predate and perhaps anticipate Butler's shifts in perspective, we nonetheless argue that is productive to see them as part of the continuous and iterative development of performative thinking.

Butler's recent work has displayed an interesting reorientation to questions of subjectivity, performativity, relationality, freedom and ethics. In *Senses of the Subject*, Butler adopts a distinctly phenomenological language to reframe the subject of her earlier work. Specifically, she emphasizes the inseparability of linguistic and bodily interpellation (2015a, 14) and the fundamental relationality and dependency that characterize the formation of subjectivity. Issues of intention and agency, likewise, are redefined, as Butler turns away from the terms of discourse toward a phenomenological account of subjectivity:

> Something is already underway by the time we act, and we cannot act without, in some sense, being acted upon. This acting that is upon us constitutes a realm of primary impressionability so that by the time we act, we enter into the action, we resume it in our name, it is an action that has its sources only partially and belatedly in something called a subject.
>
> *2015, 62–63*

Here, Butler wrestles with the fundamental paradox of subjectivity: how we are necessarily produced through our relations with others, yet we nonetheless recognize ourselves and act as an 'I'. Rather than jettison this 'I' as a site of agency, meaning and intention, Butler emphasizes the need to constantly reflect on how the 'I' is never fully formed by these others, nor fully self-fashioning. This formulation echoes and extends Butler's earlier claims that attending to the relational production of subjects through a gendered matrix 'is not to do away with the subject, but only to ask after the conditions of its emergence and operation' (1993, xvii). Here, in this relational conception of subjectivity, we encounter new, and renewed, ethical and ontological problems.

These themes are further developed in *Notes Toward a Performative Theory of Assembly*, wherein Butler considers how political protest and assembly might be thought as performative. In doing so, she cautions that we cannot 'extract the body from its constituting relations – and those relations are always economically and historically specific' (2015b, 148). Here again, Butler offers a sensitive account of how subjectivity and space are performative and how they arise at a paradoxical site defined equally by the sedimentations of history and the possibilities of the future enacted through a yet to be defined collectivity. Responding to the question of the 'agency' of the individual subject, she offers: '[f]reedom does not come from me or from you; it can and does happen as a relation between us, or, indeed, among us' (2015b, 88). These passages only hint at a larger reorientation of Butler's thought, one that holds much interest and potential for geographers both excited and frustrated by her ideas.

These new formulations, bringing together Butler's post-structuralist, phenomenological and political thinking, offer fascinating directions for further geographic work. Butler's later work, then, presents rich resources for geographers, engaging parallel theoretical shifts that we might call a 'relational turn'. Here, we use 'relational turn' to signal a diverse and emerging cluster of distinct but related perspectives – including but not limited to feminist new materialisms (FNM), non-representational theory (NRT), actor network theory (ANT) and strands of queer theory – through which scholars have reconsidered and redrawn conceptual mappings that have structured theoretical debates since the rise and consolidation of post-structuralism in the social sciences. In this section, we present work within geography and related fields that foreshadows and engages this turn, pointing to the ongoing potential of conversations between Butler's work and geographers.

Bringing Butler into conversation with Deleuze, Dewsbury (2000) usefully redefines issues of the subject by taking a relational approach to the problem of performativity, refocusing on common human practices that produce subjects and worlds, rather than subjects as such (2000, 477). This turn toward practice as a site of performativity is shared with other approaches to geography, including NRT and ANT (see Thrift 2003). In a critical review of NRT, Nash describes that it 'is concerned with practices through which we become subjects "decentered", affective, but embodied, relational, expressive and involved with others and objects in a world continually in process' (2000, 655). Nash suggests that a critical employment of NRT may create new potentials for geographic thinking, well-illustrated through her example of the historically and culturally located movements and meanings of dance. While these paradigms present interesting potentials for cross-pollination with Butler's formulations, both ANT and NRT must be approached critically, in that they both appropriate and develop the concept of performativity without drawing from feminist theory or considering gender as a necessary analytical term.

Within feminist theory, performativity has undergone related shifts. For example, feminist scholar Horowitz takes on the thorny ontological questions of Butler's account by revisiting and rethinking the (non)distinction between performance and performativity. Drawing on FNM and queer theory, she interrogates and reworks the metaphysical commitments of performative thinking, recasting performativity as an inquiry into the interaction of heterogeneous sites, discourses, bodies and subjects. For Horowitz, then, the 'referent' of performativity is necessarily a constellation of relations, rather than a subject (2013). Other feminist scholars have further 'queered' the subject of performativity, stressing processes of becoming and unbecoming and their implications for ideas of identity, agency and freedom (Bunch 2013). Further, feminist scholars working within science and technology studies have pushed the boundaries of performativity for thinking through philosophical, methodological and ethical questions posed by posthumanism and other relational ontologies (Birke, Bryld and Lykke 2004; Hovorka 2015). These directions hold much potential for geographers, inasmuch as they are concerned with the constellations of relations that produce space and subjects in historically, culturally and geographically specific ways.

A contextual and relational approach is shared across these innovative revisions of performative thinking, including FNM, ANT, NRT and strands of queer theory. What these distinct but philosophically related approaches share is a commitment to questioning taken-for-granted philosophical categories and exploring the implications of relational thinking for long-standing philosophical problems, like agency and subjectivity. We can expect, given these philosophical overlaps with the more contemporary directions of Butler's project, that there is much potential for further articulating performativity after a 'relational turn'.

Conclusion

While these complex approaches to rethinking the problems of representation, materiality and agency present us with new theoretical and political challenges, they also open up exciting spaces for critical geography. The most tangible takeaway from these critiques of performativity may be methodological: from the start, geographers have urged practitioners of performativity to maintain a critical attention to historical, cultural and geographic specificity. Where, then, do these critiques leave us as producers of geographic knowledge?

Geographic debates on performativity over the decades have clarified the uses and qualifications of any approach to performativity. While the performative remains a powerful tool for geographic thinking, it must be finely tuned to context. Nelson's critique emphasized a reorientation toward fieldwork theory as one way of producing accounts that are geographically and historically grounded (1999). Further, Nelson and others (Benwell 2014) argue that researchers using performativity must be attentive to the ethical implications of applying it to the 'subjects' of their research and must maintain critical reflexivity in doing so. A greater attention to specific historical and empirical contexts is, indeed, necessary; Nash argues that '[p]utting theories of performativity to work in discussing specific practices makes for better theory' (Nash 2000, 661). Similarly, Dewsbury explains: 'whilst the performative, as a theoretical tool or concept, can be used in any given circumstance, its usefulness and what it uncovers and creates are fundamentally specific to the context in which it is sited' (Dewsbury 2000, 475). In terms of writing and representation, McCormack argues that, 'insofar as writing is performative, one of its keys tasks might be to elicit those qualities of difference which are always excessive of textual strategies' and that we must find 'ways of allowing the qualities of movement to animate the logics of geographical thinking' (2009, 136). Amid the proliferation of innovative uses of performative thinking in geography, it is critical that the term remain connected to its original formulation; that is, as a critique of gendered and sexed subjectivity. These requirements present complex methodological challenges, yet they provide us with exciting directions for ongoing experiments with performative thinking.

In conclusion, the diversity of approaches to performativity in feminist geographies, and geography more generally, evidences the critical impact that the term has had on the field and its potential for developing ways to think about the interconnections of the subject, space and politics. Over the last two decades, critical and timely interventions have pointed out the tensions, contradictions and potentials in performative thinking and highlighted the stakes of engaging the term. Further, geographic debates on performativity, particularly the interventions posed by feminist geographers, have consistently redirected us to critical philosophical, ethical and methodological questions: namely, the question of why and how the way we theorize (gendered and sexed) subjectivity and space come to matter. Rather than be alarmed by the resistance of the term 'performativity' to settle into a strict definition or be dismayed by the difficulty of resolving the problem of the subject into conventional wisdom, we might welcome these continuing provocations and transformations in geographic thought. After all, as Butler reminds us:

> [t]o question a term, a term like 'the subject' … is to ask how it plays, what investments it bears, what aims it achieves, what alterations it undergoes … If a term becomes questionable, does that mean it cannot be used any longer, and that we can only use terms that we *already know how to master?*
>
> *Butler 1997, 162*

As of yet, 'performativity' shows no signs of mastery; nonetheless, we encourage geographers to continue to mobilize it in creative and critical explorations of geographic thought.

Key readings

Butler, J. 1997. *Excitable Speech: A Politics of the Performative*. New York: Routledge.
Butler, J. 2015. *Notes Toward a Performative Theory of Assembly*. Cambridge, MA: Harvard University Press.
Loxley, J. 2006. *Performativity*. London: Routledge.

References

Barber, T. 2015. "Performing 'Oriental' Masculinities: Embodied Identities Among Vietnamese Men in London." *Gender, Place & Culture* 22 (3): 440–455.

Bell, D., J. Binnie, J. Cream, and Valentine G. 1994. "All Hyped Up and No Place to Go." *Gender, Place and Culture: A Journal of Feminist Geography* 1 (1): 31–47.

Bell, D., and G. Valentine, eds. 1995. *Mapping Desire: Geographies of Sexualities*. New York: Routledge.

Benwell, M.C. 2014. "Considering Nationality and Performativity: Undertaking Research Across the Geopolitical Divide in the Falkland Islands and Argentina." *Area* 46 (2): 163–169.

Berndt, C., and M. Boeckler. 2009. "Geographies of Circulation and Exchange: Constructions of Markets." *Progress in Human Geography* 33 (4): 535–551.

Birke, L., M. Bryld, and N. Lykke. 2004. "Animal Performances: An Exploration of Intersections Between Feminist Science Studies and Studies of Human/Animal Relationships." *Feminist Theory* 5 (2): 167–183.

Blomley. N. 2014. "Disentangling Property, Performing Space." In: *Performativity, Politics, and the Production of Social Space,* edited by M.R. Glass and R. Rose-Redwood, 169. New York: Routledge.

Bunch, M. 2013. "The Unbecoming Subject of Sex: Performativity, Interpellation, and the Politics of Queer Theory." *Feminist Theory* 14 (1): 39–55.

Butler, J. 1988. "Performative Acts and Gender Constitution: An Essay in Phenomenology and Feminist Theory." *Theatre Journal* 40 (4): 519–531.

Butler, J. 1990. *Gender Trouble: Feminism and the Subversion of Identity.* New York: Routledge.

Butler, J. 1993. *Bodies That Matter: On the Discursive Limits of Sex*. New York: Routledge.

Butler, J. 1997. *Excitable Speech: A Politics of the Performative*. New York: Routledge.

Butler, J. 2015a. *Senses of the Subject*. New York: Fordham University Press.

Butler, J. 2015b. *Notes Toward a Performative Theory of Assembly*. New York: Routledge.

Callon, M. 2007. "What Does It Mean to Say That Economics is Performative?" In: *Do Economists Make Markets? On the Performativity of Economics,* edited by D. MacKenzie, F. Muniesa, and L. Siu, 311–358. Princeton, NJ: Princeton University Press.

Cohen, A., and L. Harris. 2014. "Performing Scale: Watersheds as 'Natural' Governance Units in the Canadian Context." In *Performativity, Politics, and the Production of Social Space,* edited by M.R. Glass and R. Rose-Redwood, 226–252. New York: Routledge.

Dewsbury, J.D. 2000. "Performativity and the Event: Enacting a Philosophy of Difference." *Environment and Planning D: Society and Space* 18 (4): 473–496.

Gibson-Graham, J.K. 2008. "Diverse Economies: Performative Practices for Other Worlds." *Progress in Human Geography* 32 (5): 613–632.

Glass, M.R. 2014. "Becoming a Thriving Region: Performative Visions, Imaginative Geographies, and the Power of 32." In: *Performativity, Politics, and the Production of Social Space,* edited by M.R. Glass and R. Rose-Redwood, 202–225. New York: Routledge.

Gregson, N., and G. Rose. 2000. "Taking Butler Elsewhere: Performativities, Spatialities and Subjectivities." *Environment and Planning D: Society and Space* 18 (4): 433–452.

Horowitz, K.R. 2013. "The Trouble With 'Queerness': Drag and the Making of Two Cultures." *Signs: Journal of Women in Culture and Society* 38 (2): 303–326.

Hovorka, A.J. 2015. "The Gender, Place and Culture." Jan Monk Distinguished Annual Lecture: Feminism and Animals: Exploring Interspecies Relations Through Intersectionality, Performativity and Standpoint. *Gender, Place & Culture* 22 (1): 1–19.

Jones, M. 2009. "Phase Space: Geography, Relational Thinking, and Beyond." *Progress in Human Geography* 33 (4): 487–506.

Kaiser, R.J. 2014. "Performativity, Events, and Becoming-Stateless." In: *Performativity, Politics, and the Production of Social Space,* edited by M.R. Glass and R. Rose-Redwood, 121–146. New York: Routledge.

Kaiser, R., and E. Nikiforova. 2008. "The Performativity of Scale: The Social Construction of Scale Effects in Narva, Estonia." *Environment and Planning D: Society and Space* 26 (3): 537–562.

Lloyd, M. 1999. "Performativity, Parody, Politics." *Theory, Culture & Society*, 16 (2): 195–213.

Magnus, K.D. 2006. "The Unaccountable Subject: Judith Butler and the Social Conditions of Intersubjective Agency." *Hypatia* 21 (2): 81–103.
McCormack, D.P. 2009. "Performativity." In: *International Encyclopedia of Human Geography*, edited by R. Kitchin and N. Thrift, 133–136. Boston: Elsevier.
McDowell, L. 1995. "Body Work: Heterosexual Gender Performances in City Workplaces." In: *Mapping desire: Geographies of sexualities*, edited by D. Bell and G. Valentine, 75–95. New York: Routledge.
McNay, L. 1999. "Subject, Psyche and Agency: The Work of Judith Butler." *Theory Culture & Society* 16 (2): 175–193.
Nash, C. 2000. "Performativity in Practice: Some Recent Work in Cultural Geography." *Progress in Human Geography* 24 (4): 653–664.
Nelson, L. 1999. "Bodies (and Spaces) Do Matter: The Limits of Performativity." *Gender, Place and Culture: A Journal of Feminist Geography* 6 (4): 331–353.
Nelson, L. 2014. "Engaging Butler: Subjects, Cernment and Ongoing Limits of Performativity." In: *Performativity, Politics, and the Production of Social Space,* edited by M.R. Glass and R. Rose-Redwood, 62–94. New York: Routledge.
Rose-Redwood, R.S. 2014. "'Sixth Avenue is Now a Memory:" Regimes of Spatial Inscription and the Performative Limits of the Official City-text." In: *Performativity, Politics, and the Production of Social Space*, edited by M.R. Glass and R. Rose-Redwood, 176–201. New York: Routledge.
Schurr, C. 2013. "Towards An Emotional Electoral Geography: The Performativity of Emotions in Electoral Campaigning in Ecuador." *Geoforum* 49: 114–126.
Secor, A. J. 2003. "Belaboring Gender: The Spatial Practice of Work and the Politics of 'Making Do' in Istanbul." *Environment and Planning A* 35 (12): 2209–2227.
Steinfort, L., B. Hendrikx, and R. Pijpers. 2017. "Communal Performativity – A Seed For Change? The Solidarity of Thessaloniki's Social Movements in the Diverse Fights Against Neoliberalism." *Antipode* 49 (5): 1446–1463.
Sullivan, R. 2016. *Geography Speaks: Performative Aspects of Geography*. New York: Routledge.
Tate, S. 2013. "The Performativity of Black Beauty Shame in Jamaica and its Diaspora: Problematising and Transforming Beauty Iconicities." *Feminist Theory* 14 (2): 219–235.
Thomas, M.E. 2004. "Pleasure and Propriety: Teen Girls and the Practice of Straight Space." *Environment and Planning D: Society and Space* 22 (5): 773–789.
Thomas, M.E. 2005. "'I Think It's Just Natural': The Spatiality of Racial Segregation at a US High School." *Environment and Planning A* 37 (7): 1233–1248.
Thrift, N. 2003. "Performance and… ." *Environment and Planning A* 35 (11): 2019–2024.

10
POLITICS AND SPACE/TIME

Doreen Massey

'Space' is very much on the agenda these days. On the one hand, from a wide variety of sources come proclamations of the significance of the spatial in these times: 'It is space not time that hides consequences from us' (Berger); 'The difference that space makes' (Sayer); 'That new spatiality implicit in the postmodern' (Jameson); 'It is space rather than time which is the distinctively significant dimension of contemporary capitalism' (Urry); and 'All the social sciences must make room for an increasingly geographical conception of mankind' (Braudel). Even Foucault is now increasingly cited for his occasional reflections on the importance of the spatial. His 1967 Berlin lectures contain the unequivocal: 'The anxiety of our era has to do fundamentally with space, no doubt a great deal more than with time.' In other contexts the importance of the spatial, and of associated concepts, is more metaphorical. In debates around identity the terminology of space, location, positionality and place figures prominently. Homi Bhabha, in discussions of cultural identity, argues for a notion of a 'third space'. Jameson, faced with what he sees as the global confusions of postmodern times, 'the disorientation of saturated space', calls for an exercise in 'cognitive mapping'. And Laclau, in his own very different reflections on the 'new revolution of our time', uses the terms 'temporal' and 'spatial' as the major differentiators between ways of conceptualizing systems of social relations.

In some ways, all this can only be a delight to someone who has long worked as a 'geographer'. Suddenly the concerns, the concepts (or, at least, the *terms*) which have long been at the heart of our discussion are at the centre also of wider social and political debate. And yet, in the midst of this gratification I have found myself uneasy about the way in which, by some, these terms are used. Here I want to examine just one aspect of these anxieties about some of the current uses of spatial terminology: the conceptualization (often implicit) of the term 'space' itself.

In part this concern about what the term 'space' is intended to mean arises simply from the multiplicity of definitions adopted. Many authors rely heavily on the terms 'space'/'spatial', and each assumes that their meaning is clear and uncontested. Yet in fact the meaning that different authors assume (and therefore – in the case of metaphorical usage – the import of the metaphor) varies greatly. Buried in these unacknowledged disagreements is a debate that never surfaces; and it never surfaces because everyone assumes we already know what these terms mean. Henri Lefebvre, in the opening pages of his book *The Production of Space*, commented on just this phenomenon: the fact that authors who in so many ways excel in logical rigour will fail to define a term which functions crucially in their argument: 'Conspicuous by its absence from supposedly epistemological studies is … the idea … of space – the fact that "space" is mentioned on every page notwithstanding.'[1] At least there ought to be a debate about the meaning of this much-used term.

Nonetheless, had this been all that was at issue I would probably not have been exercised to write an article about it. But the problem runs more deeply than this. For among the many and conflicting definitions of space that are current in the literature there are some – and very powerful ones – which deprive it of politics and of the possibility of politics: they effectively depoliticize the realm of the spatial. By no means all authors relegate space in this way. Many, drawing on terms such as 'centre'/'periphery'/'margin', and so forth, and examining the 'politics of location' for instance, think of spatiality in a highly active and politically enabling manner. But for others space is the sphere of the lack of politics.

Precisely because the use of spatial terminology is so frequently unexamined, this latter use of the term is not always immediately evident. This dawned fully on me when I read a statement by Ernesto Laclau in his *New Reflections on the Revolution of Our Time*. 'Politics and space,' he writes on page 68, 'are antinomic terms. Politics only exist insofar as the spatial eludes us.'[2] For someone who, as a geographer, has for years been arguing, along with many others, for a dynamic and politically progressive way of conceptualizing the spatial, this was clearly provocative!

Because my own inquiries were initially stimulated by Laclau's book, and because unearthing the implicit definitions at work implies a detailed reading (which restricts the number of authors who can be considered) this discussion takes *New Reflections* as a starting point, and considers it in most detail. But, as will become clear, the implicit definition used by Laclau, and which depoliticizes space, is shared by many other authors. In its simpler forms it operates, for instance, in the debate over the nature of structuralism, and is an implicit reference point in many texts. It is, moreover, in certain of its fundamental aspects shared by authors, such as Fredric Jameson, who in other ways are making arguments very different from those of Laclau.

To summarize it rather crudely, Laclau's view of space is that it is the realm of stasis. There is, in the realm of the spatial, no true temporality and thus no possibility of politics. It is on this view, and on a critique of it, that much of my initial discussion concentrates. But in other parts of the debate about the nature of the current era, and in particular in relation to 'postmodernity', the realm of the spatial is given entirely different associations from those ascribed to it by Laclau. Thus Jameson, who sees postmodern times as being particularly characterized by the importance of spatiality, interprets it in terms of an unnerving multiplicity: space is chaotic depthlessness.[3] This is the opposite of Laclau's characterization, yet for Jameson it is – once again – a formulation which deprives the spatial of any meaningful politics.

A caveat must be entered from the start. This discussion will be addressing only one aspect of the complex realm that goes by the name of the spatial. Lefebvre, among others, insisted on the importance of considering not only what might be called 'the geometry' of space but also its lived practices and the symbolic meaning and significance of particular spaces and spatializations. Without disagreeing with that, the concentration here will nonetheless be on the view of space as what I shall provisionally call 'a dimension'. The argument is that different ways of conceptualizing this aspect of 'the spatial' themselves provide very different bases (or in some cases no basis at all) for the politicization of space. Clearly, anyway, the issue of the conceptualization of space is of more than technical interest; it is one of the axes along which we experience and conceptualize the world.

Space and time

An examination of the literature reveals, as might be expected, a variety of uses and meanings of the term 'space', but there is one characteristic of these meanings that is particularly strong and widespread. This is the view of space which, in one way or another, defines it as stasis, and as utterly opposed to time. Laclau, for whom the contrast between what he labels temporal and what he calls spatial is key to his whole argument, uses a highly complex version of this definition. For him, notions of time and space are related to contrasting methods of understanding social systems. In his *New Reflections on the Revolution of Our Time*, Laclau posits that 'any repetition that is governed by a structural law of successions is space' (41) and 'spatiality means coexistence within a structure that establishes the positive nature of all its terms' (69). Here, then, any postulated causal structure which is complete and self-determining is labelled 'spatial'. This does not mean that such a 'spatial' structure cannot change – it may do – but the essential characteristic is that all the causes of any change which may take place are internal to the structure itself. On this view, in the realm of the spatial there can be no surprises (provided we are analytically well-equipped). In contrast to the closed and self-determining systems of the spatial, Time (or temporality) for Laclau takes the form of dislocation, a dynamic which disrupts the predefined terms of any system of causality. The spatial, because it lacks dislocation, is devoid of the possibility of politics.

This is an importantly different distinction between time and space from that which simply contrasts change with an utter lack of movement. In Laclau's version, there can be movement and change within a so-called spatial system; what there cannot be is real dynamism in the sense of a change in the terms of 'the system' itself (which can therefore never be a simply coherent closed system). A distinction is postulated, in other words, between different types of what would normally be called time. On the one hand, there is the time internal to a closed system, where things may change yet without really changing. On the other hand, there is genuine dynamism, Grand Historical Time. In the former is included cyclical time, the times of reproduction, the way in which a peasantry represents to itself (says Laclau, 42) the unfolding of the cycle of the seasons, the turning of the earth. To some extent, too, there is 'embedded time', the time in which our daily lives are set.[4] These times, says Laclau, this kind of 'time' is space.

Laclau's argument here is that what we are inevitably faced with in the world are 'temporal' (by which he means dislocated) structures: dislocation is intrinsic and it is this – this essential openness – which creates the possibility of politics. Any attempt to represent the world 'spatially', including even the world of physical space, is an attempt to ignore that dislocation. Space therefore, in his terminology, is representation, is any (ideological) attempt at closure: 'Society, then, is unrepresentable: any representation – *and thus any space* – is an attempt to constitute society, not to state what it is' (82, my emphasis). Pure spatiality, in these terms, cannot exist: 'The ultimate failure of all hegemonisation [in Laclau's term, spatialization], then, means that the real – including physical space – is in the ultimate instance temporal' (42); or again: 'the mythical nature of any space' (68). This does not mean that the spatial is unimportant. This is not the point at issue, nor is it Laclau's intent. For the 'spatial' as the ideological/mythical is seen by him as itself part of the social and as constitutive of it: 'And insofar as the social is impossible without some fixation of meaning, without the discourse of closure, the ideological must be seen as constitutive of the social' (92).[5] The issue here is not the relative priority of the temporal and the spatial, but their definition. For it is through this logic, and its association of ideas with temporality and spatiality, that Laclau arrives at the depoliticization of space. 'Let us begin,' writes Laclau, 'by identifying three dimensions of the relationship of dislocation that are crucial to our analysis. The *first* is that dislocation is the very form of temporality. And temporality must be conceived as the exact opposite of space. The "spatialization" of an event consists of eliminating its temporality' (41; my emphasis).

The second and third dimensions of the relationship of dislocation (see above) take the logic further: 'The *second* dimension is that dislocation [which, remember, is the antithesis of the spatial] is the very form of possibility', and 'The *third* dimension is that dislocation is the very form of freedom. Freedom is the absence of determination' (42, 43, my emphases). This leaves the realm of the spatial looking like unpromising territory for politics. It is lacking in dislocation, the very form of possibility (the form of temporality), which is also 'the very form of freedom'. Within the spatial there is only determination, and hence no possibility of freedom or of politics.

Laclau's characterization of the spatial is, however, a relatively sophisticated version of a much more general conception of space and time (or spatiality and temporality). It is a conceptualization in which the two are opposed to each other, and in which time is the one that matters and of which History (capital h) is made. Time Marches On but space is a kind of stasis, where nothing really happens. There are a number of ways in which, it seems to me, this manner of character-izing space and the realm of the spatial is questionable. Three of them, chosen precisely because of their contrasts, because of the dis-tinct light they each throw on the problems of this view of space, will be examined here. The first draws on the debates that have taken place in 'radical geography' over the last two decades and more; the second examines the issue from the point of view of a concern with gender; and the third examines the view from physics.

Radical geography

In the 1970s the discipline of geography experienced the kinds of developments described by Anderson in 'A Culture in Contraflow'[6] for other social sciences. The previously hegemonic positivist 'spatial science' was increasingly challenged by a new generation of Marxist geographers. The argument turned intellectually on how 'the relation between space and society' should be conceptualized. To caricature the debate, the spatial scientists had posited an autonomous sphere of the spatial in which 'spatial relations' and 'spatial processes' produced spatial distributions. The geography of industry, for instance, would be interpreted as simply the result of 'geographical location factors'. Countering this, the Marxist critique was that all these so-called spatial relations and spatial processes were actually social relations taking a particular geographical form. The geography of industry, we argued, could therefore not be explained without a prior understanding of the economy and of wider social and political processes. The aphorism of the seventies was 'space is a social construct'. That is to say – though the point was perhaps not made clearly enough at the time – space is constituted through social relations and material social practices.

But this, too, was soon to seem an inadequate characterization of the social/spatial relation. For, while it is surely correct to argue that space is socially constructed, the one-sideness of that formulation implied that geographical forms and distributions were simply outcomes, the end point of social explanation. Geographers would thus be the cartographers of the social sciences, mapping the outcomes of processes which could only be explained in other disciplines – sociology, economics, and so forth. What geographers mapped – the spatial form of the social – was interesting enough, but it was simply an end product: it had no material effect. Quite apart from any demeaning disciplinary implications, this was plainly not the case. The events taking place all around us in the 1980s – the massive spatial restructuring both intranationally and internationally as an integral part of social and economic changes – made it plain that, in one way or another, 'geography matters'. And so, to the aphorism of the 1970s – that space is socially constructed – was added in the 1980s the other side of the coin: that the social is spatially constructed too, and that makes a difference. In other words, and in its broadest formulation, society is necessarily constructed spatially, and that fact – the spatial organization of society – makes a difference to how it works.

But if spatial organization makes a difference to how society works and how it changes, then far from being the realm of stasis, space and the spatial are also implicated (*contra* Laclau) in the production of history – and thus, potentially, in politics. This was not an entirely new thought. Henri Lefebvre, writing in 1974, was beginning to argue a very similar position:

> The space of capitalist accumulation thus gradually came to life, and began to be fitted out. This process of animation is admiringly referred to as history, and its motor sought in all kinds of factors: dynastic interests, ideologies, the ambitions of the mighty, the formation of nation states, demographic pressures, and so on. This is the road to a ceaseless analysing of, and searching for, dates and chains of events. Inasmuch as space is the locus of all such chronologies, might it not constitute a principle of explanation at least as acceptable as any other?[7]

This broad position – that the social and the spatial are inseparable and that the spatial form of the social has causal effectivity – is now accepted increasingly widely, especially in geography and sociology,[8] though there are still those who would disagree, and beyond certain groups even the fact of a debate over the issue seems to have remained unrecognized (Anderson, for example, does not pick it up in his survey).[9] For those familiar with the debate, and who saw in it an essential step towards the politicization of the spatial, formulations of space as a static resultant without any effect – whether the simplistic versions or the more complex definitions such as Laclau's – seem to be very much a retrograde step. However, in retrospect, even the debates within radical geography have still fully to take on board the implications of our own arguments for the way in which space might be conceptualized.

Issues of gender

For there are also other reservations, from completely different sources, that can be levelled against this view of space and that go beyond the debate which has so far taken place within radical geography. Some of these reservations revolve around issues of gender.

First of all, this manner of conceptualizing space and time takes the form of a dichotomous dualism. It is neither a simple statement of difference (a, b, ...) nor a dualism constructed through an analysis of the interrelations between the objects being defined (capital:labour). It is a dichotomy specified in terms of a presence and an absence; a dualism which takes the classic form of a/not-a. As was noted earlier, one of Laclau's formulations of a definition is: 'temporality must be conceived as the exact opposite of space' (41). Now, apart from any reservations which may be raised in the particular case of space and time (and which we shall come to later), the mode of thinking that relies on irreconcilable dichotomies of this sort has in general recently come in for widespread criticism. All the strings of these kinds of opposition with which we are so accustomed to work (mind–body; nature–culture; Reason–emotion; and so forth) have been argued to be at heart problematical and a hindrance to either understanding or changing the world. Much of this critique has come from feminists.[10]

The argument is two-fold. First, and less importantly here, it is argued that this way of approaching conceptualization is, in Western societies and more generally in societies where child-rearing is performed overwhelmingly by members of one sex (women), more typical of males than of females. This is an argument which generally draws on object-relations-theory approaches to identity-formation. Second, however, and of more immediate significance for the argument being constructed here, it has been contended that this kind of dichotomous thinking, together with a whole range of the sets of dualisms that take this form (we shall look at some of these in more detail below) are related to the construction of the radical distinction between genders in our society, to the characteristics assigned to each of them, and to the power relations maintained between them. Thus, Nancy Jay, in an article entitled 'Gender and Dichotomy', examines the social conditions and consequences of the use of logical dichotomy.[11] She argues not only that logical dichotomy and radical gender distinctions are associated but also, more widely, that such a mode of constructing difference works to the advantage of certain (dominant) social groups, 'that almost any ideology based on a/Not-a dichotomy is effective in resisting change. Those whose understanding of society is ruled by such ideology find it very hard to conceive of the possibility of alternative forms of social order (third possibilities). Within such thinking, the only alternative to the *one* order is disorder' (54). Genevieve Lloyd, too, in a sweeping history of 'male' and 'female' in Western philosophy, entitled *The Man of Reason*, argues that such dichotomous conceptualization, and – what we shall come to later – the prioritization of one term in the dualism over the other, is not only central to much of the formulation of concepts with which Western philosophy has worked but that it is dependent upon, and is instrumental in the conceptualization of, among other things, a particular form of radical distinction between female and male genders.[12] Jay argues that 'Hidden, taken for granted, a/Not-a distinctions are dangerous, and because of their peculiar affinity with gender distinctions, it seems important for feminist theory to be systematic in recognizing them' (47). The argument here is that the definition of 'space' and 'time' under scrutiny here is precisely of this form, and on that basis alone warrants further critical investigation.

But there is also a further point. For within this kind of conceptualization, only one of the terms (a) is defined positively. The other term (not-a) is conceived only in relation to a, and as lacking in a. A fairly thorough reading of some of the recent literature that uses the terminology of space and time, and that employs this form of conceptualization, leaves no doubt that it is Time which is conceived of as in the position of 'a', and space which is 'not-a'. Over and over again, time is defined by such things as change, movement, history, dynamism; while space, rather lamely by comparison, is simply the absence of these things. This has two aspects. First, this kind of definition means that it is time, and the characteristics associated with time, that are the primary constituents of both space and time; time is the nodal point, the privileged signifier. And second, this kind of definition means that space is defined by absence, by lack. This is clear in the simple (and often implicit) definitions (time equals change/movement, space equals the lack of these things), but it can also be argued to be the case with more complex definitions such as those put forward by Laclau. For although in a formal sense it is the spatial which in Laclau's formulation is complete and the temporal which marks the lack (the absence of representation, the impossibility of closure), in the whole tone of the argument it is in fact space that is associated with negativity and absence. Thus: 'Temporality must be conceived as the exact opposite of space. The "spatialization" of an event consists of eliminating its temporality' (41).

Now, of course, in current Western culture, or in certain of its dominant theories, woman too is defined in terms of lack. Nor, as we shall see, is it entirely a matter of coincidence that space and the feminine are frequently defined in terms of dichotomies in which each of them is most commonly defined as not-a. There is a whole set of dualisms whose terms are commonly aligned with time and space. With Time are aligned History, Progress, Civilization, Science, Politics and Reason, portentous things with gravitas and capital letters. With space on the other hand are aligned the other poles of these concepts: stasis, ('simple') reproduction, nostalgia, emotion, aesthetics, the body. All these dualisms, in the way that they are used, suffer from the criticisms made above of dichotomies of this form: the problem of mutual exclusivity and of the consequent impoverishment of both of their terms. Other dualisms could be added which also map on to that between time and space. Jameson, for instance, as do a whole line of authors before him, clearly relates the pairing to that between transcendence and immanence, with the former connotationally associated with the temporal and immanence with the spatial. Indeed, in this and in spite of their other differences, Jameson and Laclau are very similar. Laclau's distinction between the closed, cyclical time of simple reproduction (spatial) and dislocated, changing history (temporal), even if the latter has no inevitability in its progressive movement, is precisely that. Jameson who bemoans what he characterizes as the tendency towards immanence and the flight from transcendence of the contemporary period, writes of 'a world peculiarly without transcendence and without perspective … and indeed without plot in any traditional sense, since all choices would be equidistant and on the same level' (*Postmodernism*, 269), and this is a world where, he believes, a sense of the temporal is being lost and the realm of the spatial is taking over.

Now, as has been pointed out many times, these dualisms which so easily map on to each other also map on to the constructed dichotomy between female and male. From Rousseau's seeing woman as a potential source of disorder, as needing to be tamed by Reason, to Freud's famous pronouncement that woman is the enemy of civilization, to the many subsequent critics and analysts of such statements of the 'obviousness' of dualisms, of their interrelation one with another, and of their connotations of male and female, such literature is now considerable.[13] And space, in this system of interconnected dualisms, is coded female. '"Transcendence", in its origins, is a transcendence *of* the feminine', writes Lloyd (*The Man of Reason*, 101), for instance. Moreover, even where the transcodings between dualisms have an element of inconsistency, this rule still applies. Thus where time is dynamism, dislocation and History, and space is stasis, space is coded female and denigrated. But where space is chaos (which you would think was quite different from stasis; more indeed like dislocation), then time is Order … and space is *still* coded female, only in this context interpreted as threatening.

Elizabeth Wilson, in her book *The Sphinx in the City*, analyses this latter set of connotations.[14] The whole notion of city culture, she argues, has been developed as one pertaining to men. Yet within this context women present a threat, and in two ways. First, there is the fact that in the metropolis we are freer, in spite of all the also-attendant dangers, to escape the rigidity of patriarchal social controls which can be so powerful in a smaller community. Second and following from this, 'women have fared especially badly in Western visions of the metropolis because they have seemed to represent disorder. There is fear of the city as a realm of uncontrolled and chaotic sexual licence, and the rigid control of women in cities has been felt necessary to avert this danger' (157). 'Woman represented feeling, sexuality and even chaos, man was rationality and control' (87). Among male modernist writers of the early twentieth century, she argues – and with the exception of Joyce – the dominant response to the burgeoning city was to see it as threatening, while modernist women writers (Woolf, Richardson) were more likely to exult in its energy and vitality. The male response was perhaps more ambiguous than this, but it was certainly a mixture of fascination and fear. There is an interesting parallel to be drawn here with the sense of panic in the midst of exhilaration which seems to have overtaken some writers at what they see as the ungraspable (and therefore unbearable) complexity of the postmodern age. And it is an ungraspability seen persistently in spatial terms, whether through the argument that it is the new (seen-to-be-new) time-space compression, the new global-localism, the breaking down of borders, that is the cause of it all, or through the interpretation of the current period as somehow in its very character intrinsically more spatial than previous eras. In Jameson these two positions are brought together, and he displays the same ambivalence. He writes of 'the horror of multiplicity' (363), of 'all the web threads flung out beyond my "situation" into the unimaginable synchronicity of other people' (362). It is hard to resist the idea that Jameson's (and others') apparently vertiginous terror (a phrase they often use themselves) in the face of the complexity of today's world (conceived of as social but also importantly as spatial) has a lot in common with the nervousness of the male modernist, nearly a century ago, when faced with the big city.

It is important to be clear about what is being said of this relationship between space/time and gender. It is not being argued that this way of characterizing space is somehow essentially male; there is no essentialism of feminine/masculine here. Rather, the argument is that the dichotomous characterization of space and time, along with a whole range of other dualisms that have been briefly referred to, and with their connotative interrelations, may both reflect and be part of the constitution of, among other things, the masculinity and femininity of the sexist society in which we live. Nor is it being argued that space should simply be reprioritized to share an equal status with, or stand instead of, time. The latter point is important because there have been a number of contributions to the debate recently which have argued that, especially in modernist (including Marxist) accounts, it is time which has been considered the more important. Ed Soja, particularly in his book *Postmodern Geographies*, has made an extended and persuasive case to this effect (although see the critique by Gregory).[15] The story told earlier of Marxism within geography – supposedly the spatial discipline – is indicative of the same tendency. In a completely different context, Terry Eagleton has written in his introduction to Kristin Ross's *The Construction of Social Space* that 'Ross is surely right to claim that this idea [the concept of space] has proved of far less glamorous appeal to radical theorists than the apparently more dynamic, exhilarating notions of narrative and history.'[16] It is interesting to speculate on the degree to which this deprioritization might itself have been part and parcel of the system of gender connotations. Ross herself writes: 'The difficulty is also one of vocabulary, for while words like "historical" and "political" convey a dynamic of intentionality, vitality, and human motivation, "spatial", on the other hand, connotes stasis, neutrality, and passivity' (8), and in her analysis of Rimbaud's poetry and of the nature of its relation to the Paris Commune she does her best to counter that essentially negative view of spatiality. (Jameson, of course, is arguing pretty much the same point about the past prioritization of time, but his mission is precisely the opposite of Ross's and Soja's; it is to hang on to that prioritization.)

The point here, however, is not to argue for an upgrading of the status of space within the terms of the old dualism (a project which is arguably inherently difficult anyway, given the terms of that dualism), but to argue that what must be overcome is the very formulation of space/time in terms of this kind of dichotomy. The same point has frequently been made by feminists in relation to other dualisms, most particularly perhaps – because of the debate over the writings of Simone de Beauvoir – the dualism of transcendence and immanence. When de Beauvoir wrote 'Man's design is not to repeat himself in time: it is to take control of the instant and mould the future. It is male activity that in creating values has made of existence itself a value; this activity has prevailed over the confused forces of life; it has subdued Nature and Woman',[17] she was making precisely that discrimination between cyclicity and 'real change' which is not only central to the classic distinction between immanence and transcendence but is also part of the way in which Laclau distinguishes between what he calls the spatial and the temporal. De Beauvoir's argument was that women should grasp the transcendent. A later generation of feminists has argued that the problem is the nature of the distinction itself. The position here is both that the two dualisms (immanence/ transcendence and space/time) are related and that the argument about the former dualism could and should be extended to the latter. The next line of critique, the view from physics, provides some further hints about the directions which that reformulation might take.

The view from physics

The conceptualization of space and time under examination here also runs counter to notions of space and time within the natural sciences, and most particularly in physics. Now, in principle this may not be at all important; it is not clear that strict parallels can or should be drawn between the physical and the social sciences. And indeed there continue to be debates on this subject in the physical sciences. The point is, however, that the view of space and time already outlined above does have, as one of its roots at least, an interpretation drawn – if only implicitly – from the physical sciences. The problem is that it is an outmoded one.

The viewpoint, as adopted for instance by Laclau, accords with the viewpoint of classical, Newtonian, physics. In classical physics, both space and time exist in their own right, as do objects. Space is a passive arena, the setting for objects and their interaction. Objects, in turn, exist prior to their interactions and affect each other through force-fields. The observer, similarly, is detached from the observed world. In modern physics, on the other hand, the identity of things is *constituted through* interactions. In modern physics, while velocity, acceleration and so forth are defined, the basic ontological categories, such as space and time, are not. Even more significantly from the point of view of the argument here, in modern physics, physical reality is conceived of as a 'four-dimensional existence instead of … the evolution of a three-dimensional existence'.[18] Thus 'According to Einstein's theory … space and time are not to be thought of as separate entities existing in their own right – a three-dimensional space, and a one-dimensional time. Rather, the underlying reality consists of a four-dimensional space-time' (35). Moreover, the observer, too, is part of the observed world.

It is worth pausing for a moment to clarify a couple of points. The first is that the argument here is not in favour of a total collapse of the differences between something called the spatial and the temporal dimensions. Nor, indeed, would that seem to be what modern physics is arguing either. Rather, the point is that space and time are inextricably interwoven. It is not that we cannot make any distinction at all between them but that the distinction we do make needs to hold the two in tension, and to do so within an overall, and strong, concept of four-dimensionality. The second point is that the definitions of both space and time in themselves must be constructed as the result of interrelations. This means that there is no question of defining space simply as not-time. It must have a positive definition, in its own terms, just as does time. Space must not be consigned to the position of being conceptualized in terms of absence or lack. It also means, if the positive definitions of both space and time must be interrelational, that there is no absolute dimension: space. The existence of the spatial depends on the interrelations of objects: 'In order for "space" to make an appearance there needs to be at least two fundamental particles' (33). This is, in fact, saying no more than what is commonly argued, even in the social sciences – that space is not absolute, it is relational. Perhaps the problem at this point is that the implications of this position seem not to have been taken on board.

Now, in some ways all this seems to have some similarities with Laclau's use of the notion of the spatial, for his definition does refer to forms of social interaction. As we have seen, however, he designates them (or the concepts of them) as spatial only when they form a closed system, where there is a lack of dislocation that can produce a way out of the postulated (but impossible) closure. However, such use of the term is anyway surely metaphorical. What it represents is evidence of the connotations which are being attached to the terms 'space' and 'spatial'. It is not talking directly of 'the spatial' itself. Thus, to take up Laclau's usage in more detail: at a number of points, as we have seen, he presents definitions of space in terms of possible (in fact, he would argue, impossible) causal structures: 'Any repetition that is governed by a structural law of successions is space' (*New Reflections*, 41); or 'Spatiality means coexistence within a structure that establishes the positive nature of all its terms' (69). My question of these definitions and of other related ones, both elsewhere in this book and more widely – for instance in the debate over the supposed 'spatiality' of structuralism – is 'says who?' Is not this appellation in fact pure assertion? Laclau agrees in rejecting the possibility of the actual existence of pure spatiality in the sense of undislocated stasis. A further question must therefore be: why postulate it? Or, more precisely, why postulate it as 'space'? As we have just seen, an answer that proposes an absolute spatial dimension will not do. An alternative answer might be that this ideal pure spatiality, which only exists as discourse/myth/ideology, is in fact a (misjudged) metaphor. In this case it is indeed defined by interrelations – this is certainly not 'absolute space', the independently existing dimension – and the interrelations are those of a closed system of social relations, a system outside of which there is nothing and in which nothing will dislocate (temporalize) its internally regulated functioning. But then my question is: why call it 'space'? The use of the term 'spatial' here would seem to be purely metaphorical. Insofar as such systems do exist – and even insofar as they are merely postulated as an ideal – they can in no sense *be* simply spatial nor exist only in space. In themselves they *constitute a* particular form of space-time.[19]

Moreover, as metaphors the sense of Laclau's formulations goes against what I understand by – and shall argue below would be more helpful to understand by – space/the spatial. 'Any repetition that is governed by a structural law of successions'? – but *is* space so governed? As was argued above, radical geographers reacted strongly in the 1970s precisely against a view of 'a spatial realm', a realm, posited implicitly or explicitly by a wide range of then-dominant practitioners, from mathematicized 'regional scientists' to data-bashers armed with ferociously high regression coefficients, in which there were spatial processes, spatial laws and purely spatial explanations. In terms of causality, what was being argued by those of us who attacked this view was that the spatial is externally determined. A formulation like the one above, because of the connotations it attaches to the words 'space'/'spatial' in terms of the nature of causality, thus takes us back a good two decades. Or again, what of the second of Laclau's definitions given above? – that the spatial is the 'coexistence within a structure that establishes the positive nature of all its terms'? What then of the paradox of simultaneity and the causal chaos of happenstance juxtaposition which are, as we shall argue below (and as Jameson sees), integral characteristics of relational space?

In this procedure, any sort of stasis (for instance a self-regulating structural coherence which cannot lead to any transformation outside of its own terms) gets called 'space'/'spatial'. But there is no reason for this save the prior definition of space as lacking in (this kind of) transformative dynamic *and*, equally importantly, an assumption that anything lacking in (this kind of) dynamism is spatial. Instead, therefore, of using the terms 'space' (and 'time') in this metaphorical way to refer to such structures, why do we not remain with definitions (such as 'dislocated'/'undislocated') that refer to the nature of the causal structures themselves? Apart from its greater clarity, this would have the considerable advantage of leaving us free to retain (or maybe, rather, to develop) a more positive concept of space.

Indeed, conceptualizing space and time more in the manner of modern physics would seem to be consistent with Laclau's general argument. His whole point about radical historicity is this: 'Any effort to spatialize time ultimately fails and space itself becomes an event' (84). Spatiality in this sense is agreed to be impossible. '"Articulation" ... is the primary ontological level of the constitution of the real', writes Laclau (184). This is a fundamentally important statement, and one with which I agree. The argument here is thus not opposed to Laclau; rather it is that exactly the same reasoning, and manner of conceptualization, that he applies to the rest of the world, should be applied to space and time as well. It is not that the interrelations between objects occur *in* space and time; it is these relationships themselves which *create/define* space and time.[20]

It is not of course necessary for the social sciences simply to follow the natural sciences in such matters of conceptualization.[21] In fact, however, the conceptions of space and time that are being examined here do, if only implicitly, tend to lean on versions of the world derived from the physical sciences; but the view they rely on is one which has been superseded theoretically. Even so, it is still the case that even in the natural sciences it is possible to use different concepts/theories for different purposes. Newtonian physics is still perfectly adequate for building a bridge. Moreover, there continue to be debates between different parts of physics. What is being argued here is that the social issues that we currently need to understand, whether they be the hightech postmodern world or questions of cultural identity, require something that would look more like the 'modern physics' view of space. It would, moreover, precisely by introducing into the concept of space that element of dislocation/freedom/possibility, enable the politicization of space/space-time.

An alternative view of space

A first requirement of developing an alternative view of space is that we should try to get away from a notion of society as a kind of 3-d (and indeed more usually 2-d) slice which moves through time. Such a view is often, even usually, implicit rather than explicit, but it is remarkably pervasive. It shows up in the way people phrase things, in the analogies they use. Thus, just briefly to cite two of the authors who have been referred to earlier, Foucault writes 'We are at a moment, I believe, when our experience of the world is less that of a long life developing through time than that of a network that connects points and intersects with its own skein',[22] and Jameson contrasts 'historiographic deep space or perspectival temporality' with a (spatial) set of connections which 'lights up like a nodal circuit in a slot machine'.[23] The aim here is not to disagree in total with these formulations, but to indicate what they imply. What they both point to is a contrast between temporal movement on the one hand, and on the other a notion of space as instantaneous connections between things at one moment. For Jameson, the latter type of (inadequate) history-telling has replaced the former. And if this is true then it is indeed inadequate. But while the contrast – the shift in balance – to which both authors are drawing attention is a valid one, in the end the notion of space as *only* systems of simultaneous relations, the flashing of a pinball machine, is inadequate. For, of course, the temporal movement is also spatial; the moving elements have spatial relations to each other. And the 'spatial' interconnections which flash across can only be constituted temporally as well. Instead of linear process counterposed to flat surface (which anyway reduces space from three to two dimensions), it is necessary to insist on the irrefutable four-dimensionality (indeed, n-dimensionality) of things. Space is not static, nor time spaceless. Of course spatiality and temporality are different from each other, but neither can be conceptualized as the absence of the other. The full implications of this will be elaborated below, but for the moment the point is to try to think in terms of all the dimensions of space-time. It is a lot more difficult than at first it might seem.

Second, we need to conceptualize space as constructed out of interrelations, as the simultaneous coexistence of social interrelations and interactions at all spatial scales, from the most local level to the most global. Earlier it was reported how, in human geography, the recognition that the spatial is socially constituted was followed by the perhaps even more powerful (in the sense of the breadth of its implications) recognition that the social is necessarily spatially constituted too. Both points (though perhaps in reverse order) need to be grasped at this moment. On the one hand, all social (and indeed physical) phenomena/activities/relations have a spatial form and a relative spatial location. The relations which bind communities, whether they be 'local' societies or worldwide organizations; the relations within an industrial corporation; the debt relations between the South and the North; the relations which result in the current popularity in European cities of music from Mali. The spatial spread of social relations can be intimately local or expansively global, or anything in between. Their spatial extent and form also changes over time (and there is considerable debate about what is happening to the spatial form of social relations at the moment). But, whichever way it is, there is no getting away from the fact that the social is inexorably also spatial.

The proposition here is that this fact be used to define the spatial. Thus, the spatial is socially constituted. 'Space' is created out of the vast intricacies, the incredible complexities, of the interlocking and the non-interlocking, and the networks of relations at every scale from local to global. What makes a particular view of these social relations specifically spatial is their simultaneity. It is a simultaneity, also, which has extension and configuration. But simultaneity is absolutely not stasis. Seeing space as a moment in the intersection of configured social relations (rather than as an absolute dimension) means that it cannot be seen as static. There is no choice between flow (time) and a flat surface of instantaneous relations (space). Space is not a 'flat' surface in that sense because the social relations which create it are themselves dynamic by their very nature. It is a question of a manner of thinking. It is not the 'slice through time' which should be the dominant thought but the simultaneous coexistence of social relations that cannot be conceptualized as other than dynamic. Moreover, and again as a result of the fact that it is conceptualized as created out of social relations, space is by its very nature full of power and symbolism, a complex web of relations of domination and subordination, of solidarity and cooperation. This aspect of space has been referred to elsewhere as a kind of 'power-geometry'.[24]

Third, this in turn means that the spatial has *both* an element of order *and* an element of chaos (or maybe it is the case that we should question that dichotomy also). It cannot be defined on one side or the other of the mutually exclusive dichotomies discussed earlier. Space has order in two senses. First, it has order because all spatial locations of phenomena are caused; they can in principle be explained. Second, it has order because there are indeed spatial systems, in the sense of sets of social phenomena in which spatial arrangement (that is, mutual relative positioning rather than 'absolute' location) itself is part of the constitution of the system. The spatial organization of a communications network, or of a supermarket chain with its warehousing and distribution points and retail outlets, would both be examples of this, as would the activity space of a multinational company. There is an integral spatial coherence here, which constitutes the geographical distributions and the geographical form of the social relations. The spatial form was socially 'planned', in itself directly socially caused, that way. But there is also an element of 'chaos' which is intrinsic to the spatial. For although the location of each (or a set) of a number of phenomena may be directly caused (we know why x is here and y is there), the spatial positioning of one in relation to the other (x's location in relation to y) may not be directly caused. Such relative locations are produced out of the independent operation of separate determinations. They are in that sense 'unintended consequences'. Thus, the chaos of the spatial results from die happenstance juxtapositions, the accidental separations, the often paradoxical nature of the spatial arrangements that result from the operation of all these causalities. Both Mike Davis and Ed Soja, for instance, point to the paradoxical mixtures, the unexpected land-uses side by side, within Los Angeles. Thus, the relation between social relations and spatiality may vary between that of a fairly coherent system (where social and spatial form are mutually determinant) and that where the particular spatial form is not directly socially caused at all.

This has a number of significant implications. To begin with, it takes further the debate with Ernesto Laclau. For in this conceptualization space is essentially disrupted. It is, indeed, 'dislocated' and necessarily so. The simultaneity of space as defined here in no way implies the internally coherent closed system of causality which is dubbed spatial' in his *New Reflections*. There is no way that 'spatiality' in this sense 'means coexistence within a structure that establishes the positive nature of all its terms' (69). The spatial, in fact, precisely *cannot* be so. And this means, in turn, that the spatial too is open to politics.

But, further, neither does this view of space accord with that of Fredric Jameson, which, at first sight, might seem to be the opposite of Laclau's. In Jameson's view the spatial does indeed, as we have seen, have a lot to do with the chaotic. While for Laclau spatial discourses are the attempt to represent (to pin down the essentially unmappable), for Jameson the spatial is precisely unrepresentable – which is why he calls for an exercise in 'mapping' (though he acknowledges the procedure will be far more complex than cartography as we have known it so far). In this sense, Laclau and Jameson, both of whom use the terms 'space'/'spatiality', and so on, with great frequency, and for both of whom the concepts perform an important function in their overall schemas, have diametrically opposed interpretations of what the terms actually mean. Yet for both of them their concepts of spatiality work against politics. While for Laclau it is the essential orderliness of the spatial (as he defines it) that means the death of history and politics, for Jameson it is the chaos (precisely, the dislocation) of (his definition of) the spatial that apparently causes him to panic, and to call for a map.

So this difference between the two authors does not imply that, since the view of the spatial proposed here is in disagreement with that of Laclau, it concords with that of Jameson. Jameson's view is in fact equally problematical for politics, although in a different way. Jameson labels as 'space' what he sees as unrepresentable (thus the 'crisis of representation' and the 'increasing spatialization' are to him inextricably associated elements of postmodern society). In this, he perhaps unknowingly recalls an old debate within geography that goes by the name of 'the problem of geographical description'.[25] Thus, thirty years ago H.C. Darby, an eminent figure in the geography of his day, ruminated that 'A series of geographical facts is much more difficult to present than a sequence of historical facts. Events follow one another in time in an inherently dramatic fashion that makes juxtaposition in time easier to convey through the written word than juxtaposition in space. Geographical description is inevitably more difficult to achieve successfully than is historical narrative.'[26] Such a view, however, depends on the notion that the difficulty of geographical description (as opposed to temporal storytelling) arises in part because in space you can go off in any direction and in part because in space things which are next to each other are not necessarily connected. However, not only does this reduce space to unrepresentable chaos, it is also extremely problematical in what it implies for the notion of *time*. And this would seem on occasions to be the case for Jameson too. For, while space is posed as the unrepresentable, time is thereby, at least implicitly and at those moments, *counterposed* as the comforting security of a story it is possible to tell. This of course clearly reflects a notion of the difference between time and space in which time has a coherence and logic to its telling, while space does not. It is the view of time which Jameson might, according to some of his writings, like to see restored: time/History in the form of the Grand Narrative.[27]

However, this is also a view of temporality, as sequential coherence, that has come in for much questioning. The historical in fact can pose similar problems of representation to the geographical. *Moreover*, and ironically, it is precisely this view of history that Laclau would term spatial:

> ... with inexorable logic it then follows that there can be no dislocation possible in this process. If everything that happens can be explained *internally* to this world, nothing can be a mere event (which entails a radical temporality, as we have seen) and everything acquires an absolute intelligibility within the grandiose scheme of a pure spatiality. This is the Hegelian-Marxist moment.
>
> *New Reflections, 75*

Further still, what is crucially wrong with both these views is that they are simply opposing space and time. For both Laclau and Jameson, time and space are causal closure/representability on the one hand and unrepresentability on the other. They simply differ as to which is which! What unites them, and what I argue should be questioned, is the very counterposition in this way of space and time. It is a counterposition which makes it difficult to think the social in terms of the real multiplicities of space-time. This is an argument that is being made forcefully in debates over cultural identity. '[E]thnic identity and difference are socially produced in the here and now, not archeologically salvaged from the disappearing past';[28] and Homi Bhabha enquires 'Can I just clarify that what to me is problematic about the understanding of the "fundamentalist" position in the Rushdie case is that it is *represented* as archaic, almost medieval. It may sound very strange to us, it may sound absolutely absurd to some people, but the point is that the demands over *The Satanic Verses* are being made *now*, out of a particular political state that is functioning very much in our time.'[29] Those who focus on what they see as the terrifying simultaneity of today would presumably find such a view of the world problematical, and would long for such 'ethnic identities' and 'fundamentalisms' to be (re)placed in the past so that one story of progression between differences, rather than an account of the production of a number of different differences at one moment in time, could be told. That this cannot be done is the real meaning of the contrast between thinking in terms of three dimensions plus one, and recognizing fully the inextricability of the four dimensions together. What used to be thought of as 'the problem of geographical description' is actually the more general difficulty of dealing with a world which is 4-d.

But all this leads to a fourth characteristic of an alternative view of space, as part of space-time. For precisely that element of the chaotic, or dislocated, which is intrinsic to the spatial has effects on the social phenomena that constitute it. Spatial form as 'outcome' (the happenstance juxtapositions and so forth) has emergent powers which can have effects on subsequent events. Spatial form can alter the future course of the very histories that have produced it. In relation to Laclau, what this means, ironically, is that one of the sources of the dislocation, on the existence of which he (in my view correctly) insists, is precisely the spatial. The spatial (in my terms) is precisely one of the sources of the temporal (in his terms). In relation to Jameson, the (at least partial) chaos of the spatial (which he recognizes) is precisely one of the reasons why the temporal is not, and cannot be, so tidy and monolithic a tale as he might wish. One way of thinking about all this is to say that the spatial is integral to the production of history, and thus to the possibility of politics, just as the temporal is to geography. Another way is to insist on the inseparability of time and space, on their joint constitution through the interrelations between phenomena; on the necessity of thinking in terms of space-time.

Key readings

Massey, D. 1991. Flexible Sexism. *Environment and Planning D: Society and Space* 9 (1): 31–57.

Massey, D. 1994. *Space, Place and Gender.* Cambridge: Polity Press.

Massey, D. 1995. Masculinity, Dualisms and High Technology. *Transactions of the Institute of British Geographers* 20: 487–499.

Bibliography

1 H. Lefebvre, *The Production of Space*, Oxford 1991, p.3.
2 E. Laclau, *New Reflections on the Revolution of Our Time*, London 1990. Thanks to Ernesto Laclau for many long discussions during the writing of this article.
3 F. Jameson, *Postmodernism, or, the Cultural Logic of Late Capitalism*, London 1991.
4 See, for instance, the discussion in M. Rustin, 'Place and Time in Socialist Theory', *Radical Philosophy*, no. 47, 1987: 30–36.
5 And in this sense, of course, it could be said that Laclau's space is 'political' because any representation is political. But this is the case only in the sense that *different* spaces, different 'cognitive mappings', to borrow Jameson's terminology, can express different political stances. It still leaves each space – and thus the concept of space – as characterized by closure and immobility, as containing no sense of the open, creative possibilities for political action/effectivity. Space is the realm of the discourse of closure, of the fixation of meaning.
6 P. Anderson, 'A Culture in Contraflow', nlr 180, March–April 1990: 41–78 and nlr 182, July–August 1990: 85–137.
7 Lefebvre, p. 275.
8 See, for instance, D. Massey, *Spatial Divisions of Labour: Social Structures and the Geography of Production*, Basingstoke 1984; D. Gregory, and J. Urry, eds., *Social Relations and Spatial Structures*, Basingstoke 1985; and E. Soja, *Postmodern Geographies: The Reassertion of Space in Critical Social Theory*, London 1989.
9 It should be noted that the argument that 'the spatial' is particularly important in the current era is a different one from that being made here. The argument about the nature of postmodernity is an empirical one about the characteristics of these times. The argument developed within geography was an in-principle position concerning the nature of explanation, and the role of the spatial within this.
10 See, for instance, J. Flax, 'Political Philosophy and the Patriarchal Unconscious: A Psychoanalytic Perspective on Epistemology and Metaphysics', in S. Harding and M.B. Hintikka, eds., *Discovering Reality: Feminist Perspectives on Epistemology, Metaphysics, Methodology, and Philosophy of Science*, Dordrecht 1983, pp. 245–81; and in the same volume, the 'Introduction' by Harding and Hintikka (pp. ix–xix), and L. Lange, 'Woman is Not a Rational Animal: On Aristotle's Biology of Reproduction', pp. 1–15; also J. Flax, 'Postmodernism and Gender Relations in Feminist Theory', in L.J. Nicholson, ed., *Feminism/Postmodernism*, London 1990, pp. 39–62, and N. Hartsock, 'Foucault on Power: A Theory for Women?' in the same volume, pp. 157–175.
11 N. Jay, 'Gender and Dichotomy', *Feminist Studies*, vol. 7, no. 1, Spring 1981: 38–56.
12 G. Lloyd, *The Man of Reason: 'Male' and 'Female' in Western Philosophy*, London 1984.
13 See, for instance, D. Dinnerstein, *The Rocking of the Cradle and the Ruling of the World*, London 1987; M. le Doeuff, *Hipparchia's Choice: An Essay Concerning Women, Philosophy, Etc.*, Oxford 1991; and Lloyd.
14 E. Wilson, *The Sphinx in the City: Urban Life, the Control of Disorder, and Women*, London 1991.
15 Soja and D. Gregory, 'Chinatown, Part Three? Soja and the Missing Spaces of Social Theory', *Strategies*, no. 3, 1990.
16 K. Ross, *The Emergence of Social Space: Rimbaud and the Paris Commune*, Basingstoke 1988; Eagleton's *Foreword*, p. xii.
17 S. de Beauvoir, *The Second Sex* (1949), trans. H.M. Parshley, Harmondsworth 1972: 97.
18 R. Stannard, *Grounds for Reasonable Belief*, Edinburgh 1989. Page references are given in parenthesis in the text.
19 An alternative explanation of why such structures are labelled 'spatial' is available. Moreover, it is an explanation which relates also to the much wider question (although in fact it is rarely questioned) of why structuralist thought, or certain forms of it, has so often been dubbed spatial. This is that, since such structures are seen to be non-dynamic systems, they are argued to be non-temporal. They are static, and thus lacking in a time dimension. So, by a knee-jerk response they are called spatial. Similarly with the distinction between diachrony and synchrony. Because the former is sometimes seen as temporal, its 'opposite' is automatically characterized as spatial (although in fact not by Laclau, for whom certain forms of diachrony may also be 'spatial' – see p. 42). This, however, returns us to the critique of a conceptualization of space simply and only in terms of a lack of temporality. A-temporality is not a sufficient, or satisfactory, definition of the spatial. Things can be static without being spatial – the assumption, noted earlier, that anything lacking a transformative dynamic is spatial cannot be maintained in positive terms; it is simply the (unsustainable) result of associating transformation solely with time. Moreover, while a particular synchrony (synchronic form) may have spatial characteristics,

in its extension and configuration, that does not mean that it is a sufficient definition of space/spatial itself.

20 Stannard, p. 33.

21 However, the social sciences deal with physical space too. All material phenomena, including social phenomena, are spatial. Any definition of space must include reference to its characteristics of extension, exclusivity, juxtaposition, and so on. Moreover, not only do the relationships between these phenomena create/define space-time; the spacing (and timing) of phenomena also enables and constrains the relationships themselves. Thus, it is necessary for social science to be at least consistent with concepts of physical space, although a social-science concept could also have additional features. The implications for the analysis of 'natural' space – of physical geography – are similar. Indeed, as Laclau argues, even physical space is temporal and therefore in his own lexicon not spatial: 'the real – including physical space – is in the ultimate instance temporal' (41–2). While I disagree with the labelling as spatial and temporal, I agree with the sense of this – but why only 'in the ultimate instance'?!

22 M. Foucault, 'Of Other Spaces', *Diacritics*, Spring 1986: 22.

23 Jameson, p. 374.

24 D. Massey, 'Power-Geometry and a Progressive Sense of Place' 08:42 AM 12/18/03, in Bird et al., eds., *Mapping the Futures*, London forthcoming.

25 H.C. Darby, 'The Problem of Geographical Description', *Transactions of the Institute of British Geographers*, vol. 30, 1962: 1–14.

26 Ibid., p. 2.

27 I am hesitant here in interpreting Jameson because, inevitably, his position has developed over the course of his work. I am sure that he would not in fact see narrative as unproblematic. Yet the counterposition of it to his concept of spatiality, and the way in which he formulates that concept, does lead, in those parts of his argument, to that impression being given.

28 M.P. Smith, 'Postmodernism, Urban Ethnography, and the New Social Space of Ethnic Identity', forthcoming in *Theory and Society*.

29 In 'Interview with Homi Bhabha' in J. Rutherford, ed., *Identity: Community, Culture, Difference*, London 1990, p. 215. At this point, as at a number of others, the argument links up with the discussion by Peter Osborne in his 'Modernity is a Qualitative, Not a Chronological, Category', nlr 192, March–April 1992, pp. 65–84.

11

FEMINIST ENGAGEMENT WITH THE ECONOMY

Spaces of resistance and transformation

Jessa M. Loomis and Ann M. Oberhauser

Introduction

This chapter explores the multiple and contested meanings of 'the economy' from a feminist geographic perspective. Research in this field has challenged conventional, often masculinist, approaches to studying the economy by instead examining it as a set of processes that are interconnected with social, political and cultural practices. Our discussion engages with the theoretical foundations of, and empirical work in, feminist geography regarding gender and the economy from multiple and cross-scalar perspectives. It includes scholarship that highlights the relationships among social identities, such as race, class, gender and sexuality, that produce and are constitutive of geographically and temporally diverse and alternative economies. As highlighted below, these identities are evident in global, community-based and individual economic strategies, such as livelihoods of migrant domestic workers in the Middle East, fair-trade farming among Honduran coffee growers and digital media start-ups in San Francisco.

Our analysis also examines how conventional or mainstream discourses of capital accumulation and social reproduction are challenged in feminist scholarship and praxis. We argue that 'the economy' is redefined through activities such as community-based co-ops, urban agriculture systems, microcredit schemes and other alternative systems or practices of exchange that place human relationships at their centre. These alternative approaches, in turn, create socially and geographically sustainable and just societies. This chapter thus provides a feminist framework to analyse, critique and transform the socio-political processes that drive and constitute the economy.

Feminist perspectives on labour and work

Feminist economic geographers have produced rich and nuanced scholarship on the processes, practices and subjects of economic transformations, including globalization (Nagar et al. 2002; Roberts 2004), neoliberalism (Larner 2003) and financialization (Pollard 2013; Rankin 2013). Leading up to the critical turn in economic geography, this field tended to focus on the role of the state in economic restructuring, the geography of firms, class-based labour markets and global production networks. Feminist economic geography challenges these traditional sites of economic knowledge production by shifting the focus of analysis to diverse social relations and the corporal and affective aspects of economic processes within varied scales and spaces of the economy (Gibson-Graham 2006; Oberhauser 2000; Werner et al. 2017). For example, feminist geographers have expanded research on care work, such as nursing, eldercare and other professions, where the work is devalued as a result of women's naturalized identity as caregivers (Atkinson, Lawson and Wiles 2011). These analyses consider the ethics of solidarity and revaluing labour from standpoints that are different from market-based activities and outcomes (Lawson 2007; Pratt 2012).

Critical insights to the nuanced and fluid dimensions of capital and labour are key aspects of feminist economic geography. Cindi Katz's (2001a) work on vagabond capitalism has been particularly inspiring for this field, as it underscores the social relations of capitalism that benefit those who are already resourced while dispossessing certain people and places based on class, racialized and national components. According to Katz (2001a, 709), 'vagabond capitalism puts the vagrancy and dereliction where it belongs – on capitalism, that unsettled, dissolute, irresponsible stalker of the world'. By highlighting the temporary and fleeting commitments of capital, she provides a relational framework for understanding the workings of capitalism writ large.

Feminist economic geography also analyses embodied experiences of labouring and the demands made on bodies that are performing work in particular sectors of the economy. Exploring these workspaces and the creation of worker subjectivities, Daniel Cockayne (2016) examines how workers invest in creative forms of entrepreneurial labour in San Francisco's digital media sector. This research highlights the deleterious effects of precaritization, which includes a shift from salaried work to consultant and contract forms of labour, as well as the ongoing 'flexibilization' of the workforce and working conditions (MacLeavy 2011; Richardson 2016). While 'flexible' aspects of work are often celebrated by employers, in practice the changing nature of work often erodes worker protections and devalues the skills required to perform the work. As Kim England and Caitlin Henry (2013) demonstrate in their research on international nurses in the UK, this type of work is often performed by women, immigrants and racialized minorities. Indeed, feminist scholarship has shown that contemporary economic conditions and their historical inequalities render some bodies, workforces and communities more precarious and subject to exploitation than others (Meehan and Strauss 2015; Pulido 2016).

Feminist analyses of labour relations and exploitative work conditions have recently been applied to academic institutions. In these contexts, critical scholars document an increasing reliance on casual labour (Berg et al. 2016), the politics of citation (Mott and Cockayne 2017), the speed-up of research demands (Mountz et al. 2015) and the overall neoliberal logics that influence higher education (Mott et al. 2015). In response, feminist geographers have offered slow scholarship (Mountz et al. 2015) and collective biography (Kern et al. 2014) as ways to counter neoliberal and metric-oriented narratives of success in the academy. These alternative approaches provide a bold vision of a more collaborative and sustainable version of academic life that reimagines the university through a feminist sensibility and ethics of care.

Geographies of social reproduction and scale

While the economy is often depicted in the form of hegemonic capitalist production, feminists point to the equally important role of social reproduction in maintaining the social, cultural and material subsistence of individuals, families, households and communities. In this vein, the framing of distinct public and private domains is critiqued by feminists who view these socially and spatially embedded spheres as fluid and overlapping, especially in the context of gendered labour practices. For example, Katie Meehan and Kendra Strauss (2015, 1) critique the mainstream focus on public spaces, production and the market economy, and instead provide a 'framework for examining the interaction of paid labour and unpaid work in the reproduction of bodies, households, communities, societies, and environments'. Furthermore, the 'fleshy, messy, and indeterminate stuff of everyday life' (Katz 2001a, 711) that comprises social reproduction is an important aspect of production, from the standpoint of both labour and the means of production. Katherine Mitchell, Sallie Marston and Cindi Katz's (2004, 23) analysis of social reproduction contributes to these discussions through the concept of 'life's work' or the 'various ways in which life is made outside of work ... and how differentiated subjects of transnational capitalism are produced in the course of everyday practices'. Feminist economic geography thus dismantles dichotomous categories, such as the public and private, which have been forwarded by mainstream economics in order to emphasize the contested and dynamic nature of labour and work.

Feminist geographers also trace the fluidity of production and reproduction within and across scales through empirically and conceptually rich cross-cultural research. Katz (2001a, 2001b) suggests that we harness the metaphorical power of contour lines to develop 'counter-topographies' that draw analytical connections between distinct places that are similarly situated in relation to global processes. Counter-topographies are thus a tool for building a politics of resistance against the imperial, patriarchal and racist operations of globalization. Katz's rich analysis of Howa village in South Sudan and East Harlem in New York City illustrates how these strategies work across intersecting scales. Members of these communities work within and develop resistance to structural adjustment, civil war and the expanding scope of agricultural and late-industrial economic activities (2004). Ann Oberhauser (2010) applies the analytical concept of scale in her research on spaces of resistance in gendered livelihood strategies of women in Accra, Ghana. Here, informal markets serve as important sites of economic activity and social dynamics in the context of Ghana's structural adjustment and neoliberal globalization. These examples illustrate how globalization and, by extension, scale appear to shrink or compress differences at the household and local levels, while also expanding the global reach of production and social reproduction.

Reconceptualizing the economy through the analytic of scale involves exploring the relationship between the global and the intimate. The situated nature of the global and the intimate is often studied by feminist scholars at 'local' scales, such as individuals, households and communities, which are also sites of social reproduction. Geraldine Pratt and Victoria Rosner (2006) suggest that focusing on the intimate, rather than the global/local binary, reconfigures our understanding of spatial relationships by politicizing their meaning through incorporating sensory experiences such as sound, smell, taste and touch. In addition, Sara Ahmed (2004) engages with a feminist re/construction of scale through her work on affect and geographies of fear, hate, love and other intimacies. According to Ahmed, gender and sexuality involve emotions that stem from the body through the international politics of asylum and migration, terrorism and reconciliation. In this section, we argue that feminist economic geography has disrupted conventional meanings and practices of social reproduction in a way that inserts the fluid and dynamic nature of multiple contexts and scales of work. The following discussion positions these social processes with the dynamics of globalization and the mobility of labour.

Globalization and migration

Feminist economic geography engages with transnational and global frameworks of migration and mobility in ways that highlight the dynamic and multi-scalar aspects of these economic processes within and through diverse social identities (Silvey 2013). Recent studies focus on the circular and transformative movement of migrant labour and the multiple labour markets and social contexts that they occupy (Gidwani and Sivaramakrishnan 2003). Furthermore, globalization and the migration of labour influence capital investment, finances and the embodied practices of workers in distinctively gendered ways. Saskia Sassen's (2005) foundational work on global cities and finance capital analyses the large proportion of low-paid service-sector jobs that are disproportionately occupied by female immigrants in places such as New York, London and Tokyo. These cities are built on and intensify social and spatial disparities, as the wealthy professional class exploits the labour of workers who are poorly paid and who experience insecurity in the workplace, including citizenship status (Kern and Mullings 2013). Labourers experiencing these conditions have found opportunities for resistance, including contesting unfair and unsafe working conditions. Brenda Yeoh and Kamalini Ramdas (2014, 1198) address the contested experiences of migrants who 'straddle the multiple places of being "here" and "there" simultaneously', thus revealing both 'the emancipatory and constraining nature of gendered migrant spatialities and identity politics'. Feminist geography offers critical analyses of how these social and spatial aspects of transnational migration affect the economic and political spheres of migrants' lives.

Moreover, feminist analysis of labour has shown that racialised and gendered identities are actively constructed through economic activities, both within and across national borders. For example, Rachel Silvey's (2006) analysis of Indonesian migrant domestic workers in Saudi Arabia examines how the status of temporary workers excludes them from formal citizenship. Repressive legal systems such as these position them as racialized migrant workers, whose cultures are perceived as immoral, disruptive and generally threatening to civil society. In addition, McDowell's (2013, 231) research on migrant female labour in Britain during the post-World War II era shows that these women made significant contributions to the nation even as they were continually excluded from the 'imaginary version of Britishness'. This geographical research shows how embodied experiences of labouring are highly racialized and gendered across industrial sectors, categories and conditions of employment, as well as national and historical contexts.

Feminist scholarship examining the transnational lives of migrants and their diasporic connections to home expands how we understand migration. In these contexts, experiences of migration are often gendered and defined by strong material and emotional connections, including ties to family and community elsewhere through, for example, the payment of remittances, which are often crucial to sustain households and family members in the home country. Beverley Mullings's (2014, 56) research on the Jamaican diaspora demonstrates that:

> While women's migration … has been instrumental to the alleviation of poverty in their households and communities, women's ability to financially support their families and communities has often come at the expense of their ability to be a present source of emotional support and protection … for their most vulnerable members.

Araby Smyth (2017) approaches remittances from a different perspective, arguing that transnational sharing of wealth through remittances is an alternative, or even an anti-capitalist, economic practice. While the dominant narrative of *remesas* is one of development and state-guided contributions, her research on Mexican hometown associations in New York City shows that this practice of sending money home is one of solidarity. In many instances, migrants and labour movements have stood up to the denial of access to healthcare, social services and basic worker rights through sanctuary cities, immigrant reform policies and other advocacy groups. Thus, the economic and social dynamics of labour mobility are contested arenas and processes that align with feminist analyses of these transnational and global networks.

Intersectional approaches to the economy

The feminist concept of intersectionality is attentive to how privilege and power operate through multiple intersecting forms of social difference that compound and transform experiences of oppression and exploitation (Crenshaw 1991; Peake 2010; Valentine 2007). The theory of intersectionality has been mobilized in feminist scholarship on the economy to show how capitalism relies on classed, gendered and racialized hierarchies to reproduce the dominant power relations and systems of privilege, such as hetero-patriarchy or racial capitalism. Recently, Werner et al. (2017, 1–2) argue that the production of social difference is 'integral to the functioning of political-economic systems and knowledge production processes'. An intersectional analysis, therefore, shows how socially produced categories of difference cannot be analysed in isolation if scholars seek to attend to and dismantle oppressive and exploitative systems of power.

In the tradition of feminist political economy, feminist economic geographers examine how social difference is used to fracture the labour force and discourage collective organizing. As part of this, feminist economic geographers have examined how worker remuneration depends on where the labour is performed and who performs it. Melissa Wright's (2006) research on *maquiladoras* and female labourers along the US–Mexico border shows how a woman is made 'disposable' through gendered and geographical (Third-World) discourses that devalue her life and her labour. Similarly, in research on the restructuring of firms in the Dominican Republic, Marion Werner (2010, 2012) explores how firms upgrade by devaluing labour through defining work as skilled or unskilled, based on interlocking forms of social difference including, but not always, gender. By attending to how social difference is used to create exploitable workers, these scholars offer a feminist analysis of value that is attentive to the intersectional production of the perceived worth of labour.

While class has been the traditional object of inquiry for economic geographers and while gender has been a long-standing focus of feminist economic geography, recent work in this field draws attention to the significance of race and racism to the history of capitalism. Anne Bonds (2013, 399) argues that economic geography should examine race 'not just as an effect or product of capital accumulation but rather as a systemic presence that is thoroughly embedded in economic paradigms, institutions, practices, and actors'. Recent events in the US have underscored the need for this type of analysis. Following Donald Trump's election, Roy (2016) and Gökarıksel and Smith (2017) have employed an intersectional approach, including identifying how discourses of masculinity and Whiteness work in tandem with narratives of economic abandonment to deconstruct Trump's reactionary cultural politics. In particular, Roy (2016) called on academics in their roles as scholars and educators to challenge the normalization of 'white supremacy, misogyny, and virulent nationalism'.

In recent years, scholars have examined how anti-Black violence articulates with other forms of oppression. Deborah Cowen and Nemoy Lewis (2016) offer specific examples of the 'shifting geographies of black dispossession', citing gentrification, subprime mortgage lending and the 'entrepreneurial racism of the Ferguson police', who disproportionately levied traffic fines on Black residents in order to generate revenue for the city. Furthermore, in her research focus on the murder of Michael Brown at the hands of police in Ferguson, Missouri, Kate Derickson (2017) calls for geographers to examine the racialized legacy of US urban growth and development. These examples clearly illustrate ongoing racialized violence in many parts of the US.

Feminists thus argue that an intersectional approach is vital to addressing inequality at its root and connecting the struggles across populations for transformational change. As feminist economic geography learns from colleagues who embrace Black geographies and postcolonial approaches, scholarship on the economy must account for the co-production of social difference through an intersectional approach to understanding the economy.

Gender, financial subjectivity and financial inclusion

Since the 1980s, economic geographers have been particularly interested in studying how globalization and neoliberal economic change manifests across space in uneven and contested ways. More recently, and especially since the 2008 financial crisis, geographers have studied the growing significance of finance and new forms of exclusion and marginalization. Feminists studying the institutions and cultures of financial services incorporate the spread of financial logics into daily life through gendered practices and values. For example, Linda McDowell's (1997, 2010) research focuses on the distinctly masculinist cultures of banking and financial institutions, while Caitlin Zaloom's (2006) *Out of the Pits* and Karen Ho's (2009) *Liquidated* provide rich ethnographic work in this area. Feminists have also examined investor subjectivity and the ideology of shareholder value, including the gendered notions of risk and responsibility that inform underwriting practices and financial regimes more broadly (Amoore 2011; Joseph 2014).

In recent years, feminist scholarship has been attentive to how finance operates beyond these institutional spaces, including the way that financial logics are a part of everyday life, thus replicating and sometimes exacerbating experiences of exclusion, dispossession and marginalization. In the wake of the Great Recession, feminist economic geographers paid attention to the gendered causes and consequences of the financial crisis. This scholarship examines how social relations and familial responsibilities are reconfigured alongside changes to the welfare state (Elwood and Lawson 2013; MacLeavy 2011; Pollard 2013). The austerity measures taken after the financial crisis disproportionately impacted on those households and communities living 'on the edge' in ways that are strongly gendered, racialized, classed and militarized (Ettlinger 2007; Waite 2009). Everyday experiences of austerity (Hall 2015) are also embodied as people find strategies to cope with scarcity and uncertainty, including crafting, homesteading and food sharing (Hall 2011; Parker and Morrow 2017). This work shows how the lives of individuals and families are altered as they adjust to the economic recession and the ongoing financialization of the economy.

Development projects have also taken a decidedly financial turn, as they aim to promote social and economic empowerment by expanding access to credit. Microfinance programmes extend small lines of credit to women, primarily in the Global South, in order to encourage entrepreneurship (Aladuwaka and Oberhauser 2014; Roy 2010). These programmes target women who, based on essentialized views of their role as responsible stewards of the family's finances and moral debtors, are believed to use the funds more responsibly than men (Maclean 2013). Similar gendered assumptions inform conditional cash transfer (CCT) programmes that provide cash assistance to women who meet certain conditions, such as consistently sending children to school or bringing children to the doctor for regular check-ups. Examining the effects of the CCT programme *Oportunidades*, Claudia Radel, Birgit Schmook, Nora Haenn and Lisa Green (2016) report optimistic outcomes in the areas of land control and tenure for women engaged in the agricultural sector in Mexico, yet they suggest that the programme's benefits have been successful despite, not because of, the neoliberal programme's requirements. Similarly, in her examination of Peru's CCT programme *Juntos*, Tara Cookson (2018) argues that well-meaning CCT programmes do not adequately address the legacies of exclusion that underlie low-income rural women's experiences and, as a result, replicate inequality and reproduce poverty.

In the US context, Jessa Loomis's (2018) research examines how nonprofit financial literacy and capability programmes are teaching previously unfit market actors, such as women living on limited incomes, to manage their debt, monitor their credit scores, avoid predatory lending and invest using mainstream financial products. Her research illustrates how financial empowerment programmes are helping financial institutions to expand their reach into new consumer markets as they encourage participants to be responsible debtors. In general, programmes of financial inclusion and empowerment, such as CCT, microfinance and financial capability programmes, have been criticized for making women responsible for their own poverty alleviation (Rankin 2013; Roy 2010). In contrast, programmes that encourage individuals to be accountable to and responsible for their debts have received widespread political support, even as regulation of the financial industry remains contentious. Thus, feminist geographic analyses of finance provide critical dimensions to our understanding of both empowerment and further marginalization within the economy.

Strategies of resistance – transforming economic livelihoods

In their efforts to create a more capacious definition of the economy, feminist economic geographers have argued for new representations and metaphors of economic activity and have imagined alternatives to exploitative and oppressive relations of economic production and consumption. Included in this reimagining are efforts to articulate and enact a feminist politics of the economy beyond capitalism (Cameron and Gibson-Graham 2003). The work of feminist scholars J.K. Gibson-Graham theorizes economic exchange beyond a capitalist framing and imagines a post-capitalist politics (2006). This work has since inspired research on diverse economies (2008), or the varieties of economic activity, as well as how community economies envision a different economic practice, which involves ethics, values and interdependence. These alternative and diverse economies shift the economic focus from market-based capitalism to household and community needs and resources. For instance, Oona Morrow and Kelly Dombroski's (2015) research on canning, gardening and self-provision identifies a range of communal responses to the most recent downturn and shows the coping strategies that families and communities use to survive in times of economic contraction. As Fickey and Hanrahan (2014) suggest in their review of scholarship on diverse economies and alternative economic spaces, there remains a need for scholars to take seriously the inequalities and questions of power within these hopeful examples.

While some resistance arises in opposition to capitalism, other strategies for transforming livelihoods arise from changing the means of economic exchange and altering the terms under which that exchange occurs. Scholarship on food justice and ethical consumption has been a particularly fertile topic for conceptualizing different modes of economic practice. For example, Sarah Lyon, Josefina Bezaury and Tad Mutersbaugh's (2010) work on coffee production in Mesoamerica reveals how women gain greater control over their participation in fair-trade organic coffee organizations than in smallholder agricultural production. Of course, not all attempts to create fair-trade and ethical consumption have produced the expected or desired results. Feminist approaches recognize that these results are sometimes mixed, in that women can be empowered yet still experience restrictions. Lyon et al.'s (2017) recent scholarship complicates earlier findings by suggesting that the women who gained agency and power through their coffee production still contend with domestic obligations; this 'time poverty' is a particularly gendered burden that is an all-too-common experience when economic rights do not coincide with advances in the social status of women.

Strategies of resistance often arise in attempts to engage in more ethical or collective economic practice, suggesting moves towards social and economic justice. For example, the movement #GiveYourMoneytoWomen, started by feminist Lauren Chief Elk and colleagues, calls for a direct transfer of wealth from men to women as a form of gender justice. This initiative acknowledges the long-standing structural inequality that has undervalued and refused to pay women for their labour (Hanson and Pratt 1995), including work in the home and emotional labour and care work (West 2016), and advances a radical politics of wealth redistribution from men to women. Diprose's (2017) research examining the practice known as 'timebanking' shows how the inequalities of waged work can be challenged by rethinking the value of one's own and others' labour. While often not understood in political terms, Diprose argues that these localized and embodied experiences of labouring beyond the wage can be an antidote to the powerlessness that people feel in uncertain and precarious times.

Reimagining feminist alternatives for the economy

In this chapter, we argue that feminist scholars have enhanced geographical approaches to understanding the economy by broadening the scope of analysis to diverse economic processes and practices. Our discussion highlights how feminist economic geography considers gendered, racialized, classed and militarized forms of insecurity, and provides critical spaces for activism and resistance to the dominant forms and discourses of capitalism, including worker-owned collectives, alternative economies and different ways of measuring and performing academic success. We also examine how the erosion of social support systems, austerity measures and reactionary movements in many areas of both the Global North and South are embedded in hetero-patriarchal, racialized regimes of political economy. These processes underscore the need for an intersectional approach to understanding economic marginalization, exploitation and dispossession. Drawing on feminist scholars of finance, we also suggest that financial inclusion under terms that remain predatory or discriminatory are inadequate responses to ongoing and insidious economic exclusion.

By drawing on empirical examples and reviewing major themes in the history of the subdiscipline, including work, migration, social reproduction, finance and strategies of resistance, we have shown that feminist economic geography innovates the theories, methods and objects traditionally employed in economic geography. These approaches offer alternative narratives that complicate the well-worn explanations that circumscribe economic analysis to categories such as labour, production, consumption and markets. Feminist economic analyses of contemporary events such as Brexit, the election of Trump and the rise of populist movements in the Global North, for example, illustrate what feminist theory of the economy offers for analysing political economic issues. These examples demonstrate that feminist scholarship on the economy has not only augmented economic geography writ large but also offers a capacious understanding of the workings of the economy that should be understood as more than a mere complement to mainstream economic geography. Indeed, feminist approaches to the economy have been foundational to capturing the depth and complexity of economic life. Through these approaches, feminist economic geography has reframed conventional notions of globalization and neoliberalization to instead highlight spaces of radical possibility and to offer alternative paths for ethical and productive alliances in the economy. Here, we find solidarity and resolve to reimagine these economic spaces and to create more just approaches to pressing economic questions, including those related to the provision of care and the distribution of wealth, now and into the future.

Key readings

Gibson-Graham, J.K. 2006. *The End of Capitalism (As We Knew It): A Feminist Critique of Political Economy*. Minneapolis: University of Minnesota Press.

Meehan, K., and K. Strauss. 2015. *Precarious World: Contested Geographies of Social Reproduction*. Athens: University of Georgia Press.

Pollard, J. 2013. "Gendering Capital: Financial Crisis, Financialization and (an Agenda for) Economic Geography." *Progress in Human Geography* 37 (3): 403–423.

Werner, M., K. Strauss, B. Parker, R. Orzeck, K. Derickson, and A. Bonds. 2017. "Feminist Political Economy in Geography: Why Now, What is Different, and What For?" *Geoforum* 79: 1–4.

References

Ahmed, S. 2004. "Affective Economies." *Social Text* 22 (2): 117–139.

Aladuwaka, S., and A.M. Oberhauser. 2014. "'Out of the Kitchen': Gender, Empowerment and Microfinance Programs in Sri Lanka." In: *Global Perspectives on Gender and Space: Engaging Feminism and Development*, edited by A.M. Oberhauser and I. Johnston-Anumonwo, 35–52. London: Routledge.

Amoore, L. 2011. "Data Derivatives on the Emergence of a Security Risk Calculus for Our Times." *Theory, Culture & Society* 28 (6): 24–43.

Atkinson, S., V. Lawson, and J. Wiles. 2011. "Care of the Body: Spaces of Practice." *Social and Cultural Geography* 12 (6): 563–572.

Berg, L.D., E.H. Huijbens, and H.G. Larsen. 2016. "Producing Anxiety in the Neoliberal University." *Canadian Geographer/le géographe canadien* 60 (2): 168–180.

Bonds, A. 2013. "Racing Economic Geography: The Place of Race in Economic Geography." *Geography Compass* 7 (6): 398–411.

Cameron, J., and J.K. Gibson-Graham. 2003. "Feminising the Economy: Metaphors, Strategies, Politics." *Gender, Place and Culture* 10 (2): 145–157.

Cockayne, D.G. 2016. "Entrepreneurial Affect: Attachment to Work Practice in San Francisco's Digital Media Sector." *Environment and Planning D: Society and Space* 34 (3): 456–473.

Cookson, T.P. 2018. *Unjust Conditions: Women's Work and the Hidden Cost of Cash Transfer Programs*. Oakland, CA: University of California Press.

Cowen, D., and N. Lewis. 2016. "Anti-Blackness and Urban Geopolitical Economy: Reflections on Ferguson and the Surburbanization of the Internal Colony." *Society and Space Open Site*. http://societyandspace.org/2016/08/02/anti-blackness-and-urban-geopolitical-economy-deborah-cowen-and-nemoy-lewis/. Accessed 14 May 2018.

Crenshaw, K. 1991. "Mapping the Margins: Intersectionality, Identity Politics, and Violence against Women of Color." *Stanford Law Review* 43 (6): 1241–1299.

Derickson, K.D. 2017. "Urban Geography II: Urban Geography in the Age of Ferguson." *Progress in Human Geography* 41 (2): 230–244.

Diprose, G. 2017. "Radical Equality, Care and Labour in a Community Economy." *Gender, Place & Culture* 24 (6): 834–850.

Elwood, S., and V. Lawson. 2013. "Whose Crisis? Spatial Imaginaries of Class, Poverty, and Vulnerability." *Environment and Planning A* 45 (1):103–108.

England, K., and C. Henry. 2013. "Care, Migration, and Citizenship: International Nurses in the UK." *Social and Cultural Geography* 14 (5): 558–574.

Ettlinger, N. 2007. "Precarity Unbound." *Alternatives* 32 (3): 319–340.

Fickey, A., and K.B. Hanrahan. 2014. "Moving Beyond Neverland: Reflecting Upon the State of the Diverse Economies Research Program and the Study of Alternative Economic Spaces." *ACME: An International Journal for Critical Geographies* 13 (2): 394–403.

Gibson-Graham, J.K. 2008. "Diverse Economies: Performative Practices for Other Worlds." *Progress in Human Geography* 32 (5): 613–632.

Gidwani, V., and K. Sivaramakrishnan. 2003. "Circular Migration and the Spaces of Cultural Assertion." *Annals of the Association of American Geographers* 93 (1): 186–213.

Gökarıksel, B., and S. Smith. 2017. "Intersectional Feminism Beyond US Flag Hijab and Pussy Hats in Trump's America." *Gender, Place & Culture* 2 (5): 628–644.

Hall, S.M. 2011. "Exploring the 'Ethical Everyday': An Ethnography of the Ethics of Family Consumption." *Geoforum* 42 (6): 627–637.

Hall, S.M. 2015. "Everyday Family Experiences of the Financial Crisis: Getting by in the Recent Economic Recession." *Journal of Economic Geography* 16 (2): 305–330.
Hanson, S., and G. Pratt. 1995. *Gender, Work, and Space*. New York: Routledge.
Ho, K. 2009. *Liquidated: An Ethnography of Wall Street*. Durham, NC: Duke University Press.
Joseph, M. 2014. *Debt to Society: Accounting for Life under Capitalism*. Minneapolis: University of Minnesota Press.
Katz, C. 2001a. "Vagabond Capitalism and the Necessity of Social Reproduction." *Antipode* 33 (4): 709–728.
Katz, C. 2001b. "On the Grounds of Globalization: A Topography for Feminist Political Engagement." *Signs* 26 (4): 1213–1234.
Katz, C. 2004. *Growing up Global: Economic Restructuring and Children's Everyday Lives*. Minneapolis: University of Minnesota Press.
Kern, L., R. Hawkins, K.F. Al-Hindi, and P. Moss. 2014. "A Collective Biography of Joy in Academic Practice." *Social & Cultural Geography* 15 (7): 834–851.
Kern, L., and B. Mullings. 2013. "Urban Neoliberalism, Urban Insecurity and Urban Violence: Exploring the Gender Dimensions. In: *Rethinking Feminist Interventions into the Urban*, edited by L. Peake and M. Rieker, 23–40. New York: Routledge.
Larner, W. 2003. "Neoliberalism?" *Environment and Planning D: Society and Space* 21 (5): 509–512.
Lawson, V. 2007. "Presidential Address: Geographies of Care and Responsibility." *Annals of the Association of American Geographers* 97 (1): 1–11.
Loomis, J.M. 2018. "Rescaling and Reframing Poverty: Financial Coaching and the Pedagogical Spaces of Financial Inclusion in Boston, Massachusetts." *Geoforum* 95: 143–152.
Lyon, S., J.A. Bezaury, and T. Mutersbaugh. 2010. "Gender Equity in Fairtrade – Organic Coffee Producer Organizations: Cases from Mesoamerica." *Geoforum* 41 (1): 93–103.
Lyon, S., T. Mutersbaugh, and H. Worthen. 2017. "The Triple Burden: The Impact of Time Poverty on Women's Participation in Coffee Producer Organizational Governance in Mexico." *Agriculture and Human Values* 34 (2): 317–331.
Maclean, K. 2013. "Gender, Risk and Micro-Financial Subjectivities." *Antipode* 45 (2): 455–473.
MacLeavy, J. 2011. "A 'New Politics' of Austerity, Workfare and Gender? The UK Coalition Government's Welfare Reform Proposals." *Cambridge Journal of Regions, Economy and Society* 4 (3): 355–367.
McDowell, L. 1997. *Capital Culture: Gender at Work in the City*. Oxford: Blackwell.
McDowell, L. 2010. "Capital Culture Revisited: Sex, Testosterone and the City." *International Journal of Urban and Regional Research* 34 (3): 652–658.
McDowell, L. 2013. *Working Lives: Gender, Migration and Employment in Britain, 1945–2007*. Oxford: Wiley-Blackwell.
Meehan, K., and K. Strauss. 2015. *Precarious World: Contested Geographies of Social Reproduction*. Athens, GA: The University of Georgia Press.
Mitchell, K., S.A. Marston, and C. Katz, eds. 2004. *Life's Work: Geographies of Social Reproduction*. Malden, MA: Blackwell.
Morrow, O., and K. Dombroski. 2015. "Enacting a Post-Capitalist Politics Through the Sites and Practices of Life's Work." In: *Precarious Worlds: New Geographies of Social Reproduction*, edited by K. Strauss and K. Meehan, 82–96. Athens: University of Georgia Press.
Mott, C., and D. Cockayne. 2017. "Citation Matters: Mobilizing the Politics of Citation Toward a Practice of 'Conscientious Engagement'." *Gender, Place & Culture* 24 (7): 1–20.
Mott, C., S. Zupan, A.-M. Debbane, R.L.*, and University of Kentucky Critical Pedagogy Working Group. 2015. "Making Space for Critical Pedagogy in the Neoliberal University: Struggles and Possibilities." *ACME: An International Journal for Critical Geographies* 14 (4): 1260–1282.
Mountz, A., A. Bonds, B. Mansfield, J. Loyd, J. Hyndman, M. Walton-Roberts, R. Basu, R. Whitson, R. Hawkins, T. Hamilton, and W. Curran. 2015. "For Slow Scholarship: A Feminist Politics of Resistance through Collective Action in the Neoliberal University." *ACME: An International E-Journal for Critical Geographies* 14 (4): 1235–1259.
Mullings, B. 2014. "Neoliberalization, Gender and the Rise of the Diaspora Option in Jamaica." In: *Global Perspectives on Gender and Space: Engaging Feminism and Development*, edited by A.M. Oberhauser and I. Johnston-Anumonwo, 53–70. London: Routledge.
Nagar, R., V. Lawson, L. McDowell, and S. Hanson. 2002. "Locating Globalization: Feminist (Re)readings of the Subjects and Spaces of Globalization." *Economic Geography* 78 (3): 257–284.
Oberhauser, A.M. 2000. "Feminism and Economic Geography: Gendering Work and Working Gender." In: *A Companion to Economic Geography*, edited by E. Sheppard and T.J. Barnes, 60–76. Oxford: Blackwell.

Oberhauser, A.M. 2010. "(Re)Scaling Gender and Globalization: Livelihood Strategies in Accra, Ghana." *ACME: An International E-Journal for Critical Geographies* 9 (2): 221–244.
Parker, B., and O. Morrow. 2017. "Urban Homesteading and Intensive Mothering: (Re) Gendering Care and Environmental Responsibility in Boston and Chicago." *Gender, Place & Culture* 24 (2): 247–259.
Peake, L. 2010. "Gender, Race, Sexuality." In: *The SAGE Handbook of Social Geographies*, edited by S.J. Smith, R. Pain, S. Marston, and J.P. Jones III, 55–77. London: SAGE.
Pollard, J. 2013. "Gendering Capital: Financial Crisis, Financialization and (an Agenda for) Economic Geography." *Progress in Human Geography* 37 (3): 403–423.
Pratt, G. 2012. *Families Apart: Migrant Mothers and the Conflicts of Labor and Love*. Minneapolis: University of Minnesota Press.
Pratt, G., and V. Rosner. 2006. *The Global and the Intimate: Feminism in Our Time*. New York: Columbia University Press.
Pulido, L. 2016. "Geographies of Race and Ethnicity II: Environmental Racism, Racial Capitalism and State-Sanctioned Violence." *Progress in Human Geography* 41 (4): 1–10.
Radel, C., B. Schmook, N. Haenn, and L. Green. 2016. "The Gender Dynamics of Conditional Cash Transfers and Smallholder Farming in Calakmul, Mexico." *Women's Studies International Forum* 65: 17–27.
Rankin, K.N. 2013. "A Critical Geography of Poverty Finance." *Third World Quarterly* 34 (4): 547–568.
Richardson, L. 2016. "Feminist Geographies of Digital Work." *Progress in Human Geography* 42 (2): 244–263.
Roberts, S.M. 2004. "Gendered Globalization." In: *Mapping Women, Making Politics: Feminist Perspectives on Political Geography*, edited by L. Staeheli, E. Kofman, and L. Peake, 127–140. New York: Routledge.
Roy, A. 2010. *Poverty Capital: Microfinance and the Making of Development*. New York: Routledge.
Roy, A. 2016. "Divesting from whiteness: The University in the Age of Trumpism." *Society and Space.org* 28 (November) http://societyandspace.org/ 2016/11/28/ divesting-from-whiteness-the-university-in-the-age-of-trumpism/. Accessed 14 May 2018.
Sassen, S. 2005. "The Global City: Introducing a Concept." *Brown Journal of World Affairs* 11 (2): 27–43.
Silvey, R. 2006. "Consuming the Transnational Family: Indonesian Migrant Domestic Workers to Saudi Arabia." *Global Networks* 6 (1): 23–40.
Silvey, R. 2013. "Political Moves: Cultural Geographies of Migration and Difference." In: *The Wiley-Blackwell Companion to Cultural Geography*, edited by N.C. Johnson, R.H. Schein and J. Winders, 409–422. New York: John Wiley & Sons.
Smyth, A. 2017. "Re-reading Remittances through Solidarity: Mexican Hometown Associations in New York City." *Geoforum* 85: 12–19.
Valentine, G. 2007. "Theorizing and Researching Intersectionality: A Challenge for Feminist Geography." *Professional Geographer* 59 (1): 10–21.
Waite, L. 2009. "A Place and a Space for a Critical Geography of Precarity?" *Geography Compass* 3 (1): 412–433.
Werner, M. 2010. "Embodied Negotiations: Identity, Space and Livelihood after Trade Zones in the Dominican Republic." *Gender, Place & Culture* 17 (6): 725–741.
Werner, M. 2012. "Beyond Upgrading: Gendered Labour and the Restructuring of Firms in the Dominican Republic." *Economic Geography* 88 (4): 403–422.
Werner, M., K. Strauss, B. Parker, R. Orzeck, K. Derickson, and A. Bonds. 2017. "Feminist Political Economy in Geography: Why Now, What Is Different, and What For?" *Geoforum* 79: 1–4.
West, E. 2016. "'Give Women Your Money': Radical Feminism in an Age of Choice." Blog post. https://medium.com/@ericawest/give-women-your-money-radical-feminism-in-an-age-of-choice-2d4a0bb40dae. Accessed 10 December 2017.
Wright, M.W. 2006. *Disposable Women and Other Myths of Global Capitalism*. London: Routledge.
Yeoh, B.S.A., and K. Ramdas. 2014. "Gender, Migration, Mobility and Transnationalism." *Gender, Place & Culture* 21 (10): 1197–1213.
Zaloom, C. 2006. *Out of the Pits: Traders and Technology from Chicago to London*. Chicago, IL: University of Chicago Press.

12

DISENTANGLING GLOBALIZATION

Towards a feminist geography of hair and beauty

Caroline Faria and Bisola Falola

Introduction

Each year in May, Dubai hosts Beautyworld Middle East, a trade fair bringing together 2,000 exhibitors and 35,000 producers, distributors and cosmetologists from over 60 countries. It is now the world's largest trade fair for the hair and beauty industry. The lucrative markets of Iran, Saudi Arabia and the United Arab Emirates (UAE) partly explain its dramatic growth and success. Also, increasingly attractive to Beautyworld visitors are the markets of East, West and Southern Africa and their cosmopolitan centres of Lagos and Accra, Nairobi and Kampala, Johannesburg and Cape Town. This success signals a growing desire in these places for cosmopolitan beauty, one that is rooted in a history of colonial and postcolonial international exchange. Africans are positioned not only as consumers in these commodity circuits; African entrepreneurs extend the beauty trade onto and within the continent, with African 'saloonists' adapting imported styles for their clients and innovative manufacturers producing hair weaves, skin lotions, fragrances and beauty technologies. Indeed, beauty is big business in the region, powerfully shaping and shaped by people and places.

Beauty is necessarily, then, deeply spatial. Its (im)material affects and objects, its ideologies and its imagined and embodied forms are powerful in producing, connecting and transforming our world. And yet, curiously, very little *geographic* attention has been paid to beauty (but see Fluri 2009; Wrigley-Asante, Agyei-Mensah, and Obeng 2017). In this chapter, we demonstrate some of the insights offered by a *feminist* and intersectional geographic approach to globalization, one centring hair and beauty. We do so through an analysis of the synthetic and human hair trade between the Gulf and East Africa. In particular, we demonstrate the methodological insights of studying hair and beauty via a *feminist commodity-chain analysis*. Building on the work of Priti Ramamurthy (2003, 2011) we centre three insights of this approach here: its postcolonial disruption of narratives of people and places in the Global South living with, affected by and driving globalization; its attention to the connected ideologies of gender, race, class and sexual power that reproduce and reinvent the industry; and its insistence on a 'global–intimate', relational understanding of scale. As we show, this approach reveals new insights into the embodied nature of globalization and the innovation, creative labour, trading ties and consumption politics that drive it.

Intimate geographies of globalization: feminist interventions

There is now an extensive, complex and often contradictory body of scholarship on globalization within the discipline of geography and across the social sciences. Scholars have argued that it should be understood both as a set of material processes of heightened spatial economic, political and socio-cultural integration (and in some cases, dislocation) and a powerfully dominant ideological discourse (Gibson-Graham 1996; Massey 2005; Yeoh 1999). Globalization has produced new kinds of mobilities and fixities, inclusions and exclusions and reconfigured circuits of commodities, as well as the entrenchment, recomposition and decomposition of values, norms and ideologies. Amid proclamations of an inevitable flattening of difference and a borderless world (Friedman 2005), critical scholars have pushed for grounded theorizations that attend to the 'entanglements' (Sheppard 2012) of global trade with a host of socio-cultural, economic and political spatialities that are, at once, material, discursive, emotional and power-laden (Domosh 2010; Gibson-Graham 1996; Katz 2001; Rankin 2003; Tsing 2005). One central area of interest is that of neoliberal globalization, a far more narrowly defined phenomenon concerning the emergent dominance of liberal political-economic ideology and practice around the world in the last 30 years (Carmody 2013; Peck and Tickell 2002; Power 2005). Political geographers have interrogated the biopolitical governing practices (Agnew 2005; Dodds 1998; Sparke 2006) of this form of globalization, how its logics are reinforced and reproduced (Gibson-Graham 1996), and how it has been resisted and reworked in multiple contexts (Hart 2002). Cultural engagements with economic geography, long concerned with uneven economic development and its place-based particularities, in developing and/or complicating spatialized economic models have contributed to deeper understandings of contemporary neoliberal globalization (Barnes 2001; Flew 2010; James, Martin and Sunley 2006; Warf 2012).

Feminist scholars have productively engaged with and extended political and economic studies of globalization. Centrally, such work has critiqued the tendency in both subdisciplines to focus (albeit often critically) on: formal and 'public' spheres of politics and economics and their associated spaces, places (e.g. firms and institutions, 'global cities', advanced economies of the Global North) and scales (e.g. the supranational and national); and on particular subjects and actors (e.g. supranational bodies, elite global managers) (Nagar et al. 2002; Staeheli and Kofman 2004). Such attends to 'peripheral' subjects and spaces, arguing that these are, in fact, central to understanding global political and economic relations (Oberhauser and Hanson 2008; and see Mbembe 2001). Here, feminist and cultural geographers have extended the insights gained by attending to the cultural workings of globalizations as they intersect with and co-produce those of the economic and political (Gökarıksel and McLarney 2010; Rankin 2003). Feminist scholars here have brought historical archival and contemporary ethnographic methods to geography to interrogate the power-laden processes of meaning-making that both underpin and are produced through global circuits of production, trade and consumption (Domosh 2006, 2010; Rankin 2003).

Global commodities through a feminist geographic lens

Commodity chain analyses were first developed by Hopkins and Wallerstein to document the respatialization of capital accumulation in 'core' nations at the expense of the 'periphery' through the journey of a commodity (Hopkins and Wallerstein 1986; Leslie and Reimer 1999; Wallerstein 2009). Studying processes of capital accumulation from production through to the point of sale has been the hallmark of realist commodity chain analyses undertaken by world systems scholars since the rise of the new international division of labour. Such analyses have focused on the macro-economic – valuably tracing new geographies of commodities, the relations that they formed between the Global North and South, and the parallel shifting of sovereignty between nation states and multinational corporations. With heightened economic integration, the heuristic was taken up by Gereffi, Korzeniewicz and Korzeniewicz (1994) and reconceptualized as 'global commodity chain analysis'. A host of conceptual variants have been developed since, with shifts towards an understanding of commodity 'networks', which usefully moves beyond the linearity of the original chain conceptualization (Blair and Werner 2011). Put simply, Leslie describes its most contemporary formulation as a way to 'trace the entire trajectory of a product across time and space, including the movement of material, resources, value, finance, and knowledge, as well as signs and symbols' (2012, 65).

In parallel with those broader feminist developments described above, studies of commodities in economic geography have undergone productive critical interrogation. The global commodity chain analysis has been criticized for its privileging of primary commodity production or extraction (and elision of distribution and consumption, for example) and its interest only in those 'nodes' where flows of commodities touch down momentarily (Leslie and Reimer 1999), with little interrogation of the grounded spatialities of commodity chains, the agency of participants or the varied place-based meanings, values and norms attached to commodities and consumption practices (Bell and Valentine 1997; Cook and Crang 1996; Fine and Leopold 1993; Miller 1995; Slater 1997). This work is richly complemented by Appadurai's call for an interrogation of the 'social life' of commodities (1986), one that traces their journey through space from production to consumption to better understand how seemingly abstract global circuits produce, shape and remake embodied relations of power (Bassett 2002; Cook 2004; Gökarıksel and Secor 2010; Hollander 2008; Leslie 2012; Ramamurthy 2004). Of particular import to this project is the work of feminist scholars such as Bair (2005, 2009) and Leslie (2002, 2012), who have called attention to the interwoven classed and *gendered* relations bound up with commodity networks (Ansell, Tsoeu and Hajdu 2014; Oberhauser 2010; Oberhauser and Johnston-Anumonwo 2011; Oberhauser and Yeboah 2011). Most recently, these critical insights have been complicated by postcolonial interventions in economic geography and in the study of commodities (Maclean 2013; Pollard et al. 2009; Pollard, McEwan and Hughes 2011; Poon and Yeung 2009; Participants of the 2010 Economic Geography Workshop 2011). This work demands that attention be paid to the colonial legacies of race and racism that underpin contemporary economic structures.

A *feminist* commodity chain analysis interweaves these feminist and postcolonial interventions. Ramamurthy is a leading scholar of this approach (e.g. 2004, 2011, and see Schmidt 2018). Although not a geographer, the theoretical reframing and methodological intervention that she poses in her approach is deeply spatial and thus incredibly instructive for geographic work on globalization. We put this approach into conversation with insights from feminist geography around the 'global intimate' (Pratt and Rosner 2006; Smith 2012), demonstrating new and innovative ways to understand globalization and the social, political, economic and geographical lives of global commodities. In doing so, we centre three connected interventions for work on globalization: first, the *disruption of dominant and orientalist narratives of people and places viewed as peripheral to and marginalized by globalization*; second, a feminist commodity chain relies on a *performance-based theorization of power: racial, classed, gendered, and sexualized power*; and, last, the use of a *relational understanding of scale* that refuses to disconnect supranational, national and regional policy and practice from the household and the body. What emerges is an intersectional feminist geographic approach that attends to the embodied and emotional workings of commodity circuits and that reveals new stories of globalization and ways of engaging with and understanding it. We turn to each intervention below, demonstrating their insights by grounding them in moments from the global hair and beauty trade operating through and beyond the Gulf–East African region.

Putting a feminist commodity chain analysis to work: snapshots from the Gulf–East African hair trade

Insight 1: disruptive stories of globalization: Uganda and Dubai as drivers of global beauty

To begin, feminist analyses of commodities, and their lives, aim to disrupt orientalist narratives of people and places in the Global South. A snapshot of the beauty industries in, and connecting, the Gulf and East Africa dramatically upturns the 'single story' (Adichie 2009, and see Owusu 2012) of Africa, of globalization and of beauty. While most scholarly and popular attention centres the Global North or emphasizes Africa's marginalization through global economic exchange, a look at Gulf–East African beauty industry through the lens of a feminist commodity chain analysis offers another story.

The African trade in beauty products has a long history, rooted in global circuits of colonial and postcolonial international exchange (Horton 2004; Nicolini 2009; Thomas 2008), with long-standing ties to Europe and the Americas and also to the Gulf, East and South Asian regions. From the early 1990s on, East Africa underwent rapid economic growth. Uganda, along with Kenya and Tanzania, are particularly vibrant centres for the East Africa trade in beauty products, influenced continentally by the innovative fashion centres of the Congo, Ghana, South Africa, Nigeria and further afield, through African diasporic circuits (Balogun 2012; Thomas 2012). Focusing on Uganda, between 1992 and 2011 the country saw an average rise in GDP of 7% per year (AEO 2013), albeit unevenly centred on the south, particularly the capital, Kampala. In connection, Uganda has a growing middle class, estimated to be around 6.1 million people, mirroring that of Kenya and Tanzania (ADB 2011; Tentena 2012). While the consumer base of the hair and beauty trade extends beyond the middle class, these economic shifts have helped to position Uganda as an increasingly important site for the trade in and consumption of beauty products. Kampala, the main economic centre of Uganda, is at the heart of this vibrant global connection. The city hosts some of the largest markets for clothing and beauty goods, including Gazaland Mall, known popularly as 'The House of Hair', and Luwum Street, a long-established centre for bridal shopping across the region (Whitesell and Faria 2018). The city also holds year-round beauty-related events, including a range of pageants such as Miss Uganda and Miss Tourism-Uganda, and since 2013 has hosted the internationally recognized Kampala Fashion Week (Elledge and Faria 2018).

Uganda is also the home of an industrial base for the *manufacturing* of beauty goods. In the hair industry, Darling Ltd is the longstanding leader. Until its recent buy-out by giant Indian conglomerate Godrej, Darling was owned and operated by an extended Lebanese family based in Africa for over 20 years, with operations in 22 countries across the continent, according to a personal communication (in June 2012). The Ugandan branch is a leading producer for the East African region, exporting products to the Democratic Republic of the Congo, the Republic of Rwanda and the new Republic of South Sudan. The branch designs, tests and releases 50 new styles a year, drawing on inspiration locally, across the region and internationally and expanding its consumer base (which spans low- to high-income levels from as young as 8 to 10 years old) through fashion-show advertising, 'saloonist' training, social media and mobile salons that travel to rural areas to introduce Ugandans to the latest in synthetic hair fashions. In 2014, Darling Ltd was named 'Superbrand of the Year' in Uganda, and it continues to be one of the most successful and influential hair manufacturers on the continent.

While Darling Ltd is a major supplier of synthetic hair products across the continent, beauty products are also imported, and Dubai is a vital trading point. Many of the products flowing into this region from Europe, China, the US and elsewhere arrive via Dubai, transforming the city into one of the largest trading hubs, wholesale and logistical sites and centres for cosmetological connection via the Beautyworld Middle East trade fair. Dubai is also an influential arbiter of style, being home to Huda Kattan, a renowned beauty influencer. Although a large volume of goods now arrives in East Africa directly from places such as China, Dubai remains influential. This is in part due to its proximity to Africa and the neoliberalizing imperatives that continue to ease travel and business operations.

Distributors like Mohammad and Lee, operating in the wholesale district of Deira, have worked in Dubai for over a decade. They moved from Iran and the Philippines, respectively, in search of employment and found work in the city's 'old town' in the lucrative hair and beauty trade. Small- and mid-scale entrepreneurs dominate the emergent African trade, purchasing less than 100 kilogrammes of hair on each visit to the city's downtown wholesale hair shops, a number of which are also owned and run by Africans. These traders are highly mobile, often (but not always) middle class and include members of the African diaspora. Rather than exporting via the Jebel Ali free zone, these traders fly smaller quantities via suitcases and/or plane cargo packages to the cosmopolitan market centres of Lagos, Accra, Nairobi and Kampala. Traders making larger purchases move their orders through cargo companies, which transport their goods by container from the Gulf to the port of Mombasa in Kenya. Traders like Lubega, a Ugandan man in his forties, buy hair in bulk that has been freshly shipped into Mombasa, filling the markets of Nairobi. He hired porters to help him to pack and move five or more bags filled with hair onto a bus back to Kampala. When he began his business in the early 1990s, he was one of the first to import synthetic hair into the country, and it was lucrative. He ran three shops in the city under his business name of Lubex Hair and Beauty, pioneering a trade in a commodity that would become big business across the region.

Insight 2: sigh ... 'Russians have the best hair': ideologies of power in the global hair trade

A feminist commodity chain analysis secondly insists that we attend to power: the connected and complex ways that racial, gendered, sexualized and classed norms, ideals and geometries of power operate over the life course of the commodity. A look at hair, hair care and hair management through this lens is instructive.

I (Author 1) overheard the words in the quote above at a 2015 Beautyworld Middle East trade fair. The comment gave me the chills: a simple articulation of the power-laden work of hair and beauty. Indeed, critical race scholars have long argued that hair – and hair technologies like wigs, weaves and extensions – is highly visible and meaning-laden (Hill Collins 1990; Wingfield 2009) and is powerfully used to promote, reinforce and culturally diffuse gender, race and class-based norms and hierarchies. In one way, hair beauty norms rely on and exploit historical precedents and beliefs about race, class and gender, providing products of corporeal alteration to meet these ideals (Thomas 2012). In turn, hair – and those traders, stylists and clients who work with and wear it – both troubles and reworks these norms (Balogun 2012).

A feminist analysis of the commodity reveals the ways that these seemingly contradictory processes are at work together. Interviews with traders and consumers and an archival analysis of decades of fashion coverage in the country suggest that, until 15 to 20 years ago, only the elites bought and used hair additions. At that time, these purchases acted as a marker of class, urban, gendered and cosmopolitan distinction (Faria 2013; Weiss 2009). However, this symbolic power of hair extensions heightened demand for them. In response, the material object itself has been modified, with a range of qualities, styles, lengths and modes of attachment, to expand the socio-economic consumer base. This raises new and interesting questions about a 'subaltern', 'vernacular', 'working-class' or otherwise non-elite cosmopolitanism (Gidwani 2006; Kurasawa 2004; Mohan 2006; Oza 2006), expressed through consumption. Young students in Uganda who regularly purchase hair extensions often described the way it made them feel: up to date; worldly; modern; successful. Women of the times. Of course, these modern hairstyles and textures create anxieties about threats to 'Ugandan' national or 'African' continental ideals of womanhood, as evidenced elsewhere on the continent (Balogun 2012; Faria 2013; Hackspiel 2008; Thomas 2008, 2012; Weiss 2009).

Direct marketing techniques by Darling Ltd and competitor companies, and well-travelled traders like Lubega, have introduced hair extensions into markets across Uganda, building consumers even in more rural and poorer communities. Here newer, synthetic alternatives and cost-effective designs have sharply increased the number of those who can afford these products in Uganda, according to a personal communication in 2014 (Faria and Jones, in review). On the international market, a similar widening of price and product ranges has driven the growth of hair additions over the last two decades. Here, price is determined by 'quality', a notion that reflects, and reproduces, a longstanding and *racialized* ideal of hair beauty. For example, one trader at the Beautyworld trade fair described how first the ranking is determined on the 'authenticity' of the hair (whether it is human or synthetic). Then it is based on phenotypical categorizations, such as the shape of individual strands of hair and their straightness. These thinly veiled racial hierarchies centre on their likeness to Indo–European hair. For this reason, Indian 'Virgin Remy' hair (i.e. that which has not been chemically treated and where individual strands all run in the same direction) is among the most expensive hair on the global market.

This ranking system also speaks to a very embodied classed, sexualized, gendered and *racialized* politics of hair that has been interrogated in the wider (primarily North American) African diaspora (Banks 2000) yet remains largely unaddressed in continental Africa (but see Asante 2016; Fritsch 2017; Thomas 2008, 2012 regarding skin lighteners). A feminist commodity chain analysis pushes us to pay attention to the racialized complexities of the sale and profitability of long, straightened, European-styled hair that mimic those of White bodies and are linked with the cultural practices and structural advantages of Whiteness.

However, Ramamurthy's framework also prompts deeper interrogation. A *feminist commodity chain analysis* insists that racial power is not fixed but performed, socially and historically rooted and malleable. Even as they live within the racial logics described above, the women who buy and style hair also do so *in defiance or outside* of these logics. For example, women consumers of synthetic hair in Kampala spoke about how ease, fashionability and playfulness guide their decision to buy and style synthetic hair. In turn, the use of hair weaves and the popularity of styles named after cultural icons such as Beyoncé and Rhianna demonstrate how the women who sport them identify with a wider pan-African diaspora, connecting them to global cultural circuits of Blackness (Rahier, Hintzen and Smith 2010) while also refashioning local styles. There is much more at work than simply an effort to mirror Whiteness. The complexities of these relationships – between consumers and their products, and between the people and places of the beauty trade and wider processes of globalization – are revealed in these moments through the lens of an intersectional feminist commodity chain analysis. We see, for example, how commodified hair and the wider beauty trade are produced by, perpetuate and disrupt the logics of capitalism. And that these logics are braided with, and rely upon, others – those perhaps of nationalism, colonialism, racism, heterosexism and ableism – to produce this powerful ideal, practice and industry of beauty.

Insight 3: scalar innovations: the global intimacies of hair and beauty

Lastly, a feminist commodity chain analysis challenges the macro-economic and political emphases of previous commodity chain studies, insisting upon a *relational understanding of scale* that understands seemingly abstract supranational, national and regional policy, as always bound up with the body, as themselves *corporeal*. Attention to the embodied and emotional workings of commodity circuits here relies productively on the conceptual geographic work of the 'global intimate' (Pratt and Rosner 2006). This is a particularly elegant conceptualization of power, space and scale that we use in our research to better understand globalization. Centrally, it troubles understandings of 'global' processes as masculine, distanced and disembodied, and the simultaneous rendering of the 'local', the 'domestic' and the 'bodily' as essentialized, feminized and characterized by the minutiae of the everyday (Mountz and Hyndman 2006, 446).

Through this framework, seemingly abstract processes like neoliberal globalization can be rethought by examining the intimate, the familiar, the sensory and the embodied experiences of 'living and knowing the global' (Mountz and Hyndman 2006, 448). The global intimate frame also draws upon Katz's (2001) methodological and epistemological call to construct 'counter topographies' by recognizing connections produced through global processes to other places, people and times. In doing so, the global intimate framework 'trac[es] lines across places to show how they are connected by the same processes, [while] simultaneously embedding these processes within the specifics of fully contextualized, three dimensional places' (Pratt and Yeoh 2003, 163). By moving away from the conflated binary of the global and local, where scales are rendered discrete, oppositional and in a moral- or power-based hierarchy, the global and/as intimate framing foregrounds the embodied, sensual, emotional, grounded operations of global capital. In this move, it is 'undoing' (Pratt and Rosner 2006, 20) the grand narratives of neo-liberal and postcolonial globalization.

This lens is particularly instructive to think through a particular encounter in our research on the global hair and beauty trade. At one Beautyworld Middle East booth, a Russian sales-woman woos a group of Iranian visitors. They are part of a contingent of about fifty female salon owners, distributers and clients who arrive together via a high-end chartered bus.

She hopes her product, a high-quality 'Virgin Remy' Indian hair, dyed blonde, available in 8, 10, 12 and 14 inch lengths, will appeal to them. Demonstrating how to attach and style the pieces, she encourages the women to touch them – to feel the softness and attest to the hair's purity, its lack of prior chemical treatment, its cuticles that run in the same direction and, above all, to its authenticity as human. They comply, teasing strands of hair between their manicured fingers, sighing, laughing with pleasure and nodding in agreement, while others clamour to photograph the stylists' fashionings so they can replicate them back at home. Indian hair, dyed, processed and sold by Russian workers and sales representatives in Dubai, for the high-end markets of Iran: the varied travels of purity, authenticity, beauty, glamour.

In this moment, we see the power-laden nature of the global hair and beauty trade and of this particular expression of economic globalization. Global intimate scale-thinking pushes us to tell this new geographic story: of the always and already embodied and emotional nature of international commodity exchange and globalized political decision-making: its sensory experience, its grounding in the tired limbs of travel-weary traders, the aching fingers of stylists, the relief, excitement, satisfaction of representatives who make that sale; the pleasure and anxie-ties, too, around consumption, around dressing-up.

Figure 12.1 Showcasing hair at the Beautyworld Middle East trade fair, 2015.
Source: C. Faria.

Conclusion: towards a feminist geography of hair and beauty

Just a few weeks before the 2014 Beautyworld Middle East opened its doors, the first African Global Business Forum was held in Dubai. Organized by the Dubai Chamber of Commerce and the Common Market for Eastern and Southern Africa (COMESA) (Geronimo 2013) regional investment agency, the event was designed for African countries to 'display their main attractions, facilities and business opportunities' and to 'highlight Dubai's strategic link between the continent and the rest of the world' (Africa Global Business Forum 2013). The Ugandan Prime Minister Amama Mbabaza and the Ugandan Minister of Trade Hon. Amelia Kyambadde were key speakers at the event, demonstrating the central role that Uganda has begun to play in promoting more open trade between the Middle East and East Africa. This vibrant, emergent and multi-billion-dollar beauty industry thus reflects, yet has also driven, the rising economic integration of Africa and the Middle East over the past 15 to 20 years. Hair is a significant part of this trade. African entrepreneurs are extending hair beauty products onto and within that continent, and local manufacturers are developing their own styles through the labour of women braiders and stylists. In connection, the hair trade is driven by the magic of the material object itself: the complexly gendered, classed and racialized desire evoked by commodified hair and its promises of class distinction and cosmopolitan style, its visage of modernity.

But there are many other geographic stories of beauty that should also be told. Beauty is a powerful ideology, a set of material objects and practices that dramatically shapes lives and places. Ideologies of beauty, their material objects and effects, and the people and places brought into the webs of beauty are many and varied. These stories are produced across our field for a deeper understanding of health and well-being, migration and mobility, urban change, national development and tourism, the reproduction of violent sexual, gendered and able – its norms in our everyday environments, to name a few. There is a long-established work outside of our field on beauty, such as that by Priti Ramamurthy, led by scholars of colour, critical gender and race theorists and Crip-studies theorists (Baggio and Moretti 2018; Balogun 2012; Cepeda 2018; Clare 2017; Dolan and Johnstone-Louis 2011; Ford 2016; Gentles-Peart 2018; Oza 2006; Özdemir 2016; Wingfield 2009, to name just a few examples). As geographers, we can valuably build on and extend this scholarship with our own spatial insights.

We close, then, with a call for a *feminist geography of hair and beauty*, one that is postcolonial and intersectional: attentive to the connected work of racial, gendered, sexualized, classed power and its grounding in, and escape from, the past–presents of colonialism.

Key readings

Mountz, A., and J. Hyndman. 2006. "Feminist Approaches to the Global Intimate." *Women's Studies Quarterly* 34: 446–463.

Nagar, R., V. Lawson, L. McDowell, and L. Hanson. 2002. "Locating Globalization: Feminist (Re)readings of the Subjects and Spaces of Globalization." *Economic Geography* 78: 257–284.

Ramamurthy, P. 2004. "Why is Buying a 'Madras' Cotton Shirt a Political Act? A Feminist Commodity Chain Analysis." *Feminist Studies* 30 (3): 734–769.

References

ADB. 2011. "Market Brief." *The Middle of the Pyramid: Dynamics of the Middle Class in Africa*. 20 April. Available at: www.afdb.org (accessed 20 August 2014).

Adichie, C.N. 2009. "The Power of the Single Story." TED Talk, July. Available at: www.ted.com/talks/chimamanda_adichie_the_danger_of_a_single_story.

Africa Global Business Forum. 2013. Program African Global Business Forum. Available at: www.africaglobalbusinessforum.com/ (accessed 20 August 2014).

African Economic Outlook (AEO). 2013. *African Economic Outlook Report*. Available at: www.africaneconomicoutlook.org/en/ (accessed 20 August 2014).

Agnew, J. 2005. *Hegemony: The New Shape of Global Power*. Philadelphia, PA: Temple University Press.

Ansell, N., S. Tsoeu, and F. Hajdu. 2014. "Women's Changing Domestic Responsibilities in Neoliberal Africa: A Relational Time–Space Analysis of Lesotho's Garment Industry." *Gender, Place & Culture* 22 (3): 363–382.

Appadurai, A. 1986. *Introduction: Commodities and the Politics of Value*, 3–63. Cambridge: Cambridge University Press.

Asante, G. 2016. "Glocalized Whiteness: Sustaining and Reproducing Whiteness Through 'Skin Toning' in Post-colonial Ghana." *Journal of International and Intercultural Communication* 9 (2): 87–203.

Baggio, R., and V. Moretti. 2018. "Beauty as a Factor of Economic and Social Development." *Tourism Review* 73 (1): 68–81.

Bair, J. 2005. "Global Capitalism and Commodity Chains: Looking Back, Going Forward." *Competition and Change* 9 (2): 153–180.

Bair, J. 2009. "Global Commodity Chains: Genealogy and Review." *Frontiers of Commodity Chain Research*, 1–34. Stanford, CA: Stanford University Press.

Balogun, O.M. 2012. "Cultural and Cosmopolitan: Idealized Femininity and Embodied Nationalism in Nigerian Beauty Pageants." *Gender and Society* 26 (3): 357–381.

Banks, I. 2000. *Hair Matters: Beauty, Power, and Black Women's Consciousness*. New York: New York University Press.

Barnes, T. 2001. "Retheorizing Economic Geography: From the Quantitative Revolution to the 'Cultural Turn'." *Annals of the Association of American Geography* 91 (3): 546–565.

Bassett, T.J. 2002. "Women's Cotton and the Spaces of Gender Politics in Northern Cote d'Ivoire." *Gender, Place and Culture* 9 (4): 351–370.

Bell, D., and G. Valentine. 1997. *Consuming Geographies: We Are Where We Eat*. London: Routledge.
Blair, J., and M. Werner. 2011. "Commodity Chains and the Uneven Geographies of Global Capitalism: A Disarticulations Perspective." *Environment and Planning* 43: 988–997.
Carmody, P. 2013. "A Global Enclosure: The Geo-Logics of Indian Agro-Investments in Africa." *Capitalism, Nature, Socialism* 24 (1): 84–103.
Cepeda, M.E. 2018. "Putting a 'Good Face on the Nation': Beauty, Memes, and the Gendered Rebranding of Global *Colombianidad*." *Women's Studies Quarterly* 46 (1/2): 121–138.
Clare, E. 2017. *Brilliant Imperfection: Grappling with Cure*. Durham, NC: Duke University Press.
Cook, I. 2004. "Follow the Thing: Papaya." *Antipode: A Journal of Radical Geography* 36 (4): 642–664.
Cook, I., and P. Crang. 1996. "The World on a Plate: Culinary Culture, Displacement and Geographical Knowledges." *Journal of Material Culture* 1: 131–153.
Dodds, K. 1998. "Political Geography I: The Globalization of World Politics." *Progress in Human Geography* 22 (4): 595–606.
Dolan, C., and M. Johnstone-Louis. 2011. "Re-Siting Corporate Responsibility: The Making of South Africa's Avon Entrepreneurs." *Focaal: Journal of Global and Historical Anthropology* 60 (1): 21–33.
Domosh, M. 2006. *American Commodities in an Age of Empire*. New York: Routledge.
Domosh, M. 2010. "The World was Never Flat: Early Global Encounters and the Messiness of Empire." *Progress in Human Geography* 34 (4): 419–435.
Elledge, A., and C. Faria. 2018. "'I Want to be a Part of Anything That Will Let my Country Shine': Towards a Geography of Beauty in Development Studies." Paper presented at the SWAAG Annual Meeting, Baton Rouge, October.
Faria, C. 2013. "Styling the Nation: Fear and Desire in the South Sudanese Beauty Trade." *Transactions of the Institute of British Geographers* 39 (2): 318–330.
Fine, B., and E. Leopold. 1993. *The World of Consumption*. London: Routledge.
Flew, T. 2010. "Toward a Cultural Economic Geography of Creative Industries and Urban Development: Introduction to the Special Issue on Creative Industries and Urban Development." *Information Society* 26: 85–91.
Fluri, J. 2009. "The Beautiful 'Other': A Critical Examination of 'Western' Representations of Afghan Feminine Corporeal Modernity." *Gender, Place & Culture* 16 (3): 241–257.
Ford, T. 2016. *Liberated Threads: Black Women, Style, and the Global Politics of Soul*. UNC Press, NC: Chapel Hill.
Friedman, T. L. 2005. *The World is Flat: A Brief History of the Twenty-First Century*. New York: Farrar, Straus & Giroux.
Fritsch, K. 2017. "'Trans-skin': Analyzing the Practice of Skin Bleaching among Middle-class Women in Dar es Salaam." *Ethnicities* 17 (6): 749–770.
Gentles-Peart, K. 2018. "Controlling Beauty Ideals: Caribbean Women, Thick Bodies, and White Supremacist Discourse." *Women's Studies Quarterly* 46 (1/2): 199–214.
Gereffi, G., M. Korzeniewicz, and P. Korzeniewicz, eds. 1994. "Introduction: Global Commodity Chains." In: *Commodity Chains and Global Capitalism*, 1–14. Westport, CT: Praeger.
Geronimo, A. 2013. "Africa Global Business Forum to Become an Annual Event." *Middle East Trade and Export Regional News*, 2 May. Available at: www.tradeandexportme.com/2013/05/agbf-to-become-an-annual-event/.
Gibson-Graham, J.K. 1996. *The End of Capitalism (As We Knew It): A Feminist Critique of Political Economy*. Oxford: Blackwell.
Gidwani, V. 2006. "Subaltern Cosmopolitanism as Politics." *Antipode* 38: 7–21.
Gökarıksel, B. 2012. "The Intimate Politics of Secularism and the Headscarf: The Mall, the Neighborhood, and the Public Square in Istanbul." *Gender, Place and Culture: A Journal of Feminist Geography* 19 (1): 1–20.
Gökarıksel, B., and E. McLarney, eds. 2010. "Muslim Women, Consumer Capitalism and Islamic Culture Industry." *Journal of Middle East Women's Studies* 6 (3): 1–18. Available at: https://doi.org/10.2979/MEW.2010.6.3.1.
Gökarıksel, B., and A. Secor. 2010. "'Even I Was Tempted': The Moral Ambivalence and Ethical Practice of Veiling-Fashion in Turkey." *Annals of the Association of American Geographers* 102 (4): 847–862.
Hackspiel, E. 2008. "Modernity and Tradition in a Global World: Fashion in Africa." *African Arts* 41 (2): 90–91.
Hart, G. 2002. "Disabling Globalization: Places of Power in Post-Apartheid South Africa." Berkeley: University of California Press.

Hill Collins, P. 1990. *Black Feminist Thought: Knowledge, Consciousness, and the Politics of Empowerment.* Boston: Unwin Hyman.
Hollander, G. 2008. *Raising Cane in the 'Glades': The Global Sugar Trade and the Transformation of Florida.* Chicago, IL: University of Chicago Press.
Hopkins, T., and E. Wallerstein. 1986. "Commodity Chains in the World Economy Prior to 1800." *Review X* 10 (1): 157–170.
Horton, M. 2004. "Artisans, Communities, and Commodities: Medieval Exchanges between Northwestern India and East Africa." *Ars Orientalis* 34: 62–80.
James, A., R.L. Martin, and P. Sunley. 2006. "The Rise of Cultural Economic Geography." In: *Critical Concepts in Economic Geography: vol. IV, Cultural Economy,* edited by R.L. Martin and P. Sunley, 3–18. London: Routledge.
Katz, C. 2001. "On the Grounds of Globalization: A Topography for Feminist Political Engagement." *Signs: Journal of Women in Culture and Society* 26: 1213–1234.
Kurasawa, F. 2004. "Cosmopolitanism from below: Alternative Globalization and the Creation of a Solidarity Without Bounds." *European Journal of Sociology* 45: 233–255.
Leslie, D. 2002. "Gender, Retail Employment and the Clothing Commodity Chain." *Gender, Place and Culture: A Journal of Feminist Geography* 9 (1): 61–76.
Leslie, D. 2012. "Gender, Commodities and Everyday Life." *Encounters and Engagements between Economic and Cultural Geography,* edited by B. Warf, 65–78. Dordrecht: Springer.
Leslie, D., and S. Reimer. 1999. "Spatializing Commodity Chains." *Progress in Human Geography* 23 (3): 401–420.
Maclean, K. 2013. "Evo's Jumper: Identity and the Used Clothes Trade in 'Post-Neoliberal' and 'Pluri-Cultural' Bolivia." *Gender, Place & Culture* 21 (8): 963–978.
Massey, D. 2005. *For Space.* London: SAGE.
Mbembe, A. 2001. *On the Postcolony.* Berkeley, CA: University of California Press.
Miller, D. 1995. "Consumption as the Vanguard of History: A Polemic by Way of Introduction." In: *Acknowledging Consumption: A Review of New Studies,* edited by D. Miller, 1–57. London: Routledge.
Mohan, G. 2006. "Embedded Cosmopolitanism and the Politics of Obligation: The Ghanaian Diaspora and Development." *Environment and Planning A: Society and Space* 38: 867–883.
Mountz, A., and J. Hyndman. 2006. "Feminist Approaches to the Global Intimate." *Women's Studies Quarterly* 34: 446–463.
Nagar, R., V. Lawson, L. McDowell, and S. Hanson. 2002. "Locating Globalization: Feminist (Re)readings of the Subjects and Spaces of Globalization." *Economic Geography* 78: 257–284.
Nicolini, B. 2009. "The Myth of the Sultans in the Western Indian Ocean During the Nineteenth Century: A New Hypothesis." *African & Asian Studies* 8 (3): 239–267.
Oberhauser, A.M. 2010. "(Re)Scaling Gender and Globalization: Livelihood Strategies in Accra, Ghana." *ACME: An International E-Journal for Critical Geographies* 9 (2): 221–244.
Oberhauser, A.M., and K.T. Hanson. 2008. "Negotiating Livelihoods and Scale in the Context of Neoliberal Globalization: Perspectives from Accra, Ghana." *African Geographical Review* 26: 11–36.
Oberhauser A.M., and I. Johnston-Anumonwo, I. 2011. "Globalization and Gendered Livelihoods in Sub-Saharan Africa: Introduction." *Singapore Journal of Tropical Geography* 32 (1): 4–7.
Oberhauser, A.M., and M. Yeboah. 2011. "Heavy Burdens: Gendered Livelihood Strategies of Porters in Accra, Ghana." *Singapore Journal of Tropical Geography* 32 (1): 22–37
Owusu, F. 2012. "The Geography of Globalization in Sub-Saharan Africa: Implications for Africa's Globalization Narratives." Paper presented at the Association of American Geographers Annual Meeting, New York.
Oza, R. 2006. *The Making of Neoliberal India: Nationalism, Gender, and the Paradoxes of Globalization.* New York: Routledge/Women Unlimited?
Özdemir, B.P. 2016. "Building a 'Modern' and 'Western' Image: Miss Turkey Beauty Contests from 1929 to 1933." *Public Relations Review* 42: 759–765.
Participants of the Economic Geography 2010 Workshop 2011. "Emerging Themes in Economic Geography: Outcomes of the Economic Geography 2010 Workshop." *Economic Geography* 87 (2): 111–126.
Peck, J., and A. Tickell. 2002. "Neoliberalizing Space." *Antipode* 34 (3): 380–404.
Pollard, J., C. McEwan, and A. Hughes, eds. 2011. *Postcolonial Economies.* London: Zed Books.
Pollard, J., C. McEwan, N. Laurie, and A. Stenning. 2009. Economic Geography Under Postcolonial Scrutiny." *Transactions of the Institute of British Geographers* 34 (2): 137–142.

Poon, J.P.H., and H.W. Yeung. 2009. "SJTG Special Forum – Continental Drift? Development Issues in Asia, Latin America and Africa." *Singapore Journal of Tropical Geography* 30: 3–6.
Power, M. 2005. "Working the Spaces of Neoliberalism." *Antipode* 37 (3): 605–612.
Pratt, G., and V. Rosner. 2006. "Introduction: The Global and the Intimate." *Women's Studies Quarterly* 34: 13–24.
Pratt, G., and B. Yeoh. 2003. "Transnational (Counter)topographies." *Gender, Place and Culture: A Journal of Feminist Geography* 10 (2): 159–166.
Rahier, J., P. Hintzen, and F. Smith. 2010. *Global Circuits of Blackness: Interrogating the African Diaspora.* PhD dissertation, University of Illinois.
Ramamurthy, P. 2003. "Material Consumers, Fabricating Subjects: Perplexity, Global Connectivity Discourses, and Transnational Feminist Research." *Cultural Anthropology* 18: 524–550.
Ramamurthy, P. 2004. "Why is Buying a 'Madras' Cotton Shirt a Political Act? A Feminist Commodity Chain Analysis." *Feminist Studies* 30 (3): 734–769.
Ramamurthy, P. 2011. "Value, Affect, and Feminist Commodity Chain Analysis: A Conceptual Framework." Paper presented at the conference: *Gendered Commodity Chains: Bringing Households and Women into Global Commodity Chain Analysis.* Binghamton University, October.
Rankin, K. 2003. "Anthropologies and Geographies of Globalization." *Progress in Human Geography* 27 (6): 708–734.
Schmidt, S. 2018. "Precarious Craft: A Feminist Commodity Chain Analysis." *Migration and Development* online: 1–18. doi.org/10.1080/21632324.2018.1489363.
Sheppard, E. 2012. "Trade, Globalization and Uneven Development: Entanglements of Geographical Political Economy." *Progress in Human Geography* 36 (1): 44–71.
Slater, D. 1997. *Consumer Culture and Modernity.* Cambridge: Polity Press.
Smith, S. 2012. "Intimate Geopolitics: Religion, Marriage, and Reproductive Bodies in Leh, Ladakh." *Annals of the Association of American Geographers* 102 (6): 1511–1528.
Sparke, M. 2006. "Political Geographies of Globalization II: Governance." *Progress in Human Geography* 30 (3): 357–372.
Staeheli, L., and E. Kofman, eds. 2004. *Mapping Women, Making Politics: Feminist Perspectives on Political Geography.* New York: Routledge.
Tentena, P. 2012. "East Africa: Uganda's Middle Class Grows as Poverty Dips." *AllAfrica*, 13 May. Available at: http://allafrica.com/stories/201205141224.html.
Thomas, L. 2008. "The Modern Girl and Racial Respectability in 1930s South Africa." In: *The Modern Girl Around the World: Consumption, Modernity and Globalization,* edited by A.E., Weinbaum, L.M., Thomas, P. Ramamurthy, U. G. Poiger, Y.M. Dong, and T.Ed. Barlow, 96–119. Durham, NC: Duke University Press.
Thomas, L. 2012. "Skin Lighteners, Black Consumers and Jewish Entrepreneurs in South Africa." *History Workshop Journal* 73: 259–283.
Tsing, A. 2005. *Friction: An Ethnography of Global Connection.* Princeton, NJ: University of Princeton Press.
Wallerstein, I. 2009. "Protection Networks and Commodity Chains in the Capitalist World-economy." In: *Frontiers of Commodity Chain Research,* edited by J. Bair, 83–89. Stanford, CA: Stanford University Press.
Warf, B. eds. 2012. "Encounters and Engagements between Economic and Cultural Geography." *Geojournal Library* 104: 65–78.
Weiss, B. 2009. *Street Dreams and Hip-hop Barbershops: Global Fantasy in Urban Tanzania.* Bloomington: Indiana University Press.
Whitesell, D. and C. Faria . 2018. "Gowns, Globalization, and 'Global Intimate Mapping': Geovisualizing Uganda's Wedding Industry." *Environment and Planning C: Politics and Space.* 0(0): 1–15.
Wingfield, A. 2009. *Doing Business with Beauty: Black Women, Hair Salons, and the Racial Enclave Economy.* Plymouth: Rowman and Littlefield.
Wrigley-Asante, C., S. Agyei-Mensah, and F.A. Obeng. 2017. "It's Not All about Wealth and Beauty: Changing Perceptions of Fatness Among Makola Market Women of Accra, Ghana." *Singapore Journal of Tropical Geography* 38 (3): 414–428.
Yeoh, B. 1999. "Global/Globalizing Cities." *Progress in Human Geography* 23 (4): 607–616.

PART 2

Placing feminist geographies

13
EMBODIMENT
Lesbians, space, sperm and reproductive technologies

Robyn Longhurst and Lisa Melville

Introduction

The subject of embodiment has now been on gender and feminist geographers' agendas for more than three decades. Over this period, scholarship in this area has undergone significant change. Work has continued on people's embodied interactions with space and place, but now there is also research that reflects on the interiority of bodies, for example on how the placenta and uterus can be used to prompt different kinds of geographical thinking. Another area of interest in the body that has emerged in recent years is the way in which the gendered and sexed dimensions of embodiment are fluid. Now, there is also research that moves beyond the gendered and sexual dimensions of embodiment. Intersectional approaches have led to a focus on other axes of embodied subjectivity, not just ethnicity and class but, also, for example, sexuality, age and health. How technology is shaping bodies has made its way onto gender and feminist geographers' agendas. Also on their agenda is how bodies are imbued with emotion and affect. This chapter examines these trends by turning attention to a recent project that one of us is carrying out on lesbians, space, sperm and reproductive technologies. The project addresses themes such as mothering, social relations, policies, home and public places. Embodiment is a useful concept in helping to cast light on these themes. In turn, these themes help to deepen the thinking on embodiment.

Embodiment, corporeality, the visceral or whatever one chooses to describe this 'thing' we commonly call 'the body', as stated above has now been on geographers' agendas for more than three decades. Sarah Nettleton and Jonathan Watson (1998, 1) note:

> If one thing is certain, it is that we all have a body. Everything we do we do with our bodies – when we think, speak, listen, eat, sleep, walk, relax, work and play – we 'use' our bodies. Every aspect of our lives is therefore embodied.

Every aspect of our lives is, however, not only embodied but also embedded in place. As Heidi Nast and Steve Pile (1998, 1) argue: 'since we have bodies, we must be some place'. It was important, therefore, that in the mid-1990s geographers began to attempt to understand more about how bodies create and occupy spaces and places. It was important also to consider how places create bodies with particular desires and capacities (see Grosz 1992 on the mutually constitutive relationship between bodies and places).

In particular, it was gender and feminist geographers (e.g. Longhurst 1995; Rose 1993; Teather 1999) who began to pay attention to the various theoretical and empirical ways in which bodies are a locus for gendered and sexed subjectivities. Bodies, it was argued, are a site of power relations (e.g. see McDowell and Court 1994 on how bodies 'perform work') that are always located within a specific time and place.

In addition, many gender and feminist geographers paid attention to how bodies, on account of their long association in Western thought with women, femininity, materiality and irrationality, might offer a way of dismantling the masculinist structures of knowledge production (Rose 1993), which have long been based on men, masculinity, immateriality and rationality. Phil Hubbard et al. (2002, 98) point out that since the time of the philosopher Rene Descartes in the seventeenth century there has been 'theorizing across the social sciences which privileged the mind as the seat of truth, knowledge and humanity, with the body rejected as an explicit theme in social, spatial and historical analysis'.

Since the 1990s, much of the work on 'bodies and spaces', but certainly not all, was published in the journal *Gender, Place and Culture: A Journal of Feminist Geography* (*GPC*) (see Longhurst and Johnston 2014 for a review of work on embodiment published in *GPC* over a 20-year period). We do not recite this literature here, except to say that some of the common themes addressed in the 1990s and 2000s were: deconstructing binary thinking especially sex and gender (Gatens 1991); sexuality and space (Bell and Valentine 1995); 'the gaze' and self-disciplining of bodies (Johnston 1996); sexed and gendered performativity (Bell et al. 1994); and, gendered and embodied structures of power (Krenske and McKay 2000; Nairn 2003; Prorok 2000).

Over this period – 1990s and 2000s – scholarship in the area of 'bodies and spaces' gained momentum. Terms such as 'the body' or bod*ies* (remembering that there is no *one* or *the* body; rather, bod*ies* are always plural), embodiment, and corporeality appeared regularly not just in the pages of *Gender, Place and Culture* but also in other geography journals, such as *Environment and Planning D: Society and Space*, *Progress in Human Geography*, *Social and Cultural Geography* and *Antipode*. Also, over the aforementioned two decades, these terms came to be used within a wider array of theoretical frameworks than just feminism. A range of scholars, especially but not only those with an interest in feminist theory, began to pay attention to embodiment, for example those interested in non-representational theory (Thrift 2008), theories of performativity (Nelson 1999) and theories of emotion and affect (Bondi 2005; Davidson and Milligan 2004; Pile 2010).

Having begun this chapter by providing some background information on (feminist) geographical work on embodiment, we go on to examine what we think are some of the most inspiring, challenging and interesting ways that bodies are currently being researched by feminist geographers and others. We do this by considering trends in recent work and by thinking through how these might be useful for Lisa's doctoral research on lesbian mothers in Aotearoa New Zealand (for the most part, throughout this chapter we use the signifier 'mother', but we acknowledge that this term is gendered and that it is important to trouble the binary between mothers/fathers (see later in this chapter, under the heading: *From presuming man/woman and male/female to trans bodies and geographies*)).

For both of us, Lisa as a PhD student and Robyn as one of her supervisors (the other being Lynda Johnston), writing this chapter was beneficial. It expanded the supervisory relationship into a broader mentoring relationship (see Hawkins 2018 on 'Reflections on Academic Caring as a Feminist Practice') that worked beyond the typical PhD student–supervisor boundary.

Lisa and Robyn engaged in lively discussions via email (Lisa lives in Wellington and Robyn in Hamilton) on how themes such as intersectionality, fluidity of gendered dimensions of bodies, emotion and affect can help to deepen understanding of the embodiment of particular sexed and gendered bodies, especially lesbian mothers. As a team, we were able to progress our individual and collective thinking on this topic more than if we were considering this theme in relation to Lisa's doctoral research questions alone. Our hope is that the chapter will also assist other students and researchers to think through recent trends on embodiment in relation to their own research, and we also hope, given the mutual benefits it brought about for us, that it will encourage other students and supervisors to write and publish together.

By way of positioning ourselves in this research (see England 2017 on how the concept of positionality initially emerged as a critique of omniscient researchers, producing seemingly 'objective' research), Lisa has completed the 'fieldwork' for her PhD and is now at the stage of analysing data and writing chapters. Her topic emerged out of her life experience, as she was keen to add to the limited pool of information available to her when she began to create her own family. Robyn also did her doctoral research on pregnancy, namely how women were often excluded from public space, but this was completed more than 20 years ago. For both of us, being able to bring past and present discourses about maternal bodies, subjectivities and power relations into conversation has been useful.

The chapter unfolds first with a short introduction to Lisa's doctoral project. Second, we discuss five trends in work on 'bodies and spaces' that we have identified from our reading, as emerging over the past decade. The first trend is a move from considering how the surfaces of bodies are inscribed to how bodily *interiors* can deepen understanding of sexed and gendered embodiment. Following this, we examine a trend towards understanding more about *trans bodies and geographies*. Then we examine a move from focusing on sex and gender to focusing on multiple and *intersecting subjectivities*. This section is followed by a discussion of a recent trend to focus not just on bodies per se, but on how *technology* is shaping bodies. Finally, we examine how more geographers are now paying explicit attention to the important role played by *emotion and affect* in people's lives. In discussing each of these trends, we alert readers to some important research being carried out and also to how thinking about them (and the researchers working in these areas) has helped Lisa to progress her doctoral work. We begin by briefly introducing Lisa's project.

'You've just got to make shit up sometimes': an overview of Lisa's project

Lisa's PhD examines how lesbians' experiences of conceiving, being pregnant, birthing (and often *not* conceiving, being pregnant, birthing for one partner) and mothering both reinforce and trouble the normative gendering of bodies and spaces. The research is grounded in feminist poststructuralist and queer theory, both of which recognize the power of discourse and representation to shape reality. A number of Lisa's participants noted that many of the words, descriptors, forms and legal documents encountered during and after the time that they were creating their families did not capture adequately who they are, their roles and their relationships. For example, Danielle (pseudonyms are used throughout this chapter), who identifies as lesbian, and is in her early forties with one child, says: 'The terminology can get quite, it's, you've just got to make shit up sometimes.' For her the language, labels and boxes, or lack of boxes, on forms and legal documents did not align with her experience. This was similar for Hayley, a lesbian mother of one child in her early forties, who says: 'Even on the antenatal forms there is nothing for [a] partner. I had to keep correcting it.' Given the prevalence of comments such as these on discourse, language, representation and power during the interviews, feminist poststructuralist and queer theory provided a useful lens through which to examine the data.

Lisa conducted 27 face-to-face semi-structured interviews, 16 with single interviewees (either a sole parent or one partner of a couple) and 11 with couples. The youngest participants were in the 30 to 34 age bracket, and the oldest in the 55 to 59 bracket. More than three-quarters (82%), however, were aged between 30 and 44 years. Over half of respondents (52%) identified as lesbian, while another quarter (24%) identified as 'gay'. The remainder (24%) identified as 'queer', 'mostly lesbian' and 'queer/lesbian'. Nearly all of the participants (87%) identified as *Pākehā*, 'NZ European', 'White' or similar. The remainder identified as *Māori* or of other countries. The interview questions focused on how people started their families, how they decided who the donor was going to be and their experiences with fertility and maternity services.

In order to analyse in more depth lesbians' experiences of conception, pregnancy, birth and mothering/fathering, Lisa has been thinking about recent work on bodies, including that which considers in more detail the interiority of bodies, troubles the presumption of man/woman and male/female, sees subjectivities as intersecting, understands the interaction of bodies with recent reproductive technologies and highlights emotion and affect. We address each of these themes in turn in relation to 'lesbians, space, sperm and reproductive technology'. We offer five possibilities for examining these themes in more detail.

Recent trends in research on bodies

From inscribing the surfaces of bodies to bodily interiors

While some researchers in the past have tended to focus on bodies as blank pages, unmarked texts (*tabula rasae*) or surfaces that are etched or inscribed by power relations (e.g. Lingis 1984), others (e.g. Grosz 1994; Kirby 1997) have argued that bodies are not just surfaces but a co-mingling of surfaces and depths that are constituted (externally and internally) by gendered, social and spatial power relations. For the most part, however, the interiority or internal spaces of bodies has tended to be ignored, by geographers at least. Rachel Colls and Maria Fannin (2013, 1087) argue:

> Within geographical research on 'the body', a focus on the surfaces of bodies has been useful for considering how body boundaries, most often implied to begin and end at the skin, (de)limit, (de)regulate, and (de)stabilize what we come to know as 'a body'.

Colls and Fannin (2013, 1087) continue by explaining that, while they think there is value in this work, there is also potential for 'thinking geographically about bodily interiors'. Colls and Fannin do so by focusing on the placenta, and Lisa found there is also value in doing this by considering the movement of sperm out of bodies and into bodies, and sperm without a body.

Contemplating where the boundary is that separates research 'on the body' from research 'in the body' demonstrates fluidity in thinking about embodiment. The skin is often regarded as the delineator between bodies and the outside world, but contemplating sperm problematizes these bodily boundaries. Sperm is ejaculated through an orifice, not through a layer of skin. In a vagina, the skin potentially demarcating the outside of the body is inside the body – so, is sperm inside or outside that body? Sperm can move into a specimen jar, it can die or transform. Just like bodily boundaries, interiors are also unstable, temporal and mobile.

Throughout the course of the doctoral project, Lisa has come to think that in relation to heterosexual sex, when trying to get pregnant sperm maintains an association of masculinity and potency. However, with lesbian bodies, sperm tends to be more disembodied, disassociated from the male body from which it came and, perhaps somewhat ironically, is more associated with femininity (e.g. gentleness, warm flannels and juice) than masculinity. Stacey, who identifies as lesbian and is in her early thirties with one child, comments: 'it's all about being gentle with the sperm.' Kerry, Stacey's partner, notes 'we started calling it, just baby-making juice or something'. Tracey, who is in her early forties with one child and identifies as mostly lesbian, says that the donor 'would sort of run out with this nice warm flannel [laughs] with the syringe in it'. These processes of insemination, which happened at home, can seem a long way from idealized heterosexual notions of getting pregnant through romantic sex and mutual orgasms. They are much more ordinary, mundane and, as Donna, who identifies as gay and is in her early forties with two children, suggests, 'just clinical'. This is something that Lisa plans to examine further in the thesis (also, for how lesbian couples perceive family resemblances and negotiate the involvement of a sperm donor, see Nordqvist 2011).

Gordon Waitt and Elyse Stanes (2015) pay attention to another bodily fluid – sweat – to deepen theoretical understandings of how gender is lived. Sweat, like sperm, makes its way from inside to outside the body but, unlike sweat, in the case of lesbians who want to get pregnant (but not to have sex with a man), it is then necessary to reinsert this fluid into another body. Sperm as a bodily fluid is highly sexed and gendered, arguably more so than sweat (although, as Waitt and Stanes 2015 argue in relation to men and masculinity, sweat is also highly gendered; for example, it is often considered unladylike to sweat). Paying attention to the visceral fluids and solids (especially in relation to birth) that cross the thresholds that demarcate the inside from the outside of bodies opens up possibilities for thinking geographically about bodies in ways that go well beyond bodily surfaces.

From presuming man/woman and male/female to trans bodies and geographies

Another area of interest in embodiment that has emerged in recent years is the way in which the gendered and sexed dimensions of embodiment are 'fluid' or cross sexed and gendered binaries. A useful example of this is trans geographies. In 2010, *Gender, Place and Culture* published a series of themed papers on trans geographies (for the introduction to these papers, see Browne, Nash and Hines 2015). The editors explain: 'This collection spans a range of theoretical fields in this context, including trans theories, queer engagement, feminist geographies, gender geographies and sexualities geographies' (Browne, Nash and Hines 2015, 573). The series of themed papers also presents empirical studies of trans lives.

The question of how people construct and (re)present complex and shifting subjectivities arose at the outset of Lisa's project, when she put out a call for lesbian participants. Several women who identified as bisexual immediately contacted her, claiming that the research was exclusionary and continued the trend of making bisexual women invisible. One woman mentioned her sadness at, once again, being too queer for the straight crowd and too straight for the queer crowd. This is one example of how work on trans geographies has been useful in thinking about bodies in relation to the fixing and unfixing of subject positions.

Trans geographies also facilitate thinking around queer families. Queer families are often regarded as disruptive to heteronormative family structures. However, attention is now being drawn to the ways in which lesbian families also conform to traditional notions of family and therefore are implicated in supporting the very structures that can function to erase them (e.g. Nordqvist 2010). Trans geographies, and the attention that they pay to the fluidity of sex and gender roles and relations, are useful for examining the ways in which lesbian families both disrupt and conform to heteronormative family structures.

The other aspect of Lisa's project that was highlighted by thinking about recent work on trans geographies is the unfixing of maternal bodies themselves. Claire Madge and Henrietta O'Connor (2005, 94) explain: 'There is no preconstituted "body" on to which motherhood is inscribed; what it means to be a mother is constantly produced and reproduced through varying and competing discourse and practices' over time and space. This means, if the subjectivity of 'mother' is not pre-given, then anyone can 'mother' (nurture, foster, take care of children; including a woman [mother] who has not given birth, a man or someone whom society deems to be an 'inappropriate mother'). Lesbians can 'mother' and be mothers. They can also 'father' and be fathers (Aitken 2000). Trans geographies offer ways to trace diversity, to contest and trouble binaries, to trace lines of queer kinship and parenthood. The questioning of hegemonic conventions that link sexed and gendered bodies to mothering and fathering is useful for Lisa's project.

From sex and gender to multiple and intersecting embodied subjectivities

Peter Hopkins explains:

> Intersectionality is an approach to research that focuses upon mutually constitutive forms of oppression … [it] is not only about multiple identities but is about relationality, social context, power relations, complexity, social justice and inequalities.
>
> *2017, abstract*

It is most certainly the case in Lisa's project that different axes of embodied subjectivity intersect to give rise to a wide range of experiences for lesbian mothers. Jacqui Gabb (2004) has also found this to be the case and calls for more research on how the intersectionality of ethnicity, socio-economics and geographical location affect the choices available to lesbians when trying to get pregnant. Lisa, in her research, found that some of her participants were highly aware of the way in which particular axes of their subjectivities afforded them privilege in Aotearoa New Zealand. Danielle, for example, in discussing her experience of being pregnant and using maternity services notes: 'It helped that I'm a middle-class, educated White lady, so I kind of just shuffled along … and I'm a New Zealander.' Another participant, Kitty, a lesbian mother of one and in her late thirties, comments that she had a positive experience with the fertility clinic but reflects that this is probably because she is 'reasonably feminine and, yeah, could pass as straight'. She continues explaining that 'butch friends' of hers were having trouble and that 'maybe that's part of … of the cold shoulder you're receiving from [the fertility clinic]. You don't look like [you fit] the mother role.'

In Lisa's project, it quickly became apparent that it was not just the axes of subjectivity of being a lesbian mother that mattered, but also other axes such as being feminine, being a New Zealander rather than a migrant, being read by healthcare workers and others as middle class and having White skin; that is, being European or *Pākehā*. For example, in relation to ethnicity, in the context of Aotearoa New Zealand, not only do birthing spaces and other spaces tend to be heteronormative and patriarchal but also deeply colonial (see Longhurst 2008 on 'colonizing and reclaiming birthing space for *Māori* women'). Also, for *Māori*, women's pregnant and birthing bodies tend to be constructed differently from those of *Pākehā;* that is, they are constructed as *tapu,* or under the influence of the spirits or gods (August 2005). It is not enough to simply consider lesbian mothers without also considering these other intersecting axes of embodied subjectivity and the power relations that they generate (Hopkins 2017).

From human to technology shaping bodies

In recent years it has become apparent that numerous and varied technologies, both digital and non-digital, are shaping contemporary spaces and bodies. Geography, technology and bodies are intimately linked (Warf 2017). Bodies are interpellated through technology within specific spaces. It is worth noting, however, that this link between geography, technology and bodies is not entirely new. As Stanley Brunn, Susan Cutter and J.W. Harrington Jr (2004) point out, new technologies have long been important in advancing geographic understanding. The point we are making here, though, is that over the past decade the transformation of society through rapidly changing technologies, especially digital technologies, has arguably been profound, prompting geographers to rethink categories such as bodies, matter, technologies and objects (see Ash 2013).

In the case of Lisa's research, this has become evident in the assisted reproductive technologies currently being used by some lesbians to conceive. Assisted reproductive technologies can be examined at different geographical spatial scales, which reveals some interesting ways in which these technologies are gendered. So-called 'high-end' technologies, such as in vitro fertilization (IVF), are often represented within the discourse of 'rational science'. IVF is a process of fertilization in which an egg and sperm are combined outside the body. This process takes place in highly medicalized and clinical spaces (often available only in cities) and tends to be carried out by male doctors and fertility specialists. So called 'low-end' technologies, such as home ovulation tests or predictor kits (TPKs), are often used in domestic spaces and are predominantly the domain of women. These kits, unlike IVF, are relatively cheap to purchase.

In thinking about bodies and assisted reproductive technologies, it becomes apparent that different technologies are not neutral in the way that they are incorporated into bodies, but reinforce the dominant power structures by being more readily available to particular people, for example IVF being more readily available to those who can afford it and those whose Body Mass Index (BMI) (BMI is a measure of body fat, based on height and weight) is 'acceptable' to healthcare and medical professionals. Participants in Lisa's study also reported that they feel that the interactions that happen in the course of undergoing IVF are easier for those who look like a 'mother' (read: feminine).

Assisted reproductive technologies blur the boundaries between body and matter, and bodies that matter. The construct of 'family' (bounded by blood, two parents, children created through sex) is troubled in and by lesbian families, where not all family members are genetically related, where there may be more than two parents and where insemination is not a private, romantic event but a clinical process that may involve a number of people. Interestingly, in Lisa's research with 27 lesbian families, there were 17 different ways used to create families, and yet their choices were often overshadowed by hegemonic definitions of family represented through laws, regulations and forms that retain archaic notions of heteronormative and patriarchal power.

From rationality to emotion and affect

Geographers over the past decade have turned their attention to emotion and affect and how these are lived in and through bodies (see Bondi 2005; Davidson and Milligan 2004; Pile 2010 as examples of the 'emotional turn' in geography). It could be argued that this is not exactly a new perspective, given that humanistic geographers in the 1970s and 1980s adopted approaches that rejected mechanistic models of spatial sciences, instead putting people at the centre of their work. However, much of the research undertaken by humanistic geographers during this time did not differentiate carefully between individuals and groups. Instead, it attempted to capture the experiences of everyone under the umbrella of 'Man and his environment' (Tuan 1974). Embodied power relations were often ignored. Recent work on emotion and affect, especially by feminist geographers, has tended to focus more squarely on what different bodies can do and feel, and how this provides insights into the gendered, sexed, raced, (dis)abled, and so on, lives of people as they move through different spaces.

In relation to Lisa's research, considering emotion and affect is a useful reminder that medical and legal practices, and changes to these, can prompt a range of feelings for lesbian mothers. For example, in 2006 it became legal in Aotearoa New Zealand to list two mothers as parents on the birth certificate. Nicola and Noni (in separate interviews) both expressed feeling very pleased about this change, yet had different reactions. Nicola recalls: 'I sobbed while filling out the birth certificate … that whole "other parent", I get to be "other parent". I'm still utterly wrapped with that. It makes me so happy.' Noni, who identifies as queer, is in her late thirties and has two children, says:

> When [the birth certificate] arrived … [it was] one of my … proudest moments as a New Zealander. To be, like, our country actually recognizes, in the law, our family, and we have the protections, the rights and protections, that come with that. And I'm pretty, sort of, suspicious of patriotism, but at that point I was, like, this is precious.

Conclusion

In this chapter we have highlighted that research on bodies in the last 10 years has moved in new and interesting directions:

> Twenty years ago the mere mention of bodies, especially their messy materiality, could prompt a sense of dis-ease amongst geographers (at least some!). Bodies were largely ignored in the discipline of geography.
>
> *Longhurst and Johnston 2014, 273*

Today, however, terms such as the body, bodies, embodiment and corporeality are used within a wide range of theoretical frameworks and empirical studies, often seemingly without a second thought. They now sit more easily within the frame of acceptability, although some of the new areas of scholarship discussed in this chapter, such as the interiority of bodies and trans bodies – especially when this involves invoking the fleshy materiality of bodies – may still prompt some question of their legitimacy as a 'proper' topic worthy of geographers' attention.

We have argued that bodies, recently, have been used as a frame of reference in relation to bodily interiors, trans bodies and geographies, intersectional approaches to embodied subjectivity, bodies being imbued with technology and embodied geographies of emotion and affect. We have touched briefly on these five areas, reflecting on what they might have to offer in relation to casting light on lesbians' experiences of becoming mothers in Aotearoa New Zealand. There are likely other areas of scholarship and activism around embodiment that we could also have mentioned, but we trust that others will add to this conversation. Bodies have now been on feminist geographers' agendas for more than three decades, and this interest does not appear to be waning. Instead, bodies are being (re)presented in new ways that are helping to develop an understanding of how gender is thought about and lived on a daily basis.

Key readings

Brown, K. and, E. Ferreira, eds. 2016. *Lesbian Geographies: Gender, Place and Power*. Abingdon: Routledge.
Hayman, B., and L. Wilkes. 2017. "De Novo Families: Lesbian Motherhood." *Journal of Homosexuality* 64 (5): 577–591. doi:10.1080/00918369.2016.1194119.
Johnston, L. 2016. "Gender and Sexuality I: Genderqueer Geographies?" *Progress in Human Geography* 40 (5): 668–678. doi:10.1177/0309132515592109.

References

Aitken, S. 2000. "Fathering and Faltering: 'Sorry but You Don't Have the Necessary Accoutrements'." *Environment and Planning A* 32: 581–598.
Ash, J. 2013. "Technologies of Captivation: Videogames and the Attunement of Affect." *Body & Society* 19: 27–51.
August, W. 2005. "Māori Women: Bodies, Spaces, Sacredness and Mana." *New Zealand Geographer* 61 (2): 117–123.
Bell, D., J. Binnie, J. Cream, and G. Valentine. 1994. "'All Hyped up and No Place to Go'." *Gender, Place and Culture* 1 (3): 31–47.
Bell, D., and G. Valentine, eds. 1995. *Mapping Desire: Geographies of Sexualities*. London: Routledge.
Bondi, L. 2005. "Making Connections and Thinking Through Emotions: Between Geography and Psychotherapy". *Transactions of the Institute of British Geographers* 30: 433–448.
Browne, K., C.J. Nash, and S. Hines. 2015. "Introduction: Towards Trans Geographies". *Gender, Place and Culture* 17 (5): 573–577.
Brunn, S., S. Cutter, and J.W. Harrington Jr, eds. 2004. *Geography and Technology*. Netherlands: Springer.
Colls, R., and M. Fannin. 2013. "Placental Surfaces and the Geographies of Bodily Interiors." *Environment and Planning A* 45: 1187–1104.
Davidson, J., and C. Milligan. 2004. "Editorial: Introducing Emotional Geographies". *Social and Cultural Geography* 5: 523–548.
England, K. 2017. "Positionality". *The International Encyclopedia of Geography: People, the Earth, Environment and Technology*. Wiley Online Library. Available at: https://doi.org/10.1002/9781118786352.wbieg0779.
Gabb, J. 2004. "Critical Differentials: Querying the Incongruities Within Research on Lesbian Parent Families". *Sexualities* 7 (2): 167–182.
Gatens, M. 1991. "A Critique of the Sex/Gender Distinction". In: *A Reader in Feminist Knowledge*, edited by S. Gunew, 139–157. New York: Routledge.
Grosz, E. 1992. "Bodies-cities." In: *Sexuality and Space*, edited by B. Colomina, 241–254. New York: Princeton Architectural Press.
Grosz, E. 1994. *Volatile Bodies: Toward a Corporeal Feminism*. St Leonards: Allen and Unwin.
Hawkins, H. 2018. "On Mentoring – Reflections of Academic Caring as a Feminist Practice." Posted by Gender and Feminist Geographies Research Group, 12 February. Available at: www.gfgrg.org/on-mentoring-reflections-on-academic-caring-as-a-feminist-practice-by-harriet-hawkins/.
Hopkins, P. 2017. "Social Geography I: Intersectionality". *Progress in Human Geography*, 1: 1–11. doi:10.1177/0309132517743677.
Hubbard, P., R. Kitchin, B. Bartley, and D. Fuller. 2002. *Thinking Geographically: Space, Theory and Contemporary Human Geography*. London: Continuum.
Johnston, L. 1996. "Flexing Femininity: Female Body-Builders Refiguring 'the Body'." *Gender, Place and Culture* 3: 327–340.
Kirby, V. 1997. *Telling Flesh: The Substance of the Corporeal*. London: Routledge.
Krenske, L., and J. McKay. 2000. "'Hard and Heavy': Gender and Power in a Heavy Metal Music Subculture." *Gender, Place and Culture* 7: 287–304.
Lingis, A. 1984. *Excesses: Eros and Culture*. New York: State University of New York.
Longhurst, R. 1995. "The Body and Geography." *Gender, Place and Culture* 2: 97–106.
Longhurst, R. 2008. *Maternities: Gender, Bodies and Space*. New York: Routledge.
Longhurst, R., and L. Johnston. 2014. "Bodies, Gender, Place and Culture: 21 Years On". *Gender, Place & Culture* 21 (3): 267–278.
McDowell, L., and G. Court. 1994. "Performing Work: Bodily Representation in Merchant Banks." *Environment and Planning D: Society and Space* 12: 727–750.

Madge, C., and H. O'Connor. 2005. "Mothers in the Making? Exploring Liminality in Cyber/space." *Transactions of the Institute of British Geographers* 30 (1): 83–97.

Nairn, K. 2003. "What has the Geography of Sleeping Arrangements Got to Do with the Geography of our Teaching Spaces?" *Gender, Place and Culture* 10: 67–81.

Nast, H., and S. Pile, eds. 1998. *Places Through the Body*. London: Routledge.

Nelson, L. 1999. "Bodies (and Spaces) Do Matter: The Limits of Performativity." *Gender, Place and Culture* 6: 331–353.

Nettleton, S., and J. Watson, eds. 1998. *The Body in Everyday Life*. London: Routledge.

Nordqvist, P. 2010. "Out of Sight, Out of Mind: Family Resemblances in Lesbian Donor Conception." *Sociology* 44 (6): 1128–1144.

Nordqvist, P. 2011. "'Dealing with Sperm': Comparing Lesbians' Clinical and Non-Clinical Donor Conception Processes." *Sociology of Health & Illness* 33 (1): 114–129.

Pile, S. 2010. "Emotions and Affect in Recent Human Geography". *Transactions of the Institute of British Geographers* 35: 5–20.

Prorok, C.V. 2000. "Boundaries are Made for Crossing: The Feminized Spatiality of Puerto Rican Espiritismo in New York City." *Gender, Place and Culture* 7: 57–79.

Rose, G. 1993. *Feminism & Geography: The Limits of Geographical Knowledge*. Cambridge: Polity Press.

Teather, E.K., ed. 1999. *Embodied Geographies: Spaces, Bodies and Rites of Passage*. London: Routledge.

Thrift, N. 2008. *Non-representational Theory: Space, Politics, Affect*. London: Routledge.

Tuan, Y.-F. 1974. *Topophilia: A Study of Environmental Perception*. New York: Columbia University Press.

Waitt, G., and E. Stanes. 2015. "Sweating Bodies: Men, Masculinities, Affect, Emotion." *Geoforum* 59: 30–38.

Warf, B., ed. 2017. *Handbook on Geographies of Technology*. Cheltenham: Edward Elgar.

14

THE INTIMATE GEOPOLITICS OF RACE AND GENDER IN THE UNITED STATES

Chris Neubert, Sara Smith and Pavithra Vasudevan

Introduction

In a small town in the US South, a 92-year old grandmother collects obituaries in a scrapbook to document those who died young after decades of long and gruelling hours working in the toxic environment of an aluminium smelting plant. In the segregated company town of Badin, North Carolina, US, the industrial plant and its lethal wastes have for generations discriminately poisoned the Black workers and the families that loved and supported them. Their stories are mostly forgotten now even by their children, who moved away from this 'sacrifice zone' (Lerner 2012) of capitalist development. It produced millions of tons of aluminium for urban infrastructure, military equipment and everyday products for 90 years. Meanwhile, in the US agricultural heartland of Iowa, families struggling with dramatic changes in their communities hold onto a cruel optimism (Berlant 2011) that they can maintain the good life of comfort and stability as rapidly intensifying global capitalism shifts the demographics and economic opportunities in their towns. Riling up embodied fears numerous threatening 'others', Donald Trump promises that he understands them and can protect them from this uncertain and threatening world. Embodying whiteness and a revanchist masculinity, he extracts their fears and distills them into an external threat that has a supposed solution: a strong defence against outsiders and America First policies to restore a golden era (Gökarıksel and Smith 2016).

In this chapter, we argue that the vignettes above teach us as much about geopolitics, or the relationship between politics, international relations and geography, as do studies of Cold War rhetoric or the 'War on Terror'. Building on work in feminist geopolitics, we develop intimate geopolitics as a framework for understanding how of gender and race function in the contemporary US. We argue that the study of geopolitics is greatly enriched by attention to how international geopolitical relations between states are manifested in the relationships between people and populations on the ground, particularly as colonial hierarchies of race and gender structure our ways of knowing and being in the world (Lugones 2016; Spillers 2003; Weheliye 2014; Wynter 2003). We trace the formation of white agricultural masculinities, represented as central to US national identity, alongside the disproportionate toxic burden borne by communities of colour to argue that relationships between the environment, race and gender can be productively understood through attention to intimate experiences and representations. The discourse of the timeless 'family farm', at the heart of US national imaginaries, centres a patriarchal and hetero-normative family structure anchored in nationalism and settler colonialism. An alternate discourse of the US as emblematic of industrial modernity is undergirded by environmental racism, where Black bodies and geographies are exposed to toxic waste. Here, we bring together these twinned embodied manifestations to push for a feminist and intimate geopolitical approach to understanding the (mis)management and representation of life in the contemporary US.

In what follows, we provide a brief overview of geopolitics as a subdiscipline of geography, then turn to recent developments in feminist geopolitics. After this review of the literature, we will engage with brief case studies based on research in Iowa and North Carolina to flesh out the ways that an embodied and intimate understanding of the geopolitical can enable us to see things that we might otherwise miss.

From geopolitics to critical geopolitics

Geopolitics, as a field of study and as a particular practice, is closely aligned with the broader study of geography and geographic thought, though the term itself has long been contested, reimagined and reworked to examine the operation of state power on populations and territories from different theoretical engagements. The earliest geopolitical thinkers in the late-nineteenth and early-twentieth centuries could also be easily identified as geographers. Geographic thought at the time was steeped in environmental determinism – the theory that the physical geography of a given place defines the biological profile of the people who live there. Deterministic thought was heavily influenced by Social Darwinism, an application of emergent Eurocentric 'objective' empirical approaches to social analysis that sought to uncover the 'natural laws' defining the essence of human societies, legitimizing racial categories for the benefit of imperial capitalist agendas (Peet 1985). An overly simplistic adaptation of Darwin's evolutionary theories by early geographers like Ellen Churchill Semple shrouded environmental determinism in a cloak of scientific objectivity and was used to justify a range of state-sanctioned oppressive acts. Semple herself was a student of Friedrich Ratzel, who is perhaps most closely associated with bringing Social Darwinism and determinism together in an imperial, territorializing strategy that became 'geopolitics'. Ratzel theorized nation states as living organisms, competing for survival, and justified imperial expansion rooted in each organism's need for *Lebensraum* (living space).

Inextricably linked with imperial practices of statecraft, the work of early geopolitical thinkers such as Ratzel in Germany, Mackinder in the UK, and Mahan in the US conceptualized the world as driven by inevitable conflict among global imperial powers (Kearns 2011) and developed geopolitics as a science of understanding how states could maximize their geographic resources to consolidate power, further their interests, and protect themselves. Ratzel's thinking, in particular, profoundly influenced Nazi-era Germany, which often justified atrocities as the necessary consequence of the expansion of German geopolitical power over territories inhabited by people whom they considered genetically inferior. While geographers eventually moved away from the more obviously racist elements of environmental determinism, the theory's influence on geopolitics was notable and came to define global interstate relations in the postwar era.

Isaiah Bowman, as the first director of the American Geographical Society and later President of Johns Hopkins University, exemplified the thinking and practice of geographers at the time. While Bowman worked to distance geography from the deterministic ideas that came to define Nazi geopolitics, a 'hidden determinist agenda' remained central to his thinking (Livingstone 1992; Peet 1985, 328). Bowman worked closely with the US State Department and influenced policymakers in the US such as Kissinger and Brzezinski, who also frequently cited Mackinder positively, ensuring that the geopolitics of the state actors in the Cold War era was indelibly linked to an imperialist agenda (Kearns 2011). The publication of Huntington's *The Clash of Civilizations* (1993) continued this trend of consolidating differences between peoples and states to create ill-defined 'civilizations', which were hierarchically ranked by arbitrary and uncritical categories like 'freedom'. While not without its critics, Huntington's work remains widely influential among geopolitical strategists in international relations, particularly in the US. However, criticism of geopolitics as a field within geography began to gain attention, and the publication of *Critical Geopolitics* (Toal 1996) exposed the deep links between geopolitics and imperialism. Engaging with poststructuralist thinkers such as Foucault and Derrida, Toal demonstrated how geopolitical theory, rather than objectively describing the existing order of the world, actively intervenes to shape the world through political practices that emerge from geopolitical discourse. The field of 'critical geopolitics' that emerged focused on demonstrating how space functions as an important tool of imperial and state power, through which conventional geopolitics prescribes strategies and trains subjects to carry them out (Hyndman 2004).

Concurrently, feminist geographers, increasingly influenced by Haraway's (1988) call for 'situated knowledge' and criticism of the 'God trick', began intervening in the conversation opened by critical geopolitics. The earliest feminist critiques demanded that critical geopolitics account for the continuation of a detached 'view from nowhere', which maintains that statecraft is a practice of all-seeing intellectuals or statesmen and thus inaccessible to ordinary people, who apparently await 'their regular injection of [geopolitical] ideas' (Sharp 2000, 362). Often, these early critiques questioned the position of those geographers in articulating a critical geopolitics, arguing that their unexamined gender analysis maintained the masculinist orientation of geopolitics and served to keep women hidden from geopolitical work. Even in critical geopolitics, feminist geographers argued, women's bodies were acted upon by states, yet were never recognized as actors in geopolitical discourse (Dowler and Sharp 2001; Staeheli 2001).

Feminist interventions

Feminist geographers embarked on an ambitious project to do more than rewrite women into geopolitical histories; they sought to articulate a perspective through which 'the everyday experiences of the disenfranchised can be made more visible' (Dowler and Sharp 2001, 169). These geographers demanded an acknowledgement that geopolitical discourses are not simply written onto bodies but produce active geopolitical subjects. Thus, early feminist geopolitical interventions highlighted how geopolitics functions at scales other than the nation state (Massaro and Williams 2013).

This work questioned how scale was deployed in geopolitical theory. While traditional and critical geopolitics emphasizes an analysis on 'public displays of power', with the state assumed to be the 'basic unit of analysis' (Staeheli 2001, 185), feminist geopolitics demonstrates that the global and local are not unilaterally ordered in a spatial hierarchy. Instead, these scales are intertwined and diffuse, 'scaffolded' and interpenetrating (Marston 2000). This shift in perspective makes alternative visions of the world possible. 'Counter-topographies', for instance, demonstrate how distinct places are analytically connected to other sites through material and social relationships (Katz 2001). Secor (2001) proposes 'counter-geopolitics' as a way to demonstrate the multiple scalar categories that 'ultimately blur, overlap and collapse into one another in the making of a political life' (201). Similarly, 'alter-geopolitics' questions the function of states as 'immutable forces', and instead offers that everyday moments 'speak back' and change these forces (Koopman 2011, 276). In each of these interventions, it becomes possible to see how geopolitical discourses about states emerge through intimate relationships, how state power focuses attention on the body, and how emotions run through the flows of power and knowledge.

As both the subfields of critical geopolitics and feminist geopolitics have developed in the last 20 years, scholars continue to work in tandem, but feminist geopolitics specifically has come to focus attention on key themes: intimacy; the body; and emotion. Examining intimacy and intimate relationships with regard to geopolitical power focuses attention on the 'lived materiality' of the bodies of geopolitical subjects (Pratt and Rosner 2012). Such work aligns with feminist interventions that have long demonstrated that supposedly 'private' realms are, in fact, intertwined with global circuits of power, and intimate geopolitics shows us how our political worlds are built around personal attachments with other bodies, materials or objects that are touched by such power. Importantly, this is not to suggest that the intimate is simply 'dripped down upon' by geopolitics but rather that, in taking the intimate as a starting point for geopolitical analysis, it becomes possible to see how the home is a site of state security and violence, as well as resistance to that violence (Fluri 2011; Pain 2015, 64–65). By focusing on intimate space, we can see that there are other forces beyond states and statesmen working to produce geopolitical realities, and that there is no neat 'spatial hierarchy' that demarcates everyday intimate spaces from global, geopolitical ones (Pain and Staeheli 2014).

Within the intimate realm, where struggles over population and territory are materially realized, feminist geopolitical scholars have focused attention on how 'the body itself becomes a geopolitical site' (Smith 2011, 456). By examining the role of the body and questioning whose bodies are able to move through space, feminist geographers have demonstrated that state borders are negotiated in bodily space (Mountz and Hyndman 2006). From the perspective of the state, the body is a key target of various security strategies, with securitization functioning to either 'protect' certain bodies or to mark them as 'a potential agent of "insecurity"' (Fluri 2014, 797). With regard to violence, Pain (2015) shows how violence is committed against bodies not just by geopolitical actors; it is already present in intimate relationships, which 'soak up and are shaped by these wider forces' (66). Thus, in contrast to classical or even critical geopolitics, feminist geopolitics shows how the unevenly articulated borders of the state permeate the most intimate spaces of everyday life, even as states continue to disregard the full impact of such embodied power. Indeed, these borders are manifest within the body, as evidenced by recent work on stem cells (Dixon 2014) and foetuses (Wang 2017). In essence, by seeing bodies both as the site of geopolitical strategy and as active, mobile and resisting agents themselves, it is possible to understand how borders and territory are created not by distant and disembodied figures of statecraft but in the relationships between people in the intimate realm, which makes them geopolitical actors (Smith, Swanson and Gökarıksel 2016, 260).

Intimate relationships, of course, are formed through emotional attachments and commitments that, from a feminist geopolitical perspective, must also be understood within this context of discourses about nationhood and the state (Smith 2012). Love, desire, fear, hope and anxiety – these and other emotions can be found in everyday moments that pin intimate interactions and relationships to geopolitical struggles for territory and the future of nations. Management of fear in the face of sexual violence, for instance, contributes to a 'banal nationalism' through which people experience the state in the intimate realm (Christian et al. 2016). Research on the politics of love during wartime destabilizes masculinist narratives about war while simultaneously demonstrating that the lived dimensions of war are complicated – about pain and death as much as about love and life (Tyner and Henkin 2015). Thus, in contrast to geopolitics from the perspective of the state, feminist geopolitics demonstrates how we experience the state as a complicated tangle of emotions, intimate moments and embodied interactions, rather than as neat borders negotiated in distant conference rooms.

Masculinity, agriculture and the good life: farmers and others in Iowa

In 2016, standing in front of a crowd of Iowans honouring bikers, veterans and farmers, Donald Trump – who made his fortune building skyscrapers and casinos – forcefully pledged to solve the problems of a rural population that feels increasingly disenfranchised in American politics. With rows of Iowa corn swaying in the summer sun behind him, Trump declared: 'Family farms are the backbone of this country. Remember that. And I know what's happening to you.'

The crowd's cheers were deafening; the people had waited a long time to hear this promise. Trump launched into a long litany of 'happenings' faced by the rural Iowan farmer: job-killing regulations; the Environmental Protection Agency's intrusions into farmers' homes; unfair trade deals; and illegal immigration. Trump told the crowd, 'We are going to end this war on the American farmer'.

To anyone unfamiliar with life in the rural regions of the US today, such a proclamation must seem particularly strange. What is this war, exactly? How can an economic sector that receives billions in federal subsidies each year still be disenfranchised? How can residents in a state that becomes the focal point of American presidential politics every four years feel ignored? Answering these questions requires a feminist geopolitical approach to rural America, to the right and to the rise of Trumpism, which jettisons popular media discourses about a white working-class voting against its own interests. On paper, at least, the agricultural economy of rural Iowa is doing quite well – unemployment rates are low; commodity prices are modest compared to earlier booms yet nowhere near the 1980s farm-crisis levels; and land prices have been at record highs for several years.

The 'war' that Trump speaks of, then, is about something deeper than simple economic anxiety. It is, as this section contends, about a perceived loss of respect for the American farmer, about a betrayal of rural values, and a desire for a return to better times. Whiteness and masculinity are central to the respect and values that make up the rural agrarian nostalgia that is thought of as lost. When Trump tells his rapt audience, 'I know what's happening to you', the unspoken happening is the perceived breakdown of the cultural borders that have protected whiteness and manhood for generations in these rural spaces. In this sense, contrary to the conventional post-election analyses in 2016, people are not voting against a narrow understanding of their own economic interests – they are, instead, supporting the first candidate in a generation who appears to share their specific sociocultural worldview.

Political consciousness in the rural US has come to be defined by a sense of resentment against cities and a belief that these cities are imposing regulations and supporting their own decadence through the stolen labour of rural white Americans (Cramer 2016). The political discourse about government regulation has 'historically been made by equating deservingness with whiteness', and conversations about whom government policies should support 'are about race even when race is not mentioned' (Cramer 2016, 86). Even as agriculture in the US remains heavily subsidized, a racist discourse about regulation ensures that white farmers do not see themselves as the recipients of government aid. To be effective, however, such a spatial hierarchy of race must appear democratic and 'immediately intelligible to the masses', thus racist codes must not appear racist but should instead act as an 'interpretive key' or a shared secret knowledge of how individuals should be rightly placed in the social world (Balibar and Wallerstein 1991). Thus, when Trump states to rural Iowans, 'I know what's happening to you', he never needs to explicitly say what it is that is going on yet the vociferous audience is clearly receiving the message.

'What's going on' can therefore be understood as a penetration of pure white space by a Brown threat, originating in the cities, and these discourses do the work of securitizing nationalist space by identifying for their audience the various 'outliers who actively work against core American beliefs' (Silva 2016, 29). Indeed, these are not merely electoral debates for hearts, minds and votes: they are embodied struggles for territory – 'geopolitical power struggles … firmly tethered to brown bodies' (Silva 2016, 6). The rise of the right in rural America in the last two decades can be connected to this power struggle. Right-wing movements that either implicitly or explicitly acknowledge the racialization of these spatial contestations offer a sense of optimism and hope that they can restore 'the good life' to white, rural America. This explains why Trump rallies, with their dark, threatening messages, are quite often jubilant affairs. There is a palpable hope that, in exchange for this faith in optimism, the individual will finally be recognized by the state as a deserving subject, even if that recognition is little more than a 'misrecognition you can bear, a transaction that affirms you without, again, necessarily feeling good or being accurate' (Berlant 2011, 43).

The influence of whiteness on the rural Iowa landscape cannot be separated from its intricate connections to masculinity. If rural communities must be protected from the Brown threats encroaching from the cities, then it is rural white men who must do the protecting. The history of perceived threats to rural space can be read as threats to the role of men as protectors and providers. Still, despite its profound influence, 'masculinity goes through great pains to hide itself and cloak its influences … it never has to acknowledge its role in organizing social, economic, and cultural life' (Shabazz 2015, 9). The fear of lost masculinity is a formative and powerful discourse crafting space in rural landscapes.

Kimmel (2015) writes that the farm crisis of the 1980s, followed by continued economic shocks, globalization and corporatization, left rural white men with the sense that 'the very people who had built America were the ones who were being pushed aside' (13). This sense of entitlement is infused into the discourse about modern farming, where the independent 'family farm' – led by a strong male head – is continually positioned as the cornerstone of agriculture. A 2013 Super Bowl advertisement for Dodge Ram vehicles perfectly summarizes this discourse. Images of men baling hay, caring for livestock, praying, waving flags, wearing dirty cowboy hats and jeans and (of course) driving Dodge Ram trucks are shown, while radio personality Paul Harvey recites a famous 1978 speech to the Future Farmers of America:

> God said, 'I need somebody … who'd bale a family together with the soft strong bonds of sharing, who would laugh and then sigh, and then reply, with smiling eyes, when his son says he wants to spend his life doing what dad does.' So God made a farmer.

The production of this particular kind of masculinity, which is indelibly linked to farming, emerged as part of a historical process of state-sponsored projects, like Future Farmers of America and 4-H, where citizenship programmes produced wholesome white youths performing 'gender-appropriate' labour on the family farm. These projects merged this 'vision of rural normalcy with American nationalism and the language of civic obligation, casting white, commercial family farmers as the backbone of the nation' (Rosenberg 2015, 8). This rhetoric persists in contemporary discourse, as exemplified by Trump's Des Moines speech. It is powerful because of its effective emotional resonance – being described as 'the backbone of the nation' links the body of the white man with the body of the nation. As Smith (2012) writes, this complex interplay between embodiment and emotion – material movement and practices, as well as love, fear, and pain – are 'not the side effects of geopolitical practice but its principal manifestation' (1524). In essence, from their perspective, white rural Iowan men are the body and soul of America – they, alone, truly belong. And when they see themselves reflected in the voice and body of Donald Trump, they envision themselves taking their rightful place at the head of the nation.

Intimate entanglements of geopolitical care: Race and waste in Badin, NC

In community meetings held in church basements, the residents of West Badin, North Carolina, lament the state of disrepair and decline that their community faces. West Badin was a residential settlement established in 1917 for Black workers at the Alcoa (formerly the Aluminum Company of America) Badin Works, an aluminium smelting plant, and their families. Since the closure of the plant in 2007, small grants from Alcoa have supported beautification efforts, a museum and historical markers in the predominantly white East Badin and attempts to rebrand the company town as a quaint tourist destination. On the other side of the plant, in West Badin, the broken-down roads and homes abandoned by residents without any economic prospects reflect a general sense of hopelessness. Those who remain are caught between nostalgic remembrances of a once-thriving Black community and contentious politics regarding ongoing contamination from nearly a century of industrial toxicity. Racism has defined the occupational, political and social history of Badin. Black workers were hired for the most physically demanding and dangerous jobs in the aluminium plant. Black residents were exposed most directly to the toxic chemicals from the plant's smokestacks and from the unlined industrial landfill, located adjacent to their residences. When Alcoa was required by environmental regulators to account for and manage the contaminated areas, the Black residents were excluded by the town council, company and state agencies.

It would be easy to see Badin as an isolated instance whose conditions are specific to a town in the twentieth-century US South. However, the disproportionate combined burden of toxicity and racialized political exclusion reflects a pattern of 'environmental racism' (Pulido 2016), which refers to the unhealthy living and working environments that result from racially driven unequal development. Sites like Badin are racialized 'sacrifice zones' (Lerner 2012) to capitalist development, where the collusion of corporate harm and state neglect creates territorial zones whose residents are excluded from state welfare and representation, despite their nominal citizenship. Badin's history is replete with the racialized sacrifice of Black workers and residents in the cause of national security, a story of twentieth-century US geopolitics driven by the profitability and promises of war.

Badin Works was one of the earliest aluminium smelting plants in the US. Described as 'the material of modernity' (Sheller 2014), aluminium was vital to the emergence of the US state as a global powerhouse in the mid-twentieth century, from kettles and cookware to portable construction equipment, to the wings and fuselages of aircraft. The materials of warfare for World Wars I and II were manufactured with this new 'light metal' that promised a more efficient, mobile, sleek and altogether modernized American military. As the sole aluminium producer for the US military in the interwar period, Alcoa consolidated power and geopolitical influence. Alcoa's vertical integration of smelting and commodity production operations and full control over government-funded materials research ensured that Alcoa held a monopoly on aluminium production through the war years. Their executives and investors gained key posts in the federal government while the company expanded its holdings across a vast geography, acquiring bauxite mines, factories, and dams, becoming one of the world's first multinational corporations

(Sheller 2014, 62). The intimate entanglements of corporate power, technological expertise, and government funding that enveloped Alcoa probably inspired President Dwight Eisenhower's cautionary note in his 1961 farewell address against the 'unwarranted influence' of government by the 'military-industrial complex'.

In January 1943, the Badin Works aluminium smelting plant in Badin, North Carolina, was honoured with the Army Navy E-Award for 'Excellence in Wartime Production'. Received by only 5 per cent of all US military producers and contractors, the award recognized the 'high and practical patriotism' of the factory's workers, in the words of Under Secretary of War Robert Patterson. The programme brochure quotes President Roosevelt's 1942 State of the Union address: 'Modern methods of warfare make it a task not only for shooting and fighting but an even more urgent one of working and producing.' The production of materials for modern warfare in Badin is quite revealing of the production of modernity itself. Looking back on the experiences of Badin's Black workers and residents, we start to see a different picture of the sacrifices demanded for US military might.

Badin was a real-world test lab, where Alcoa was able to perfect methods for producing aluminium cheaply and efficiently, while conveniently exploiting Black bodies as receptacles for lethal industrial waste. As such, these bodies become intertwined with geopolitical strategies of war (Pain 2015) that, for residents of Badin, produce a complex mess of feelings, including nostalgia for the times when the plant was open and a sense of betrayal at its closure, along with a sense of complicity in the violence enacted on their own families. Often, residents brought toxins home on their clothes and bodies, connecting the intimate space of the home and the intimate ecologies of the body (Vasudevan 2019) to global processes of security and violence through these very finely grained expressions of industrial–military power (Fluri 2011).

Yet the residents of Badin also show how geopolitical efforts to transform our homes and bodies into sites of territorial struggle are resisted. In a context where company loyalty was instilled from childhood as a patriotic duty, Black workers and residents have struggled for generations to challenge Alcoa's portrayal of itself as a caring and responsible company through creative strategies that maintain a communal history erased in the formal archives. An older generation of women who could not openly voice dissent has compiled in homemade scrapbooks the obituaries of workers who died prematurely from illnesses due to toxic exposure. The scrapbooks, a form of 'wake work' (Sharpe 2016), document the tremendous burden of caring for dying workers that was borne by the women of the community. This gives meaning in death to those who were not valued in life. We may not recognize social reproductive practices such as these as geopolitics, yet they enable both the survival and the possibility of future politics for those racialized communities who are subject to ongoing state violence through neglect and abandonment (Gilmore 2008; Pulido 2016). Today, as West Badin residents seek to hold Alcoa responsible for poisoning them, artifacts like the scrapbooks are a reminder of the generational inheritance of both oppression and resistance in their intimate lives.

Conclusions

In Iowa, rural voters who feel neglected by the state invest their hopes in a presidential candidate who promises them safety, security and a return to the good days; while in Badin, North Carolina, Black workers and their families do the work of justice and survival that is necessary for them to exist at the intersections of geopolitics and racial capitalism. In both these cases, to omit either the geopolitical framing of the case or the intimate and entangled ways that questions of geopolitics come to rest within homes and bodies would be to miss a significant component of the story at play. For white Iowans, the embodied politics of Trump speaks to fears and concerns that blur the geopolitical (unprotected borders, trade wars) with the insecurity of family economics and a future that looks increasingly uncertain. These are felt and embodied, and find resonance in Trump's outsized performances.

In the moment of affective connection between a politician at a rally and the global imaginary of a country under threat, these fears shift electoral possibilities and manifest the revanchist white masculinity in the White House, sending ripples of geopolitical anxiety across the world (even as we know that this is not a new era but an intensification or retrenchment of older eras). Likewise, in Badin, North Carolina, we find a reconfiguration of older iterations of racial capitalism: one in which the imperial drives for modernity and military might that are manifest in aluminium production now continue the devaluation of Black bodies, yet in forms that have shifted from those in earlier eras. Here, in addition to the intimately felt geopolitics that work through family histories of illness, the residents of Badin also work to destabilize these geopolitical regimes of erasure and devaluation when they mark, remember and seek justice for the lives cut short.

Both these cases resonate with an intimate geopolitics of the good life deferred, in which visions of happy family and community life in the future lead the residents of Badin and Iowa to participate in geopolitical work. In creating a good life for their families, the Black workers at the plant found themselves complicit in the company's dumping of waste in the community yet were told by their elders not to speak up, lest they should lose their economic stability. In Iowa, attachment to a particular dream of the good life, offered but never attained through neoliberalism, sets residents toward an exclusionary masculinist and nationalist politics that disallows the formation of different forms of community in a changing rural space.

Key readings

Dowler, L., and J. Sharp. 2001. "A Feminist Geopolitics?" *Space and Polity* 5 (3): 165–176.

Fluri, J. L. 2014. "States of (in)Security: Corporeal geographies and the Elsewhere War." *Environment and Planning D: Society and Space* 32 (5): 795–814.

Smith, S. 2012. "Intimate Geopolitics: Religion, Marriage, and Reproductive Bodies in Leh, Ladakh." *Annals of the Association of American Geographers* 102 (6): 1511–1528.

References

Balibar, E., and I.M. Wallerstein. 1991. *Race, Nation, Class: Ambiguous Identities.* London: Verso.

Berlant, L.G. 2011. *Cruel Optimism.* Durham, NC: Duke University Press.

Christian, J., L. Dowler, and D. Cuomo. 2016. "Fear, Feminist Geopolitics and the Hot and Banal." *Political Geography Special Issue: Banal Nationalism 20 Years On* 54: 64–72.

Cramer, K.J. 2016. *The Politics of Resentment: Rural Consciousness in Wisconsin and the Rise of Scott Walker.* Chicago, IL: University of Chicago Press.

Dixon, D.P. 2014. "The Way of the Flesh: Life, Geopolitics and the Weight of the Future." *Gender, Place & Culture* 21 (2): 136–151.

Dowler, L., and J. Sharp. 2001. "A Feminist Geopolitics?" *Space and Polity* 5 (3): 165–176.

Fluri, J.L. 2011. "Bodies, Bombs and Barricades: Geographies of Conflict and Civilian (in)security." *Transactions of the Institute of British Geographers* 36 (2): 280–296.

Fluri, J.L. 2014. "States of (in)Security: Corporeal Geographies and the Elsewhere War." *Environment and Planning D: Society and Space* 32 (5): 795–814.

Gilmore, R.W. 2008. "Forgotten Places and the Seeds of Grassroots Planning." In: *Engaging Contradictions: Theory, Politics, and Methods of Activist Scholarship*, edited by C.R. Hale, 31–61. Berkeley, Los Angeles and London: University of California Press.

Gökarıksel, B., and S. Smith. 2016. "Making America Great Again?: The Fascist Body Politics of Donald Trump." *Political Geography* 54: 79–81.
Haraway, D. 1988. "Situated Knowledges: The Science Question in Feminism and the Privilege of Partial Perspective." *Feminist Studies* 14 (3): 575–599.
Huntington, S.P. 1993. "The Clash of Civilizations?" *Foreign Affairs* 72 (3): 22–49.
Hyndman, J. 2004. "Revisiting Mackinder 1904–2004." *Geographical Journal* 170 (4): 380–383.
Katz, C. 2001. "On the Grounds of Globalization: A Topography for Feminist Political Engagement." *Signs: Journal of Women in Culture and Society* 26 (4): 1213–1234.
Kearns, G. 2011. "Geopolitics." In: *The SAGE Handbook of Geographical Knowledge*, edited by J.A. Agnew and D.N. Livingstone, 610–622. London: SAGE. Available from: http://sk.sagepub.com/reference/hdbk_geoknowledge/n47.xml (accessed 16 January 2018).
Kimmel, M. 2015. *Angry White Men: American Masculinity at the End of an Era.* New York: Nation Books.
Koopman, S. 2011. "Alter-geopolitics: Other Securities are Happening." *Geoforum, Themed Issue: Subaltern Geopolitics* 42 (3): 274–284.
Lerner, S. 2012. *Sacrifice Zones: The Front Lines of Toxic Chemical Exposure in the United States.* Cambridge, MA: MIT Press.
Livingstone, D.N. 1992. *The Geographical Tradition: Episodes in the History of a Contested Enterprise.* Oxford: John Wiley & Sons.
Lugones, M. 2016. "The Coloniality of Gender." In: *The Palgrave Handbook of Gender and Development*, edited by W. Harcourt, 13–33. London: Palgrave Macmillan.
Marston, S. 2000. "The Social Construction of Scale." *Progress in Human Geography* 24 (2): 219–242.
Massaro, V.A., and J. Williams. 2013. "Feminist Geopolitics: Redefining the Geopolitical, Complicating (In) Security." *Geography Compass* 7 (8): 567–577.
Mountz, A., and J. Hyndman. 2006. "Feminist Approaches to the Global Intimate." *Women's Studies Quarterly* 34 (1/2): 446–463.
Pain, R. 2015. "Intimate War." *Political Geography* 44 (Supplement C): 64–73.
Pain, R., and L. Staeheli. 2014. "Introduction: Intimacy-geopolitics and Violence." *Area* 46 (4): 344–347.
Peet, R. 1985. "The Social Origins of Environmental Determinism." *Annals of the Association of American Geographers* 75 (3): 309–333.
Pratt, G., and V. Rosner, eds. 2012. *The Global and the Intimate: Feminism in Our Time.* New York: Columbia University Press.
Pulido, L. 2016. "Flint, Environmental Racism, and Racial Capitalism." *Capitalism, Nature, Socialism* 27: 1–16.
Rosenberg, G.N. 2015. *The 4-H Harvest: Sexuality and the State in Rural America.* Philadelphia: University of Pennsylvania Press.
Secor, A.J. 2001. "Toward a Feminist Counter-geopolitics: Gender, Space and Islamist Politics in Istanbul." *Space and Polity* 5 (3): 191–211.
Shabazz, R. 2015. *Spatializing Blackness: Architectures of Confinement and Black Masculinity in Chicago.* Urbana: University of Illinois Press.
Sharp, J. P. 2000. "Remasculinising Geo-politics? Comments on Gearoid O'Tuathail's Critical Geopolitics." *Political Geography* 19 (3): 361–364.
Sharpe, C. 2016. *In the Wake: On Blackness and Being.* Durham, NC: Duke University Press.
Sheller, M. 2014. *Aluminum Dreams: The Making of Light Modernity.* Cambridge, MA: MIT Press.
Silva, K. 2016. *Brown Threat: Identification in the Security State.* Minneapolis: University of Minnesota Press.
Smith, S. 2011. "'She Says Herself, 'I Have No Future': Love, Fate and Territory in Leh District, India." *Gender, Place & Culture* 18 (4): 455–476.
Smith, S. 2012. "Intimate Geopolitics: Religion, Marriage, and Reproductive Bodies in Leh, Ladakh." *Annals of the Association of American Geographers* 102 (6): 1511–1528.
Smith, S., N.W. Swanson, and B. Gökarıksel 2016. "Territory, Bodies and Borders." *Area* 48 (3): 258–261.
Spillers, H.J. 2003. *Black, White, and in Color: Essays on American Literature and Culture.* Chicago, IL: University of Chicago Press.
Staeheli, L.A. 2001. "Of Possibilities, Probabilities and Political Geography." *Space and Polity* 5 (3): 177–189.
Toal, G. 1996. *Critical Geopolitics: The Politics of Writing Global Space.* Minneapolis: University of Minnesota Press.
Tyner, J., and S. Henkin. 2015. "Feminist Geopolitics, Everyday Death, and the Emotional Geographies of Dang Thuy Tram." *Gender, Place & Culture* 22 (2): 288–303.
Vasudevan, P. 2019. "An Intimate Inventory of Race and Waste." *Antipode.* Doi: 10.1111/anti.12501.

Wang, S.H. 2017. "Fetal Citizens? Birthright Citizenship, Reproductive Futurism, and the 'Panic' over Chinese Birth Tourism in Southern California." *Environment and Planning D: Society and Space* 35 (2): 263–280.
Weheliye, A. 2014. *Habeas Viscus: Racializing Assemblages, Biopolitics, and Black Feminist Theories of the Human.* Durham, NC: Duke University Press.
Wynter, S. 2003. "Unsettling the Coloniality of Being/Power/Truth/Freedom: Towards the Human, After Man, Its Overrepresentation – An Argument." *New Centennial Review* 3 (3): 257–337.

15

HOME-KEEPING IN LONG-TERM DISPLACEMENT[1]

Cathrine Brun and Anita H. Fábos

Introduction

In this chapter, we propose a set of alternative solutions to long-term displacement, using shared human concepts and experiences of 'home' as our starting point. Based on insights into what people actually do when living with displacement and the strategies they use to go about their lives in long-term exile, we explain how mobilizing the notion of home can change current thinking around long-term refugee and displacement situations and produce policies based on a shared understanding of agency. Our feminist analysis of the refugee policy discourse contends that the current thinking about repatriation, local integration and resettlement (the three 'durable solutions' to the global refugee problem) is based on a static, bounded and binary understanding of home, which drives policy. In response, we have developed a nuanced and three-dimensional concept of home in the present policy environment that takes into account home-keeping in displacement, as well as connections with a home-place, to generate more flexible ways of understanding and addressing long-term displacement (see also Brun and Fábos 2015). Beginning with the interconnections between the global and the intimate embraced by feminist geographers, we analyse how refugees and displaced populations make and keep home in interaction and negotiation with the policies formulated to manage their lives. We rely on the critical analysis of home produced by feminist scholars and geographers such as Alison Blunt (2003) and her work with Robyn Dowling (Blunt and Dowling 2006), Iris Marion Young (2005) and Ann Varley (2008). We have also conducted our own feminist analysis of the policy discussions of durable solutions represented by manuals and guidelines, and drawn upon our own long-standing ethnographic research with refugees and internally displaced populations in showing these interconnections. With a feminist geopolitics, we enable an understanding of homemaking that takes place at the meeting point between embodied experiences and geo-political tactics.

We first analyse current conceptions of home in the refugee policies and contemporary solutions formulated to solve displacement crises. Second, we present a feminist critique of home in the context of displacement to demonstrate how feminist geography enables a more refined understanding of home, which encompasses temporal and spatial dimensions of homemaking. We then exemplify three dimensions of home (home-Home-HOME) that form a constellation of ideas to show how refugees and internally displaced (IDPs)[2] make home through daily actions, relate to Home from a position of displacement and are enmeshed in a political system built on HOME. Finally, we bring the three dimensions together by proposing that policymakers can mobilize the ways in which displaced populations 'keep home' to form alternative solutions to long-term refugee displacement.

Policy conceptions of home and contemporary solutions to displacement

International policies that address population displacement have developed in tandem with last century's rise of national frameworks of belonging and homeland, often couched in terms of citizenship, in which people are believed to belong to one place, one nation state. The international political and legal structures in place to address displacement and the plight of displaced people are thus built around concepts such as 'return', 'repatriation' and 'country of origin', to put people back into place. These legal terms signal a shared understanding of how the international community deals with the problem of people 'out of place' (Malkki 1992). The terms have been presented across several compendia prepared by the key actors involved in supporting *refugee repatriation* or, if a return to a country of origin proves impossible, *local integration* in the country that had granted asylum or *resettlement* in a third country. Despite the acknowledgement of their weaknesses, these are well-recognized and commonly accepted terms, framed as the 'durable solutions' to displacement: 'durable' because they address a humanitarian need, should not require revisiting and, by implication, represent an end to displacement.

The discourse around the durable solutions may, however, be seen in stark contrast to the circumstances of the contemporary era in which long-term displacement for both refugees and IDPs has become the norm. In 2015, the United Nations High Commissioner for Refugees stated that in the 32 long-term situations of displacement 'of concern,' people had been displaced for an average of 26 years (UNHCR 2015). Prevailing approaches to solving refugee crises – the three 'durable solutions' – have largely failed to produce a meaningful end to displaced people's predicament (Brun and Fábos 2017). Moreover, the current policies for long-term refugee situations contribute to the dominant narrative that displaced people are either 'stuck' in limbo and passive, or on the move and threatening. The implied solution to this double bind is for refugees to 'go home'.

If there is a contemporary discourse that 'going home' is the best solution for refugees, the concept of home itself is notably undefined. The first edition of the *International Thesaurus of Refugee Terminology*, published in 1989,[3] offers terms related to housing, shelter, settlements, relations with host communities and family, in which specific definitions including home birth and home economics can be found, but home itself does not appear as a unique thesaurus class. In 2006, the United Nations High Commissioner for Refugees (UNHCR) formulated a master glossary of terms to guide consistency in policy and practice. The term 'home' was not part of that glossary, either. Neither does the second edition of the International Organisation for Migration's *Glossary on Migration*, published in 2011, include 'home' as an entry, nor does the UNESCO *Handbook of Selected Terms and Concepts for People on the Move* (2008). Discussing this matter with some of those involved in formulating the glossary, it is clear that home as a term of policy use was never added to the thesaurus; it is too complex and has too many meanings, while simultaneously so commonplace as to seem unnecessary to include. Despite a surge in research on home and forced migration since the early 2000s, the word is still, to a large extent, taken for granted in policy.

The concept's taken-for-grantedness influences policy in particular ways. An analysis of key humanitarian law and refugee protection instruments, such as the 1951 Geneva Convention, shows that the terminology around asylum, protection and return of refugees is clearly laid out, yet 'home' remains undefined and the implication of the geographic home-place around which refugees' flight, loss and displacement centres remains unspecified. The meaning of home that emerges from our analysis is a 'flat' home, a home that is synonymous with the nation state: an understanding that does not capture the multiple scales and dimensions of home. Home is related to *a* location; it is *either* where someone originated from *or* where they have become integrated or naturalized. It thus refers to a particular scale: the nation state. Additionally, it is an understanding of home without a temporal dimension, and is a static and bounded notion of home.

Largely missing from the policy discourse is what forced migrants themselves do. People living in long-term displacement must create and re-create home on a daily basis, while developing strategies for the future and maintaining connections to their previous lives. The majority of policy documents are devoid of any acknowledgement of refugee agency, and do not refer to homemaking as a process that takes place at several levels. In a sense, the refugee 'predicament' is rather a predicament for the international community to deal with the elephant in the room, namely the inadequacy of the current geopolitical model – where home is relegated to bounded, static and unidimensional nation-state territories for the majority of mobile people.[4] This predicament drives the long-term encampment, warehousing or permanent temporariness of refugees and IDPs (Brun and Fábos 2017).

The analysis presented here challenges the ways in which nation states and the 'inter-national' community employ encampment, minimum standards and 'don't die survival' (Cindy Horst, in Hyndman and Giles 2011) to address unending displacement. Essentialist and static notions of home continue to fix forced migrants in both place and time, depriving them of agency and the opportunities to move on and make homes in displacement. A feminist approach to the agentive work of making home helps us to unpack the gendered aspects of control inherent in policies that derive from such a static understanding of home.

Feminist critique of home in the context of displacement

Increased scholarly interest by the social sciences and humanities on home has led to a critical scrutiny of the often taken-for-granted notions that we identified in the previous section. With contributions from feminist research, geography and scholarship on migration, home is established as a more unsettled and problematic entity in which tension and conflict are replete (Brickell 2012; Brun and Fábos 2015). Home is defined as a multiscalar and multidimensional concept. It is a site or several sites, and may be understood as both material and immaterial.

Feminist scholarship on home has related to, analysed and criticized home across several orientations, from socialist feminist to post-colonial scholars (Blunt and Dowling 2006). A central feature has been the dismissal of home by some, due to the association with patriarchy and the subordination that women experience in the home (Martin and Mohanty 1986), where a woman's role is to *be* the home for the rest of the family (Young 2005). With this particular association of women to home, women became stuck in the home not only physically or geographically but symbolically, too. Women were – and still are – often associated with the home in specific ways: as the maintainers of home, as representing the identity of home as house and nation, and as the nurturers of home. Home symbolized for many women the impossibility of progressing, developing their own life and a future (de Beauvoir 1952/1988). Home came to represent a bounded place without progress. As we have pointed out elsewhere, this way of understanding home ties in closely with experiences of protracted displacement (Brun 2016; Brun and Fábos 2015).

In protracted displacement, the future is uncertain, which leads to an experience of 'stuckness' indicated by legal limbo, encampment and other securitization strategies that immobilize refugees over the long term, contributing to a 'feminization of refugees' – a depiction of displaced people as helpless, passive and static (Hyndman and Giles 2011). This feminization discourse further associates refugees and their homemaking strategies with stasis and immanence. However, feminist scholars (Ahmed 1999; Young 2005) have made a case for retaining the concept and idea of 'home' by emphasizing the actual praxis of home as homemaking.

Not all homemaking is housework (Young 2005). Homemaking is a wider term, which incorporates the ways in which people create place in a mobile world (Koraç 2009). Our work focuses on the role of homemaking practices that are pursued in displacement, which, we argue, takes place through the interconnections between different scales (Brun and Fábos 2015). With the help of feminist geopolitics, we develop a notion of home that challenges the usual scale of inquiry and policymaking by drawing attention to the everyday and embodied sites and discourses through which transnational political relations are forged and contested (Dowler and Sharp 2001; Williams and Massaro 2013). Even the most intimate and everyday aspects of life are key sites where geopolitical power is (re)produced and negotiated (Hyndman 2007; Pain and Staeheli 2014). The multiple meanings of home that are experienced and acted out in long-term displacement are a case in point.

Retaining – but reconsidering – home through a feminist geopolitical lens has expanded the range of understandings of 'home' in academic research and in politics. What promises does feminist scholarship make for 'home' (Gardey 2016)? The continuing process of creating alternative understandings of home demonstrates that the making of home in/on the margins remains a crucial political act, one that provides a blueprint for revising the gendered disparity between practices of maintaining and practices of constructing. In the context of displacement, home constitutes several locations, made and maintained by mobilizing a diverse set of resources and relations. Like Gardey (2016), we see home as a common space to be built in which the associations between home and identity and between home and our orientations in/towards the world are challenged (Ahmed 2006; hooks 1990; Wilkins 2017). We also find inspiration in Ahmed's (2017, 7) suggestion that feminism is homework:

> By homework, I am not suggesting we all feel at home in feminism in the sense of feeling safe or secure. Some of us might find a home here; some of us might not. Rather, I am suggesting feminism is homework because we have much to work out from not being at home in a world. In other words, homework is work on as well as at our homes. We do housework. Feminist housework does not simply clean and maintain a house. Feminist housework aims to transform the house, to rebuild the master's residence.

As Ahmed notes, home is not the equivalent of safety and security; because, for displaced people, 'housework', 'homework' and 'homemaking' variously indicate the ways in which home is constantly changing, made and remade according to the everyday lives led in close interaction with geopolitical governance of refugees and other mobile populations, where displaced and non-displaced bodies meet and interact. In this context and by building on the active vocabulary that accompanies 'home', we introduce and bring with us into the next section the concept of 'home-keeping' – a way to help analyse how people *do* home in protracted displacement. Our notion of home-keeping refers to the ways in which subjects make home: how home is made, remade and the ways in which different dimensions of home are held together through homemaking practices, housework and homework, and how subjects constantly negotiate the shifting demands related to the different dimensions of home and navigate their day-to-day requirements at the cross-currents of geopolitical challenges and demands.

Emphasizing these geopolitical acts of maintenance – of home-keeping – alerts us not only to people's creativity and agency but also to the ways in which the importance of these acts is diminished by policies that take for granted and thereby uphold and sustain temporariness, limbo and other static conditions. In the next section we show how, in situations of protracted displacement, people continue to organize their daily lives and think about their futures, even while their abilities to plan appear to be limited and their home-making practices shaped by hardship and uncertainty.

Constellations of home: a feminist approach to home and displacement

Displacement leads to the experience of loss of home, but simultaneously to a redefinition of home. Focusing on home and protracted displacement from a feminist perspective demonstrates that home-keeping for refugees and displaced persons – far from being static and propelled only by the need for survival in the present moment – is a masterful dialogue that spans place and time, incorporating ideal concepts of home and the homeland, aspirations to return 'home' and hopes to achieve a more stable exile by strategizing to go somewhere else. We suggest that these multiple concepts exist simultaneously, while the people who hold them move among various locations to form a very complex idea of home that we have called 'constellations of home'. This metaphor is useful to demonstrate how human beings turn points of reference into meaningful patterns, but that the same points may be imagined differently from each site of observation.

We have derived a simplified triadic constellation that helps us to think about the interconnected and multidimensional implications of homemaking in protracted circumstances of displacement. To distinguish between the various strands that make up this constellation, we visually code them as 'home', 'Home' and 'HOME'.

Beginning with 'home', we take this to mean the day-to-day practices that help to create the place of displacement as a particularly significant kind of place. Such practices involve both material and imaginative notions of home and may be improvements or even investments in temporary dwellings; they include the daily routines that people undertake in these dwellings; and they incorporate the social connections that people make in a neighbourhood, a section of a camp or other institutions formed to 'take care of' refugees and IDPs.

'Home', the second modality in our constellations of home, represents values, traditions, memories and subjective feelings of home. Discussions of home and displacement tend to concern an ideal Home; the Home that many displaced people dream of and long for. These ideas are created by the experiences that displaced people have of lost homes, past homes, and their dreams and hopes for future homes. Home articulated during protracted displacement refers to a more generalized ideal in a particular socio-cultural context and influences domestic practices in temporary dwellings. Emerging from the ideal Home are the material standards that a dwelling must have for it to be inhabitable; while some minimum standards may be commonly shared across socio-cultural contexts, certain aspects such as what constitutes privacy may vary widely. The ideal Home for forced migrants in protracted situations is then reflected in the dwelling, but is also expressed at different scales. For example, numerous studies on home and diaspora analyse the ways in which nostalgia and longing for the homeland nurture an ideal, idealized or even invented Home.

Finally, grappling with homemaking in protracted displacement requires engaging with the dominant meaning and institutionalization of home for the current global order. While we recognize that the notion of homeland is highly politicized for forced migrants idealizing their Home, our focus on the modality coded here as 'HOME' refers to the broader social, political, economic and historical context in which it is understood and experienced by displaced people, and also by the perpetrators of nationalist exclusion and violence and the policymakers addressing protracted displacement through the optic of'durable solutions'. HOME refers to the geopolitics of nation and homeland that contribute to situations of protracted displacement, the ways in which politics of home are necessarily implicated in the causes of displacement and how displaced populations negotiate these conditions in their day-to-day lives. Including HOME in our constellation makes the rift more visible between assumptions about displaced people in a (largely) fixed global order and the fluid conditions of precariousness and unsettledness.

The constellations of home framework enable an open and dynamic understanding of home, one that incorporates the interaction between the intimate and the global and that captures power relations and gendered dynamics of home-keeping. Our own work suggests that this framework captures what refugees and displaced populations actually do and how they go about their lives in a far more nuanced and realistic way than current policies for protracted displacement can account for. In the final pages of this chapter, we will show examples from our own research how the framework can help to adjust policies to be more attentive to what people do.

home

In long-term displacement, daily home-keeping practices are a window into people's fundamental need to control, organize and share their domestic space. For displaced ethnic Georgians in Tbilisi, these routine practices, established under duress and uncertainty, have helped to delineate their communal identity, and they illustrate the key dimension of home that is established through daily living. The war between Georgia and Abkhazia forced nearly 250,000 ethnic Georgians to leave their homes and move to Georgian-controlled areas as internally displaced persons (IDPs). They found places to live in so-called 'collective centres', established in old hotels, hospitals and dormitories for students and factory workers – buildings not meant for permanent dwelling. Since the 1990s, these collective centres were temporary homes for displaced Georgians, symbolizing the experience of permanent impermanence. Everyday life was distinguished by waiting and queueing for water, bathrooms or other facilities in these crowded spaces, where two and sometimes three generations shared one room, with shared toilet facilities and cooking in the corridor. But everyday routines were still established. Home-keeping took hold through the labour of daily life: the cooking in the corridor, getting children to school, cleaning and tidying up and organizing a liveable space in these unlovable conditions.

The emotional and physical labour of crafting a home through daily activities constituted homemaking in what was initially not people's home but gradually became more homely. People made their dwelling spaces beautiful. In particular, one interview stands out, in which two artist brothers described their struggle to generate resources to beautify their surroundings. These daily practices made home – the first dimension of our constellation of home – a particularly significant place, established through housework and maintaining social relations among the displaced in the collective centres. Through these housekeeping practices, people connected the past and the future through the few objects, photos and other items that they may have brought or sourced from Abkhazia. This measure of home-keeping can also engage the production of Home or HOME in a place.

Home

The homemaking practices of the large, displaced population of Arabic-speaking Muslim Sudanese nationals living in Cairo in the 1990s illustrates Home, our second dimension of 'constellations of home'. While daily practices to set up and maintain homely spaces in large, scattered apartment blocks in this mega-city were taking place, individuals and families spent a great deal of time and resources travelling to visit other Sudanese in their temporary rented accommodations or in places of employment. Since Sudanese did not cluster in a specific quarter of Cairo, conducting visits involved a variety of modes of transport (underground, micro-buses, taxis and on foot), and usually included more than one household in an afternoon of visiting, with some visits as short as 15 minutes before the visitors moved on to the next stop. Notably, visits were made not only to nearby relatives, friends and colleagues, as had been a normal part of Sudanese social life back in Sudan, but rather to a variety of far-flung households in a shifting network of Sudanese nationals with various residence statuses in Egypt. Hosts – whether resident for a number of years or newly arrived – would offer tea, juices and biscuits, along with comforting traditions from Home, such as burning incense and playing Sudanese music cassettes.

Social relations for displaced Sudanese people in Cairo thus centred around a new form of visiting that helped them to articulate their cultural identity and national origin by knitting together individual dwellings across space to produce a Home in exile. Furthermore, while narrating these practices helped displaced Sudanese to remember their ideal Home in Sudan and process their loss and yearning, their visiting practices also comprised acts of re-imagining a future nation that was more inclusive, more equitable and represented the best of Sudanese culture and aspirations.

HOME

In the constellations of home negotiated by long-term displaced people, the HOME dimension comprises the current international framework itself, with accompanying rights to citizenship and temporary protections for people out of place. It should be clear from our two ethnographic examples – the Georgian internally displaced people in Tbilisi and the Sudanese refugees in Cairo – that their temporary status did not prevent the home-keeping practices that helped to sustain families, support their day-to-day lives and provide a sense of home and identity. Nevertheless, the current policy environment of exclusionary legal membership, with its distinct social, political and economic constraints, prevents displaced people from becoming full members of the society to which they have moved. The Georgians and Sudanese created their networked and practical meanings of 'home' and 'Home' in order to move on, to develop their own sense of security and to deal with the unlikelihood of return and the accompanying permanent temporariness inherent in their status as forced migrants. The HOME dimension, a central condition of all constellations of home, illuminates the tension between the dominant policies governing displacement and people's struggle for recognition within the current system. While we do not dismiss or romanticize the experiences of marginalization, abuse or violence in displaced people's home-keeping practices in making 'home' and 'Home', the cases that we present here demonstrate that home-keeping has created new opportunities, new homes and new practices that have gradually changed families and communities. Grappling with HOME, however, caused heightened vulnerability, a sense of uncertainty and a lack of recognition due to displaced people's limited rights in societies. The HOME dimension underscores the reality that, for refugees and other displaced populations, unmarked belonging often lies in the past, when they were full members of a society. Nevertheless, displaced populations negotiate the limitations of HOME on a daily basis, creating new and expanded meanings that constantly erode policy assumptions of people in limbo. In fact, protracted displacement frequently leads displaced people to practise a translocal form of HOME from whence people mobilize resources both within and across particular social groups to challenge the temporality that their physical location (and in consequence their legal and political status) represents.

Conclusion: home-keeping as an alternative to 'durable solutions'

Currently, displaced people's struggle for inclusion takes place in tension with the dominant policy understanding that they are away from home, in limbo and in need of re-emplacing via one of the 'durable solutions'. This discourse restricts the possibility for people to develop all three dimensions of the constellation. While we are not the only scholar/practitioners calling for a more dynamic, fluid and flexible enactment of memberships, recognition and rights of displaced people, our conviction rests upon policymakers recognizing an alternative to the current norms of citizenship by reframing HOME as an interface. Since HOME incorporates settled citizens as well as displaced denizens and since all people, regardless of legal status, are agentive makers of home and Home, we call upon state-based humanitarian actors to realign their discourse with the real progress towards inclusive practices and thinking that is being driven by displaced persons and creative helpers.

Home-keeping is the process through which the different dimensions of home–Home–HOME come together in an articular constellation. Refugee policies and national politics are crucial components that shape a particular constellation and restrict practices and possibilities for home-keeping. Our feminist observation that displacement does not stop people from producing constellations of home also encourages new policies and practices on the part of states, inter-governmental agencies and humanitarian actors, which all contribute to shape the conditions for home-keeping in displacement. Displaced people and refugees mobilize their resources and social relations to enact an expanded notion of home, while feminist scholarship and practice help us to identify how mobilizing home may influence and change policy. First, as we have shown here, policy currently expresses a unidimensional understanding of home. Home has not been problematized as a concept and practice; by focusing on what people do and the centrality of home-keeping as part of displacement practices, home needs to be put on the agenda. Second, by placing home on the agenda by way of understanding people's practices, it is possible to expand meanings of home, Home and HOME beyond the dominant use, to refer to the nation state as well as to unsettle the taken-for-grantedness of the concept and processes that home-keeping involves. Third, with an expanded notion of home, more focus could potentially be placed on enabling people to make home in exile and thereby be recognized as active participants. Home – reconfigured as the constellations of home – then becomes a political and feminist space and home-keeping a political act that contributes to make place in a mobile world.

Notes

1 We are very grateful to the people living with displacement who have become research participants and interlocutors over the many years that we have been thinking about the concept of home. These individuals are too numerous and live in too many places to name, but we thank them for sharing their intimate experiences with us. We also appreciate the following individuals, who took part in Anita's seminar at Clark University, Displacement & Development in the Contemporary World: Angela Abdel Sater, Vonia Adams, Sophia Graybill, Donggie Hong, Ray Kane, Siphie Komwa, Alexandra Kramen,

Faith Nelson-Tibuah, Helen Perham, Ron Peterson, Stephanie Rowlett, Inge Salo, Krithi Vachaspati and Corie Welch. Finally, we are extremely grateful to Elisa Mason for exploring ideas and vocabularies of home and policy with us.

2 Internally Displaced Persons (IDPs) is a category used to describe people who have been forced to leave their homes in similar circumstances as refugees, but who have not crossed an internationally recognized nation state border. While they are still in the same country as the home-place they left, they may be in an equally vulnerable situation compared to refugees, as the state that is supposed to protect them is unable or unwilling to do so. Similar to refugees, the long-term situations of displacement have become the norm as conflicts are unresolved, and much of the same policy language and principles are shared between agencies working with refugees and the internally displaced. In this entry, we thus consider the two policy categories together.

3 'The Thesaurus grew out of the *Draft Thesaurus of Refugee Terminology* compiled in English for the UNHCR by Piers Campbell in 1986. In 1988, the then chief of the UNHCR Centre for Documentation on Refugees (CDR), Hans Thoolen, invited Jean Aitchison, a respected expert in the field of thesauri, to provide the technical expertise for the revision and restructuring of this draft. She worked closely with the CDR and with an international working group, and the result of their collaboration was the first edition of the International Thesaurus of Refugee Terminology, in three separate language editions, published in 1989'. UNHCR 2006.

4 We are reminded of the select cosmopolitan few who are able to traverse national boundaries with ease, often with access to more than one national affiliation. This is not the case for the majority of people excluded from these transnational pathways as a function of racialized and sexualized bordering.

Key readings

Blunt, A., and R. Dowling. 2006. *Home*. London: Routledge.

Brun, C., and A.H. Fábos. 2015. "Homemaking in Limbo? A Conceptual Framework." *Refuge* 31 (1): 5–18.

Young, I.M. 2005. "House and Home: Feminist Variations on a Theme." *On Female Body Experience. Throwing Like a Girl and Other Essays*, chapter 7, 27–45. New York: Oxford University Press.

References

Ahmed, S. 1999. "Home and Away. Narratives of Home and Estrangement." *International Journal of Cultural Studies* 2 (3): 330–331.

Ahmed, S. 2006. "Orientations. Towards a Queer Phenomenology." *GLQ: A Journal of Lesbian and Gay Studies* 12 (4): 543–574.

Ahmed, S. 2017. *Living a Feminist Life*. Durham, NC: Duke University Press.

Blunt, A. 2003. "Collective Memory and Productive Nostalgia: Anglo-Indian Diaspora at McCluskieganj." *Environment and Planning D: Society and Space* 21: 717–738.

Blunt, A., and R. Dowling. 2006. *Home*. London: Routledge.

Brickell, K. 2012. "'Mapping' and 'Doing' Critical Geographies of Home." *Progress in Human Geography* 36 (2): 225–244.

Brun, C. 2016. "Dwelling in the Temporary: The Involuntary Mobility of Displaced Georgians in Rented Accommodation." *Cultural Studies* 30 (3): 421–440.

Brun, C., and A.H. Fábos. 2015. "Homemaking in Limbo? A Conceptual Framework." *Refuge* 31 (1): 5–18

Brun, C., and A.H. Fábos. 2017. "Mobilising Home for Long-Term Displacement: A Critical Reflection on the Durable Solutions." *Journal of Human Rights Practice* (special issue: *The End of International Refugee Protection?)* 9 (2): 177–183.

De Beauvoir, S. 1952/1988. *Second Sex*, reprint. London: Cape.

Dowler, L., and J. Sharp. 2001. "A Feminist Geopolitics?" *Space and Polity* 5 (3): 165–176.

Gardey, D. 2016. "'Territory Trouble': Feminist Studies and (the Question of) Hospitality." *Differences* 27 (2): 125–152.

hooks, b. 1990. *Yearning. Race, Gender and Cultural Politics*. Southend: Southend Press.

Hyndman, J. 2007. "Feminist Geopolitics Revisited: Body Counts in Iraq." *Professional Geographer* 59 (1): 35–46.

Hyndman, J., and W. Giles. 2011. "Waiting for What? The Feminization of Asylum in Protracted Situations." *Gender, Place & Culture* 18 (3): 361–379.

Koraç, M. 2009. *Remaking Home: Reconstructing Life, Place and Identity in Rome and Amsterdam.* Oxford: Berghahn Books.
Malkki, L. 1992. "National Geographic: The Rooting of People and the Territorialization of National Identity Among Scholars and Refugees." *Cultural Anthropology* 7 (1): 22–44.
Martin, B., and C.T. Mohanty. 1986. "Feminist Politics: What's Home Got to Do With It?" In: *Feminist Studies/Cultural Studies,* edited by T. de Lauretis, 191–212. Bloomington: Indiana University Press.
Organisation for Migration (IOM). 2011. *Glossary on Migration.* International Migration Law No. 25. Geneva: IOM.
Pain, R., and L. Staeheli. 2014. "Introduction: Intimacy – Geopolitics and Violence." *Area* 46 (4): 344–360.
UNESCO. 2008. *People on the Move. Handbook of Selected Terms and Concepts,* version 1.0. The Hague/Paris: THP Foundation/UNESCO.
UNHCR. 2006. "About the International Thesaurus of Refugee Terminology." In: *Master Glossary of Terms,* Rev.1. Geneva: UNHCR. Available at: www.refworld.org/cgi-bin/texis/vtx/rwmain?docid=42ce7d444 (accessed 3 December 2017).
UNHCR. 2015. *Global Trends. Forced Displacement in 2015.* Geneva: UNHCR.
Varley, A. 2008. "A Place Like This? Stories of Dementia, Home, and the Self." *Environment and Planning D: Society and Space* 26: 47–67.
Wilkins, A. 2017. "Gender, Migration and Intimate Geopolitics: Shifting Senses of Home among Women on the Myanmar–Thailand Border." *Gender, Place & Culture* 24 (11): 1549–1568.
Williams, J., and V. Massaro. 2013. "Feminist Geopolitics: Unpacking (In)Security, Animating Social Change." *Geopolitics* 18 (4): 751–758.
Young, I.M. 2005. "House and Home: Feminist Variations on a Theme." *On Female Body Experience. Throwing Like a Girl and Other Essays,* chapter 7, 123–154. New York: Oxford University Press.

16

ENVIRONMENTAL POLITICS IN THE EVERYDAY

Jam, red meat and showers

Gordon Waitt and Rebecca Campbell

Introduction

The everyday human environment interface has been pivotal to the discipline of geography. Take, for example, the work of Carl Sauer (1963, 343) and his notion of 'cultural landscape', in which 'culture is the agent [and] the natural area the medium'. Of concern to us in this chapter is how feminist research has approached the taken-for-granted division of the everyday worlds that we encounter and study in two categories: 'culture' and 'nature'. Until the 1980s, this division shaped the understanding of the human environment interface within the discipline of geography, underscored by the problematic separation of 'physical' and 'human geography'. This chapter endeavours to outline some of the ways in which feminist thinking reconfigured and troubled humanist assumptions underpinning the culture–nature antinomy. To do so, the chapter discusses various feminist theoretical and methodological strands that have shaped geographical analysis of the human environment interface.

Our chapter is divided into three sections. In the first, we outline how the ecofeminist politics that sought to make everyday lives more environmentally sustainable were contested. The second examines how post-Marxist feminist approaches to science studies took account of bodies, devices and codes in the knowledge practice of scientists. Post-Marxist feminism addressed the gendered cultural politics of nature to trouble the culture–nature binary that rendered everyday social worlds as solely human achievements. The third and most substantial section discusses feminist work on embodiment that takes a visceral approach and attends to corporeal geographies by drawing on assemblage concepts to think outside the nature–culture divide and take up questions of nonhuman agency. To do this, we draw on a series of different ethnographic research projects that examine mundane practices in a range of everyday and research contexts: eating jam; plating up kangaroo; and washing oneself. These examples illustrate how domestic lives and subjectivities are assembled through relationships that blur the boundaries between the natural and cultural and are bound up in the affective intensities through which bodies' capacities to act become either diminished or enhanced. In the context of questions that focus on sustainability, liveability and climate change, we outline how a visceral approach that is alive to the notion of affective intensities has potential practical and political applications to help to rethink what mobilizes people to change their everyday choices.

Ecofeminisms

Ecofeminists of the 1970s, 1980s and early 1990s (see Gaard 1993; Plumwood 1991, 1993; Seagar 1993) posited a close connection between gender/women and nature/environmental conservation/sustainability, specifically addressing the oppression of both nature and women within Cartesian–Kantian understandings of knowledge. Ecofeminism is a contested concept and there is no single agreed form. Two strands of ecofeminist thinking characterize this intellectual and practical movement; cultural or spiritual ecofeminism; and social ecofeminism. Both are entrenched in the binary thinking that assumes that things were assigned to either the categories of 'culture' or 'nature' and thus rendered the world an exclusively human achievement.

Central to cultural or spiritual ecofeminism is the appropriation of 'Mother Nature' as an ideal. This allowed spiritual ecofeminists to advocate for women-as-nature. One of the key tenets of spiritual ecofeminism is that women have a special relationship with nature, because of their reproductive role. Ecofeminists for whom 'Mother Nature' held spiritual resonance embraced an essentialist position and advocated that women's qualities (caring, mothering and nurturing) were intrinsic biological attributes. In addition, spiritual ecofeminists argued that these biological women–nature connections were undermined by patriarchal conditions that inform environmental management policies. Rather than challenging the knowledge that took for granted the association of women and nature, spiritual ecofeminists positioned women as stewards whose interests aligned with nature and thus sought to reverse the social hierarchies of patriarchal societies.

Social ecofeminists rejected the essentialist arguments of spiritual ecofeminism that reduced women to their biology. Instead, this ecofeminist position advocated for rethinking the women–nature connection as socially constructed and, thus, a political rather than a biological category. This strand of ecofeminism centred on the dominant representation of women as carers and nurturers. Socialist feminists critiqued the masculinist view of environmental knowledge born of a Cartesian–Kantian framework that naturalized the dualist thinking that aligns women with nature and men to culture. Such a divided worldview set up nature as 'out there' and separate from 'man', and therefore could be objectively controlled, studied and manipulated, whereas, following this Western masculinist tradition, women were positioned as closer to nature and as both irrational and emotional.

This realization allowed for the recognition of how masculinist thinking operates as a system of opposition and exclusion. How masculinist thinking shaped human–environment relationships is demonstrated in taken-for-granted assumptions that nature is 'out there', still waiting to be protected, remade or reclaimed by men. By contrast, women are thought to become more concerned about the environment through naturalized traditional gendered divisions of labour that position women through their labour as nurturers and carers (Plumwood 1993). For example, Emel (1995) showed how a version of Western frontier masculinity arose in the late nineteenth century in the American west through wolf-eradication programmes. The portrayal of men-who-hunted wolves as chivalrous, virile, moral and civilizing relied upon understandings of wolves as savage, cowardly and with a pack mentality. Emel's (1995) study shows not only how the wolves were destroyed but also how social hierarchies were generated around a White colonial frontier masculinity. Here, nature–culture is conceived as necessarily entangled and spatial.

In summary, social ecofeminism is primarily a social constructivist approach that critiqued the systems of power and cultural politics that have privileged dominant models of Western masculinist thought. The social ecofeminist paradigm posited that all oppressions (deriving not only from gender, but also from race, class, sexuality, and so on) are set up by hierarchical dualisms from which the subjugation and exploitation of both women and nature stem (Gaard 1993). By this token, it is therefore impossible to truly emancipate any oppressed group if equal efforts to do the same for nature itself are ignored. Social ecofeminsm destabilized the positivist, rationalist, essentialist and deeply entrenched Western habitus of environmental history that enabled the oppression of both women and nature by holding the concepts of nature and culture apart. The arguments of social ecofeminism are imperative to destabilizing taken-for-granted gendered categories, identities and discourses.

Post-Marxist feminisms

Post-Marxist feminisms advocate for notions of nature–culture conjoining or coupling. Hence, culture should be conceptualized as being 'in' or 'folded' through the natural. There are at least two strands of post-Marxist feminist explanations of the gendered dimensions of the human–environment relationship: feminist science studies; and feminist political ecology. These feminist approaches emphasize how the social and ideological, alongside technologies, shape gendered knowledge, modes and relationships of production and political activism.

Feminist science studies

Feminist critiques of science provide an important avenue to rethink the nature–culture divide. Here, we focus on two contributions: first, the knowledge practice of science as situated; and, second, human–machine interdependencies with respect to the notion of 'cyborg'. The feminist approach to the question of knowledge practice in science articulated by Haraway (1988) and Rose (1997) positions scientific knowledge as embedded within, rather than divorced from, gendered and classed social worlds. Feminist science studies therefore identified as important the interrogation of the social and ideological construction of science, which is predicated on the idea that nature, as feminized Other, is a site for scientific experimentation and discovery (Merchant 1980). Hence, feminist writers like Haraway advocated for situated knowledges; that is, an epistemology grounded in a reflexive process that reveals the researcher's subject position in relation to the uneven social power relationships that underpin their knowledge claims. In doing so, feminist science studies critiqued how environmental knowledge is not immune to the 'subjective' forces of culture by arguing that scientific communities are embedded in whole series of interconnections that are personal, economic, political and technical (including Haraway 1988; Merchant 1980; Plumwood 1993). Feminist science scholars argued that stronger forms of objectivity stemmed from the positioned rather than idealized humanist subject. These scholars illustrated the strong tendency of scientists to render nonhuman nature as knowable and predictable, rather than chaotic, unpredictable and coupled or conjoining natural–human system. The notion of situated knowledge transcends the gendered dimensions of Eurocentric knowledge by acknowledging that objectivity and disinterest in the subject is the 'god trick of seeing everything from nowhere' (Haraway 1988, 581).

Consequently, feminist postcolonial science studies acknowledge Indigenous knowledges that are generally marginalized through colonialism, inequality and power relations. For example, Carey et al. (2016) advocate for a feminist postcolonial glaciology that analyses not only situated knowledge and gender dynamics but also folk glaciology, which is usually silenced through the operations of patriarchy, colonialism and Western science. While glaciologists may try to understand glacial ice through measurement, a folk glaciology, as offered by Cruikshank (2005), challenges scientific understandings of glaciers by engaging all the senses and the narratives of how the lives of women and glaciers are intimately connected.

Haraway's (1996) motif of the cyborg helps feminists to think beyond the separation of culture and nature with the objective of being alive to how technology is blurring the boundaries of bodies and conventional notions of a discrete and autonomous human subject. The possibilities of a cyborg folding of flesh and technology acknowledge human–machine–animal interdependencies, including the activities of using hearing aids, wearing glasses, driving a car, talking on a mobile phone and working on a computer. Haraway acknowledges that our minded bodies are engaged in scientific innovations and political decisions. Such ideas seek a more inclusive ecological politics that envisages power as located neither in the environment or in humans but as produced through interrelationships between both together. For example, a car, a driver and driving infrastructure (roads, petrol, parking) become a new entity that has power, space and everyday life-making potentials. The scientific interventions of battery technologies and the driverless electric car, for example, create a whole new set of challenges, from how we organize our everyday lives to how we think of ourselves in the car. The cyborg motif has provided inspiration for feminist geographical research to explore the different ways in which science is reshaping the human–nature relationship, including through modifying foodstuffs, receiving blood and mapping DNA (Nash 2005).

Feminist political ecology

Post-Marxist feminist concepts of nature found in political ecology build upon post-colonial science studies' and ecofeminism's interrogations of the dominant Western, masculinist paradigm to offer the concept of 'socio-natures' to address the culture–nature binary. A feminist approach to political ecology addresses how patriarchal systems of power are founded upon the capitalist production of goods (Littig 2001). Feminist political ecology, for the most part, adopted a macro-sociological approach that looks to the structural, ideological and discursive to better understand the relationships between gender and the environment. It argues that men and women experience environmental issues differently and have different environmental attitudes and politics, because of the social construction of gendered roles and difference. There are arguably parallels between the capitalist subjugation of the natural world and women's oppression (Bennholdt-Thomsen and Mies 1997, cited in Littig 2001; Mies et al. 1988; Mies and Shiva 1993). As Seagar (1993) points out, the capitalist wealth upon which state power is predicated is built not only on the exploitation of resources available domestically but also on the claiming of the resources of Others. Control over natural resources is therefore key to performances of power. This is made possible by systems of ownership and privatization that stem from the Western, masculinist view of nature as something that is Other to humanity and thus an object to be dominated (Seagar 1993). Therefore, some feminist political ecology scholars argue that 'a liberation from the constraints of capitalist material consumption and the preservation and restoration of subsistent economic activity is the only successful feminist counter strategy' (Littig 2001, 43) to oppressions of nature and of socially marginalized groups.

Feminist political ecology researchers call for studies to move beyond gender to include analysis of power and justice. Hence, political ecology scholars challenge concepts like capitalism and class as taken for granted. As researchers in feminist political ecology argue, that the capitalist enterprise goes largely unquestioned and unchallenged is the product of various discursive ideologies (shaped by essentialist, phallocentric and binary thinking) permeating

Western discourse (Gibson-Graham 1996). Gibson-Graham (1996) sought to rethink capitalism as a 'driverless juggernaut' to instead envisage it as comprised of diverse economies and economic difference. Thus, feminist engagements with economy and uneven social power relations challenge dominant, capitalist relations with nature as a resource, and call for alternative economic narratives that acknowledge marginalized, hidden, non-dominant knowledges and practices (Gibson-Graham 2008). The work of Gibson-Graham (1996, 2008), for example, offers alternative conceptual lenses to understand the capitalist regime and its consumption of nature and its resources. If, as suggested above, the critique and dismantling of hegemonic capitalist ideology is the only method by which to liberate nature and women, Gibson-Graham's (1996, 2005, 2008) insights into Other spaces and performances of economy offer possible sites of resistance and change.

A visceral approach to environmental politics in the everyday

Aligned with broader discussions and conceptualization in feminist scholarship on corporeality (see Alaimo 2010; Neimanis 2017), the recent turn of feminist geographies to the most intimately lived geography of the body to address power, justice and knowledge production offers opportunities for scholars to engage with environmental issues through people's everyday lived encounters, visceral experiences, emotions, affects and embodied realities. A visceral approach offers much to feminist studies of the human environment interface by de-essentializing the corporeal body and acknowledging it as a complex nexus of practices, materialities, discourses and social relations that disrupts the masculinist binaries through which the body is traditionally viewed (Longhurst and Johnston 2014; Valentine 1999). A visceral approach allows for insights into culturally diverse environmental knowledges that are not necessarily based on Western, rationalist epistemologies (Waitt and Welland 2019). The philosophy of Deleuze and Guattari (1988) is a significant resource for feminist geographers' rethinking of the human environment interface, through their insistence on analysis that begins in the middle of things and, in our understanding of the world, on attention to forces that are both material (bodies, things, technology, plants, water) and expressive (affects, emotions, feelings, sensations, ideas).

Following Hayes-Conroy and Hayes-Conroy (2008, 462), the visceral is 'the realm of internally-felt sensations, moods and states of being, which are born from the sensory engagement with the material world'. A visceral approach brings bodies to the fore by attending to affect. In this regard, drawing on the work of Deleuze, and of Probyn's (2000) translation of his ideas, affect is a visceral way of behaving that involves the senses – touch, sounds, taste, sight and smell. That is, affect is conceived as the push in the world, an outcome of embodied knowledge that is at once non-cognitive and cognitive that may circulate between and through bodies and may be stored in places, things and ideas. Attention turns to emotions and feelings articulated through everyday encounters or convergences that either increase or decrease the body's capacity to act. In this regard, affects may be conceptualized as a relational force. That is, when bodies and embodied knowledge are brought to the fore, before we signify affective forces as emotions or feelings such as shame, pride or disgust, we acknowledge the force that pushes us towards or away from that which we encounter in the world. Shame or pride are thus registered at a visceral level as blushes or euphoria in a way that is attuned to 'going with your gut response'. In turn, affective forces are conceived to be intensified by discursive framings that align words with specific performances and places. As Ahmed argues (2004, 119), affective relations 'involve subjects and objects, but without residing positively within them', for example football fans who get goose bumps of pride at the winning performances of a sports team and who then may shout or sing with elation. The affective capacity, at some molecular level in the fans' bodies, produces an enhanced capacity to act in the presence of a winning team and supporters. It is the relationship between the supporters' bodies and the football team that presents the affect. Furthermore, affective relations have the capacity momentarily to mobilize people to forge a collective, as demonstrated by the euphoria of a winning team.

In this regard, following the feminist philosopher Elspeth Probyn (2000), a visceral approach is a way to conceptualize the body as actively participating in the unfolding of human behaviour, subjectivities and social difference, alongside other processes more often recognized as sociological, psychological and physiological. This attention to the body therefore invites a shift in theoretical approaches to embodiment and so conceptualizes emotions as circulating as much by affective communication in and through moments of proximity or contact between human and non-human bodies as by more discursive modes of interaction. To illustrate the implications of feminist geographers embracing the concepts of Deleuze and Guattari (1988) to advance a visceral approach, our chapter turns to jam, kangaroos and showers.

Jam – sticky encounters with a difference

Picking up and extending Probyn's work, feminist geographers Hayes-Conroy and Hayes-Conroy (2008, 2010), Hayes-Conroy and Martin (2010) and Longhurst et al. (2009) have addressed how the deeply visceral attributes of eating make food a particularly compelling entry point for exploring the relationship between subjectivity and place in our accounts of the politics of eating. This work shows how what we eat is shaped not only by sets of ideas but also through multi-sensory engagements with the materiality of things that become food, on an everyday basis (Hayes-Conroy and Martin 2010). It pays close attention to the physical body's role in social life and, specifically, to how particular oppressive regimes may become embodied. Furthermore, the visceral approach has equipped scholars with the conceptual tools to envisage how the body's affective capacities could be forces to open possibilities for future change. For example, these feminist scholars conceptualized taste as within the visceral, rather than as attributable solely to the biological realm. In this task, they paid close attention to the affective intensities of individual eating practices, including the foodways of homemade and commercial jams.

The example of tasting jam provided Hayes-Conroy and Hayes-Conroy (2008) with an applied entry point to illustrate how a visceral approach can offer new insights not only into 'taste' but also into the politics of eating. They deployed a visceral approach to critique the politics of the 'slow food' movement, which advocates for the politics of sustainability based on the biological taste of organic foods and knowing where and how food was produced. Hayes-Conroy and Hayes-Conroy (2008) drew on an education programme with primary schoolchildren run by 'slow-food' leaders in Australia. The leaders prepared the students for a blind tasting of homemade and commercial strawberry jams. When asked about which jam they thought would taste better, most opted for the homemade jams made by their mothers. Yet, after the blind tasting, an overwhelming majority of participants selected commercial jam as their preferred taste; indeed, most of them had grown up eating commercial jams.

Through this example, Hayes-Conroy and Hayes-Conroy (2008) alerted us to the possible classed and gendered assumptions of the 'slow food' educators. They argued that taste is always more than a biochemical process. When the taste of jam is conceived through the visceral realm, it points to how the affective intensities of taste are not only aligned to particular social groups along class and ethnicity but are also tied to memories and places. The work of Hayes-Conroy and Martin (2010) is also attentive to how the body is implicated in the unfolding politics of the slow-food movement for people to eat more sustainably. Rather than limiting taste to biological and chemical processes, Hayes-Conroy and Martin (2010) recognize that the molecular and chemical dimensions of taste are connected to the ideas, beliefs and social categories that we inhabit. In doing so, they illustrate how the taste of jam opens up connections between its materiality and particular ideas and identities. Their point is that taste is materially valid yet is complicated by how social difference and affective intensities enter the material realm and shape our everyday food choices. Hence, this work points to wider implications arising from the visceral experience of food for social movements seeking to influence eating habits. Without close attention to the social–spatial circumstances that trigger negative visceral responses, food programmes designed to change eating practices will continue to fail.

Kangaroo

Building on the corporeal feminist scholarship of Probyn (2000) and Hayes-Conroy and Hayes-Conroy (2008), Waitt and Appleby (2014) conceive the human body as relationally produced and materially affected to better understand the resistance to eating more kangaroo meat in Australia. Environmental scientists (Morrison et al. 2007; Wilson and Edwards 2008) and economists (Garnaut 2008) disseminate information that asserts the benefits of regularly eating kangaroo – including reducing the greenhouse gas emissions associated with what we eat, particularly from the increased consumption of meat and dairy products in Western diets. Here, the logic is that kangaroos produce less methane than cattle due to differences in their digestive tract. Yet, most Australian households seem disinterested in eating kangaroo regularly. This is despite supplies of kangaroo meat being available through leading supermarket chains since 2000. By attending to the affective intensities of produced through seeing, smelling, touching and tasting kangaroo meat, Waitt and Appleby (2014) show how food preparation and eating contribute to the resistance to practise ethnic identities in new ways. Through a discussion of visceral disgust, they illustrate the body's material agency in arranging social patterns and categories of settler-Australian society. Like the work of Hayes-Conroy and Hayes-Conroy (2008) and Slocum (2008), Waitt and Appleby (2014) illustrate the way that bodily judgements of plating up and eating kangaroo arise from the specific structural circumstances of settler Australian society that position it as holiday 'bushtucker' rather than an everyday meal. Overall, the kangaroo tastes (strong, gamey) and textures (tough, stringy) did not connect viscerally with the thoroughly embodied set of culinary practices, labour, skills and social relations that comprise everyday home life and the subjectivities of a 'good' grandparent or parent. As Probyn (2000, 9) notes, 'eating … becomes a visceral reminder of how we variously inhabit the axes of economy, intimate relations, gender, sexuality, history, ethnicity, and class'. In the case of eating kangaroo meat, visceral disgust serves to fix cultural and familial food traditions through the uncomfortable feeling of coming too close to something already rejected as inedible when categorized as wormed or, at best, pet food. However, as Slocum illustrates (2007, 2008), thinking through the visceral and the relational body reveals that affective forces do have the capacity to make way for imagining new attachments and desires, and unsettling cultural and familial food traditions. A visceral approach is committed to the uncertainty of outcomes and the possibility of affective intensities for the world being otherwise.

Showers

Our final example draws on Burmese refugees' and migrants' experiences of washing themselves in Australia. This project attended to the visceral experiences of the touch of tap water on the skin. Concentrating on the visceral experiences of bathing, insights were offered into participants' gut feelings about washing themselves. In this instance, the visceral experiences were about cleanliness, freshness, identity, belonging and longing and place. Insights were offered into how the performances, alongside the touch of water, mobilized bodies to act in particular ways. For those participants who continued to scoop water over themselves from a bucket, this habitual practice reconnected them with their home country, age, faith, freshness and personal ethics to minimize water use. For others, the affective intensity of flowing tap water from the shower was often connected to transitioning between identities and feelings of being 'clean' and 'at home' in Australia. The affective geographies of showering for these participants were bound up with ideas of water security, and the experiences of the shower became a device to help to constitute professional identities and public selves alongside managing hectic schedules and spatially fragmented lives. These participants illustrate how discourses around cleanliness, convenience and abundance combined with bathing materials (taps, pipes, showerheads, hot-water tanks) in the visceral body. The discourses of the convenience and comfort of showers are not separate from their material design and the sensuous touch of water. Furthermore, this research underscored that bathing is never 'just washing'. Education programmes designed to reduce water consumption alone may prove ineffective because of how subjectivities and daily lives are entangled with the touch of the shower water. As Waitt and Welland (2019, 39) argue:

> the scenario of washing skin from the touch of cool water splashed from a bucket, rather than the flow of warm water for a shower, may raise a series of alarm bells precisely because it challenges habits and routines of deeply embedded racialised, classed, aged, gendered and ethical norms that stabilise white, healthy, classed bodies and western public spaces as deodorised.

To inform household sustainability policy, we also need to understand how the experiences associated with practices – being touched by water, touching water and experiencing the feelings associated with the touch of water – are important components of ethnic, gendered and religious subjectivities and places. Hence, Waitt and Welland (2019) point to the importance of environmental programmes that engage, particularly, through the unpredictability of bodily relationality. Visceral political action could involve, for example, enrolling the felt intensity of a 'bucket-bathing challenge' once a year to encourage people to reflect on the social norms of bathing to excess and our connections to river systems. Bound up in the bodily and affective intensities of washing skin are possibilities for embodied social change.

Conclusions: gender, power and the human environment interface

In this chapter, we have outlined how different feminist concepts that engage with the human–environment interface expose how patriarchal structures of domination have worked historically to supress women through a romanticized affinity with nature. The concepts serve to open up corporeal knowledge to rethink what mobilizes everyday choices. Our review highlights how feminist scholarship offers various theoretical portals to understand the human environment interface, each underpinned by a shared commitment to the gendered politics of sustainability/environment. The politics of spiritual ecofeminism was theorized by women as having essential biological qualities and, as such, were used to organize events that sought to celebrate this romanticized connection with nature. In contrast, the politics of social ecofeminism departed from the binary Western hierarchical thinking of man: women, culture: nature, urban: rural, and civilized: wild. The concepts of social ecofeminists enabled a cultural politics that became a technique for challenging the normativity that aligned gendered human environment relationships to their social, spatial and representational dimension.

Post-Marxist feminisms have built upon such a way of thinking that for too long imprisoned women and the environment within analytical divisions of a masculinist knowledge of the world. Post-Marxist feminist thinking – specifically, the concept of 'situated knowledge', where knowledge is set in its specific socio-economic context – became a vehicle for an environmental politics that raised questions of how scientific knowledge itself was embedded in gendered and class interests that then legitimated certain 'truths' about nature. Furthermore, post-Marxist feminist thinking through the concept of 'cyborg' entailed a challenge to the human–nature binary by highlighting the ontological existence of things alongside their powers, capacities and consequences. Feminist political ecology set research agendas embedded in economic relations, institutional sites and state policies that were attuned to how dominant gendered and class agendas were integral to shaping paid and unpaid work, and thus relationships with the environment.

The aim of a visceral approach is to better understand the non-cognitive and cognitive elements that comprise our experiences of everyday life and situate these within political processes, including sustainability, liveability and climate change. For many decision-makers, everyday life is now at the fore of liveable futures. The question of how policymakers and activists can mobilize everyday habits (including bathing, diets and transport) is thus highly germane. The body is the entry point for the visceral approach of corporeal feminism. Corporeal feminists argue that the body should be treated as a relational concept in ways that do not predetermine our analysis of either outcomes or experiences. Thus, a visceral approach requires attention to the interconnections between thinking and being, paying specific attention to the role of affect and emotions in how the dualisms of our social–material world are made, remade or undone. Gender, together with other social categories, is therefore conceived as emerging within the spatially embedded relationships between ideas, actions, things and experiences. According to corporeal feminists, affective intensities that are expressed as emotions, such as disgust, pride and shame, are crucial to our situated subjectivities and our everyday choices and actions. Attention therefore focuses on the multi-sensual dimensions of everyday life encounters with people, things, ideas and places. Corporeal feminists offer important insights to the politics of everyday life and social movements by illustrating the unpredictability of bodily relationality and its implications for being political and changing everyday life.

Key readings

Hayes-Conroy, A., and J. Hayes-Conroy. 2008. "Taking Back Taste: Feminism, Food and Visceral Politics." *Gender, Place and Culture* 15 (5): 461–473.

Waitt, G., and L. Welland. 2019. "Water, Skin and Touch: Migrant Bathing Assemblages." *Social and Cultural Geography*, 20 (1): 24–42.

Waitt, G., and B. Appleby. 2014. "It Smells Disgusting: Plating up Kangaroo for a Changing Climate." *Continuum: Journal of Media and Cultural Studies* 28 (1): 88–100.

References

Ahmed, S. 2004. "Affective Economies." *Social Text* 22 (2): 117–139.

Alaimo, S. 2010. *Bodily Natures: Science, Environment and the Material Self.* Bloomington: Indiana University Press.

Carey, M., M. Jackson, A. Antonello, and J. Rushing. 2016. "Glaciers, Gender and Science: A Feminist Glaciology Framework for Global Environmental Change Research." *Progress in Human Geography* 40 (6): 770–793.

Cruikshank, J. 2005. *Do Glaciers Listen? Local Knowledge, Colonial Encounters, and Social Imagination.* Vancouver: University of British Columbia Press.

Deleuze, G., and F. Guattari. 1988. *A Thousand Plateaus: Capitalism and Schizophrenia*. Minneapolis: University of Minnesota Press.
Emel, J. 1995. "Are You Man Enough, big and Bad Enough? Ecofeminism and Wolf Eradication in the USA." *Environment and Planning D* 13 (6): 707–734.
Gaard, G. 1993. "Living Interconnections with Animals and Nature." In: *Ecofeminism: Women, Animals, Nature,* edited by G. Gaard, 1–12. Philadelphia, PA: Temple University Press.
Garnaut, R. 2008. *Climate Change Review*. Canberra: Australian Government Publishing Service.
Gibson-Graham, J.K. 1996. *End of Capitalism (As We Knew It): A Feminist Critique of Political Economy.* Minneapolis: University of Minnesota Press.
Gibson-Graham, J.K. 2005. "Surplus Possibilities: Postdevelopment and Community Economies." *Singapore Journal of Tropical Geography* 26 (1): 4–26.
Gibson-Graham, J.K. 2008. "Diverse Economies: Performative Practices for 'Other Worlds'." *Progress in Human Geography* 32 (5): 613–632.
Haraway, D. 1988. "Situated Knowledges: The Science Question in Feminism and the Privilege of Partial Perspective." *Feminist Studies* 14 (3): 575–599.
Haraway, D. 1996. *Simians, Cyborgs and Women: The Reinvention of Nature*. London: Free Association Books.
Hayes-Conroy, A., and J. Hayes-Conroy. 2008. "Taking Back Taste: Feminism, Food and Visceral Politics." *Gender, Place and Culture* 15 (5): 461–473.
Hayes-Conroy, A., and J. Hayes-Conroy. 2010. "Visceral Difference: Variations in Feeling (Slow) Food." *Environment and Planning A: Economy and Society* 42 (12): 2956–2971.
Hayes-Conroy, A., and D.G. Martin. 2010. "Mobilising Bodies: Visceral Identification in the Slow Food Movement." *Transactions of the Institute of British Geographers* 35 (2): 269–281.
Littig, B. 2001. *Feminist Perspectives on Environment and Society.* Abingdon: Routledge.
Longhurst, R., and L. Johnston. 2014. "Bodies, Gender, Place and Culture: 21 Years On." *Gender, Place and Culture* 21 (3): 267–278.
Longhurst, R., L. Johnston, and E. Ho. 2009. "A Visceral Approach: Cooking 'at Home' with Migrant Women in Hamilton, New Zealand." *Transactions of the Institute of British Geographers* 34 (3): 333–345.
Merchant, C. 1980. *The Death of Nature: Women, Ecology and the Scientific Revolution*. San Francisco, CA: HarperCollins.
Mies, M., and V. Shiva. 1993. *Ecofeminism*. London: Zed Press.
Mies, M., V. Bennholdt-Thomsen, and C. Werlhof. 1988. *Women: The Last Colony.* London: Zed Press.
Morrison, M., C.S. McSweeney, and A-D.G. Wright. 2007. "The Vertebrate Animal Gut in Context, Microbiomes, Metagenomes and Methane." *Microbiology Australia* 28 (3): 107–110.
Nash, C. 2005. "Geographies of Relatedness." *Progress in Human Geography* 30 (4): 449–462.
Neimanis, A. 2017. *Bodies of Water.* London: Bloomsbury.
Plumwood, V. 1991. "Nature, Self and Gender: Feminism, Environmental Philosophy and the Critique of Rationalism." *Hypatia* 6 (1): 3–27.
Plumwood, V. 1993. *Feminism and the Mastery of Nature.* London: Routledge.
Probyn, E. 2000. *Carnal Appetites: Food Sex Identities*. London: Routledge.
Rose, G. 1997. "Situating Knowledges: Positionality, Reflexivities and Other Tactics." *Progress in Human Geography* 21 (3): 305–320.
Sauer, C.O. 1963. "The Morphology of Landscape." In: *Land and Life: A Selection from the Writings of Carl Ortwin Sauer,* edited by J. Leighly, 315–350. Berkeley: University of California Press.
Seagar, J. 1993. *Earth Follies.* London: Earthscan.
Slocum, R. 2007. "Whiteness, Space, and Alternative Food Practice." *Geoforum* 38 (3): 520–533.
Slocum, R. 2008. "Thinking Race Through Corporeal Feminist Theory: Divisions and Intimacies at the Minneapolis Farmers' Market." *Social & Cultural Geography* 9 (8): 849–869.
Valentine, G. 1999. "A Corporeal Consumption of Geography." *Environment & Planning D: Society and Space* 17 (3): 329–351.
Waitt, G., and B. Appleby. 2014. "It Smells Disgusting: Plating up Kangaroo for a Changing Climate." *Continuum: Journal of Media and Cultural Studies* 28 (1): 88–100.
Waitt, G., and L. Welland. 2019. "Water, Skin and Touch: Migrant Bathing Assemblages." *Social and Cultural Geography,* 20 (1): 24–42.
Wilson, G.R. and Edwards, M.J. 2008. "Native Wildlife on Rangeland to Minimise Methane and Produce Lower-emission Meat: Kangaroos Versus Livestock." *Conservation Letters* 1 (3): 119–128.

17
GENDER AND URBAN NEOLIBERALIZATION

Carina Listerborn

Introduction

Feminist movements have been claiming public space and asserting rights through both strategic action and everyday practices in cities for over a century. In the Global North, feminist movements have been essential to the development of welfare societies and especially since the 1970s, feminist scholars, activists, architects and planners have supported egalitarian and gender equality ideals in urban development (Hayden and Wright 1976; Wekerle 1980). Based on a critique of male dominance in both the planning professions and academia, a feminist urban research agenda has emerged for a 'non-sexist city' (Hayden 1980). This field of research brings together disciplines from urban and architectural history, geography, sociology, ethnology, economics, political science, planning, architecture and the growing field of urban studies. Originally, much attention was given to rewriting the role of women in the history of cities (Hayden 1981; Spain 2001; Wilson 1992), the dichotomy between private (home) and public spheres under industrial capitalism (Bondi and Domosh 1998), gendered spaces (Domosh and Seager 2001) and women's right to equal access to public services such as transportation, housing and social services (Fainstein and Servon 2005; Weisman 1994). The home as a site of resistance was articulated by Black feminists (hooks 1981), challenging the dominant idea of home as merely imprisoning women. In Sweden and Denmark, collective housing became an important technique for breaking up traditional domestic gender roles in the 1970s and 1980s (Vestbro 2010).

In the early 1990s, the gendered experiences of women in public spaces, from the scale of the street to the nation and beyond, gained recognition within geography and the growing research field of gender studies. Women's fear of sexual assaults in public spaces emerged as a powerful statement on the 'geography of fear' (Pain 1991, 2001; Valentine 1989). As a result, safety planning was initiated by women's groups (Wekerle and Whitzman 1995; Whitzman 1992). When social sustainability was introduced at the 1992 UN conference in Rio de Janeiro, participatory planning processes from a gender perspective became part of Agenda 21. Feminist scholars had been advocating for a more inclusive planning process, which suited this agenda

(Sandercock 1998). Community-based work has always been a strong base within feminist urban research and activism, going back to Jane Addams Hull House in Chicago. Today, there is a strong focus on women as a category in both research and policy, while aspects of class, race and sexuality are much less articulated. Less attention still is paid to masculinity and urban space, even though different forms of masculinities are formed in relation to public spaces (Kimmel 2008).

Integrating a feminist perspective in academia, policy and planning is not without challenges. Feminist theory and policy development draw on political traditions from both the left and the social liberals, which leads to different understanding of power relations in society. Feminist interventions have been fundamental to public planning in welfare societies, and today gender equality and gender mainstreaming in planning have become integral to the process, though to various degrees in different places. At the United Nations Conference on Women in Beijing 1995, and at Habitat II in Istanbul 1996, women's rights in relation to housing, transport and safe cities were put on the agenda. In Europe, the Council of Europe (1998) was central to the conceptualization of gender mainstreaming, which is defined as the process of changing policy routines to promote equality between women and men and to combat discrimination. Today, several municipalities have signed the EU charter for Equality of Women and Men in Local Life by the Council of European Municipalities and Regions (CEMR), which is a tool for integrating a gender perspective into political decision-making and practical activities. The impact on policymaking by such tools, of course, varies. The Nordic countries are often reported to be the most gender equal in Europe (see, for example, European Institute for Gender Equality, EIGE 2013). Rankings on gender equality are commonly used in branding strategies, and gender mainstreaming tools are given much attention by governments and authorities. Gender mainstreaming is meant to advance women's interests yet has difficulties in offering radical transformations of gendered power relations in urban spaces.

The concept of gender mainstreaming is not unproblematic, either in terms of the utility of the mainstreaming strategy or in the meaning of 'gender' (Eveline and Bacchi 2005), and raises questions about what is included in or excluded from the strategy. Its consensualist approach, leaving out the oppositional politics and the feminist movement, and its technocratic understanding of knowledge and tools risk hindering actual empowerment. If gender equality is reduced to 'inclusion of women' in a continuation of previous policies, without transforming or challenging existing hierarchies, it risks being an empty signifier (Verloo 2005). Officials possessing gender expertise develop courses and checklists, turning gender equality into an administrative and profit-making task, following the logic of neoliberal new public management.

Gender mainstreaming is usually organized within the public sector, while private investments or public–private partnerships rarely engage in gender mainstreaming strategies. This means that large-scale urban development, renewal projects, event architecture and new residential areas often lack gender perspectives. The role of private construction companies and large property owners has been largely ignored by the feminist urban research agenda, with a few exceptions (see, for example, Fainstein 2001; Parker 2017).

Acknowledging the importance of early urban feminist research and activism, this chapter will now turn to the emerging contemporary feminist critique of neoliberal planning practices to illustrate that a feminist agenda for egalitarian urban life is in constant need of revisions in relation to new forms of gendered spatial relations. After an introduction of urban neoliberalization, two examples will be given of how a feminist agenda, gender equality and the category of women are being used to promote economic growth rather than to create a more just society.

A gendered urban neoliberalization

Feminist urban research has only recently engaged with the vast research on urban neoliberalism (for example, Brenner, Peck and Theodore 2010; Hackworth 2007; Leitner et al. 2008). A gendered critique of urban neoliberalism, intersecting with class and race, highlights the reproduction of power relations in new spatial orders. Neoliberal policies reshape cities and seek to improve the economic effectiveness of urban spaces. Neoliberalism assumes that there is a 'market' as a solution to all problems (Peck and Tickell 2002), including social policies. Economic growth is supposed to solve social crises and, in this way, 'economic growth *is* the state's social policy' (Brown 2015, 63–64). The state becomes more like a private company, and private companies start to act more like the public sector through, for example, adopting commitments to certain concepts of sustainability.

In many cities, the entry points for neoliberal policies have been times of crisis, unemployment and de-industrialization. In Western cities, by contrast, planning is located in the public sector and acts in regard to long-term public interests to organize land use, neoliberalism and market mechanisms (Baeten 2012). Sager (2015) describes neoliberal ideologies as a matter of how 'the market should discipline politics'. This goes firmly against a traditional Social-Democratic view that 'politics should discipline the market'. In the process, the role and influence of urban planning is effectively transformed into a process of neoliberalization, even in traditionally strong welfare societies (Listerborn 2017; Sager 2011). The anticipated economic boost is expected to 'trickle down naturally' and also be of benefit to the lower classes and be sufficient to maintain a welfare society. This seldom works (Holgersen and Baeten 2017). Today, private stakeholders are financing and building the majority of new housing and services, while the public sector remains responsible for public spaces, where democracy is given symbolic value. 'Gender-equal' spaces, or safety measures, are examples of such municipal investments. Brown's (2015) thesis that the logics of economic growth has integrated democracy into capitalism in a way that has deprived democracy of its critical potential seems appropriate, in this context. Neoliberal planning and urban development risk intensifying and/or creating injustices and inequality, not least through the privatization of planning and dismantling the public sector, in which the majority of women are employed. New planning paradigms, like smart cities for example, are promoted by large private companies like IBM and Siemens, and are not particularly attentive to gender inequalities (Sangiuliano 2014).

Gendered, racialized and classed power relations existed before neoliberalization processes. Patriarchal, colonial, racist and capitalist practices and dominations follow historical paths. The feminist challenge is to follow and trace such paths and to be attentive to continuity, change, contradictions and contestations. In the process of a neoliberal urban development – from industrial to de-industrial, from a working city to a consumption and leisure city, for attracting visitors, businesses and the creative class – changing gender roles play a central role. In times when women are no longer primarily associated with the domestic sphere but have gained space in the labour market, gender will affect urban development. This should not be conflated with gender equality because, as I discuss in the remainder of this chapter, it can lead to women being used to gentrify the city and attract the creative class, and to women's movements being co-opted in the name of urban safety.

Gender and urban renewal

Studies of urban neoliberalism have been surprisingly inattentive to gender, even though dominating urban agendas like attracting the 'creative class' feature gendered dimensions. In her study of Milwaukee, Brenda Parker concludes that 'creative-class discourses bolstered neoliberal rationalities and were embodied and embraced by the elite, in part because they aligned with and affirmed extant raced, gendered and classed power relations' (Parker 2017, 167). The ideal creative type is a mobile, autonomous, flexible and hypercapitalist worker and, to conform with this ideal, men and women have to accept neoliberal ideals and new forms of insecurities. 'The highest paying industries remained most easily occupied by white, male, elite heterosexual subjects with limited social reproduction responsibilities' (Parker 2017, 168). Care work is seldom part of such ideologies of creativity, which indicates both a limited interpretation of the concept of creativity and the low appreciation of reproduction labour.

Regardless of its actual outcomes, the celebration of diversity is a large part of the creative class discourse. To increase the role of diversity, Richard Florida, the main promotor of the creative class, made use of the Gay Index introduced by demographer Gary Gates in the early 2000s to tout the prediction of the regional success of high-tech industries. In contemporary urban development,

> one can celebrate diversity and cast tolerance as a new investment strategy at the same time as one assails those very features by naming the acceptance of people of color, transgender women, and people of low income as 'liabilities' of a neighborhood.
>
> *Hanhardt 2013, 187*

Women, overtly or implicitly functioning as a category representing the middle classes and family, can be part of such urban renewal strategies. Van den Berg (2017) introduces the concept of 'genderfication', resonating with gentrification, as the production of space for post-Fordist gender relations in which women and families with children are the new attractive urbanites for cities. Her case study from Rotterdam shows how the city has used femininity as its marketing strategy. In 2013, the city's alderman stated that 'Rotterdam needs tits', building on the previous mayor's 2008 plea for a more 'round' and 'breasted' city Rotterdam (Van den Berg 2017, 31). The urban transformation process focused on making the city more attractive to women and families, replacing the traditional (migrant) male working class. Through urban programmes like the City Lounge, women could join in public yoga, and at the same time a strict 'ban on gathering' was introduced to displace any loitering men. These strategies accompany plans to replace 20,000 affordable homes with 36,000 properties for middle- and upper-income households. In short, 'women' and the marketing of heterosexual, reproductive femininity were used to cleanse the city of its working-class history and reinforce a gentrification process.

Though rooted in places like Rotterdam, gentrification is a global urban process that produces an unequal city in relation to its resources. As Curran (2018) states, such processes are gendered yet largely under-researched, thus reinforce patriarchal practices. Historically, women were early urbanites and experienced the city as a site of freedom of expression that provided empowering employment opportunities (Wilson 1992). At the same time, women and single parents are often victims of poverty in these gentrifying urban processes, which increase the cost of living, narrow the housing choice, make social reproduction more expensive and limit the scope of democratic influence (Curran 2018). Consequences of gentrification, such as evictions and displacement or 'pressure of displacement' (Marcuse 1985) reinforce the existing power relations. The perspectives of tenants who stay put throughout urban renewal processes remain understudied (Shaw and Hagemans 2015). The social cost of both displacement and remaining are high, not least for the elderly and single parents. Social networks, child support and friendships in the local communities are broken (Pull and Richard 2019).

The neglect, lack of maintenance, increased territorial stigmatization and threat of displacement due to increased rents have been part of a growing local movement of resistance among residents impacted by gentrification. bell hooks (1981) pinpointed the importance of understanding such *sites* of resistance by conceptualizing the 'homeplaces' as sources of self-dignity, agency and solidarity in which – and from which – resistance can be organized and theorized. In hooks' writings, the private spheres transgresses into the public. Through private coordination and communication, resistance movements are formed. The concept of homeplaces indicates a site of comfort, safety and grounding, but also of dignity. The homeplace, in the case of housing, is the meaningful site of resistance and space appropriation, while the claims that the movements raise go beyond neighbourhood territories to create alliances between different areas and cities. To claim the right to dignity is a resistance to the territorial and bodily stigma that people experience in so-called 'problem areas'.

Feminist urban research maintains that neoliberalization is reasserting masculinity, while women's burdens of home, care and justice work have increased. Nonetheless, women and femininity may be used in changing a city's branding. The processes of urban revanchism and gentrification are closely related to neoliberal safety work, the topic that I turn to next, and where a long-standing feminist claim has become a tool in the hand of urban developers.

The neoliberalization of safety work

The fear of sexual assault is an important issue for women's groups around the world, which also has had an impact on authorities at all levels. Safety work is organized on supra-national levels, like the Safe Cities global initiative, which is a partnership of municipal governments, local communities, organisations and United Nations (UN) focused on the situation for women and girls. Several cities around the world are setting up safety work through conducting diagnostic data studies, engaging community members and improving the lighting and design of streets and buildings, training and sensitizing police, and recruiting more women police officers. UN Women is partnering Microsoft, for example, to find ways to use mobile technology to stop sexual harassment and violence in public spaces by training women to use their smartphones to map safety risks, such as 'faulty infrastructure or services, obscured walking routes, and lack of lighting' ('Making Safe Cities for Women and Girls', 2013). In many cities, safety work is part of gender mainstreaming planning.

When safety work becomes a part of the mainstream agenda, it also risks being depoliticized. In a neoliberal urban context, safety and freedom from violence may become a commodity when issues of fear and safety are co-opted by authorities, property companies and other stakeholders. Political and activist history is replaced by technocratic solutions that do not unsettle any gendered power relations. Feminist groups have organized marches in Western cities to 'Take Back the Night' in protests against sexual threats in public spaces since the 1970s. Inspired by the radical feminist Andrea Dworkin, the first march was initiated in 1978 in the US, and similar marches are still being organized all around the world, representing women's claims to space and their assertions of a right to the city. Global estimates published by the World Health Organization indicate that about one in three (35%) of women worldwide have experienced either physical and/or sexual intimate partner violence or non-partner sexual violence in their lifetime. The #metoo movement confirms such statistics. But, transformed from an issue of freedom from sexual harassment and a cry for empowerment, safety is now a tool for business (not the least for smart-city related businesses) or a way to increase social control, surveillance and security. Within this new 'safety discourse' that has gained an important status in planning and urban renewal, women and their fear of sexual violence are used as an argument for urban renewal projects. The category of 'women' is limited to White middle-class women living in urban areas, excluding the fear of violence experienced by women outside of this privileged category (Listerborn 2016).

One example of how safety work has been transformed from a feminist intervention to a neoliberal branding strategy is the case of Toronto, which in the early 1980s enhanced its status by focusing on urban safety from a gender perspective. A committed women's movement, including feminist researchers, put the issue of fear of sexual violence on the political agenda. In the aftermath of the killing of 14 female engineering students in Montreal in 1989,[1] a debate arose on the lack of a gender perspective in contemporary crime prevention. This galvanized the safety work already initiated, further boosting public support (Whitzman 1992) and spurring the analysis and rethinking of urban design, transport systems, public spaces and education programmes for the police. Safety audits were introduced, an example of a grassroots-based planning tool (Wekerle and Whitzman 1995). The ambition of the women's groups was to empower women and to engage the most vulnerable groups for safety efforts, and the activists and the researchers involved were experienced in community development work. In the run-up to elections, the women's group organized hearings where, for example, sexual violence was discussed with politicians (Whitzman 1992). This ambitious and strategic work was undoubtedly based on an understanding of gendered power relations, and it inspired other women's groups, including in Europe. Overall, safety issues acquired a political resonance and became a marketing tool for the City of Toronto.

In the early 2000s, safety work was evident in the urban design and transport system in Toronto, based on the work by feminist activists and researchers, but the new municipal government had cut its budget for work against sexual violence, reduced the women's groups' influence on safety work and developed a parallel strategy focusing on traditional situational crime prevention and a 'hard-target' approach (Listerborn 2002). Embedded in these new political and economic programmes, the gender aspects had become merely cosmetic. In the following year, surveillance and hard-target security were strengthened in most Western countries. In the aftermath of the World Trade Center attacks in 2001 and in the context of the 'War on Terrorism', 'the local level, the focus on community crime prevention in urban areas that emerged in the 1990s … has also been appropriated for the antiterrorism project' (Wekerle and Jackson 2005, 43). Safety also became a convenient tool to upgrade areas and market new developments. With the turnaround of New York City from its criminal reputation in the 1980s to a tourist attraction today, such safety work was essential. The format of BIDs (Business Improvement Districts) and police measures reshaped the social fabric of the city, leading to gentrification (Smith 1996).

To return to Toronto, Leslie Kearn and Beverly Mullings (2013) analysed a wave of apartment block development in the mid-1990s that lasted until 2008. Based on the global, all-encompassing idea of inevitable progress towards a culture- and knowledge-based economy, the City of Toronto sought to boost urban density through high-rise buildings and investment in 'disinvested' areas to include them in 'better' use. Apartment blocks in this project were 'aggressively marketed towards women, particularly young and single women' (Kearn and Mullings 2013, 32), framed within a discourse of women's liberation and urban space as affording emancipation for women. Safety aspects were highlighted as selling points, as some of the new buildings were located in low-income and racially stigmatized neighbourhoods. Their locations were presented as attractive financial investments that would increase in value over the years, and the new buildings were lent a sense of urban authenticity through the presence of marginalized people. Fear and danger were mitigated by private security mechanisms and potential wealth accumulation. For the property companies:

> the extensive marketing of security features is not simply a way to attract women buyers; it is also a way for developers to colonise areas of the city that would have been deemed too risky for such investment because of social stigma and fear of crime.
>
> *Kearn and Mullings 2013, 33*

Borrowing the rhetoric from earlier feminist work on women's fear and independent access to the city, property companies earned new credentials yet simultaneously contributed to the displacement of marginalized groups from various city zones to the periphery. Low-income women, single mothers, recent immigrants and people with disabilities do not belong to this group of young and mobile urban professionals.

Safety work, initiated by women's groups, was transformed from a feminist strategy that challenged gendered power relations to a marketing and branding tool for real estate companies and the city. The underlying power relations of fear identified by feminists are still relevant but are used for purposes of capital investments. The claim for safety and security; that is, freedom from sexual and racist violence, has been repackaged and has entered the urban development market to become a central selling point, and sold to women as a market group. Comparable investments in transit, services and safety are not made in the districts where the marginalized people live; poverty-stricken residents of marginalized neighbourhoods both are exposed to more crime and experience more fear in their everyday lives than others (Estrada and Nilsson 2011). In addition, social movements and urban activists working against violence and fear are themselves subjected to violence and fear (Wekerle and Jackson 2005). When pronounced inequality is based on gender, race or religion, activists are often subjected to violent threats from antagonists, and anti-feminist agendas act upon notions of safety and connect these to a nationalist and racist agenda (Sager and Mulinari 2017). Pinpointing the unequal spatial distribution of fear and sexist and racist violence must be part of an overall notion of the right to the city and radical equality, but this needs to be revitalized and redefined in times of neoliberal urban development.

Conclusion

Feminist urbanites, activists, politicians, planners, architects and researchers have made progress, at the same time recurrently facing comparable challenges. Gender mainstreaming has been important to gain gender awareness in policy development, while contributing to depoliticize and incorporate gender issues into an agenda of technocratic new public management. As such, an increased gender awareness and focus on women in urban development may have negative long-term consequences. As Van den Berg writes:

> genderfication may seem to be cause for celebration. After decades of feminist urban studies, Jane Jacobs is the hero of the day in planning discourse, planners are thinking about more gender-equal cities and child-friendly cities are increasingly a popular adagio in public policy circles.
>
> *2017, 107*

But instead of leading to a less sexist and ageist city, Van den Berg wants to shed light on the risk of the use of gender equality and femininity as an instrument for class upgrading and the creation of profit. When gender equality and women's safety become a non-threatening entity, they are easily co-opted by commercial interests and there is a risk of reinforcing the existing power relations of class, ethnicity and ability. A radical transformation to a more gender-equal city cannot be attentive only to White middle-class women.

For contemporary feminist urban scholars, this is a highly important warning and at the same time a call for more research to understand more deeply how gender (both femininities and masculinities) is manifested in planning and urban development at different times and in different spaces of urban restructuring. Gendered issues, like safety, are being highlighted, while female poverty and racialized issues are being downplayed. The visions for the good city, based on a capitalist, patriarchal and colonial ground, will not be able to include the feminist call for a non-sexist and non-racist city. Remember the words of Audre Lord: 'The master's tools will never dismantle the master's house' (Lorde 2001).

Note

1 In the 'Montreal Massacre' of 1989, 14 women died when a gunman, Marc Lépine, staged his attack at Montreal's École Polytechnique on 6 December. Armed with a rifle, he stormed into a classroom, separated the men from the women and declared that he hated feminists before opening fire on the women. The shooting sparked a national debate about gun control and violence against women. In 1991, Parliament declared 6 December the National Day of Action and Remembrance on Violence Against Women (Source: www.cbc.ca/news/canada/montreal/montreal-massacre-events-mark-1989-shootings-1.791903).

Key readings

Curran, W. 2018. *Gender and Gentrification*. New York: Routledge.
Domosh, M., and J. Seager. 2001. *Putting Women in Place*. New York: Guilford Press.
Peake, L., and M. Rieker, eds. 2013. *Rethinking Feminist Interventions into the Urban*. London: Routledge.

References

Baeten, G. 2012. "Neo-liberal Planning: Does it Really Exist?" In: *Contradictions of Neo-liberal Planning. Cities, Policies, and Politics*, edited by T. Tasan-Kok and G. Baeten. Heidelberg: Springer.
Brenner, N., J. Peck, and N. Theodore. 2010. "Variegated Neoliberalization: Geographies, Modalities, Pathways." *Global Networks* 10: 182–222. doi: 10.1111/j.1471-0374.2009.00277.x.
Bondi, L., and Domosh, M. 1998. "On the Contours of Public Space: The Tale of Three Women." *Antipode* 30 (3): 279–289.
Brown, W. 2015. *Undoing the Demos*. New York: Zone Books.
Council of Europe. 1998. "What is Gender Mainstreaming?" Gender Equality. Available at: www.coe.int/en/web/genderequality/what-is-gender-mainstreaming.
Curran, W. 2018. *Gender and Gentrification*. London: Routledge.
Domosh, M., and J. Seager. 2001. *Putting Women in Place*. New York: Guilford Press.
Estrada F., and A. Nilsson. 2011. "Fattigdom, segregation och brott" ("Poverty, Segregation and Crime." In: *Utanförskap* (*Outsiderness*), edited by S. Alm et al. Stockholm: Dialogos and Institute for Futures Studies.
European Institute for Gender Equality, EIGE 2013. "Annual Report." Available at: www.eige.europa.eu.
Eveline, J, and C. Bacchi. 2005. "What Are We Mainstreaming When We Mainstream Gender?" *International Feminist Journal of Politics* 7 (4): 496–512.
Fainstein, S. 2001. *The City Builders: Property Development in New York and London, 1980–2000*. Lawrence: University Press of Kansas.
Fainstein, S., and L. Servon, eds. 2005. *Gender and Planning: A Reader*. New Brundwick, New Jersey and London: Rutgers University Press.
Hackworth, J. 2007. *The Neoliberal City: Governance, Ideology, and Development in American Urbanism*. New York: Cornell University Press.
Hanhardt, C. 2013. *Safe City: Gay Neighborhood History and the Politics of Violence*. Durham, NC: Duke University Press.
Hayden, D. 1980. "What Would a Nonsexist City Be Like?" *Signs* 5 (3).
Hayden, D. 1981. *The Grand Domestic Revolution: A History of Feminist Designs for American Homes, Neighborhoods, and Cities*. Cambridge, MA: MIT Press.

Hayden, D., and G. Wright. 1976. "Architecture and Urban Planning." *Signs* 1 (4): 923–933.
Holgersen, S., and G. Baeten. 2017. "Beyond a Liberal Critique of 'Trickle Down': Urban Planning in the City of Malmö." *International Journal of Urban and Regional Research* 40: 1170–1185.
hooks, b. 1981. *Ain't I a Woman?: Black Women and Feminism*. London: Pluto Classics.
Kearn, L, and B. Mullings. 2013. "Urban Neoliberalism and Violence." In: *Rethinking Feminist Interventions into the Urban*, edited by L. Peake and M. Rieker. London: Routledge.
Kimmel, M. 2008. *Guyland: The Perilous World Where Boys Become Men*. New York: Harper.
Leitner, H., E. Sheppard, and K.M. Sziarto. 2008. "The Spatialities of Contentious Politics." *Transactions of the Institute of British Geographers* 33: 157–172.
Listerborn, C. 2002. "Trygg stad. Diskurser om kvinnors rädsla i forskning, policyutveckling och lokal praktik" ("Safe City: Discourses on Women's Fear in Research, Policy Development and Local Practice." Dissertation, Chalmers University of Technology, Gothenburg.
Listerborn, C. 2016. "Feminist Struggle over Urban Safety and the Politics of Space." *European Journal of Women Studies* 23 (3): 251–264.
Listerborn, C. 2017. "The Flagship Concept of 'the 4th Urban Environment'. Branding and Visioning in Malmö, Sweden." *Planning, Theory and Practice* 18 (1): 11–33.
Lorde, A. 2001. "The Master's Tool Will Never Dismantle the Master's House." In: *Feminism and Race*, edited by K.K. Bhavnani, 89–92. New York: Oxford University Press.
"Making Safe Cities for Women and Girls." 2013. *The Guardian*, 21 February. Available at: www.theguardian.com/global-development/2013/feb/21/making-cities-safe-women-girls (accessed 7 December 2017).
Marcuse, P. 1985. "Gentrification, Abandonment, and Displacement: Connections, Causes, and Policy Responses in New York City." *Washington University Journal of Urban and Contemporary Law* 28: 195.
Pain, R. 1991. "Space, Sexual Violence and Social Control: Integrating Geographical and Feminist Analyses of Women's Fear of Crime." *Progress in Human Geography* 15: 415–431.
Pain, R. 2001. "Gender, Race, Age and Fear in the City." *Urban Studies* 38: 899–913.
Parker, B. 2017. *Masculinities and Markets: Raced and Gendered Urban Politics in Milwaukee*. Geographies of Justice and Social Transformation Service. Athens: University of Georgia Press.
Peck, J., and Tickell, A. 2002. "Neo-liberalizing Space." In: *Spaces of Neo-liberalism. Urban Restructuring in North America and Western Europe,* edited by N. Brenner and N. Theodore, 33–57. Malden, MA: Blackwell.
Pull, E., and Å. Richard. 2019. "Domicide: Dis*placement* and Dispossessions in Uppsala, Sweden." *Social and Cultural Geography*. doi: 10.1080/14649365.2019.1601245.
Sager, T. 2011. Neo-liberal Urban Planning Policies: A Literature Survey 1990–2010." *Progress in Planning* 76, 147–199.
Sager, T. 2015. "Ideological Traces in Plans for Compact Cities: Is Neo-Liberalism Hegemonic?" *Planning Theory* 14: 268–295.
Sager, M., and D. Mulinari. 2017. "Safety for Whom? Exploring Femonationalism and Care-racism in Sweden." *Women's Studies International Forum*. Available at: https://doi.org/10.1016/j.wsif.2017.12.002.
Sandercock, L. 1998. *Towards Cosmopolis: Planning for Multicultural Cities*. New York: Wiley.
Sangiuliano, M. 2014. "Smart Cities and Gender: Main Arguments and Dimensions for a Promising Research and Policy Development Area." Working Paper. WRGS. Available at: www.ohchr.org/Documents/Issues/Women/WRGS/GenderDigital/MariaSangiuliano.pdf.
Shaw, K.S., and I.W. Hagemans. 2015. "'Gentrification Without Displacement' and the Consequent Loss of Place: The Effects of Class Transition on Low income Residents of Secure Housing in Gentrifying Areas." *International Journal of Urban and Regional Research*, 1–19. Available at: https://doi.org/10.1111/1468-2427.12164.
Smith, N. 1996. *New Urban Frontier: Gentrification and the Revanchist City*. London: Routledge.
Spain, D. 2001. *How Women Saved the City*. Minneapolis: University of Minnesota Press.
Valentine, G. 1989. "The Geography of Women's Fear." *Area* 21: 385–390.
Van den Berg, M. 2017. *Gender in the Post-Fordist Urban. The Gender Revolution in Planning and Public Policy*. Cham: Palgrave Macmillan.
Verloo, M. 2005. "Displacement and Empowerment: Reflections on the Concept and Practice of the Council of Europe Approach to Gender Mainstreaming and Gender Equality." *Social Politics: International Studies in Gender, State, and Society* 12 (3): 344–365.
Vestbro, D.U. 2010. *Living Together – Cohousing Ideas and Realities Around the World*. Stockholm: KTH.
Weisman, L.K. 1994. *Discrimination by Design: A Feminist Critique of the Man-made Environment*. Chicago: University of Illinois Press.

Wekerle, G.R. 1980. "Women in the Urban Environment." *Signs* 5 (3): 188–214.
Wekerle, G.R., and C. Whitzman. 1995. *Safer Cities: Guidelines for Planning, Design and Management.* New York: Van Nostrand-Reinhold.
Wekerle, G.R., and P.S.B. Jackson. 2005. "Urbanizing the Security Agenda: Analysis of Urban Trends, Culture, Theory, Policy, Action." *City* 9 (1): 33–49. doi:10.1080/13604810500050228.
Whitzman, C. 1992. "Taking Back Planning: Promoting Women's Safety in Public Places – The Toronto Experience." *Journal of Architectural and Planning Research* 9 (2): 169–179.
Wilson, E. 1992. *The Sphinx in the City: Urban Life, the Control of Disorder and Women.* London: University of California.

18

GENDER AND SEXUALITY IN PARTICIPATORY PLANNING IN ISRAEL

A journey between discourses

Tovi Fenster and Chen Misgav

Introduction

The chapter presents the discourse changes in Israeli geography research on gender and sexuality. Moving over three decades (1990s–2018), the chapter shows how the theoretical concepts, methodologies and experiences of participatory planning processes have changed both in identity-related focus (from women to gender and later to LGBTQ identities) and in geographical focus (from the peripheries to the city).

This change in discourse in feminist geographies takes place within the Israeli socio-political context of a (Jewish) hegemonic nation state and dominant Zionist–security/militarism circumstances, which shapes gender relations and gender writings in Israeli academia (Berkovitch 1999; Herzog 2011). Together, these growing neoliberal and neo-colonial influences create new gender subjectivities and citizenry patterns and reorganize social and cultural relations (Sasson-Levy and Misgav 2017).

In reviewing the change in discourse in Israeli geography and especially in participatory planning, we use three theoretical concepts: *women's* forbidden and permitted spaces; *gendered* civic capacity; and *LGBTQ* safe urban spaces. We will work on three case studies. Each represents a different time/space gendered and feminist planning discourse and different identity perceptions.

The first, *women's* forbidden and permitted spaces, represents the gendered and geographies discourses in Israeli participatory planning of the 1990s regarding patriarchal perspectives of planning Bedouin settlements in the Negev, in the south of Israel, and their effects on Bedouin *women* only. The second, *gendered* civic capacity, represents later gendered geographies discourses and identities developed in the mid 2000s, referring to lower-class Jewish *women's and men's* participation in neighbourhood regeneration project in Bat-Yam, a suburb in southern Tel Aviv. The gendered civic capacity discourse represents a more egalitarian approach to participatory planning, viewing *women and men* as part of the participatory planning process. The last discourse change is represented by the notion of *LGBTQ* safe urban spaces. It points to the shift in Israeli gendered geographies of the late 2000s, introducing LGBTQ identities, as reflected in the case of the Tel Aviv's LGBTQ community and the municipal project of the Tel Aviv's gay centre. The

LGBTQ safe urban spaces show how the 'participatory planning' process around the opening of the gay centre in Tel Aviv might seem to be a product of a new LGBTQ-friendly politics adopted by the city authorities and mayor for mainly economic and touristic reasons, yet in fact this change followed long and persistent LGBTQ grassroots activism, tangible changes in the power relations and a discourse between the community and the political levels. The discussion on LGBTQ people and geographies helps to expose the discourse changes within Israeli gender scholarship and the planning paradigm in including sexuality issues as a main category.

This is followed by a long discussion on the relationships between feminist and queer geographies and the ways in which queer geographies refer to issues of intersectionality (Binnie and Valentine 1999; Knopp 2007; Misgav 2015a). As Brown (2012, 2) argues, it is 'unsurprising that sexualities' intersection with gender has been the most consistent trend in the literature'. Some scholars place the connection between gender and queer scholarship within the participatory planning paradigm (Fincher and Iveson 2008; Fincher et al. 2014).

As we argue in this chapter, the three discourses are not totally separate and, in fact, can be interlinked and are relevant to the others as well. However, as we argue, they do represent a *discourse change* over time in Israeli gendered geographies regarding Israeli participatory planning in their different identity perceptions and geographical scales.

Forbidden and permitted gendered spaces: gendered blinded urban planning approach

Forbidden and permitted spaces (Fenster 1999) emphasize the cultural and gendered construction of spaces that were exposed when researching Bedouin women's everyday life in the then-new Bedouin towns planned for them by the Israeli government. Those forbidden and permitted spaces were constructed as a result of cultural gendered norms of women's bodies, modesty, clothing and freedom of movement. As the findings revealed (Fenster 1999), patriarchal perceptions of space as modest/permitted or immodest/forbidden became crucial in terms of reducing women's movement. This created what seemed at the outset to be a paradox in which the modernity project that, in principle, promotes equality and free access to resources turned out to be more restrictive than their traditional (or 'primitive') settlements, without services and infrastructure, before they moved to the modernized planned towns (Fenster 1997). This is because the project failed to recognize the need to treat non-Western cultures differently and to formulate different planning solutions for such cultures.

There was no participation in this project during the planning process, especially not for the women. It emphasizes how the procedural, modernized planning process did not consider cultural constructions of space at all but looked at space only as a material, physical component with no connection to its social and cultural constructions as forbidden or permitted, especially for the Bedouin women. During the research, Bedouin women were interviewed and shared their experiences on forbidden and permitted spaces at the domestic level, the only space in which they felt that inner design is planned and controlled by women, to the neighbourhood level, most of which was planned as cul-de-sacs to avoid undesirable interaction. However, most of the women felt suffocated and lonely because they could not move out of their own neighbourhood. Moreover, the high residential density at the town level made the town, as a whole, a forbidden space for most women. Thus, the large modern parks, like in Rahat, one of the Bedouin towns, which was planned to be located away from the residential areas, were impossible for Bedouin women to use because they were located near the main road and thus they could be spotted when using them.

At the regional level, a fascinating spatial mobility has been explored in which Bedouin women's mobility is very limited within the town, yet other cities, which are large and anonymous, are considered less public in terms of codes of modesty than the neighbourhoods of the town itself. This interestingly demonstrates that, for Bedouin women, there is no direct relationship between distance and boundary formulations of forbidden and permitted spaces. A more distant place may be considered permitted, and therefore allow women more mobility, than a place in close proximity (Fenster 1997, 1999).

This research, focusing on women rather than on gender issues, has been one of the first in Israel to highlight the necessity to engage with women's issues in urban planning. It is true that the research dealt with ethnic minorities' (Bedouin) women's issues, but this has later become a widespread phenomenon of engaging with women's issues in urban planning (see also Fenster 1998 for Ethiopian women; and Fenster 2007 for Jewish religious/secular women). Later, concepts such as comfort, belonging and the commitment of both men and women to their homes and cities have been researched, proposing a better understanding of the meaning of life in the city (Fenster 2004a, 2004b, 2005, 2012, 2013).

Gendered civic capacity in urban regeneration projects: paradoxical moments

Civic capacity is commonly defined as general participation in the public sphere (from voting in elections to volunteering), rendering it as a set of individualistic actions and choices. Several components have been identified as playing a key role in the development of civic capacity: *direct action or the exercise of power toward a weighty change in the community*. Such action is made possible through *increased community organization and sufficient development of coalitions and leadership* (Saegert 2006; Stone 2001, 2006). As already mentioned, it is also promoted by and relies upon *high levels of community social capital*, most notability the reciprocity, trustworthiness and commitment it involves (Saegert 2006). Civic capacity as a strategic outcome of the planning with communities' approach becomes the dominant factor in socially concerned planning processes, especially of disadvantaged communities. In our research (Fenster and Eisenberg 2016), we have suggested adding two components to the articulation of civic capacity: professional knowledge transmission (Fenster and Kulka 2016) and its effects on community/individual power; and opening communication channels with private and municipal agencies to promote community and individual civic capacity. Those components are manifested in the case study project of urban regeneration in Bat Yam.

The participatory research in Bat Yam (Fenster and Eisenberg 2016; Fenster and Misgav 2014) engaged lower-class Jewish *women and men* living in a neighbourhood undergoing a regeneration project. It explores the central role that gender plays across a variety of scales in the development of civic capacity among the residents of a neighbourhood. In this research, a new, gendered reading of civic capacity is proposed, which involves transforming women's and men's ways of thinking and acting out of the ordinary in this regeneration project. The aim has been to update and revise the term 'civic capacity' as related to planning with communities approaches.

Using a performative understanding of gender based on feminist poststructuralist analysis to identity knowledge/power and place, we combined the analysis of community and personal scales, looking at 'paradoxical moments' to understand how the transformation in power relations has taken place and how civic capacity is developed. We did it by conducting a critical analysis of bi-weekly meeting minutes and in-depth interviews held over a three-year period (2010–2013). Our understanding is that this civic capacity development is particularly important for women, who managed to enhance and increase their social capital throughout those years.

It is important to emphasize that our planning with communities work in Meonot-Yam was not gendered from the start. During the first year, we became aware of the growing intensity of women's involvement in the project and thus placed more attention on the gender aspects of the process. Such subsequent introduction of gender aspects is not unique. We became increasingly aware of the gender differences, especially after noticing the growing role that women play in 'conflictive moments', when discussions are harsh and when there is considerable anger and frustration among the residents. We encountered them as paradoxical, complicated, multidimensional, nuanced moments embodying 'paradoxical spaces' of women's survival and resistance in masculinized field of knowledge (Rose 1993). Paradoxical spaces provide a helpful lens for understanding the complex politics surrounding social life (Longhurst 2006). In such complex masculinized environments, women develop 'a political-personal strategy of survival and resistance that is, at the same time, a critical practice and a mode of knowledge' (de Lauretis 1986, 9, in Rose 1993, 138). Paradoxical space is a significant term for participatory planning, as the processes of engaging community members in planning includes many 'paradoxical moments', which result from the different interests and encounters of the women and men involved. Rose (1993) associates paradoxical spaces with the acknowledgment of the Other's differences and explains the confinement of space that women experience in daily life (Mahtani 2001). She examines the particular paradox of occupying these spaces as resulting from specific circumstances of power and identity; thus, the sense of paradoxical space can challenge the exclusions caused by masculinist geography and knowledge (Desbiens 1999). The paradox is, in fact, that of being both included in and excluded from the same space and oscillating back and forth between the two positions. We argue that such moments occurred in the Bat Yam project when women experienced themselves and others as being or becoming insiders/outsiders; included/excluded enabled the development of civic capacity among women, characterized by transformative change of thinking by women and men.

We have investigated these paradoxical moments of gendered civic capacity at the community scale in terms of knowledge power changes, looking at the following planning themes: cleanliness and hygiene; parking; regeneration alternatives; and re-parceling options. We also looked at gendered civic capacity at the personal level, focusing on the construction of social capital by the women involved in the project. Our conclusions related to the community scale, at which we concluded that civic capacity has been developed mainly by transforming women's and men's thinking, especially in paradoxical moments where women acted as insiders/outsiders or the inclusionary/exclusionary at the same time. Women were both the centre and margin, both the suppressed and the leaders, and this paradox is what enhanced the transformation of women's and men's ways of thinking. This transformation is reflected at the personal level, where women reported to have undergone a more significant transformation in their public role and in their self-growth, thus enhancing their civic capacity to the point of being able to engage seriously with various issues concerning regeneration. This was mainly evident in their knowledge of planning matters and their ability to build cooperative relations as part of their increasing social capital.

LGBTQ safe urban spaces: the complex relations between community and state authorities

Since the 1990s, geographers have offered important insights into the relation between queer theory, spatial and geographical theories (Bell and Valentine 1995; Browne, Lim and Brown 2007; Hubbard 2012; Johnston and Longhurst 2010). While there is an extensive literature on geographies of sexuality that draws primarily on feminist and queer theory, urban planning is by comparison notably lacking. After Foucault, pioneering scholars pointed to the ways in which planning is embedded in idea of heterosexuality that operate to silence sexual diversity and to the ways that planning preserves and promotes the heterosexual order and social and family values, while ignoring the needs of LGBTs and how they use the urban space (Forsyth 2001;

Frisch 2002). Others show how planners disregard the possible effects of their decisions on LGBT communities (Doan 2011). With this background in geography and planning scholarship, more attention was paid to urban politics (Andrucki and Elder 2007; Bell and Binnie 2004; Brown 2008; Kenney 2001) and to the creations and constructions of LGBTQ[1] safe spaces, mainly within cities. While some scholars focused on the gender nature of queer safe spaces in different geographical contexts (in North America, see Doan 2008; Doan and Higgins 2011; in Israel, see Hartal 2017), our focus here is on the participatory planning process that aims to create queer space. We discuss the case study of the participatory planning process led by the Tel Aviv municipality and its planning authorities with the LGBTQ community.

This participatory process started at the beginning of the previous decade, with two large-scale surveys that were carried out by the municipality and planning units to identify the community's needs. These two surveys reflect the first formal recognition of the community needs in the city and the resulting redistribution of resources (Misgav 2015b). They also relate to the globalization processes that have weakened the state and its institutions and strengthened municipal autonomy in certain areas, sometimes to the point of acting against national policies, for example regarding the LGBT community (Alfasi and Fenster 2005).

The process of negotiation between the community and the municipality, and the latter's growing autonomy, culminated in the decision to establish a community centre, at the municipality's initiative and with municipal funds. The socio-political context for this process has to do with the significant changes that the Israeli LGBT community has undergone since the late 1980s. Before that time, Israel was a conservative place for LGBTs and, in many ways, excluded them from the public sphere. The turning point was the amendment in 1988 of the penal code that prohibited homosexual intercourse. LGBT identity began being recognized not only legally but also in the media and in other symbolic spaces in Israeli society (Kama 2011; Gross 2015). In the 1990s, Tel Aviv became Israel's gay capital and, unlike the official capital, Jerusalem, it began supporting the LGBT community (Alfasi and Fenster 2005; Kama 2011). The local gay scene flourished and began attracting growing numbers of gay tourists and venues. Tel Aviv also led the way in LGBT municipal representation: in 1998 the first lesbian politician, Michal Eden, was elected to the municipality. Since then, LGBTs were elected to all city councils in Tel Aviv. This development changed both the public visibility and legal status of LGBTs in Israel.

The final outcome of these process the opening of the Tel Aviv's municipal gay centre in early 2008 as a safe urban space for the community (Misgav 2015a, 2015b, 2016b, 2019). The centre is located in a three-storey former school building owned by the municipality at the edge of the park in the heart of the city. It includes an information centre, activity rooms, a performance hall, a coffee shop and a clinic for the LGBT community. It also serves as the headquarters of many community organizations that used to be scattered in different places. About half of the funding comes from the centre's revenues (office rentals, the coffee shop license and minimal participation fees for activities); the other half is provided by the municipality. The model of the Gay Center is unique and innovative on a global scale as, although there are social community centres for the LGBT in many cities in the West, they are mostly operated exclusively by NGOs, with no direct connection to the municipality or any other authority (Andrucki and Elder 2007), even though they often receive some financial support from the municipality. In the Israeli case, the connection between the municipality and the centre is not limited to the use of the municipal building. The budget has been transferred to joint management, which also pays the salaries of the centre's employees. Moreover, the centre was founded by the municipality to serve the community all over the city and even beyond it, unlike some that serve a geographical community (a neighbourhood, for example).

Studying the gay centre as an urban phenomenon, a hub for activism and a safe space for LGBTQ people enables us to question the complex connections between urban planning, urban politics towards LGBT people and the notion of recognition (Andrucki and Elder 2007). In previous research (Misgav 2015b, 2019), the central argument was that, via a 'queer analysis', it is possible to observe how the recognition of the LGBT community is partial and serves only certain privileged groups who can benefit from the new (municipally funded) queer safe space, while excluding and alienating others. This political and urban reality is important to understand since it happens in the most liberal and 'gay-friendly' city in Israel (Kama 2011). This argument joins a growing literature that places more emphasis on categorizing specific groups and why urban planning policies lack the awareness of diversity complexity (Doan 2011, 2015; Misgav 2015b, 2016a, 2016b, 2019; Muller-Myrdahl 2011). It also shows how official recognition of certain LGBTQ activities make them legitimate and deserving of institutional support and excludes activities considered less normative, while distinguishing former queer spaces that served for (mainly sexual) meetings.

At the same time, it reveals the success of Tel Aviv and its Gay Center in reaching tangible results such as providing recognition and safe space in the city centre for various groups within the community, such as transgender people (Misgav 2015a), gay elders (Misgav 2016b), lesbians (Hartal 2015) or youth and, as such, adds value to our understanding of the pros and cons of targeting specific groups in the urban policy process.

The case study of the Tel Aviv Gay Center and the participatory planning process that preceded its opening demonstrates that establishing the centre promoted both distributive and procedural justice and created a new safe space for the LGBTQ community (Misgav 2015b, 2019). Yet it also gave rise to a greater complexity, given the institutionalization dilemma or, in this case, the queer dilemma. This was mainly because the centre, while transforming the relationship between the local (and even national) establishment/state/planning authorities and the LGBTQ community, promoted recognition of an essentialist, mainstream national(ist) and consumer(ist) LGBTQ identity, to the exclusion of other sectors of the community and the safe spaces that served them.

Conclusion

To conclude, reviewing the discourse changes in Israeli feminist geographies exposes the significant contribution of Israeli scholars to the international academic research on this topic. This is reflected specifically in the formulation of new analytical concepts, such as 'forbidden and permitted spaces', gendered comfort/belonging and commitment in urban planning, gendered civic capacity and the contribution of new understanding based on local experience to concepts such as queer safe spaces/queerying planning. These concepts enable the use of different theoretical perspectives and methodological approaches to propose more nuanced and sensitive gendered participatory planning. They also help to expand the discussion around participatory planning to be more inclusive of marginalized people. These tendencies reflect not only changes within the academic discourse but also a deeper social and cultural change within Israeli society and geography.

Note

1 By LGBTQ we refer to anti-normative subjects' position with respect to sexuality. As an inclusive term, it has gradually replaced earlier ones that focused on gays, or gays and lesbians, and refer both as an umbrella term akin to LGBT, and (as the Q for Queer means) as an anti-identity and anti-essentialist political position in the spirit of the queer theory that strives to deconstruct conventional identity definitions and to emphasize the fluidity and subversion of the concept of identities, based on sexual tendency or gender (see Gross 2015; Jagose 1996).

Key readings

Doan, P., ed. 2011. *Queerying Planning: Challenging Heteronormative Assumptions and Reframing Planning Practice.* Farnham: Ashgate.
Fenster, T. 2005. "The Right to the Gendered City: Different Formations of Belonging in Everyday Life." *Journal of Gender Studies* 14 (3): 217–231.
Fincher, R., and K. Ivesont. 2008. *Planning and Diversity in the City.* New York: Palgrave Macmillan.

References

Alfasi, N., and T. Fenster .2005. "A Tale of Two Cities: Jerusalem and Tel Aviv–Jaffa in an age of Globalization." *Cities* 22: 351–363.
Andrucki, M., and G. S. Elder. 2007. "Locating the State in Queer Space: GLBT Non-profit Organizations in Vermont, USA." *Social and Cultural Geography* 8 (1): 89–104.
Bell, D., and J. Binnie. 2004. "Authenticating Queer Space: Citizenship, Urbanism and Governance." *Urban Studies* 41 (9): 1807–1820.
Bell, D., and G. Valentine, eds. 1995. *Mapping Desire: Geographies of Sexualities.* London: Routledge.
Berkovitch, N. (1999). *From Motherhood to Citizenship: Women's Rights and International Organizations.* Baltimore, MD: Johns Hopkins University Press.
Binnie, Jon, and G. Valentine. 1999. "Geographies of Sexuality – a Review of Progress." *Progress in Human Geography* 23 (2):175–187.
Brown, M. 2008. "Working Political Geography Through Social Movements Theory: The Case of Gay and Lesbian Seattle." In: *The SAGE Handbook of Political Geography*, edited by K. Cox, M. Low, and J. Robinson, 285–304. Thousand Oaks, CA: Sage.
Brown, M. 2012. "Gender and Sexuality I: Intersectional Anxieties." *Progress in Human Geography* 36: 541–550.
Browne, K., J. Lim, and G. Brown, eds. 2007. *Geographies of Sexualities: Theory, Practices and Politics.* Farnham: Ashgate.
Desbiens, C. 1999. "Feminism 'in' Geography. Elsewhere, Beyond and the Politics of Paradoxical Space." *Gender, Place and Culture* 6 (2): 179–185.
Doan, P. 2008. "Queers in the American City: Transgendered Perceptions of Urban Space." *Gender, Place and Culture* 14 (1): 57–74.
Doan, P., ed. 2011. *Queerying Planning: Challenging Heteronormative Assumptions and Reframing Planning Practice.* Farnham: Ashgate.
Doan, P., ed. 2015. *Planning and LGBTQ Communities: The Need for Inclusive Queer Spaces.* New York and London: Routledge.
Doan, P.L., and H. Higgins. 2011. "The Demise of Queer Space? Resurgent Gentrification and LGBT Neighborhoods." *Journal of Planning Education and Research* 31 (1): 6–25.
Fenster, T. 1997. "Spaces of Citizenship for the Bedouin in the Israeli Negev." *Progress in Planning* 47 (4): 291–306.
Fenster, T. 1998. "Ethnicity, Citizenship and Gender: Expressions in Space and Planning." *Gender, Place, Culture* 5 (2): 177–189.
Fenster, T. 1999. "Space for Gender: Cultural Roles of the Forbidden and the Permitted." *Environment and Planning D: Society and Space* 17: 227–246.
Fenster, T. 2004a. "Belonging, Memory and the Politics of Planning in Israel." *Social and Cultural Geography* 5 (3): 403–417.
Fenster, T. 2004b. *The Global City and the Holy City – Narratives on Knowledge, Diversity and Planning.* London: Pearson.
Fenster, T. 2005. "The Right to the Gendered City: Different Formations of Belonging in Everyday Life." *Journal of Gender Studies* 14 (3): 217–231.
Fenster, T. 2007. "Between Identity and Urban Governance – Citizenship Dilemmas in the Global City." *Alpaim* 31: 253–272 (in Hebrew).
Fenster, T. 2012. *Whose City Is It? Planning, Knowledge and Everyday life***.** Tel Aviv: Hakibutz Hameuhad Publishers (in Hebrew).
Fenster, T. 2013. "Moving between Addresses: Home and Belonging for Israeli Migrant and Palestinian Indigenous Women over 70." *Home Cultures* 10 (2): 159–188.
Fenster, T., and E. Eisenberg. 2016. "Planning with Communities: Towards Gendered Civic Capacity?" *Gender, Place and Culture* 23 (9): 1254–1269.

Fenster, T., and T. Kulka. 2016. "Whose Knowledge, Whose Power in Urban Regeneration Projects with Communities?" *Geografiska Annaler B* 98: 221–238.
Fenster, T., and C. Misgav. 2014. "Memory and place in participatory planning'" *Journal of Planning Theory and Practice* 15 (3): 349–369.
Fincher, R., and K. Iveson. 2008. *Planning and Diversity in the City.* New York: Palgrave Macmillan.
Fincher, R., K. Iveson, H. Leitner, and V. Preston. 2014. "Planning in the Multicultural City: Celebrating Diversity or Reinforcing Difference?" *Progress in Planning* 92: 1–55.
Forsyth, A. 2001. "Sexuality and Space: Nonconformist Populations and Planning Practice." *Journal of Planning Literature* 15 (3): 339–358.
Frisch, M. 2002. "Planning as a Heterosexist Project." *Journal of Planning Education and Research* 21 (3): 254–266.
Gross, A. 2015. "The Politics of LGBT Rights in Israel and Beyond: Nationality, Normality, and Queer Politics." *Columbia Human Rights Law Review* 46 (2): 81–152.
Hartal, G. 2015. "The Gendered Politics of Absence: Homonationalism and Gendered Power Relations in Tel Aviv's Gay-Center." In: *Lesbian Geographies: Gender, Place and Power,* edited by K. Browne and E. Ferreira, 91–112. London: Ashgate/Gower.
Hartal, G. 2017. "Fragile Subjectivities: Constructing Queer Safe Spaces." *Social & Cultural Geography*: doi 10.1080/14649365.2017.1335877.
Herzog, H. 2011. "NGOization of the Israeli Feminist Movement: Depoliticizing or Redefining Political Spaces?" In: *The Contradictions of Israeli Citizenship: Land, Religion and State,* edited by G. Ben-Porat and B. Turner, 175–197. New York: Routledge.
Hubbard, P. 2012. *Cities and Sexualities.* London: Routledge.
Jagose, A. 1996. *Queer Theory: An Introduction.* New York: New York University Press.
Johnston, L., and R. Longhurst. 2010. *Space, Place and Sex, Geographies of Sexualities.* Ashgate: Rowman & Littlefield.
Kama, A. 2011. "Parading Pridefully into the Mainstream: Gay and Lesbian Immersion in the Civil Core." | In: *The Contradictions of Israeli Citizenship: Land, Religion and State,* edited by G. Ben-Porat and B. Turner, 181–200. London: Routledge.
Kenney, M.R. 2001. *Mapping Gay L.A: The Interaction of Place and Politics.* Philadelphia: Temple University Press.
Knopp, L. 2007. "On the Relationship Between Queer and Feminist Geographies." *Professional Geographer* 59 (1): 47–55.
Longhurst, R. 2006. "Plots, Plants and Paradoxes: Contemporary Domestic Gardens in Aotearoa/New Zealand." *Social & Cultural Geography* 7 (4): 581–593.
Mahtani, M. 2001. "Racial ReMappings: The Potential of Paradoxical Space." *Gender, Place and Culture* 8 (3): 299–305.
Misgav, C. 2015a. "With the Current, Against the Wind: Constructing Spatial Activism and Radical Politics in Tel Aviv's LGBT Community Center." *ACME: An International E-Journal for Critical Geography* 14 (4): 1208–1234.
Misgav, C. 2015b. "Planning, Justice and LGBT Urban Politics: The Case of Tel Aviv's Gay-Center." *Tichnon – Journal of the Israeli Planning Association* (special issue on planning and justice edited by Oren Yiftachel and Rani Mendelbaum 12 (1): 180–195 (in Hebrew).
Misgav, C. 2016a. "Gay-Riatrics: Spatial Politics and Activism of Elderly Gay Men in Tel-Aviv's Gay Community Center." *Gender, Place and Culture: A Journal of Feminist Geography* 23 (11): 1519–1534.
Misgav, C. 2016b. "Some Spatial Politics of Queer-Feminist Research: Personal Reflections from the Field." *Journal of Homosexuality* 63 (5): 719–739.
Misgav, C. 2019. "Planning, Justice and LGBT Urban Politics in Tel-Aviv: A Queer Dilemma." *Documents d'Anàlisi Geogràfica* 65 (3): 1–22.
Muller-Myrdahl, T. 2011. "Queerying Creative Cities." In: *Queerying Planning: Challenging Heteronormative Assumptions a Reframing Planning Practice* edited by P. Doan, 157–168. Farnham: Ashgate.
Rose, G. 1993. *Feminism and Geography: The Limits of Geographical Knowledge.* London: Polity Press.
Saegert, S. 2006. "Building Civic Capacity in Urban Neighborhoods: An Empirically Grounded Anatomy." *Journal of Urban Affairs* 28 (3): 275–294.
Sasson-Levy, O., and C. Misgav. 2017. "Gender Research in Israel: Between Neo-Liberalism and Neo-Colonialism" (in Hebrew). *Megamot* 51 (2): 165–2016.
Stone, C.N. 2001. "Civic Capacity and Urban Education." *Urban Affairs Review* 36 (5): 595–619.
Stone, C.N. 2006. "Power, Reform, and Urban Regime Analysis." *City and Community* 5 (1): 23–38.

19

RURALITY, GEOGRAPHY AND FEMINISM

Troubling relationships

Barbara Pini, Robyn Mayes and Laura Rodriguez Castro

Introduction

In this chapter, we provide a selective rather than a comprehensive review of the literature. The latter has been ably undertaken numerous times (e.g. Bryant and Pini 2011; Grace and Lennie 1998; Gorman-Murray et al. 2013; Ibarra García and Escamilla-Herrera 2016; Little and Panelli 2003; Pini and Whelehan 2015). We examine the literature through the lens of the troubled relationship between rurality, geography and feminism. We give particular emphasis to more recent scholarship, cognizant that it constitutes a dialogue with foundational feminist rural geographies (e.g. Brandth 1994, 1995; Liepins 1998, 2000; Little 2002: Pini 2002, 2003). In mobilizing Judith Butler's (1990) metaphor of trouble, we seek to highlight fractures and resonances among rurality, geography and feminism. Our aim is to show each of these as multifaceted and complex theories, experiences and perspectives that should trouble the others.

The chapter is divided into four parts. In the first, we detail how feminist scholarship has troubled the subdiscipline of rural geography. Feminist incursions into rural geography remain highly circumscribed, with gender typically corralled as a separate line of inquiry. However, we demonstrate the often-unacknowledged ways in which feminism has broadened and deepened rural geography epistemologically and methodologically. In the second part of the chapter, we trouble the relationship between rural feminist geography and other feminist interventions in geography. We detail how feminist rural geographers have challenged urban-centric feminist theorizations and one-dimensional representations of rural women. At the same time, we identify gaps in rural feminist geographic scholarship compared to the empirical foci of broader spatialized feminist research. Following this, in the third part of the chapter we trouble the privileging of White, Western neoliberal feminism in rural geography. In troubling the colonial gaze of feminist rural geography, we introduce decolonial feminisms in the final section of the chapter, demonstrating its political and theoretical potential. Decolonial feminisms are part of the body of work emerging from Latin America on decoloniality that destabilizes the existing colonial matrix of power embedded in the specific histories of ongoing colonization of the Americas (see Mignolo and Walsh 2018). Thus, we show how it has troubled universalist constructions of feminism and rurality, and how its engagement may assist in addressing the troubling relationships between rurality, geography and feminism.

Feminist rural geography: troubling rural geography epistemologically and methodologically

Feminist scholarship – as a multiplicity of intertwined theoretical and methodological projects variously emphasizing and seeking to address unequal relations of power – has broadened and enriched the subject of rural geography in a range of compelling ways. In terms of extending the empirical foci in the study of rural farming families, Haugen, Brandth and Follo (2015), for example, attend to the intimate gender dynamics of family break-up from the perspective of women who have divorced farming partners in the past three years. This study moves empirical work away from ideal(ized) heteronormative rural families to encompass diversity and change, while recognizing women's evolving roles and agency in Western farm life. Importantly, the research identifies a tendency among participants to privilege the farm's survival in the divorce process. This is attributed to the patriarchal structure and discourse of the 'family farm', along with farming community norms, which, in turn, highlights the way that 'local structures and cultures of care and obligation to others' are central to rural Norwegian women's senses of self and the protection of their reputation and good conscience (Haugen, Brandth and Follo 2015: 47). Haugen and Brandth (2015) further explore divorce in farming couples in relation to gendered moralities.

A nuanced illustration of the way in which feminist infused scholarship has enriched rural geography is Riley's (2009) work on the spatial and temporal dynamicity of farm gender relations in the UK. Through methodologically innovative farm life histories, he demonstrates the shifting nature and temporalities of the gendering of farm practices and spaces. In doing so, he adds nuance to our understandings of men's and women's farm identities and their interrelationships. Perhaps most importantly, this work demonstrates women's agency in the production of hybrid feminine identities. Similarly, Asher (2004) takes a territorial approach to explore Afro-Colombian women's activism by reflecting on everyday lived experiences, cultural traditions and ancestral medicine practices. By spatially tracing how women organize along the Caucan River, Asher (2004) explores resistances to neoliberal development, government abandonment and racism in their localities. This study is based on doing geographies of place that unveil spatially local resistances.

Alongside this troubling of understandings of 'the farm' as a stable, if not neutral, focus of rural geography, feminist scholarship has highlighted other elisions deriving from White masculinist privilege in scholarly representations of rurality. Drawing on Whiteness theory and engaging with the work of Australian Indigenous feminist scholars, including the highly influential scholarship of Moreton-Robinson (2000), Ramzan, Pini and Bryant (2009, 442) acknowledge the 'power and privilege' underpinning the subject position of 'white academic feminists', along with the inherently 'situated and partial' nature of knowledge production. In doing so, they demonstrate the inadequacies of understandings of rurality marked by a lack of Indigenous voices. This is evidenced, in one example, in the primacy given to White farmers 'in constructions of rural Australia, despite the integral role that Indigenous people played in the establishment and development of pastoral industries' (Ramzan, Pini and Bryant 2009, 441).

Just as Ramzan, Pini and Bryant's (2009) research deployed a feminist methodology grounded in the notion of 'conversation', Bryant and Livholts (2015) use of a 'memory work' approach is a powerful exemplar of how innovative methods grounded in feminist theorizing have extended the subject of inquiry in rural geography. Drawing on Donna Haraway's notions of situatedness and embodied knowledge, Bryant and Livholts (2015) trouble the taken-for-granted separation of subject and object underpinning rural scholarship (and, indeed, scholarship more widely). Specifically, they draw on their own situated memories to show the 'interconnectivity of self and landscape' (2015, 183) opening up to scrutiny the dynamic interrelations of gendered bodies and rural landscapes. In adopting a memory work approach, they thus highlight the active processes through which embodied memories inform interactions with place. As they argue, adopting such a method 'enables the connecting of personal narratives and experiences with social structures bringing to the fore relations of power and how they impact on body and place' (Bryant and Livholts 2015, 193).

In rural geography, feminist scholarship has further foregrounded complex dynamics of power and place in addressing intersections of normative heterosexuality and senses and experiences of rural locales. Research by Pini, Mayes and Boyer (2013) examining the privileging of rural heterosexuality and the local geographies of containment of non-normative or 'scary' heterosexuality in rural communities has both expanded and troubled understandings of the interrelations of rurality and heterosexuality. Such work has demonstrated how the women employed in Australian rural hotels as 'skimpie' barmaids – those who wear just lingerie while serving behind the bar – are seen to alleviate the risks locally associated with frontier hypermasculine heterosexuality. At the same time, skimpie barmaids can be seen to destabilize gendered stereotypes around fixity and mobility (Boyer, Mayes and Pini 2017). Attention to the contemporary cultural management of rural sexualities illuminates tensions and contradictions informing the ongoing (re)production of contemporary ruralities.

These tensions and contradictions are taken up by Abelson (2016) in interviews with 45 transgender men living in rural areas of the US. As Abelson notes, transgender people are often imagined to reside only in urban areas, and those in rural locations are excluded. However, through an intersectional analysis she demonstrates that transgender men can mobilize their Whiteness as 'a key constitutive element of rural sameness' (2016, 1453), along with working-class heterosexual identities, to forge connections and a sense of belonging. This is enabled in significant part by economic change and rural decline, along with demographic change, so that the boundaries of local inclusion and sameness are extended to maintain racial hierarchies; that is, the terms by which transgender men are included in the rural communities are grounded in the othering of people of colour.

As demonstrated by the above, contemporary feminist rural geography has enriched mainstream rural geography by highlighting embodied dimensions of gendered rural lives. This is also advanced by research on the politics of embodied care work undertaken by Fullagar and O'Brien (2018) and Bryant and Garnhams (2017). In the former, the authors explore rural women's experiences of depression and recovery, departing from individualized biomedical models to employ instead a relational understanding. Fullagar and O'Brien (2018, 15) highlight the stigma that attaches to the bodies of depressed rural women, who are positioned as 'less than good mothers, self-sacrificing wives, and stalwart community volunteers'. Formal spaces of care, if they exist, are often avoided and instead women employ a range of embodied self-care practices, such as gardening and swimming, to facilitate their recovery. In a similar respect to Fullagar and O'Brien (2018), Bryant and Garnham (2017) report on the centrality of discourses of the 'good mother' in shaping how rural women enact care. In this instance, mothers caring for adult disabled children are viewed as embodying special knowledge and intuition in relation to their children, which creates difficulties in accessing the limited formal social care in rural communities so that care continues in the home. The investigations by Fullagar and O'Brien (2018) and Bryant and Garnham (2017) foreground topics of inquiry in feminist geography, namely care work and motherhood. However, as we discuss in the following section, in the main this has not resulted in the inclusion of rural perspectives in the development of these literatures.

Feminist rural geography: troubling urban geography

Rural feminist geographers have troubled feminist theorizing outside of its subdisciplinary field. They have done so by seeking to understand and illuminate what Brandth (2002, 115) has described as the 'poor fit' between rural women and feminism. As Brandth (2002) rightly observes, anxieties about rural women's relationship with feminism are embedded in a number of assumptions, including, for example, a modernist belief in progress and a second-wave feminist view that there is a commonality to women's experience. To address the question of why rural women have not adopted feminism, Brandth (2002, 115) counsels that there is a need to refute these assumptions and, importantly, to undertake empirical investigations that privilege rural women's voices on feminism as potentially 'individual and collective, heterogeneous, contingent, ambivalent, complex, and embedded within various specificities of rural contexts'. This is a task undertaken by Egge and Devine (2015) in an historical analysis of how women in the rural mid-west of the US responded to the national suffrage amendment in the 1970s. The authors reveal that, while rural women have often been portrayed as failed political subjects, they have often been actively involved politically, as they were in this campaign. However, their activism has occurred outside of mainstream and formal spaces, and so has been overlooked. In another examination of more recent feminist activism in rural spaces in the Global North, Leach (2015) charts the fate of women's service organizations in rural Canada in recent decades. Interviews with feminist employees in these services reveal experiences of 'mockery, minimization and misunderstanding' in some conservative rural communities, but interviewees forge connections and alliances with each other as well as with urban-based feminists (Leach 2015, 92). Their very presence in and commitment to their rural clients provides a useful corrective to urban imaginaries of feminism.

Troubling the relationship between feminist rural geography and other feminist geographical scholarship should not be thought of as unidirectional. There is much that feminist rural geographers can learn from engaging in spatialized feminist research outside the subdiscipline. In preparing this review, for example, we have noted the limited engagement with a diversity of feminist theorizing in rural geography. There are some notable exceptions, such as Kimura and Katano's (2014) insightful feminist political ecological study of the experiences of organic farmers in Japan, following the Fukushima Nuclear reactor accident in 2011, and Tuitjer's (2018) new materialist research examining the role of housebuilding in the lives of rural German women.

Kath Browne's (2011) research on 'festigoers' at the Michigan Womyn's Music Festival moves away from the widely used lens of the rural idyll to engage with rural utopias as a means to enable consideration of the 'rural possibilities' afforded to non-hegemonic sexualities. Through a detailed focus on lesbian rural utopias, Browne (2011) shows how rural others create rural spaces as sites of empowerment and resistance. Also indicative of a more diverse engagement in theory is Bjorklund's (2013) use of Sara Ahmed's theory of affect to explore embodiment and belonging for rural young lesbians in contemporary Swedish youth fiction. In another theoretically innovative and important study, Rachael E. Sullivan (2009) uses feminist and queer theory to de-tether the conflation of masculinity and male bodies. She examines how some straight rural women are (mis)read as queer because of their investment in masculinity, while some queer women in the same space are (mis)read as heterosexual. Importantly, she asserts that that masculine femininity is a 'highly valued quality within the "macho" landscape' of the rural area of her study, so it is understandable why some women have considerable investment in such a gender performance (Sullivan 2009, 3). What she offers is not only an insightful analysis of how rural lesbian women creatively read their social space to locate and identify queer women but a nuanced account of enactments of rural masculinity.

Despite the above studies, overall feminist studies in rural geography have not fully exploited the diverse theoretical terrain in gender studies to understand rural gendered lives more comprehensively. This limitation can be illustrated through reference to postfeminism. At present, we know little, if anything, about the constitution of rural masculinities and femininities in the context of postfeminism, a period associated with both a backlash against feminism and a popular(ized) belief that feminist goals have been achieved (McRobbie 2009). The few studies that have been undertaken in this area suggest that there is much to be gained by redressing the urban bias of postfeminist scholarship. While she does not specifically name postfeminism, Little's (2013, 2015) work on women's consumption of fitness holidays and spa retreats in rural spaces explores a number of post-feminist sensibilities. These include women's self-surveillance and monitoring, personal transformation and intensified consumption. She reveals the connection between these postfeminist practices and the imagined natural environment of the rural setting. Elsewhere, Whelehan and Pini (2017) examine postfeminist discourses in three recent women's memoirs about finding romance in the countryside. They reveal the ambivalences and contradictions inherent in postfeminism as it is manifest in the rural setting. The writers present themselves as strong and capable women with rich professional lives in the city. However, they simultaneously construct the urban as a place to be left behind, sometimes with regret, and the move to the country as a cleansing experience, signifying a reframing of life priorities.

Troubling White, Western feminism in rural geography

A foundational critique in the literature emerging from global rural feminisms problematizes its White, Western bias (Gargallo Celentani 2014; Hernández Castillo 2013; Lozano 2016; Marcos 2005; Moreton-Robinson 2000). Notably, these neoliberal feminist narratives and actions have resulted in discourses that position the majority of rural women, around the globe, as 'other', 'exotic' and 'powerless'. Simultaneously, they ignore women's liberation and progress that do not align with a colonial and neoliberal logic. Moreover, within the colonial gaze of feminist rural geography, there is an idea that poor women, rural women and/or women of colour need 'rescuing', by prioritizing the global over the local and legitimizing racism. Therefore, there is a commonality between neoliberal globalization and hegemonic feminist discourses, as both are premised on a notion of a linear pathway to 'progress'.

This commonality is taken up in research on the daily work of *campesina*[1] women in Colombia by Rodriguez Castro et al. (2016). The authors note that a Western reading of the women participants in their study would typically be that they are failed subjects of both feminism and neoliberal globalization. However, they caution against such a reading, despite the gendered division of labour and the challenges of forging an income from small-scale farming. They highlight participants' creativity in resisting agricultural globalization through strategies such as selling farm produce in local networks, their pride in the description of being a *campesina* and their attachment to their rural community and natural surrounds. As such, rural women are viewed as agentic and knowledgeable rather than as in need of saving, as would be the case if seen through the purview of modernist feminist and capitalist/neoliberal orthodoxies.

Despite the sustained and compelling challenges to hegemonic White feminism, there is still a tendency in rural feminist geography to universalize the experience of women. In such work, the 'we' that is invoked is typically unstated but taken for granted to be White, middle-class feminists, and the differences between women (although sometimes acknowledged) are then conveniently obscured or muted. This is evident in a celebratory auto-ethnographic review in January 2017 by Garrett Graddy-Lovelace (2017) of the relationship between agrarian feminism and the so-called unproblematized 'Women's March' in Washington. Despite repeatedly using the language of intersectionality, Graddy-Lovelace (2017, 691) concludes by providing a call to action by asking that 'women rise up – together'. In summoning this undifferentiated social category, she erases oppressions of class, race, disability and sexuality and ignores the material reality that many women benefit from these oppressions. In appropriating the Women's March as the impetus for global feminist action, she simultaneously re-entrenches the centrality not only of White, urban, Western feminism but a particularly US version of this feminism. This highlights the need, which we take up in the following section, for feminist rural geographers to engage more fully with decolonial feminisms.

Decolonial feminisms

White Western neoliberal feminism has been challenged in academic and activist spaces via the emergence of what is termed 'decolonial feminism', as expressed in the work of the *Red de Feminismos Descoloniales* in Mexico (Méndez Torres et al. 2013; Millán et al. 2014). Decolonial feminists seeks to destabilize truths and question the above perspectives on progress and well-being through an engagement with women's localized resistances in their territories and communities. Equally, they call for a feminist agenda that denounces racism and that does not silence or exclude indigenous, *campesina* and Black women (see also Cabnal 2015; Hernández Castillo 2013; Rivera Cusicanqui 2010).

Across the literature that has emerged from decolonial thought, we have identified two main claims about its potential in relation to rural feminist geography. First, it has the capacity to strengthen territorial struggles as a way to contest colonialism and neoliberalism (including patriarchy and racism) by understanding that territory, body and land are intimately related (Gargallo Celentani 2014; Lozano 2016; Suárez Navaz and Hérnandez Castillo 2008). This is the concept of *territorio cuerpo-tierra* (Cabnal 2015). The political importance of this concept lies in the reality that territorial struggles are central to the localized resistances of rural communities to colonial practices and neoliberal globalization (Teubal and Ortega Breña 2009). As land is understood as more than just a physical space, so is the territory, and therefore these spaces have history, feelings, life and memory. Cabnal (2015) argues that recognizing the relationship of territory body-land is essential to resist violence against women's bodies. She notes that 'there are a number of testimonies of resistance: from the grandmothers to the great grandmothers against the different forms of colonial domination, to the contemporary women who place their bodies in the frontline to defend life' (Cabnal 2015, np).

Feminist rural geographers have already shown that the home may be differently experienced by rural women in the Global South and, doing so, reveal the efficacy of a conceptualization of territory body-land. In White, Western feminist geography, the positive connotations of home have often been ignored, yet the home is often where rural women, through their gardens, engage in food sovereignty struggles (Espino et al. 2012). These results have been resonated in the work of feminist rural geographers based in other countries in the Global South (Christie 2006; Robson 2006; Wardrop 2006).

A second way in which decolonial thought may be usefully engaged in feminist rural geography is methodologically. In rural geography, Carolan (2008) and Woods (2010) have called for an engagement in methodologies that explore the spatial relationships of neoliberal globalization by focusing on place, both the everyday and embodied lived experiences. However, decolonial work takes this further, advocating a feeling-thinking approach to research (*sentipensando*) in which the heart and thinking processes are not separate but, rather, entangled in our subjectivities as researchers (Fals Borda 2015; López Intzín 2013; Rodriguez Castro 2017). In the context of feminist rural geography, this involves embodied research in which the study of place and its spatial relationships (both human and non-human) are essential to understanding the countryside. It also requires more communitarian research practices to be committed to render visible the rural actors who have been historically invisible in the mainstream, especially rural women (Espino et al. 2012; Méndez Torres et al. 2013; Riley 2010; Rodriguez Castro 2017). By being attuned to the lived experiences of feeling-thinking, we privilege everyday knowledge in research (Fals Borda 2013, 2015). As a consequence, our emotions, feelings and politics become a site of negotiation, which is a necessary prerequisite to contesting the colonial gaze and to enabling global dialogue.

Conclusion

In this chapter, we have taken up Judith Butler's (1990) notion of trouble to explore the troubling relationship between rurality, geography and feminism. We began by considering the positive ways in which feminists have troubled rural geography epistemologically and methodologically. Despite these achievements, feminist scholarship continues to operate on the margins in rural geography, in that the work by feminists is seldom cited by mainstream geographers. This means that the radical potential of a feminist epistemic lens has not been fully realized by rural geographers. Indeed, a significant body of writing in more recent years has reverted to (or, perhaps, has never moved on from) addressing gender simply by adding it as a variable. Such research is not engaged with the critical political project of feminism that goes beyond the academy in its concern with addressing inequalities of power and advocating for social change.

In the second part of the paper we examined the ways in which feminist rural geography has troubled feminist theorizing more broadly, and vice versa. Again, we concentrated in this review on the constructive conversations that have occurred and should continue to occur as a result of this troubling. What is problematic however, is that rural feminist geography is often overlooked outside of our own subdisciplinary area. It does not seem to find an audience in the larger body of feminist geographical writing, perhaps because it is seen as too niche or particular. This means that the work done by feminists in rural geography is labelled peripheral not just by mainstream rural geographers but also by feminist geographers more broadly.

In the final two sections of the chapter, we troubled the colonial and racial perspective of rural feminist geography. In addition to identifying the criticisms that have been levelled against rural geography by women outside of the narrow confines of the Western world, we engaged with the notion of decolonialism. We contend that this perspective could usefully inform rural feminist geography on two counts. That is, it can both facilitate a deeper and more critically reflexive understanding of the intricate relationship between land, body and territory in women's lives and encourage more embodied, place-based and communitarian research practices.

Note

1 *Campesina* is the widely used and culturally appropriate word to refer to the peasant women of Colombia. It has important historical and political implications in the history of rural resistance in the country.

Key readings

Millán, M., et al., eds. 2014. *Mas Allá del Feminismo: Caminos para Andar.* México, D.F.: Red de Feminismos Descoloniales.

Moreton-Robinson, A. 2000. *Talkin' Up to the White Woman: Aboriginal Women and Feminism.* Queensland: University of Queensland Press.

Pini, B., B. Brandth, and J. Little. 2015. *Feminisms and Ruralities.* Lanham, MD: Lexington.

References

Abelson, M.J. 2016. "'You Aren't From Around Here': Race, Masculinity, and Rural Transgender Men." *Gender, Place and Culture* 23 (11): 1535–1546.

Asher, K. 2004. "Texts in Context: Afro-Colombian Women's Activism in the Pacific Lowlands of Colombia." *Feminist Review* 78: 38–55.

Bjorklund, J. 2013. "Coming Out, Coming In: Geographies of Lesbian Existence in Contemporary Swedish Youth Novels." In: *Sexuality, Rurality and Geography,* edited by A. Gorman-Murray, B. Pini, and L. Bryant, 159–171. Lanham, MD: Lexington Books.

Boyer, K., R. Mayes, and B. Pini. 2017. "Narrations and Practices of Mobility and Immobility in the Maintenance of Gender Dualisms." *Mobilities* 12 (6): 847–860.

Brandth, B. 1994. "Changing Femininity: The Social Construction of Women Farmers in Norway." *Sociologia Ruralis* 34 (2/3): 127–149.

Brandth, B. 1995. "Rural Masculinity in Transition: Gender Images in Tractor Advertisements." *Journal of Rural Studies* 11 (2): 123–133.

Brandth, B. 2002. "On the Relationship between Feminism and Farm Women." *Agriculture and Human Values* 19 (2): 107–117.

Browne, K. 2011. "Beyond Rural Idylls: Imperfect Lesbian Utopias at Michigan Womyn's Music Festival." *Journal of Rural Studies* 27: 13–23.

Bryant, L., and B. Garnham. 2017. "Bounded Choices: The Problematisation of Longterm Care for People Ageing With an Intellectual Disability in Rural Communities." *Journal of Rural Studies* 51: 259–266.

Bryant, L., and M. Livholts. 2015. "Memory Work and Reflexive Gendered Bodies: Examining Rural Landscapes in the Making." In: *Feminisms and Ruralities*, edited by B. Pini, B. Berit, and J. Little, 181–193. Lanham, MD: Lexington.

Bryant, L., and B. Pini, 2011. *Gender and Rurality.* London: Routledge.

Butler, J. 1990. *Gender Trouble.* London: Routledge.

Cabnal, L. 2015. *De las opresiones a las emancipaciones: Mujeres indígenas en defensa del territorio cuerpo-tierra.* Revista Pueblos. Available at: www.revistapueblos.org/?p=18835 (accessed 20 March 2017).

Carolan, M.S. 2008. "More-than-Representational Knowledge/s of the Countryside: How We Think as Bodies." *Sociologia Ruralis* 48 (4): 408–422.

Christie, M.E. 2006. "Kitchenspace: Gendered Territory in Central Mexico." *Gender, Place & Culture* 13 (6): 653–661.

Egge, S., and J. Devine. 2015. "Putting the Community First: Feminism and American Women's Activism in the Twentieth Century." In: *Feminisms and Ruralities,* edited by B. Pini, B. Berit, and J. Little, 15–20. Lanham, MD: Lexington.

Espino, I.W.A., et al. 2012. "Alternatives under Construction in Latin America." *Development* 55 (3): 338–351.

Fals Borda, O. 2013. "Action Research in the Convergence of Disciplines." *Journal of Action Research* 9 (2): 155–167.

Fals Borda, O. 2015. *Una sociología sentipensante para América Latina.* Buenos Aires: CLACSO.

Fullagar, S., and W. O'Brien. 2018. "Rethinking Women's Experiences of Depression and Recovery as Emplacement: Spatiality, Care and Gender Relations in Rural Australia." *Journal of Rural Studies* 58: 12–19.

Gargallo Celentani, F. 2014. *Feminismos desde Abya Yala: Ideas y Propocisiones de las Mujeres de 607 Pueblos en Nuestra América.* Mexico: Editorial Corte y Confección.
Garrett Graddy-Lovelace, G. 2017. "Latent Alliances: The Women's March and Agrarian Feminism as Opportunities of and for Political Ecology." *Gender, Place and Culture* 24 (5): 674–695.
Gorman-Murray, A., B. Pini, and L. Bryant, eds. 2013. *Sexuality, Rurality and Geography.* Lanham, MD: Lexington.
Grace, M., and J. Lennie. 1998. "Constructing and Reconstructing Rural Women in Australia: The Politics of Change, Diversity and Identity." *Sociologia Ruralis* 38 (3): 351–370.
Haugen, M.S., and B. Brandth. 2015. "When Farm Couples Break Up: Gendered Moralities, Gossip and the Fear of Stigmatization in Rural Communities." *Sociologia Ruralis* 55 (2): 228–242.
Haugen, M.S., B. Brandth, and G. Follo. 2015. "Farm, Family and Myself: Farm Women Dealing with Family Break-up." *Gender, Place and Culture* 22 (1): 37–49.
Hernández Castillo, R.A. 2013. "Comentarios a: Mujeres Mayas-Kichwas en la apuesta por la descolonización de los pensamientos y corazones." In: *Senti-pensar el género: perspectivas desde los pueblos originarios*, edited by G. Méndez Torres et al., 16. Guadalajara, Mexico: Red Interdisciplinaria de Investigadores de los Pueblos Indios de México; Red de Feminismos Descoloniales.
Ibarra García, M.V., and I. Escamilla-Herrera, eds. 2016. *Geografías Feministas de Diversas Latitudes: Orígenes, Desarrollo y Temáticas Contemporáneas.* Mexico DF: Universidad Nacional Autónoma de México.
Kimura, A.H., and Y. Katano. 2014. "Farming After the Fukushima Accident: A Feminist Political Ecology Analysis of Organic Agriculture." *Journal of Rural Studies* 34: 108–116.
Leach, B., 2015. "Feminist Connections in and Beyond the Rural." In: *Feminisms and Ruralities*, edited by B. Pini, B. Berit, and J. Little, 81–94. Lanham, MD: Lexington.
Liepins, R. 1998. "The Gendering of Farming and Agricultural Politics: A Matter of Discourse and Power." *Australian Geographer* 29 (3): 371–388.
Liepins, R. 2000. "Making Men: The Construction and Representation of Agriculture-based Masculinities in Australia and New Zealand." *Rural Sociology* 65 (4): 605–620.
Little, J. 2002. *Gender and Rural Geography: Identity, Sexuality and Powering the Countryside.* London: Pearson.
Little, J. 2013. "Pampering, Well-being and Women's Bodies in the Therapeutic Spaces of the Spa." *Social and Cultural Geography* 14 (1): 41–58.
Little, J. 2015. "Nature, Wellbeing and the Transformational Self." *Geographical Journal* 181 (2): 121–128.
Little, J., and R. Panelli. 2003. "Gender Research in Rural Geography." *Gender, Place and Culture* 10 (3): 281–289.
López Intzín, S. 2013. *Ich'el ta muk'*: La trama en la construcción del *Lekil kuxlejal* (vida plena-digna-justa). In: *Senti-pensar el género: perspectivas desde los pueblos originarios,* edited by G. Méndez Torres et al., 18–40. Guadalajara, México: Red Interdisciplinaria de Investigadores de los Pueblos Indios de México; Red de Feminismos Descoloniales.
Lozano, B.R. 2016. "Feminismo Negro – Afrocolombiano: Ancestral, insurgente y cimarrón, un feminismo en – lugar." *Insterticios de la Política y la Cultura* 5 (9): 23–48.
Marcos, S., 2005. "The Borders Within: The Indigenous Women's Movement and Feminism in México." In: *Dialogue and Difference: Feminisms Challenge Globalization*, edited by M. Waller and S. Marcos, 81–113. New York: Palgrave Macmillan.
McRobbie, A. 2009. *The Aftermath of Feminism: Gender, Culture and Social Change.* London: Sage.
Méndez Torres, G., et al. 2013. *Senti-pensar el género: perspectivas desde los pueblos originarios.* Guadalajara, México: Red Interdisciplinaria de Investigadores de los Pueblos Indios de México; Red de Feminismos Descoloniales.
Mignolo, W.D., and C. Walsh. 2018. *On Decoloniality: Concepts, Analytics and Praxis.* Durham, NC: Duke University Press.
Millán, M., et al., eds. 2014. *Mas Allá del Feminismo: Caminos Para Andar.* Mexico, DF: Red de Feminismos Descoloniales.
Moreton-Robinson, A. 2000. *Talkin' Up to the White Woman: Aboriginal Women and Feminism.* Queensland: University of Queensland Press.
Pini, B. 2002. "Focus Groups, Feminist Research and Farm Women: Opportunities for Empowerment in Rural Social Research." *Journal of Rural Studies* 18 (3): 330–351.
Pini, B. 2003. "Feminist Methodology and Rural Research: Reflections on a Study of an Australian Agricultural Organisation." *Sociologia Ruralis* 43 (4): 418–433.
Pini, B., R. Mayes, and K. Boyer. 2013. "'Scary' Heterosexualities in a Rural Australian Mining Town." *Journal of Rural Studies* 32, 168–176.

Pini, B., and I. Whelehan. 2015. "The Feminist and the Cowboy: Reading 'An Unlikely Love Story'." In: *Feminisms and Ruralities*, edited by B. Pini, B. Berit, and J. Little, 95–104. Lanham, MD: Lexington.

Ramzan, B., B. Pini, and L. Bryant. 2009. "Experiencing and Writing Indigeneity, Rurality and Gender: Australian Reflections." *Journal of Rural Studies,* 25 (4): 435–443.

Riley, M. 2009. "Bringing the 'Invisible Farmer' into Sharper Focus: Gender Relations and Agricultural Practices in the Peak District (UK)." *Gender, Place and Culture* 16 (6): 665–682.

Riley, M. 2010. "Emplacing the Research Encounter: Exploring Life Histories." *Qualitative Inquiry* 16 (8): 651–662.

Rivera Cusicanqui, S. 2010. *Ch'ixinakax Utxiwa: Una Reflexión sobre prácticas y discursos descolonizadores.* Buenos Aires: Editorial Retazos/Tinta Limón.

Robson, E. 2006. "The 'Kitchen' as Women's Space in Rural Hausaland, Northern Nigeria." *Gender, Place & Culture* 13 (6): 669–676.

Rodriguez Castro, L. 2017. "The Embodied Countryside: Methodological Reflections in Place." *Sociologia Ruralis* 58 (2): 293–231.

Rodriguez Castro, L., B. Pini, and S. Baker. 2016. "The Global Countryside: Peasant Women Negotiating, Recalibrating and Resisting Rural Change in Colombia." *Gender, Place & Culture* 23 (11): 1547–1559.

Suárez Navaz, L., and R.A. Hernández Castillo, eds. 2008. *Descolonizando el Feminismo: Teorías y prácticas desde los márgenes.* Madrid: Catedra.

Sullivan, R.E. 2009. "The (Mis)translation of Masculine Femininity in Rural Space: (Re)reading 'Queer' Women in Northern Ontario, Canada." *Thirdspace: A Journal of Feminist Theory and Culture* 8 (2): 1–19.

Teubal, M., and M. Ortega Breña. 2009. "Agrarian Reform and Social Movements in the Age of Globalization: Latin America at the Dawn of the Twenty-first Century." *Latin American Perspectives* 36 (4): 9–20.

Tuitjer, G. 2018. "'A House of One's Own': The *Eigenheim* Within Rural Women's Biographies." *Journal of Rural Studies* 63: 156–163.

Wardrop, J. 2006. "Private Cooking, Public Eating: Women Street Vendors in South Durban" *Gender, Place & Culture* 13 (6): 677–683.

Whelehan, I. and B. Pini . 2017. "Farm Lit: Reading Narratives of Love on the Land." In: *Cultural Sustainability in Rural Communities: Rethinking Australian Country Towns,* edited by K. Driscoll, K. Darian-Smith, and D. Nichols, 68–83. Abington: Ashgate.

Woods, M. 2010. "Performing Rurality and Practising Rural Geography." *Progress in Human Geography* 34 (6): 835–846.

20
NATIONHOOD
Feminist approaches, emancipatory processes and intersecting identities

Maria Rodó-Zárate

Introduction

Feminist geopolitics, since its inception in early 2000s, has provided an analytical lens that has redefined what counts as geopolitics and has considered the effects of power at multiple scales: from the body to home and nation state (Massaro and Williams 2013). This field of inquiry, as Dixon (2015) argues, draws attention to how the conditions of people's lives across the globe are shaped by political, economic, cultural and environmental factors. Despite the relevance of feminist perspectives in fields such as social, cultural or economic geography, their impacts on political geography have been 'surprisingly few' (Hyndman 2003), influencing the lack of feminist perspectives on geopolitics in general and specifically on issues such as nationhood and nationalism.

Feminist geographers have explored the gendering of nations in a variety of contexts, showing different perspectives built upon work that is empirically grounded. Some examples include the exploration of gendering of cultural nationalism in South Sudan (Faria 2013), the analysis of how a gendered spatiality underlies discourses of nationhood in Ecuador (Radcliffe 1996), how nation state and diaspora are interdependent and shaped by gender ideologies in Singapore and China (Yeoh and Willis 1999), the practice of identity and the Irish nationalist sense of place (Martin 1997), militarism and the 'culture of war' as productive of gendered national identities in Uzbekistan (Koch 2011), hierarchical domains of nation and masculinity as builders of gendered national subjects in India (Sabhlok 2017) and how family reunification redefines the relation of kinship to nation through cross-border marriages between Chinese and Taiwanese (Friedman 2017). Beyond geographical studies, even though 'all nations depend on powerful constructions of gender' (McClintock 1993, 61), it was not until the 1990s that feminist perspectives were incorporated to studies on nations, nationhood and nationalism (Palau 2015). Since then, feminist approaches have been central to understanding the configurations of nations.

In the following sections, I present some feminist perspectives on nation and nationalism, pointing to the sexual and gender dimension of their construction, representation and configuration. Next, I analyse the relations between feminist movements and national projects. Finally,

I focus on the Catalan case in order to show how feminist pro-independence groups can offer new frameworks for thinking about nations and nationalism.

Gendering the nation

Feminist authors have largely analysed the gendering of the nation and of nationalism, showing how women's bodies and movements are controlled, policed and regulated in the nation's interests and maintenance (Rodó-de-Zárate 2019). Nira Yuval-Davis and Floya Anthias, in their work *Women-Nation-State* (1989), focus on five ways in which women are placed in relation to nations, which are generally described as an imagined, socially constructed, political communities that create bonds of belonging and use a wide range of symbols to project a national past and present (Anderson 1983). As the authors contend, first, women have been seen as *biological reproducers* of the nation. This idea has been materialized in the form of pronatalist policies to force the sterilization of certain groups and also systematic rape and sexual enslavement, where women's bodies are taken as a battleground in wars. Thus, being the biological reproducers of nations implies direct regulations and restrictions of their sexual and reproductive rights. Second, women have also been conceived as *social reproducers* of group members and cultural forms, including cultural transmission. Usually, this aspect takes the form of legislation on property inheritance, marriage or the 'battle of the nursery' (Peterson 1994), which also involves the ideological reproduction of group members. Third, women have been conceptualized as *symbolic markers* of the nation, portraying the nation-as-woman metaphor and implying that their bodies are the battleground for group struggles and group identity. Fourth, and beyond the idea of being mere symbols, women are active in conflicts that belong to their communities and are also seen as *participants* in political identity struggles. Yuval-Davis and Anthias (1989) note in this way that women have had relevant roles in their participation in public and political activities, even though they have traditionally been denied participation in them. Fifth, and finally, women are viewed as *members* of society in general. However, nationalist projects treat different groups of women differently, incorporating some of them and excluding others (Yuval-Davis 1993). This makes it also evident that women are not a homogeneous category, given the interrelations of gender with other social divisions such as ethnicity, religion, class, age and sexual orientation. The intersection of these multiple axes of oppression should also be seen as mutually constitutive, not as additive layers of inequality, in order to understand the various positionings of women in every context (Crenshaw 1991).

Feminists have also largely analysed the gendering of the state and of nationalism, rendering visible how women's bodies and movements are controlled, policed and regulated for the nation's interests and maintenance. Nationalism is problematic in relation to conflict between groups in general but also for those within the nation who share least in elite privileges and political representation (Peterson 1994, 77). In this sense, control over access to the benefits of belonging to a nation is invariably gendered (Mayer 2000). As Peterson (2014) argues, nations constitute heteropatriarchal family and household formations as the foundational socio-economic unit of the state and the basis of inheriting property and citizenship claims, showing that modern state-making processes normalize binaries of gender, sexuality and race, and that the regulation of sexuality, property, membership and intimacy has been key to imperial projects.

The places in which women have been situated in relation to nationhood are symptoms of a more profound distinction between the public and private spheres. Nation and nationalism are normally discussed as part of the public political sphere, from which women are excluded (Yuval-Davis 1993). This distinction is grounded on the social contract and the idea that women are located in the private domain, which is not seen as political (Pateman 1988). This conceptualization has relevant implications for the role of women in the different dimensions of public and political action, and thus the dynamics of nationhood, and also for conducting research on this question.

It is also important to notice that the relation between gender and nation is not expressed only through women's situations but in relation to the positioning of masculinity and heteronormativity. As Cynthia Enloe (1990, 45) argues, 'nationalism has typically sprung from masculinized memory, masculinized humiliation and masculinized hope'. Nagel (1998) also shows the strong links that relate nationalism to masculinity, patriotism, militarism and imperialism, and explores the gender gap regarding men and women's different goals and agendas for the nation. Usually, the imagined community has been based on the notion of 'brotherhood' (Anderson 1983), which in turn is fundamental to the symbolical construction of the idealized mother and the rejection of homosexuality, thus allowing the creation of fraternal relations (Ugalde 1996). Actually, as Peterson (1999) defends, the process of state-making and its centralization of political authority and coercive power is inextricably linked to a heterosexist ideology. The author defends the view that heterosexism and gender binary are constitutive of the early and modern Western state-making, given that the heterosexist state denies male homosocial sexuality in favour of male homosocial politics while it denies women's homosexual and political bonding through the public-private dichotomy (Peterson 1999).

Sexuality plays a central role in nation-building and national identity, making it impossible to think about nation as sexless (Mayer 2000). This axis of sexuality is not only fundamental to the creation of national identity as the reinforcement of the heteronormative state but also to the creation of certain dynamics of inclusion of groups of sexual dissidents. While sexual dissidents have not been portrayed as symbols of the nation, there have been movements of an assimilation of their claims. In this sense, there has been a growing body of work that shows how LGBT claims are accommodated in a process of integration and acceptance of the 'proper homosexuals' into neoliberal citizenship, at the expenses of rising Islamophobia and exclusion of the racial 'others'. Jasbir Puar (2007) has defined these dynamics as homonationalism, and it has proved a crucial notion for understanding the relation between the LGBT community, citizenship and the construction of the nation state. The approach from homonationalism allows an analysis of inclusion and exclusion from the nation in relation to LGBT discourses, identifying that Whiteness is reinforced in national projects through the assimilation of the LGBT collective. Puar (2007) also puts forward the idea of sexual exceptionalism as the construction of the nation as better than others in relation to LGBT rights, thus putting in opposition any given nation as Other, often represented as 'Arab' and uncivilized.

Feminisms and national projects

We have seen how nationhood and nationalist projects have used women's bodies and work for their interest, situating women and other minoritized groups in very specific imaginaries, directly excluding them or instrumentalizing these groups for further national purposes. However, nationalism, beyond being state-led when there is assimilation of all within a state, can be state-seeking when there is mobilization for recognition as an independent state (Peterson 1999). In a context of 'globalization, changing sovereignties, proliferation of actors, deterritorialization and space/time compression' (Peterson 1999, 57), the definition of nationalism from a state-centric perspective appears to be problematic. Regarding this, Smith's (1971, 1986) influential typology distinguishes between 'ethnic-genealogical' and 'civic-territorial' nationalisms, as well as between 'ethnocentric' and 'polycentric' nationalism. Ethnocentric nationalists would consider their own nation as superior to that of others, associated with imperialism and the desire to conquer and dominate other nations. Polycentric nationalism does not regard its own nation as superior and would be associated with nationalist movements of colonized nations. Gregory Gleason (1991), in turn, identifies three faces of nationalism: liberation (related to self-determination and democratization); exclusivity (promotion of group homogeneity); and domination (suppressing difference within the group).

The relation between nationalism and feminism is generally seen as problematic with respect to a broader debate in the social sciences regarding the nature of individual as opposed to collective rights and identities within nation state formations (Jacoby 1999). To be concrete, most feminists consider nationalism as detrimental to women and feminism (Seodu Herr 2003). It has been argued by Enloe (1990) that, in nationalist movements and conflict, women are relegated to minor or symbolic roles, and this can be seen as a way of rendering invisible women's actions and contributions to the field. Moreover, the complexity between feminism and nationalism has been obscured by the tendency to subordinate gender justice to national priorities (Vickers 2006). The main tensions and conflicts that have been identified are a permanent accusation that they generate division in the nationalist struggle, the problems facing patriarchal assumptions of male colleagues in such movements or 'now is not the time' as a way of not incorporating feminist demands in the agenda (Vickers 2000).

In a globalized context, Western feminists have generally rejected the potential of nationalism as an emancipatory framework and regarded women participating in nationalist projects as not following the 'authentic' feminist parameters (Hasso 1998, in Jacoby 1999). This results in a situation in which '[f]eminism and nationalism are almost always incompatible ideological positions within the European context' (Kaplan 1997, 3). However, this tendency has been contested by feminist nationalist movements in a variety of socio-economic and political contexts that simultaneously seek rights for women and for nationalists, thus reconstructing the meanings of both nationalism and feminism from their perspective (West 1997). Another way in which feminism and nationalism might approach is through women being more likely to organize to insert feminist goals into national projects is if the project is open, pluralistic and advocates a more 'positive' sex/gender regime than if it is committed to militarism/fascism (Walby 1997). Moreover, women of minority nations experience a greater sense of responsibility to reproduce their national identity. This may predispose them to struggle to liberate 'their' nations and to join such national projects, even if women may support a *different* national project from that of men (Walby 1997). Along the same lines, Amurrio Vélez (2002) argues that the structuring of gender is influenced by nation and shows that women who participate in national projects often direct their commitments in a different way. On the other side of the coin, Lee and Cardinal (1998) argue that those who are part of majority nations cannot see themselves as 'nationalistic', precisely because members of dominant nations within a nation state understand their nationality in terms of common citizenship – a logic that McRoberts (2004) describes as collective amnesia.

In relation to 'Third World feminists', Ranjoo Seodu Herr (2003) states that, even if most of them also consider nationalism as detrimental to their goals, there have also been collaborations between feminists and nationalists in their pursuit of nation independence from a colonial power. In this sense, Jayawardena (1986) shows that in Asia and in other colonized nations feminism was compatible with the modernizing dynamic of anti-imperialist national liberation movements. Jacoby (1999) argues that nationalist women in non-Western contexts become politically active through their participation and collaboration with their male counterparts in struggles against colonial/post-colonial power or external threat (Jacoby 1999). The possibility of such collaboration in political mobilization is more probable when 'nation' is understood as a 'large community whose members differentiate themselves from others through their possession of a common "pervasive" or "societal" culture' (Seodu Herr 2003, 148). In this definition, there is no reified essence of a nation; it can accommodate foreign ideas and people to create a 'hybrid' culture. In this respect, Seodu Herr (2003) argues that 'polycentric' nationalism has potential for advocating feminist causes, as they may consider women's liberation to be possible only when the sovereignty of their nation is achieved. The author defends that grassroots, bottom-up nationalism in the Third World, reconceptualized in a nonessentialist way, may have genuine potential for promoting democracy and feminist causes. Also, nationalism should be conceived in this sense as involving an external goal (self-determination and recognition) and an internal one (equality among members of the nation). In this sense, the goal of 'polycentric' nationalism is the attainment and maintenance of national self-determination.

Finally, in relation to certain compatibility between gender and nation yet going beyond this dyad, Yuval-Davis (2011) has developed the concept of 'politics of belonging' to comprise specific political projects aimed at constructing belonging to particular collectivity/ies, within specified boundaries, and tightly related to the question of citizenship. The author defends that, while solidarity based on shared history, language and culture can be an important tool of mobilization, self-determination and liberation, any political project of belonging that uses the boundaries of the collectivity to exclude or delegitimize the full individual and collective rights of all other members of society can end up constructing its own version of racialized autochthony, which reproduces exclusion and discrimination.

The Catalan case: feminist perspectives on independence

As shown above, feminism has tended to regard nationalism from a distance, mostly due to the negative consequences that nationalism and the state have had for women in relation to sexual, reproductive, civic and political rights and the violence suffered in the name of national projects. Moreover, the recent rise of far-right and supremacist movements and politicians, especially in the US and in some countries in Europe and South America that function under explicit expressions of xenophobia and mobilizing around hate-centred nationalist frames, reinforces the idea that nationalism is dangerous, ethnicist and racist. However, Third World feminists have criticized Western feminists for the Eurocentrism that goes unnoticed in their claims and that reproduces the discrimination of colonized women, thus making visible that nationalistic expressions are not only materialized as xenophobia but that these can also be expressed through liberatory frameworks.

In this section, I want to show how, within the context of the intense struggle for independence in Catalonia, there are pro-independence leftist feminists who have been working for national liberation from critical perspectives. I argue that it is important to focus on feminist movements' discourses and practices to understand why some feminists join the struggle for independence and how they challenge hegemonic conceptualizations of what national identity is and how independence could favour women's and other minoritized group's interests. The current shifting situation makes it impossible to analyse the Catalan case from a distant perspective, as this chapter has been written during autumn 2017, even as events are radically changed the known frameworks. My own participation in Catalan feminist politics situates my perspective as an insider, a position I do not consider as a limitation but one with potential to render visible other perspectives that are generally misrepresented. Given that situated knowledge is also part of feminist epistemologies, in the following section I want to provide a reflection of the current situation with the aim of highlighting feminist proposals on the Catalan situation.

It is important to note that the Catalan case has specifics worth highlighting. Even though the national issue has been present in Catalan politics since the eighteenth century, it was during the later years of Francoism (1939–1976) when the Catalan nationalist movement began to re-emerge after the defeat of the legitimate government of the Second Republic in 1939 (see Guibernau 2004). However, the pro-independence will was not a majority until a few years ago. Huge uprisings started in 2010 due to Spanish government's limitation of sovereignty through cuts to the *Estatut de Catalunya* (the 'Catalan Constitution', approved by a vast majority in the Catalan Parliament). Since then, on 11 September (National Day in Catalonia) massive demonstrations demanding independence have been held in Barcelona and other towns. In the 2015 elections, pro-independence parties won the elections with a majority in the Catalan Parliament, with the objective of celebrating a referendum of self-determination that the Spanish government had never agreed to negotiate. The referendum was organized and celebrated on 1 October 2017, with the participation of more than two million voters: more than 90 per cent voted for independence. The images of repression by the Spanish police were seen around the world, showing the anti-democratic and violent response by the Spanish government that caused more than a thousand people to be injured. The declaration of independence took place on 27 October and, after that, part of the elected Catalan government was imprisoned. The president and other members of the executive were exiled to Belgium (2 November 2017). Two activists of cultural and political associations, who had been organizing the demonstrations, were also imprisoned. The Spanish government subsequently applied Article 155 of the constitution, which gives the power to intervene in Catalan institutions, to relieve the government and also to call for elections.

The role of the Catalan feminist movement in this process has been made invisible by general media and is absent from academic work. However, feminist developments on the Catalan nation were already present in the 1970s. One of the main contributions that relate the feminist developments to the Catalan situation was actually the conception of the interrelation of oppressions. The following text, written by feminist pro-independence radical leftist activists, shows a different perspective on nation and gender that is rarely taken into account:

> if feminism and nationalism have a separate space in everybody's minds, the area in which both fields interact is a kind of no one's land … We should give all alternative as women, being conscious that feminism must be a tool for global freedom, one that will break the chain of interrelated and inseparable oppressions in which our society is built upon … Our specific experience as women that feel oppressed because of being members of the feminine sex and because of belonging to a dominated nation, objects of a cultural and linguistic genocide, has also been very complex.
>
> *Olivares et al. 1982, 99*

The authors argue that, even though in any political struggle there is a tendency to separate various oppressions, what causes the constant tension and discomfort is the experience of oppression; they argue that, for women, there must be a project of total liberation against gender, class and national oppression. They argue the necessity of struggling from the space where oppressions are interrelated and against those political groups that 'deny that one and the other [oppressions] intersect' (Olivares et al. 1982, 99).

These kinds of statements resonate directly with the intersectional theorizations that were developed during the 1980s and 1990s in the US by authors such as hooks (1981), Collins (1990) and Crenshaw (1991) in relation to the 'inseparability of oppressions', the 'specific experience' of women crossed by different oppressions or the 'space where oppressions intersect' as the starting point for transformative reflection and practice (see Rodó-de-Zárate 2019).

The first challenge for the conceptualization of gender and nation is therefore the intersectional perspective, one that is based on the lived experience of oppression and on the national and gender identities as sources of this experience and also of knowledge production. This relates to what Yuval-Davis (2011) has called *situated intersectionality*, a perspective that acknowledges the role of context and particular social and historical configurations in order to examine the complexity of social relations. For the authors, national identity is seen as a source of oppression because of the belonging to a dominated nation, so 'nation' and 'gender' do not appear as contradictory but as constituting the intersectional experience, a perspective that is generally missing from accounts on women and nation. It is rare to find 'nationality' as an oppressive intersectional oppression (at least in the European context), yet conceptualizing it in this way helps to show why feminists engage in national struggles: not only as a strategy to pursue gender justice but as an aim in itself, to end lived oppression.

This experience of oppression is based on cultural, linguistic and political dimensions, and this is what feminists relate to national struggle. Such claims have historically been labelled as nationalistic, obscuring that the position of Spanish identity was also nationalistic. For instance, as Palau (2015) explains, the programme of the *Asociación Nacional de Mujeres Españolas* (National Association of Spanish Women), founded in 1918 by Maria Espinosa with the aim of struggling for women's civil and political rights, stated from the outset that it opposed 'by all means at the disposal of the association, any purpose, act, or event which might jeopardize the integrity of the national territory'. The association worked to 'ensure that all Spanish mothers work in perfect harmony with teachers to instil in their children love of one indivisible fatherland from the earliest childhood'. This Spanish nationalism, found in a feminist group linked to Spanish colonialism and an imperialist past, shows that feminism and nationalism are different axes to which people are variously oriented, as intersectional theory rightly points out. Even if nowadays feminists develop their position in other terms, this quote relates to the current collective amnesia (McRoberts 2004) that hides Spanish nationalism as 'nationalistic', situating the national identity only in relation to those that struggle for secession and not to those that want to maintain, also through physical violence, the unity of the nation state.

Violence, here, is also an important issue, as the Catalonia struggle for independence has been peaceful, with massive and permanent non-violent mobilizations. It is a rare case, given that secessionist claims have historically escalated into violence and war (Duffy Toft 2012), with exceptions such as Gandhi's policy of non-violence and civil disobedience in the Indian independence movement. The anti-militarist tradition in Catalonia and the feminist pacifist groups have contributed to situate non-violence as a central element in the struggle for independence, which has promoted the creation of organizations like *En peu de pau* (www.enpeudepau.cat), which seeks to extend and promote pacifist civil and non-violent resistance in the current situation.

Intersectionality and a non-violent strategy are both important elements that may challenge conceptualizations of nationalism from a feminist perspective. However, at the centre of the debate is the question of identity. As has been shown, there are multiple interpretations and classifications of types of nationalism. The celebration of the 1 October referendum, the defence of the ballot boxes, the popular organization the days before and during the referendum, the general strike on 3 October and the demonstrations against the gaoling of political prisoners showed that the popular struggle was constructed around something other than national identity. The leftist pro-independence party in the Catalan Parliament (Candidatura d'Unitat Popular – CUP), which identifies itself as feminist, has been explicit to avoid ethnicist discourse. One of its members of parliament, during his speech in the Chamber, said: 'It is not about taking out a flag to rise another one. What we want is a Republic to be able to build everything … we will never defend an ethnic or identitarian project' (Benet Salellas, MP). The arguments to defend the right to hold a referendum have been articulated in relation to the right of self-determination, the right of peoples to determine freely their own political status. As Hacket (1995) argues, self-determination has important connotations for feminists in relation to the right to decide about women's rights, which relates to the right to decide what kind of society people want. This claim is also strongly related to the demand of sovereignty and the ability of the legitimate institutions to legislate. An example of this is the denouncement of the suspension by the Constitutional Court in Madrid of the laws approved by the Catalan Parliament, like the law on banks taxes, against evictions, the agency for social protection, against energetic poverty, for gender equality and against fracking, among others.

These kinds of cuts to sovereignty, together with police brutality and the imprisonment of the elected government, have shifted from an articulation based on self-determination and national claims to a struggle against the anti-democratic, authoritarian and violent nature of the Spanish state. This relation might be well illustrated by the popular quote from a musician and poet, Ovidi Montllor: 'There's people who don't like people to speak, write or think in Catalan. It's the same people who don't like people to speak, write or think.' The repression of Catalan has historically been linked to a repression of political freedom and democracy, and this is nowadays evidence that turns the struggle for independence into a democratic struggle, framing a national struggle beyond identities and culture. The creation of Republic Defence Committees, a network that coordinates assemblies in every neighbourhood, village and city, which has been crucial in the organization of strikes and demonstrations, is an example of the popular organization to defend the institutions beyond national discourses. Feminists have also organized themselves within these groups in order to promote gender equality in the organization, actions and discourses of these horizontal assemblies.

This coincides with what Walby (1997) argues in relation to the probability that women engage in national projects if they are open and pluralistic. Much evidence may be found in the Catalan case of feminist activists being active in various fields, raising their voices to situated gender and LGBT issues in the political agenda and organizing through different levels in order to develop strategies to build the Feminist Catalan Republic (see *Feministes per la Independència*).

However, although important efforts have been made by different social movements and political parties in order to break with ethnic discourses on Catalan identity, there have also been important critiques from feminist and post-colonial activists in relation to citizenship policies in Catalonia. For instance, as both Ortiz (2017) and Aatar (2017) denounce in their chapters in the recently published book *No One's Land: Feminist Perspectives on Independence* (Gatamaula 2017), the Law for Juridical Transitory (a foundational law for the Catalan Republic, approved on September 2017). This specifies that children of migrant parents will not have the right to Catalan nationality, even if they are born in Catalonia. This implies the maintenance of Spanish (and European) current migratory laws and has been seen as a racist policy that excludes people from Catalan citizenship on the basis on origin. This is also an alert regarding the possibilities for a radical change if a supposed Catalan Republic was part of the European Union, with its racist and capitalist policies. In the same edited-collection there are various views from feminists and LGBT activists that alert us to homonationalist discourses in Catalonia (Sadurní 2017). Others propose to build a queer Catalan identity based on diversity and fluidity (Olid 2017), that nationalism and belonging can be constructed as an idea of community against individualism (Forcades 2017) or that the transition to a republic could be thought as the transition to a gender identity, from a trans position (Preciado 2017). Internationalist perspectives on independence that highlight the historical feminist solidarities between Catalonia and Kurdistan (Çiçek 2017) and the Basque Country (Eizaguirre 2017) also show a tendency towards 'polycentric' nationalism, in this case through feminist organizations from several nations. All these contributions show how feminists engage with the national struggle with innovative perspectives that link gender, sexuality, ethnicity and nation from a situated point of view. Nation and nationalism can be approached from various perspectives, and organized feminist movements in the struggle for national liberation provide new views that may contest the hegemonic conceptions.

The role of place in recent mobilizations has also been crucial, and feminist geographies could certainly provide new insights on the political processes of democratization, as in the Catalan case. For instance, the body as a site of resistance against state violence has been a central issue in relation to the 1 October referendum. Pro-independence politicians, such as the Catalan President Carles Puigdemont, to defend the legitimacy of the referendum referred to 'people who used their bodies to defend democracy'. The defence of the polling stations as symbols of resistance or the squares and streets as the permanent way of expressing political opinions have also been central spatial metaphors.

Concluding remarks

The feminist perspective on nation and nationalism shows that there is a complex relation between gender and nation that has been central to understanding women's experiences and geopolitics in general. There is a tendency to focus on nation states when analysing nationalism, and there is also an invisibilization of feminist voices engaged in national projects that comes from different sides: invisibilization by feminist state-nationalists, by national liberation movements that silence women's voices, and by academia and mass media, which tend to focus on the hegemonic positions and hide actions and proposals that come from social movements and women and LGBT groups. This obscures the theories and practices developed by feminists in various contexts to understand the national struggle from perspectives that may challenge definitions of nation or nationalism. There is a great diversity of examples, and here I have tried to show how feminist groups have been engaged in the struggle for independence, in the Catalan context, from an intersectional and critical perspective that works for national freedom while it criticizes and contradicts the patriarchal discourses that persist in national movements. Feminist geographies could shed light on such processes of democratization from a situated perspective that renders visible that place matters, and feminist activists are voices to take into account when researching geopolitics.

Key readings

Dixon, D.P. 2015. *Feminist Geopolitics: Material States.* Farnham: Ashgate.

Gatamaula feminista. 2017. *Terra de Ningú: perspectives feministes sobre la independència.* Barcelona: Pol·len edicion. (Translation from Spanish forthcoming 2018).

Yuval-Davis, N., and F. Anthias. 1989. *Woman-Nation-State.* Houndsmills, Basingstoke: Palgrave Macmillan.

References

Aatar, F. 2017. "Quan la desobediència es racialitza." In: *Terra de Ningú: perspectives feministes sobre la independencia,* edited by Gatamaula feminista. Barcelona: Pol·len edicions.

Amurrio Vélez, M. 2002. "El proceso de construcción generizada de la nación vasca." *Euskonews & Media* 192: 13–20.

Anderson, B. 1983. *Imagined Communities.* London: Verso

Çiçek, M. 2017. "Amb Kurdistan: Reivindiquem l'era de la revolució de les dones mitjançant aliances democràtiques" In: *Terra de Ningú: perspectives feministes sobre la independencia,* edited by Gatamaula feminista. Barcelona: Pol·len edicions.

Collins, P.H. 1990. *Black Feminist Thought: Knowledge, Power and the Politics of Empowerment.* Boston, MA: Unwin Hyman.

Crenshaw, K. 1991. "Mapping the Margins: Intersectionality, Identity Politics, and Violence against Women of Colour." *Stanford Law Review* 43: 1241–1299.
Dixon, D.P. 2015. *Feminist Geopolitics: Material States.* Farnham: Ashgate.
Duffy Toft, M. 2012. "Self-determination, Secession, and Civil War." *Terrorism and Political Violence* 24 (4): 581–600.
Eizaguirre, N. 2017. "Amb Euskal Herria: El nacimiento de la República Catalana. La acción internacionalista del aquí y el ahora." In: *Terra de Ningú: perspectives feministes sobre la independencia,* edited by Gatamaula feminista. Barcelona: Pol·len edicions.
Enloe, C. 1990. *Bananas, Beaches, and Bases: Making Feminist Sense of International Politics.* Berkeley: University of California Press
Faria, C. 2013. "Staging a New South Sudan in the USA: Men, Masculinities and Nationalist Performance at a Diasporic Beauty Pageant" *Gender, Place & Culture* 20: (1).
Forcades i Vila, T. 2017. "Identitat nacional i feminisme: pertinença, comunitat i llibertat." In: *Terra de Ningú: perspectives feministes sobre la independencia,* edited by Gatamaula feminista. Barcelona: Pol·len edicions.
Friedman. S.L. 2017. "Men who 'Marry Out': Unsettling Masculinity, Kinship, and Nation Through Migration across the Taiwan Strait." *Gender, Place & Culture* 24 (9).
Gatamaula feminista. 2017. *Terra de Ningú: perspectives feministes sobre la independència.* Barcelona: Pol·len edicions.
Gleason, G. 1991. "Nationalism in Our Time." *Current World Leaders* 34 (2): 213–234.
Guibernau, M. 2004. *Catalan Nationalism: Francoism, Transition and Democracy.* London: Routledge.
Hackett, C. 1995. "Self-determination: The Republican Feminist Agenda". *Feminist Review* 50: 111–116.
hooks, b. 1981). *Ain't I a Woman?: Black Women and Feminism.* Boston, MA: South End Press.
Hyndman, J. 2003. "Mind the Gap: Bridging Feminist and Political Geography Through Geopolitics" *Political Geography* 23 (3): 307–322.
Jacoby, T.A. 1999. "Gendered Nation: A History of the Interface of Women's Protest and Jewish Nationalism in Israel." *International Feminist Journal of Politics* 1 (3): 382–402.
Jayawardena, K. 1986. *Feminism and Nationalism in the Third World.* London: Zed Books.
Kaplan, G. 1997. "Feminism and Nationalism: The European Case." In: *Feminist Nationalism,* edited by L. West, 3–40. London: Routledge.
Koch, N. 2011. "Security and Gendered National Identity in Uzbekistan." *Gender, Place & Culture* 18 (4).
Lee, J., and L. Cardinal. 1998. "Hegemonic Nationalism and the Politics of Feminism and Multiculturalism in Canada." In: *Painting the Maple: Essays on Race, Gender and the Construction of Canada,* edited by: V. Strong-Boag, S. E. Grace, A. Eisenberg, and J. Anderson, 215–241. Vancouver: UBC Press.
Martin, A.K. 1997. "The Practice of Identity and an Irish Sense of Place." *Gender, Place & Culture* 4 (1).
Massaro, V.A., and J. Williams. 2013. "Feminist Geopolitics." *Geography Compass* 7/8: 567–577.
Mayer, T., ed. 2000. *Gender Ironies of Nationalism. Sexing the Nation.* London: Routledge.
McClintock, A. 1993. "Family Feuds: Gender, Nationalism and the Family." *Feminist Review* 44: 61–80.
McRoberts, K. 2004. "The Future of the Nation-State and Québec Canada Relations." In: *The Fate of the Nation-State,* edited by M. Seymour, 390–402. Montreal: McGill-Queen's University Press.
Nagel, J. 1998. "Masculinity and Nationalism: Gender and Sexuality in the Making of Nations." *Ethnic and Racial Studies* 21 (2): 242–269.
Olid, B. 2017. "República queer de Catalunya." In: *Terra de Ningú: perspectives feministes sobre la independència,* edited by Gatamaula feminista. Barcelona: Pol·len edicions.
Olivares, M., C. Maso, E. Duran, T. Cusí, M.M. Marçal, and S. Sentis. 1982. "Dona i Nació: Feminisme i Nacionalisme." *Segones Jornades Catalanes de la Dona.*
Ortiz, D. 2017. "Cataluña, colonialidad y racismo institucional." In: *Terra de Ningú: perspectives feministes sobre la independencia,* edited by Gatamaula feminista. Barcelona: Pol·len edicions.
Palau, M. 2015. "Gender and Nation: Conditioned Identities." In: *Conditioned Identities. Wished-for and Unwished-for Identities.* Berna: Peter Lang.
Pateman, C. 1988. *The Sexual Contract.* Stanford, CA: Stanford University Press.
Peterson, V.S. 1994. "Gendered nationalism." *Peace Review: A Journal of Social Justice* 6 (1): 77–83.
Peterson, V.S. 1999. "Political Identities/Nationalism as Heterosexism." *International Feminist Journal of Politics* 1 (1): 34–65.
Peterson, V.S. 2014. "Family matters: How queering the intimate queers the international." *International Studies Review* 16 (4): 604–608.
Preciado, P.B. 2017. "Catalunya Trans." In: *Terra de Ningú: perspectives feministes sobre la independencia,* edited by Gatamaula feminista. Barcelona: Pol·len edicions.

Puar, J. 2007. *Terrorist Assemblages: Homonationalism in Queer Times*. Durham, NC: Duke University Press.
Radcliffe, S. 1996. "Gendered Nations: Nostalgia, Development and Territory in Ecuador." *Gender, Place & Culture* 3 (1).
Rodó-de-Zárate, M. 2019. "Gender, Nation and Situated Intersectionality: The Case of Catalan Pro-independence Feminism." *Politics and Gender.*
Sabhlok, A. 2017. "'Main Bhi to Hindostaan Hoon': Gender and Nation-state in India's Border Roads Organisation." *Gender, Place & Culture* 24 (12).
Sadurní Balcells, N. 2017. "LGTBI+, racisme i colonialitat. Una aproximació a la independència des del concepte d'homonacionalisme." In: *Terra de Ningú: perspectives feministes sobre la independencia,* edited by Gatamaula feminista. Barcelona: Pol·len edicions.
Seodu Herr, R. 2003. "The Possibility of Nationalist Feminism." *Hypatia* 18 (3): 135–60.
Smith, A.D. 1971. *Theories of Nationalism*. London: Duckworth.
Smith, A.D. 1986. *The Ethnic Origins of Nations.* Oxford: Basil Blackwell.
Ugalde, M. 1996. "Notas para una historiogafía sobre la nación y la diferencia sexual." *Arenal: Revista de Historia de las Mujeres* 3/2.
Vickers, J. 2000. "Feminisms and Nationalisms in English Canada." *Journal of Canadian Studies* 35 (2): 128–148.
Vickers, J. 2006. "Bringing Nations In: Some Methodological and Conceptual Issues in Connecting Feminisms with Nationhood and Nationalisms." *International Feminist Journal of Politics* 8 (1).
Walby, S. 1997. *Gender Transformations*. London: Routledge.
West, L. 1997. *Feminist Nationalism*, London: Routledge.
Yeoh, B.S.A., and K. Willis. 1999. "'Heart' and 'Wing', Nation and Diaspora: Gendered Discourses in Singapore's Regionalisation Process." *Gender, Place & Culture* 6 (4).
Yuval-Davis, N. 1993. "Gender and Nation." *Ethnic and Racial Studies* 16 (4): 621–632.
Yuval-Davis, N. 2011. *Politics of Belonging: Intersectional Contestations*. Sage.
Yuval-Davis, N., and F. Anthias. 1989. *Woman-Nation-State.* Houndsmills, Basingstoke: Palgrave Macmillan.

21

UNSETTLING GENDER AND SEXUALITY ACROSS NATIONS

Transnationalism within and between nations

May Farrales and Geraldine Pratt

Introduction

Transnationalism holds within it the promise and contradiction of movement and stasis. While much popular rhetoric celebrates the crossings and connections enabled by the *trans* in transnationalism, the *national(ism)* weighs down the possibility of crossings and connections at the same time as it facilitates the movement of bodies, goods and capital. It is thus not surprising that geographers in general, and feminist geographers in particular, have paid attention to the spatialities of transnationalism. Feminist geography has brought to transnationalism questions of how and where gender and sexuality intersect with flows of migration, goods and capital and how to create knowledge communities and solidarities across national contexts and relations of inequality. There are now many excellent reviews and accounts of feminist geographical approaches to transnationalism (see Katz 2001; Nagar 2014; Silvey 2006; Swarr and Nagar 2010). In this chapter, we aim to follow feminist, anti-racist, decolonial and queer approaches to transnationalism that apprehend the transnational as a space and site of relations of power. In these relations of power (that lie at the core of the contradictory nature of transnationalism), conditions of possibility for particular subjectivities and subjects both emerge in new ways and consolidate old patterns and certainties.

To this end, we consider what is most useful about a transnational approach to gender relations and what might be productively rethought. First, we work through a range of examples, focusing mostly on the Filipino diaspora, to consider the ways that gender relations and sexualities become reworked through migration experiences and how norms in one place are taken up and reused in another to both destabilize or restabilize gender norms. Our focus on the Filipino diaspora emerges from our lived experience and research (May Farrales) and extended research collaborations (Geraldine Pratt) and the fact that so much research has focused on the Filipino labour diaspora, one of the largest globally. Second, we examine how gender norms are resistant to change, and how racial difference, racism and nationalisms are enacted through hetero- and homonormativities. Finally, we seek to set in motion or unsettle what is typically taken to be the relatively static part of transnationalism: the nation. We do this by considering Indigenous critiques of normative notions and dominant forms of nation in settler colonial contexts; that is, we consider the transnational already embedded within the settler colonial state. We pay particular attention to how the non-Indigenous and non-Black racialized diasporic subject figures in settler colonial relations. By bringing these critiques into conversation with transnational approaches, we leave the chapter with questions of what building a transnational approach into a settler nation brings to studies of transnationalism.

Gender, sexuality and race in transit

In the opening chapter of his book *Filipino Crosscurrents: Oceanographies of Seafaring, Masculinities, and Globalization* (2011), Kale Fajardo dwells on the photograph of a forlorn Filipina nurse adorning the cover of the *New York Times* containing a magazine article about Global South migration. Fajardo describes the woman's image: she is alone, barefoot on the sand, dressed in hospital scrubs and looking into the distance over an expanse of water. She stands as the quintessential image of the modern-day overseas Filipino migrant, Fajardo argues. Over the past few decades, the Filipina overseas worker, in multiple iterations as a domestic helper, entertainer, nurse or nanny, has come to figure large in the imaginary of transnational migration. Worldwide, almost half of temporary workers – who are by definition in transit between nations – are now women, often migrating alone to work in feminized, racialized and sexualized jobs (Global Commission on International Migration 2005). Of the new recruits from the Philippines in 2010, for example, 68 per cent were employed as domestic helpers or housekeepers (Philippine Overseas Employment Administration 2010). Both in sheer numbers and in the gendered, sexual and racialized nature of the care work that they perform across the globe, Filipinas have come to function, as Rachel Parreñas (2001) put it, as the 'servants of globalization', who are slotted into insecure, low-paid jobs through 'capitalist scripts' that operate through persistent gendered and racialized inequalities. Transnational labour migration, therefore, often works within, extends and hardens existing gendered and other power relations on a global scale.

Transnational migration also can ossify and exacerbate gender binaries at home. Migrant mothers (but not migrant fathers) are often stigmatized in their home countries. Mexican migrant mothers, Joanna Dreby argues, 'bear the moral burdens of transnational parenting' (2010, 4; see also Parreñas 2005a), exposing how gender norms can be resistant to change. Migration calls upon traditional gendered scripts, in most cases by passing care work to other female kin. In her study of children of migrant mothers living in Manila, Parreñas found that these same scripts led children to feel abandoned. 'Again and again children describe the nurturing provided by migrant mothers as "not enough"' (Parreñas 2005b, 33). Parreñas argues that children's inability to recognize the care that they receive from their mothers and their extended female kin, and their feelings of abandonment and longing for greater intimacy with their mothers who are working abroad are instilled in the Philippines by the norms of patriarchal gender relations in the heteronormative family, promulgated in the media and literally taught in the state-regulated Values Formation curriculum at school (Parreñas 2010). She calls for an expansion of the ideology of mothering in the Philippines to include economic provisioning as a respected maternal role, as well as a recalibration of the gender division of the labour of social reproduction, such that fathers assume a greater caregiving role. But she sees no evidence that migration or transnationalism, in and of themselves, have altered gender relations in a country – the Philippines – where it is estimated that one in ten citizens is at any time working abroad (but see Aguilar 2013 and McKay 2007 for accounts that view family separation as much less damaging).

Fajardo complicates this picture of the feminization of transnational labour migration, by going to what he calls the 'backwater geographies and oceanographies of globalization': to the men who live and work on the container ships, moving goods through global transportation networks. The Philippine state champions these Filipino seamen's masculinity as counter to neocolonial and orientalist narratives that feminize the nation (Tadiar 2004). Filipino men are cast as heroic hyper-masculine figures, manning the ships of globalization. Through their remittances, they are heralded as heroes in a 'national struggle (not against colonialism) but against [foreign] debt' (85). Fajardo enriches this understanding of the seafarers' masculinity by seeing it as produced at the cross-current of multiple trajectories of power: class; race; sexuality; and citizenship. Their experience of masculinity is shaped through their experiences on the ship, including their sense of being entrapped in space and suspended in time in a 'queer' space–time. Fajardo suggests that there is a loosening of gender scripts and expectations in the spaces of transit and transportation; for example, seafarers' masculinities shift in relation to the figure of the *tomboy*. Counter to Western narratives that posit bodies from the Global South as wholly repressed or excessively sexualized, the Filipino *tomboy* occupies a particular identity in the Philippine imagination as a form of female masculinity or 'an embodiment of female manhood' with specific working-class roots. Instead of falling in line with the dominant state narrative of the Filipino seaman working on foreign-owned container ships to support family and nation, the men tend to embody queer masculinities in finding common ground with the *tomboy* figure's masculinity and working-class mythology. For Fajardo, these queer encounters provide opportunities to hold together feminine masculinities and masculine femininities along a continuum of sexualities instead of cleaving the masculine and feminine apart.

In this thread of work, of which Fajardo's work is illustrative, the transnational is understood as a space and site of power relations in which subjectivities emerge in negotiation with normative narratives and structures. Subjects and subjectivities, reworked through migration experiences, are afforded a degree of mutability and distance from dominant gender relations and sexualities (see, for example, Cruz-Malave and Manalansan 2002; Manlansan 2003; Pratt and Yeoh 2003; Yeoh and Ramdas 2014; Thangaraj 2015). Gayatri Gopinath (2005), for example, theorizes the concept of a queer South Asian diaspora as a translation. Gopinath captures the multiple registers through which transpire 'the formation of sexual subjectivity within transnational flows of culture, capital, bodies, desire, and labor' (13). In this formulation, Gopinath disrupts a linear causal link between the concepts of 'queer' and 'heterosexuality', and 'diaspora' and 'nation', where heterosexuality and nation act as the original, and queer and diaspora are its mere twin copy. She proposes instead that 'queer' and 'diaspora' are translations of 'heterosexuality' and 'nation', and vice versa: 'Translation here cannot be seen as a mimetic reflection of a prior text but rather as a productive activity that instantiates new regimes of sexual subjectivity even as it effaces earlier erotic arrangements' (14). In other words, for Gopinath, the transnational or diaspora is a process of translation, not a monolithic site attached to nation (both home and host) as 'mimetic reflection'. Instead, it is a productive and generative space in which different subjectivities emerge. Gopinath argues that with migration, then, diasporic subjects cannot be easily captured by homonormative gender and sexual paradigms in the host countries of the Global North.

The idea that the queer racialized diasporic subject confounds normative sexual scripts, because their gender, sexual and racial subjectivities are reworked through transnational migration, is echoed in Martin Manalasan's (1995, 2003) ethnography of Filipino gay men in New York:

> Filipino gay men are not typical immigrants who 'move' from tradition to modernity; rather, they rewrite the static notions of tradition as modern or as strategies with which to negotiate American culture. Immigration, therefore, does not always end in an assimilative process but rather in contestation and reformation of identities.
>
> *2003, 14*

Stories of migrating from Global South to Global North, he notes, are often constructed within a teleological movement from tradition to modernity, in the case of gay Filipino men from *bakla* to global gay (see Benedicto 2014 for the ways in which elite gay men in Manila deploy this teleological narrative). In contrast, in diasporic spaces such as New York, *bakla* is not a 'premodern antecedent to gay' but rather is 'recuperated and becomes an alternative form of modernity' (2003, 21). Practices of cross-dressing in the Filipino diaspora are poorly understood, he argues, within the White conventions of drag, underlining the point that queer and feminist theories in the Global North likely miss the complexity of transnational lives if they work within existing, largely Eurocentric concepts and categories.

While Manalansan appreciates that the rewriting and reworking of subjectivities are part of a process of negotiation with dominant structures, at the same time he signals the stubbornness of the forces involved in creating the diasporic queer's conditions of possibility. The tenacity of certain gender, sexual and racial norms makes these resistant to change. Feminist scholarship that pays attention to the stubbornness of gender norms both highlights the work that gender and power do in transnational processes and investigates how normative and dominant gender, sexual and racial regimes are often recuperated and redeployed in the transnational. The promise of gender equity, or a reworking of gender relations, sits as a promise in much of the literature on transnational lives (e.g. Mahler and Pessar 2006); but, as we explore further in what follows, it is a promise that needs close interrogation.

The 'national' in 'transnationalism'

An August 2017 media report of a planned anti-immigrant rally that was thwarted by almost a thousand anti-racist protestors in Vancouver incorporates a photograph of an older White man in the midst of the rally-goers.[1] The man is wearing a baseball cap emblazoned with the Canadian flag. He is pointing a finger, as if scolding someone in the crowd, and is holding a sign that reads 'Population Reduction. Zero Immigration'. While it might be an extreme position in a country that prides itself on its liberal multicultural messaging of tolerance, the sign nonetheless captures a sentiment that points to the stasis inherent in transnationalism. In particular, the refrain 'Population Reduction. Zero Immigration' evokes the nation and its racialized and gendered borders. To explore that which weighs down the potential of crossings and connections, in this section we focus on the *national(ism)* in *transnationalism*. Following the work of critical race and feminist scholars, we propose that the 'nation', itself a rendering of colonial, imperial and capitalist designs, aligns transnational migrations to its own normative formations. This alignment, we suggest, points to enduring or stubborn hegemonic gender, sexual and racial norms and relations that set the conditions of possibility when transnational migration is anchored in dominant ideas and workings of nation.

Harsha Walia (2013), for example, theorizes the concept of 'border imperialism' to situate contemporary transnational migrations in the broader geopolitical historical and present-day context of colonialism, imperialism and capitalism. She explains:

> Border imperialism is characterized by the entrenchment and retrenchment of controls against migrants, who are displaced as a result of the violences of capitalism and empire, and subsequently forced into precarious labor as a result of state illegalization and systemic social hierarchies.
>
> *2013, 38*

Walia forwards an analytical framework for transnational migrations from the Global South to North America that simultaneously critiques the imperial and colonial workings of nation states that dispossess, displace and force migration at the same time as they centre the ways that states limit the inclusion of migrant bodies through established hierarchies of race, class and gender. Or, as she puts it, 'Western states thus are the major arbiters in determining if and under what conditions people migrate' (39). The concept of border imperialism invites closer scrutiny of the work that nations and states do in transnationalism. It makes clear that material conditions limit the possibilities for transnational migrants.

Nation states (and national citizens) not only repurpose hierarchies of gender, race, class and uneven development; in some cases, hard-won struggles for gender and sexual equality are being used to similar effect, to harden the borders of the Global North and mobilize narratives of life in the Global South as traditional, 'backward' and a threat to liberties in the Global North. In many countries in the Global North, gender equality and tolerance of sexual diversity are now taken as a measure of liberal moral progress and secular modernity (and of the implicit backwardness and threat of Islamic societies where such tolerance is not exhibited). Jasbir Puar (2007) has identified the melding of sexual liberation with nationalism as 'homonationalism'. Sexual politics has become, in Puar's words, an 'optic and operative technology' in the production and disciplining of Muslims (2007, xiii; see also Butler 2008; Mepschen, Duyvendak and Tonkens 2010). Gender equality can serve the same end, and accepting a diversity of gender norms (e.g. female circumcision; the *hijab*) has proved challenging for many Western feminists. It is perhaps not entirely surprising that those who carry out the everyday hate crimes towards Muslim women in Malmo in Sweden, such as pushing, spitting on, verbally abusing or attempting to pull off their veils, are often found to be older non-Muslim Swedish women (Listerborn 2015). Sara Ahmed (2000) argues that such acts of exclusion both differentiate and include 'the stranger' (i.e. racialized migrants) to secure the dominant citizen-subject in Western liberal nations:

> The recognition of others [is] … central to the constitution of the subject. The very act through which the subject differentiates between others is the moment that the subject comes to inhabit or dwell in the world. The subject is not, then, simply differentiated from the (its) others, but comes into being by learning how to differentiate between others.
>
> *2000, 24*

These racialized, sexual and gendered forms of inclusion/exclusion are important to understand if transnationalism is to be apprehended as a process that is mired in a contradiction of movement and stasis. The tension between movement and stasis operates in a number of registers. Devaluing the feminized work of social reproduction (whether uncommodified or commodified) has been essential to profitability in capitalist societies (Federici 2004; Fraser 2016; Katz 2001), as has the devaluation of racialized labour (Gilmore 2007; Pulido 2018; Robinson 1983). One need not resolve debates on whether gender and racial hierarchies are internal or external to the workings of capitalism to notice the utility of such hierarchies for profitability in such societies. Global movements of labour from the Global South are drawn into and are part of such processes. Such hierarchies, as we have suggested, are not simply imposed by the nation state: they enter into our formation as subjects; and seemingly unrelated cultural politics (such as legalized gay marriage) can become technologies for racial differentiation and border control.

This is not to say that transnational mobility is simply, or always, or easily absorbed into the sameness of the repetition of racial and gender hierarchies. It is important as well to disrupt the assumption that North America and Western Europe are the inevitable and prefered destinations for transnational migration. In the last 15 years, more and more Filipino *balikbayans* (members of the overseas Filipino immigrant diaspora, in contrast to migrant overseas Filipino workers) have been returning to retire in the Philippines, where they are creating new processes and forms of inequality and difference (Pido 2017). The Philippine state is enthusiastic about the return of those who had previously migrated to North America, viewing their return and assets as an economic development opportunity. This enthusiasm is captured by President Benigno Aquino III's Proclamation 181, which designated the years 2011 to 2016 as the 'Pinoy Homecoming Years' (Pido 2017, 148).

Alternatively and additionally, in some cases migrants actively resist their instrumentalization by the nation state. Filipino migrant worker activists in Vancouver in Canada, for instance, note how experiencing gender and racial hierarchy transnationally can itself be radicalizing. In trying to understand the relative difficulty of mobilizing Filipino male migrant domestic workers, they note that men and women come to Canada as domestic workers through different migration trajectories: male domestic workers tend to come directly from the Philippines, whereas many women come to Canada after working as foreign domestic workers elsewhere. This pattern reflects differential (gendered) access to resources. Women are more likely to be what Paul (2011, 2015, 2017) refers to as 'capital constrained', thus must work their way to their desired migration destination through 'step-wise migration'. Filipinas coming to Canada as migrant domestic workers often work first as domestic servants in Hong Kong, Singapore, Taiwan or the Middle East, working in at least one other destination country for several years before securing the necessary economic, human, social and cultural resources to come to Canada. Activists assess that this step-wise migration experience, in and of itself, has a politicizing effect: 'And for them, given their experience before, there's really no worry because, you know, either you die fighting or you just accept it.' Step-wise migration is an education in itself in the commodification of Filipina labour around the globe, an education that has had radicalizing effects (Pratt and Migrante 2018).

Transnational mobility reaches not only across space but through families and over time. In an early collaboration (Pratt, in collaboration with the Filipino-Canadian Youth Alliance 2003/04), we explored the need on the part of many second-generation Filipino–Canadian youths in Vancouver in Canada to claim their transnational identity. Parents' experiences of dislocation and relocation touched the lives of their children in what Katherine Sugg (writing about the children of exiles from Cuba, 2003) has termed a 'generational legacy'. Youths came to identify both with their parents' wounds of dislocation – including their experiences of systematic racism in Canada – and the Philippines. Learning about, visiting and becoming involved in political activism in the Philippines then shaped and radicalized Filipino–Canadian youths' analyses and activism in Canada. Critical race, feminist and queer scholarship thus provides a sustained critique of dominant forms of nation and its racial, gendered and sexual workings vis-à-vis the transnational subject and suggests some powerful oppositional impulses that can emerge from within it.

Transnationalism in a settler colonial nation

Writing from the Canadian context, as a second-generation Filipinx and a fourth-generation White settler, respectively, we raise a further contradiction *within* the nation to question how approaches to transnationalism can be and have been reinforcing the enduring colonial violence of settler colonialism. Often, scholarship on transnationalism, while attuned to how colonial forms of gender, racial and sexual regimes operate in transnationalism, falls short of critiquing the role of settler colonialism in the transnational flows of bodies, capital, goods and politics, and the position of racialized arrivants within it.

Chickasaw scholar Jodi Byrd's (2011) work on Indigeneity as a transit for US empire-building provides a way to think about diaspora differently in ways that connect, for example, the Philippines and Canada as two different colonial and capitalist enterprises. Byrd, along with other Indigenous scholars in Turtle Island (the place now known as North America), reminds us that Indigenous nations, systems of governance and relations to their lands and each other continue to exist despite the onset of settler colonialism. Even the Canadian nation state now recognizes nation-to-nation relationships with Indigenous peoples in Canada.[2]

For Byrd, centring this perspective asks that:

> [The] settler, native, and arrivant each acknowledge their own positions within empire and then reconceptualize space and history to make visible what imperialism and its resultant settler colonialisms and diasporas have sought to obscure. Within the continental United States, it means imagining an entirely different map and understanding of territory and space: a map constituted by over 565 sovereign Indigenous nations, with their own borders and boundaries, that transgress what has been naturalized as contiguous territory divided into 48 states.
>
> *2011, xxx*

Byrd offers important points that are imperative for conceptualizing how transnationalism works in and with settler colonialism. First, she regards diasporas and settler colonialisms as the twinned results of imperialism but refuses to conflate processes of racialization with the logics of colonial dispossession. As many Indigenous and allied scholars have noted, dispossession of land and resources (in the case of Indigenous peoples) are not the same as racialized labour exploitation, and equating the racialization of Indigenous and migrant peoples renders Indigenous peoples, Byrd argues, as 'unactionable in the present as their colonization is deferred along the transits that seek new lands, resources, and peoples to feed capitalistic consumption' (221). Second, she argues that ideas of Indian and Indianness have functioned as the 'transit' of US empire, the 'ontological ground through which US settler colonialism enacts itself as settler imperialism' (2011, xix). In the case of the US, transnational exchanges of bodies, ideas and legal precedents have taken place between settler colonialism and imperialist ventures elsewhere. For example, the modes of warfare and the actual high-ranking US army officers from the wars fought with the Plains Nations in North America were literally reemployed in the Philippine–American War of 1899 to 1902 (see also Miller 1982). Neferti Tadiar (2015, 143) documents how the incorporation of the Philippines into US domestic space as a colonial acquisition 'bore the legal memory' of earlier landmark legal cases, framing citizenship for African and Native Americans in the US. In not extending the Fourteenth Amendment of the US Constitution to the Philippines as an unincorporated territory in 1901, 'we could say Filipino is racialized as *not* black, and *like* Indian' at one defining moment in its national formation (143). Third, recognizing relationships made possible by 'imperialism and its resultant settler colonialisms and diasporas' allow for an alternative understanding of territory and space that potentially denaturalizes the authority of settler nations and states.

The need to denaturalize settler states and their presumed right to Indigenous lands and territories remains a persistent call from Indigenous scholars and communities. Kwagiulth scholar Sarah Hunt (2014) theorizes 'colonialscapes' as 'a way of seeing that naturalizes the relations of domination and dehumanization inherent in colonial relations' (7), including ongoing violence against Indigenous peoples and communities. One way of disrupting the assumed authority of settler states and nations, Hunt and other Indigenous scholars argue, is to re-centre concepts of land, intimate relations, territory and nation that flow from Indigenous forms of sovereignty and knowledges (see, for example, Byrd 2011; Coulthard 2014; Daigle 2016; Hunt 2014; Million 2014; Simpson 2014; Tuck and Yang 2012). This central tenet of Indigenous communities and scholars in settler colonial situations puts pressure on transnational scholarship to consider the nation in different ways, reconsidering how diasporic peoples of colour come to be in settler colonial relations (within geography, see Farrales 2017; Johnston and Pratt 2017; Pulido 2018; beyond geography, see Day 2016; Lawrence and Dua 2005; Razack et al. 2010; Saranilllio 2013; Walia 2013).

Alongside the interwoven histories that create the ground for solidarities, immigration can and often does reinforce the colonial and multicultural nation state. As Patrick Wolfe (2006) has argued, there is nothing preventing even colonized natives from one region becoming settlers in another, if their actions support the dispossession of Indigenous peoples in the new locale. 'Power does not simply target historically oppressed communities,' Filipino scholar, Dean Saranillio, notes, 'but also operates through their practices, ambitions, narratives and silences' (2013, 286). Walia (2013) concedes that many mainstream immigrant groups on Coast Salish territories (Vancouver) perpetuate the dispossession of Indigenous lands and peoples by framing immigrant struggles as 'integration issues'. However, she insists that while:

> migrants of color are inevitably implicated in settler colonialism and have a responsibility to ally with Indigenous struggles, [I] do not believe that migration as a process in and of itself, especially in this late period of capitalist globalization and global neo-colonialism, can inherently be understood as a form of settler colonialism.
>
> *2013, 129*

The migration of Filipinos to Canada, for example, is entangled and implicated in the political economy and relations of settler colonialism by way of the Canadian state's active recruitment of Filipino (im)migrant labour and by Filipinos' presence in Canada. As part of this migration process, settler colonial logics make particular demands on racialized (im)migrants. In a study on the sexual, gendered and racialized performances of Filipino men at community-organized basketball games on Coast Salish territories, the game of basketball reperformed in Canada helps to realign and discipline Filipino racialized masculinities to the settler colonial project (Farrales 2018, 2017). Such racialized heteromasculinities fit into the demands and dynamics of the colonialscapes that posit Filipinos as cheap labourers, striving to be proper citizens in ways that naturalize the authority of settler state.

There is clearly considerable room for research and debate, and Indigenous critiques of liberal settler nations invite scholars of transnationalism to ask different questions, for example, how narratives of immigrant integration and success might reinforce and further naturalize the legitimacy of the settler-colonial state. Equally, what are the varying and resonant ways that the racial state has systematically disrupted and devalued Indigenous and migrant families? How has the colonial state in a variety of colonial contexts disciplined sexual and gender multiplicity as a technology of colonial governance and control (Hunt and Holmes 2015)? How have, and how might, solidarities and alliances be effectively forged across different experiences of colonialism and transnational displacement (within and beyond the nation state), so as to imagine other ways of living and relating (Lowe 2015)? These are the types of questions that open up new avenues of thinking and research within and beyond feminist transnational studies.

Conclusion

In this chapter, we have attended to contradictions and tensions in the concept and processes of transnationalism, in the first instance by noticing the tension between mobility and fixedness inherent in the term transnationalism. Focusing on the Filipino diaspora in particular, we have examined how transnational scholarship theorizes gender relations and sexualities in migration as a process of subject formation. We have outlined transnational approaches that show how processes of subject formation can both disrupt and reproduce dominant and normative paradigms and power relations in countries of origin and destination. We conclude by suggesting that transnational scholarship in geography should consider Indigenous critiques of liberal settler nations and states. In particular, in order to explore more fully the possibilities of movement and mobility, we highlight those Indigenous critiques that bring into question the presumed supremacy and normalcy of the settler state and nation. By bringing them into conversation with transnational approaches, transnational scholarship registers the ways in which immigration can function to both stabilize and destabilize settler colonialism. It extends and deepens the questions that we might ask about place, territory and nation to open possibilities for new relations of gender, sexuality and race.

Notes

1 http://vancouversun.com/news/local-news/anti-immigration-rally-at-vancouver-city-hall
2 For a critique of the Canadian state's recognition of nation-to-nation relationships with Indigenous peoples, see Coulthard (2014) and Simpson (2014).

Key readings

Farrales, M. 2018. "Repurposing Beauty Pageants: The Colonial Geographies of Filipina Pageants in Canada." *Environment and Planning D: Society and Space* 37: 46–64.

Hunt, S.E., and C. Holmes. 2015. "Everyday Decolonization: Living a Decolonizing Queer Politics." *Journal of Lesbian Studies* 19: 154–172.

Katz, C. 2001. "On the Grounds of Globalization: A Topography for Feminist Political Engagement." *Signs: Journal of Women in Culture and Society* 26: 1213–1234.

References

Aguilar, F. 2013. "Brother's Keeper? Siblingship, Overseas Migration, and Centripetal Ethnography in a Philippine Village." *Ethnography* 14 (3): 346–368.

Ahmed, S. 2000. *Strange Encounters: Embodied Others in Post-coloniality*. London: Routledge.

Benedicto, B. 2014. *Under Bright Lights: Gay Manila and the Global Scene*. Minneapolis: University of Minnesota Press.

Butler, J. 2008. "Sexual Politics, Torture, and Secular Time." *British Journal of Sociology* 59 (1): 1–23.

Byrd, J.A. 2011. *The Transit of Empire: Indigenous Critiques of Colonialism*. Minneapolis: University of Minnesota Press.

Coulthard, G.S. 2014. *Red Skin, White Masks: Rejecting the Colonial Politics of Recognition*. Minneapolis: University of Minnesota Press.

Cruz-Malave, A., and M. Manalansan. 2002. *Queer Globalizations: Citizenship and the Afterlife of Colonialism*. New York: NYU Press.

Daigle, M. 2016. "Awawanenitakik: The Spatial Politics of Recognition and Relational Geographies of Indigenous Self-Determination." *Canadian Geographer / Le Géographe Canadien* 60 (2): 259–269. Available at: https://doi.org/10.1111/cag.12260.

Day, I. 2016. *Alien Capital: Asian Racialization and the Logic of Settler Colonial Capitalism*. Durham, NC: Duke University Press.

Dreby, J. 2010. *Divided by Borders: Mexican Migrants and their Children*. Berkeley: University of California Press.

Fajardo, K.B. 2011. *Filipino Crosscurrents: Oceanographies of Seafaring, Masculinities, and Globalization*. Minneapolis: University of Minnesota Press.

Farrales, M. 2017. "Gendered Sexualities in Migration: Play, Pageantry, and the Politics of Performing Filipino-ness in Settler Colonial Canada." Dissertation, University of British Columbia. Available at: https://doi.org/10.14288/1.0354499.

Farrales, M. 2018. "Colonial, Settler Colonial Tactics and Filipino Canadian Heteronormativities at Play on the Basketball Court." In: *Diasporic Intimacies: Queer Filipinos and Canadian Imaginaries, Critical Insurgencies*, edited by R. Diaz, M. Largo, and F. Pino, 183–198. Evanston, IL: Northwestern University Press.

Federici, S. 2004. *Caliban and the Witch: Women, the Body and Primitive Accumulation*. Brooklyn, NY: Autonomedia.

Fraser, N. 2016. "Contradictions of Capital and Care." *New Left Review* 100: 99–117.

Gilmore, R.W. 2007. *Golden Gulag: Prisons, Surplus, Crisis and Opposition in Globalizing California*. Berkeley: University of California Press.

Global Commission on International Migration. 2005. *Global Migration for an Interconnected World: New Directions for Action*. Available at: www.gcim.org.

Gopinath, G. 2005. *Impossible Desires: Queer Diasporas and South Asian Public Cultures*. Durham, NC: Duke University Press.

Hunt, S.E. 2014. "Witnessing the Colonialscape: Lighting the Intimate Fires of Indigenous Legal Pluralism." Dissertation, Environment: Department of Geography. Available at: http://summit.sfu.ca/item/14145%23310.

Hunt, S.E., and C. Holmes. 2015. "Everyday Decolonization: Living a Decolonizing Queer Politics." *Journal of Lesbian Studies* 19: 154–172.

Johnston, C., and G. Pratt. 2017. "Tlingipino Bingo, Settler Colonialism and Other Futures." *Environment and Planning D: Society and Space* 35 (6): 971–993. Available at: https://doi.org/10.1177/0263775817730699.

Katz, C. 2001. "On the Grounds of Globalization: A Topography for Feminist Political Engagement." *Signs: Journal of Women in Culture and Society* 26: 1213–1234.

Lawrence, B., and E. Dua. 2005. "Decolonizing Antiracism." *Social Justice* 32: 4(102): 120–143.

Listerborn, C. 2015. "Geographies of the Veil: Violent Encounters in Urban Public Spaces in Malmo, Sweden." *Social and Cultural Geography* 16 (1): 95–115.

Lowe, L. 2015. *The Intimacies of Four Continents*. Durham, NC: Duke University Press.

Mahler, S.J., and P. Pessar. 2006. "Gender Matters: Ethnographers Bring Gender from the Periphery toward the Core of Migration Studies." *International Migration Review* 40 (1): 27–63.

Manalansan, M.F. 1995. "In the Shadows of Stonewall: Examining Gay Transnational Politics and the Diasporic Dilemma." *GLQ: A Journal of Lesbian and Gay Studies* 2 (4): 425–438. Available at: https://doi.org/10.1215/10642684-2-4-425.

Manalansan, M.F. 2003. *Global Divas: Filipino Gay Men in the Diaspora*. Durham, NC: Duke University Press.

McKay, D. 2007. "Sending Dollars Shows Feelings: Emotions and Economies in Filipino Migration." *Mobilities* 2(2): 175–194.

Mepschen, P., J.W. Duyvendak, and E. Tonkens. 2010. "Sexual Politics, Orientalism and Multicultural Citizenship in the Netherlands." *Sociology* 44 (5): 962–979.

Miller, S.C. 1982. *Benevolent Assimilation: The American Conquest of the Philippines, 1899–1903*. New Haven, CT: Yale University Press.

Million, D. 2014. "There is a River in Me: Theory from Life." In: *Theorizing Native Studies*, edited by A. Simpson and A. Smith, 31–42. Durham, NC: Duke University Press.

Nagar, R. 2014. *Muddying the Waters: Coauthoring Feminisms across Scholarship and Activism*. Champaign: University of Illinois Press.

Parreñas, R.S. 2001. *Servants of Globalization: Women, Migration and Domestic Work*. Stanford, CA: Stanford University Press.

Parreñas, R.S. 2005a. *Children of Global Migration*. Stanford: Stanford University Press.

Parreñas, R.S. 2005b. "Long Distance Intimacy: Class, Gender and Intergenerational Relations Between Mothers and Children in Filipino Transnational Families." *Global Networks* 5: 317–333.

Parreñas, R.S. 2010. "Transnational Mothering: A Source of Gender Conflicts in the Family." *North Carolina Law Review* 88: 1825–1856.

Paul, A.M. 2011. "Stepwise International Migration: A multi-stage migration pattern for the aspiring migrant." *American Journal of Sociology* 116 (6): 1842–1886. Available at: doi:10.1080/09663699725422.

Paul, A.M. 2015. "Capital and Mobility in the Stepwise International Migrations of Filipino Migrant Domestic Workers." *Migration Studies* 3 (3): 438–459. Available at: doi:10.1093/migration/mnv014.

Paul, A.M. 2017. *Multinational Maids: Stepwise Migration in a Global Labor Market*. Cambridge: Cambridge University Press.

Philippine Overseas Employment Administration. 2010. *Overseas OFW Deployment by Skill and Country for the Year.* Available at: www.poea.gov.ph/ofwstat/depperskill/2010.pdf.

Pido, E.J. 2017. *Migrant Returns: Manila, Development and Transnational Connectivity*. Durham, NC: Duke University Press.

Pratt, G. in collaboration with the Filipino-Canadian Youth Alliance. 2003/2004. "Between Homes: Displacement and Belonging for Second-Generation Filipino-Canadian Youths." *BC Studies* 140: 41–68.

Pratt, G. and B.C. Migrante. 2018. "Organising Filipina Domestic Workers in Vancouver, Canada: Gendered Geographies and Community Mobilization." *Political Power and Social Theory* 35: 99–119.

Pratt, G., and B.S.A. Yeoh. 2003. "Transnational (Counter) Topographies." *Gender, Place & Culture* 10 (2): 159–166. Available at: https://doi.org/10.1080/0966369032000079541.

Puar, J. 2007. *Terrorist Assemblages: Homonationalism in Queer Times*. Durham, NC: Duke University Press.

Pulido, L. 2018. "Geographies of Race and Ethnicity III: Settler Colonialism and Nonnative People of Color." *Progress in Human Geography* 42 (2): 309–318. Available at: https://doi.org/10.1177/0309132516686011.

Razack, S., M. Smith, and S. Thobani, eds. 2010. *States of Race: Critical Race Feminism for the 21st Century*. Toronto: Between the Lines.

Robinson, C. 1983. *Black Marxism: The Making of the Black Radical Tradition*. Chapel Hill: University of South Carolina Press.

Saranillio, D.I. 2013. "Why Asian Settler Colonialism Matters: A Thought Piece on Critiques, Debates, and Indigenous Difference." *Settler Colonial Studies* 3 (3–4): 280–294. Available at: https://doi.org/10.1080/2201473X.2013.810697.

Silvey, R. 2006. "Geographies of Gender and Migration: Spatializing Social Difference." *International Migration Review* 40 (1): 64–81.

Simpson, A. 2014. *Mohawk Interruptus: Political Life Across the Borders of Settler States*. Durham, NC: Duke University Press.

Sugg, K. 2003. "Suspended Migrations: Sexuality and Cultural Memory in Achy Obejas and Carmelita Tropicana." *Environment and Planning D: Society and Space* 21: 461–477.

Swarr, A. and R. Nagar, eds. 2010. *Critical Transnational Feminist Praxis*. Albany: SUNY Press.

Tadiar, N.X.M. 2004. *Fantasy Production: Sexual Economies and Other Philippine Consequences for the New World Order*. Hong Kong: Hong Kong University Press.

Tadiar, N.X.M. 2015. "Decolonization, 'Race', and Remaindered Life Under Empire." *Qui Parle* 23 (2): 135–160.

Thangaraj, S.I. 2015. *Desi Hoop Dreams: Pickup Basketball and the Making of Asian American Masculinity*. London and New York: NYU Press.

Tuck, E., and K.W. Yang. 2012. "Decolonization Is Not a Metaphor." *Decolonization: Indigeneity, Education & Society* 1 (1). Available at: www.decolonization.org/index.php/des/article/download/18630.

Walia, H. 2013. *Undoing Border Imperialism*. Oakland, CA: AK Press.

Wolfe, P. 2006. "Settler Colonialism and the Elimination of the Native." *Journal of Genocide Research* 8 (4): 387–409.

Yeoh, B.S.A., and K. Rambas. 2014. "Gender, Migration, Mobility and Transnationalism." *Gender, Place & Culture* 21 (10): 1197–1213. Available at: https://doi.org/10.1080/0966369X.2014.969686.

22
MOBILITIES AND CITIZENSHIP

Tamir Arviv and Symon James-Wilson

Introduction

Citizenship has commonly been conceptualized as the rights and responsibilities of people organized in and through a single, territorially bound nation state (Purcell 2003). Within this framework, citizenship refers to a person's individual legal status and their social identification with a state's specific cultural narrative. Citizens are imagined as political actors who agree to a 'social contract' with the state. They revoke elements of their autonomy, and they consent to being ruled by the state in exchange for access to certain privileges, protections and mobilities.

British sociologist T.H. Marshall's 1950 definition of citizenship, which remains influential to this day, conceptualizes citizenship as 'a status bestowed on those who are full members of a community. All who possess the status are equal with respect to the rights and duties with which the status is endowed' (28–29). In Marshall's view, the realization of the state and its citizens' 'social contract' hinges on whether members have full, unprejudiced access to civil, political and social rights conferred by the state. When imagined this way, citizenship and belonging are predicated on individuals' collective association with a sense of shared national past and unilateral cultural history. A citizen's primary political community and loyalty must lie at the scale of the nation state, even if they might be members of, and active participants in, other subordinated or subaltern political communities (Purcell 2003).

While these traditional understandings of what constitutes citizenship have been hugely impactful and enduring in many cases, important critical scholarship began to emerge in the 1970s to challenge liberal theorists' narrow comprehension of citizenship as a universally equitable and inclusive vehicle of social ordering. Feminist and critical geographers were among the first radical scholars to push for alternative conceptions of citizenship that were less contingent on authoritative discourses of nationalism and state-centrism. As a result, there has been increasingly thoughtful and robust engagement with the social and geographic unevenness of citizenship, particularly in terms of how gender and race-based power hierarchies operate in contradiction to the mainstream discourse of universal equality.

Actively interrogating the societal strategies designed to systematically exclude women and marginalized communities from full citizenship, this pivotal turn in citizenship and mobilities studies challenged the assumed homogeneity and universality within the normative conception of citizenship. This scholarship charted a new course for the field by attending to the politics of identity, place and scale among diverse gendered, racialized, sexualized and classed actors' enactments of citizenship more deliberately and meaningfully (see Bell 1995; Fenster 2005; Mahtani 2002a).

Central to this intellectual paradigm shift was the move away from methodological nationalism; the idea that social life logically and automatically takes place within the nation state framework (Wimmer and Glick Schiller 2002). Post-colonial scholars such as Stuart Hall (1990), Paul Gilroy (1993) and Homi Bhabha (1994) offered important critiques of methodological nationalism's racist and exclusionary notion of a so-called 'pure' national identity. Their scholarship importantly troubled methodological nationalism's claims to be apolitical by drawing attention to how the theory was centrally hinged on historic inaccuracies and over-simplifications. Methodological nationalism's assertion that people live their lives in one place, according to a single set of national and cultural norms bounded by impermeable territorial borders, not only ignores historic routes and global flows of goods, ideas and people (e.g. the Silk Road and the transatlantic slave trade) but also more recent circulations that were catalyzed by decolonizing movements in the former Third World, transnational migration of post-colonial racialized communities and the explosive rise in urbanization globally. Hall, Gilroy and Bhabha's work considered the myriad techniques that diverse groups of people have used to make claims to multiple forms of citizenship and belonging across space and time.

Attending to these new geographies, feminist scholars and critical race theorists have increasingly studied how citizenship shapes everyday life and the ways individuals constitute their membership to local, national and transnational publics. Building on anti-essentialist critiques, their studies of citizenship have continued to shift away from state-centred perspectives towards more careful examinations of individuals' and groups' claims to, and enactments of, citizenship. In particular, careful attention has been paid to individuals and groups who belong to socio-political spaces in more than one nation state simultaneously (Basch, Glick Schiller and Szanton Blanc 1994). They are commonly referred to as 'transnational migrants' or 'transmigrants'. With growing numbers in large urban centres generally, and global cities especially, migrants with undocumented, 'illegal' or irregular status have been the focus of numerous scholarly articles and international policy reports (Holston and Appadurai 1999).

In conjunction with these advancements in citizenship theory, several incipient socio-spatial processes began to redefine the meaning of citizenship in the twenty-first century, including but not limited to: a) economic globalization in late-stage capitalism; b) rising rates of transnational labour migration; c) the global expansion of information and communication technologies (ICTs); and d) the burgeoning influence of multinational corporations and NGOs. Primarily, these interrelated forces of change have amplified the number of transnational communities that are based on economic interests, cultural exchanges, social relations and political affiliations that operate at scales beyond a single city or nation state (Basch, Glick Schiller and Szanton Blanc 1994). Additionally, these processes have complicated social actors' experiences of time–space compression and uneven power geometries, as multiple axes of difference, such as race, class and gender, have enhanced the physical and social mobility of some at the expense of others (Massey 1994).

Citizenship is a socially and spatially negotiated, emotionally saturated and boundary-breaking process. As diverse populations – including transmigrants, ethnic minorities and other historically marginalized subjects – continue to migrate to cities around the world, enactments of citizenship occur not only at border crossings but in everyday spaces like private homes, public schools and neighbourhood recreation centres. In this chapter, case studies of urban diasporic community-building in several global cities are presented to provide contemporary examples of the current tensions and alternative futures in citizenship and mobility studies.

Citizenship and mobility studies in the twenty-first century global city

Since the mid- to late-twentieth century, global cities have emerged as a new kind of 'command and control' centre from which the world economy is managed and serviced (Sassen 1991). Global cities – such as Hong Kong, Mumbai, London and New York – have increasingly relied on labour migrants to fulfil precarious service sector jobs. Narratives that emphasize global cities' culturally progressive image, ample economic opportunities and ethnic social networks have enticed a growing number of migrants, despite the intense race, class and gender-based discrimination that they are met with.

Feminist geographers have used qualitative research methodologies – including interviews, participant observations and content analysis – to explore the contested negotiation processes in which migrants in global cities engage to advocate for their often-denied human and legal rights. This research has inspired other critical geographers and migration scholars to investigate the politics of identity and scale. Specifically, the resulting body of literature has collectively worked to unsettle spatial entrapment theses, paying greater attention to the politics of representation, identity formation and multiple human territorial strategies across geographic scales (Fenster 2005, 218).

Considering new geographies of citizenship within and beyond the nation state, these studies have examined the interplay between mobility and confinement in diverse urban contexts more thoroughly. They have notably rejected historic narratives of static marginality, which insisted that there are always clear winners and clear losers in socio-spatial negotiations. Exploring the possibility of 'in-betweeness', these much-needed conceptual developments have contributed a more nuanced understanding of human agency and place-making practices. This has, in turn, highlighted the more transformative potentials for citizenship and belonging in global cities.

Turning to an example from the East and Southeast Asian context, recent studies on the socio-political organizing and place-making practices of female migrant domestic workers in East and Southeast Asian global cities have investigated transnational geographies of agency and knowledge production with a greater degree of intricacy (Lyons and Yee 2009). In Singapore, female migrant domestic workers' occupation of state and private-interest controlled spaces – including Singapore's Botanic Gardens, the Lucky Plaza shopping centre and places of worship – on their weekly rest day (mandated by Singaporean law since January 2013) have provided robust evidence of the ways in which domestic workers' place-making practices empower their individual and collective agency (Lyons and Yee 2009; Rahman 2005; Yeoh and Annadhurai 2008). In Hong Kong and Taipei, migrant women have politically organized around a wide range of topics, including human rights abuses, low wages and poor working in conditions (Constable 2007). Non-governmental groups such as the Hong Kong Confederation of Trade Unions (HKCTU) have made important legal claims to space for domestic workers and created spaces for women to advocate for change on their own terms (Constable 2009).

This literature on migrant domestic workers' place-making processes has advanced discourse on the relationship between cities and citizenship ('urban citizenship') in numerous ways. Principally, it has demanded that greater attention be paid to the gendered, cultural, ethnic and racial contours and local specificities of place-making practices. Further, feminist geographers and sociologists in this topical field have concurrently detailed the need for a greater distinction between citizenship in the formal, legal sense (in this case, citizenship as a legal status that binds individuals to a nation state by conferring them with certain rights) and citizenship in an informal or substantive sense (meaning a more complex and expressive sense of citizenship, primarily defined by both material and symbolic belonging to one or more political communities) (Staeheli 2003).

Citizenship, migration, and the 'right to the city'

Activists and scholars who have explored the links between citizenship, migration and cities have drawn considerable insight from French Marxist philosopher Henri Lefebvre's conceptualization of the 'right to the city'. Lefebvre challenged traditional notions of 'the citizen' by offering a definition of citizenship that centres more on abstract imaginations of belonging rather than on formal political and/or legal status (Lefebvre [1968] 1996). Lefebvre did not define citizenship as membership of a nation state, but instead contended that anyone who inhabits the city is eligible for citizenship and the right to the city, thereby entitling all urban dwellers to a role in the decision-making processes within the city that they inhabit (Purcell 2003). This more inclusive idea of citizenship includes individuals' rights to 'full and complete usage' and participation in the social, cultural, political and economic spaces they create in the city over the course of their everyday lives (Lefebvre 1996/1968). Lefebvre's radical retheorization of political community and citizenship beyond the state is particularly relevant to migrant workers, who have traditionally been denied nationalist forms of belonging in the global city era – usually due to their precarious legal status.

David Harvey, who expanded on Lefebvre's 'right to the city' framework (Harvey 2008), has further illustrated how cities are dynamic places for gendered, racialized and classed actors to mobilize (Harvey 2012). Citizens have the right to use urban environments not solely in service of producing economic growth but also as spaces of vibrant geographic imaginations and realizations of social and political justice (Harvey 2012). The urbanization of collective action in contemporary cities – especially global cities – has rendered the city a central arena for citizenship claim-making and identity formation (Holston and Appadurai 1999). As Isin (2002) notes, the modern city has become 'the battleground through which groups define their identities, stake their claims, wage their battles, and articulate citizenship rights and obligations … the city as a difference machine relentlessly provokes, differentiates, positions, mobilizes, immobilizes, oppresses, liberates' (50). The city is no longer a passive, contained arena within the state polity status quo.

In this view, citizenship is actively produced through relational, cultural and spatial acts that enable subjects to constitute themselves as citizens. There is a range of tactics and strategies that migrant groups and other marginalized groups have used to challenge inherited, normative conceptions of liberal democracies and public urban space. Using strategic sites throughout the city, these citizens have carved out spaces to express themselves, make demands and call on the state for more protective regulations.

Literature from the queer geographies subfield, in particular, has explored gender and sexuality's evolving influence on the politics of citizenship and belonging in urban spaces today. Positioning queernesses' refusal to be 'located' within a singular spatial, temporal or material reality, queer geographies have engaged human agency by moving between and beyond urban, global and transnational scales (Duff 2010). Destabilizing the notion that sexualized social actors are static space-occupiers of hetero-/homo-normative cityscapes, the evolution of queer geographies has motioned new analytical demands that push Lefebvre and Harvey's 'right to the city' discussion even further.

As the global city discourse has gained momentum, contested struggles for the 'right to the city' have catalyzed queer geography researchers to develop new methodological techniques.

Shifting away from a Cartesian planar analysis, place-based case study models quickly became the emblematic method of the subdiscipline (Binnie and Valentine 1999; Brown 2014). Beginning with particularly public health-focused agendas (following the proliferation of HIV/AIDS and problematic substance use in the 1980s), cities such as New York, Miami Beach and London became important sites for local, regional, and increasingly international comparisons (Gandy 2012; Kanai and Kenttamaa-Squires 2015). Informed by the relational geographies' turn, more-embodied and ethnographic approaches were gradually adopted to further ground multiple 'city-makings' beyond the 'ghetto of community studies' (Binnie and Valentine 1999, 181). Survey, interview and participant-as-observer methods have frequently been employed to explore the real and imagined peripatetic urban im/mobilities of sexualized citizens (Binnie and Valentine 1999).

Queer geographies' contributions to the 'right to the city' literature has not only challenged the notion that homosexual territoriality is confined to homonormative spaces but has gone a step further to argue that it is, in fact, through contested spatial transgressions, infiltrations and connections that multiple sexualized human and nonhuman agents relate to and perform 'queer' (see Hubbard 2008; Kanai and Kenttamaa-Squires 2015). Now situated as politically, socially and economically pervasive – rather than spatially bound or marginal – the field of queer geographies has importantly expanded the kinds of sites that geographers look to when investigating gender's and sexualities' influence on contemporary forms of citizenship and belonging in urban spaces (Binnie and Valentine 1999).

Citizenship, migration and (trans)national identities in the global city: case studies from Toronto, Canada

For Isin (2002), urban space is a condition for *becoming and being political* (43–45). Individuals and groups explore their identities both in and through public spaces. The city is a site in which: a) the lives of people are organized, assembled together and rendered meaningful; b) there is socialization into various identities; and c) individuals develop both their unique sense of self and their sense of belonging to broader community formations beyond themselves (266).

For transnational migrants in particular, citizenship in the contemporary global city involves a variety of complex processes. This includes having to navigate multiple negotiations between conflicting national identities as they make human and legal rights claims. Isin and Simiatycki's work (2002) on the struggles of recent Muslim migrants to build mosques in Toronto, Canada, concludes that 'citizenship is about making a place and about identifying with boundaries, markers, and identities of place' (208). Their research uncovered notable insights into the racialized structure of nationality and citizenship in contemporary Canada. In particular, Isin and Simiatycki's (2002) findings assert that these intercultural struggles over place-making were about more than just establishing houses of worship where newcomers could practise their religious freedoms and faith. These contestations also involved Muslims actively seeking a material articulation that could represent their presence in the Western cosmopolis of Toronto, both symbolically and spatially.

Louisa Veronis' study (2007) of Barrio Latino – a Latino ethnic neighbourhood in Toronto – demonstrates that ethnic urban spaces often serve a complex duality in the contemporary global city. On the one hand, ethnic urban spaces reinforce racialized logics of neoliberal multiculturalism. These neighbourhoods are often formed on the basis of shared experience of 'othering' and exclusion and, as a result, they are characteristically both fixed and flexible. This was true of the Barrio Latino community centre, 'Casa', that Veronis (2007) investigated. On the other hand, ethnic urban spaces also operate as critical sites for resisting race, class and gender-based discrimination. They can create vital opportunities for political mobilization and participation in the city, especially for recent migrants. These spaces not only amplify individual voices from the community but simultaneously support the re/production of a shared local identity. In this way, community spaces in ethnic neighbourhoods play an integral role in minoritized groups' individual and collective urban identity formation. Consequently, residents and stakeholders develop a sense of community that is territorialized yet not necessarily tokenistic or essentialized.

Public spaces in the contemporary global city – such as the street and the square – are essential sites for social groups to make visible their demands and constitute themselves as 'public' citizens (Mitchell 2003, 129). Urban public spaces have tremendous social value, in that they provide considerable opportunities for interpersonal exchange. They are crucial to the formation of a healthy civic culture (Amin 2008; Young 1990) and an inclusive public sphere (Kohn 2004). In addition to creating new ethnic spaces, recent migrants' temporary appropriations of urban public space for transnational public events – such as patriotic demonstrations, national cultural festivals and religious rituals – have had transformative impacts on city life. This has become a central focus of recent literature (see Arviv 2017; Ashutosh 2013; Cohen 2011; Oosterbaan 2014; Veronis 2006).

For example, in her study on the 'Canadian Hispanic Day Parade' (CHDP) in Toronto, Veronis (2006) argues that Latin Americans and other minoritized groups use transnational public events to assert their membership of the Canadian polity and lay claim to full and equal citizenship rights. Importantly, Veronis (2006) notes that the organizers' choice to host CHDP in Toronto's Jane and Finch neighbourhood was highly deliberate. Jane Street – which features numerous privatized spaces that were purposefully designed to preclude spontaneous social encounters – is creatively refashioned into a democratic public space for meaningful representation and participation in city life through CHDP's place-making process.

Other feminist and critical race theory (CRT) geographers have added to these conversations by launching in-depth inquiries into the role that gendered and racialized power hierarchies play in urban citizens' encounters and participation in public space. Scholars like Ruddick (1996) have found that cities often function as localities where people begin to understand themselves and their sense of place through other people's points of view (Ruddick 1996). In urban public space, as Isin (2002) explains, racialized groups tactfully decide whether to emphasize solidarity or differentiation with other social groups. These choices are negotiated differently, depending on whether the intention is to emphasize solidarity or distinctiveness. This has meant that individuals and social groups have had to learn how to nuance their performances of race, class, gender and sexuality strategically in ways that complicate – and in some cases directly challenge – established power hierarchies (Mahtani 2002b).

Ashutosh (2013), for example, explores the concurrent protests led by the Sri Lankan Tamil diaspora in Toronto, London and Oslo, Norway that arose in response to the escalation of violence in Sri Lanka between 2008 and 2009. Ashutosh (2013) argues that these protests represent *transnational acts of citizenship.* These acts were positioned as being both against the limits of national membership in Sri Lanka and as being in search for new forms of belonging in the Tamil diaspora. Notably, the protests reject the narrow categorization of 'citizens' and 'outsiders'. This choice was a direct refusal of binaristic nationalism and, instead, reflected a commitment to placing citizenship in a larger, transnational context.

Ashutosh (2013) concludes that immigrants' protests are acts against the limits of political membership and (un)belonging that underwrite citizenship. These protests represent grounded and concrete acts of citizenship through which transmigrant groups can 're-imagine belonging beyond the territories of the nation state' (Ashutosh 2013, 198). They allow for assertions and performances of multiple political memberships and expressions of belonging that move in counter-rhythm to dominant narratives of race and nation.

Arviv (2017) conceptualizes pro-Israeli public events in Toronto as *racialized spatial acts of citizenship* in which Jewish–Israeli activists imagine, perform and negotiate their multiple political memberships and loyalties – to Canada, Israel, to the local Jewish community and to 'the West'. By temporarily aligning themselves with Canadian-born Jews (i.e. strategically suspending intra-Jewish cultural, political, racial and sexual differences), these migrants perform their belonging to the idea of a unified White, Western and Zionist Jewish collective. By carrying Israeli and Canadian flags, they express their political loyalties to both Israel and Canada.

These Jewish–Israeli activists' actions represent racialized spatial performances of 'Western citizenship'. While their imaginations and representations of citizenship subvert the conventional conceptions of citizenship (as an identification with a single and territorially defined nation state), they nevertheless confirm the emotional and racialized rhetoric of citizenship and nationhood in the sending state (Israel), the receiving state (Canada) and across the West (Pain and Smith 2008). As the Jewish–Israeli activists perform the role of defenders of Israel and 'Western civilization', they aim not only to critique local and transnational pro-Palestinian groups, who they claim are anti-Western and anti-Semitic radical Islam, but also to demonstrate their alliances with non-Jewish Whites in Canada.

Beyond celebration

Studies by feminist geographers, grounded in ethnographic methods, have provided evidence of urban transnational and diasporic communities' multiple allegiances and solidarities and, in turn, challenged the presumed sameness of citizenship. Notably, these findings have illuminated alternative ways of conceptualizing local, national and global forms of belonging, while also cautioning uncritical celebrations of these non-national forms of citizenship as universally progressive and inclusive. This is especially significant, given contemporary global city contexts where movements for cultural protectionism, nativism and popularism have proliferated amid new types of neoliberal and geopolitical alliances. As migrants and religious minorities have increasingly been positioned as cultural others within national communities, even if those nation states are considered officially multicultural (Hopkins 2016; Razack 2007), the urban scale has gained considerable relevance to critical scholarship.

While the powerful new forces of twenty-first century globalization and transnationalism have somewhat decoupled citizenship, identity and political loyalty from the space of the nation state, they have not diminished the nation state's power. The proliferation of new transnational and supranational institutions has allowed some nation states to maintain power and expand their bordering practices. Borders, in this era of globalization, have become progressively more porous for financial goods and transnational business elites while barring refugee claimants and low-waged service workers. For example, in global cities like Hong Kong expatriate migrants from countries such as the UK, the US, Australia and Canada can enjoy citizenship rights and participate fully in political practices (such as voting in federal elections) in more than one nation state, while there are no pathways for formal citizenship for migrant domestic workers or construction workers from countries like Bangladesh and the Philippines. Ong (1999) conceptualizes this ease of mobility for some as 'flexible citizenship'.

Moreover, sending states are still playing an important role in transmigrants' lives. Michael Peter Smith (2003) points to the paradox whereby the growth in flexible transnational connections has, in fact, entrenched the links between nation states and their citizens in other parts of the world. In the last few decades, many sending states have transnationalized citizenship and nationhood in an attempt to 'recapture' the investments, remittances and loyalties of their citizens living abroad. This political formation has been called the 'deterritorialized' nation state (Basch, Glick Schiller and Szanton Blanc 1994).

The deterritorialization process includes creating formal channels for communication across national borders. For example, since the 1990s, the Israeli state has launched a large number of outreach programmes and provided new services in several emigrant destinations through 'the Israeli house' project (Cohen 2007). These initiatives were designed to encourage its Jewish–Israeli citizens living abroad to return 'home', or at the very least to reassert control over their citizens' political identifications, attachments and participations (Cohen 2007). As a consequence of deterritorialization, transmigrants have increasingly had to counterbalance their affiliations in multiple, often competing, communities (Guarnizo and Smith 1998).

Finally, recent empirical studies in geography that highlight transmigrants' personal perspectives on their lived experiences demonstrate that the polyvalence of political identities, allegiances and practices does not erase national-territorial identifications and meaning systems. Migrants are often involved in complex negotiations between multiple and sometimes conflicting national identities, loyalties and commitments. They express a desire to become formal national citizens in their new country of residence while maintaining their 'old' citizenship status – thereby participating in more than one national community. Interestingly, national identity and commitment to the sending state figure prominently, even as migrants plan for a future in their host country (Arviv 2017; Cohen 2011; Leitner and Ehrkamp 2006).

Key readings

Fenster, T. 2005. "The Right to the Gendered City: Different Formations of Belonging in Everyday Life." *Journal of Gender Studies* 14 (3): 217–231.

Isin, E.F., and G.M. Nielsen, eds. 2013. *Acts of Citizenship*. London: Zed Books.

Young, I.M. 2011. *Justice and the Politics of Difference*. Princeton, NJ: Princeton University Press.

References

Amin, A. 2008. "Collective Culture and Urban Public Space." *City* 12 (1): 5–24.

Arviv, T. 2017. "Fear, Imagination, and Public Representations of 'Western Citizenship' Among Jewish–Israeli Activists in Toronto." *ACME–An International E-Journal for Critical Geographies* 16 (3): 383–404.

Ashutosh, I. 2013. "Immigrant Protests in Toronto Diaspora and Sri Lanka's Civil War." *Citizenship Studies* 17 (2): 197–210.

Basch, L., N. Glick Schiller, and C. Szanton Blanc. 1994. *Nations Unbound: Transnational Projects, Postcolonial Predicaments, and Deterritorialized Nation-States*. London: Routledge.

Bell, D. 1995. "Pleasure and Danger: The Paradoxical Spaces of Sexual Citizenship." *Political Geography* 14 (2): 139–153.

Bhabha, H.K. 1994. *The Location of Culture*. London: Routledge.

Binnie, J., and G. Valentine. 1999. "Geographies of Sexuality – A Review of Progress." *Progress in Human Geography* 23 (2): 175–187.

Brown, M. 2014. "Gender and Sexuality II: There Goes the Gayborhood?" *Progress in Human Geography* 38 (3): 457–465.

Cohen, N. 2007. "From Overt Rejection to Enthusiastic Embracement: Changing State Discourses On Israeli Emigration." *Geojournal* 68 (2–3): 267–278.

Cohen, N. 2011. "Rights Beyond Borders: Everyday Politics of Citizenship in the Israeli Diaspora." *Journal of Ethnic and Migration Studies* 37 (7): 1137–1153.

Constable, N. 2007. *Maid to Order in Hong Kong: Stories of Migrant Workers,* 2nd edition. Ithaca, NY: Cornell University Press.

Constable, N. 2009. "Migrant Workers and Many States of Protest in Hong Kong." *Critical Asian Studies* 41 (1): 143–164.

Duff, C. 2010. "On the Role of Affect and Practice in the Production of Place." *Environment and Planning D: Society and Space* 28 (5): 881–895.
Fenster, T. 2005. "The Right to the Gendered City: Different Formations of Belonging in Everyday Life." *Journal of Gender Studies* 14: 217–231.
Gandy, M. 2012. "Queer Ecology: Nature, Sexuality, and Heterotopic Alliances." *Environment and Planning D: Society and Space* 30 (4): 727–747.
Gilroy, P. 1993. *The Black Atlantic: Modernity and Double Consciousness*. Cambridge, MA: Harvard University Press.
Guarnizo, L.E., and M.P. Smith. 1998. "The Locations of Transnationalism." In: *Transnationalism from Below*, edited by M.P. Smith and L.E. Guarnizo, 3–34. New Brunswick, NJ: Transaction Books.
Hall, S. 1990. "Cultural Identity and Diaspora." In: *Identity: Community, Culture, Difference*, edited by J. Rutherford, 222–237. Lawrence and Wishart, London.
Harvey, D. 2008. "The Right to the City." *New Left Review* 53: 23–39.
Harvey, D. 2012. *Rebel Cities: From the Right to the City to the Urban Revolution*. New York: Verso.
Holston, J., and A. Appadurai. 1999. "Introduction: Cities and Citizenship." In: *Cities and Citizenship*, edited by J. Holston, 1–18. Durham, NC: Duke University Press.
Hopkins, P. 2016. "Gendering Islamophobia, Racism and White Supremacy: Gendered Violence Against Those Who Look Muslim." *Dialogues in Human Geography* 6 (2): 186–189.
Hubbard, P. 2008. "Here, There, Everywhere: The Ubiquitous Geographies of Heteronormativity." *Geography Compass* 2 (3): 640–658.
Isin, E.F. 2002. *Being Political: Genealogies of Citizenship*. Minneapolis: University of Minneapolis Press.
Isin, E.F., and M. Simiatycki. 2002. "Making Space for Mosques: Struggles for Urban Citizenship in Diasporic Toronto." In: *Race, Space and the Law: Unmapping a White Settler Society*, edited by S. Razack, 185–209. Ottawa: Between the Lines Press.
Kanai, J.M., and K. Kenttamaa-Squires. 2015. "Remaking South Beach: Metropolitan Gayborhood Trajectories Under Homonormative Entrepreneurialism." *Urban Geography* 36 (3): 385–402.
Kohn, M. 2004. *Brave New Neighbourhoods: The Privatization of Public Space*. New York: Routledge.
Lefebvre, H. [1968]1996. "Right to the City." In: *Writings on Cities: Henri Lefebvre*, edited by E. Kofman and E. Lebas, translated by E. Kofman and E. Lebas, 61–181. Oxford: Blackwell.
Leitner, H., and P. Ehrkamp. 2006. "Transnationalism and Migrants' Imaginings of Citizenship." *Environment and Planning A* 38 (9): 1615–1632.
Lyons, L., and Yee, Y.C. 2009. "Migrant Rights in Singapore: Political Claims and Strategies in Human Rights Struggles in Singapore." *Critical Asian Studies* 41 (4): 575–604.
Mahtani, M. 2002a. "Interrogating the Hyphen-Nation: Canadian Multicultural Policy and 'Mixed Race' Identities." *Social Identities* 8 (1): 67–90.
Mahtani, M. 2002b. "Tricking the Border Guards: Performing Race." *Environment and Planning D* 20 (4): 425–440.
Marshall, T.H. 1950. *Citizenship and Social Class*. Cambridge: Cambridge University Press.
Massey, D. 1994. *Space, Place, and Gender*. Minneapolis: University of Minnesota Press.
Mitchell, D. 2003. *The Right to the City: Social Justice and the Fight for Public Space*. New York: Guilford Press.
Ong, A. 1999. *Flexible Citizenship: The Cultural Logics of Transnationality*. Durham, NC: Duke University Press.
Oosterbaan, M. 2014. "Public Religion and Urban Space in Europe." *Social and Cultural Geography* 15 (6): 591–602.
Pain, R., and Smith, S.J. 2008. *Fear: Critical Geo-Politics and Everyday Life*. Aldershot: Ashgate.
Purcell, M. 2003. "Citizenship and the Right to the Global City: Reimagining the Capitalist World Order." *International Journal of Urban and Regional Research* 27 (3): 564–590.
Rahman, N.A. 2005. "Shaping the Migrant Institution: The Agency of Indonesian Domestic Workers in Singapore." In: *The Agency of Women in Asia*, edited by L. Parker, 182–216. Singapore: Marshall Cavendish Academic.
Razack, S. 2007. *Casting Out: The Eviction of Muslims From Western Law and Politics*. Toronto: University of Toronto Press.
Ruddick, S. 1996. "Constructing Difference in Public Spaces: Race, Class, and Gender as Interlocking Systems." *Urban Geography* 17 (2): 132–151.
Sassen, S. 1991. *The Global City: New York, London, Tokyo*. Princeton, NJ: Princeton University Press.
Sim, A. (2003). "Organising Discontent: NGOs for Southeast Asian Migrant Workers in Hong Kong." *Asian Journal of Social Science* 31 (3): 478–510.

Smith, M.P. 2003. "Transnationalism and Citizenship." In: *Approaching Transnationalisms*, edited by B.S. Yeoh, K.P. Lai, M.W Charney, and T.C. Kiong, 15–38. New York: Springer.
Staeheli, L.A. 2003. "Introduction: Cities and Citizenship." *Urban Geography* 24 (2): 97–102.
Veronis, L. 2006. "The Canadian Hispanic Day Parade, Or How Latin American Immigrants Practice Suburban Citizenship in Toronto." *Environment and Planning A* 38 (9): 1653–1671.
Veronis, L. 2007. "Strategic Spatial Essentialism: Latin Americans' Real and Imagined Geographies of Belonging in Toronto." *Social and Cultural Geography* 8 (3): 455–473.
Wimmer, A., and N. Glick Schiller. 2002. "Methodological Nationalism and Beyond: Nation–State Building, Migration and the Social Sciences." *Global Networks* 2 (4): 301–334.
Yeoh, B.S.A., and K. Annadhurai. 2008. "Civil Society Action and the Creation of 'Transformative' Spaces for Migrant Domestic Workers in Singapore." *Women's Studies* 37: 548–569.
Young, I.M. 1990. *Justice and the Politics of Difference*. Princeton, NJ: Princeton University Press.

23
GEOGRAPHIES OF GENDERED MIGRATION
Place as difference and connection

Eleonore Kofman and Parvati Raghuram

Introduction

In the past two decades, there has been a resurgence of interest in migration and its gendered nature as the population on the move has grown. In this chapter, we will explore the reasons for its importance, the key achievements of existing research and some future directions. Towards this end, we explore how one key geographical concept, place, is deployed in two ways: as a marker of difference; and through theorizations that connect places in different ways. In doing so, it highlights how geographies of gendered migration bring place into play in various ways.

The chapter begins with an empirical outline of gendered migration processes, including some of the key stocks and flows in contemporary migration at a time when the numbers, theorizations and research questions that female migration[1] poses are receiving increasing attention. The subsequent section identifies selected significant concerns that have shaped theoretical discussions of gendered migration and draws out how place has been viewed in these debates. The chapter ends by summarizing a research agenda for geographies of migration that encompasses both missing geographies and incomplete theorizations.

Geographies of gendered migration

In 2015 there were 244 million international migrants, 3.3 per cent of the world's population. Much of this migration was from middle-income towards high-income countries. The largest senders at that time were India (15.6 million), Mexico (12.3 million), Russia (10.6 million), China (9.5 million) and Bangladesh (7.2 million), while the largest country of immigration was the US. The other important destination countries were Germany (12.0 million), Russia (11.6 million), Saudi Arabia (10.2 million) and the UK (8.5 million) (International Organisation for Migration [IOM] 2017). Intra-regional migration also accounts for significant proportions of migrant flows. Recent conflicts, especially in the Middle East, have led to a sharp increase in refugee stocks, which in 2017 were calculated at 22.1 million people globally (United Nations High Commissioner for Refugees [UNHCR] 2017).

International migration is overshadowed by the much larger flows of internal migration, estimated at 763 million in 2013 (United Nations Department for Economic and Social Affairs [UNDESA] 2013). Data on internal migration are limited in both availability and accuracy, but in 2010 Chan (2013) estimated that there were 155 million such migrants in China alone. Moreover, the growth in urbanization globally has generally been driven by migration. Although internal migration, and especially internal displacement of people within a country due to conflict, is an important socio-economic issue, particularly in the Global South, public debates have centred on the problem of international migration. Tied in with nationalism and xenophobia, it has created political ripples in many countries. Given its importance and for reasons of brevity, we will view the question of migration through an international lens, incorporating internal migration only as it meshes with international migration.

Women are an important part of both international and internal migration. In many countries, women, because of marriage migration, are the primary movers. However, in measurements of internal and international boundary crossings (provinces or states) as well, women may move within the region or across neighbourhoods rather than over administrative boundaries. Internationally, approximately 48 per cent of migrants are women (United Nations [UN] 2015). In 2016, there were more female migrants than male in Europe and North America, in part because women migrants are the majority in family flows, which constitute the main form of permanent migration, while in Africa and Asia, particularly Western Asia, migrants were predominantly men due to their predominance in labour migration.

The differences between female and male migration patterns were first noted by Ravenstein (1885) but, at least in Europe, interest in the differences between the two was sparked particularly by concerns over labour migration in the 1970s (Morokvasic 1984). The political interest in migrant lives was underpinned by the possibility of gender equity that migration, particularly from Global South to North, seemed to promise. Europe was seen as modern and gender equitable, while sending regions were marked as traditional and patriarchal, so that migration allowed women to shake off traditional gender roles and enter modernity. Gender discrimination and inequalities and oppressive social norms are now increasingly acknowledged as reasons that lead to female international migration (Ferrant and Tuccio 2015). The right to move also remains important for women engaged in migration advocacy, as it is argued that women should be allowed to move from a less to a more gender-egalitarian society, escaping gender-based violence, for instance.

However, it is the growth of female labour migrants since the late 1990s, alongside the availability of gender disaggregated data on migration from the UN (Zlotnik 2003), that has catapulted the topic of gendered migration into academic and policy debates. The number of women crossing borders to work as domestic workers, carers, teachers and nurses in the female-dominated caring professions has increased. Men continued to migrate, working in sectors such as construction, seafaring and agricultural work. Through the 1990s and 2000s there was a sharp increase in highly skilled migration, especially in the male-dominated sectors such as information technology and finance. However, women play an important, albeit little acknowledged (Kofman 2000), part in these skilled flows, too, especially in welfare sectors such as medicine, nursing and teaching, where migration continues to be seen through the lens of brain drain (Raghuram 2009). Education enables women from poorer countries to migrate – 20 per cent of highly educated women from sub-Saharan African have emigrated, but only 0.4 per cent of the least educated have moved (Dumitru and Marfouk 2015, 40). Moreover, an increasingly important form of migration is international student mobility, which has gone up from 0.8 million in 1975 to 4.5 million in 2012 and is expected to rise to 8 million by 2025 (OECD 2014).

Women move not only to continue their reproductive roles in waged work but also within families, as family migration remains the largest source of permanent migration in OECD countries, ahead of labour and humanitarian migrations (OECD 2017).[2] Around 38 per cent of all migrants entered through this route. The largest group is of spouses entering as marriage migrants, especially where family reunification is not permitted for the lesser skilled. This is followed by children and parents (Organisation for Economic Cooperation and Development [OECD] 2017). In Asia, such migration has increased (Chinsung, Keuntae and Piper 2016) in countries like South Korea, Japan and Taiwan, where it intertwines with labour flows from poorer to richer countries. Women marry men who have a weak position on the marriage market but are in a higher socio-economic location due to labour migration. Nicole Constable (2005) calls this 'global hypergamy'. A location in the Global North or in richer countries makes men desirable partners.

Refugee flows and internal displacement also have increased with the proliferation of protracted conflicts and political instability in the Middle East and Africa (Hyndman and Giles 2017). Syria has become the largest refugee- producing country (4.2 million in mid-2015), with the largest numbers seeking safety in neighbouring countries such as Turkey (2.7 million in March 2016). Although women comprise the majority of the internally displaced, a much smaller percentage manage to cross international borders to seek asylum in the Global North, because moving a long distance requires considerable resources and frequently necessitates smugglers. Not all asylum seekers are escaping from generalized conflict but may be seeking to get away from gender-related forms of persecution, such as domestic violence, forced and early marriage and genital mutilation, as well as experiencing difficulties in openly expressing their sexual orientation and gender identity (Freedman 2017).

These patterns reflect changing migration regimes, economic conditions in the Global South and North and how they are intertwined through various forms of gendered mobility. Thus, the masculinized labour migration to the Global North, which, for instance, was required for post-war reconstruction in Europe and drew in people from the Global South (especially the ex-colonies), and the agricultural and construction sectors in countries like Canada and the US, are now accompanied by a feminized migration. However, elsewhere, as in the Gulf region, men still dominate labour migration. Families continue to shape migration through all kinds of householding strategies (Douglass 2012), generating diverse forms of transnational living. Regional migration has been enabled in some contexts by the opening up of free-movement areas, as in the Southern African Development Community (SADC) countries, the European Union (EU) and the 12 countries of South America that now form part of the Union of South American Nations (UNASUR). However, increasing nationalism accompanies this change, in some parts of the world, with fears over migration driving anti-immigration sentiments and regulations that seek to radically reduce the number of migrants. This has manifested itself in the UK's decision to exit from the EU and in the criminalization of migration. These variegated gendered migration patterns have been analysed through new theorizations, as we will go on to see.

Analysing gendered migration geographically

Migration is deeply geographical (Raghuram 2013), as migrants are ultimately moving from place to place. Focusing on these flows from place to place and the connections that precede, enable and succeed these movements has been a central part of migration theorizations. However, the focus has largely remained on the flows or, when place is investigated, it is the empirical particulars of place and how this affects the theorizations that become the focus. This section steps back from these details to see what analytical manoeuvres about place are used in such studies. This section uses place as difference and places as connected as two distinct tropes through which gendered migration has been analysed, albeit often in unacknowledged ways.

Place as difference

A dominant mode of understanding place is through specificity; what makes a place unique and what are the meanings imbued to place (Massey 1994). Place is particularly important in analysing migration, because the differences between places are, arguably, what drive migration. In this sub-section we explore various aspects of these differences: i) economic differences between places; ii) gendered variations in the ability to migrate; iii) dissimilarities in the performance of gender identities in each place; iv) the differences that migration can make to these performances; and v) how the meanings of place are themselves changed as migrants inhabit place.

Economic differences between places were central to early gendered interventions into an implicitly male-dominated analysis of labour migration. The causes for migration are captured in this analysis by the hyphen between receiving-sending countries or here-there. It was argued that there are not only differences between places but also a hierarchy of places, especially in labour market conditions that drive migration. This analysis continues to dominate narratives of female labour migration. Here, the target of feminist analysis is global inequalities and the ways in which gender is reinscribed into migrant working lives. However, gender itself continues unchanged in many of the theorizations of female migration. For instance, research has emphasized elements of femininity, such as a caring disposition, either in paid work or through the feminized and masculinized roles that women and men take up. Women may be employed as domestic workers and cleaners and sometimes in skilled work like teaching and nursing; and men find work as gardeners and odd-job men or in professions like information technology (Perrons et al. 2010; but see Raghuram 2008; and for a discussion of femininity and work see Kofman and Raghuram 2013). Similarly, research on sexuality and gendered migration has primarily emphasized sex work. Sexuality has also often been conflated with compulsion and, problematically, merged with trafficking. Sex work by choice and the ways in which sex work may be combined with heterosexual and homosexual relations have been much less considered (Walsh et al. 2008).

Second, the difference between places is not just in economic opportunities or in work but also in the diverse social contexts from which people migrate and how gender is performed differently in different places. These differences can influence who migrates, as well as what happens to them when they migrate. For instance, Bylander (2015) differentiates between the pressures that young Cambodian men feel to move in the context of a stagnant economy and the possibilities that are still open for women to stay put. Thus, mobility is inscribed with particular gender codes and what ageing and youth means can be variable and dependent not only on biology but also societal expectations and roles. Both the meaning of what constitutes being a woman or man and what it implies for mobility can vary.

Third, how women and men perform their gendered identities may also vary across place. Thus, Aija Lulle and Russell King (2016), in their study of older Latvian migrants, found that the ability to quadruple their pension by moving from Eastern to Western Europe was accompanied not only by better economic prospects but also by finding new sexual and emotional relationships, rewriting how they performed gender after migration. Similarly, younger migrants may move for a number of reasons – to take up new opportunities and for excitement. Here, too, gendered performances may be rewritten, as Sondhi and King (2017) describe in their study of Indian students in Canada. They found that both the women's clothing and the physical displays of affection, such as hugging among men – commonplace in India – had to be adjusted to fit the Canadian norms of heterosexual femininity and masculinity. Thus, what constituted appropriate gendered performances were adjusted according to place.

Although these and other studies have attempted to dismantle the 'vulnerability trope' that has haunted migration at both ends of the age spectrum (King et al. 2016), social upheavals, such as conflict or disease and economic compulsions due to the death of parents or because they are trafficked, remain important causes of migration. These examples also point to the fact that migrants move for multiple reasons at different periods of their lives. The notion of vulnerability is also commonly used in relation to refugees, which tends to represent women and children as exploited and therefore more deserving of protection and additional resources than men. The concept is attached to the person rather than the circumstances creating the vulnerability (Turner 2016).

Fourth, as in the 1970s, something that continues to fascinate academics is the question of what happens to gender identities, roles and responsibilities and relations, not only in themselves but what it means for the inequalities in gender that exist prior to, during and after migration and in these different places. Given the almost-universal nature of patriarchy, the question that arises is: is patriarchy transcended, transformed or transported during migration? The evidence is mixed. For some, migration has been empowering. Migration and gender research provides us with contradictory findings that demonstrate, on the one hand, the empowering potential of migration for women (Duda-Mikulin 2013), where women are able to transcend many of the limitations due to patriarchy, to cases where patriarchy may be reinforced due to migration (Walton-Roberts and Pratt 2006). But much more common are stories of limited changes, where some aspects alter but others do not (Kofman and Raghuram 2015). For instance, women may escape control of older patriarchs and matriarchs but, on the other hand, they also have less help with childcare and, in some cases, may be limited to the household because of it. This leads to deskilling among the highly qualified and can also cause mental health issues for migrant women. An ability to form new, satisfying social relations while maintaining old ones in their country of origin seems to be the best response.

It is not only women whose gender roles are transformed by migration but also men. Female migration can lead to men taking on more responsibility for care of the children and the household (Asis 2006). In other instances, the pressures of these responsibilities can lead men to themselves migrate, leaving the children behind with other women. The prevalence of female labour migrants means that the proportion of male migrants who come through the family reunion route is increasing. This is specially so for the spouses of women in highly skilled sectors; less-skilled migrants usually face limitations in family reunion, as difficulties come into operation concerning meeting the minimum thresholds of salary, appropriate accommodation or the nature of the visa through which they enter. Normative notions of masculinity may be reinscribed as women become the wage earners and men have limited or no access to the labour market. Moreover, the impact of being removed from male-dominated societal norms as they become separated from their kinship groups can have an effect on how masculinity is defined (Friedman 2017). As Friedman argues, migrant men no longer benefit from what Connell (1987) defines as the patriarchal dividend, or the cultural, economic and material benefits that are societally reproduced and that all men supposedly enjoy.

Fifth and finally, there is a vibrant literature on questions of home and belonging and the meanings imbued to place by migrants. For instance, Wilkins (2017) explores how women alter their attachments to home as house, as well as nation, as they traverse the Myanmar–Thailand border. She recounts the personal transitions that women make in the face of the insecurities of border crossings.

In sum, although place differences drive migration, so too does gender difference. Moreover, as people migrate, both place and the meanings of gender and how it is performed alter in place, so that place is a powerful way of understanding migration.

Connecting places

Although places may be seen as distinctive and posited against each other, the connections between places have also played a part in theorizations of gendered migration. In this section we will explore three ways in which these connections have been theorized: through causation of migration as in notions of a care chain (Dumitru 2017); through a transnational lens (Bastia 2015); and as physical connections between places through which migrants transit (Stock 2012).

The neo-Marxist take on gendered migration is encapsulated in theories of the global care chain, where the emotional and material labour of care is transferred from the Global South to the North through the movement of care workers. This arises out of poverty in the sending areas, caused in part by the flow of money and resources to the Global North. But it is also underpinned by local changes in the receiving areas. For instance, the growth of the two-wage family, prompted by a shift in welfare policies from the family wage to the adult worker model, has drawn more and more women into the workplace and has led to a care deficit in some households in the Global North. Drawing on Rhacel Parreñas's work (2001) on the migration of nannies from the Philippines to the US, Arlie Hochschild (2000) argues that such migration simply transfers the care deficit from households in the US to households in the Philippines as women from poorer countries migrate to perform paid care work in richer countries. The next link in the chain is when women move from poorer parts of the Philippines to look after the families of these international migrants, and each cascading chain of care results in its performance being less well remunerated, and sometimes unpaid. The care chain identifies a hierarchy of places based on the movement from the periphery to providing care and making good a care deficit in a Northern context. Emotional and material labour connects places.

Although this pattern was first observed in the Philippines, similar movements have been noted globally, from Indonesia, Sri Lanka and Latin American countries such as Bolivia, Ecuador and Peru. These studies show how care is redistributed from poorer to wealthier countries but, as Dumitru (2014) has argued, many of those taking up such work are nurses and teachers, so that what is seen as a care drain is also actually a brain drain. She rightly emphasizes the skills of the carers and also the methodological nationalism that underpins this analysis, whereby the nation state becomes the primary way of thinking about migration. In this instance, international migration is the arbiter of the first link in the chain. However, as Dumitru (2014) points out, such chains may also exist internally, within a country.

The focus on care as the primary activity that female migrants take up has some limitations. Care is both an important set of activities and an ethic, but it only connects loosely to the world of production and the debates therein. As a result, care can become relegated as important but unconnected to the (male) world of production. Returning to the language of social reproduction is therefore important (Kofman and Raghuram 2015). Social reproduction involves the production of people through various kinds of work – mental, manual and emotional – aimed at providing what is necessary to maintain existing life and to reproduce the next generation (Laslett and Brenner 1989). It includes 'the varying institutions within which this work is performed, the varying strategies for accomplishing these tasks, and the varying ideologies that both shape and are shaped by them' (383). Thus, social reproduction is necessary for production: for biological reproduction, for the skilling and training including the social comportments, and the generational reproduction of the next round of labour (Kofman and Raghuram 2015). It connects places, generations and the social and economic fabric of society. Moreover, migrant men, too, are involved in social reproduction, as Gallo and Scrinzi (2016) have shown in studies in the domestic sphere, where they are more likely to undertake the care of the elderly. Other activities in the international division of reproductive labour, such as gardening and handiwork, are dominated by men (Perrons et al. 2010).

A second form of spatial connection that is widely catalogued in the literature on gendered migration is transnationalism. Transnationalism highlights the interconnectivity between people in different nations. It challenges the exclusivity of belonging to one place, often ascribed to theories of assimilation, for instance. Instead, migrants, it is argued, do not simply leave behind their past attachment to the places they left behind and the social relations there, but continue to maintain them while developing social relations in the receiving context (Bastia 2015). For instance, although there is some global transfer of care and social reproductive activities, migrants often continue to fulfil their responsibilities in the country of origin. In effect, place connections are not sequential but simultaneous across places; social reproduction occurs in both places. This has been documented in a growing number of studies of transnational parenting, especially of mothers, and the impact on their children, based on research in Latin American and South East Asian migrations (Bastia 2015; Hoang and Yeoh 2015). However, much less is known about the impact of recent African migrations on the circulation of care and transnational parenting, or of fathers (Mazzucato 2015). Though most attention has been paid to children, in recent years studies have focused on elderly parents, who may both be looking after grandchildren and needing care themselves (Hoang and Yeoh 2015).

Finally, migration involves physical movement through places. These places may be varied, from ships to aeroplanes, but most commonly over land. These technologies of movement are themselves places that migrants inhabit in differentially gendered ways, although there is little literature on it thus far. The little literature that does exist shows the overwhelmingly masculine assumptions made about transit migrants and a tendency to see women in transit primarily as victims, particularly of sexual violence (Stock 2012).

Future research

Migration is always a movement of people from one place to another. As such, place specificities and connections have been an important, albeit implicit part of research on gendered migration. This chapter takes a step back and outlines how gender and place are differently configured in this analysis. It does so because these relations between place are crucial to a feminist agenda in unpacking suppositions and for achieving gender equity.

Although there is a wealth of research on gendered migration which recognizes these place relations, there are still a number of gaps in the existing research. We explore this through three questions relating to place. First is the relative absence of some places and associated mobilities, given their importance in migration patterns. Second, we argue for a recognition of the entanglements between migrants and the objects, including cash flows, that make up migrant lives. Finally, we suggest the need for further research on immobilities: of having the option of staying in one place and not moving.

Although there is considerable literature on migrants, much of it focuses on South–North mobilities. This is slowly changing in some parts of the world, with more work on intra-regional migration in East and South-East Asia, for instance, which explores key issues of care, remittances and the long-term implications of changing family life in both the sending and receiving countries (Hoang and Yeoh 2015). As Wall and Bolzman (2014, 75) comment, 'examining the changes in female migration trajectories is a key issue for understanding the emergence of new forms of transnational families'. Women are migrating more often alone (as single, divorced or widowed) and initiating migration, as well as moving with their partners and children. Their migration changes family life. Yet the impact of mobility on family dynamics has been poorly understood due, in part, to the tendency to treat the family as a secondary consideration, unlike the primacy accorded to labour migration and remittances.

Second, migrant connections between places are always entangled with flows of money, goods, ideas and knowledge but, so far, the migrant's own mobility is still prioritized. One of the main ways in which material resources circulate through the global economy and transnational families is remittances. Though they decreased in 2015 and 2016, remittances have steadily increased over the longer period and are expected to increase to $596 billion in 2017 (World Bank 2017). In the past, the interest in remittances focused on the economic aspects of transfers but, increasingly, research in remittances has turned its attention to the social and gendered practices of remittance sending and receiving (King et al. 2013). Complex processes of negotiation within households determine how the remittances are spent and who has the power to determine their usage, who benefits from them and what effects they have on the welfare of the family. They also embody values and serve to reconstruct social identities within families and communities (Hoang and Yeoh 2015). These entanglements between places that arise through money and goods transfers need more attention.

Another area that has been little studied is gendered immobilities. For many people, migration has become a necessary means of earning a living. For some, development offers the opportunity not to migrate. But for others, societal development has not produced gendered freedom and so national development has not been adequate. When do women and men have the ability not to migrate, and how can that be fostered? The increasing importance of environmental changes and natural resource depletion by international companies in shaping migration, as well as how that links to the above discussions around migrant labour, needs further attention.

Since the 1980s, and particularly in the past two decades, our understanding of gendered migrations has advanced enormously. However, there remains much to do in relation to the implications of gendered migrations in the Global South, their interaction with internal migrations, which remain significant, and the impact of changing places on family lives and gender and generational relationships. Gendered migration is driven both by differences between places and by unequal connections. Furthermore, place then interplays with gender as a variable that acts, not as a static variable, but rather as a performance that can be altered as men and women move between places. As migration continues to dominate political agendas nationally and globally, these notions of how place and gender interact will become increasingly salient.

Notes

1 This chapter, for reasons of space, has adopted a binary approach to gender and is largely focused on the insights into female migration from feminist research.

2 The OECD bases its figures on those who are given family permits rather than on those stating that they have migrated for family reasons. Entering on a family permit does not mean that one is not intending to enter the labour market immediately, as part of a family project.

Key readings

Dumitru, S. 2014. "From 'Brain Drain' to 'Care Drain': Women's Labor Migration and Methodological Sexism." *Women's Studies International Forum* 47: 203–212.

Kofman, E., and P. Raghuram. 2015. *Gendered Migration and Global Social Reproduction*. New York: Palgrave Macmillan.
Morokvasic, M. 1984. "Birds of Passage are Also Women." *International Migration Review* 18 (4): 886–907.

References

Asis, M.M. 2006. "Living With Migration: Experiences of Left-Behind Children in the Philippines." *Asian Population Studies* 2 (1): 45–67.
Bastia, T. 2015. "'Looking After Granny': A Transnational Ethic of Care and Responsibility." *Geoforum* 64: 121–129.
Bylander, M. 2015. "Contested Mobilities: Gendered Migration Pressures Among Cambodian Youth." *Gender, Place & Culture* 22 (8): 1124–1140.
Chan, K.W. 2013. "China, Internal Migration." In: *The Encyclopedia of Global Migration*, edited by I. Ness and P. Bellwood, 1–17. Chichester: Wiley Blackwell.
Chinsung, C., K. Keuntae, and N. Piper. 2016. "Preface: Marriage Migration in Southeast and East Asia Revisited through a Migration-Development Nexus Lens." *Critical Asian Studies* 48 (4): 463–472.
Connell, R. 1987. *Gender and Power: Society, The Person and Sexual Politics*. Palo Alta: University of California Press.
Constable, N. 2005. *Cross-Border Marriages: Gender and Mobility in Transnational Asia*. Philadelphia: University of Pennsylvania Press.
Douglass, M. 2012. "Global Householding and Social Reproduction: Migration Research, Dynamics and Public Policy in East and Southeast Asia." *Asia Research Institute Working Paper Series*, No. 188. Available at: www.ari.nus.edu.sg/docs/wps/wps12_188.pdf (accessed 20 November 2017).
Duda-Mikulin, E.A. 2013. "Migration as Opportunity? A Case Study of Polish Women: Migrants in the UK and Returnees in Poland." *Problemy Polityki Spolecznej* 23 (4): 105–121.
Dumitru, S. 2014. "From 'Brain Drain' To 'Care Drain': Women's Labor Migration and Methodological Sexism." *Women's Studies International Forum* 47: 203–212.
Dumitru, S. 2017. "How Neo-Marxism Creates Bias in Gender and Migration Research: Evidence from the Philippines." *Ethnic and Racial Studies*. https://doi.org/10.1080/01419870.2017.1397279.
Dumitru, S. and A. Marfouk. 2015. "Existe-t-il une féminisation de la migration internationale? Féminisation de la migration qualifiée et invisibilité des diplômes." *Hommes et Migrations* 1311: 31–41.
Ferrant, G., and M. Tuccio. 2015. *How Do Female Migration and Gender Discrimination in Social Institutions Mutually Influence Each Other?* Paris: OECD Development Centre.
Friedman, S.L. 2017. "Men Who 'Marry Out': Unsettling Masculinity, Kinship, and Nation through Migration across the Taiwan Strait." *Gender, Place & Culture* 24 (9): 1243–1262.
Gallo, E., and F. Scrinzi. 2016. *Migration, Masculinities and Reproductive Labour. Men of the Home*. Basingstoke: Palgrave Macmillan.
Hoang, L.A., and B. Yeoh, eds. 2015. *Transnational Labour Migrations, Remittances and the Changing Family in Asia*. New York: Palgrave Macmillan.
Hochschild, A. 2000. "Global Care Chains and Emotional Surplus Value." In: *On the Edge: Living with Global Capitalism*, edited by A. Giddens and W. Hutton, 130–146. London: Jonathan Cape.
Hyndman, J., and W. Giles. 2017. *Refugees in Extended Exile. Living on the Edge*. London: Routledge.
IOM. 2017. *Migration in the World*. Available at: www.iom.sk/en/about-migration/migration-in-the-world (accessed 29 November 2017).
King, R., A. Lulle, D. Sampaio, and J. Vullnetari. 2016. "Unpacking the Ageing–Migration Nexus and Challenging the Vulnerability Trope." *Journal of Ethnic and Migration Studies* 43 (2): 182–198.
King, R., D. Mata-Codesal, and J. Vullnetari. 2013. "Migration, Development, Gender and the 'Black Box' of Remittances: Comparative Findings from Albania and Ecuador." *Comparative Migration Studies* 1 (1): 69.
Kofman, E. 2000. "The Invisibility of Skilled Migrants and Gender Relations in Studies of Skilled Migration in Europe." *Population, Space and Place* 6 (1): 45–59.
Kofman, E., and P. Raghuram. 2013. "Knowledge, Gender, and Changing Mobility Regimes: Women Migrants in Europe." In: *Mobilities, Knowledge, and Social Justice,* edited by S. Ilcan, 59–75. Montreal: McGill-Queen's University Press.
Kofman, E., and P. Raghuram. 2015. *Gendered Migration and Global Social Reproduction*. New York: Palgrave Macmillan.

Laslett, B., and J. Brenner. 1989. "Gender and Social Reproduction: Historical Perspectives." *Annual Review of Sociology* 15: 381–404.
Lulle, A., and R. King. 2016. *Ageing, Gender, and Labour Migration*. New York: Palgrave Macmillan.
Massey, D. 1994. *Space, Place and Gender*. Cambridge: Polity Press.
Mazzucato, V. 2015. "Transnational Families Between Africa and Europe." *International Migration Review* 49 (1): 142–172.
Morokvasic, M. 1984. "Birds of Passage Are Also Women." *International Migration Review* 18 (4): 886–907.
OECD. 2014. *Education at a Glance 2014. OECD Indicators*. Paris: OECD Publishing.
OECD. 2017. *Trends in International Migration*. Paris: OECD Publishing.
Parreñas, R., ed. 2001. *Servants of Globalization: Women, Migration and Domestic Work*. Stanford, CA: Stanford University Press.
Perrons, D., A. Plomien, and M. Kilkey. 2010. "Migration and Uneven Development Within An Enlarged European Union. Fathering, Gender Divisions and Male Migrant Domestic Services." *European Urban and Regional Studies* 17 (2): 197–215.
Raghuram, P. 2008. "Migrant Women in Male-Dominated Sectors of the Labour Market: A Research Agenda." *Population, Space and Place* 14 (1): 43–57.
Raghuram, P. 2009. "Caring about 'Brain Drain' Migration in a Postcolonial World." *Geoforum* 40 (1): 25–33.
Raghuram, P. 2013. "Theorising the Spaces of Student Migration." *Population, Space and Place* 19 (2): 138–154.
Ravenstein, E.G. 1885. "The Laws of Migration." *Journal of the Statistical Society of London* 48 (2): 167–235.
Sondhi, G., and R. King. 2017. "Gendering International Student Migration: An Indian Case-Study." *Journal of Ethnic and Migration Studies* 43 (8): 1308–1324.
Stock, I. 2012. "Gender and The Dynamics of Mobility: Reflections on African Migrant Mothers and 'Transit Migration' in Morocco." *Ethnic and Racial Studies* 35 (9): 1577–1595.
Turner, L. 2016. *Are Syrian Men Vulnerable Too? Gendering the Syrian Refugee Response*. Middle East Institute, online 29 November. Available at: www.mei.edu/content/map/are-syrian-men-vulnerable-too-gendering-syria- refugee-response (accessed 2 December 2017).
UNDESA. 2013. "Cross-National Comparisons of Internal Migration: An Update on Global Patterns and Trends." Technical Paper No. 2013/1. Available at: www.un.org/en/development/desa/population/publications/pdf/technical/tp2013-1.pdf (accessed 29 November 2017).
UNHCR. 2017. *Figures at a Glance*. Available at www.unhcr.org/uk/figures-at-a-glance.html (accessed 29 November 2017).
United Nations (UN). 2015. "Trends in International Migrant Stock: Migrants by Age and Sex." *United Nations Database*. New York: UN.
Wall, K., and C. Bolzman. 2014. "Mapping the New Plurality of Transnational Families: A Life Course Perspective." In: *Transnational Families, Migration and The Circulation of Care. Understanding Mobility and Absence in Family Life*, edited by L. Baldassar and L. Merla, 61–78. New York: Routledge.
Walsh, K., H. Shen, and K. Willis. 2008. "Introduction To Special Issue: Heterosexuality and Migration in Asia." *Gender, Place and Culture* 15 (6): 575–579.
Walton-Roberts, M., and G. Pratt. 2006. "Mobile Modernities: A South Asian Family Negotiates Immigration, Gender and Class in Canada." *Gender, Place & Culture* 12 (2): 173–195.
Wilkins, A. 2017. "Gender, Migration and Intimate Geopolitics: Shifting Senses of Home Among Women on the Myanmar–Thailand Border." *Gender, Place and Culture* 24 (11): 1549–1568.
World Bank. 2017. *Migration and Remittances: Recent Developments and Outlook*. Washington, DC: World Bank.
Zlotnik, H. 2003. "The Global Dimensions of Female Migration." *Migration Information Source*, March.

24
REPRESENTING WOMEN AND GENDER IN MEMORY LANDSCAPES

Danielle Drozdzewski and Janice Monk

Introduction

Representations of memory in the public landscape – monuments, memorials, street names – communicate how, what and, often, who are considered important to conceptions of a nation's past. Such conceptions are always mediated by power, most commonly by the government and/or regime in power. Thus, representations of memory also have a politics. Choices are made about what and who has a place in the public memory landscape; commonly, they toe a political line, supporting and/or commemorating individuals, places and events that have resonance for a particular (and most commonly a government's) interpretation of identity (national, community or religious, for example). Memory of the nation is 'used' in public landscapes, then, as a tool to reinforce certain aspects of the nation's identity. The events in 2017 in Charlottesville in the US show how the past, and its representation in the public landscape, matter; how they intersect into our everyday lives; and how they have a longevity that also subjects them to changing interpretative contexts. Clearly, public memory is complex, and it matters to representations of the self in the present (whether that self is the nation, individual and/or community). We use this contention – that memory matters – as a pivot point to consider how women and gender are represented in memory landscapes, asking what kind of commemorative atmospheres these representations generate, contribute and (re)produce about the identity of/in those places.

Representations of memory tell us stories about identity. Concomitantly, they tell us stories about whom a nation thinks it 'should' remember. As critical feminist geographers, we also ask who is silenced and/or absent from these representations, how power is attributed (perhaps unevenly) and by whom. Such critical attendance to intersectionality in the memory landscape follows Rose-Redwood, Alderman and Azaryahu's (2010, 454) assertion that 'a critical analysis of the politics of spatial inscription remains one of the most effective strategies for challenging essentialist claims to affixing stable identities to particular spaces'. In this vein, here, we turn our focus towards representations of women and gender in memory landscapes.

In writing this chapter, we each reflect our own experiences and perspectives on landscapes and memories. One of us, Janice, has a long-term interest in gender and feminist geography. Her awareness that monuments in the landscape reflect political and cultural priorities of those with power emerged in her youth. As an undergraduate at the University of Sydney, she took breaks from reading at the State Library in the nearby Royal Botanic Gardens. Among classical monuments there were *The Boxers* – nude male figures whose pugilist poses conveyed physical power. Nearby were demure female figures of *Spring*, *Summer* and *Autumn* and of a weary old male, *Winter*. In her subsequent years as an academic in the US and as feminist scholarship emerged, she recalled those early images and turned some of her interests to gender representations in the landscape, locally and during her wider travels. Danielle's research – on memory, place and identity – is often allied to her cultural heritage and has resulted in a sustained engagement with diverse discourses of Polish cultural identity and cultural memories, both with the Polish diaspora in Australia and in terms of public articulations of nationalism in Poland. Indeed, a conversation between Janice and Danielle was the starting point for this chapter, initiated on a tram in Kraków, Poland, after a conference session sponsored by the International Geographical Union Geography and Gender Commission.

Our focus here is twofold: to examine how women and gender are portrayed in articulations of public memory and to consider what these representations say about the identity and memory politics of the nation. The chapter is divided into two key discussion sections. In the first we detail a feminist approach to memorialization and identity maintenance, including an explication of how women and gender are represented in memory landscapes and what we can read about these representations. Integral to this section is a discussion of absence and allegory, showing how the feminine form, when actually portrayed in the memory landscape, often takes on 'other' forms. In the second section, we highlight examples from memory landscapes in the context of women in politics and women of war. We describe notable exceptions of women in memory landscapes and provide an analysis of how these examples 'bear traces of deeper stories about how they were created, by whom, and for what ideological purpose' (Dwyer and Alderman 2008, 168). To contextualize these two conversations, next we provide an overview of the thematics that are the key to our discussions of memory landscapes.

Memory landscapes

To begin an analysis of the representations of women and gender in public memory landscapes, we must first understand why public memorialization plays a key role in expressing, cementing and maintaining specific narratives of a nation's past. Memories shape national identity and conceptions of belonging to the nation (Finney 2002; Gillis 1994). Rolston (2018, 5) calls those responsible for such shaping 'memory entrepreneurs', stating that it is their role to 'articulate an interpretation of the past which enables a society to pull together and build a common identity'. In policy and politics, reinforcing collective narratives of national identity is a state imperative, because it creates consensus through collectivity and shared notions of what citizens of the nation remember as part of the nation's past. Hoelscher and Alderman (2004, 350) have contended that in spatializing these collective notions, public memory landscapes, then, are: 'explicitly designed to impart certain elements of the past – and, by definition, to forget others – such lieux de memoire are the sites where, as Nora (1989, 7) puts it, "memory crystallizes and secretes itself"'.

This idea of crystallizing a certain view/version of the past in public space fixes that narrative to place and makes it public, so that passing audiences read and view that memory in that place. It can also mean that that memory is celebrated as part of a commemorative and/or calendar ritual, such that it transcends material space and becomes both performances *and* practices of memory. In discussing the role that monuments play in public memory, Abousnnouga and Machin (2010, 132) have argued that 'visual elements and features combine and operate as part of a system or visual grammar'; they create 'symbolic capital' (Rose-Redwood 2008, 431).

Alderman (2008, 196) speaks specifically of the symbolic capital generated through toponymic naming. Placing memory in public landscapes in monumental form also brings 'distinction and status to landscapes and the people associated with them'. Representations of memory form links to place identities, which increase in familiarity when embedded into everyday routines.

These representations of memory in the everyday have impact; such impact is, of course, not always straightforward. It is contested, and its reception is not necessarily assured. Yet much work by memory scholars and geographers has discussed the relationship between what is portrayed publicly about a nation's past and its reception – that is, between what the public knows about a nation's past and what it sees, hears and participates in. For example, Hoelscher and Alderman (2004) have discussed the critical relationship between memory and place; Rolston (2018) has investigated the role of propaganda posters in urban Iran; Verovsek (2017) champions the power of the past as a resource for political change; Winter (2014) has explored public perceptions of war memorials in regional Australia; and Bevan (2006) has discussed the destruction of built memory forms and the ensuing loss to local and national communities. Gender features only on the fringes of these discussions, if at all. Indeed, as Rolston (2018, 15) noted, 'representations of women are rare'. This silence is not only context and culturally contingent (as in the case of Rolston's research in Iran) but speaks to the heteropatriarchal dominance of scholarship about public memory landscapes. Furthermore, amid this corpus of work, and much more, the role of memory landscapes in influencing public discourse about the nation reveals that 'a politics of memory operates in, on and with (re)productions of places and identities' (Drozdzewski et al. 2016, 1). Dwyer and Alderman (2008, 168) note 'that memorials typically reflect the values and world views of government leaders and members of the dominant class'; their comments point us in the direction of why this discussion of the relationship between public memory and its reception/perception is pertinent to our chapter on gendered memory landscapes. They continue by stating that, in choosing who and what to memorialize, 'they [government leaders and the dominant class] tend to exclude the histories of minority and subaltern groups or appropriate these histories for elite purposes' (Dwyer and Alderman 2008, 68). Thus, what is chosen for representation in the public memory landscape (or not), and how these representations, or lack thereof, have an impact and the potential to shape public opinion.

Women and gender in memory landscapes: absence and allegory

When we critically examine public memory landscapes, representations of women are few and far between. As Puwar (2004, 6) notes, 'women's inclusion into the nation has been quite specific'. Women often do not appear at all, or they appear in allegorical and/or mythical form. Further, and in reference to women on war monuments, Abousnnouga and Machin (2013, 105) found that they are often depicted in 'the form of protected wives and daughters, or mythical figures personifying nations, place and values'. In this section, we delve into these two facets: the absence of women in memory landscapes and representation of femininity in non-human forms. To explain both, we draw from examples of two women in the Planty, the parkland that encircles the Old Town of Kraków, Poland. The remainder of this section is structured by first outlining each of the two aforementioned facets, followed by a worked example of the gendered representation of women in the Planty.

First, to speak of an absence of women in memory landscapes, it is necessary to flag up that men dominate memory landscapes (Puwar 2004). Men also often appear as themselves rather than as transmuted mythological creatures. This domination and form of depiction tells an important story about the types of masculinities, memories, achievements and losses that are validated and valorized in society, as well as those who receive less, or no, attention. As

Weidenmuller et al. (2015, 438) have argued: 'this gendered nature of statues and monuments of people is important because their psychical solidity contributes directly to the seeming finality of these landscapes', and thus to 'the ongoing gendering of space, and thereby perpetuate, normalize, and naturalize a gendered society through landscape.' The paucity of monuments commemorating women is indicative of how society (and the governments and regimes that make choices about public memory) values women's achievements. Monk's (1992, 129) assertion that 'power over the built environment remains in male hands and women have limited control over its form' means that it is of little surprise that 'the normative figure of leadership and especially in battle has been masculine' (Puwar 2004, 6).

A lack of power leads to a diminished and gendered assignation of the value of women's contributions to a nation's history and its identity, *and* normalizes women's absence in public landscapes. For example, a list of public art in the City of Sydney catalogues 102 items of public statuary.[1] There are 13 women depicted in these public monuments. However, there are only five representations of 'real' women, and two of these are of Queen Victoria. Of the remaining eight, seven represent women allegorically – for example, in the *Venus Fountain* – and one depicts religious iconography.

In a more recent example, these narratives of absence and their outcomes have featured in popular media. For example, in *Time* magazine, Rhodan (2017) discussed that in San Francisco the lack of female representation in the city's memory landscape sends the message that 'women did not participate and they do not deserve the respect that men do who are portrayed across the country'. Recently, however, 'comfort women' statues are now appearing in cities in the US. These statues depict young teenage girls – usually from Korea, China and the Philippines – as symbols of sexual slavery. They commemorate the tens of thousands who were detained in 'rape camps' by the Japanese Imperial Army in World War II (McGrane 2017).

Across the Atlantic Ocean, Schwartz (2017) outlines that 'among the UK's 925 public statues, 158 are women, 29 of those depict Queen Victoria'. Capps (2016), writing for the CityLab blog, referred to the lack of female representation in public landscapes as the 'Gender Gap in Public Sculpture'. She also notes that women 'hardly ever [appear] as real women from lived history, with first and last names'. In the news website CNN, Peled (2017) quotes the Smithsonian American Art Museum's online inventories catalogue to state that 'of the 5,575 outdoor sculpture portraits of historical figures in the United States, 559 portray women'. These examples substantiate a growing interest in the absence of women in the memory landscape. Yet it was more than three decades ago that Marion Warner (1985), in *Monuments and Maidens*, used feminist theory to discuss gendered representations in memory and public landscapes. Warner (1985, 331) drew attention to the fact that 'men often appear as themselves, as individuals, but women attest the identity and value of someone or something else'. It is with this contention that we turn to the example of the women represented in the Planty in Kraków, Poland.

The Planty is the former moat that surrounds the Old Town of Kraków. Represented in the Planty are two of the three bards of Polish literature, Adam Mickiewicz and Juliusz Słowacki. The works of both poets remain pivotal to Polish nationalism and are symbolic of the Polish struggle for freedom from the oppression of foreign occupation (Krzyzanowski 1968, 1978; Miłosz 1969). Exceptionally, both poets are represented in the Planty through the leading female characters of two of their poems, *Grażyna* and *Lilla Weneda*, respectively. Their representation attests to the allegorical significance of the two characters or, in Warner's (1985) words, the 'something else'. By using these two examples, we will show how, even when women are represented in the landscape, their representation rarely relates to an individual 'real' woman's achievement but instead to women's supportive role, to the femininity of their form and/or to some other moral virtue.

Figure 24.1 *Grażyna and Litawor* (1823), Planty, Kraków.
Source: 36 108mm/digital/exp.auto/DD.

In Mickiewicz's poem, Grażyna upholds her husband Litawor's honour by foiling a disreputable pact. Under this pact, Litawor was to help the Teutonic knights to overthrow the ruler of his homeland, the Grand Duke of Lithuania. However, Grażyna took up arms against the Teutonic knights and died in battle protecting her homeland, while saving him from the dishonour of an alliance with the enemy. Grażyna appears in the Planty as a warrior – she is dressed in her husband's armour, carrying a sword and wearing a helmet. She is ready for battle. Warner (1985, 147) has argued that armed 'maidens' are symbols of strength and virtue: they 'work magic on the side of good against the bad'. Furthermore, Warner (1985, 258) has contended that, to represent ideals of justice, virtue and fortitude – ideals upheld by Grażyna in Mickiewicz's poem – a woman's body 'must have its surface reinforced', literally in armour, to strengthen woman's inherent physical and metaphorical weakness. Bulbeck (1992, 2) has contended that the 'female form is often sealed or strengthened with armoury', because 'real females are fragile "leaky" vessels' (see also Longhurst 2001). Grażyna, like other allegorical women warriors, is armed, and her armour makes her body masculine though the 'buckler, breastplate, helmet and spear' (Warner 1985, 124). Such reinforcements, Warner (1985, 124) argues, invoke a sense of women's 'law-abiding chastity, [and] their virtuous consent to patriarchal monogamy'. The intersectionality of Grażyna's representation is telling of the unevenness power and devaluing of her contribution based on her gender. Grażyna's character is strong and heroic. Her strength is emphasized through her willingness to champion the plight of her country. Yet, her heroism towards and for the nation is not enough; and this standpoint is signalled by the fact that her defence of her husband's morality is a key element of the narrative. As Weidenmuller et al. (2015, 452) have contended, such characters of mythical femininity, while providing examples of

> women with a sense of power and strength ... only [do so] in a mythical or fictional sense, as if women can only be heroic and strong in myth or as if strength and courageousness can only be something of legend.

The monument of *Lilla Weneda* presents a more evocative image of the female body. A snake extends up from the ground around Lilla's naked torso, towards the harp held aloft in her hands. Lilla's eyes are closed and her demure composure presents a familiar representation of the feminine figure in public statutory. Bulbeck (1992, 2) suggests that 'the exposed breast' is congruent with 'the abundance of nature, mediated by the "Motherland" or the state'. Similarly, Warner (1985, 324) states that feminine nakedness expresses a close association with nature outside the realm of our 'flawed and fallen world'. Słowacki, the author of the poem *Lilla Weneda,* wrote about the rise and fall of a nation through using two mythical and ancient 'tribes' pitted against each other. One tribe, the Weneds, were the original owners of the land who fought the invading Lechs. The interference of a trivial character, who disrupts the destiny of and ultimately victory for the Weneds, is integral to the poem's narrative of struggle. Słowacki's external world was flawed, because it was a world in which his homeland Poland no longer existed.

The means of regaining independence for the Weneds was the harp held aloft in Lilla's arms. Despite its closeness, she was unable to secure victory and freedom for her family and tribe. The proximity of a snake to the harp represents the external threat closing in on their autonomy – the ever-present vulnerability of a nation's autonomy – yet also the vulnerability of the female form and female gender (Bulbeck 1992). Abousnnouga and Machin (2013, 111) proffer that the use of feminine nudity can be read as representing 'women not as the vulnerable, but as the spiritual ideal'. Warner (1985) lambastes the use of nudity in public statutory, equating it with sexual desire. She cautions that the use of nude females straddles a complex narrative line. One foot is aligned with the sanctity and virtue of the naked human form, and the other foot with the capacity of that body to incite carnal sin. In both points, the female body represents 'something else': virtue, desire, fragility, rapture. These values are what Warner (1985, 12) also identified as 'generic and universal' characteristics – ones 'with symbolic overtones'.

Figure 24.2 *Lilla Weneda* (1839), Planty, Kraków.
Source: 36-1088mm/digital/exp.auto/DD.

Women in memory landscapes: women of politics and women of war

Exceptionally, and perhaps with increasing incidence, real women do appear in public memory landscapes. Most commonly, representations of women, as women, commemorate the achievements and/or notoriety of women involved in politics, including female heads of state and political activists, *and* women's roles during war (though this is not an exhaustive list). In what follows, we point to these two categories to draw the distinction on their 'exceptional' character.

Women of politics

In choosing who, what and how to remember, a concomitant choice is made about what *not* to remember. Dwyer (2004, 423) suggests that 'forgetting is intrinsic to the act of commemoration', primarily because it directs attention away from the event being forgotten towards that event being remembered. The dualism of remembering and forgetting is important when observing the gendering of memory landscapes, because the near absence of women renders the message that their contributions are not only unrecorded but undervalued and apparently non-existent. In the context of monuments commemorating political contributions, women's formal involvement in politics is relatively recent, given that women's legal right of involvement began only around 130 years ago. Nonetheless, there are still few women chosen for public commemoration and, indeed, few monuments denoting political movements involving women, such as suffrage. For example, 'out of nearly 150 public historical statues in New York City, only five of them are historical female leaders' (Gross 2018, np). The act of selectively 'remembering otherwise' (Esbenshade 1995, 87) is undeniably gendered; reading the memory landscape, then, tells us an important, though more obfuscate, story about the politics of memory, power and identity of the nation.

The examples that follow explicate what it takes to memorialize actual women in public. Of note in the two cases (one in New York, the other in Washington, DC) is that the monuments were not instigated by the state but rather through private endeavours and sponsorship. While private sponsorship of monuments may be common in different geographical contexts, the necessity of private money – and, with it, individual momentum – to commemorate women indeed provides revealing clues about the politics of the memory landscape. The first example centres on Central Park in New York City (NYC). Currently, of the 23 statues in Central Park, none represents an actual woman (Blakemore 2015). Substantiating our claims in the first sections of this chapter, the two female characters that do appear in the park are allegorical and fictional – Alice in Wonderland and Shakespeare's Juliet. Seeking the representation of a woman in the park, a non-profit, all-volunteer organization, the Elizabeth Cady Stanton and Susan B. Public Anthony Statue Fund, Inc., sought permission from the NYC Parks Department to donate a work of art. To do so, the organization began a long, bureaucratic process alongside a substantial fundraising campaign 'to bring the first statue of a woman to Central Park' (MonumentalWomen 2018). With the hashtag #MonumentalWomen, the organization has been successful in raising over USD 1.5 million to contribute to a statue to be located in New York's Central Park that would commemorate America's women's suffragist movement. It has also received the necessary approval from NYC Park's Department.[2] While the forthcoming statues of the Elizabeth Cady Stanton and Susan B. Anthony represent a step towards public recognition of women's political achievements, it remains that there is 'no women of color represented in the Park' (Carlson 2018, np).

This example from Central Park demonstrates that monuments honouring women's political achievements (may) take decades to achieve significant placement. In another example from the US, the *Portrait Monument* located in the Capitol Rotunda in Washington DC celebrates three leaders of the suffrage movement of the late-nineteenth century – Susan B. Anthony, Lucretia Mott and Elizabeth Cady Stanton. The monument – originally called the *Woman's Movement Monument* – was gifted to the United States Congress in 1921 by the National Woman's Party. Two days after receiving this 'gift' on behalf of Congress, the 'all-male Joint Library Committee' relegated it to the basement (known as the crypt) of the building (Weber 2016, 3). It remained there for 75 years. Further, the inscription on the monument, 'Woman first denied a soul, then called mindless, now arisen, declaring herself an entity to be reckoned' was removed by Congress (Boissoneault 2017, np). It was not until private donors and a founder of the National Women's History Museum became aware of its banishment that discussion about moving the monument began. Organizers from the Museum worked (initially with sympathetic members of Congress, including Senator Ted Stevens [Alaska], whose mother had been an active suffragette), to have it moved. By raising significant private funds, they managed to arrange the arduous and complex task of its removal in 1996 to the Rotunda. Since then, more than 39 million visitors have viewed the monument (Stone 2017). In both cases, the suffragist monuments have been initiated by private (and predominately female) interest groups. Women have lobbied for their representation, funded the provision of this representation and have still, in the case of the *Portrait Monument,* struggled to hold the same space in the public memory landscape as men. Clearly, what the suffragist movement achieved – the right for women to vote and participate in the country's politics – is still stymied by uneven and gendered politics. If #MonumentalWomen or the National Women's History Museum had not taken issue with whether and how women were represented, there would still likely be no statue in Central Park and one still gathering dust in the crypt on Capitol Hill.

Other international examples commemorating suffrage evince the significant lag time to commemorate women in public memory landscapes. In New Zealand, for example, women first gained the franchise in 1893, with rights accorded to all (including Indigenous Māori) women over 21years of age. Only on the centenary of this date was a celebratory monument established in Christchurch. That monument features life-size sculptures of Kate Sheppard and other women, including Helen Nicol (women's suffrage campaigner from Dunedin), Ada Wells (who worked for education of girls), Harriet Monson (in support of employed women), Meri Te Tai Mangakāhia (of Te Rarawa, Ngati Te Reinga, Ngati Manawa, Te Kaitutae, who was, in 1868, the first woman to speak in any New Zealand parliament) and Amey Daldy of the Auckland Women's Christian Temperance Union (Grimshaw 1972). They are carrying a petition for women's suffrage to Parliament. Surrounding the figures are scenes of women's everyday lives.

In the case of suffrage monuments in Britain, it was only in 2018, also a century after suffrage was achieved, that a monument to the leader, Millicent Fawcett, was unveiled in London's Parliament Square. The campaign to place the monument was initiated by feminist activist and journalist Caroline Criado Perez, originally from Latin America. The significance of this monument is augmented by the fact that only 2 per cent of statues recognizing women in Britain are of women other than members of the British Royal Family.[3] The location of the British suffrage monument, not just its establishment, is worthy of further explication. Stevens (2015, 40) has argued that meaning arises not only from the design and form of a monument but also from the 'spatial relationships to other buildings and memorials, and public activities around it'. Location, thus, affects who sees a monument. Location influences the temporality of the viewing (for example, would the monument be sited on an everyday route to work or shopping or located in a crypt), the viewing vantage point (Abousnnouga and Machin 2010) and also the context (for example, located independently or amid other monuments) (Stevens 2015). The suffrage

Figure 24.3 *Monumental Women Belong Here* (2018).
Permission granted by Stanton and Anthony Statue Fund.

monument's location shares space with both British and international political figures, including Winston Churchill and Nelson Mandela. Its location at London's Parliament Square relates the tenor of the monument to the social and cultural context of the location.

Women of war

Women's wartime contributions, roles and actions in the public memory landscape are often commemorated in line with specific social and cultural norms. For example, Abousnnouga and Machin (2010, 105) note about World War I (WWI) monuments that female 'representations played an important part in the decontextualization of social practice … [thereby] deleting women's actual roles during the war'. Similarly, Bulbeck (1992, 8) has suggested that, as depicted in war memorials, women often 'mimic the memorials to men' as 'patriot and martyr'. Yet women's wartime contributions extended well beyond the battlefield; they played a pivotal role in wartime in arms factories, in other arenas where jobs had traditionally been held by men, and, for some, at the front as nurses. However, these more 'active' roles do not commonly feature in the memory landscape (Abousnnouga and Machin 2010, 105). Rather, women were often depicted as 'protected wives and daughters', as caregivers and as recipients of the courage and bravery of male combatants (Abousnnouga and Machin 2010). Moreover, and as we have discussed, Abousnnouga and Machin (2010) also connect women in war memory landscapes with their representation as mythical creatures alongside male war heroes (represented as themselves). For example, it was in the early 1990s when Janice first saw a statue of a US Civil War hero, General Sherman (located at the south-eastern corner of Central Park). In glistening gold, Sherman is positioned astride a majestic horse, preceded – not led – by a winged female figure holding a palm frond. She is a delicate, virginal figure, keeping pace with the forward movement of horse and rider (Rosenblum 1984, cited in Monk 1992). This *Winged Victory* (cf Inglis 1987, 41) portrays women as synonymous with peace yet also with a moral virtue, passivity and beauty. This idyll is far from what the nurses on the home front felt during both world wars or, indeed, what the women working in factories and at home embodied.

In a small number of cases, women have achieved comparable representation in war monuments. Inglis (1987) identified where women have been named collectively or individually on WWI monuments in Australia. For example, he noted that women's names – often those of the 2,300 nurses who served overseas – appear alongside men's names in war memorial roll calls. Further, he specified that many a 'local memorial lists one or more nurses with the soldiers, and thereby classifies them as equals' (Inglis 1987, 37). Yet in other examples, such as in Wallsend, New South Wales, nurses' names have been added afterwards to those existing at the time of the initial unveiling.

In Washington DC, the Vietnam War monuments are indicative of incremental change in the normative status quo of war commemoration. The *Vietnam Veterans Memorial*, designed by the then 21-year-old woman architecture student, Maya Lin (it was selected from a competition in which the names of entrants were not identified) depicts women and men in the same format, as names listed in the wall. The nearby *Vietnam Women's Memorial* depicts three nurses: two support a wounded male soldier, the third scans the skies. While the memorials demonstrate movement towards acknowledging women's roles in armed conflict, that war shared temporal space with progressions in the women's rights movement, and thus differs from older memorials established shortly after World Wars I and II. Yet the broader spectrum of women's war commemoration in the US tells a different and equally intriguing story. The *National Women's Memorial* is 'the only major national memorial honouring all servicewomen – past, present and future' (Women's Memorial 2017, np). Unveiled in 1997, the remit of this memorial encompasses the commemoration of '*3 million women* who have served or are serving in or with the US Armed Forces, starting with the American Revolution' (Women's Memorial 2017, np, our emphasis). At the memorial, women who served can register their names as part of the memorial's database. In 2016 and as a ramification of the cessation of Congressional funds to support the memorial five years earlier, a fundraising call was made for private donations to keep the memorial open (see www.armytimes.com/military-honor/saluteveterans/2016/11/19/america-s-only-memorial-to-military-women-needs-your-help/). The campaign, coordinated by the AcademyWomen MilitaryWomen eMentor Community, stated that the cost of running the memorial without public funding jeopardized its future. Its website included the following passage:

> The Women's Memorial would not exist today except for the vision and efforts of military women who 20 years ago began fundraising to build and open this memorial. It's up to military women now, to ensure the memorial stays open for another 20 years.
>
> *AcademyWomen 2018*

While the outcome of this campaign has been somewhat difficult to track,[4] the more pertinent point for this chapter is that it involves women asking other (military) women to help to support their own memorial. The idea of mobilizing like-minded support tells us something about what is required to sustain such a monument in public space and who might 'care' about the longevity of the monument (or not). It also ties this example back to the case studies noted in our section on 'Women of politics' – non-public interest and funding both helped to initiate the monument and is required to sustain it.

Like the *National Women's Memorial*, the *Monument to the Women of World War II* was unveiled on Whitehall, London, in 2005, as a statement of national recognition of female contribution to WWII. Almost half the funding for the monument was generated by a national campaign initiated by Baroness Betty Boothroyd, who designated it as the beneficiary of her winnings in a television programme (*Who Wants to be A Millionaire?)*. The monument does not specifically name individual women; rather the inscription states: 'This memorial was raised to commemorate the vital work done by nearly seven million women in World War II' (IWM 2018, np). The implication of anonymity in the inscription and the absence of specific individual inscriptions – in contrast with (more) traditional war monuments, which show roll calls of the deceased – is confirmed in the monument's design.

While this is a monument to women, it does not feature actual women. Rather, uniforms and female attire hang from hooks concealed by hats. These articles of clothing are sculpted as if they are being worn, yet the figures have no legs, arms or heads. The sculptures take the form of a female body, but the articles of clothing are hung up as if signifying the hanging up of uniform at the end of a day's work. In representing women's contribution under the auspices of 'a day's work', a certain level of ordinariness of contribution is implied. In *The Unwomanly Face of War*, Svetlana Alexievich (2017) has argued that:

> Women's war has its own colours, its own smells, its own lighting, and its own range of feelings. Its own words. There are no heroes and incredible feats, there are simply people who are busy doing inhumanly human things.

Alexievich (2017) sought to capture the everyday missing details of women's involvement in the Soviet military campaign of WWII. Further, she sought to tell these stories to counter the louder heroism consistent in male histories of the war. National war memorials commemorating women raise intriguing questions about how women of war are represented in the public memory landscape. As our examples have shown, often these memorials do not depict actual women or name them, but use the female form and 'represent' the achievements/contributions of women.

Conclusions and aims for movements forward

Returning to Warner's (1985, 331) comments that women are often represented as 'something else' gives us pause to consider the examples presented here. In not using actual women, such memorials do a disservice to the contributions of the women whom they represent. Do they (and their discursive narratives) embody the actual character of these contributions, as per the everyday remit that Alexievich (2017) details and the significant mobilization of female support that the *National Women's Memorial* and its campaign demonstrates? We need to consider that women's wartime contributions usually take the heteronormative form and shape of male monuments to war. But, if women are not portrayed this way (as real women in heroic or active pose, for example), is there a risk of a broader scale (and) continued undervaluation of their contributions, especially to audiences for whom the representation of women of war already presents an anomaly in the memory landscape?

Many more monuments and works of commemorative public art feature women – too many to feature in one chapter.[5] For those who do not have the tools to read these monuments outside of their orthodox contexts, how can we (and should we at all) challenge the normative depiction of courage, strength and heroism in the action that we see in monuments depicting combatants and chivalrous infantry? How then can we show that women's contributions were as valuable to the overall cause – but different? What these questions demonstrate is that the geographies of remembrance are complex (Drozdzewski et al. 2016). Overlaying a critical feminist lens to the memory landscape raises more questions. It highlights the uneven and gendered politics and make us think, judiciously, about how these representations matter, how they tell important stories and possibly also how we can lay foundation stones for more equal representation, on our own terms.

We end this chapter with a call to action, to push beyond the pervasive images of women and genders in public places of memory. What are the opportunities to enhance the range of representations of women in public monuments, to recognize their achievements? We recall that as feminist research and teaching developed in the early 1980s it critiqued the existing practices of historians' histories for their focus on men's lives and orientation to 'kings, wars, and politics'. The endurance of public monuments perpetuates similar masculinist perspectives to those that these feminist scholars were challenging. Following their lead, we thus advocate research and teaching that also identify the diversity of women's achievements and offer some examples from other arenas of life in the arts and literature. We can start to think about pushing these agendas by asking, for example, students to look around and analyse the monuments, sculptures, streets and parks of their university and home towns, following the model of Weidenmuller et al. (2015). Who is represented there, to what effect, what are the silences, who sponsored the items and what are the students' hopes for the types of public memory markers in the future? Further, recognition of unevenly constructed memory landscapes and providing the tools to make that assessment are as important. For example, coordinating walking tours with students or joining existing tours that use critical lenses on the gendering of public space are useful starting points for change. Can attentiveness to public places, and the tutelage of these critical geographic skills, generate a capacity to look and think differently about the memory landscape?

In some places, action has been taken and change initiated. For example, Lynette Long established the EVE (Equal Visibility Everywhere) organization, with the mandate of 'Changing the face of America, one symbol at a time' (EVE 2018). Long has campaigned for a statue to Amelia Earhart in National Statuary Hall. New York City's 'She Built NYC' is to 'commission public monuments and sculptures that properly recognize women's history' ('Tired of Monuments' 2018, npn). The 'Spark Movement' established a campaign in cooperation with Google 'To Put Women on the Map'. They have mapped '100 amazing and impressive women, and connected their stories with landmarks and locations significant to their lives in dozens of cities and 28 countries around the world' (Spark 2015). To cite just a few examples of monuments commemorating women who attained international and national reputations in arenas other than war, we note the statue recently erected in Wellington, New Zealand, to celebrate the locally born writer of the early twentieth century, Katherine Mansfield.

The *Woman of Words* statue was sponsored collaboratively by three groups: the Katherine Mansfield Society (a literary group); the private Wellington Sculpture Trust; and Wellington City Council. In Bergen, Norway, tour guides draw attention to the statue of feminist novelist and political activist Amalie Skram, whose writings in the early-twentieth century drew attention to the implications of marriage and family relations for women. In Barcelona, a bust of the distinguished early-twentieth-century Catalan painter, Pepita Teixidor, a visible acknowledgement of woman's creativity, is located in the key city park of La Ciutadella. In Sydney,

Figure 24.4 *Woman of Words.* Katherine Mansfield statue, Wellington, Aotearoa New Zealand. Source: Lynda Johnston.

Australia, the 'Jessie Street' Gardens on Loftus Street honour her activism for human rights, including those of Aboriginal people, also featuring two sculptures, one honouring pioneer women settlers, the other the women military of WWII. A nearby plaque honours the woman founder of a local theatrical endeavour.

On the completion of their 'feminist walk of London's monuments to women', Rosie Martin and Louise Rondel (2017) stated: 'We feel that we are occupying a different city to other people and through this occupation, we are making the city differently'. In the early 1990s, and in an Australian context, Bulbeck (1992, 3) urged us to look more closely for monuments of women: because there are 'fewer signposts to women's memorials', they are often not written about in guidebooks or given central stage in parks or public squares. Of course, we maintain that there are fewer representations of women per se, and even fewer of real women who are commemorated for their actions, heroism and civic duties. We urge scholars to look for them, note them and tell others about them.

Notes

1 This catalogue is located at: https://en.wikipedia.org/wiki/List_of_public_art_in_the_City_of_Sydney.
2 See http://docs.wixstatic.com/ugd/a1edeb_1fbb6fe825ec4897a4e4fa898bf10e76.pdf.
3 A commitment has been made, however, to complete another statue of suffragist Emmeline Pankhurst in Manchester in 2019. See www.bbc.com/news/uk-england-manchester-35360244.
4 The fundraising webpage states that the campaign ended on 4 February 2018, with only $114,508 raised (approx. 8%) of the total $1.5M goal. Yet, on the official Women's Memorial webpage, further information about this shortfall is non-existent (see for example: https://fundrazr.com/SaveTheWomensMemorial?ref=ab_2Ib79qf8pFX2Ib79qf8pFX and www.womensmemorial.org/).
5 Of these examples, we would like to point readers not only to the accomplishments but also to other recurring challenges. These have accompanied aspects of the design of the South African *Women's Monument* in Pretoria (Marschall 2004). McDowell's (2008) article analyses the protracted and frustrating struggles of addressing male dominance in 'gendering the past and present in post-conflict Northern Ireland'. Davidson's (2018) recent article, 'Three Stories about a Statue', in discussing the commemoration of Asian 'comfort women' provided for military men, notes how these are complicated by both the dynamics of ethnic diversity in US sites and of contemporary international relations with the regions from which the women originated.

Key readings

Drozdzewski, D., S. De Nardi, and W. Waterton. 2016. "The Significance of Memory in the Present." In: *Memory, Place and Identity: Commemoration and Remembrance of War and Conflict*, edited by D. Drozdzewski, S. De Nardi and E. Waterton, 1–16. Abingdon: Routledge.
Monk, J. 1992. "Gender in the Landscape: Expressions of Power and Meaning." In: *Inventing Places: Studies in Cultural Geography,* edited by K. Anderson and F. Gale, 123–138. Melbourne: Longman Cheshire.
Warner, M. 1985. "Monuments and Maidens: The Allegory of the Female Form." London: Butler and Tanner.
Weidenmuller, E., T. Williamson, C. Leistensnider, and J. C. Finn. 2015. "History Written in Stone: Gender and the Naturalizing Power of Monuments in Southeastern Virginia." *Southeastern Geographer* 55 (4): 434–458.

References

Abousnnouga, G., and D. Machin. 2010. "Analysing the Language of War Monuments." *Visual Communication* 9 (2): 131–149.
Abousnnouga, G., and D. Machin. 2013. *Language of War Monuments.* London: Bloomsbury.
Academy Women. 2018. "Save the Memorial that Honors Military Women." Available at: https://fundrazr.com/SaveTheWomensMemorial?ref=ab_634mHbcI9pa634mHbcI9pa (accessed 15 August 2018).

Alderman, D.H. 2008. "Place, Naming, and the Interpretation of Cultural Landscapes." In: *The Ashgate Research Companion to Heritage and Identity,* edited by B. Graham and P. Howard, 195–213. Aldershot: Ashgate.
Alexievich, S. 2017. *The Unwomanly Face of War: An Oral History of Women in World War II.* New York: Random House.
Bevan, R. 2006. *The Destruction of Memory: Architecture at War.* Wiltshire: Reaktion.
Blakemore, E. 2015. "Central Park Has 22 Statues of Historical Figures. Every Single One is a Man." *Smithsonian,* 22 July. Available at: www.smithsonianmag.com/smart-news/central-park-has-no-statues-real-women-180955973/#7vHZ3kyvfl8heUXw.99 (accessed 24 September 2018).
Boissoneault, L. 2017. "The Suffragist Statue Trapped in a Broom Closet for 75 Years." *The Smithsonian,* Available at: www.smithsonianmag.com/history/suffragist-statue-trapped-broom-closet-75-years-180963274/#3y14GwvzPLf4Ha1o.99 (accessed 16 August 2018).
Bulbeck, C. 1992. "Women of Substance: The Depiction of Women in Australian Monuments." *Hecate* 18 (2): 8–24.
Capps, K. 2016. "The Gender Gap in Public Sculpture." CityLab blog. Available at: www.citylab.com/design/2016/02/the-gender-gap-in-public-sculpture/463170/ (accessed January 2018).
Carlson, J. 2018. "Here's the First Statue Depicting Real Women in Central Park, Coming in 2020." *Gothamist,* 20 July. Available at: http://gothamist.com/2018/07/20/central_park_first_female_statue.php (accessed 23 September 2018).
Davidson, R. 2018. "Three Stories about a Statue." *Yearbook of the Association of Pacific Coast Geographers* 80: 41–65.
Drozdzewski, D., S. De Nardi, and E. Waterton. 2016. "The Significance of Memory in the Present." In: *Memory, Place and Identity: Commemoration and Remembrance of War and Conflict,* edited by D. Drozdzewski, S. De Nardi, and E. Waterton, 1–16. Abingdon: Routledge.
Dwyer, O.J. 2004. "Symbolic Accretion and Commemoration." *Social & Cultural Geography* 5 (3): 419–435.
Dwyer, O.J., and D.H. Alderman. 2008. "Memorial Landscapes: Analytic Questions and Metaphors." *Geojournal* 73 (3): 165–178.
Esbenshade, R.S. 1995. "Remembering to Forget: Memory, History, National Identity in Postwar East-Central Europe." *Representations* 49: 72–96.
EVE. 2018. "Who We Are". Equal Visibility Everywhere. Available at: http://equalvisibilityeverywhere.org/who-we-are/ (accessed 24 September 2018).
Finney, P. 2002. "On Memory, Identity and War." *Rethinking History* 6: 1–13.
Gillis, J.R. 1994. "Memory and Identity: The History of a Relationship." In: *Commemorations the Politics of National Identity,* edited by J.R. Gillis, 3–24. Princeton, NJ: Princeton University Press.
Grimshaw, P. 1972. *Women's Suffrage in New Zealand.* Auckland: Auckland University Press.
Gross, E.L. 2018. "The Five Female Historical Statues in New York City Are Decorated For International Women's Day." *Forbes,* 3 August. Available at: www.forbes.com/sites/elanagross/2018/03/08/the-five-female-historical-statues-in-new-york-city-are-decorated-for-international-womens-day/#37a09bd57c26 (accessed 15 August 2018).
Hoelscher, S., and D.H. Alderman. 2004. "Memory and Place: Geographies of a Critical Relationship." *Social and Cultural Geography* 5: 347–355.
Imperial War Museum (IWM). 2018. "Women of World War Two." Available at: www.iwm.org.uk/memorials/item/memorial/51288 (accessed 15 August 2018).
Inglis, K. 1987. "Men, Women, and War Memorials: Anzac Australia." *Daedalus* 116 (4): 35–59.
Krzyzanowski, J. 1968. *Polish Romantic Literature.* New York: Books for Libraries.
Krzyzanowski, J. 1978. *A History of Polish Literature.* Poland: PWN–Polish Scientific.
Longhurst, R. 2001. *Bodies: Exploring Fluid Boundaries.* London: Routledge.
Marschall, S. 2004. Serving Male Agendas: Two National Women's Monuments in South Africa. *Women's Studies* 33 (8): 1009–1033.
Martin, R., and L. Rondel. 2017. "Re-figuring the City: A Feminist Walk of London's Monuments to Women." *Institute of British Geographers Gender and Feminist Geographies Research Group Blog,* Available at: www.gfgrg.org/re-figuring-the-city-a-feminist-walk-of-londons-monuments-to-women/ (accessed 23 September 2018).
McDowell, S. 2008. "Commemorating Dead 'Men': Gendering the Past and Present in Post-conflict Northern Ireland." *Gender, Place and Culture: A Journal of Feminist Geography* 15 (4): 335–354.
McGrane, S. 2017. "An Important Statue for 'Comfort Women' in San Francisco." *New Yorker.* Available at: www.newyorker.com/culture/culture-desk/an-important-statue-for-comfort-women-in-san-francisco (accessed 17 October 2018).

Monk, J. 1992. "Gender in the Landscape: Expressions of Power and Meaning." In: *Inventing Places: Studies in Cultural Geography*, edited by K. Anderson and F. Gale, 123–138. Melbourne: Longman Cheshire.
Monumental Women. 2018. Monumental Women. Available at: www.monumentalwomen.org/ (accessed 15 August 2018).
Miłosz, C. 1969. *The History of Polish Literature*. London: Macmillan.
Nora, P. 1989. "Between Memory and History: Les lieux de memoire." *Representations* 26: 7–24.
Peled, S. 2017. "Where Are the Women? New Effort to Give Them Just Due on Monuments, Street Names." *CNN*. Available at: http://edition.cnn.com/2017/03/08/us/womens-monument-project-trnd/index.html (accessed January 2018).
Puwar, N. 2004. *Space Invaders: Race, Gender and Bodies Out of Place*. London: Berg.
Rhodan, M. 2017. "Inside the Push for More Public Statues of Notable Women." *Time*, 3 August. http://time.com/4903612/women-statues-san-francisco/ (accessed January 2018).
Rolston, B. 2018. "When Everywhere is Karbala: Murals, Martyrdom and Propaganda in Iran." *Memory Studies*, September. Online First. doi:1750698017730870.
Rose-Redwood, R.S. 2008. "From Number to Name: Symbolic Capital, Places of Memory and the Politics of Street Renaming in New York City." *Social & Cultural Geography* 9 (4): 431–452.
Rose-Redwood, R., Alderman, D. and Azaryahu, M. 2010. "Geographies of Toponymic Inscription: New Directions in Critical Place-name Studies." *Progress in Human Geography* 34: 453–470.
Schwartz, E. 2017. "Why We Need Female Monuments." *Econlife*. Available at: https://econlife.com/2017/09/need-for-female-monuments/ (accessed January 2018).
"Spark Puts Women on the Map!" 2015. "Spark Puts Women on the Map!" Spark Movement, online, 2 March. Available at: www.sparkmovement.org/ 2015/03/02/spark-puts-women-on-the-map/ (accessed 24 September 2018).
Stevens, Q. 2015. "Masterplanning Public Memorials: An Historical Comparison of Washington, Ottawa and Canberra." *Planning Perspectives* 30 (1): 39–66.
Stone, A.E.W. 2017. "Moving the Women into the Light." National Women's History Museum. Available at: www.womenshistory.org/articles/moving-women-light (accessed 15 August 2018).
"Tired of Monuments." 2018. "Tired of Monuments Ignoring Women's History? Now You Can Nominate Female Trailblazers for Statues in NYC." *Frieze*, 21 June. Available at: https://frieze.com/article/tired-monuments-ignoring-womens-history-now-you-can-nominate-female-trailblazers-statues-nyc (accessed 15 August 2018).
Verovsek, P.J. 2017. "Memory, Narrative, and Rupture: The Power of the Past as a Resource for Political Change." *Memory Studies*. ISSN 1750-6980.
Warner, M. 1985. *Monuments and Maidens: The Allegory of the Female Form*. London: Butler and Tanner.
Weber, S. 2016. *The Woman Suffrage Statue: A History of Adelaide Johnson's Portrait Monument to Lucretia Mott, Elizabeth Cady Stanton and Susan B. Anthony at the United States Capitol*. Jefferson, NC: McFarland.
Weidenmuller, E., T. Williamson, C. Leistensnider, and J.C. Finn. 2015. "History Written in Stone: Gender and the Naturalizing Power of Monuments in Southeastern Virginia." *Southeastern Geographer* 55 (4), 434–458.
Winter, C. 2014. "Public Perceptions of War Memorials: A Study in Ballarat." *Australasian Journal of Regional Studies* 20 (1) 210–230.
Women's Memorial. 2017. Women's Memorial. Available at: www.womensmemorial.org/ (accessed 15 September 2018).

25

FEMINIST POLITICAL ECOLOGIES

Race, bodies and the human

Sharlene Mollett, Laura Vaz-Jones and Lydia Delicado-Moratalla

Feminist political ecologies

Feminist political ecology (FPE) is an interdisciplinary subfield that blends feminist theories and concepts with the wider field of political ecology, a conceptual framework linking political economy with nature and the environment (Neumann 2005; Peet and Watts 2004). In the subfield's seminal text, *Feminist Political Ecology: Global Issues and Local Experiences* (1996), Rocheleau, Thomas-Slayter and Wangari maintain that gender is 'a critical variable in shaping resource access and control interacting with class, caste, race, culture, and ethnicity to shape processes of ecological change' (Rocheleau et al. 1996, 4; see also Carney 1993; Jarosz 1992). FPE scholarship was shaped by a need to complicate the singularity of the 'land manager' category, once dominant in political ecology and frequently imagined as male. Feminist environmental movements around reproductive rights, food consumption and agricultural work also influenced policy shifts in women in development (and later gender and development), affirming women's relationship to nature and the environment through their daily interactions. Such foci required attention to more than class dynamics (Gururani 2002). Overall, since the early 1990s, the feminist movement and scholarship have brought meaningful questions to FPE with regards to various kinds of oppression and the degradation of nature (Sundberg 2017).

In this chapter, we highlight how FPE advances from three scholarly angles: ecofeminism; feminist critiques of science; and feminist critiques of development. Ecofeminists argue that the oppression of women by men and the exploitation of nature by humans are linked. Such dualisms are believed to be a legacy of Western philosophical traditions, whereby:

> the human capacity for reason and abstract thought as the grounds for transcendence and domination of nature … is framed as masculine through its opposition to and domination of all that is associated with nature, the body, reproduction, emotion and ultimately the feminine.
>
> *Sundberg 2017, 2*

Second, feminist critiques of science maintain that gender and patriarchal norms influence knowledge production and shape whose knowledge(s) have value. In this vein, feminist scholars challenge the concept of objectivity and instead argue that all knowledge is partial, rendering objectivity mythic and comprised of the incomplete perspectives of Euro-American, White, heterosexual, property-owning men (Haraway 1991; Harding 1986; Sundberg 2017). Feminist epistemologies and methodologies align diverse interests in FPE, with a collective view that argues that all 'knowledge is partial, situated, and emerges from embodied social locations' (Resurrecciòn 2017, 80). This perspective contributes to the commitments embedded in feminist knowledge production, which challenge latent relationships of domination within objective knowledge claims (Haraway 1988).

Feminist epistemologies also influence the ways in which feminist political ecology understands how power operates within the seemingly mundane spaces and ecologies of everyday life. According to Sundberg (2017, 5):

> The scale of the everyday is where social reproduction takes place, where subject identities and social orders are brought into being and contested. Attending to daily life allows FPE to shed light on otherwise neglected dimensions of environmental engagements.

The socially uneven terrain upon which natural resource struggles occur is constituted in large part through the everyday negotiations of ecological processes. Wangui (2014), for example, examines how the changing pastoral livelihoods in Kenya are shifting the labour demands on women and men. She draws attention to the everyday negotiations and performances that women partake in as they withdraw and contribute their labour to the household as a challenge to patriarchal relations inside the household and vis-à-vis the market. FPE thereby takes stock of the ways that scales are linked and both shape and are shaped by the everyday and intimate spaces of the globe, nation, region community, home, household and the body (Elmhirst 2011; Hawkins et al. 2011).

A third research direction employs FPE as a conceptual framework for international development critique. In this vein, FPE scholarship illustrates how women are consistently marginalized and exploited by sustainable development, population control, land titling and conservation projects, even when such projects are conducted in the name of gender and development and/or gender and poverty programmes (Carney 1992; Jarosz 1992). Moreover, this focus offers a feminist critique of development that is strongly influenced by the work of Chandra Mohanty and her critiques of development's 'Third World difference', a representation of women in the Global South as 'needy' recipients of Western aid programmes. These images elide the plurality of women's multiple social locations, experiences and knowledges that Mohanty charges are evidence of:

> assumptions of privilege and ethnocentric universality on the one hand, and inadequate self-consciousness about the effect of western scholarship on the 'third world' in the context of a world system dominated by the west on the other.
>
> *1988, 63*

This produces

> an analysis of 'sexual difference' in the form of a cross-culturally singular, monolithic notion of patriarchy or male dominance [that] leads to the construction of a similarly

> reductive and homogeneous notion of what I call the 'third world difference' – that stable ahistorical something that apparently oppresses most if not all the women in these countries.
>
> *Mohanty 1988, 63*

FPE critiques of development are also shaped by material concerns raised by feminist environmentalism, built on a global movement that challenges a society and world economy organized for the profit of a small number of:

> white men [which] has created the conditions for widespread unemployment, violence at home and in the streets, oppression of third world peoples, racists attacks, inadequate food, housing and health care, and finally the ecological devastation of the earth.
>
> *Women and Life on Earth 1979, cited in Seager 2003, 948*

Bina Agarwal's contribution to feminist environmentalism is best seen in her text *A Field of One's Own* (1994). Her work attends to the multiple relations of power that control women in private and public spaces. She maintains that, in the context of enduring environmental degradation, women's share in the division of labour and the time for daily domestic chores in the household and in agricultural fields increases. This is particularly problematic because the fields are owned exclusively by men. Relatedly, Agarwal argues for women's land rights. With such rights, women are better protected against food insecurity, because they have more say in decisions such as crop choices. They are also empowered within their households, making them less dependent on their spouses (1994). Land rights for women thus mitigate the patriarchal dominance and the embodied consequences of environmental degradation.

During the first 20 years of FPE, gender and gender inequalities preoccupied feminist political ecology development critiques. However, more recently, the subfield has re-emerged as 'new feminist political ecologies' with a profound commitment to Mohanty's challenge and the intersectional nature of domination and privilege (Elmhirst 2011; Mollett 2010). In the challenge to '"mess" with gender by "doing race"', FPE takes seriously Sundberg's argument that 'processes of racialization articulate in and through the environmental formations and vice versa' (2008, 579, cited in Mollett and Faria 2013, 118). For Mollett and Faria (2013, 118):

> this means more than simply working in or writing about communities of colour. It necessitates recognition of the power inequities between global north and global south, shaped by the legacies of colonial racisms, as well as (colonial) patriarchies.

A call to multiple forms of power is bolstered through the growing influence of the concept of intersectionality in FPE scholarship. Borrowing from Black feminist thinkers, intersectionality as both conceptual framework and epistemology understands race, gender, class and ethnicity as interlocking forms of oppression and privilege that shape the multiple dimensions of people's lives (Collins 2000; Crenshaw 1991). The concept offers us a way to understand multiple power dynamics as mutually constituted, and it helps to advance our knowledge of the daily lives of people across the globe and in the context of development programmes. For example, Mollett and Faria (2013) explore race as constitutive of gendered subjectivities and challenge the limited theorization of race in political ecology writ large. They write, 'postcolonial intersectionality acknowledges the way patriarchy and racialized processes are consistently bound in a postcolonial genealogy that embeds race and gender ideologies within nation-building and international development processes' (Mollett and Faria, 2013, 120). Leila Harris insists that 'FPE has the potential to unsettle and challenge dominant assumptions ... Yet, to do so, the very understanding of feminism itself must be problematized and unsettled, moving towards intersectional understandings' (2015, xx). Indeed, in the recent engagements with post-colonial and decolonial scholarship, FPE scholars are energizing forms of thought that challenge Western understandings of gender and development.

FPE and postcolonial and decolonial influences

Political ecology's engagements with more-than-human nature is part of a decolonial aim to critique dualist constructions of culture and nature that underpin Euro-American knowledge production. Challenging dualist conceptions of nature and society helps to demonstrate how human and nonhuman natures are interconnected (Chagani 2014; Collard 2012). Feminist geographers contribute to these critiques through an array of approaches, such as animal and species geographies (Gillespie 2018; Hovorka 2012), and post-humanism and race (Anderson 2007) and Indigenous ontologies (Daigle 2018; Hunt 2014). For example, Sundberg explores the political ecology of the US–Mexico border in relation to how nonhuman-natures contend with boundary enforcement practices that then influence the ways that humans, technology and money are deployed in the interest of security (2011). Mullaney's (2014) work on maize varieties in Mexico brings together post-humanism, anticolonial geopolitics and feminist political ecology to examine how crop seeds are linked to knowledge systems, cultural identities and labour structures in ways that challenge state and corporate control over agrarian and food systems. In a study of forest management in India, Münster (2016) highlights how the intimate working relationship between Indigenous forest labourers and captive elephants unfolds through both intimate and violent interaction. This work highlights how ethnography can move beyond 'human' observation to reveal that human and more-than-human relationships share collective histories of region and space that shape and are shaped by conservation management.

Feminist political ecologists aptly suggest that, while complicating the dualist notions of nature and culture, post-humanist approaches (and more-than-human critiques) do not go far enough to decolonize knowledge production in the academy and the daily lives of people who are dependent on natural resources (Mollett 2017; Sundberg 2014). As such, feminist political ecological research is increasingly engaging with decolonial theorists and the field of Indigenous geographies. For instance, Indigenous geographies scholarship illuminates the practice of 'epistemic ignorance' and 'epistemic violence' in geography in the ways in which Indigenous knowledges are elided from post-humanist approaches and how 'epistemologies or ontologies embedded in other worlds are not alive to us as resources for critical thought today' (Sundberg 2014, 38; see also Hunt 2014). Feminist engagement with decolonial thinking influences the way that feminist political ecology calls for decolonial processes in the fields both out there and inside our academic and epistemic worlds (Asher 2017; Radcliffe 2015). Such a call aligns with post-colonial feminist thinkers, challenging Western feminism to dispel presuppositions about 'third-world woman' categories in both development thought and practice (Mohanty 1988). Post-colonial and decolonial insights increasingly shape feminist geographic discussions of nature and natural resource struggles (Mollett 2017; Zaragocin 2017). One of the ways in which these lines of thought merge is through a focus on the body.

In feminist geography, broadly defined, the body figures prominent in shaping understandings of subjectivity to 'place the embodied subject and engage the body to better understand difference through a range of approaches' (Mountz 2017, 3). Highlighting the immaterial, embodied and affective dimensions of environmental relations and struggles discloses how the body is a site and scale of analysis within environmental politics. In FPE, the body provides a scale and location upon and through which ideas, ideologies and politics unfold and are made visible, linking the global with the intimate (Mountz 2017; Pratt and Rosner 2006). The body is also a site where autonomy and sovereignty are manifest, a site of articulation of claims and resistance to the violent legacies of colonialism and transatlantic slavery (Goeman 2013; McKittrick 2006).

For many FPE scholars, the body figures prominently because struggles over resources and ecological processes are not simply economistic or 'rational' choices; they are influenced by the physicalities of bodies, imaginaries and conceptions of belonging, territory and ancestry (Doshi 2017; Harris 2015). Environmental politics and processes are visible in the ways in which women's bodies are at risk of violence when engaging the most natural of activities, including urination, defecation and menstruation (Truelove 2011). Sultana (2011), for example, illustrates that suffering stemming from arsenic contamination of water in rural Bangladesh is unevenly distributed, and argues that emotions and embodiment crucially shape the reasons and conditions under which people access and use resources. According to Sultana:

> Abstractions of 'resource struggles' and 'resource conflicts' are thereby grounded in embodied emotional geographies of place, peoples, and resources, enabling us to enhance our comprehension of the complex ways resources, bodies and emotions come to matter in survival strategies and everyday resource management practices.
>
> *2011, 163*

Feminist political ecology also concerns itself with how ecological processes shape and extend through bodies (Hayes-Conroy and Hayes-Conroy 2013; Sultana 2011). The body is imbricated in complex relationships with land and resources – and their life-sustaining qualities. It is often the foundation upon which environmental struggles are waged (Doshi 2017; Harris 2015; Sundberg 2008, cited in Mollett and Faria 2013).

We argue here that a focus on the entanglements of bodies, difference and environmental struggles not only lays bare hierarchical relations between human and nonhuman nature but makes evident the human-to-human domination, subjugation and resistance. Relevant to the Sustainable Development Goals (SDGs), we are particularly interested in how racial difference and the processes of racialization shape subjectivity as an intersectional and embodied process. We argue for more profound attention to what it means to be human (see McKittrick 2015). Thus, in the space below, we place feminist political ecology in conversation with Black Feminist Thought (BFT) to engage with the historically and spatially situated ways that race, patriarchy, sexuality and capitalism are entangled and classify some people as *human* and others as not. Embedded in our analysis is the historical and spatial fact that 'the *Human* (and its meanings) comes through a system that has rendered Africans (and people of African descent) outside of humanity' (McKittrick 2015; Mollett 2017, 13; Weheliye 2014). To illustrate, we flesh out how racialization, gender and sexuality are mutually constituted in the dehumanization of sex-trafficked Nigerian women within neoliberal tourism development in Spain, as part of a global patriarchal social-racial order.

Black feminist thought: embodied geographies of slavery

Black feminist thought (BFT) emerges from the intellectual and activist work of Black women and radical women of colour (Collins 2000; Crenshaw 1989; Spillers 1987). BFT challenges hegemonic images of Black women that justify their economic, political and sexual exploitation and highlights the rich legacy of Black women's struggles for social justice (Collins

2000; Crenshaw 1989). The entanglement of BFT with human geography offers insights into how space and place are imbued with meaning through Black women's everyday struggles over livelihoods and belonging, while interrogating the hegemonic imaginaries that seek their subjugation (King 2016; McKittrick 2006). Such a focus both works against an epistemological tradition that has often 'displaced, rendered ungeographic' Black women and resists disciplinary tendencies to place race, as a form of power, and critical race scholarship at the periphery of knowledge production (McKittrick 2006, x). Black feminist thinking has long shaped feminist geographic scholarship, particularly with regards to the concept of intersectionality, since 'intersectionality was, at its inception, already a deeply spatial theoretical concept, process and epistemology' (Mollett and Faria 2018, 2; see also Kobayashi and Peake 1994).

The influence of BFT on geographic scholarship both challenges and makes visible the various ways in which Black people are not simply disavowed for their Blackness in a White supremacist patriarchal global social order but are misrecognized as *less than human* (Gilroy 2015; King 2016; McKittrick 2006; Mollett 2017). Black feminist understandings of the transatlantic slave trade illuminate how slavery was legitimated and sustained through the objectification of the captive Black body (Hartman 1996; McKittrick 2006; Spillers 1987). Women's bodies and reproductive labour, in particular, represent the material, sociospatial and symbolic sites for the reproduction of slavery (Hartman 2016; Morgan 2004). Slavery transformed bodies into commodities in ways that illuminate both gendering and (un)genderings (Spillers 1987). Namely,

> when it was profitable to exploit [Black women] as men, they were regarded, in effect, as genderless, but [Black women] they could be exploited, punished and repressed in ways suited only for women, they were locked into their exclusively female roles.
>
> *Davis 1983, 6*

Labouring in other spaces of the plantation as housekeepers, caretakers, nannies and wet nurses did not spare women of the distinct forms of violence and cruelty and domestic expectations of owners who reduced Black women's value to their wombs, as their sexual reproduction created future slaves and wealth creation for slave owners (Davis 1983; Hartman 2016). White male consumption of Black slave women's labour and bodies was not driven simply by economic rationale; as Davis powerfully argues, '[o]ne of racism's salient historical features has always been the assumption that white men – especially those who wield economic power – possess an incontestable right of access to Black women's bodies' (1983, 175) that was 'profitable' and 'pleasurable' (McKittrick 2006, 71). The systemic sexual exploitation of Black women and the denial of their humanity is a lasting legacy of the past within a global imaginary of Blackness and Black femininity, manifest in the 'afterlife' of slavery (Hartman 2016). The mutual constitution of race, gender, capitalism and colonialism imbue spatial imaginaries and reproduce Black female representations as disposable, violable and subject to 'multilayered and routinized forms of domination' in their everyday lives (Crenshaw 1991, 1245). By bringing together FPE and Black feminist theorizing, we advance a feminist political ecology of race that simultaneously questions what it means to be human and clarifies the dehumanizing processes that place Black women outside of humanity. To illustrate, we briefly reflect on the forced migration and labour of Nigerian women in Spain. In doing so, we interrogate how bodies are unevenly situated in social landscapes through which environmental degradation and racial-sexual disposability are woven together.

Sex trafficking and prostitution: from the Niger Delta, Nigeria to Alicante, Spain

The oil and gas sector in Nigeria accounts for 80 per cent of the state's income, 40 per cent of the country's GDP and yet employs only 4 per cent of the population (Oluduro 2014). Despite its expansive energy section, the (mal)distribution of oil wealth has meant that 62.6 per cent of Nigerians continue to live in poverty (IMF 2018). Economic disparities have been exacerbated by the devastation of the socio-environmental integrity of the Niger Delta (Ugor 2013). Export-oriented agriculture managed by international agribusinesses uses vast amounts of land and water. Monocropping and oil extraction have degraded the land and polluted the water and soil with disastrous effects on those with farming- and fishing-based livelihoods (Oluduro 2014). Resistance to disenfranchisement and environmental destruction in the region is countered by state and non-state violence, which has left millions of people displaced from their homes and many dead in the Niger Delta (Egharevba and Osunde 2001; Nixon 2011; Watts 2005).

Environmental degradation and income disparities exacerbate the existing ethnic, class and gender hierarchies (Watts 2005). For example, unemployment is drastically high in the region and is particularly stark among women, who have low rates of employment in the formal sector and are not considered for work in the oil sector (Ikelegbe 2005; Ukeje 2004). Women in the Niger Delta suffer the gravest consequences of water and land pollution, because they are responsible for agricultural labour, household water provision and all domestic work (Egharevba and Iweze 2004). Oil spill-related harms are linked to increased neonatal mortalities in the region, speaking to the ways in which the political ecologies of oil extraction are gendered and embodied (Hodal 2017).

The economic and ecological harms of resource extraction faced by women in the Niger Delta have also increased their vulnerability to other kinds of exploitation. In Benin City, the capital of Edo State, which has the highest unemployment in the country (Omorodion 2009), one in every three girls/women between 15 and 20 years old has been contacted to travel to European countries by a recruiter working in organized crime (IOM 2006). Young girls are targeted for forced sex work, often by the deception of promises of jobs in domestic work such as cleaning and nannying Poverty, ethnicity, illiteracy and high unemployment rates among youth make Edo women vulnerable as targets of transnational sex trafficking (Omorodion 2009). This has led some to deduce that '[Nigerian] women have become the new natural resource for exportation' (Elabor-Idemudia 2003, 104). But, such trafficking of bodies is not new, as Nigeria's Atlantic coast was a crucial site of transatlantic slave capture in the colonial period and remains a significant location from which Nigeria women are trafficked to Europe for sex work in the twenty-first century (UNESCO 2006; see also Delicado-Moratalla 2017).

The Global Slavery Index (2018) reports that most Nigerian women are trafficked from Nigeria not knowing that they will work as prostitutes in European cities. Many travel in the expectation of working in domestic service. Of those who make it to Spain, many left their rural areas of the Niger Delta voluntarily for Benin City, where they learned that they had been deceived by their family members and neighbours, who had recruited them under the pretence of working in domestic service in Europe to secure a better life (Benavides 2018; Delicado-Moratalla 2017). However, the women quickly find that their lives are in danger. Many young ethnic Yoruba women, after being kidnapped in Benin City, undergo a *juju* ritual, sometimes undertaken first in their home village, where their bodies, in part and in full, are doused in animal blood. As part of the ritual, traffickers obtain their pubic hair, fingernails and/or menstrual blood, which are stored and recorded under their name (UNESCO 2006). According to the well-known Spanish publication *El Pais*, this ritual gives traffickers both an embodied and psychological hold over young women and girls to increase their submission (Domínguez and Gálvez 2017). After being forcibly taken by crime networks in Edo State, women and their traffickers cross the Mediterranean Sea in small, clandestine ships from northern Morocco to Spanish cities like Barcelona and Alicante. During this crossing, they are frequently raped and often arrive pregnant, and are forced to give up their babies (Women's Link Worldwide 2014). This journey leaves the women locked into enormous debts to the traffickers, supposedly for their travel costs – a discursive justification that the traffickers use to coerce them into prostitution (Euroweekly.com). Intensifying women's vulnerability, traffickers steal and retain the women's passports upon arrival (UNESCO 2006). Their precarious immigration status makes it extremely difficult for trafficked women to secure protection and situates them as even more vulnerable to violence and coercion within their forced participation in commercial sex.

The sex trafficking of Yoruba women in Spain, as well as the desperation that drives family and friends to recruit these women for the traffickers in exchange for compensation, is linked to the environmental harms and impoverishment tied to extraction production in the Niger Delta; their economic and environmental precarity makes possible their captivity. Such transactions are followed by more-than-human rituals; dousing bodies in animal blood and drawing upon religion and faith in ways that compel these women to acquiesce to their subjugation. Such is reinforced by invented debt bondage that further strengthens the traffickers' supposed ownership of these women and their bodies. Furthermore, the dehumanization of the women in Nigeria transforms them into a recognizable sexual symbol on the streets of Spain's cities 'that turns Blackness into an ideological currency that moves beyond the [original] moment of sale' (McKittrick 2006, 78).

La Calle de las Negras: 'This is not my way, I don't like it, I want to leave'

Spain is a vital hub for prostitution in Europe (Cacho 2010). In recent years, the increasing number of brothels, escort services, dance clubs and massage parlours has expanded sexual services throughout the country. Spain is also a top sex-tourism destination, generating roughly $22 billion dollars a year and rivalling the 'drug and arms trade in terms of prevalence and profit, according to Spain's National Rapporteur on Human Trafficking' (Domínguez and Gálvez 2017). In 1995, commercial sex was decriminalized, creating an environment in which prostitution is tolerated and unregulated. While there are some exceptions that fall under municipal jurisprudence, such as the *Ley Orgánica 4/2015 de Protección de la Seguridad Ciudadana* (Constitutional Law for the Protection of Citizen's Security), which prohibits engagement with prostitution in specific public spaces, a lack of regulation bolsters sex trafficking. Journalist Lucia Benavides of *Bright Magazine* argues that '90% of sex workers could be under the control of organized crime networks' (2018). The change in the racial composition of sex workers over the last 30 years has transformed the landscape of commercial sex. Where once European and Spanish women dominated the trade, today immigrants from Latin America, former Eastern Europe and Sub-Saharan Africa comprise the majority of women working as prostitutes in Spain.

On the Mediterranean coast of Spain lies the south-east port city of Alicante (population of 300,000). Situated on the Costa Blanca, the city is known for its warm waters, Mediterranean climate and decadent beach vacations frequented by international and domestic tourists. Alicante is a key destination in a global sex-trafficking network that kidnaps and smuggles poor, ethnic Yoruba-Nigerian women, who are then forced to work in the sex trade as street prostitutes. For them, the region is a dangerous place for prostitutes: between 2010 and 2015, Costa Blanca had the highest incidence of prostitute homicide in the country (Feminicidio.Net 2016). In 2017, ten people were arrested and suspected of forcibly trafficking 'dozens' of Nigerian women, coerced into prostitution for as little as €5 per transaction and 'under conditions of absolute slavery' (Shedrofsky 2017). Despite the many brothels in Spain, in Alicante the Nigerian prostitutes work on the street. The Nigerian women occupy a street near the centre of the Alicantian sex industry, *La Calle de las Negras* (Black Women's Street). According to Maria, an Edo girl from Niger working on *La Calle*: 'I need money, I need a job ... this is not my way' (Delicado-Moratalla interview 2017). Another Nigerian woman, Eugenie, similarly states, 'This is not my way, I don't like it ... I want to leave'. Eugenie describes how she was very ill. Eventually, the Red Cross unit found her homeless, starving and infected with HIV (Delicado-Moratalla interviews 2017).[1] According to social workers at the Red Cross, which often provides services to prostitutes and sex workers in Alicante, the women of *La Calle de las Negras* fear violence and certain death if they try to break from their pimps and traffickers. Many of the women working on *La Calle* are girls in their mid-teens, forced to provide sex for particular men on a regular basis. While *La Calle de las Negras* is a site of commercial sex and forced prostitution, their services are also sold on the internet very cheaply (Ranea Triviño 2017). According to the women, the 'johns' (mostly White male tourists) do not wear condoms and, indeed, stories of HIV infection among the women are common (Delicado-Moratalla interviews 2017).

This brief example of trafficked Nigerian women and girls exemplifies more than the racial and carnal forms of dehumanization. In fact, as McKittrick (2015) explains in the context of nineteenth-century US slavery and the auction block, the 'moment of sale' satisfies more than economic aspirations. It also speaks to the role of sexual desire imbued with 'the repetitive and sometimes mundane economic exchanges' that we argue give meaning to the street name itself, *La Calle de las Negras,* it involves sexual desire. *La Calle* is a place of multi-scaled violence, where the sex trafficking of Nigerian women and girls extends beyond the 'moment of sale' (McKittrick 2006, 71). Like the violence that comes after slave markets, the 'moment of sale' 'obscures Black humanity by violently transforming human beings into commodity objects through the act of economic exchange', intensifying racial sexual difference and gendered disposability (McKittrick 2006, 73; see also Razack 2016). A racialized, patriarchal system materializes a process of dehumanization through a variety of practices that, so often repeated, may go unnoticed: the kidnap and theft of women's bodies, including hair and blood, raping as a form of terror and subjugation in transit, the separation of mothers from their children and placing women and girls on the street and internet to be 'rented' for sex by a largely White, male, tourist market. Such a process is aided by a patriarchal elite state and its neo-colonial foreign and national investors, who destroy homelands and displace and make vulnerable the ethnic minority populations once economically and culturally tied to this environment (Nixon 2011).

Final thoughts

Since its emergence in the early 1990s, feminist political ecology (FPE) has foregrounded how environmental processes and subject formation are co-constituted. We do so here in a way that makes clear the embodied consequences of environmental devastation and poverty. The ways in which young, poor and ethnic minority women in Nigeria are disproportionately vulnerable to sex slavery and the violence of human trafficking are an extension of the environmental degradation of their homelands. Our work here responds to Doshi's (2017) insistence on the embodied attributes of uneven development within political ecology and brings modern slavery to feminist political ecological critiques of development. As is captured by the United Nations' adoption of the eradication of human trafficking as a Sustainable Development Goal (SDG) (Global Slavery Index 2016), recognizing sexual slavery as an environment-development challenge advances our understanding of the connections between environmental degradation and bodily harm. Our contribution reveals how control over the land's natural resources not only means embodied consequences but demands attention to how the control over natural resources often extends to the control over the very bodies, constructed as different, that depend on them. FPE is a promising corrective to the technocratic and positivist approaches that pervade environmental policies and sustainable development agendas.

Thus, we draw insights from the overlapping contributions of FPE, decolonial and postcolonial thought and Black feminist thinking about the body and slavery. These scholarly conversations suggest that future entanglements between FPE and Black feminist thinking will prove instructive to deepen our work and justice-oriented scholarship around human rights. That the category of 'human' is contested means that feminist political ecologists are well placed to interrogate how some humans are misrecognized as less-than-human, offering a novel path for advancing feminist critiques of international development in intersectional and embodied ways. Such a focus is at the centre of our critique and speaks directly to the complex challenges of human lives.

Note

1 Fieldwork in Alicante was conducted using participant observation while volunteering at the local Red Cross unit, which provides basic healthcare support and language-learning skills to women involved in street prostitution. During fieldwork, Delicado-Moratalla interviewed ten Nigerian women involved in street prostitution. Most of them were originally from Benin City.

Key readings

Doshi, S. 2017. "Embodied Urban Political Ecology: Five Propositions." *Area* 49 (1): 125–128.

McKittrick, K. 2006. *Demonic Grounds: Black Women and the Cartographies of Struggle.* Minneapolis: University of Minnesota Press.

Mollett, S., and C. Faria. 2013. "Messing with Gender in Feminist Political Ecology." *Geoforum* 45: 116–125.

References

Agarwal, B. 1994. *A Field of One's Own: Gender and Land Rights in South Asia.* Cambridge: Cambridge University Press.

Anderson, K. 2007. *Race and the Crisis of Humanism.* London: Routledge.

Asher, K. 2017. "Spivak and Rivera Cusicanqui on the Dilemmas of Representation in Postcolonial and Decolonial Feminisms." *Feminist Studies* 43 (3): 512–524.

Benavides, L. 2018. "Decriminalizing Sex Work in Spain Made it Safer for Women—and Traffickers." *Bright Magazine* 11 April. Available at: https://brightthemag.com/legalizing-sex-work-spain-prostitution-human-rights-trafficking-immigration-gender-78b96c05e6fa

British Council Nigeria. 2012. *Gender in Nigeria Report 2012: Improving the Lives of Girls and Women in Nigeria.* Nigeria: Department for International Development.

Cacho, L. 2010. *Esclavas del poder: un viaje al corazón de la trata sexual de mujeres y niñas en el mundo.* Mexico City: Grijalbo.

Carney, J.A. 1992. "Peasant Women and Economic Transformation in The Gambia." *Development and Change* 23 (2): 67–90.

Carney, J. 1993. "Converting the Wetlands, Engendering the Environment: The Intersection of Gender with Agrarian Change in The Gambia." *Development* 69 (4): 329–348.

Chagani, F. 2014. "Critical Political Ecology and the Seductions of Posthumanism." *Journal of Political Ecology* 21 (1): 424–436.

Collard, R.-C. 2012. "Cougar – Human Entanglements and the Biopolitical Un/Making of Safe Space." *Environment and Planning D: Society and Space* 30 (1): 23–42.

Collins, P.H. 2000. *Black Feminist Thought: Knowledge, Consciousness, and the Politics of Empowerment.* New York: Routledge.

Crenshaw, K. 1989. "Demarginalizing the Intersection of Race and Sex: A Black Feminist Critique of Antidiscrimination Doctrine, Feminist Theory and Antiracist Politics." *University of Chicago Legal Forum* 140: 139–167.
Crenshaw, K. 1991. "Mapping the Margins: Intersectionality, Identity Politics, and Violence against Women of Color." *Stanford Law Review* 43 (6): 1241–1299.
Daigle, M. 2018. "Embodying Relations of Accountability in Settler Colonial Contexts." In: *Interventions: Bringing the Decolonial to Political Geography,* edited by L. Naylor, M. Daigle, S. Zaragocin, M.M. Ramírez, and M. Gilmartin. *Political Geography* 66: 199–209.
Davis, A.Y. 1983. *Women, Race & Class*. New York: Vintage Books.
Delicado-Moratalla, L. 2017. "Esclavitud, Género y Racialización en Alicante: La Colonización de los Cuerpos Femeninos (ss. XVII-XVIII)." *Géneros: Multidiscipliary Journal of Gender Studies* 6 (2): 1334–1360.
Domínguez, I., and J.J. Gálvez. 2017. "5.600 víctimas de esclavitud afloran en España tras los cambios legales." *El País* 17 April. Available at: https://elpais.com/politica/2017/04/14/actualidad/1492152357_266303.html
Doshi, S. 2017. "Embodied Urban Political Ecology: Five Propositions." *Area* 49 (1): 125–128.
Egharevba, R.K., and F.A. Iweze. 2004. "Sustainable Agriculture and Rural Women: Crop Production and Accompanied Health Hazards on Women Farmers in Six Rural Communities in Edo State Nigeria." *Journal of Sustainable Agriculture* 24 (1): 39–51.
Egharevba, R.K., and D.O. Osunde. 2001. "The Effect of Crude Oil on Seedling Growth of Two Forest Fruit Trees: *Chrysophyllum Albidum (Gambaya Albida)* and *Dacryodes Edulis* G. Don." *Journal of Sustainable Agriculture* 18 (2–3): 25–35.
Elabor-Idemudia, P. 2003. "Migration, Trafficking and the African Woman." *Agenda* 58: 101–116.
Elmhirst, R. 2011. "Introducing New Feminist Political Ecologies." *Geoforum* 42 (2): 129–132.
Feminicidio.Net. 2016. *Feminicidio en el sistema prostitucional del Estado español. Víctimas 2010-2015: 31 mujeres asesinadas*. Available at: www.feminicidio.net/articulo/feminicidio-sistema-prostitucional-del-estado-espa%C3%B1olv%C3%ADctimas-2010-2015-31-mujeres.
Gillespie, K. 2018. "Placing Angola: Racialisation, Anthropocentrism, and Settler Colonialism at the Louisiana State Penitentiary's Angola Rodeo" *Antipode* 50 (5): 1267–1289.
Gilroy, P. 2015. "Offshore Humanism." *Antipode RGS-IBG Lecture*. Available at: https://antipodefoundation.org/2015/12/10/paul-gilroyoffshore-humanism/ (accessed 29 April 2016).
Global Slavery Index. 2016. *Measuring Progress*. Available at: www.globalslaveryindex.org/2019/findings/measuring-progress/ (accessed 6 February 2020).
Global Slavery Index. 2018. *Africa Region Report*. Perth: Walk Free Foundation.
Global Slavery Index. 2019. *Findings: Measuring Progress*. Perth: Walk Free Foundation.
Goeman, M. 2013. *Mark My Words: Native Women Mapping our Nations*. Minneapolis: University of Minnesota Press.
Gururani, S. 2002. "Forests of Pleasure and Pain: Gendered Practices of Labor and Livelihood in the Forests of the Kumaon Himalayas, India." *Gender, Place & Culture* 9 (3): 229–243.
Haraway, D. 1988. "Situated Knowledges: The Science Question in Feminism and the Privilege of Partial Perspective." *Feminist Studies* 14 (3): 575–599.
Haraway, D.J. 1991. *Simians, Cyborgs, and Women: The Reinvention of Nature*. New York: Routledge.
Harding, S.G. 1986. *The Science Question in Feminism*. Ithaca, NY: Cornell University Press.
Harris, L. 2015. "Foreword: A Quarter Century of Knowledge and Change: Pushing Feminism, Politics, and Ecology in New Directions with Feminist Political Ecology." In: *A Political Ecology of Women, Water and Global Environmental Change*, edited by S. Buechler and A. Hanson, x–xxiii. London: Routledge.
Hartman, S. 1996. "Seduction and the Ruses of Power." *Callaloo* 19 (2): 537–560.
Hartman, S. 2016. "The Belly of the World: A Note on Black Women's Labors." *Souls* 18 (1): 166–173.
Hawkins, R., D. Ojeda, K. Asher, B. Baptiste, L. Harris, S. Mollett, A. Nightingale, D. Rocheleau, J. Seager, and F. Sultana. 2011. "A Discussion." *Environment and Planning D: Society and Space* 29 (2): 237–253.
Hayes-Conroy, J., and A. Hayes-Conroy. 2013. "Veggies and Visceralities: A Political Ecology of Food and Feeling." *Emotion, Space and Society* 6: 81–90.
Hodal, K. 2017. "'Absolutely Shocking': Niger Delta Oil Spills Linked with Infant Deaths." *The Guardian* 6 November. Available at: www.theguardian.com/global-development/2017/nov/06/niger-delta-oil-spills-linked-infant-deaths.
Hovorka, A.J. 2012. "Women/Chickens vs. Men/Cattle: Insights on Gender–Species Intersectionality." *Geoforum* 43 (4): 875–884.
Hunt, S. 2014. "Ontologies of Indigeneity: The Politics of Embodying a Concept." *cultural geographies* 21 (1): 27–32.

Ikelegbe, A. 2005. "The Economy of Conflict in the Oil Rich Niger Delta Region of Nigeria." *Nordic Journal of African Studies* 14 (2): 208–234.
International Monetary Fund. 2018. *Nigeria: Selected Issues.* Washington, DC: International Monetary Fund.
International Organization for Migration. 2006. *Migration, Human Smuggling and Trafficking from Nigeria to Europe*. Geneva: International Organization for Migration.
Jarosz, L. 1992. "Constructing the Dark Continent: Metaphor as Geographic Representation of Africa." *Geografiska Annaler. Series B, Human Geography* 74 (2): 105–115.
King, T.L. 2016. "The Labor of (Re)Reading Plantation Landscapes Fungible(ly)." *Antipode* 48 (4): 1022–1039.
Kobayashi, A., and L. Peake. 1994 "Unnatural Discourse: 'Race' and Gender in Geography." *Gender, Place & Culture* 1 (2): 225–243.
McKittrick, K. 2006. *Demonic Grounds*. Minneapolis: University of Minnesota Press.
McKittrick, K. 2015. *Sylvia Wynter: On Being Human as Praxis*. Durham, NC: Duke University Press.
Mohanty, C.T. 1988. "Under Western Eyes: Feminist Scholarship and Colonial Discourses." *Feminist Review* 30: 61–88.
Mohanty, C.T. 2003. *Feminism without Borders: Decolonizing Theory, Practicing Solidarity*. New Delhi: Zubaan.
Mollett, S., and C. Faria. 2013. "Messing with Gender in Feminist Political Ecology." *Geoforum* 45 (March): 116–125.
Mollett, S., and C. Faria. 2018. "The Spatialities of Intersectional Thinking: Fashioning Feminist Geographic Futures." *Gender, Place & Culture* 25 (4): 565–577.
Mollett, S. 2010. "Está Listo (Are You Ready)? Gender, Race and Land Registration in the Río Plátano Biosphere Reserve" *Gender, Place & Culture* 17 (3): 357–375.
Mollett, S. 2017. "Irreconcilable Differences? A Postcolonial Intersectional Reading of Gender, Development and *Human* Rights in Latin America" *Gender, Place & Culture* 24 (1): 1–17.
Morgan, J. 2004. *Laboring Women: Reproduction and Gender in New World Slavery*. Philadelphia: University of Pennsylvania Press.
Mountz, A. 2017. "Political Geography III: Bodies." *Progress in Human Geography* 42 (5): 759–769.
Mullaney, E.G. 2014. "Geopolitical Maize: Peasant Seeds, Everyday Practices, and Food Security in Mexico." *Geopolitics* 19 (2): 406–430.
Münster, U. 2016. "Working for the Forest: The Ambivalent Intimacies of Human–Elephant Collaboration in South Indian Wildlife Conservation." *Ethnos* 81 (3): 425–47.
Neumann, R. 2005. *Making Political Ecology*. London: Routledge.
Nixon, R. 2011. *Slow Violence and the Environmentalism of the Poor.* Cambridge, MA: Harvard University Press.
Oluduro, O. 2014. *Oil Exploitation and Human Rights Violations in Nigeria's Oil Producing Communities.* Cambridge: Intersentia.
Omorodion, F. 2009. "Vulnerability of Nigerian Secondary School to Human Sex Trafficking in Nigeria." *African Journal of Reproductive Health* 13 (2): 33–48.
Peet, R., and M. Watts. 2004. *Liberation Ecologies: Environment, Development, Social Movements*. London: Routledge.
Pratt, G., and V. Rosner. 2006. "Introduction: The Global & the Intimate." *Women's Studies Quarterly* 34 (1): 13–24.
Radcliffe, S.A. 2015. *Dilemmas of Difference: Indigenous Women and the Limits of Postcolonial Development Policy*. Durham, NC: Duke University Press.
Ranea Triviño, B. 2017. "(Re)pensar la prostitución desde el análisis crítico de la masculinidad." In: *Elementos para una teoría crítica del sistema prostitucional,* edited by N. Gómez and M. Álvarez, pp. 135–142. Editorial Comares.
Razack, S.H. 2016. "Gendering Disposability." *Canadian Journal of Women and the Law* 28 (2): 285–307.
Resurrección, B.P. 2017. "Gender and Environment in the Global South: From 'Women, Environment, and Development' to Feminist Political Ecology." In: *Routledge Handbook of Gender and Environment*, edited by S. MacGregor, 71–85. Abingdon: Routledge.
Rocheleau, D., B. Thomas-Slayter, and E. Wangari. 1996. *Feminist Political Ecology: Global Issues and Local Experience.* London: Routledge.
Seager, J. 2003. "Rachel Carson Died of Breast Cancer: The Coming of Age of Feminist Environmentalism." *Signs: Journal of Women in Culture and Society* 28 (3): 945–972.
Shedrofsky, K. 2017. "Morocco & Spain Arrest 10 for Trafficking and Sex Slavery." *Organized Crime and Corruption Reporting Project*, 15 February.
Spillers, H.J. 1987. "Mama's Baby, Papa's Maybe: An American Grammar Book." *Diacritics* 17 (2): 65–81.

Sultana, F. 2011. "Suffering for Water, Suffering from Water: Emotional Geographies of Resource Access, Control and Conflict." *Geoforum* 42 (2): 163–172.
Sundberg, J. 2011. "Diabolic Caminos in the Desert and Cat Fights on the Río: A Posthumanist Political Ecology of Boundary Enforcement in the United States–Mexico Borderlands." *Annals of the Association of American Geographers* 101 (2): 318–336.
Sundberg, J. 2014 "Decolonizing Posthumanist Geographies." *cultural geographies* 21 (1): 33–47.
Sundberg, J. 2017. "Feminist Political Ecology." In: *International Encyclopedia of Geography: People, The Earth, Environment, and Technology*, edited by R. Kitchin and N. Thrift, 1–12. Hoboken, NJ: Wiley.
Truelove, Y. 2011. "(Re-)Conceptualizing Water Inequality in Delhi, India through a Feminist Political Ecology Framework" *Geoforum* 42 (2): 143–152.
Ugor, P. 2013. "Survival Strategies and Citizenship Claims: Youth and the Underground Oil Economy in Post-Amnesty Niger Delta." *Africa* 83 (2): 270–292.
Ukeje, C.U. 2004. *Oil Capital, Ethnic Nationalism and Civil Conflicts in the Niger Delta of Nigeria.* Doctoral Dissertation, Department of International Relations, Obafemi Awolowo University, lle-lfe, Nigeria.
UNESCO. 2006. *Policy Paper Poverty Series - Human Trafficking in Nigeria: Root Causes and Recommendations.* Paris: United Nations Educational, Scientific and Cultural Organizations.
Wangui, E.E. 2014. "Livelihood Shifts and Gender Performances: Space and the Negotiation for Labor among East Africa's Pastoralists." *Annals of the Association of American Geographers* 104 (5): 1068–1081.
Watts, M.J. 2005. "Righteous Oil? Human Rights, the Oil Complex, and Corporate Social Responsibility." *Annual Review of Environment and Resources* 30 (1): 373–407.
Weheliye, A.G. 2014. *Habeas Viscus: Racializing Assemblages, Biopolitics, and Black Feminist Theories of the Human.* Durham, NC: Duke University Press.
Women's Link Worldwide. 2014. *La trata de mujeres y niñas nigerianas: esclavitud entre fronteras y prejuicios.* Available at: www.occrp.org/en/daily/6079-morocco-and-spain-10-arrested-for-trafficking-and-turning-nigerian-women-into-sex-slaves.
Zaragocin, S. 2017. "Decolonized Feminist Geopolitics: Coloniality of Gender and Sexuality at the Center of Critical Geopolitics." In: *Interventions: Bringing the Decolonial to Political Geography*, edited by L. Naylor, M. Daigle, S. Zaragocin, M.M. Ramírez, and M. Gilmartin. *Political Geography* 66: 199–209.

PART 3

Engaging feminist geographies

26

TRAUMA, GENDER AND SPACE

Insights from Bangladesh, Malaysia and the UK

Rachel Pain, Nahid Rezwana and Zuriatunfadzliah Sahdan

Introduction

In recent years, trauma has escalated as an issue for public and scholarly attention. It is evident throughout history and across space that how we understand it is shaped by the cultural and political contexts in which trauma, and we, are located. As a psychiatric disorder, it was formally identified in the 1980s as 'the response to an unexpected or overwhelming violent event or events that are not fully grasped as they occur, but return later in repeated flashbacks, nightmares, and other repetitive phenomena' (Caruth 1996, 91). As we show in this chapter, trauma is also understood as a social and spatial condition with broader causes and consequences.

While single-instance events (such as car accidents) may lead to debilitating symptoms of post-traumatic stress disorder (PTSD), theorizing trauma beyond an individual ailment is central to feminist understandings. Trauma may manifest as a collective condition that affects communities, social groups and even whole societies (Schwab 2010). This recognizes the scale and reach of the after-effects of violence, which are repeated over a longer timeframe and are open to structural analysis: systemic violence relating to uneven distributions of power, such as slavery or racism. Our focus in this chapter is gender-based violence (GBV), 'violence that occurs as a result of the normative role expectations associated with each gender, along with the unequal power relationships' (Bloom 2008, 14), an especially common cause of trauma. We examine some of its various forms in Bangladesh, Malaysia and the UK. While men and boys can be its targets, it is most common against women and girls, and our discussion reflects this.

Feminist researchers and activists have played an important role in challenging medical and popular understandings of trauma. GBV is marked by its 'everywhereness' and paradoxical invisibility (Pain 2014). In the late-twentieth century, feminists questioned which traumas are taken seriously, given that survivors of war and GBV experience similar rates of PTSD (Herman 1997). Before then, the history of trauma from GBV is harder to locate, and social status and geographical location still profoundly influence whose traumas are recognized and responded to. The psychiatrist Judith Herman (1997) played a key role in identifying 'chronic' trauma arising from entrapment and repeat victimization. Intersectional analyses view trauma as folded into inequalities of gender, sexuality, race, class and colonial oppression, producing uneven experiences; and we try to illuminate these differences across and within the three case studies that we refer to in this chapter. There is also irregularity in who gets to narrate trauma, reflected in debates over its history, meaning and treatment. Trauma, then, is not a fixed or objective medical state but culturally constructed and enmeshed in social politics (Nguyen 2011; Tamas 2011).

In particular, Western perspectives on trauma have had a disproportionate influence on theories of causation and treatment. Post-colonial and Indigenous analysts argue that this dominant Western trauma paradigm, heavily influenced by Freud, tends to reflect single-event trauma and privileges the suffering of White Europeans (Fassin and Rechtman 2009). This paradigm depoliticizes and dehistoricizes others' trauma (Visser 2015), especially connected to historical colonial violence and its contemporary traces. Fanon (1963) famously observed that the causes of psychological trauma in soldiers and civilians during the French/Algerian war were far deeper than conventional psychiatry could address, and Brave Heart (2000) identified historical trauma among Indigenous Native Americans as the direct result of colonization and racism. For Black feminist bell hooks (2003), slavery and racism underpin contemporary trauma for African–Americans, and contemporary re-enactments retraumatize the population. There is deep suspicion among some communities that medical science can provide resolution, given its historical role in oppression: 'until the legacy of remembered and re-enacted trauma is taken seriously, black America cannot heal' (hooks 2003, 24). The diagnosis of PTSD, so widespread in the West, is sometimes little known in the regions where traumatic events are most frequent. Yet the many traditional methods of healing and rebuilding were disrupted by colonialism, globalization and the export of Western medical practices, and there are now movements to reconnect and establish them. We return to the issue of recovery at the end of the chapter.

Our first aim in this chapter is to highlight these structural roots of trauma. Our second aim is to signal the importance of trauma's geographies. As an immediate experience, trauma is inherently geographical, involving spatial and temporal psychological dislocation and triggered by aspects of place or particular sites where memory, pain or ongoing violence linger. Place thus becomes hardwired in trauma, and trauma can become hardwired in place, affecting the possibilities for recovery. For example, and as feminist geographers have long argued, the home may become a traumatic space when violence is experienced there, holding negative memories. Trauma is also present across scales, as our discussion above of collective trauma shows, and produces scales. And trauma is mobile and multidirectional, as Pratt et al. (2015) describe in trauma narratives that travel between places with specific historical politics. As Coddington and Micieli-Voutsinas (2017, 52) put it, 'trauma has a productively complex relationship to space … it is both rooted in place, yet defies geospatial logics'.

In the rest of the chapter, we discuss trauma arising from a number of forms of GBV in Bangladesh, Malaysia and the UK. In each context, trauma is privately and publicly recognized, contested, treated and reinforced to varying degrees. Our intention is to show something of the situated everyday experiences of trauma, the combined significance of trauma's effects and the importance of spatial setting in offering or closing down possibilities for survivors' empowerment and recovery.

Chronic trauma: multiple forms of gender-based violence in Bangladesh

Gender-based violence is a major social problem in Bangladesh. Domestic violence, rape, sexual abuse and verbal public sexual harassment are common forms, and early/forced marriage and wife abandonment take place at alarming rates. Many women are subject to multiple forms of GBV. In 2016, the country was in fifth position for the rate of child marriage (UNICEF 2016a); 66 per cent of girls are married before the age of 18 (UNICEF 2016b), and Bangladesh has the highest rate of child marriage for girls below 15 years (UNCF 2014). This usually ends female education and increases economic dependency. A high number of Bangladeshi women experience domestic violence. According to a Government report in 2016, 72.6 per cent of married women are abused at least once in their lifetime, including physical, sexual, economic and emotional abuse as well as controlling behaviour. Around 41.7 per cent of those experiencing physical or sexual violence suffer injuries (BBS 2016), but most face psychological impacts that are not usually recorded.

Recent qualitative research on GBV was undertaken in the coastal region of Barguna, a remote, cyclone-prone coastal district of Bangladesh. Four group discussions, 45 individual interviews and observations were conducted to identify and explore GBV with female and male key informants and survivors.

The research revealed very common signs of psychological trauma (Rezwana and Pain 2018); feelings of intense fear, deep sadness, helplessness, loss of control, hyperarousal and intrusion. One survivor describes her experience:

> My body … my blood might have become black … I was beaten so many times after I have been married to him. Everyone beats me, my husband, my mother-in-law, father-in-law and even sister-in-laws. I cannot show all the bruises, there are many more … I was working with the sticks, she (mother-in-law) took them and poked me … I could not get any treatment. It was so painful … I wanted to die … I cannot go back to my parents.

The reasons for domestic abuse vary from disagreements over dowry payments and poverty to extra-marital relationships, extra marriages and keeping discipline in the family in the name of tradition. Violence, however, has profound psychological impacts. According to Pain (2014), domestic abuse can be considered a form of everyday terrorism, creating long-lasting fear and chronic trauma that reinforce the perpetrator's control over the abused person. Survivors in this study were in constant fear that their husband would become violent, which intensifies trauma symptoms. They abided by all their husbands' rules and orders in order to keep them calm in an effort to avoid violence: 'I do whatever he says, never disobey him now … he is like my master'; 'I never want anything from him to buy, I do it myself, I want him to be calm'. Being beaten in one's own house not only creates immense fear but decreases self-confidence and changes self-perception.

Wife abandonment is another significant fear for these women. Being abandoned means losing the shelter, economic security and social status that come with living in their husband's house. Abandoned married women are not welcomed back to their parents' house due to cultural traditions, and this increases their social insecurity; they often become the target of local miscreants who may threaten or abuse them further. This woman describes her feelings of trauma and intrusion after her husband left her:

> They roamed around outside my house at night, I could hear them whispering, they wanted to come near to the door … I called loudly … they went away … I could not sleep for many nights. It became a trauma for me. I left that house, which was near the bazaar. Now, I stay in my rural home. However, I cannot visit the bazaar now. I feel frightened.

In Barguna, survivors of GBV are trapped by overlapping layers of location (due to the remoteness of the region and poor level of service provision), socio-economic conditions, gender identity and responsibilities, social attitudes, culture and traditions. Geographical location increases the vulnerability of women in Barguna to cyclones and GBV following disasters; domestic violence, forced marriage, rape, sexual assault and trafficking increase during and after each cyclone (Rezwana 2018; Rezwana and Pain 2018). Women are victimized in the name of traditions of early/forced marriage, family discipline, dowry and the priority given to sons, and their complaints are considered through the lens of these traditions. They are often advised to tolerate the violence due to social stigma and fear of the perpetrator, to avoid the further embarrassment of abandonment or divorce and to maintain tradition and social status ('Bangladesh' 2016; Hossain and Suman 2013).

All of these factors increase and intensify chronic trauma. Most harmful, as the respondents describe it, is being unable to express their feelings and share their experiences with anyone in a position of responsibility, such as law enforcement, doctors or psychologists, or religious or community leaders. This increases their vulnerability to mental health problems, as well as leaving the violence unreported (Hossain and Suman 2013). Some 72.7 per cent of women who have been abused in the home do not share their experiences and only 2.6 per cent seek legal support ('Bangladesh' 2016; BBS 2016). Family members are the first with whom GBV is shared and then, sometimes, local leaders are informed. Many women do not wish to report violence, as this can aggravate trauma. Social stigma, economic dependency and the lack of victim support facilities, healthcare centres and knowledge about government helplines are also key reasons for not seeking formal support. Thus, too many Barguna women are trapped within a cycle of GBV, trauma, economic dependency and lack of empowerment – four cornerstones that compound each other.

Understandings of trauma: domestic abuse in Malaysia

In Malaysia, there is very little understanding of GBV or associated trauma. Furthermore, domestic violence, despite its major contribution to women's mental and physical ill health, is understudied. Women become more vulnerable to violence if they marry without their family's approval, enter into arranged marriages or are involved in deception ('marriages' set up for human trafficking). Their vulnerability is reinforced by the resulting social and spatial isolation from their families, homelessness and economic exploitation by their partner. Malaysian law does not recognize informal partnerships, including dating relationships and unmarried sexual relationships, and thus does not offer legal protection to those women.

A recent study with survivors of domestic violence in a women's refuge in a large urban area in Malaysia (the location of which cannot be disclosed) has documented the significant effects of trauma (Sahdan 2018). The research with ten women involved interviews, storytelling, photovoice and the production of a mural. Violent events such as the one recalled by Rekha (below) have a long-term impact:

REKHA: When the clock hit 12, I was petrified. I went to my room, and kept quiet.
LIEYA: You mean every day?
REKHA: Yes … I'm scared out of my wits.
LIEYA: Did he drink every day at 12?
REKHA: Yes, at night. At 12 … Or one in the morning … Then, he would break into the room. Smashing it. When I see the clock reached 12 … there was a sound, 'ting, ting, ting, ting', I quickly ran into the house, went into the next room, locked myself up and 20 minutes later, he would show up.

LIEYA: So he'd break into the room?

REKHA: Yes.

LIEYA: Are you still living in fear?

REKHA: Yes, I can still sense it (shaking in fear). I can't let it go, I won't forget it till my death (sighing).

Trauma is compounded by the spatial settings of the abuse experienced, first at home and later in wider society. However, its relationship to public and private spaces depends on the cultural, ethnic and religious background of the perpetrator. Perpetrators negotiate and use the spaces in which the abuse occurs, and their power is expressed in different ways. The spatial dynamics of trauma centre on the politics of home or 'private' space in multicultural societies such as Malaysia. As Herman (1997, 74) argues, 'a man's home is his castle; rarely is it understood that the same home may be a prison for women and children'. This invisible prison, as other scholars have pointed out, involves stretching the practices and times of domination, before and after women leave the abusive relationship (Warrington 2001). It is multi-sited, beyond the limits of the visible space of the house and bounded by the physical and invisible barriers created by psychological, economic, social and legal subordination (Stark 2004). The home setting is therefore flexible but under the perpetrator's intimate control, as is the woman's psyche and the body it contains. Intimate settings may promote resistance as well as domination (Pain and Staeheli 2014), but every action or sign of resistance is countered with tactics by perpetrators. This prevents the abused women from fleeing, deepening the coercive relationship and rendering them prisoners. Being entrapped in this way results in post-traumatic stress symptoms such as contradictory thoughts, trauma-bonding, hyper-alertness, flashbacks and re-enactment (Herman 1997).

These symptoms of trauma are very close to societal beliefs about demonic possession. In many societies, there is a widespread cultural belief that spirits may inhabit the body. In Malaysia, this idea of demonic possession becomes part of the widespread culture that sustains domestic violence. Symptoms of trauma, and the cumulative effects on personality and identity of those experiencing long-term abuse, were identified as either the causes or consequences of demonic possession by many of the survivors and perpetrators in the research (Sahdan 2018) just as, in Western contexts, cultural beliefs about women's roles and culpability for violence against them are often mobilized. The result is to divert attention from the structural explanations for violence that highlight societal gender inequality. Meanwhile, patriarchy and religious beliefs are used to justify abuse (in this study, Hinduism and Islam, but other religions are used in this way elsewhere), based on the perpetrators' assumed position in traditional culture in Malaysia. Some are seen to use religion as a ploy to dominate their wife, on the grounds of solidifying the religion. As in many other contexts, the wife is pressured into believing that it is her fault that her husband becomes abusive, and those around her may reinforce this belief.

Yet, to the outside, perpetrators appear to be 'normal husbands', making it hard for women to seek help. The government and NGOs in Malaysia are engaged in combating violence against women, yet without any culturally specific understanding. Formal assistance is the only way out for abused women in some ethnic groups, but the police do not always treat domestic violence as a serious issue, siding with the perpetrators or assuming that it is a private family matter. Meanwhile, the abused women do not consider healthcare facilities as independent agencies that might intervene and help. Some women do not seek treatment because their condition is too critical, and they feel trapped and under the control of perpetrators. This is also the case for women in the study who were pregnant, in confinement or suffering from other diseases such as cancer. Effective physical liberation, for many abused women, only comes when they are able to escape to a secret place not in the intimate knowledge of the perpetrator, such as an NGO-run refuge.

Refuges, however, are a short-term intervention, and resettlement services for domestic violence are not yet established in Malaysia. After the three-month period of protection at the shelter, women are expected to live independently. Because other formal interventions are less effective, the majority of women then face difficulties, often with no financial backup, safe shelter or reliable childcare. Many survivors come out of the shelter to return to perpetrators or to live in their family's house, where the perpetrator finds them. They sometimes hand over their children to their perpetrators because they are unable to support them alone, or they lose custody battles or the children are forcibly taken by the perpetrators. Living alone after leaving the shelter involves a magnification of trauma and fear. A space that would promote stable recovery is difficult to secure.

Contested trauma: sexual assault and UK universities

In the UK, the nature of the spaces in which GBV occurs, and contested public understandings of its different forms, also influence the societal and institutional impetus for intervention. This section examines GBV in a particular institutional space, rather than a domestic space. The issue of sexual assault on the university campus has become visible only recently, and many would argue that inadequate responses still largely fail to protect survivors. The problem is widespread. In the US one in five women is sexually assaulted at college (Krebs et al. 2016) and in the UK one in seven female students suffers a serious physical or sexual assault (NUS 2010). Perpetrators are usually students known to those whom they victimize. In the UK, the reported rates are higher at elite institutions. For example, Durham University, which promoted itself in 2014 as one of the top five safest universities in the UK (Durham University 2014), went on to reveal in 2017 that it has the highest level of reported sexual assault among students in the UK, jointly with Oxford (The Times 2017).

Trauma is shaped by the setting of the campus, and it profoundly changes the experience of this space. Prolonged or repeated exposure to a perpetrator compounds the effects of trauma (Herman 1997). Half of survivors reported mental health issues, and two-thirds said their relationships had been affected after serious sexual assault on campus (NUS 2010). To date, survivors have been more likely than perpetrators to leave university after an assault, either pausing their studies or dropping out. Living and working on campus means exposure to the contextual and sensory triggers associated with the attack, and risks seeing the perpetrator, keeping present a space in which the double-bind of the PTSD symptoms intrusion/constriction can flourish and where managing trauma is difficult. Despite this, the responses of government and the institutions involved have been extremely slow. An inquiry into campus violence against women in 2015 found that fewer than half of the elite universities monitored sexual violence and that one in six had no procedure for reporting it. As Dowler, Cuomo and Laliberte (2014) argue in their analysis of the long-term cover-up of campus sexual assaults in the US, the neoliberal cultural economies of universities lead them to prioritize institutional reputation over individual welfare.

Since recent publicity, UK universities have made efforts to address sexual assault on campus. Unlike the US, there has been no legislative change, but the umbrella organization Universities UK has established a taskforce that produced guidelines in 2016. These guidelines spell out universities' duty of care to students since the 2010 equality legislation, which had previously been overlooked. The responses of individual universities have varied, but new policies and practices have included clear procedures for reporting and training and guidance for staff and students, the provision of specialist counselling and peer support, and joint working with local police forces and rape crisis centres. While, on one hand, the changes are dramatic, on the other they are often in the early stages of development, rarely involving trauma-informed care or environments addressing the range of ways that trauma inflects the campus (and vice versa). Nor do they commonly acknowledge intersectional differences or the structural roots of discrimination and violence on campus. While sexual assault targets mostly young women, there are higher risks for students of colour and LGBTQ students. Micro-aggressions – 'brief and commonplace daily verbal, behavioral, or environmental indignities, whether intentional or unintentional, that communicate hostile, derogatory, or negative racial slights and insults' (Sue et al. 2007) – may also compound trauma. Trauma is most common for students who come to university with previous experiences of violence and more limited access to healthcare, social and financial resources. In this way, as Carter (2015) argues, inequalities of race, class, gender, sexuality and citizenship determine who lives with trauma and where.

Collective traumas may also be etched into the space of the university campus, given elite Western universities' role in historical and contemporary forms of social and political violence. Many have historical connections to imperialism, built on slavery or functioning as seats of science in the service of colonial power and subjugation (Chatterjee and Maira 2014). Today, universities function within the logics of global capitalism and its uneven development, and globalized higher education markets have not been accompanied by significant action on campus racism. Similarly, as Phipps and Young (2015) have argued, the neoliberalization of higher education encourages a competitive and masculinized 'lad culture' among students, which many students experience as hostile.

Efforts to bring campus trauma to wider attention have been met by a forceful backlash. The political contestation that often accompanies public uncovering of private trauma is illustrated by recent debates on trigger warnings, the content alerts on educational materials designed so that people who have experienced trauma can prepare themselves and manage the symptoms. Trauma destroys the sense of time, and sensory cues in the present can cause time-space slippage so that our bodies take pre-cognitive action ('fight, flight or freeze') in the present. This experience can prevent us from engaging with the present: in the classroom, for some students this affects their access to the education they have the right to.

In a notorious critique of the use of trigger warnings, Lukianoff and Haidt (2018) contend that these are more harmful than the supposed traumas they protect from, infantilizing students, restricting their education and worsening their mental health. Such an argument is ignorant of the scientific and social realities of PTSD and the long-term effects of intimate and racist violence. For Carter (2015), it is an ableist and White supremacist critique that further silences already marginalized groups; it misconstrues 'what students are actually requesting: recognition of their lived experiences and institutional support regarding how those experiences influence their education'. Some feminists express concern that trigger warnings may unintentionally set up a hierarchy of trauma, further marking out certain groups as vulnerable. From a geographical perspective, however, we might argue that the spaces of education are distinct, heavily imbued with particular relations of power – students have little choice but to engage in classrooms that are captive as well as potentially empowering. Carter (2015) calls, instead, for classroom pedagogies that incorporate and teach trauma as a justice issue.

Researching trauma

Researching trauma also requires that we adopt caring professional practices. As researchers working with survivors, we have experienced secondary trauma through prolonged contact with stories of violence and their close analysis, as well as retraumatization when memories of past experiences are brought back to the surface (see Coddington 2017; Tamas 2011). These impacts, while less severe than the trauma that our respondents tell us about, are important to recognize in research that embodies a feminist ethics of care (Drozdzewski and Dominey-Howes 2015).

Recovery, rebuilding, resistance

Our three case studies illustrate the influence of physical, social and political dimensions of space on trauma following GBV. All the survivors we have spoken to work to resist violence and trauma in different ways, drawing on available individual and collective resources. Most formal trauma interventions that are provided by the state, NGOs or the private sector are premised on a notion of 'recovery' that implies a return to previous life. We suggest that the term 'rebuilding' captures better the challenges of moving forward into the future: as these case studies illustrate, there is no 'before' or 'after', rather a series of interconnected experiences of violence and abuse. Tamas (2011) critiques the recovery paradigm in Western responses to trauma, which assumes that psychotherapeutic interventions effect a linear process of recovery. Not only is this a poor fit with complex everyday experiences but it can hold responsible those whose post-violence path differs. For Herman (1997), individual treatment must be tied to social movements that challenge oppression, and many other feminist and post-colonial scholars also direct attention to the structural underpinnings of ongoing violence that survivors continue to deal with (e.g. Brave Heart 2000; Fanon 1963). Before feminists brought to light the nature of GBV-related trauma, the psychiatric profession considered to be symptomatic the way that the survivors told their stories; nightmares, intrusive thoughts, dissociation and self-doubt were even seen as failings or illness that caused, rather than resulted from, the abuse (Alcoff and Gray 1993). More recently, in the West, trauma discourses have moved further away from survivors' empowerment, acquiring neoliberal and neocolonial logics as they are co-opted by the state and mainstream medicine (Tseris 2013). Yet, in some countries, as we see in Bangladesh, trauma is rarely reported or treated because of the negative social consequences of making it known.

The wider social setting also plays an important role in rebuilding after trauma. For Herman (1997), on the one hand, healing requires strong and caring connections with others, and the sense that the wider community condemns the violence experienced. On the other hand, the social and institutional dissociation that we have seen in the examples above tends to compound the psychological effects on individuals. In the case of collective trauma, predominant social narratives may deny the wrong committed or blame the survivors, as in many instances of genocide and oppression of Indigenous people. Collective experience, however, has traditionally provided social support and mechanisms for challenging further violence (hooks 2003; Marshall 2014).

In our three case studies, we see that survivors take various paths of resistance and rebuilding, working within and also against wider cultural framings of trauma. This may include simply surviving or coping with the everyday effects of trauma. In the coastal communities of Bangladesh, contesting entrapment with the extremely limited economic resources that are available is challenging. There is little or no provision within healthcare systems for the psychological effects of GBV, even where women can access healthcare, and often limited family support and societal recognition of victimization (Rezwana 2018; Rezwana and Pain 2018). Local NGOs are aware of the extent of trauma, yet the limited treatment that they provide does not always work within the precarious contexts of many women's lives (Rezwana 2018). In Malaysia, NGOs working with domestic abuse survivors also employ psychotherapeutic treatments based on Western models of trauma. However, the research suggests that this often conflicts with lived experiences and beliefs about trauma (Sahdan 2018). There are no local culturally sensitive alternatives within formal healthcare or the voluntary sector (see also Marshall 2014), and many people seek treatment instead from spirit healers. Elsewhere, Western secularism, hand-in-hand with colonialism, has led to a loss of the Indigenous belief systems and practices that were previously used to treat trauma (Visser 2015). In the UK, survivors of campus sexual assault may receive therapeutic support, although this is limited by cuts under recent austerity and the tighter rationing of mental health services. Therapy helps some survivors to build resilience and rebuild their lives yet rarely acknowledges the structural and everyday contexts of GBV, which, as we have shown, may continue to re-traumatize.

Fassin and Rechtman (2009) state that trauma is now such a dominant discourse in the West that there is widespread public support for survivors but, for many GBV survivors, the everyday reality depends who and where you are (Pain 2014). As we have seen from our three examples, the existence of trauma among marginalized social groups may be denied or disputed, yet the intersectional nature of trauma is centrally important, producing very uneven experiences. Attention to the everyday environments that retraumatize (see Nguyen 2011) and to societal trauma narratives is an important task for geographers. Trauma is best understood as a social, spatial and political condition, and so addressing its social, spatial and political contexts has a key role in prevention and healing.

Acknowledgements

We would like to thank our research respondents for sharing their stories so generously. We are also very grateful to Lynda Johnston and Betsy Olson for their superb editing and advice.

Key readings

Coddington, K., and J. Micieli-Voutsinas. 2017. "On Trauma, Geography, and Mobility: Towards Geographies of Trauma." *Emotion, Space and Society* 24: 52–56.

Herman J. 1997. *Trauma and Recovery: The Aftermath of Violence from Domestic Abuse to Political Terror.* New York: Basic Books.

Marshall, D.J. 2014. "Save (Us From) the Children: Trauma, Palestinian Childhood, and the Production of Governable Subjects." *Children's Geographies* 12 (3): 281–296.

References

Alcoff, A.M., and L. Gray. 1993. "Survivor Discourse: Transgression or Recuperation?" *Signs* 18: 260–290.

"Bangladesh" 2016. "Bangladesh." *Daily Star*, 3 October 2016.

BBS. 2016. *Report on Violence Against Women Survey 2015.* Bangladesh: Bangladesh Bureau of Statistics.

Bloom, S. 2008. *Violence against Women and Girls.* UNC Chapel Hill: Carolina Population Center.

Brave Heart, M.Y.H. 2000. "Wakiksuyapi: Carrying the Historical Trauma of the Lakota." *Tulane Studies in Social Welfare* 21–22: 245–266.

Carter, A.M. 2015. "Teaching with Trauma: Disability Pedagogy, Feminism, and the Trigger Warnings Debate." *Disability Studies Quarterly* 15 (35): 1–18.

Caruth, C. 1996. *Unclaimed Experience: Trauma, Narrative, and History.* Baltimore, MD: Johns Hopkins University Press.

Chatterjee, P., and S. Maira. 2014. *The Imperial University: Academic Repression and Scholarly Dissent.* Minneapolis: University of Minnesota Press.

Coddington, K. 2017. "Contagious Trauma: Reforming the Spatial Mobility of Trauma Within Advocacy Work." *Emotion, Space and Society* 24: 66–73.
Coddington, K., and J. Micieli-Voutsinas. 2017. "On Trauma, Geography, and Mobility: Towards Geographies of Trauma." *Emotion, Space and Society* 24: 1–5.
Dowler, L., D. Cuomo, and N. Laliberte. 2014. "Challenging 'The Penn State Way': A Feminist Response to Institutional Violence in Higher Education." *Gender, Place and Culture* 21 (3): 387–394.
Drozdzewski, D., and D. Dominey-Howes. 2015. "Research and Trauma: Understanding the Impact of Traumatic Content and Places on the Researcher." *Emotion, Space and Society* 17: 17–21.
Durham University. 2014. "Durham is one of the safest universities, according to new report." Available at: www.dur.ac.uk/news/newsitem/?itemno=21541 (accessed 5 February 2020).
Fanon, F. 1963. *The Wretched of the Earth*. New York: Grove Weidenfeld.
Fassin, D., and R. Rechtman. 2009. *The Empire of Trauma*. Princeton, NJ: Princeton University Press.
Herman, J. 1997. *Trauma and Recovery*. New York: Basic Books.
hooks, b. 2003. "Lasting Trauma." In: *Rock My Soul: Black People and Self Esteem*, edited by b. hooks, 21–34. New York: Atria Books.
Hossain, K.T., and S.R. Suman. 2013. "Violence against Women: Nature, Causes and Dimensions in Contemporary Bangladesh." *Bangladesh e-Journal of Sociology* 10 (1): 79–81.
Krebs, C., C. Lindquist, M. Berzofsky, B. Shook-Sa, K. Peterson, M. Planty, L. Langton, and J. Stroop. 2016. *Campus Climate Survey*. Washington, DC: Bureau of Justice Statistics Research and Development.
Lukianoff, G., and J. Haidt. 2018. *The Coddling of the American Mind*. New York: Penguin.
Marshall, D.J. 2014. "Save (Us From) the Children: Trauma, Palestinian Childhood, and the Production of Governable subjects." *Children's Geographies* 12 (3): 281–296.
National Union of Students (NUS). 2010. *Hidden Marks: A Study of Women Students' Experiences of Harassment, Stalking, Violence and Sexual Assault*. London: NUS.
Nguyen, L. 2011. "The Ethics of Trauma: Re-Traumatization in Society's Approach to the Traumatized Subject." *International Journal of Group Psychotherapy* 61: 26–47.
Pain, R. 2014. "Everyday Terrorism: Connecting Domestic Violence and Global Terrorism." *Progress in Human Geography* 38: 531–550.
Pain, R., and L. Staeheli. 2014. "Introduction: Intimacy – Geopolitics and Violence." *Area* 46: 344–347.
Phipps, A., and I. Young. 2015. "Neoliberalisation and 'Lad Cultures' in Higher Education." *Sociology* 49 (2): 305–322.
Pratt. G., Johnston C., and V. Banta. 2015. "Filipino Migrant Stories and Trauma in the Transnational Field." *Emotion, Space and Society* 24: 83–92.
Rezwana, N. 2018. *Disasters, Gender and Healthcare Access: Women in Coastal Bangladesh*. London: Routledge.
Rezwana, N., and R. Pain. 2018. "Understanding Gender-based Violence During Disasters in the Coastal Region of Bangladesh." Unpublished paper.
Sahdan, Z. 2018. "Demonic Possession: Spatial and Cultural Accounts of Domestic Violence in Malaysia." Unpublished PhD dissertation, Durham University.
Schwab, G. 2010. *Haunting Legacies: Violent Histories and Transgenerational Trauma*. New York: Columbia University Press.
Stark, E. 2004. *Coercive Control*. New York: Oxford University Press.
Sue, D. et al. 2007. "Racial Microaggressions in Everyday Life: Implications for Clinical Practice." *American Psychologist* 62: 4.
Tamas, S. 2011. *Life after Leaving*. London: Routledge.
The Times. 2017. "Durham's conspiracy of silence over rape." January 21. Available at: www.thetimes.co.uk/article/durhams-conspiracy-of-silence-over-rape-2frc75d6q (accessed 5 February 2020).
Tseris, E. J. 2013. "Trauma Theory Without Feminism? Evaluating Contemporary Understandings of Traumatized Women." *Affilia* 28: 153–164.
UNCF. 2014. *Ending Child Marriage: Progress and Prospects*. New York: UNICEF.
UNICEF. 2016a. *State of World Children 2016*. New York: UNICEF.
UNICEF. 2016b. "Child Protection from Violence, Exploitation and Abuse." Available at: www.unicef.org (accessed 20 October 2016).
Visser, I. 2015. Decolonizing Trauma Theory: Retrospect and Prospects. *Humanities* 4 (2): 250–265.
Warrington, M. 2001. "'I Must Get Out': The Geographies of Domestic Violence." *Transactions of the Institute of British Geographers* 26: 365–382.

27
GEOGRAPHIES OF VIOLENCE
Feminist geopolitical approaches

Katherine Brickell and Dana Cuomo

Introduction

With some exceptions, domestic violence has often served as an ancillary example for feminist geographers when complicating the public/private binary or affirming the feminist mantra that the 'personal is political'. More recently however, domestic violence has emerged as a distinct and sustained focus of feminist geographic research, a promising trend in response to directed calls for such work (Basu 2016; Brickell 2015; Brickell and Maddrell 2016a, 2016b; Pain 2014; Tyner 2012, 2016; Warrington 2001). Notably, much of this recent scholarship on domestic violence situates itself within a feminist geopolitical framework that emphasizes the multi-sited and multi-scalar links between intimate and global violence (Pain and Staeheli 2014). While we do not suggest that all work in geographies on violence necessarily takes feminist geopolitics as its epistemological and methodological grounding, in this chapter we centre on feminist geopolitical approaches and their continued influence on the field of gender and feminist geography, including our own work on domestic violence.

Feminist geopolitics 'embodies an approach that advocates a finer scale of "security" accountable to people, as individual and groups, and analyses the spaces of violence that traverse public/private distinctions' (Hyndman 2001, 219). Its emphasis on the everyday politics of (in)security and fear (for a review, see Williams and Massaro 2013) have served as an entryway for feminist geographers studying domestic violence, including Pain's (2014) understanding of domestic violence as 'everyday terrorism'. By tracing fear as a foundational component of domestic violence and global terrorism, Pain illustrates how both forms of political violence 'operate across scales rather than being restricted to global or everyday securities' (ibid., 535). Similarly, Cuomo (2013) traces how masculinist security discourses feature in both the local policing response to domestic violence in the US and global military interventions to 'protect' vulnerable women around the world. This work illustrates how security interventions can paradoxically increase the fears of those whom these actions purport to protect. Likewise, Brickell's work on domestic violence in Cambodia shows the overlap between types of violence utilized in war and the home, and the uneven consequences of post-conflict peace and reconciliation processes for security and safety at both global and intimate scales (2015, 2016, 2017, 2020). Despite the warlike dynamics within abusive relationships and the institutional responses to it, Pain (2015) resists reducing domestic violence to simply everyday militarism. Rather, she envisions domestic violence and international warfare as a single complex of violence, given their 'common gendered, psychological, and emotion-laden foundations of power', and reasserts their intimate-geopolitical intertwining as 'intimate war' (64).

In the first half of the chapter, we explore research focused explicitly on domestic violence from a feminist geopolitical perspective, and include reference to connected work in gender and feminist geography more broadly. The second half of the chapter hones in on research on the intimate violences inherent, but often ignored, in the military and militarization. Despite our organizational separation of these halves, both are attentive to the ways in which masculinized 'hot' geopolitics (e.g. war) and feminized 'banal', emotional and intimate violences (e.g. sexual assault in the military and on college campuses) are inseparable (Christian et al. 2016).

Indeed, the writing of this chapter and its timing render painfully urgent what is at stake when such connections fail to be recognized or are obscured. In February 2018, one of the world's worst mass school shootings took place in Parkland, Florida. Seventeen people – 14 high school students and three staff – were killed. It has been argued that media portrayals of these rampages are designed to shock the public and, in doing so, obscure debate and action on how the culture of hegemonic masculinity in the US creates a sense of aggrieved entitlement conducive to violence (Kalish and Kimmel 2010). 'Toxic masculinity', which frames masculinity as constituted through violence in patriarchal culture, is one such 'underlying ailment' (Datta 2018; Haider 2016) that warrants further research and unpacking in feminist and gender geography. A growing number of news reports in the US and internationally, for example, are bringing into public view the records of male mass shooters who have histories of domestic violence (e.g. Filipovic 2017; 'In Texas and Beyond' 2017; 'In Orlando, as Usual' 2016). Likewise, the gun control advocacy group 'Everytown For Gun Safety' (2018) contends that, 'Despite impressions from media coverage, mass shootings in which at least four people were killed with a gun are also typically acts of domestic or family violence', to the extent that 54 per cent of mass shootings between 2009 and 2016 were committed by intimate partners of family violence. INCITE! (2006), a national activist organization of US radical feminists of colour, reinforces the connections between intimate and state violence in its anthology *Color of Violence,* which identifies links between gender-based violence, militarism, reproductive and economic violence, prisons and policing, colonialism and war.

In the weeks following the Parkland School shooting, a group of student survivors powerfully advocated on social media for gun control, with the hashtag #NeverAgain, taking inspiration from the #MeToo campaign and its focus on the prevalence of sexual assault and harassment of women. In March 2018, the student survivors went on to lead the 'March for Our Lives' in Washington DC, attended by tens of thousands, with another 800 sibling marches organized around the world (https://marchforourlives.com). Notably, while the student survivors of the mass shooting in Parkland are mostly affluent and White, the march in Washington DC centred and elevated African–American voices, whose sustained activism against gun violence has often gone ignored by the media and larger public. These public platforms for speaking out call attention to the connective tissue of resistance and do so through intersectional frameworks that are foundational to a feminist strategy that contests violence in all its forms.

Across the chapter, we pay close and synergistic attention to the methodologies and methods that feminist geographers employ to study such violence, which tends to prioritize collaboration with research participants in the design, implementation and distribution of research findings (Sharp 2004). In sum, then, our aim is to provide a holistic, albeit non-exhaustive, sense of current directions of work on geographies of violence in gender and feminist geography by drawing on ideas from feminist geopolitics.

Researching domestic violence

In the last five years, feminist geographers have begun using the conceptual frameworks of intimate war, intimate terrorism and everyday terrorism to analyse variously scaled projects on domestic violence. Faria (2017) deploys a 'countertopography of intimate war' to examine the domestic violence homicide of a South Sudanese refugee after fleeing to the US, alongside the multiple forms of state and structural violence that the victim and other refugee women experienced in the diaspora. In the same year, Little (2017) published her study of the spatiality of rural domestic violence as intimate terrorism to examine how constructions of rural masculinity and femininity shape experiences of, and responses to, such violence. Meanwhile, Laliberte (2016) shows how post-war Northern Ugandan peacebuilding programmes rely on Orientalist narratives to target violence in the home as a site for development. Her work is also an important reminder that domestic violence conveniently emerges from the shadows of international warfare and global terrorism to take centre stage when racialized and gendered tropes of violence associated with certain populations serve a strategic purpose. As Faria (2017) notes, with its recognition of the inseparability of everyday and global violence, the impact of Pain's 'intimate war' extends beyond domestic violence, as feminist geographers apply this framework to research on state-sanctioned violence against racialized, gendered and classed bodies more broadly (Massaro 2015).

Feminist geographic research on domestic violence also emphasizes the links between state, structural and intimate violence, showing how multiple 'forms of interwoven violence' (Piedalue 2017) make some bodies disproportionately vulnerable (Sweet 2016; Tyner 2012). This includes work that argues that the analysis of domestic violence must occur in conversation with other systems of oppression, such as race, class and sexuality (see also special issues in *Dialogues in Human Geography* by Brickell and Maddrell 2016a, 2016b; *Gender Place and Culture* by Fluri and Piedalue 2017). Smith (2016) analyses the multi-scalar relationship between state violence and domestic violence among low-income residents of Cairo, Egypt, showing how the violence of the state shapes and occurs simultaneously with violence in the home. Cuomo (2017) reinforces, too, the connections between different forms of political violence by showing how patriarchal ideology and coercive control work in tandem with neoliberal citizenship to normalize political recognition as the primary citizenship right for domestic violence survivors. Understanding domestic violence as indistinguishable from other institutional and structural violences, including economic vulnerability, also provides insight into the challenges that survivors encounter when attempting to flee domestic violence.

Arguing that women's experiences of fleeing domestic violence have been under-recognized in UK policy and practice, Bowstead (2015a, 2017) theorizes survivors' escape from abusive relationships as forced migration to illustrate the connections between different types of gendered migrations and the (dis)empowering elements that accompany such journeys. Feminist geographers addressing such experiences also emphasize service provision and the impact of austerity measures on refuge and shelter funding (Bowstead 2015b; Coy et al. 2011; Graham and Brickell 2019; Pain and Scottish Women's Aid 2012), further illustrating the 'entangled inequalities' (Fluri and Piedalue 2017) within experiences of state and intimate violence.

The research undertaken to explore the experiences of survivors in this work commonly turns to participatory action research (PAR) approaches that aspire to participants collaborating equitably in all stages of the research process to address a problem or issue within a community (Kindon et al. 2010). This is an approach that complements the philosophical tenets of service providers of community-based domestic violence agencies, who emphasize self-determination and recognize that survivors are the experts of their own lives (Schetcher 1982). Pain's research partnership with Scottish Women's Aid, a non-profit organization working to prevent domestic violence, for example, illustrates the participatory, reflexive and politically active components of PAR as the project's findings resulted in a co-authored report (Pain/Scottish Women's Aid 2012) and a BBC Scotland television programme broadcast in 2014 to raise public awareness of the harm that domestic violence causes. In addition to employing a feminist methodology attentive to power and ethics in the research, for this project Pain trained in counselling skills before interviewing survivors of domestic violence, and then worked with a counsellor throughout the research for both supervision and therapy.

Such methodological strategies are shared widely among feminist geographers and are used to prevent secondary trauma and to support emotional health while conducting research on violence. This includes the specialized approaches adopted by scholars studying violence from outside the academy. Influenced by the training that she received while working as a victim advocate, for example, Cuomo employs the methodological practice of boundary-making when conducting research with domestic violence survivors (Cuomo and Massaro 2016). Boundary-making, when coupled with a self-care practice that relies on space and time to consciously process the grief associated with trauma exposure, has supported Cuomo's long-term professional, academic and activist work on interpersonal violence (Cuomo 2019). These approaches work to address not only the ethical issues that arise when conducting research with survivors of violence (Meth and Malaza 2003) but also to prioritize protecting the emotional health of the researcher. As we echo calls for further feminist geographic research on interpersonal violence, we encourage a methodological praxis that responds to the emotional implications of studying violence and trauma and prioritizes self-care.

Researching militarized violence

Since the 2010s, interest in gender and feminist geography on militarized violence and its gendered experiences, impacts, meanings and dynamics has intensified. From work examining police-linked death squads and Black women's organizations in Brazil (Alves 2013), to the police-led murder and disappearance of four dozen student teachers in Mexico (Wright 2018) and to the unregulated violence of state-hired private security companies against female forced eviction activism in Cambodia (Brickell 2014), the militarization of everyday and urban life takes on a range of guises in various sites and spaces of study. Militarization in feminist scholarship, Dowler (2012), writes, 'acknowledges that subjective forms of violence, such as wars, always reach deeper into societies than conventional reports would portray' (492). In this section of the chapter, we adopt a conventional focus on war and military violence, enacted both during and after its formal end, to demonstrate the importance of going beyond tidy distinctions between war and peace, domestic and foreign, military and civil society (Loyd 2011).

The gendered politics of military enlistment is particularly interesting in this regard and brings into view the targeting of civilian lives for military mobilization and violence. Cowen and Siciliano (2011) argue that militaries 'are not simply warehouses for surplus populations but are themselves increasingly means and sites of accumulation' sourced through men and boys in inner-city poor and deindustrialized small towns. In her research on military enlistment, Christian (forthcoming) uses art-based methods to explore race and citizenship among youth in Houston, Texas. This includes visual fieldnotes, comic vignettes, activist art and graphic narrative to theorize visually how youth pathways into the military interlock with local, state and global forms of violence. In her work with anti-violence activist organizations in Houston and as part of her feminist, anti-racist praxis, Christian donates posters for protests and sells her original designs, with all proceeds supporting local social movements. Such visual and art-based methods that focus on engaging in solidary work with research participants are increasingly popular in geography. Indeed, Hawkins (2015) notes that these creative methods lend to 'embodied and practiced-based doings' (248), which we see as especially complementary when studying experiences of violence.

That these creative 'doings' are being brought into practice by feminist geographers serves as a segway to discussion in political geography on the creative industry that designs and markets video games that engender hyper-masculinized identities and virtual war spaces of performance and consumption. In the last three years, for example, several publications have arisen from the observed rise of militarism 'in the living room' through video-gaming (Robinson 2016). The online communities that form through this game-playing, Bos (2018, 163) argues, are also 'highly gendered, heteronormative, and promote and reinforce national identities', and thus asks scholars 'to think further about the private, public and virtual spaces which constitute and influence particular popular geopolitical encounters and practices' (see also Woodward 2014 on future directions on this 'military industrial-media-entertainment').

The intersections of militarized violence, visual technologies and gender are also becoming more evident in research in geography on drones. Accompanying the proliferation of scholarship and interest in political geography on the international legality of surveillance and killing via drone warfare (Gregory 2011; Shaw 2016), recent work has looked to the targeting of drone strikes, including 'geographies of legal terror' against military-aged men (known by US soldiers as MAMs) in Muslim countries such as Pakistan (Wall 2016). Interdisciplinary feminist scholarship has also tended to focus its concerns on the violent trajectories of drone militarism (Feigenbaum 2015) and describes drone warfare as 'the intermediation of algorithmic, visual, and affective modes of embodiment', which 'reproduces gendered and racialized bodies that enable a necropolitics of massacre' (Wilcox 2017, 11). 'Drone stalking' is a particularly cogent example of how such military technologies are further expanding into civilian life the opportunities for violence.

The domestication of military technologies means that the phenomenon of 'drone peeping toms' has become a growing yet unexplored threat in academic work. The US government report *Integration of Drones Into Domestic Airspace: Selected Legal Issues* (Dolan and Thompson 2013) highlights that, as drones become more readily available to private citizens, the technology will be likely used to commit various offences, including drone stalking of women (see Gallagher 2013 and 'Drone Stalking Several Women' 2017 for media coverage of these cases). Such examples align with the wider need for vigilance to the threats of digital technologies in the realm of corporeal security, including technology-enabled violence against women (Brickell and Cuomo 2019). Intimate war thus 'gains its devastating potential precisely because it does not concern strangers, but people in relationships that are often long term' (Pain 2015, 67).

Domestic violence in the military itself is also an apposite example of an intimate yet often-unseen war that is taking place. Gray (2016) writes, for example, about the prioritization of operational effectiveness in the British military, which can override the self-defined needs of civilian women's experiences of domestic abuse in marriages to servicemen. This work, Gray argues, contributes to feminist geography scholarship that makes the connections 'between the intimate spaces of the home and the public spaces of geopolitics' and emphasizes 'the role that gendered interactions within military families play in the enactment of militarism writ large' (921). The inattention to these militarized marriages, Enloe (2016, 321) argues, is problematic – 'they have been and continue to be shaped by the interactions of militarized elite strategies, militarized popular cultures, assorted dynamic patriarchies, myriad racisms and ethno-centrisms, as well as diverse heterosexisms'.

As such, these patriarchal marriages are part of a broader panorama of military violence that extends to other types of intimate violence enacted against girls and women in times of war, during peace-keeping missions and while women actively serve in the military. Dowler's (2011) research on female soldiers who experience sexual violence from their male cohort illustrates the multiple risks that women experience while serving in a military institution embodied by masculinity and virility. Reiz and O'Lear's (2016) study examines cases of civilian rape by UN military personnel and police in Haiti. Using a 'critical legal geography of rape' as their guiding framework, the authors show how 'governance of these cases and the corresponding jurisdictional logics effectively silence and marginalize the survivors of rape and sexual assault by international security forces' (456). Both Gray's research with military wives and this latter work on civilian rape show how survivors' intimate securities are commonly subordinated below those of a 'higher' security order. That Reiz and O'Lear's research focuses on long-term military deployment and the creation of spaces where violence and impunity burgeon is particularly timely. Such violence, they note, 'is inexorably linked and entangled in a much larger, complex context of history, colonialism, geopolitics, governance practices, and globalization' (456).

Described as the 'the largest coordinated effort of wartime sex trafficking and forced sexual labor in human history, despite the near ubiquity of sex work in the environs of modern militaries' (Pilzer 2014, 3), the long-ignored problem of 'comfort women' is especially resonant here. There are believed to have been around 200,000 women from occupied countries, including Korea, China and the Philippines, who were forced to work by the Japanese military in military brothels during the Asia–Pacific War (1931–1945). The 2017 *Column of Strength* memorial (Figure 27.1) in San Francisco joins others around the world, including the first, sited outside Japan's embassy in Seoul, which led to a diplomatic fall-out between Japan and South Korea (Shepherd 2016). This example speaks to issues of the gendered geopolitics of memorialization and the occlusion of women's everyday experiences of war in comparison to men's (Tyner and Henkin 2015).

In order to bring the voices and experiences of those speaking historically from the margins into public view, testimonials have proved an important means by which feminist geographers have drawn on individual recollections to speak to collective witnessing (Reyes and Curry Rodriquez 2016). This methodology has been used in Truth Commissions and human rights work, including Rigoberta Menchu's testimony detailing the state violence and terror that Mayan women experienced from the US-supported Guatemalan army (Hanlon and Shankar 2000). The potential for testimonial to capture the interwoven experience of intimate and state violence is evidenced in Pratt's work detailing Filipino women's experience of family separation and state violence (Pratt 2012; Pratt and Johnson 2014) and Valencia's (2017) research analysing Mexican migrant women's experiences of risk and insecurity in relation to the militarization of the border and immigration policies. Hanlon and Shankar explain that testimonial 'does not simply shed light on past abuses, but puts a spotlight on the abusers and the mechanisms of state terror and genocide' (2000, 267). In other words, testimonials hold potential for individual and collective healing by documenting marginalization as a shared, rather than individualized, experience (Johnston and Pratt 2010). Diaries too, Tyner and Henkin (2015) argue, are an important means to explore (deceased) women's narratives, 'given they are imbued with a multitude of emotional responses and relations that provide an alternative, embodied account of war in contrast to detached, emotionless military histories, or heroic and overly virile glorifications of war' (289).

Figure 27.1 San Francisco's *Column of Strength* memorial.
Photograph: K. Brickell, April 2018.

Conclusion

> [G]ender relations are like a linking thread, a kind of fuse, along which violence runs. They run through every field (home, city, nation-state, international relations) and every moment (protest, law enforcement, militarization), adding to the explosive charge of violence in them. If most, if not all, violence has a gender component, violence reduction calls for a feminist gendered strategy.
>
> *Cockburn 2004, 44*

In this Handbook chapter we have sought to review and inspire work on geographies of violence that takes feminist geopolitics seriously as an explanatory framework. Mapping the spatial linkages between the hurt and the hurter (Philo 2005) means following, connecting and understanding all the intersecting systems of oppression that contribute to the violent fuse of violence that Cockburn refers to in the above quote. Geographical research on violence that takes a feminist geopolitical approach has an important role to play, therefore, in supporting the analytical and political visibility of these intersecting violences and injustices.

Key readings

INCITE! 2016. *Color of Violence*. Durham, NC: Duke University Press.
Pain, R. 2015. "Intimate War." *Political Geography* 44: 64–73.

References

Alves, J.A. 2013. "From Necropolis to Blackpolis: Necropolitical Governance and Black Spatial Praxis in São Paulo, Brazil." *Antipode* 46 (2): 323–339.

Basu, S. 2016. "Strategizing Spaces: Role of Women's Courts in Addressing Domestic Violence in a Low-income Neighbourhood of Delhi." Conference paper, Royal Geographical Society-Institute of British Geographers Annual Conference London, September. Available at: http://conference.rgs.org/AC2016/312

Bos, D. 2018. "Answering the Call of Duty: Everyday Encounters with the Popular Geopolitics of Military-themed Videogames." *Political Geography*. Online before print. https://doi.org.10.1016/j.polgeo.2018.01.001.

Bowstead, J. 2015a. "Forced Migration in the United Kingdom: Women's Journey's to Escape Domestic Violence." *Transactions of the Institute of British Geographers* 40 (3): 307–320.

Bowstead, J. 2015b. "Why Women's Domestic Violence Refuges Are Not Local Services." *Critical Social Policy* 35 (3): 327–349.

Bowstead, J. 2017. "Women on the Move: Theorising the Geographies of Domestic Violence Journeys in England." *Gender, Place & Culture* 24 (1): 108–121.

Brickell, K. 2014. "'The Whole World is Watching': Intimate Geopolitics of Forced Eviction and Women's Activism in Cambodia." *Annals of the Association of American Geographers* 104 (6): 1256–1272.

Brickell, K. 2015. "Towards Intimate Geographies of Peace? Local Reconciliation of Domestic Violence in Cambodia." *Transactions of the Institute of British Geographers* 40 (3): 321–333.

Brickell, K. 2016. "Gendered Violences and Rule of/by Law in Cambodia." *Dialogues in Human Geography.* 6 (2): 182–185.

Brickell, K. 2017. "Clouding the Judgment of Domestic Violence Law: Victim Blaming by Institutional Stakeholders in Cambodia." *Journal of Interpersonal Violence* 32 (9): 1358–1378.

Brickell, K. 2020. *Home SOS: Gender, Violence and Survival in Crisis Ordinary Cambodia*. London: Wiley RGS-IBG Series.

Brickell, K., and A. Maddrell. 2016a. "Geographical Frontiers of Gendered Violence." *Dialogues in Human Geography* 6 (2): 170–172.

Brickell, K., and A. Maddrell. 2016b. "Gendered Violences: The Elephant in the Room and Moving Beyond the Elephantine." *Dialogues in Human Geography* 6 (2): 206–208.

Brickell, K., and D. Cuomo. 2019. "Feminist Geolegality." *Progress in Human Geography* 43 (1): 104–122.

Christian, J. Forthcoming. "Creative Geopolitics: Art and Solidarity as Feminist Geopolitical Praxis." *Environment and Planning C.*
Christian, J., L. Dowler, and D. Cuomo. 2016. "Learning to Fear: Feminist Geopolitics and the Hot Banal." *Political Geography* 54: 64–72.
Cockburn, C. 2004. "The Continuum of Violence: A Gender Perspective on War and Peace." In: *Sites of Violence: Gender and Conflict Zones,* edited by W. Giles and J. Hyndman, 24–44. Berkeley: University of California Press.
Cowen, D., and A. Siciliano. 2011. "Surplus Masculinities and Security." *Antipode* 43 (5): 1516–1541.
Coy, M., L. Kelly, J. Foord, and J. Bowstead. 2011. "Roads to Nowhere? Mapping Violence Against Women Services." *Violence Against Women* 17 (3): 404–425.
Cuomo, D. 2013. "Security and Fear: The Geopolitics of Intimate Partner Violence Policing." *Geopolitics* 18 (4): 856–874.
Cuomo, D. 2019. "Calling 911: Intimate Partner Violence and Responsible Citizenship in a Neoliberal Era." In: *Social and Cultural Geography* 20 (7): 879–898.
Cuomo, D. 2019. "Self-care and Trauma: Locating the Time and Space to Grieve." In: *Vulnerable Witness: The Politics of Grief in the Field,* edited by K. Gillespie and P. Lopez, 174–186. Berkeley: University of California Press.
Cuomo, D., and V.A. Massaro. 2016. "Boundary-Making in Feminist Research: New Methodologies for 'Intimate Insiders'." *Gender, Place & Geography* 23 (1): 94–106.
Datta, A. 2018. "Why So Many Rapes? Sexual Violence against Women in India." *Geography and You* 18 (112): 38–42.
Dolan, A.M., and R.M. Thompson. 2013. "Integration of Drones into Domestic Airspace: Selected Legal Issues. Congressional Research Service R42940. Available from: https://fas.org/sgp/crs/natsec/R42940.pdf (accessed 4 August 2019).
Dowler, L. 2011. "The Hidden War: The 'Risk' to female soldiers in the U.S. Military." In: *Reconstructing Conflict: Integrating War and Post-War Geographies,* edited by S. Kirsch and C. Flint, 295–314. Aldershot: Ashgate.
Dowler, L. 2012. "Gender, Militarization and Sovereignty." *Geography Compass* 6 (8): 490–499.
"Drone Stalking Several Women." 2017. "Drone Stalking Several Women in Rural Port Lincoln Community Part of Growing List of UAV Concerns." *ABC News,* 3 November. Available at: www.abc.net.au/news/2017-11-02/drone-stalking-several-women-in-port-lincoln-rural-community/9112926 (accessed 5 August 2019).
Enloe, C. 2016. "Flick of the Skirt: A Feminist Challenge to IR's Coherent Narrative." *International Political Sociology* 10: 320–331.
"Everytown for Gun Safety." 2018. "Domestic Violence." Available from: https://everytownresearch.org/issue/domestic-violence/ (accessed 5 August 2019).
Faria, C. 2017. "Towards a Countertopography of Intimate War: Contouring Violence and Resistance in a South Sudanese Diaspora." *Gender, Place & Culture* 24 (4): 575–593.
Feigenbaum, A. 2015. "From Cyborg Feminism to Drone Feminism: Remembering Women's Anti-Nuclear Activisms." *Feminist Theory* 16 (3): 265–288.
Filipovic, J. 2017. "Guns Don't Kill People, Domestic Abusers Do. Take Their Guns Away." *Guardian.* Available from: www.theguardian.com/commentisfree/2017/nov/07/guns-domestic-abusers-texas-shooting (accessed 5 August 2019).
Fluri, J.L., and A. Piedalue. 2017. "Introduction: Embodying Violence: Critical Geographies of Gender, Race, and Culture." *Gender, Place & Culture* 24 (4): 534–544.
Gallagher, R. 2013. "Drones Could Be Used for Stalking, Voyeurism." *Slate,* 1 February. Available from: www.slate.com/blogs/future_tense/2013/02/01/drones_could_be_used_for_stalking_voyeurism_says_congressional_research.html (accessed 5 August 2019).
Graham, N., and K. Brickell. 2019. "Sheltering from Domestic Violence: Women's Experiences of Punitive Safety and Unfreedom in Cambodian Safe Shelters." *Gender, Place and Culture* 26 (1): 111–127. doi: 10.1080/0966369X,2018.1557603.
Gray, H. 2016. "Domestic Abuse and the Public/Private Divide in the British Military." *Gender, Place & Culture* 23 (6): 912–925.
Gregory, D. 2011. "From a View to a Kill: Drones and Late Modern War." *Theory, Culture & Society* 28 (7–8): 188–215.
Haider, S. 2016. "The Shooting in Orlando, Terrorism or Toxic Masculinity (or Both?)." *Men and Masculinities* 19 (5): 555–565.

Hanlon, C.N., and F. Shankar. 2000. "Gendered Spaces of Terror and Assault: The Testimonio of REMHI and the Commission for Historical Clarification in Guatemala." *Gender, Place & Culture* 7 (3): 265–286.
Hawkins, H. 2015. "Creative Geographic Methods: Knowing, Representing, Intervening. On Composing Place and Page." *Cultural Geographies* 22 (2): 247–268.
Hyndman, J. 2001. "Towards a Feminist Geopolitics." *Canadian Geographer.* 45 (2): 210–222.
INCITE! 2016. *Color of Violence.* Durham, NC: Duke University Press.
"In Orlando, as Usual." 2016. "In Orlando, as Usual, Domestic Violence Was Ignored Red Flag." *Rolling Stone* 13 June. Available from: www.rollingstone.com/politics/news/in-orlando-as-usual-domestic-violence-was-ignored-red-flag-20160613.
"In Texas and Beyond." 2017. "In Texas and Beyond, Mass Shootings Have Roots in Domestic Violence." NPR, 7 November. Available at: www.npr.org/sections/health-shots/2017/11/07/562387350/in-texas-and-beyond-mass-shootings-have-roots-in-domestic-violence (accessed August 2019).
Johnston, C., and G. Pratt. 2010. "Nanay (Mother): A Testimonial Play." *Cultural Geographies* 17 (1): 123–133.
Kalish, R., and M. Kimmel. 2010. "Suicide by Mass Murder: Masculinity, Aggrieved Entitlement, and Rampage School Shootings." *Health Sociology Review* 19 (4): 451–464.
Kindon, S., R. Pain, and M. Kesby, eds. 2010. *Participatory Action Research Approaches and Methods: Connecting People, Participation and Place.* London: Routledge.
Laliberte, N. 2016. "'Peace Begins at Home': Geographic Imaginaries of Violence and Peacebuilding in Northern Uganda." *Political Geography.* 52: 24–33.
Little, J. 2017. "Understanding Domestic Violence in Rural Spaces." *Progress in Human Geography.* 41: 472–488.
Loyd, J. 2011. "'Peace is Our Only Shelter': Questioning Domesticities of Militarization and White Privilege." *Antipode* 43 (3): 845–873.
Massaro, V.A. 2015. "A Response to Rachel Pain: Intimate War and the Politics of Men." *Political Geography.* 44: 77–78.
Meth, P., and K. Malaza. 2003. "Violent Research: The Ethics and Emotions of Doing Research with Women in South Africa." *Ethics, Place & Environment: A Journal of Philosophy & Geography* 6 (2): 143–159.
Pain, R. 2014. "Everyday Terrorism: Connecting Domestic Violence and Global Terrorism." *Progress in Human Geography* 38 (4): 531–550.
Pain, R. 2015. "Intimate War." *Political Geography* 44: 64–73.
Pain, R. and Scottish Women's Aid. 2012. *Everyday Terrorism: How Fear Works in Domestic Abuse.* Durham University and Scottish Women's Aid. Available at: www.dur.ac.uk/resources/geography/downloads/EverydayTerrorism.pdf
Pain, R., and L.A. Staeheli. 2014. "Introduction: Intimacy-Geopolitics and Violence." *Area* 46 (4): 344–347.
Philo, C. 2005. "The Geographies That Wound." *Population, Space and Place* 11: 441–454.
Piedalue, A. 2017. "Beyond 'Culture' as an Explanation for Intimate Violence: The Politics and Possibilities of Plural Resistance." *Gender, Place & Culture* 24 (4): 563–574.
Pilzer, J.D. 2014. "Music and Dance in the Japanese Military 'Comfort Women' System: A Case Study in the Performing Arts, War, and Sexual Violence." *Women and Music: A Journal of Gender and Culture* 18: 1–23.
Pratt, G. 2012. *Families Apart: Migrant Mothers and the Conflicts of Labor and Love.* Minneapolis: University of Minnesota Press.
Pratt, G., and C. Johnston. 2014. "Filipina Domestic Workers, Violent Insecurity, Testimonial Theatre and Transnational Ambivalence." *Area* 46 (4): 358–360.
Reiz, N., and S. O'Lear. 2016. "Spaces of Violence and (In)justice in Haiti: A Critical Legal Geography Perspective on Rape, UN peacekeeping, and the United Nations Status of Forces Agreement." *Territory, Politics, Governance* 4 (4): 453–471.
Reyes, K.B., and J.E. Curry Rodriquez. 2016. "Testimonio: Origins, Terms and Resources." *Equity & Excellence in Education* 45 (3), 525–538.
Robinson, N. 2016. "Militarism and Opposition in the Living Room: The Case of Military Videogames." *Critical Studies on Security* 4 (3): 255–275.
Schetcher, S. 1982. *Women and Male Violence: The Visions and Struggles of the Battered Women's Movement.* Cambridge: South End Press.
Sharp, J. 2004. "Doing Feminist Political Geographies." In: *Mapping Women, Making Politics: Feminist Perspectives on Political Geography*, edited by L.A. Staeheli, E. Kofman, and L.J. Peake, 87–98. New York: Routledge.

Shaw, I.G.R. 2016. "Scorched Atmospheres: The Violent Geographies of the Vietnam War and the Rise of Drone Warfare." *Annals of the Association of American Geographers* 106 (3): 688–704.
Shepherd, R. 2016. "Cosmopolitanism Nationalism and the Heritage of Shame: Comfort Women Memorials and the Legacy of Slavery in the United States." *International Journal of Cultural Policy* 25 (2): 125–139.
Smith, C.E. 2016. "Caring Practices: The Connection Between Logics of State and Domestic Violence in Cairo, Egypt." *Gender, Place & Culture* 23 (9): 1227–1239.
Sweet, E. 2016. "Carceral Feminism: Linking the State, Intersectional Bodies, and the Dichotomy of Place." *Dialogues in Human Geography* 6 (2): 202–205.
Tyner, J. 2012. *Space, Place and Violence*. New York: Routledge.
Tyner, J. 2016. "Herding Elephants: Geographic Perspectives on Gendered Violence." *Dialogues in Human Geography* 6 (2): 190–197.
Tyner, J.A., and S. Henkin. 2015. "Feminist Geopolitics, Everyday Death, and the Emotional Geographies of Dang Thuy Tram." *Gender, Place & Culture* 22 (2), 288–303.
Wall, T. 2016. "Ordinary Emergency: Drones, Police, and Geographies of Legal Terror." *Antipode: A Radical Journal of Geography* 48 (4): 1122–1139.
Warrington, M. 2001. "'I Must Get Out': The Geographies of Domestic Violence." *Transactions of the Institute of British Geographers* 26: 365–382.
Wilcox, L. 2017. "Embodying Algorithmic War: Gender, Race, and the Posthuman in Drone Warfare." *Security Dialogue* 48 (1): 11–28.
Williams, J., and V.A. Massaro. 2013. "Feminist Geopolitics: Unpacking (In)Security, Animating Social Change." *Geopolitics* 18 (4): 751–758.
Woodward, R. 2014. "Military Landscapes: Agendas and Approaches for Future Research." *Progress in Human Geography*. 38(1) 40–61.
Wright, M. W. 2018. "Against the Evils of Democracy: Fighting Forced Disappearance and Neoliberal Terror in Mexico." *Annals of the American Association of Geographers*. 108 (2): 327–336.
Valencia, Y. 2017. "Risk and Security on the Mexico-to-US Migrant Journey: Women's Testimonios of Violence." *Gender Place & Culture* 24 (11): 1530–1548.

28

SCALING A SURVIVOR-CENTRIC APPROACH FOR SURVIVORS OF SEXUAL VIOLENCE

The case of an action-based research project in India

Andréanne Martel and Margaret Walton-Roberts

Introduction

Through challenging and changing gendered and other forms of power inequality, scholarship in feminist geography contributes to and simultaneously challenges the process of constructing knowledge. For feminist geographers, the research process, specifically the scale at which research occurs, and how the benefits of such research are shared are central and important issues of concern. In this chapter, we examine these factors in more detail by exploring a project based on scaling access to justice for survivors of sexual violence. This case illustrates the value of scaling the results of effective action-based research in order to increase its transformative potential.

Feminist geography and praxis

The work of feminist geographers can be interpreted as concerned not only with theorizing the operation of power in relation to gendered and other inequalities but also with the development of tools to transform those relations and make power more accountable for the reproduction and maintenance of such inequalities (Hyndman 2004). Staeheli and Lawson (1995, 321) laid out the methodological implications of such feminist geographical analysis: 'The goal of this more inclusive research and knowledge is the transformation of gendered power relations.' The power- and process-focused dimension of feminist geography thus compels researchers to be accountable for how they engage in research and with those whose lives are being researched. The physical and relational location of the researcher and those researched speaks to issues of power relations and space-time. Doreen Massey (1993) termed this 'power geometries', to indicate how our social location informs how we experience, control, change and access resources; this structures the particular constellation of social relations that frame people's lives.

Research, for feminist geographers, thus entails a significant amount of reflection on why the researcher is engaging with certain topics, why those topics matter, how the researched population is to be included in the formation of the research issue and data collection and how that population will benefit from the knowledge produced (Moss, Al-Hindi and Kawabata 2002). Critical and feminist geographers have addressed such concerns by highlighting the importance of community-based research or participant action research – where researched populations become partners in the process of constructing new knowledge through non-hierarchal partnerships that contribute to change – which represents an important methodological approach for feminist geographers (Pain 2004; Pain et al. in this volume, Chapter 26). The type of methodological approach and research methods used is centrally important when researching and collaborating with vulnerable populations facing significant social inequality (Brickell and Cuomo, in this volume, Chapter 27). Deeply qualitative research approaches can result in powerful yet potentially personal and politically incendiary research outcomes. Collaborative research can contribute to feminist analysis, but the difficulties and dangers of interrogating and exposing the roots of deep-seated inequality are also evident, as the Sangtin Writers Collective and Richa Nagar (2006) reveal in their critical intersectional analysis of caste and gender and the NGO sector in India.

Another debate linked to research relevance and effectiveness is that of scale. The issue of scale is central to research methodology and outcome; understanding the scale at which gender-based violence originates and is occurring is vital to determining how to prevent and address it (Pain et al. and Brickell and Cuomo, this volume, Chapters 26 and 27). Anyone who wants to have an impact on policy must understand the importance of their research being persuasive in terms of sample size or alignment with a larger body of literature with similar findings and recommendations. This element of scale helps to convince government agencies and other institutional bodies of the value and relevance of qualitative research in developing more equitable policies, and can contribute to meaningful policy engagement (Mountz and Walton-Roberts 2006).

Feminist geography has contributed to important theoretical debates regarding scale. This includes the political power of connecting issues across and beyond different scales that are seen as oppositional or mutually exclusive. This theorization is evident in the idea of the global and the intimate (Pratt and Rosner 2006), which elaborates on the political potential of feminist work to 'disrupt traditional organizations of space, to forge productive dislocations, reconfigure conventions of scale' (Pratt and Rosner 2012, 1). Also, Katz's (2001) counter-topographies, where political potential can be forged through collaborative engagements that upend or query scalar or spatial distances and find commonalities, are evident in human struggles around production and social reproduction (see also Walton-Roberts 2010). Significant debates regarding scale as a social construction have been informed by feminist arguments to take seriously social reproduction and consumption (Marston 2000), which in turn has informed rich debates about the politics of scale and new approaches such as rescaling (Swyngedouw 2004), as well as the need to focus on scales such as those of the body (Mountz 2004) and the home (Blunt 2011).

Thinking about scale together with community collaborative research can yield valuable examples that speak to the power and ideals of feminist geography and its focus on identifying and challenging all forms of social inequality. This chapter engages with these themes and provides an illustration of research with vulnerable populations: survivors of sexual violence and their access to justice.

In this chapter, we explore the outcomes of a project that took effect in 2012, a collaboration between Majlis – a Mumbai-based, women-led legal organization committed to ensuring access to justice for women in India – and the local Department of Women and Child Development, with financial support from the International Development Research Center (IDRC) in Canada. The memorandum of understanding between these organizations led to the establishment of RAHAT, a survivor-support programme that provides socio-legal assistance to survivors of sexual violence and their families. To further the transformative potential of this work, the Mumbai-based organization continued to build its partnerships after the end of the grant from the Canadian funder in order to bring the model to scale, in this case to other locations in the State of Maharashtra and, eventually, to other sites in India.

Before we start into the example, it is important to be explicit about the politics of space and scale that this chapter represents. By focusing on India and gendered violence, we are not suggesting that gendered violence does not exist in other contexts; it does (see also Pain et al. and Brickell and Cuomo, this volume, Chapters 26 and 27). Our chapter could have easily reported on how the Trump administration has rescaled gendered violence beyond the US to women globally through re-imposition of the 'global gag rule', which limits funding to organizations that counsel or provide information on abortion, a policy that will lead to increased child and maternal mortality in countries that are dependent on donor funds (Singh and Karim 2017). Likewise, we could detail how in the UK Theresa May's government-imposed two-child welfare cap on low-income households will cause disproportionate harm both to low-income women who already face structural marginalization and to their children (Machin 2017). Moreover, at the time of writing this chapter, Canada has experienced its second largest mass killing since the 1989 École Polytechnique massacre in which 14 women were murdered by a man who claimed to be fighting feminism. In April 2018, a man drove a van for 2 kilometres along a busy pavement in North Toronto, killing 10 people, eight of them women. The attacker claimed he was a 'foot soldier' for Incel, an online community of involuntary celibate men who hold women, and the feminist movement, responsible for their inability to form intimate relationships with the opposite sex. Thus, there is no shortage of examples of forms of gender-based violence enacted by governments, social groups and individuals across countries, regardless of their economic, cultural, religious or political context.

We examine the situation of India in this chapter from our shared interest in that country, and also because India is the site of important community-based innovations that target entrenched forms of gendered-based violence through transformative policy and practice. The work of many Indian NGOs is highly perceptive of the structural dimensions of gendered violence and the need to enact system-wide changes to empower survivors of violence, as well as to make states, agencies and individuals accountable for their role in perpetrating such violence. Such groups demand that organizations engage in transformation change in order to prevent the continuation of gender-based violence, but their work is highly context-sensitive, aimed at transformative change that leads to material improvements in the lives of women and girls even when faced with entrenched gendered inequality. As an example, we observed in India other similar interventions involving the potential of scaling feminist transformative work into the legal system. The 'women's courts' (*mahila adalat* or *mahila mandal*) in low-income neighbourhoods in Delhi and the recent proliferation of these alternative courts in various cities in India provide an interesting context to discuss the idea of scaling access to justice for women in this country (Vatuk 2013). As in the dissemination of 'women courts', we believe there may be important lessons from the Rahat case to share.

A survivor-centric approach for social and legal support for survivors of sexual violence: the RAHAT initiative

A few months before the highly publicized gang rape on a bus in Delhi in December 2012 made the headlines, a local journal in Mumbai had reported that a four-year-old girl had been raped by her school watchman. Lawyers and social workers at Majlis decided to offer socio-legal assistance to the victim-survivor's family. For Majlis, this case highlighted the existing flaws in the criminal justice system in addressing sexual offence cases and dealing with victim-survivors. When this horrific crime happened, data on sexual offence occurrences from the National Crime Bureau of India indicated that rape was occurring every 26 minutes, molestation every 14 minutes and dowry death every 63 minutes. Impunity for these crimes was the norm, as only 27 per cent of the perpetrators were convicted (IDRC 2016).

The purpose of the research was to build a survivor-centric approach to provide greater access to justice. To implement this approach, Majlis created the RAHAT unit. Lawyers and social workers working in this unit – the RAHAT team – launched an action-based research project that aimed systematically to follow and document the High Courts' and the Supreme Court's ruling in cases of violence against women and children and to identify the gaps in the criminal justice system (IDRC 2016, 1). This included tracking the poor implementation of the existing law[1] and led to the development of a holistic, survivor-centric approach to address those lacunae. This unit came from a collaborative project between the Department of Women and Child Development, Government of Maharashtra, and the Majlis Centre – a women's legal organization created in 1991 in Mumbai, funded by IDRC for three years. The action-based research project, entitled 'Interrogating Sexual and Domestic Violence and Evolving Protocols for State Agencies', came into effect in August 2012.

The Majlis survivor-centric approach focuses on minimizing trauma, to transform the victim into a survivor. In order to achieve that process, the RAHAT team focuses on rehabilitation and prioritizes the needs and rights of victim-survivors in its intervention (RAHAT 2015, 20). The approach prescribes joint collaboration and coordinated actions among multiple stakeholders to provide victims with the appropriate legal, medical, social or other support required, throughout the litigation process and beyond. The model of support is not limited to assistance while in hospital, but can also go directly to the victim's residence as soon as the case is reported (RAHAT 2015, 20).

Between 2012 and 2015, the RAHAT team documented the positive rulings by the High Courts and the Supreme Court and analysed them. In actual terms, it has followed up on 490 cases, having analysed 140 judgments of trials in 2011 and 2012 to ascertain trends in rape trials, and has followed almost 500 cases of sexual violence. The team has developed tools, protocols and training to help the stakeholders such as police and protection officers involved in the justice system to fulfil their roles more effectively. The RAHAT team has developed standard operating procedures (SOPs) for justice officials' use in cases of sexual offence against women and children. To date, these SOPs are being followed by over 2,000 police officers in Mumbai. The team has also designed training programmes for key stakeholders such as police, public prosecutors, judges and others in the criminal justice system, and is currently monitoring progress towards the institutionalization of these practices.

IDRC support to Majlis ended in 2015, but the RAHAT team continues to support women and children victims of violence. The tools and protocols developed by the team have been used in 93 police stations, as well as by several judges and public prosecutors in Mumbai City and suburbs. In the process of its work it began to ask how it could make its survivor-centric approach sustainable and expand it to other communities. As the RAHAT team was achieving success in Mumbai City in 2014 and 2015, it decided to explore the potential for scaling up its approach to other cities and districts in India.

Scaling a survivor-centric approach

Following the success of the survivor-centric approach in Mumbai City and suburbs, the RAHAT team is now pursuing both horizontal and vertical scaling strategies (Naqvi and Mehta 2015). 'Horizontal scaling involves the replication or expansion of an innovation in a different community; while vertical scaling implies the institutionalization of an innovation through policy, political, legal, regulatory, budgetary, or other [systems changes]' (in Rottach et al. 2012, 2; Rottach et al. 2012).

The RAHAT team has pursued three strategies in scaling its approach. A first horizontal strategy involved replication of the approach to additional sites across the 36 districts of Maharashtra. The following two involved a vertical strategy. It is scaling its approach through its adoption by the state government or other NGOs.[2] The final strategy focused on promoting behaviour change at various levels of the legal jurisdiction to ensure sustainability, as well as to improve accountability of state agencies and reduce impunity. The development of SOPs, training programmes and monitoring the institutionalization of procedures were the main pathways to scale the impact of the survivor-centric approach. This combination of mutually reinforcing scaling strategies focused on enhancing the capacities of state agencies and other public stakeholders to abide by and apply the law. It is also in line with the fundamental mandate of Majlis, which, at ground level, focuses on supporting women victims of violence and advocating for policy change. The RAHAT team is improving access to justice for survivors, holding state agencies accountable and ultimately decreasing the impunity for perpetrators of violence against women. From its perspective, a critical element for the sustainability of this approach is monitoring the behavioural changes of officials and strengthening state capacity. Each scaling strategy has encountered challenges in terms of feasibility, the quality of the model and the coordination with other stakeholders.

Replicating the survivor-centric approach in a new community: Navi Mumbai

The initial uptake of the survivor-centric approach in Mumbai and the suburbs was effective and efficient, because it was built on a solid network of organizations that believed in the approach and trusted the RAHAT team (Martel 2019). This close collaboration is the key to its success, yet is also the main obstacle to replicating the model. The scaling approach involved mapping potential new locations to identify relevant networks and organizations in each district; this was important in order to avoid competition and duplication. Even more important was identifying feminist organizations, feminist lawyers and social workers with a strong belief in the approach and a willingness to form a social network with shared values. Efforts to replicate the approach demanded the RAHAT team maintain the model's fidelity while understanding that adaptation to new contexts was critical. To stay relevant to victims' needs, the researchers had continuously to assess and adapt their approach to various types of vulnerability contexts, such as rural/urban, class, caste, ethnicity, religion, and so on. One of the cornerstones of the RAHAT survivor-centric approach's success is the Mumbai police commissioner's support, and seeking this level of support in other settings was an important feature of scaling.

As seen, a scaling process is not the achievement of a single organization but, rather, requires the participation of many partners and collaborators at several levels of the scaling pathway. In considering the replication of its approach in the city of Navi Mumbai, the RAHAT team first mapped all the organizations involved in providing social and legal support for victims of violence. This phase was critical to providing a better understanding of the context in which sexual offences against women and children were occurring and the actors involved in tackling such issues. Implementing the survivor-centric approach at the original site also involved a high number of official partners and collaborators. For example, the memorandum of understanding between the Department of Women and Child Development (DWCD) and Majlis Legal Centre represents one of the key factors in the success of this project. As noted by Naqvi and Mehta (2015, 12), this strategic collaboration 'allowed mutual needs of the DWCD and Majlis to merge toward shared goals'. The Government of Maharashtra was trying to gain public acknowledgement of its efforts to fight crime against women and obtain the trust of the public. Through this partnership, the RAHAT team had access to all key stakeholders in the judicial system, including the courts. The partnership gave RAHAT the credibility and legitimacy it needed to access the First Information Reports (FIR) recorded by the Mumbai police and the opportunity to train public prosecutors and judges. Over the course of the project, the RAHAT team became, in a sense, part of the justice system and a significant advocate for survivors of violence. Its expanding role in supporting victims of sexual violence allowed it to build a close relationship with major stakeholders within the justice system, starting with the police commissioners, public prosecutors and judges. This level of partnership was essential to its ability to scale its approach, and RAHAT leveraged its official partnerships with the DWCD and the police commissioner in Mumbai to access police stations and state agencies in its new implementing site, Navi Mumbai.

In scaling its approach horizontally to sites outside of Mumbai (Navi Mumbai), RAHAT's formal partnership with the DWCD enabled it to create a trusting environment and build relationships in other districts within the State of Maharashtra, where it was not so well known. This partnership was also a key element in the vertical adoption of the approach and its institutionalization within systems of regulatory oversight. Even though the replication of the approach in the State of Maharashtra was facilitated by its having the same laws and legal system, the actual application of the law remained highly contextual in each community. The scaling process sought to systemize the survivor-centric approach and expand official commitments to supporting it.

Concurrent strategies to reach the optimal scale

Identifying the optimal scale to replicate an approach should result from a dynamic and challenging process of assessing the resources available and the proper replication context (McLean and Gargani 2019). The survivor-centric approach illustrates some of these challenges.

In the new site of Navi Mumbai, the RAHAT team undertook a needs assessment to ensure that its approach was as context-specific as possible. Even though it had followed a high number of cases in Mumbai to inform the development of its model, the RAHAT team still needed to understand better the contextual factors in each new setting. Questions covered by the assessment varied from 'What is the socio-cultural context?' and 'What types of violence are women facing?' to 'What are the vulnerability factors?' and 'Which organizations are working in the same sector?' All these questions helped the team to make a decision regarding its scaling strategy in Navi Mumbai. Its experience in Mumbai City and suburbs allowed it to understand the multiple and complex contextual challenges that victims of violence face. The diversity of socio-economic conditions in the city represented a microcosm of the multiple facets of violence. Nevertheless, the specificity of operating in a smaller city or rural area still needed to be further assessed, as well as the types and roles of organizations already operating there.

Another important component that informed the scaling strategy was the effective capacity of the organization itself to scale the approach. The time and expertise demands that the scaling process involved might have stretched the Majlis staff beyond their limits. In this regard, it is about the capacity to deliver on a model built on their expertise. One scaling challenge of this strategy for a grassroots organization like Majlis was that it required it to allow other organizations to implement the model, running the risk that they would diminish its quality (fidelity) or transformative potential. A decentralized scaling strategy combined with deliberate partnerships with similar feminist organizations provided a means for Majlis to expand the model without compromising it (Naqvi and Mehta 2015, 5).

The third scaling strategy, changing the behaviour of stakeholders, involved many training events and rigorous monitoring of court rulings. The latter is one of the critical challenges of this model. To expand this scaling strategy, the team requires more human resources or will need to rely on partner organizations. Given the limited numbers of lawyers using a feminist perspective and the years that it took to build the specific and unique capacity of the RAHAT team, this is one of the key challenges of the various types of scaling strategy.

For this specific project, the goal was not policy change, because new regulations were already in place (e.g. the 2012 Amendment of Criminal Law and the 2012 Protection of Children from Sexual Offences Act). Instead, it was a matter of ensuring proper application of the law in a manner that supported the intentions of the survivor-centric model. Considering these needs, an optimal approach to scaling the model might involve simultaneously targeting behavioural change at the policy level by training state representatives on the existing laws and replicating the approach at a community level. These two concurrent strategies help to build cohesion around the approach and help the different stakeholders to adopt the model progressively. Ideally, improved application of the law would lead to the model's expansion rather than deliberate NGO-led scaling strategies.

The success of scaling *up* based on behavioural change might mean scaling *down* the role of the NGO, which might then be an indicator of success. Ideally, a successful scaling up strategy allows the implementing organization eventually to step back and let others (NGOs through adoption, state agencies through behavioural change) standardize the new approach.

Gender transformative strategies while scaling

A gender transformative paradigm aims to change power dynamics and empower marginalized groups or individuals. A transformative strategy also seeks to challenge gender norms and attitudes by improving the status of women in society (Kabeer 1994). Gender accommodating approaches, on the other hand, acknowledge gender norms and inequalities, and develop actions that adjust to, and often compensate, for these norms. Martel (2019) demonstrates that both gender transformative and accommodative strategies could be used in a scaling process in order to enhance the ownership of a particular approach. Though the general mandate of the survivor-centric approach was gender transformative, as it seeks to challenge how the justice system and society look at women and children victims of sexual violence, the RAHAT team decided to use both gender accommodative and transformative strategies in scaling its approach, after analysis of the gender barriers inherent to the scaling process. For instance, RAHAT needed to consider gender barriers and to articulate the role of gender at each stage of the process, from the design of the approach to the scaling strategy:

> [Their approach] is not universal and they need to be very context-specific if they want to be able to replicate it in different districts with an appropriate uptake by the communities. [...] The overall Survivor Centric Approach is deeply gender

> transformative as it seeks to challenge the way the justice system and society look at victims of sexual violence and help the victim become a survivor. To achieve that mandate, the RAHAT team had to challenge, for example, a popular misperception that all rape cases were false.
>
> *Martel 2019, 123*

The direct impact of this misperception was the normalization or dismissal of violence by the police and subsequent official inertia when addressing sexual violence complaints. To challenge this, RAHAT conducted a massive sensitization campaign for police officers called 'Zero FIR'. This campaign focused on raising awareness around the importance of filing a First Information Report (FIR) within the first 24 hours following the report of a rape. The socio-legal support that Majlis provided to its clients through legal counselling, skills training or job re-entry programmes aimed to counteract the dismissal of victims by empowering women to go through a transformation that positioned them as survivors, rather than victims, of sexual violence.

In the RAHAT project, accommodative strategies were used to enhance the ownership of the approach by key stakeholders and consequently ensure its sustainability. One of the main accommodative components was the skills training programme for police officers, which had a twofold aim. It included a sensitization component aimed at educating them about their roles in addressing cases of sexual violence (e.g. the importance of filling out FIRs, following SOP protocols, women officers recording the complaints, etc.) (Martel 2019, 122). The training was deliberately non gender-sensitive,[3] which means that it was not trying to change the mindset of police officers about sexual violence and the role of women in society but, instead, to create a rote form of standardized practice. This decision aimed to avoid resistance and gain buy-in. In this case, a pragmatic, gender accommodative approach was more appropriate than a transformative one, and stakeholder analysis was useful in identifying this more-effective approach. For the scaling purpose, RAHAT needed to collaborate with stakeholders who have very different and sometimes opposing perspectives on gender roles and norms. Yet the skills training programme also had a gender transformative component, as it included training exclusively for female police officers, and this aimed to empower them and reinforce their skills. This non-mixed training sought to make the female police officers aware of their role in recording female victim statements.

Conclusion

This chapter explored an example of a Canadian-funded Indian NGO-led development project in which strategies of scaling were used effectively to create access to justice for vicitims of sexual violence in a gender transformative manner – transforming victims into survivors and making agencies of the state support and advance that transformation. The RAHAT team challenged the geometries of power that precluded women from accessing justice, and rescaled its survivor-centric model horizontally to other communities, as well as vertically to police and legal systems. Various scales and strategies of intervention were employed to bring solutions to communities in need of justice. In some cases, these approaches were transformative (working with women police officers), while in others they were accommodating, in terms of implementing changes that were not explicitly presented as gender transformative but rather as systematic operating procedures or rote bureacratic requirements of officials.

The example of RAHAT reveals how gender transformative research needs to be highly context-specific; geographical sensitivity is evident in how the RAHAT team understood the need to operate with exisiting hierarchies of government, justice and police structures.

RAHAT understood how *transformative* change could be advanced through gender *accommodating* initiatives, such as developing new standard operating procedures and embedding this change in the rote official norms of police responsibility. Assessment of the capabilities of relevant NGOs and institutions in the new scaling site was vital to the success of a scaling strategy.

The project demonstrates how feminist-inspired research, in combination with multiple state agencies, can widen the scope of gender transformation research to materially transform structures of gendered inequality. Scaling these programmes entailed working within existing systems of legislation to make them more accountable to survivors of gendered violence, thereby transforming power structures through both gender transformative and accomodating strategies.

Notes

1 This chapter is limited to the RAHAT initiative. The overall project funded by IDRC includes a second initiative, Monitoring of Hinsa (PWDV) Act in Maharashtra (MOHIM), focusing on domestic violence issues and the application and implementation of the Protection of Women from Domestic Violence Act (PWDVA). One of the major outputs of this component of the project was the publication and implementation of a handbook published by the Maharashtra government in August 2014, entitled *The Protection of Women from Domestic Violence Act, 2005: Maharashtra State Handbook on Protocols, Best Practices and Reporting Formats* (RAHAT 2015).

2 During a field visit in September 2016 by Martel, a major international NGO was evaluating the possibility of replicating the model in Pune.

3 However, they were also advocating with several strategies to sensitize police officers. A pledge by Mumbai police to treat women and children with dignity was put up at the entrance of every police station in Mumbai.

Key readings

Nagar, R. 2006. *Playing with Fire: Feminist Thought and Activism through Seven Lives in India*. Minneapolis: University of Minnesota Press.

Pratt, G., and V. Rosner. 2006. "Introduction: The Global & The Intimate." *Women's Studies Quarterly* 34: 13–24.

Walton-Roberts, M. 2010. "The Family and Field Work: Intimate Geographies and Counter Topographies." In: *Family Geographies: The Spatiality of Families and Family Life*, edited by B. Hallman, 170–183. Oxford: Oxford University Press.

References

Blunt, A. 2011. *Domicile and Diaspora: Anglo-Indian Women and the Spatial Politics of Home*. London: John Wiley & Sons.

IDRC. 2015. "A Strategy for Gender in Agriculture and Food Security at IDRC." IDRC Internal Document.

IDRC. 2016. "Project Completion Report (PCR). Sexual and Domestic Violence: Policy Protocols." IDRC Internal Document.

Hyndman, J. 2004. "Mind the Gap: Bridging Feminist and Political Geography through Geopolitics." *Political Geography* 23: 307–322.

Kabeer, N. 1994. *Reversed Realities: Gender Hierarchies in Development Thought*. London: Verso.

Katz, C. 2001. "Vagabond Capitalism and the Necessity of Social Reproduction." *Antipode* 33 (4): 709–728.

Machin, R. 2017. "The Professional and Ethical Dilemmas of the Two-child Limit for Child Tax Credit and Universal Credit." *Ethics and Social Welfare* 11 (4): 1–8.

Marston, S.A. 2000. "The Social Construction of Scale." *Progress in Human Geography* 24 (2): 219–242.

Martel, A. 2019. "Scaling Access to Justice for Survivors of Sexual Violence." In: *Scaling Impact: Innovation for the Public Good*, edited by R. McLean and J. Gargani, 112–130. Abingdon: Routledge.

Massey, D. 1993. "Power-Geometry and a Progressive Sense of Place." *Mapping the Futures: Local Cultures, Global Change* 1: 59–69.

McLean, R., and J. Gargani. 2019. *Scaling Impact: Innovation for the Public Good.* London: CRC/Routledge.

Moss, P., K.F. Al-Hindi, and H. Kawabata. 2002. *Feminist Geography in Practice: Research and Methods.* Oxford, MA: Wiley-Blackwell.

Mountz, A. 2004. "Embodying the Nation-State: Canada's Response to Human Smuggling." *Political Geography* 23 (3): 323–345.

Mountz, A., and M. Walton-Roberts. 2006. "Gender, Geography and Policy: The Conundrums of Engagement." *GeoJournal* 65 (4): 263–273.

Naqvi, F., and N. Mehta. 2015. "Walking the Legal Talk. An Evaluation of Majlis (RAHAT and Mohim Initiatives)." Project Sexual and Domestic Violence: Policy Protocols." Evaluation Report. Available at: http://majlislaw.com/file/Final_Evaluation_Report_Majlis_IDRC_Sept20_2015.pdf

Pain, R. 2004. "Social Geography: Participatory Research." *Progress in Human Geography* 28 (5): 652–663.

Pratt, G., and V. Rosner, eds. 2012. *The Global and the Intimate: Feminism in Our Time.* New York: Columbia University Press.

Pratt, G., and V. Rosner. 2006. "Introduction: The Global and the Intimate." *Women's Studies Quarterly* 34: 13–24.

RAHAT. 2015. "Pursuing This Thing Called Justice. A Survivor Centric Approach towards Victims of Sexual Violence." Final Technical Report. Majlis Legal Centre. Available at: http://ic.idrc.ca/sites/projects/project/107101%20Sexual%20and%20Domestic%20Violence%20-%20Policy%20Protocols/Majlis%20-Final%20Report.pdf.

Rottach, E., K. Hardee, R. Jolivet, and R. Kiesel. 2012. "Integrating Gender into the Scale-Up of Family Planning and Maternal, Neonatal, and Child Health Programs." Working Paper No. 1. Health Policy Project, USAID. Available at: www.healthpolicyproject.com/pubs/51_ScaleupofGenderintoFPMCHprogramsJuly.pdf.

Sangtin Writers Collective, and R. Nagar. 2006. *Playing with Fire: Feminist Thought and Activism through Seven Lives in India.* Minneapolis: University of Minnesota Press.

Singh, J.A., and S.S.A. Karim. 2017. "Trump's 'Global Gag Rule': Implications for Human Rights and Global Health." *Lancet Global Health* 5 (4): e387–e389.

Staeheli, L.A., and V.A. Lawson. 1995. "Feminism, Praxis, and Human Geography." *Geographical Analysis* 27 (4): 321–338.

Swyngedouw, E. 2004. "Globalisation or 'Glocalisation'? Networks, Territories and Rescaling." *Cambridge Review of International Affairs* 17 (1): 25–48.

Vatuk, S. 2013. "The 'Women's Court' in India: An Alternative Dispute Resolution Body for Women in Distress." *Journal of Legal Pluralism and Unofficial Law* 45 (1): 76–103.

Walton-Roberts, M. 2010. "The Family and Field Work: Intimate Geographies and Countertopographies." In: *Family Geographies: The Spatiality of Families and Family Life*, edited by B. Hallman, 170–183. Oxford: Oxford University Press.

29

MOTHERHOOD IN FEMINIST GEOGRAPHY

Current trends and themes

Kate Boyer

Introduction

Feminist geography now boasts a rich history of scholarship on issues relating to the lived practices of motherhood. This chapter offers a (necessarily selective) overview of key trends in research in feminist geography on mothering and space, focusing on recent scholarship, together with a selection of key theoretical influences in the Anglophone context. Themes that will be considered include: the political economy of mothering; motherhood, identity and difference; mothering and transnationalism; activist mothering; motherhood and embodiment; and mothering with the more-than-human.

The political economy of mothering

Contemporary understandings of motherhood in feminist geography have been strongly influenced by the work of Adrienne Rich on how motherhood is shaped and constrained by patriarchy (Rich 1995) and Sharon Hays' work on how mothers are expected to be both endlessly selfless toward their children while also being intensively engaged in the wage-labour market under capitalism ('intensive' mothering) (Hays 1998). Building on this, one of the key ways in which feminist geographers have engaged the concept of motherhood has been by looking at the gendered spatial dynamics of how wage-work and care-work relate to one another; together with women's lived experiences of mothering under patriarchy and capitalism. Scholarship in this vein has explored the structural changes under post-Fordism, which have led to much greater numbers of mothers entering the wage-labour market with the dissolution of the Fordist gender contract since the late-twentieth century (McDowell 1991), as well as the spatial confinement and isolation of mothers over the life course across different cultural contexts (Bowlby 1990; Katz 2014; Katz and Monk 1993). It has encompassed the experiences of mothers in the wage-labour market as they seek to engage in both wage-work and care-work (Perrons et al. 2006), including the impact of childcare responsibilities, such as the school run, on labour-market participation and career progression (England 1996; Schwanen et al. 2008).

This scholarship has also produced analyses of the ways in which contemporary understandings of 'good motherhood' are underwritten by neoliberal expectations that mothers be responsible for the health of their children, leading both to practices such as 'helicopter parenting', intended to mitigate risk, and feelings of shame or guilt over failing to meet impossible standards as a working mother (Holloway 1998; McDowell et al. 2005; Pain 2006). Following on in this vein, scholarship has also considered experiences of folding embodied care-work within the time/space of wage labour (Boyer and Spinney 2014) and the interface of motherhood with the sharing economy in the US (Parker and Morrow 2017).

In light of current political realities, recent scholarship is beginning to examine the experiences of economically marginalized mothers and families under austerity budgeting in the UK. Austerity has brought with it cuts and closures to a range of programmes serving low-income parents, such as Sure Start centres, together with new forms of tax (such as the bedroom tax), which hit low-income families hardest. Feminist geographers are beginning to explore how UK mothers are coping in the context of these regressive changes (Jupp 2017).

Motherhood, difference, intersectionality and identity

A second key line of scholarship within feminist geography on mothering has explored the variation in experiences of motherhood across time and space, as well as within various cultural contexts. This field of scholarship has highlighted the ways in which practices and understandings of motherhood are shaped by intersecting factors of class, race/ethnicity, age, sexual orientation, gender identity and other forms of social differentiation that structure advantage and disadvantage.[1] As this work has amply shown, experiences of motherhood are highly differentiated, both in their cultural context and in the social position/location within that culture.[2] As Patricia Collins noted in her classic essay 'Shifting the Center: Race, Class and Feminist Theorizing about Motherhood' (1994) about motherhood in the 1990s US: 'For women of color, the subjective experience of mothering/motherhood is inextricably linked to the sociocultural concern of racial ethnic communities – one does not exist without the other' (Collins 1994, 47). I would suggest that this observation holds a wider truth for the power of culture and social location to shape maternal experience (though, of course, *how* experience is shaped will depend on the particulars of a given mother's intersectional social position, and particularly how she is situated relative to geometries of social power and discrimination).

Considerations of motherhood, identity difference within geography have included exploration of the experiences (and sometimes struggles) of lesbian mothers in the UK (Gabb 2005) and Australia (Luzia 2010, 2013), including by exploring how maternal identities emerge in and through embodied spatial practice at various scales of home, neighbourhood and city (Luzia 2010). Meanwhile McDowell et al. (2005) and Holloway (1999) have called attention to the ways in which class structures how mothers are viewed by the state (with more disciplining regimes typically being trained on working-class mothers) and in which normative understandings of what constitutes 'good mothering' vary by class and neighbourhood.

The role of the internet in the formation of maternal identities has also emerged as an important line of inquiry in recent years. Through research from the UK and Hong Kong, this scholarship has noted how mothering chat rooms and online bulletin boards can constitute an important source of companionship and emotional support, providing mothers with an opportunity to express uncertainty, anger, frustration and other feelings that they might not disclose to friends offline (Chan 2008; Madge and O'Connor 2005). Though highlighting that such fora typically cater to heterosexual, tech-savvy middle-class mothers, these authors nevertheless note the value of online interactions as a means for new mothers to move between maternal and other, more familiar, identities (see also Longhurst 2008). Further to the theme of conceptual understandings of maternal identity, Gregson and Rose (2000) suggest that maternal subjectivity can usefully be understood as indeterminate and ambiguous, while Longhurst (2000) observes that within any one mother there are multiple maternal (and other kind of) identities. Extending understanding of the role of new media to the spatial practice of mothering, Longhurst has also identified the role that Skype can play in mothering at a distance (Longhurst 2013).

Mothering and transnationalism

Feminist geographers have also been instrumental in instigating a broader turn within the discipline to attend to the gendered politics, spatialities and geo-politics of care-work. In addition to exploring the phenomenon of birth 'tourism' through the case of Turkish women giving birth in the US for citizenship purposes (Balta and Altan-Olcay 2017), this scholarship has explored the complex issue of mothering at a distance in the context of women who travel from the Global South (including the Philippines and Latin America) or Eastern Europe. They are economic migrants who travel to work as nannies to look after the babies of middle-class women in the Global North (Cox 2006; Hondagneu-Sotelo and Avila 1997; Lam and Yeoh 2018; Pratt 2012). Moreover, this scholarship investigates the hardships faced by co-resident migrant mothers (Gilmartin and Migge 2016). This scholarship has raised critical issues relating to diasporic motherhood spanning the legal and economic frameworks that enable this phenomenon as well as the profound ethical issues, body politics and forms of discrimination and micro-aggressions that it entails and enables.

Activist mothering

Another important strand of scholarship in feminist geography is the work of mothers endeavouring to affect social and political change. This work has ranged from mothers' efforts to claim public space to efforts to shine light on – and change – forms of injustice against mothers and children. Building on scholarship on activist mothering from beyond geography (such as Naples 1998, 2014), this scholarship has ranged in scale from the intimate scale of the body to that of the nation state and across different cultural contexts. Scholarship in this vein has explored mother-activism in protesting against violence to women and femicide in Northern Mexico (Wright 2007) and the work of the Revolutionary Association of the Women of Afghanistan (Fluri 2008). It has shone light on mothers' efforts to challenge unjust and racist patterns of mass incarceration of African–American young people (Gilmore 1999) and mothers' activism in challenging welfare reform in the US (Gilbert 2001). Work has also taken in the activist work of socially marginalized mothers in Bolivia (Berckmans et al. 2016) and breastfeeding activism in the UK (Boyer 2011). Moving into the realm of praxis, feminist geographers have proposed mothering as a modality through which to approach their professional practice in terms of how to interact with students and colleagues (Datta and Lund 2017) and, through testimonial theatre (Pratt and Johnston 2013), engaged in participatory action research to raise consciousness about the experiences of trans-national mothering in the case of Filipina nannies working in North America.

Mothering and embodiment

Recent years have seen an increasing number of geographers engaging with feminist theory as a way to analyse the spatial politics of embodiment. This line of inquiry has produced new theorizations of motherhood as an embodied practice, drawing on the work of Elisabeth Grosz (1994, 1998, 2005); Rosi Braidotti (1994, 2002, 2003); Luce Irigaray (2004, 1985); Judith Butler

(1993); and others. Scholars working in this vein have sought to give more analytical attention to how motherhood is experienced bodily, as well as to sharpen analysis of how practice and representation relate to one another, along the lines of what feminist theoretician Hannah Stark has called the need to 'focus on the lived practices that reveal the various ways in which subjects are embodied, located and connected' (Stark 2017, 66).

This literature has considered how the pregnant body has come under increasing medical and state surveillance and oversight (Fox et al. 2009), as well as in the context of sexualization and pornography (Longhurst 2001). Clement and Waitt have posited how walking with one's baby can function as form of maternal becoming (Clement and Waitt 2017), while Louise Holt has drawn attention to the concept of *inter-embodiment* as a means to highlight the way that maternal subjectivity can emerge relationally through (often) significant physical contact with one's baby (Holt 2013). Along similar lines, Kelly Dombroski has explored the affective aspects of mother–baby relations in the form of elimination communication (Dombroski 2017).

Scholarship in this vein has, moreover, included analyses of breastfeeding as a particular form of embodied practice of motherhood. This work has considered the physical pain that can accompany breastfeeding (and sometimes feelings of shame for stopping breastfeeding before planned) (Robinson 2016); the ways that discourses of 'discretion' govern and discipline breastfeeding in public in the US (Lane 2014); the embodied experiences and affective environments that can emerge from breastfeeding outside the home in the UK (Boyer 2012); and cross-species breastfeeding (Longhurst 2001).

Mothering with the more-than-human

Building on this, the final key area of scholarship to be considered here is that which has begun to explore mothering in the context of post-humanism. This line of inquiry has emerged alongside the increasing interest that has emerged in the late-twentieth and early-twenty-first century within the discipline in Anthropocene and human–nonhuman relations. Conceptually, this work often draws on theory that seeks to conceptualize practice, emotions, affect and the nonhuman in ways that do not ignore embodied power relations (such as those of gender, race, class and sexual orientation). A body of theory that has proven particularly useful in this regard is that of the new materialism. Composed of scholarship cutting across the humanities, philosophy and the natural sciences, new materialist social theory is concerned with creating analyses that take seriously both discourse and other forms of representation as well the non-representational, including matter, affect and emotion (Alaimo and Hekman 2008; Barad 2007, 2008; Braidotti 1994, 2002, 2003; Colebrook 2008; Coole and Frost 2010; van der Tuin and Dolphijn 2012). Drawing on a long tradition of feminist scholarship, scholars working in the new materialist tradition seek to give more analytical attention to the politics of materiality and embodied practice, as well as to sharpen analysis of how practice and representation relate to one another.

New materialism also draws on feminist and Deleuzoguattarian conceptualizations of subjectivity as an ongoing process of becoming, highlighting the role of matter in those processes (Braidotti 2002). Within this approach subjects appear, after Braidotti, as: 'embodied, embedded, assembled of agentic sub-materials within; and through encounters with the material and more-than-human world' (Braidotti 2002, 62). In the tradition of Haraway (2013), Whatmore (2002), Colls and Fannin (2013) and others, new materialism seeks to destabilize firm conceptual boundaries between bodies and matter by attending to what Karan Barad terms 'agential intra-action', referring to how meanings are produced *relationally*, in and through the relations between phenomena (Barad 2007, 33). As Stark puts it, invoking a Deleuzian frame: 'the body cannot be thought of as individual, bounded or coherent because it is constituted fundamentally by the connections it enters into' (Stark 2017, 75), noting further that 'these connections do not discriminate between the human and the non-human' (Stark 2017, 75).

Within geography, scholars have considered the role of the more-than-human in parenting practice in a range of ways. These have included: mothers' use of family photos to mediate their evolving/unfolding identities (Rose 2003, 2004); the family car as a parenting tool (Dowling 2000); and the place of prams (Boyer and Spinney 2016) and 'baby things' (Waight 2014) in the embodied practices of early mothering. Lucilla Newell has usefully proposed the concept of breastfeeding *assemblages*, which include both human as well as more-than-human components (such as nursing pads, breastpumps, bottles, etc.) as a way of conceptualizing infant feeding that does not ignore the nonhuman (Newell 2013). Colls and Fannin have explored the politics of intra-corporeal matter in their work through an approach that synthesizes both feminist and non-representational approaches to analyse placentas as a mediating force between the body of the mother and that of the foetus (Colls and Fannin 2013); and, along similar lines, Boyer has explored the agentic role of breastmilk within breastfeeding assemblages (Boyer 2018).

Conclusion

While necessarily selective, this short chapter has sought to trace some of the key trends and themes that characterize current scholarship in feminist geography on motherhood. I hope to have shown how this field is both socially and politically engaged, while at the same time also engaging with (and helping to define) the cutting edge of conceptual work. At the same time, despite the scope and range of extant scholarship as cultural, political and theoretical landscapes change, there will always be myriad avenues for further work. In particular, more work is needed to understand the spaces, politics and experiences of motherhood outside the Anglosphere and the Global North. Building on extant work, I will conclude by suggesting just a few of the many areas where more scholarship might be done:

The ongoing efforts of mothers to challenge oppressions emerging from intersecting systems of racism, sexism, heterosexism and economic marginalization: Building on feminist geography's strong tradition of politically engaged and scholar/activist research, more work is needed to better understand mothers' myriad work as political actors and agents in naming and challenging discrimination, oppression and intersectional inequality across all cultural contexts.

Mothering under the pressures and challenges of neoliberalism and austerity: How have specific political–economic landscapes, as they have emerged over recent years, changed experiences of motherhood? What new challenges do mothers today face in coping under conditions of (in many places) less state support, and what strategies have they devised for coping with these challenges?

The struggles and strategies of mothers and families coping with diaspora and climate change: What does mothering look like in the context of broader forces of global warming, intensifying weather and changing patterns of water scarcity and coastline change? What does it mean to mother as a climate (or other kind of) refugee?

Technological mediation of the spaces and embodied practices of conception and motherhood: The last ten years have witnessed a massive expansion in technologies relating to conception, pregnancy and motherhood, from fertility and pregnancy apps to in vitro fertilization (IVF) and egg extraction/freezing to devices for at-home pregnancy monitoring to the myriad social media outlets in which mothers participate. How have these changes affected how motherhood is understood and experienced? In what ways have they made motherhood easier (or harder?)

More work is also needed on *the experiences of mothering (as well as fathering) trans and non-binary children:* As transgender rights and visibility increase, how can parents best support gender non-conforming children in (often hostile) wider worlds and public spaces? And finally, more work is needed on broader questions of *how the gender of care-work generally and the body- and emotion-work of parenting specifically is itself changing:* How, where and to what extent are understandings of the gender of care-work changing? What are they changing to? What new negotiations of 'who does what' in the home are emerging across various cultural contexts, and what does this mean for experiences of motherhood, fatherhood and parenthood? What are the experiences of male primary carers, and how can families challenge and break down binarized, normative scripts relating to the work of parenting?

Notes

1 It should be recognized both that the meanings of these intersecting forms of social differentiation change over time and that within any one (intersectional) social location lie infinite forms of difference. Van der Tuin and Dolphijn refer to this as the 'thousand tiny intersections' (Van der Tuin and Dolphijn, 2012, 140) that lie beneath any identity, bringing a Deleuzian lens to the concept of intersectionality. This insight serves as an important reminder of the fact that social markers such race, class, gender, and so on can never fully capture the myriad difference within any given identity.
2 Noting also that social contexts are themselves in continual states of flux.

Key readings

Boyer, K. 2018. *Spaces and Politics of Motherhood.* London: Rowman and Littlefield.
Longhurst, R. 2008. *Maternities: Gender, Bodies, and Spaces.* New York: Routledge.
Pratt, G. 2012. *Families Apart: Migrant Mothers and the Conflicts of Labour and Love.* Minneapolis: University of Minnesota Press.

References

Alaimo, S., and S.J. Hekman. 2008. *Material Feminisms.* Bloomington: Indiana University Press.
Balta, E., and Ö. Altan-Olcay. 2017. "Born in the USA: Citizenship Acquisition and Transnational Mothering in Turkey." *Gender, Place & Culture* 24 (8): 1204–1223.
Barad, K. 2007. *Meeting the Universe Halfway: Quantum Physics and the Entanglement of Matter and Meaning.* Durham, NC: Duke University Press.
Barad, K. 2008. "Posthumanist Performativity: Toward an Understanding of How Matter Comes to Matter." In: *Material Feminisms,* edited by S. Alaimo and S. Hekman, 120–156. Bloomington: Indiana University Press.
Berckmans I., M.L. Velasco, and G. Loots. 2016. "Breaking Silence: Exploring Motherhood and Social Transformation in a Participatory Action Research with Alteñan Mothers." *Gender, Place & Culture* 23 (7): 1017–1032.
Bowlby, S. 1990. "Women, Work and the Family: Control and Constraints." *Geography* 76: 17–26.
Boyer, K. 2018. "Natureculture in the Nursery: Lively Breast Milk, Vibrant Matter and the Distributed Agencies of Infant Feeding." *Spaces and Politics of Motherhood.* London: Rowman and Littlefield.
Boyer, K. 2011. "The Way to Break the Taboo is to Do the Taboo Thing" Breastfeeding in Public and Citizen-activism in the UK." *Health and Place* 17 (2): 430–437.
Boyer, K. 2012. "Affect, Corporeality and the Limits of Belonging: Breastfeeding in Public in the Contemporary UK." *Health and Place* 18 (3): 552–560.
Boyer, K. 2014. "Neoliberal Motherhood: Workplace Lactation and Changing Conceptions of Working Motherhood in the Contemporary US." *Feminist Theory* 15 (3): 269–288.
Boyer, K., and J. Spinney. 2016. "Motherhood, Mobility and Materiality: Material Entanglements, Journey-making and the Process of 'Becoming Mother'." *Environment and Planning D: Society and Space* 34 (6): 1113–1131.

Braidotti, R. 1994. *Nomadic Subjects: Embodiment and Sexual Difference in Contemporary Feminist Theory*. New York: Columbia University Press.
Braidotti, R. 2002. *Metamorphoses: Towards a Materialist Theory of Becoming*. Cambridge: Polity Press.
Braidotti, R. 2003. "Becoming Woman: Or Sexual Difference Revisited." *Theory, Culture & Society* 20 (3): 43–64.
Butler, J. 1993. *Bodies that Matter: On the Discursive Limits of Sex*. London: Routledge.
Chan, A.H.N. 2008. "Life in Happy Land: Using Virtual Space and Doing Motherhood in Hong Kong." *Gender, Place & Culture* 15 (2): 169–188.
Clement, S., and G. Waitt. 2017. "Walking, Mothering and Care: A Sensory Ethnography of Journeying on Foot with Children in Wollongong, Australia." *Gender, Place & Culture* 24 (8): 1185–1203.
Colebrook, C. 2008. "On Not Becoming Man: The Materialist Politics of Unactualized Potential." In: *Material Feminisms*, edited by S. Alaimo and S.J. Hekman, 52–84. Bloomington: Indiana University Press.
Collins, P.H. 1994. "Shifting the Center: Race, Class, and Feminist Theorizing about Motherhood." In: *Mothering: Ideology, Experience, and Agency*, edited by E. N. Glenn, G. Chang ,and L. R. Forcey, 45–65. London: Routledge.
Colls, R., and M. Fannin. 2013. "Placental Surfaces and the Geographies of Bodily Interiors." *Environment and Planning A* 45 (5): 1087–1104.
Coole, D., and S. Frost, eds. 2010. *New Materialisms: Ontology, Agency, and Politics.* Durham, NC: Duke University Press.
Cox, R. 2006. *The Servant Problem: The Home Life of a Global Economy*. London: IB Tauris.
Datta, A., and R. Lund. 2018. "Mothering, Mentoring and Journeys Towards Inspiring Spaces." *Emotion, Space and Society* 26: 64–71.
Dombroski, K. 2018. "Learning to be Affected: Maternal Connection, Intuition and 'Elimination Communication'." *Emotion, Space and Society* 26: 72–79.
Dowling, R. 2000. "Cultures of Mothering and Car Use in Suburban Sydney: A Preliminary Investigation." *Geoforum* 31 (3): 345–353.
England, K. 1996. *Who Will Mind the Baby?: Geographies of Child Care and Working Mothers.* London: Psychology Press.
Fluri, J.L. 2008. "Feminist-nation Building in Afghanistan: An Examination of the Revolutionary Association of the Women of Afghanistan (RAWA)." *Feminist Review* 89 (1): 34–54.
Fox, R., K. Heffernan, and P. Nicolson. 2009. "'I Don't Think It Was Such an Issue Back Then': Changing Experiences of Pregnancy Across Two Generations of Women in South-east England." *Gender, Place and Culture* 16 (5): 553–568.
Gabb, J. 2005. "Locating Lesbian Parent Families: Everyday Negotiations of Lesbian Motherhood in Britain." *Gender, Place & Culture* 12 (4): 419–432.
Gilbert, M.R. 2001. "From the 'Walk for Adequate Welfare' to the 'March for Our Lives': Welfare Rights Organizing in the 1960s and 1990s." *Urban Geography* 22 (5): 440–456.
Gilmartin, M., and B. Migge. 2016. "Migrant Mothers and the Geographies of Belonging." *Gender, Place & Culture* 23 (2): 147–161.
Gilmore, R.W. 1999. "You Have Dislodged a Boulder: Mothers and Prisoners in the Post-Keynesian California Landscape." *Transforming Anthropology* 8 (1–2): 12–38.
Gregson, N., and G. Rose. 2000. "Taking Butler Elsewhere: Performativities, Spatialities and Subjectivities." *Environment and Planning D: Society and Space* 18 (4): 433–452.
Grosz, E. 1994. *Volatile Bodies: Towards A Corporeal Feminism*. Bloomington: Indiana University Press.
Grosz, E. 1998. "Bodies-Cities." In: *Places Through the Body*, edited by S. Pile and H. Nast, 42–51. London: Routledge.
Grosz, E. 2005. *Time Travels: Feminism, Nature, Power*. Durham, NC: Duke University Press.
Haraway, D. 2013. *Simians, Cyborgs, and Women: The Reinvention of Nature*. New York: Routledge.
Hays, S. 1998. *The Cultural Contradictions of Motherhood*. London: Yale University Press.
Holloway, S. 1998. "Local Childcare Cultures: Moral Geographies of Mothering and the Social Organisation of Pre-school Education." *Gender, Place and Culture: A Journal of Feminist Geography* 5 (1): 29–53.
Holloway, S. 1999. "Mother and Worker?: The Negotiation of Motherhood and Paid Employment in Two Urban Neighbourhoods." *Urban Geography* 20 (5): 438–460.
Holt, L. 2013. "Exploring the Emergence of the Subject in Power: Infant Geographies." *Environment and Planning D: Society and Space* 31 (4): 645–663.
Hondagneu-Sotelo, P., and E. Avila. 1997. "'I'm Here, But I'm There': The Meanings of Latina Transnational Motherhood." *Gender & Society* 11 (5): 548–571.

Irigaray, L. 1985. *Speculum of the Other Woman*. Ithica, NY: Cornell University Press.
Irigaray, L. 2004. *Way of Love*. London: A&C Black.
Jupp, E. 2017. "Families, Policies and Place in Times of Austerity." *Area* 49 (3): 266–272.
Katz, C. 2014. *Full Circles: Geographies of Women over the Life Course*. London: Routledge.
Katz, C., and J. Monk, eds. 1993. *Full Circles: Geographies of Women over the Life Course*, London: Routledge.
Lam, T., and B. Yeoh. 2018. "Migrant Mothers, Left-behind Fathers: The Negotiation of Gender Subjectivities in Indonesia and the Philippines." *Gender, Place & Culture* 25 (1): 104–117.
Longhurst, R. 2000. "Corporeographies of Pregnancy: 'Bikini Babes'." *Environment and Planning D: Society and Space* 18 (4): 453–472.
Longhurst, R. 2001. *Bodies: Exploring Fluid Boundaries*. New York: Routledge.
Longhurst, R. 2008. *Maternities: Gender, Bodies, and Spaces*. New York: Routledge.
Longhurst, R. 2013. "Using Skype to Mother: Bodies, Emotions, Visuality, and Screens." *Environment and Planning D: Society and Space* 31 (4): 664–679.
Lane, R. 2014. "Healthy Discretion? Breastfeeding and the Mutual Maintenance of Motherhood and Public Space." *Gender, Place & Culture* 21 (2): 195–210.
Luzia, K. 2010. "Travelling in Your Backyard: The Unfamiliar Places of Parenting." *Social & Cultural Geography* 11 (4): 359–375.
Luzia, K. 2013. "'Beautiful but Tough Terrain': The Uneasy Geographies of Same-sex Parenting." *Children's Geographies* 11 (2): 243–255.
Madge, C., and H. O'Connor. 2005. "Mothers in the Making? Exploring Liminality in Cyber/space." *Transactions of the Institute of British Geographers* 30 (1): 83–97.
McDowell, L. 1991. "Life Without Father and Ford: The New Gender Order of Post-Fordism." *Transactions of the Institute of British Geographers* 16 (4): 400–419.
McDowell, L., K. Ray, D. Perrons, C. Fagan, and K. Ward. 2005. Women's Paid Work and Moral Economies of Care." *Social & Cultural Geography* 6 (2): 219–235.
Naples, N. 1998. *Community Activism and Feminist Politics*. New York: Routledge.
Naples, N. 2014. *Grassroots Warriors: Activist Mothering, Community Work, and the War on Poverty*. New York: Routledge.
Newell, L. 2013. "Disentangling the Politics of Breastfeeding." *Children's Geographies* 11 (2): 256–261.
Pain, R. 2006. "Paranoid Parenting? Rematerializing Risk and fear for Children." *Social & Cultural Geography* 7 (2): 221–243.
Parker, B., and O. Morrow. 2017. "Urban Homesteading and Intensive Mothering: (Re) Gendering Care and Environmental Responsibility in Boston and Chicago." *Gender, Place & Culture* 24 (2): 247–259.
Perrons, D., C. Fagan, L. McDowell, K. Ray, and K. Ward. 2006. *Gender Divisions and Working Time in the New Economy*. Cheltenham: Edward Elgar.
Pratt, G. 2012. *Families Apart: Migrant Mothers and the Conflicts of Labor and Love*. Minneapolis: University of Minnesota Press.
Pratt, G., and C. Johnston. 2013. "Staging Testimony in Nanay." *Geographical Review* 103 (2): 288–303.
Rich, A. 1995. *Of Woman Born: Motherhood as Experience and Institution*. London: WW Norton.
Robinson, C. 2016. "Misshapen Motherhood: Placing Breastfeeding Distress." *Emotion Space and Society*. ePub ahead of print: https://doi.org/10.1016/j.emospa.2016.09.008.
Rose, G. 2003. "Family Photographs and Domestic Spacings: A Case Study." *Transactions of the Institute of British Geographers* 28 (1): 5–18.
Rose, G. 2004. "'Everyone's Cuddled Up and It Just Looks Really Nice': An Emotional Geography of Some Mums and Their Family Photos." *Social & Cultural Geography* 5 (4): 549–564.
Schwanen, T., M. Kwan, and F. Ren. 2008. "How Fixed is Fixed? Gender Rigidity of Space-time Constraints and Geographies of Everyday Activities." *Geoforum* 39 (6): 2109–2121.
Stark, H. 2017. *Feminist Theory After Deleuze*. London: Bloomsbury.
Van der Tuin, I., and R. Dolphijn. 2012. *New Materialism: Interviews & Cartographies*. London: Open Humanities Press.
Waight, E. 2014. "Second-hand Consumption Among Middle-class Mothers in the UK: Thrift, Distinction and Risk." *Families, Relationships and Societies* 3 (1): 159–162.
Whatmore, S. 2002. *Hybrid Geographies: Natures Cultures Spaces*. London: Sage.
Wright, M.W. 2007. "Urban Geography Plenary Lecture – Femicide, Mother-Activism, and the Geography of Protest in Northern Mexico." *Urban Geography* 28 (5): 401–425.

30

EMBODIED LABOUR IN THE BIOECONOMY

Maria Fannin

Introduction

This chapter focuses on feminist work on the political economy of the biosciences. New technologies aimed at extending, altering, isolating or accumulating the body's living vitality are reworking traditional economic concepts such as labour, work and value. Drawing on an empirical investigation of surrogacy, tissue donation and other forms of embodied participation in the bioeconomy, this chapter seeks to extend and deepen feminist geographers' analysis of new forms of embodied labour in what has been termed the 'bioeconomy'.

In policy parlance, the bioeconomy refers to the 'set of economic activities relating to the invention, development, production and use of biological products and processes' and 'a world where biotechnology contributes to a significant share of economic output' (OECD 2009). This policy agenda is grounded in the view that biological entities possess a latent and untapped value, to be extracted and transformed as engines of future economic growth and development (Birch 2012; Waldby 2002). The late-twentieth-century emergence of a bioeconomy has moved from a 'niche interest to political mainstream' as national governments and private investors seek to capitalize on the potential of biotechnology to develop alternatives to fossil fuels and innovate in the sectors of food and materials production (OECD 2018, 11). Although recent public policy interventions concentrate on the industrial biomass and biofuel sector (see OECD 2018), the bioeconomy sector's concern with technological innovation in the life sciences and medicine more generally provides the background for this chapter's examination of embodied labour.

From the bioeconomy perspective, life itself is a potential untapped natural resource to be explored and exploited. Biological entities can be disaggregated into their constituent parts and made amenable to manipulation and transformation to generate new products and services. Technologies aimed at modifying forms of life are said to offer solutions to some of the globe's most intractable problems, from genetically modifying wheat strains to be more resilient to climate change to harnessing the digestive processes of microbes to generate alternatives to fossil fuel. As this chapter will explore, these processes involve tapping into the latent potential for growth and development inside the cells of living beings, marking a new phase in the capitalist search for new resources and new markets. In this new phase of what one critical commentator calls 'biocapital':

> life itself has been made amenable in these new economic relations, as vitality is decomposed into a series of distinct and discrete objects – that can be isolated, delimited, stored, accumulated, accorded a discrete value, traded across time, space, species, contexts, enterprises – in the service of many distinct objectives.
>
> *Rose 2006, 6–7*

Geographical analysis of the political economy of this new biology, in which technologies such as gene editing and other means of manipulating and engineering organisms demonstrate the mutability and indeterminacy of biological life, emphasizes the speculative and financialized dimensions of biocapital (Birch 2016; Rajan 2006). In this literature, the bioeconomy is underwritten by two imperatives: capitalist processes of profit-making through market-based and speculative forms of investment; and biopolitical imperatives for health and the promotion of bodily well-being, from the personal(ized) to the population.

Economic geographers have studied the bioeconomy through the workings of the biotechnology sector for biofuels, pharmaceuticals and the agro-food industry (e.g. Calvert et al. 2017; Horner 2014; McDonagh 2015). The hype surrounding biotechnology in the late-twentieth century spurred a range of studies interested in assessing the sector's capacities for growth, its institutional dynamics and legal infrastructures, and the blurring of public and private finance that characterized research and development in the sector. The following discussion explores some of the key debates and themes in alternative approaches to the bioeconomy informed by feminist and post-colonial perspectives. Feminist contributions to analysing this new terrain include closer examination of how the health and vitality of some has been made fungible or commodifiable in order to benefit others. This chapter situates the discussion of these phenomena in relationship to what feminist economic geographers and others have identified as patterns of 'stratified labour', putting this conceptual frame into conversation with empirical work on surrogacy, human tissue vending and participation in clinical trials.

One of the significant contributions of feminist scholars to theories of labour in the bioeconomy is the development of the concept of clinical labour (Cooper and Waldby 2014). Like other geographical explorations of dimensions of new forms of labour (see for example, work on affective labour), clinical labour has joined feminist economic geographers' repertoire of critical concepts to situate the harnessing of vitality for another's use or benefit. In this chapter, I consider how work in clinical labour is characterized by compliance to treatment regimes, the provision of access to one's bodily (cellular, molecular, metabolic) processes of growth or development, as well as by its marginal status in relationship to other forms of labour in the bioeconomy. Clinical labour is, as Catherine Waldby and Melinda Cooper (2008, 24) write, 'labour that is "peripheral in terms of rights, but central in terms of the … value produced"'. Although the concept of clinical labour might seem abstract, I argue that it is a powerful way to capture the emergent and unorganized forms of labour central to the bioeconomy. Feminist geographers and others are increasingly engaging with these and other notions of labour that trouble notions of agency in the bioeconomy, offering new directions for research on clinical and other forms of embodied labour.

Clinical labour

Feminist geographical work on gendered labour has elaborated on the concept of stratified labour to critically examine the dynamics of how labour is differentially valued, based on workers' migration status, gender, race or other axes of difference (Batnitzky and McDowell 2011). Geographical work on the economies of domestic and caring labour in cosmopolitan cities such as New York and Vancouver also considers how such stratification stretches across space (see England 2015; Pratt and Rosner 2012). For example, feminist geographical work considers how women who migrate to the US or Canada carry out caring tasks for others, working outside the home, often leaving their own children to be cared for by relatives. These geographical mappings of global migrant trajectories provide fruitful insights into the dependencies and vulnerabilities of those who perform the gendered labour of caring for others.

One recognized lacuna in feminist scholarship on the political economy of gendered labour and feminist work on the social reproductive sphere of economic activity is work on the body and the 'materiality' of bodily processes (see Meehan and Strauss 2015). The reference to materiality in this context opens up research on social reproduction to the recent writings of feminist scholars who seek to bring the materiality of the body to the fore, interrogating how bodily processes and other dimensions of lively matter can be fruitful and generative sites for new modes of theory and politics. This new materialist feminism seeks to return feminist analysis to the problem posed by the fleshy and differentiated body, and refuses the notion of the body and bodily processes as solely socially constructed (see Colls 2012; Mansfield and Guthman 2015).

These concerns are relevant for new conceptualizations of embodied labour when practices such as surrogacy, tissue donation and participation in clinical trials continue to be ambivalently constituted *as* labour. Recent feminist scholarship on these practices suggests that labour is a more apt concept for analysing who and what produces value in the bioeconomy, distinguishing concerns over new forms of labour in the bioeconomy from the extensive bioethical and legal debates over property in the body that informed feminist responses to reproductive technologies in the 1980s and 1990s. Property-oriented critiques of surrogacy or the use of bodily tissues arising from fertility therapy, pregnancy and birth (e.g. eggs, cord blood, etc.) centred on the problem – and the potential – of making property claims to the body. These studies highlighted how specific forms of property were granted legal status in ways that property in the body was and is not. Intellectual property is the primary means by which to generate surplus in the knowledge economy and the benefits of patents on new technologies, such as stem cell lines, accrue to those who developed the technology, with the contributors of tissue rendered invisible and denied any property claims to the resulting materials derived from their biological materials (Dickenson 2007). However, the call to assert property claims of ownership or stewardship over one's body or body parts was deemed by some sympathetic critics to simply mirror intellectual property claims that reward notions of private ownership, rather than to escape them.

More recently, feminist critiques of the production of value in the bioeconomy have centred attention on theorizing women's bodily participation – as egg donors, surrogates and contributors of biological materials – as a form of labour. One of the most powerful concepts emerging from this scholarship on the gendered dimensions of the bioeconomy is the notion of clinical labour, offering a way to explore the dynamics of bodily participation in the bioeconomy that is obscured by the emphasis on an individual's claims to a right to property in the body.

Clinical labour as a concept brings to the fore an analysis of the conditions of work for surrogates, experimental subjects and tissue donors. Catherine Waldby and Melinda Cooper (2008) view the emergence of clinical labour as a reconfiguration of the role of reproductive labour in the First World welfare or social state, characterized by a male breadwinner receiving a family wage and supporting a full-time, stay-at-home mother (at least for the middle classes). The decline of state-funded social welfare, the privatization of industry and the deregulation and depression of wages have brought a slow end to this arrangement. The latter half of the twentieth and now the twenty-first century is characterized by an altogether different arrangement, which Waldby and Cooper refer to as the 'post-Fordist competition state', characterized by two-wage families, rising costs of housing and healthcare, the emergence of competition between localities, regions and nation states for investment and the growth of finance capital. This shift, from a post-war welfare state to a neoliberal competition state, is also congruent with the transformation from nation state-focused policies aimed at governing natality through prohibitions on contraception and abortion and the creation of social welfare programmes supporting maternity (e.g. benefits accruing directly to the mother of children, state-funded childcare, and so on), to a loosening of restrictions on birth control and abortion and the articulation of reproduction as an individual 'choice'.

These shifts, from post-war state-centred biopolitics underpinned by pro-natalist policies supporting women's childbearing and national health and social support systems to the late-twentieth-century emphasis on individual responsibility, entailed a whole set of changes to the landscape of reproduction:

> These changes dramatically increase the economic and emotional costs of reproduction, and lead women, especially middle-class women, to delay childbearing or avoid it altogether … It is evident, then, that one of the unintended consequences of neoliberalism has been the state's loss of traction over female reproductive biology and its disengagement from nation-building projects.
>
> *Waldby and Cooper 2008, 58*

The contemporary landscape of reproduction entails new forms of reproductive biopolitics in which 'the processes of reproduction have been deregulated, privatized, and made available for investment and speculative development' (ibid.). These sites of speculation and investment include commercial provision of in vitro fertilization (IVF) services, the emergence of global reproductive tourism and the tissue-intensive fields of regenerative medicine.

In the context of embodied labour in the bioeconomy, clinical labour entails 'a direct, often highly experimental, involvement of the body's biology in the creation of surplus value' (Waldby and Cooper 2008, 65). Historically aligned to other forms of feminized, sexualized or socially reproductive work, which has often been performed by women, clinical labour confounds analyses that take the industrial worker as their model. This new form of embodied labour entails compliance with often-complex medical regimes and aspects of self-monitoring, granting access to one's bodily vitality.

Clinical labour encompasses the work of surrogacy and forms of reproductive labour such as oocyte or egg donation, including new forms of embodied labour implicated in the tissue economies surrounding contemporary life sciences research and their reliance on parts of the body (organs, tissues, cells) as raw material for further therapeutic or research use. It includes these emergent forms of reproductive labour, as well as forms of experimental labour exemplified by participation in clinical trials. Clinical labour is geographically uneven and, as processes like reproduction become open to transnational flows of money and biological materials, clinical labour becomes globally stratified. For example, in the work of Carolin Schurr (2017) on transnational markets for surrogacy, mobile surrogacy service providers seek out destinations in the Global South that will accommodate a clientele composed primarily of relatively privileged consumers from the Global North. Echoing the ethos of the 'footloose' transnational corporate sector, surrogacy agencies seek out destinations where light-touch regulation and a pool of potential surrogates with low labour costs come together to enable individuals and couples to contract surrogacy services (see Map 30.1).

Schurr's work highlights how the transnationalization of reproduction and the racial hierarchies embedded in Mexico's post-colonial present meet the desires of (White) clients from the North for children, producing the gestational surrogate who contributes no genetic material

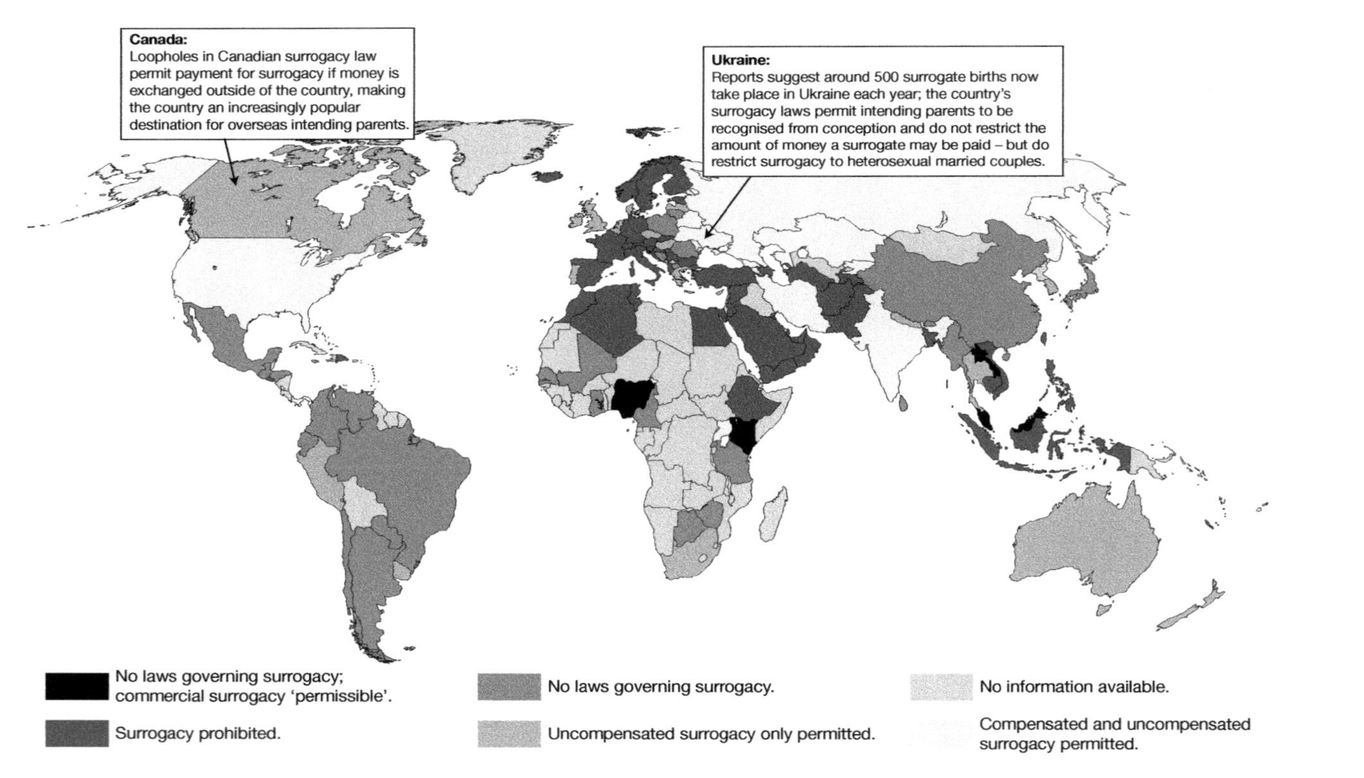

Map 30.1 Surrogacy laws by country.

Note: Surrogacy laws and surrogacy practice vary significantly from state to state. Recent media reports suggest that two new hotspots, Canada and the Ukraine, have emerged as destinations for overseas parents seeking either lower-cost surrogacy services or avoidance of restrictions on surrogacy in their home countries.

Sources: International Human Rights: Policy Advocacy Clinic, Cornell University Law School with National Law University-Delhi, 2017, 'Should Compensated Surrogacy Be Permitted or Prohibited? Policy Report Evaluating the New York Child-Parent Security Act of 2017 that Would Permit Enforceable and Compensated Surrogacy' at https://scholarship.law.cornell.edu/facpub/1551/ (accessed 12 July 2019); A. Motluk (5 October 2018), 'How Canada became an international surrogacy destination', *The Globe and Mail*, at www.theglobeandmail.com/opinion/article-how-canada-became-an-international-surrogacy-destination/ (accessed 18 February 2019); K. Ponniah (13 February 2018), 'In Search of Surrogates, Foreign Couples Descend on Ukraine', *BBC News* at: www.bbc.co.uk/news/world-europe-42845602 (accessed 13 February 2019).

of her own to conception, as a de-personalized 'womb' for the gestation and birth of a highly desired child. Schurr writes how 'the everyday practices of those who select egg donors … are deeply shared by (post)-colonial ideas of white desirability' (2017, 252). Egg donors and surrogates are stratified and racially marked by their capacities to fulfil clients' desires for White children. In the complex racial hierarchies of Mexican gestational surrogacy, egg donors are selected according to specific aesthetic criteria that privilege Whiteness, while non-White surrogates perform the relatively invisible labour of gestation and birth.

These stratifications are often left out from popular representations of commercial surrogacy, in which surrogates are positioned as rational actors making calculated choices to benefit their families, or as altruistic agents or mutual beneficiaries of transnational circuits of care, or alternatively as hyper-exploited victims of overseas consumers' desires for a child. In this context, the 'larger picture of uneven globalization in terms of international division of labour is frequently obscured or left unnoticed and undiscussed' (Lau 2018, 672). These global divisions of labour stratify surrogates as clinical labourers whose conditions of work depend on the myriad dynamics at play in particular places and times: the regulatory environment for surrogacy and its accessibility as a market to overseas clients; the particularities of indebtedness in which surrogacy becomes a means to pay off debts rather than a vehicle for class mobility; deeply entrenched notions of genetic parenthood that make surrogacy desirable; the relationship to practices and laws surrounding adoption; the intimate and affective geographies of gestation that shape surrogates' articulation of emotional attachment or detachment to their pregnancies; and the role of payment as either compensation or wage, permitting the framing of surrogacy to be bound by contract or by notions of giving (Schurr 2018; see also Parry 2015b). Like the kidney vendors in anthropologist Lawrence Cohen's ethnography of Indian transplant markets, selling and giving parts of one's body point to a complex terrain of moral hierarchies, medical technique, market structures and 'disciplinary agencies fashioning particular classes of persons' (Cohen 2003, 686).

All of these dynamics at play in the practice of surrogacy are opened up by the proposal that surrogacy is a form of labour and thus amenable to economic analysis. However, as Schurr writes, such forms of labour have only recently been given attention by economic geographers (or, I would add, by labour geographers), despite the centrality of surrogacy to the new biopolitics and economics of transnational reproduction. Geographically sensitive research on surrogacy as a form of labour highlights how the on-the-ground context of surrogacy in relation to other forms of labour and broader geopolitical dynamics matters greatly: the emergence of surrogacy 'hotspots' around the globe signals how specific locales become concentrated sites for reproductive tourism and nodes for the flows of donor gametes, surrogates and intending parents (Parry 2015b). These global dynamics invite more empirical study of the actual practices of surrogacy, building on the recent and significant contributions of geographers to this field (Bhattacharjee 2018; Lau 2018; Schurr and Militz 2018).

These dynamics also invite reflection on the conceptual frameworks through which surrogacy is framed as labour in ethical and legal debates. In Sophie Lewis' (2018, 2019) work on surrogacy, the significance of surrogate labour is expanded to encompass what Lewis describes as the paradigmatic experience of all reproductive labour. Lewis (2019, 26) explores surrogacy's affirmative potential to transform genetic notions of the family by exploding the presumptions of who children belong to, arguing for a radically expanded notion of care 'based on comradeship, a world sustained by kith and kind more than kin'.

Clinical labour also encompasses the work of tissue donors to provide material resources for therapy and research in the bioeconomy. Human tissue donation itself is a gendered practice, marked by presumptions about the value of donated tissues and by expectations and norms surrounding the motivations of donors and of recipients (Kent et al. 2018). In Rene Almeling's (2011) work on gamete markets in the US, eggs and sperm are conceived as distinctive kinds of tissues, not only because of differences in the ease of procuring these tissues but because egg and sperm donors are also perceived to have different and distinctively gendered motivations for donation and different expectations for the eventual use of their tissues. Egg vendors are viewed as desiring a connection to the intended recipients of their tissues and as receptive to the sentiment that their material contributions would create a family; by contrast, sperm vendors are presumed to be motivated by pecuniary interest in payment and to have little desire for emotional or affective connection to the recipient of their sperm.

Almeling's research illustrates that tissue donation is a gendered practice in which social hierarchies are reinforced and reproduced. Similarly, Bronwyn Parry's work on sperm vendors and the sperm export market in the US demonstrates that the high demand for American sperm is the effect of longer practices of qualifying such material through detailed profiling of donor characteristics imagined to enable the reproduction of 'elite' attributes carried through sperm itself (Parry 2015a). The designation of 'quality' also shapes the economy for human tissues as research materials, as Juliane Collard (2018) demonstrates in her study of the production of research embryos in Californian IVF clinics. Embryos designated 'abnormal' and therefore unavailable for IVF are repurposed as research materials, often gaining value in the process.

These forms of donation or sale of tissues are often not recognized as work, especially when unremunerated. In the UK, for example, receiving payment for eggs or sperm is not permitted under the Human Fertilisation and Embryology Authority regulations. Egg and sperm donors are permitted, however, to receive compensation for their costs of up to £750 per round of donation. Those undergoing or seeking IVF treatment may be enticed to donate 'surplus' materials from their own IVF treatments in order to receive discounted treatment. The reception of these incentives to exchange bodily materials for discounted treatment varies, with IVF clients in the UK much less wary of this mechanism of inducement than clients in Australia (Haimes et al. 2012; Roberts and Throsby 2008; Waldby and Carroll 2012).

Donation and sale of other tissues are also invested with gendered and racialized presumptions about the value of donated materials. Research on the promotional campaigns organized by the national blood service in the UK signals how racial categories and classifications are used to simultaneously present blood as being raced, as well as scarce and therefore valuable. These campaigns position Black, Asian and ethnic minority 'communities' as lacking awareness of this value, in turn deeming them collectively responsible for addressing the scarcity of particular blood types in the donation economy (Kierans and Cooper 2011).

Finally, clinical labour encompasses the experimental labour involved in participation in clinical drug trials, a now globalized practice in which drug testing for products aimed at markets in the Global North – and particularly the US market, which accounts for almost half of the pharmaceuticals consumed worldwide – is increasingly outsourced to the Global South (EFPIA 2017). The outsourcing of clinical trials to venues outside the largest pharmaceutical markets responds to the desire on the part of drug companies to reduce the labour costs of drug production. Little geographical research to date has focused on these forms of embodied labour, despite the important ways in which trial participation, as a form of temporally intensive but short-term, sub-contracted labour, corresponds to the growth in precarious forms of work in other sectors of the economy (see Strauss 2018).

To summarize, biomedical or clinical labour goes largely unrecognized in geographical analyses of emergent forms of contemporary labour. The work of the clinical labourer is to provide access to one's *in vivo* or *in vitro* biology. Such access may involve compliance with treatment and testing regimes or other forms of medical discipline related to diet, behaviour and other patterns of consumption. While most analysis of the bioeconomy foregrounds the role of the knowledge economy in shaping the biotech sector, and often frames discussion of intellectual property rights as *the* key problem in the bioeconomy, this chapter contends that none of these activities are possible without access to bodies and tissues. Clinical labour is enrolled in emergent forms of 'primitive accumulation', characterized by geographical unevenness and even extreme forms of bodily indebtedness. As a concept, it allows discussions of the performance of surrogacy, donation and trial participation to avoid romanticizing participants as passive victims or reproducing the narrative of the selfless and generous donor. Rather, clinical labour suggests that the dynamics of the non-clinical world of labour are also potentially relevant to emergent forms of embodied participation in reproductive, experimental and donation economies.

New directions for research on embodied labour in the bioeconomy

This chapter has reviewed the state of research to date by feminist geographers and other spatially attuned scholars in the social sciences and humanities. Scholarship by feminist geographers on the bioeconomy is still an emerging field in economic geography and the geographies of science and technology. Yet, the research to date demonstrates how emergent and evolving forms of embodied labour are critical to the bioeconomy and to new forms of biopolitics. Like other capitalist formations, the bioeconomy is characterized by uneven development and thus raises questions about the inequitable distribution of risks and benefits from new technologies, particularly for those who make bodily contributions to the services and commodities produced by these technologies. Where do the benefits of these technologies accrue in the context of speculative research programmes, for-profit drug development efforts and privatized healthcare systems? Geographical analysis of neoliberal economies and state-centred biopolitics reveals much about the general tendencies of these processes but less about the differentiated and differentiating nature of their effects on reproductive technologies, drug testing and tissue donation – to name just a few. These are just some of the areas that need further research.

To echo Carolyn Schurr's (2018) call for more research on reproductive economic geographies, this chapter would add a call for studies of the many different new forms of embodied labour, including reproductive labour, that involve rendering the body's vitality open to processes of value creation. This could encompass user-generated health data, for example the collection of information from smart technologies that monitor vital statistics or from social media and apps that use algorithms to scan users' posts for signs of mental ill-health.

It remains to be seen whether those who perform clinical labour will organize as an emerging class of workers. Will the possibilities for collective organization embedded in the concept of labour improve the working conditions of those performing such labour or result in the more equitable sharing of benefits from tissue donation, for example, with those who donate? More detailed, context-rich accounts of the actual existing practices of clinical labour are needed to discern how clinical labour practices – and the affirmative potentials of such labour as a collective form of identity – vary in situ (see Parry 2015b). This is precisely the kind of work that geographers are well placed to do. Geographical approaches can do more than track and map the sites of these transformations, although that is also important work; geographers can also engage in more spatially nuanced theorization and gather the rich empirical evidence needed to study these emergent forms of embodied labour.

In addition, studies of embodied human labour could engage in more productive cross-cutting conversations with animal geographers on the processes by which the living biology of humans *and* non-humans becomes incorporated into capitalist (and other) economies. Indeed, political ecologists and labour geographers alike could begin to consider the ways in which concepts like 'metabolic' or 'nonhuman' labour also encompass many of the characteristics of new forms of human labour in the bioeconomy (Barua 2017; Beldo 2017). All of these new means of monetizing and generating value from the body's living vitality require careful empirical investigation and the refinement of critical concepts, such as clinical labour, in order to assess their significant yet often under-recognized role in the bioeconomy.

Key readings

Cooper, M., and C. Waldby. 2014. *Clinical Labor: Tissue Donors and Research Subjects in the Global Bioeconomy.* Durham, NC: Duke University Press.

Lau, L. 2018. "A Postcolonial Framing of Indian Commercial Surrogacy: Issues, Representations, and Orientalisms." *Gender, Place & Culture* 25 (5): 666–685. doi:10.1080/0966369X.2018.1471047.

Schurr, C. 2017. "From Biopolitics to Bioeconomies: The ART of (Re-)producing White Futures in Mexico's Surrogacy Market." *Environment and Planning D: Society and Space* 35 (2): 241–262.

References

Almeling, R. 2011. *Sex Cells: The Medical Market for Eggs and Sperm.* Berkeley: University of California Press.

Barua, M. 2017. "Nonhuman Labour, Encounter Value, Spectacular Accumulation: The Geographies of a Lively Commodity." *Transactions of the Institute of British Geographers* 42 (2): 274–288.

Batnitzky, A., and L. McDowell. 2011. "Migration, Nursing, Institutional Discrimination and Emotional/Affective Labour: Ethnicity and Labour Stratification in the UK National Health Service." *Social & Cultural Geography* 12 (2): 181–201.

Beldo, L. 2017. "Metabolic Labor: Broiler Chickens and the Exploitation of Vitality." *Environmental Humanities* 9 (1): 108–128.

Bhattacharjee, D. 2018. "'It is a Jail Which Does Not Let Us Be...:' Negotiating Spaces of Commercial Surrogacy by Reproductive Labourers in India." In: *Reproductive Geographies*, edited by M.R. England, M. Fannin, and H. Hazen, 169–194. London: Routledge.

Birch, K. 2012. "Knowledge, Place, and Power: Geographies of Value in the Bioeconomy." *New Genetics and Society* 31 (2): 183–201.

Birch, K. 2016. "Rethinking Value in the Bio-economy: Finance, Assetization, and the Management of Value." *Science, Technology, & Human Values* 42 (3): 460–490.

Calvert, K.E., P. Kedron, J. Baka, and K. Birch. 2017. "Geographical Perspectives on Sociotechnical Transitions and Emerging Bio-economies: Introduction to a Special Issue." *Technology Analysis & Strategic Management* 29 (5): 477–485. doi: 10.1080/09537325.2017.1300643.

Cohen, L. 2003. "Where It Hurts: Indian Material for an Ethics of Organ Transplantation." *Zygon* 38 (3): 663–688.

Collard, J. 2018 "Biological Reproduction, Respatialised: Conceiving Abnormality in a Biotech Age." In: *Reproductive Geographies*, edited by M.R. England, M. Fannin, and H. Hazen, 50–80. London: Routledge.

Colls, R. 2012. "Feminism, Bodily Difference and Non-representational Geographies." *Transactions of the Institute of British Geographers* 37 (3): 430–445.

Cooper. M., and C. Waldby. 2014. *Clinical Labor: Tissue Donors and Research Subjects in the Global Bioeconomy.* Durham, NC: Duke University Press.

Dickenson, D. 2007. *Property in the Body: Feminist Perspectives.* Cambridge: Cambridge University Press.

EFPIA (European Federation of Pharmaceutical Industries and Associations). 2017. *The Pharmaceutical Industry in Figures: Key Data 2017.* Brussels: EFPIA.

England, K. 2015. "Nurses across Borders: Global Migration of Registered Nurses to the US." *Gender, Place & Culture* 22 (1): 143–156. doi:10.1080/0966369X.2013.832658.

Haimes, E., K. Taylor, and I. Turkmendag. 2012. "Eggs, Ethics and Exploitation? Investigating Women's Experiences of an Egg Sharing Scheme." *Sociology of Health & Illness* 34 (8): 1199–1214.

Horner, R. 2014. "Strategic Decoupling, Recoupling and Global Production Networks: India's Pharmaceutical Industry." *Journal of Economic Geography* 14 (6): 1117–1140.

Kent, J., M. Fannin, and S. Dowling. 2018. "Gender Dynamics in the Donation Field: Human Tissue Donation for Research, Therapy and Feeding." *Sociology of Health and Illness.* doi:10.1111/1467-9566.12803.

Kierans, C., and J. Cooper. 2011. "Organ Donation, Genetics, Race and Culture: The Making of a Medical Problem." *Anthropology Today* 27 (6): 11–14.

Lau, L. 2018. "A Postcolonial Framing of Indian Commercial Surrogacy: Issues, Representations, and Orientalisms." *Gender, Place & Culture* 25 (5): 666–685. doi: 10.1080/0966369X.2018.1471047.

Lewis, S. 2018. "International Solidarity in Reproductive Justice: Surrogacy and Gender-inclusive Polymaternalism." *Gender, Place & Culture* 25 (2): 207–227. doi: 10.1080/0966369X.2018.1425286.

Lewis, S. 2019. *Full Surrogacy Now*. New York: Verso.

Mansfield, B., and J. Guthman. 2015. "Epigenetic Life: Biological Plasticity, Abnormality, and New Configurations of Race and Reproduction." *cultural geographies* 22 (1): 3–20.

McDonagh, J. 2015. "Rural Geography III: Do We Really Have a Choice? The Bioeconomy and Future Rural Pathways." *Progress in Human Geography* 39 (5): 658–665.

Meehan, K., and K. Strauss, eds. 2015. *Precarious Worlds: Contested Geographies of Social Reproduction*. Athens: University of Georgia Press.

OECD (Organisation for Economic Co-operation and Development). 2009. *The Bioeconomy to 2030: Designing a Policy Agenda*. Paris: OECD. Available at: https://read.oecd-ilibrary.org/economics/the-bioeconomy-to-2030_9789264056886-en#page12 (accessed August 2018).

OECD (Organisation for Economic Co-operation and Development). 2018. *Meeting Policy Challenges for a Sustainable Bioeconomy*. Paris: OECD. http://dx.doi.org/10.1787/9789264292345-en, (accessed October 2018).

Parry, B. 2015a. "A Bull Market? Devices of Qualification and Singularisation in the International Marketing of US Sperm." In: *Bodies Across Borders: The Global Circulation of Body Parts, Medical Tourists and Professionals*, edited by B. Parry, B. Greenhough, T. Brown, and I. Dyck, 53–72. Farnham: Ashgate.

Parry, B. 2015b. "Narratives of Neoliberalism: 'Clinical Labour' in Context." *Medical Humanities* 41 (1): 32–37.

Pratt, G., and V. Rosner, eds. 2012. *The Global and the Intimate: Feminism in our Time*. New York: Columbia University Press.

Rajan, K.S. 2006. *Biocapital: The Constitution of Postgenomic Life*. Durham, NC: Duke University Press.

Roberts, C., and K. Throsby. 2008. "Paid to Share: IVF Patients, Eggs and Stem Cell Research." *Social Science & Medicine* 66 (1): 159–169.

Rose, N. 2006. *The Politics of Life Itself: Biomedicine, Power, and Subjectivity in the Twenty-First Century*. Princeton, NJ: Princeton University Press.

Schurr, C. 2017. "From Biopolitics to Bioeconomies: The ART of (Re-)producing White Futures in Mexico's Surrogacy Market." *Environment and Planning D: Society and Space* 35 (2): 241–262.

Schurr, C., and E. Militz. 2018. "The Affective Economy of Transnational Surrogacy." *Environment and Planning A: Economy and Space*. doi:10.1177/0308518X18769652.

Strauss, K. 2018. "Labour Geography 1: Towards a Geography of Precarity?" *Progress in Human Geography* 42 (4): 622–630.

Waldby. C. 2002. "Stem Cells, Tissue Cultures and the Production of Biovalue." *Health* 6 (3): 305–323.

Waldby, C., and K. Carroll. 2012. "Egg Donation for Stem Cell Research: Ideas of Surplus and Deficit in Australian IVF Patients' and Reproductive Donors' Accounts." *Sociology of Health & Illness* 34 (4): 513–528.

Waldby, C., and M. Cooper. 2008. "The Biopolitics of Reproduction: Post-Fordist Biotechnology and Women's Clinical Labour." *Australian Feminist Studies* 23 (55): 57–73.

31
CARE, HEALTH AND MIGRATION

Kim England, Isabel Dyck,
Iliana Ortega-Alcázar and Menah Raven-Ellison

Introduction

'Care,' Marian Barnes writes, 'is fundamental to the human condition and necessary both to survival and flourishing' (2012, 1). There is a growing interdisciplinary literature on theorizing and expanding the concept of care, including using it as a broad framework for making moral, political and policy decisions (see, for example, Folbre 2012; Held 2006; Tronto 2013). Feminist care ethics, especially when in conversation with debates about political values, social policy and citizenship, offers an alternative conceptualization of society that recognizes that care is essential for human life.

Feminist geographers have been contributing to this rich vein of scholarship. This includes reflexive interventions about the ways forward for a more caring discipline and caring academic structures (e.g. Lawson 2007; Puāwai Collective 2019). Much attention focuses on the varied spaces and places associated with the politics, processes and practices of care, including connecting the spaces of everyday life with a range of institutions and state agencies (e.g. England and Dyck 2012; Milligan and Wiles 2010; Power and Hall 2018; Radicioni and Weicht 2018). Moreover, the wide appeal of care ethics and care more broadly, to feminist geographers, is apparent in a range of research, such as the politics of farming and animals (Gillespie 2016) and rethinking justice in urban theory (Williams 2017).

In this chapter, we focus on care through the lens of healthcare, specifically addressing immigration, citizenship and belonging in the context of the provision of healthcare in the UK. Feminist geographers have a long-standing commitment to intersectional analyses, and an additional dimension of healthcare is the complexity that comes from accounting for international migration and the ethnic diversity of the contemporary UK. We explore these themes thorough three different groups of immigrants: well-established immigrant communities from India and West Africa; international nurses recruited to address the National Health Service (NHS) nurse shortage; and asylum seekers who have been released from detention centres. Their different experiences of the same healthcare system demonstrate the significance of questioning taken-for-granted categories such as health, care and 'immigrant woman'.

Care, citizenship and the welfare state

At the core of feminist care ethics is a relational ontology of connection that positions people as embodied, interdependent beings. We are all vulnerable and dependent on others at numerous points throughout our lives, and we are each enmeshed in networks of care relations (Barnes 2012; Tronto 2013). This contrasts strongly with the autonomous, independent, individualistic neoliberal subject who, discursively, inhabits the policy documents of neoliberal capitalism.

Several scholars suggest extending care ethics beyond something restricted to the domestic sphere among intimates to thinking more broadly about bringing feminist care ethics into public debates in order to reframe political issues (Barnes 2012; Tronto 2013; Williams 2018). Tronto (2013) proposes revising political values around care as a set of concrete practices and a deep commitment to equality and justice, with the goal of producing a caring democracy with equal access to good care for everyone, while Folbre (2012, 183) argues that care 'contributes to the development and maintenance of human capabilities that represent a "public good". Human capabilities have intrinsic value and also yield important positive spill-overs for living standards, quality of life and sustainable economic development'. In short, directly and indirectly, care provides individual and collective benefits. It has broad social value, and rethinking care as a public good is becoming a central aim of scholarship on feminist care ethics.

Actual healthcare policies reflect government decisions regarding the cost, quality of care, accessibility, delivery and programme evaluation. The UK has a publicly funded universal healthcare system – the NHS – premised on being comprehensive, universal and accessible to all citizens (and administered by the state, rather than the market). It was introduced in 1948, at a time when the public supported welfare measures to address social and economic inequalities with the goal of constructing a more equitable society. The UK welfare state was built around social liberalism, with state–market relations configured around greater state intervention to regulate and alter market forces to meet the goal of social equality. This is achieved by minimizing the risks inherent in unemployment, ill health and old age through the provision of some level of income security for individuals and households. Social ills, such as poverty, were seen as structural, and thus social programmes were built around a consensus of social rights and collective responsibility to provide for the basic social needs and economic security of the population. This is closely tied to the emergence of social citizenship – 'the whole range from the right to a modicum of economic welfare and security to the right to share to the full in the social heritage and to live the life of a civilized being according to the standards prevailing in society', as T.H. Marshall (1949/1992) famously defined them. Social citizenship rights allow citizens to make claims on the state for particular benefits and services, including the NHS for public healthcare. More broadly, citizenship is a multifaceted concept, and recent scholarship emphasizes that the same person or group might be privileged in some aspects of citizenship yet not others (Staeheli et al. 2012). For instance, legal/judicial citizenship entitles an immigrant to health services, but everyday exclusionary experiences, such as casual race-based discrimination, erode the sense of membership of, and belonging to, a community.

The contemporary restructuring of the welfare state, including healthcare systems, involves reconfigured roles, responsibilities and governance arrangements between civil society, the market and the state in the name of neoliberal-flavoured fiscal austerity. As a state form, neoliberalism involves an agenda aimed at increasing economic efficiency and competitiveness through the rhetoric of free markets, privatization and marketization while reducing government spending, especially on social welfare programmes (MacLeavy 2012). Discourses of citizenship were formerly configured around collective responsibility, ameliorating social risk and social entitlements, and public healthcare became its cornerstone. As citizenship shifts towards the neoliberal values of possessive individualism, consumerism and individual responsibility, social problems are recast as failures of the individual rather than the result of structural inequalities, and the 'good citizen' is an atomized market player, self-reliant, who does not look to the government for help (England and Ward 2016).

Health, place and policy

Geographers draw attention to the intertwining materialities and sets of socio-spatial relations that are dynamic and constitute the meanings and experiences of health. They make key interventions into understandings of the spatiality of the processes and practices of health and the difference that space, scale and sites make in the processes and practices of care (e.g. Brown et al. 2018; Crooks, Andrews and Pearce 2018). In addition, feminist geographers have brought other approaches into understanding health through the lenses of embodiment, ways of knowing, care theory and affect (e.g. Davidson and Bondi 2004; Greenhough et al. 2015). The meanings of health and the parameters of healthcare provision shift and have different consequences over time and cross-culturally, with political economy, social policy and migration regimes having profound effects on the border between health and ill health (Dyck 2006).

The specific form that a healthcare system takes at any moment reflects a nation's underlying values regarding the balance between equity (providing access to necessary medical services to everyone) versus efficiency (minimizing costs and maximizing revenues), and it is generally the most complex system of a country (Rosenberg 2017). The NHS is touted as one of the world's most comprehensive public healthcare systems. In the contemporary moment, the NHS seems in a constant state of crisis, yet it remains central to British national identity and it is still frequently ranked as the most popular institution in the UK (Burki 2018). Aneurin Bevan, the post-war Minister of Health in the Labour Government, oversaw the creation of the NHS. Funded by general taxation, it was premised on being comprehensive, universal and accessible to all citizens (and administered by the state). It was intended to provide healthcare on a uniform basis throughout the UK, based on regional health authorities (Burki 2018; Mossialos et al. 2018). However, current healthcare policy focuses increasingly on concerns about the escalating costs of healthcare and how to 'contain' and minimize public expenditures and make healthcare more 'cost effective' and 'efficient'. In many instances, this has meant the erosion of healthcare as a public good.

In the rest of the chapter, we address the themes of care, citizenship and belonging in healthcare practices and policy though studies of the three migrant groups. These three instances also allow us to examine themes of inclusion and exclusion, access to care and self-care.

Health and well-being among migrants in London

Healthy lifestyle promotion is a core issue in the NHS debates over the use of finite resources, and an emerging question is how to engage populations effectively in taking control of their own health management. Such debate is situated within a political and social climate that increasingly places responsibility for health on individuals and families, a move noted in Nikolas Rose's (2007) work on the emergence of a new 'biological citizenship' that discounts biological destiny. Despite a breadth of scholarship that demonstrates the complex embeddedness of health knowledge and practices in political, social and cultural processes, an uncritical, medicalized view of health usually prevails in health science and health promotion discourse (Dyck 2006). In such medicalized discourse an echo of cultural determinism lingers in essentializing stereotypes of particular immigrant or minority ethnic groups, which deserves to be challenged if policy initiatives are to reach a diverse population. Certainly, there is growing recognition of the value of listening to those whom the policies aim to target and of integrating their perspective into the policy-making process (Cowden and Singh 2014). Briefing notes for NHS research and policy programme staff now explicitly encourage the involvement of members of the public as active partners in research. This recognizes that if NHS research reflects the needs and views of service users, it is more likely to produce results leading to improved health and social service (Hanley et al. 2004).

As part of a larger study examining changes and continuities in health practices in the context of international migration, Iliana Ortega-Alcázar and Isabel Dyck (2012) explored the complex understandings of the health practices, beliefs and needs of two contrasting and long-established migrant groups (from Nigeria and India) in London, and revealed how notions of health promotion are taken up in everyday lives. The study is based on 40 photo-elicitation interviews of migrants living in one of London's poorer, ethnically diverse boroughs. They were asked to take photos of anything that they considered important to their health and well-being: objects, places, people, activities and other events were given as examples. After the participants brought their photographs, prints were made and the photographs returned to the participant. An interview was then conducted in a location of their choice and, beforehand, the participants were asked to group the prints that they felt should go together. Each group of individual prints was displayed on a table and the participants were invited to talk about them.

The participants' photographs indicated how both physical and social spaces in the local area were important to their health and demonstrated transnational dimensions as part of their health practices. For instance, several individuals took photographs of the African and Indian shops in the local area, including the foodstuffs and products bought there, and explained that being able to buy these things contributes to their well-being. Sometimes, the products were directly related to their physical health, such as items for the preparation of natural remedies or what were considered healthy meals. Others were described as having cultural significance and related to well-being in a different way; one example was an African broom, which enabled the continuity of a particular everyday practice and played a significant role in countering the owner's feelings of displacement.

The significance of place to the participants' accounts of health and well-being was also shown in terms of the physical environment of their neighbourhoods. Their photographs included shots of gardens, parks and buildings such as libraries and community centres. These prompted discussions in interviews of the healthiness of a particular neighbourhood and how the location of a participant's housing might facilitate a healthy lifestyle, whether by its relation to local amenities or by the physical structure of the building. One older Indian woman photographed the view from her balcony, explaining that rather than using the lift (elevator), she uses the 'stairs because in that way it is exercise' and walks to the shops. Her other indoor photographs depicted particular institutional spaces within the neighbourhood that facilitated health, through both the activities and the provision of health information. These included the Age Concern Centre and the Indian Cultural Society, where she meets friends and attends exercise classes and health information classes run by NHS workers. In talking about the photographs, she wove an account of her everyday life that indicates how housing, urban structures and the presence of a local Asian community all contribute to her health.

The active configuration of place for health and well-being is further illustrated in other participants' comments concerning links between neighbourhood involvement and belonging in wider society. A retired Nigerian woman's narrative of her photograph choices of various activities and local groups showed that engaging with her community was pivotal to her well-being. She actively resisted views of immigrants as burdens on their host society and holding no sense of attachment. About her voluntary work with the local Primary Care Trust, she said: 'It makes me feel fulfilled. I'm contributing something to the neighbourhood and to society.' She provides visible proof of the construction of herself as a healthy citizen, emphasizing her identity performance as a contributing citizen, which she interlinked with health or well-being practices.

Some participants moved beyond descriptions of particular health practices and provided interpretations that challenged the stereotyping discourses on health and migration framed in culturalist terms. For example, a young Nigerian woman actively resisted an essentialized and stereotyped view of Nigerian people being outside the dominant global health discourse and practice. She showed photographs of traditional ways of keeping healthy – for instance, she took a picture of a Nigerian dish, pepper soup, talking about its medicinal properties and use by women to recover after giving birth. Yet, throughout her narrative, she rejected a view of fixed 'Nigerian' health practices and beliefs that diverge from those of the UK as a result of 'cultural difference'. There may be Nigerian perspectives on health practices, but not a single, unified Nigerian approach to practising health. She drew attention to the historical colonial links between the two countries, further questioning a common view in the dominant discourse on health and migration that places migrants' health practices as inherently 'other'.

Thus, a care ethics approach draws attention to social and cultural well-being as important dimensions of health, as well as the more usual conceptualization of physical and mental health. A sensitivity to social and cultural differences in health practices is critical for designing effective health policies, and those practices may vary across space.

Importing international nurses

Healthcare is a labour-intensive industry, and nurses make up the largest group of health professionals. In a context of publicly funded healthcare systems, nurses are an important part of healthcare as a public good and for supporting social citizenship rights. Thus, the availability of nursing personnel is fundamental to achieving effective health service delivery and maintaining an adequate level of overall public health. More and more countries are facing nurse shortages, and talk of a global crisis in the healthcare workforce has become common. The increased transnational migration of nurses and other healthcare professionals (particularly doctors) over the last decade has brought the idea of a 'global health care economy' (Kingma 2006) into common parlance.

The migration of internationally trained nurses into the UK began to increase substantially in the 1990s (Hardill and MacDonald 2000; Kingma 2006). It is difficult to track over time the nationalities of foreign-trained nurses coming to work in the UK. However, nurses are governed through the Nursing and Midwifery Council (NMC) and all nurses are required to register in order to practise in the UK. The NMC register captures the country of their initial registration and annually publishes summary data on registrants admitted (to 31 March each year). Kim England used these initial admissions data, which include information on the country where a nurse first registered (this is not necessarily their country of birth).[1] The NMC breaks down the public data into those who first registered in the UK, the European Union (EU) or 'Overseas' (classified as those who initially registered outside the EU).

Figure 31.1 indicates that the numbers admitted to the NMC register have varied over time, with rapid growth from 1999 through the early 2000s, then a drop off to 2010, followed by another period of growth and apparent slowdown from about 2015. The UK numbers are less variable than the non-UK admissions. This suggests periodic nurse shortages and cyclical 'booms and busts' in the UK nursing labour market. Breaking down the overall data by EU and

'Overseas' indicates a growing portion of overseas admissions to 2004, followed by a drop off, with indications of another upward swing beginning in 2015. EU nurses comprised a smaller proportion in the early 2000s, followed by an uptick from the mid-2000s, with a noticeable increase in the mid-2010s.

Nurse migration patterns tend to follow paths of historical economic and colonial relationships, and the UK has historically relied on Commonwealth countries to supply nurses to fill its periodic nursing shortages. Reflecting this, the NMC data (not shown) indicate that in 1999 the top sending regions were (in rank order) Australia, New Zealand, South Africa, New Zealand, the Caribbean and Canada. They accounted for 16 per cent of all new registrants and 80 per cent of those from overseas: by far the largest group were of nurses initially registered in Australia, followed distantly by South Africa. In the early 2000s, Australia and South Africa remained important, with a new trend of increased numbers from elsewhere in Sub-Saharan Africa accounting for about 25 per cent of each year's overseas initial registrations. In addition, admissions to the register from the Philippines and India picked up markedly through the 2000s and soon became largest group of initial registrations; by 2012, those two countries accounted for 23 per cent of overseas registrants. The expansion of the EU over the past two decades caused an increase in numbers from the EU accession states, and there has also been active recruitment by some NHS Trusts from long-standing EU countries, such as Spain and Portugal. However, new registrants from the EU have dropped since the 2016 EU Referendum, and an uptick in 'Overseas' nurses is again evident.

There is a range of explanations for these trends. The UK is faced with an ageing nurse population, exacerbated by early retirement and turnover. Stress from inadequate staffing, increased workloads (sometimes without pay) and stagnating pay means that nurses leave the field or retire early, and retention has become a major issue (Marangozov et al. 2017). Some of the fluctuation is linked to institutional strategies to address concerns about patient care, nurse shortages and healthcare policy changes. In the late 1990s, the Labour government sought to 'modernize' the NHS, with plans and funding to expand the NHS workforce. Some funding went into increasing training opportunities for domestic student nurses. However, between 1998 and 2005

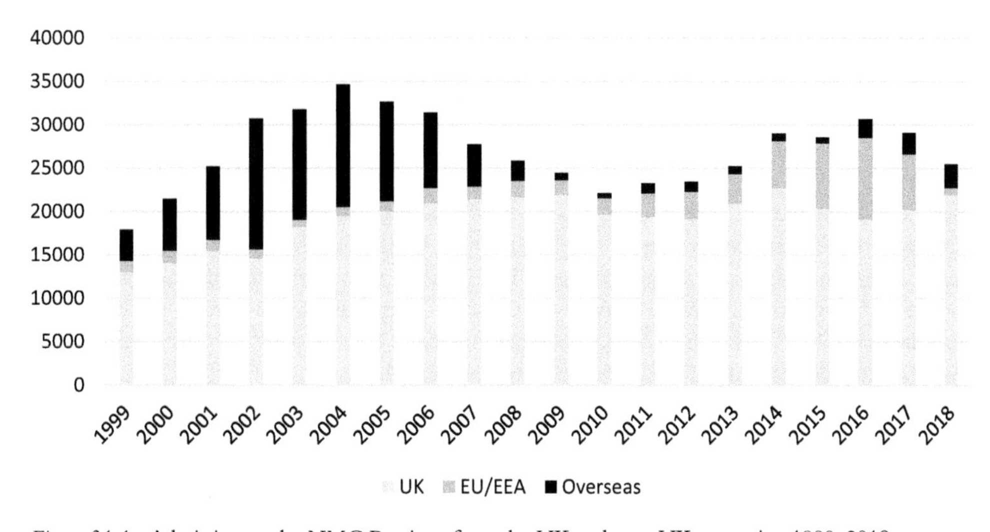

Figure 31.1 Admission to the NMC Register from the UK and non-UK countries, 1999–2018.
Source: NMC data, various countries.

there was also an explicit and coordinated policy to ramp up the international recruitment of nurses. Then, starting in 2005, the NMC introduced additional requirements aimed at overseas nurses (i.e. from outside the EU), including a period of supervision (to reskill nurses, relative to the UK healthcare system, and to safeguard patient safety); following this, in 2007 the NMC introduced more stringent English language requirements.

At the level of central government, in 2006 the Home Office removed many lower-grade nursing and general nursing occupations from the Shortage Occupation list for work permits. The 2008 shift to a points-based immigration system and the 2012 work permit immigration rules further impacted employers' ability to recruit overseas nurses. More recent austerity politics and the post-2008 recession impacted on NHS funding, with neoliberal-flavoured policies not only constraining public sector expenditures but causing a stagnation in pay, recruitment freezes, fewer training places and even redundancies. The 2013 Francis Report pinpointed the inadequate staffing levels as a reason behind the failings at the Mid Staffordshire NHS Trust (the subsequent review also recommended better nurse–patient ratios). In response, safer staffing levels were introduced, increasing the demand for nurses (Marangozov et al. 2017). By the mid-2010s, there were calls to return all nurses to the Shortage Occupation list, which the Home Office subsequently did.

While these policy decisions are reflected in the overall trends in the NMC's admissions data, a closer look reveals additional consequential geographies. Importing nurses has long been a popular 'quick fix' solution to address shortfalls. The UK has certainly been a significant player in the global migration of nurses. Starting in the early 2000s, concern was raised about the number of nurses coming from Commonwealth countries in Sub-Saharan Africa. Policy decisions resulting in the active recruitment of international nurses to 'solve' nurse shortages were interpreted as the UK depleting other countries of their nurses, in turn impacting on the ability of those governments to provide adequate healthcare for their own citizens and, potentially, to damage their country's future social and economic development (England and Henry 2013). Other, potentially more caring, strategies are also possible. For example, the Royal College of Nursing argues that the gap can also be addressed in numerous ways domestically, including by increasing wages, improving the retention of nurses and investing in nurse education. More broadly, coordinated strategic and long-term workforce planning could be a way to move the UK in the direction of a more caring democracy.

Health after detention

Our three case studies demonstrate how health, migration and care intersect variously for contrasting groups of migrants in the UK. This section focuses on the experiences of former detainees. Immigration detention is portrayed by UK government policy as a last resort in 'extreme cases', but recently the practice has become integral to the asylum process. In 2017, some 27,346 people were detained in the UK 'for the purposes of immigration control' and placed in Immigration Removal Centres (IRC), short-term holding facilities or pre-departure accommodation (Home Office 2018). Women accounted for about 16 per cent of detainees, and about a third of women detainees go on to be released.

Detainees are often ill prepared for release. Concerns about this have even been raised by HM Chief Inspectorate of Prisons, who observed of Yarl's Wood IRC: '[t]here was no systematic approach to assessing needs and helping detainees to prepare for removal or release' (2011: 16). Indeed, the only reference to preparing detainees for release within the Detention Centre Rules (2001), the primary guidance for UK immigration detention, is about their healthcare records and returning their material items to the detainees. Klein and Williams (2012, 743) suggest that:

> Because release into the community does not correspond to policy rhetoric emphasizing an unproblematic progression through detention to removal or naturalization, it is more convenient for the UK government to airbrush the very large numbers of released migrants from view than to engage in a public discussion.

What little scholarship there is on the long-term implications of detention signals that former detainees experience marginalization, marked by fear of arrest, further detention and potential removal. Many also experience precarity, because they lack the legal entitlement to work or other means to achieve a decent quality of life. Additionally, the legacies of detention include negative effects on their mental health, social relationships, perceptions of the UK and sense of belonging (McGregor 2009; Klein and Williams 2012).

Menah Raven-Ellison (2015) focused on the ongoing legacy of detention and the consequences for the sense of belonging, social integration and the health and well-being of former detainees. She conducted in-depth narrative interviews with 16 previously detained women, all of whom had been detained in Yarl's Wood IRC, the main IRC for holding women. The women ranged in age from 19 to 60 years, the majority from nine different African countries and three from Asian countries. The complex interrelationship between health, well-being and the processes and legacies of detention emerged as a central point of concern. For two of her participants, Pru (a 42-year-old Burundian asylum seeker, detained for six weeks) and Jana (a 45year old Nigerian refugee, detained for three months), the erosive implications of uncertain status on migrants' everyday lives undermine the extent to which they see themselves as having choices and opportunities to plan for the future. However, Jana was awarded refugee status, and she identified this a turning point:

> After winning your status then it becomes different, you know? You are able to have mail in your name, have proof of address, able to open a bank account, able to have an NHS number, you are able to have an ID – you are somebody, you are recognised as a person.

Refugee status bestowed on her a legitimacy and sense of purpose that she had lacked as an asylum seeker. The fact that she identified as having an NHS number and described herself as 'recognised as a person' suggests that she is recovering coherence and stability in her life, as well as some power and control and, with it, an emerging sense of belonging.

Both Pru and Jana, along with five other participants, described themselves as torture survivors. Rule 35 (3) of the Detention Centre Rules (2001) was created to identify people who have been tortured, to prevent their detention. There is consensus that detention is damaging to mental health, not least because it can arouse memories of incarceration, thus provoking re-traumatization. NGO workers find that this safeguard is often ignored, resulting in the incarceration of vulnerable individuals for whom detention is described as a 'second torture'. All seven respondents had been detained, but only Pru had been assessed and released from detention in accordance with Rule 35. The consequences for their well-being were significant, exacerbated by the poor overall standards of healthcare in detention. Their experiences of torture occurred not only in their country of origin but, in three participants' cases, including Pru's, during lengthy periods of enslavement in the UK by their traffickers.

Pru's experience illustrates some of the health concerns amongst migrants, for she was severely traumatized by an accumulation of factors, including witnessing atrocities in her home country, experiencing torture, being separated from family and then enduring three years of captivity and sexual exploitation in the UK. She would eventually be diagnosed with post-traumatic stress disorder (PTSD). Pru experienced flashbacks that disturbingly transform her present-day experience of certain public places, seemingly converging unpredictably with past events. She described how horrific past events in Burundi merged with images of UK immigration officers chasing her, creating a distorted experience of space, rendering it unstable, unbounded and unpredictable. The distressing experiences discouraged her from engaging with her local community, increasing her social isolation and further undermining her emotional well-being. After release from detention, Pru received little support and was only eventually considered eligible for assistance from the local authority and mental health services when her behaviour became inappropriate and noticeably 'out of place'.

There are glaring disparities between the priorities outlined in UK mental health policy and the relevance of these to individuals, such as Pru, who have a mental illness yet have a legally uncertain position in society. The NHS is infused with a recovery-orientated approach that claims to strive for the meaningful inclusion of individuals with mental health problems in society. Despite an admirable, seemingly caring philosophy, the narratives of people like Pru, who remain particularly marginalized in society and who already have limited rights and questionable legal status, expose the inconsistences in the NHS's inclusion rhetoric. Indeed, these experiences highlight the limitations of healthcare systems, as they can lack the ability to accommodate difficulties that fall outside of a medicalized, Western approach to pathology, not least how the care for these individuals is coordinated and delivered.

Conclusion

Geographers have made important contributions, theoretically and empirically, that demonstrate the links between places, people and health and how the interactions among them influence health, illness, survival and flourishing. In addition, viewing health through the lens of international migration spotlights intersectional analyses. We took up these ideas and interwove them with the growing interdisciplinary literature on theorizing and expanding the concept of care, including its use as a broad framework for making moral, political and policy decisions.

Iliana Ortega-Alcázar and Isabel Dyke's work (Ortega-Alcázar and Dyck 2012) focused on the everyday practices of health citizenship of two well-established migrant groups in London to reveal how place influences their health and well-being, underscoring how appropriate engagement of social groups must play a pivotal role in the NHS's creation and implementation of successful health promotion. Kim England's analysis of NMC data showed how the UK has generally welcomed international nurse migrants to fulfil quickly its social citizenship obligation in terms of publicly funded healthcare, possibly at the expense of better long-term planning. Menah Raven-Ellison's research with formerly detained migrants demonstrated the ways that the UK's immigration policy and healthcare system serve to exclude some of the most vulnerable of migrants.

Our goal, through our case studies, is to show that feminist care ethics as practice, politics and discourse offers a possible route to reframe political debate and rethink citizenship and belonging in ways that centrally address healthcare and well-being. Care ethics is always about alerting us to relational power inequalities. An emerging trend in feminist geographers' scholarship is to think more critically about care: to take seriously the empirical and conceptual implications of the spatial diversity of care practices across the globe (Raghuram 2016) and to expose the 'uncomfortable politics of care', because caring practices do not inherently produce good care (Bartos 2018). Our three case studies speak to the ways that the UK's healthcare policies are interwoven with differential power relations and reveal unequal caring through the different boundaries of inclusion and levels of exclusion experienced by each of the groups. This, in turn, shows that the NHS is not as comprehensive as it implies. In short, viewing health provision through the lens of care ethics makes it apparent that the discourses, practices and provision of healthcare are embedded in broader cultural contexts and are situated in policies formulated in particular political-economy moments. More broadly, the trend towards a deeper engagement and a further sophistication of understandings of care brings new challenges, as well as opportunities for robust feminist geography scholarship, going forward.

Note

1 The NMC does not publish a list of specific countries. A Freedom of Information request was filed to receive these data.

Key readings

Barnes, M., T. Brannelly, L. Ward, and N. Ward, eds. 2015. *Ethics of Care: Critical Advances in International Perspective*. Bristol: Polity Press.

Brown, T., I. Dyck, B., Greenhough, M. Raven-Ellison, M. Ornstein, and S.W. Duffy. 2019. "'They Say It's More Aggressive in Black Women': Biosociality, Breast Cancer, and Becoming a Population 'At Risk'." *Transactions of the Institute of British Geographers* 44: 509–523.

Ortega-Alcázar, I., and I. Dyck. 2012. "Migrant Narratives of Health and Wellbeing: Challenging 'Othering' Processes through Photo-elicitation Interviews." *Critical Social Policy* 32 (1): 106–125.

References

Barnes, M. 2012. *Care in Everyday Life: An Ethic of Care in Practice*. Bristol: Policy Press.

Bartos, A. 2018. "The Uncomfortable Politics of Care and Conflict: Exploring Nontraditional Caring Agencies" *Geoforum* 88: 66–73.

Brown, T.G.J., S. Andrews, B. Cummins, D. Greenhough, D. Lewis, and A. Power (eds). 2018. *Health Geographies: A Critical Introduction*. New York: Wiley.

Burki, T. 2018. "From Health Service to National Identity: The NHS at 70." *The Lancet* 392 (10141): 15–17.

Cowden, S., and G. Singh. 2014. "A Critical Analysis of Service User Struggles." In: *Rethinking Anti-Discriminatory and Anti-Oppressive Theories for Social Work Practice*, edited by C. Cocker and T. Hafford-Letchfield, 93–107. Basingstoke: Palgrave Macmillan.

Crooks, V.A., G.J. Andrews, and J. Pearce, eds. 2018. *Routledge Handbook of Health Geography*. London: Routledge.

Davidson, J., and L. Bondi. 2004. "Spatialising Affect; Affecting Space: An Introduction." *Gender, Place & Culture* 11 (3): 373–374.

Detention Centre Rules. 2001. *Detention Centre Rules*. Available at: www.legislation.gov.uk/uksi/2001/238/pdfs/uksi_20010238_en.pdf (accessed January 2019).

Dyck, I. 2006. "Travelling Tales and Migratory Meanings: South Asian Migrant Women Talk of Place, Health and Healing." *Social and Cultural Geography* 7: 1–18.

England, K., and I. Dyck. 2012. "Migrant Workers in Home Care: Routes, Responsibilities, and Respect." *Annals of the Association of American Geographers* 102 (5): 1076–1083

England, K., and C. Henry. 2013. "Care Work, Migration and Citizenship: International Nurses in the UK." *Social and Cultural Geography* 14 (5): 558–574.

England, K., and K. Ward. 2016. "Theorizing Neoliberalization." In: *The Handbook of Neoliberalism,* edited by S. Springer, K. Birch, and J. MacLeavy, 50–60. London: Routledge.

Folbre, N. 2012. *For Love and Money: Care Provision in the United States.* New York: Russell Sage.

Gillespie, K. 2016. "Witnessing Animal Others: Bearing Witness, Grief, and the Political Function of Emotion." *Hypatia* 31 (3): 572–588.

Greenhough, B., B. Parry, I. Dyck, and T. Brown. 2015. "The Gendered Geographies of 'Bodies Across Borders'." *Gender, Place & Culture* 22 (1): 83–89.

Hanley, B. et al. 2004. "Involving the Public in NHS, Public Health, and Social Care Research: Briefing Notes for Researchers." 2nd edition. Eastleigh: INVOLVE.
Hardill, I., and S. MacDonald. 2000. "Skilled International Migration: The Experience of Nurses in the UK." *Regional Studies* 34 (7): 681–692.
Held, V. 2006. *The Ethics of Care: Personal, Political, Global.* Oxford: Oxford University Press.
HM Inspectorate of Prisons. 2011. "Report on an announced inspection of Yarl's Wood Immigration Removal Centre." Available at: www.justice.gov.uk/downloads/publications/inspectorate-reports/hmipris/immigration-removal-centre-inspections/yarls-wood/yarls-wood-2011.pdf (accessed November 2018).
Home Office. 2018. "Immigration Statistics, Year Ending June 2018." Online. London: Home Office. Available at: www.gov.uk/government/statistics/immigration-statistics-year-ending-june-2018-data-tables (accessed August 2018).
Kingma, M. 2006. *Nurses on the Move: Migration and the Global Health Care Economy.* Ithaca, NY: Cornell University Press.
Klein, A., and L. Williams. 2012. "Immigration Detention in the Community: Research on the Experiences of Migrants Released from Detention Centres in the UK." *Population, Space and Place* 18: 741–753.
Lawson, V. 2007. "Geographies of Care and Responsibility." *Annals of the Association of American Geographers* 97 (1): 1–11.
MacLeavy, J. 2012. "The Lore of the Jungle: Neoliberalism and Statecraft in the Global-Local Disorder." *Area* 44: 250–253.
Marangozov, R., C. Huxley, C. Manzoni, and G. Pike. 2017. "Royal College of Nursing Employment Survey (Report 513)." Royal College of Nursing. Available at: www.employment-studies.co.uk/resource/royal-college-nursing-employment-survey (accessed August 2019).
Marshall, T.H., and T. Bottomore. 1949/1992. *Citizenship and Social Class.* London: Pluto.
McGregor, J. 2009. *Narratives and Legacies of Detention.* London: UCL Migration Unit and Zimbabwe Association.
Milligan, C., and J. Wiles. 2010. "Landscapes of Care." *Progress in Human Geography* 34 (6): 736–754.
Mossialos, E., and A. McGuire, M., Anderson, A. Pitchforth, A. James, and R. Horton. 2018. "The Future of the NHS: No Longer the Envy of the World?" *The Lancet* 391 (10125): 1001–1003.
Ortega-Alcázar, I., and I. Dyck. 2012. "Migrant Narratives of Health and Wellbeing: Challenging 'Othering' Processes Through Photo-elicitation Interviews." *Critical Social Policy* 32 (1): 106–125.
Power, A., and E. Hall. 2018. "Placing Care in Times of Austerity." *Social and Cultural Geography* 19 (3): 303–313.
Puāwai Collective. 2019. "Assembling Disruptive Practice in the Neoliberal University: An Ethics of Care." *Geografiska Annaler: Series B, Human Geography* 101 (1): 33–43.
Radicioni, S., and B. Weicht. 2018. "A Place to Transform: Creating Caring Spaces by Challenging Normativity and Identity." *Gender, Place & Culture* 25 (3): 368–383.
Raghuram, P. 2016. "Locating Care Ethics Beyond the Global North." *ACME* 15 (3): 511–533.
Raven-Ellison, M. 2015. "States of Precarity: Negotiating Home(s) Beyond Detention." PhD Dissertation, Geography, Queen Mary University of London.
Rose, N. 2007. *The Politics of Life Itself: Biomedicine, Power, and Subjectivity in the Twenty-first Century.* Princeton, NJ: Princeton University Press.
Rosenberg, M.W. 2017. "Health Systems." In: *The International Encyclopedia of Geography,* edited by D. Richardson, N. Castree, M.F. Goodchild, A. Kobayashi, W. Liu, and R.A. Marston, 3295–3300. Oxford: John Wiley.
Staeheli, L.A., P. Ehrkamp, H. Leitner, and C.R. Nagel. 2012. "Dreaming the Ordinary: Daily Life and the Complex Geographies of Citizenship." *Progress in Human Geography* 36 (5): 628–644.
Tronto, J. 2013. *Caring Democracy: Markets, Equity and Justice.* New York: NYU Press.
Williams, F. 2018. "Care: Intersections of Scales, Inequalities and Crises." *Current Sociology* 66 (4): 547–561.
Williams, M.J. 2017. "Care-full Justice in the City." *Antipode* 49: 821–839.

32

CONTEXTS OF 'CARING MASCULINITIES'

The gendered and intergenerational geographies of men's care responsibilities in later life

Anna Tarrant

Introduction

This chapter documents and advances an emerging feminist geography of caring masculinities. Feminist geographers have long sought to investigate the complex spatialities and temporalities of care and caregiving, scrutinizing a range of social spaces of care (McEwan and Goodman 2010) at a number of scales (Atkinson et al. 2011). This scholarship has been instrumental in identifying inequalities in gendered experiences of care and the ways in which these play out in particular spaces and places over the lifecourse, particularly for women, who remain responsible for most unpaid labour.

Recent evidence indicates, however, that an increasing number of older men are now providing informal care (Giesbrecht et al. 2016; Milligan 2009). This has been linked in part, to the ageing population in developed countries, whereby men are living longer and are more likely to take on care responsibilities later in life. Shifts in the gendered landscapes of care have also been identified since the global recession in 2008, whereby men's un- and under-employment is thought to be creating the conditions for more men to engage in a range of care practices (Boyer et al. 2017). These broader structural processes have implications for the dynamic relationship between changing divisions of labour and the social construction of gendered identities, including masculinities (see McDowell 2004). Yet men's everyday experiences of caregiving and their interpersonal relationships are frequently overlooked.

This chapter begins by documenting developments in feminist geographical research about care and caregiving and parallel advances in the geographies of masculinities subfield in which men are increasingly recognized *as* men: that is, as gendered subjects who, like women, experience a range of dependencies and interdependencies across the lifecourse. Findings from two studies, both conducted by the author, are presented to explore the unfolding geographies of men's gendered and generational experiences of caregiving. Both were based in the UK. The first explored the everyday geographies of grandfatherhood (2007–2011) and the second the unfolding care responsibilities of men living in low-income families and localities in a northern English city (2014–2018). Both shed light on the spatial and temporal dynamics of these men's relatively invisible practices of caregiving across the lifecourse. The findings highlight the range of contexts and processes involved in producing opportunities for men to engage in what have elsewhere been conceptualized as 'caring masculinities' (Elliott 2016; Hanlon 2011), including the key spaces where these can flourish.

Geographies of care

Ten years ago, Lawson (2007) demonstrated that geographers have long been concerned with their ethical responsibility to care. Yet, as she also insightfully noted, contemporary neoliberal trends such as the extension of market relations into caring realms – xemplified by shifts to discourses of personal responsibility and the withdrawal of public provision of social supports (e.g. welfare retrenchment) – were marginalizing care as a subject of critical enquiry at a time when questions of care have never been a more pressing concern. Indeed, her arguments were presented just prior to the widespread imposition of a raft of austerity measures that have further entrenched these processes and have transformed the caring landscape. This landscape characterizes an increasingly uneven redistribution of responsibility from the state to the private and third sectors and to individual families (Lawson 2007; Power and Hall 2017). Since Lawson's address, the boundaries of this body of work have rapidly expanded (Power and Hall 2017), highlighting the ways in which care is profoundly relational, embodied, emotional and subject to transformation. Feminist geographers have continued to engage with a range of interdisciplinary discourses and perspectives around care and caregiving to carve out what might be distinctive about geographical approaches. Milligan and Wiles (2010) for example, developed the 'landscapes of care' framework to capture the complex relationships between people, places and care. The power of this ambitious analytic framework, they argue, is that it helps in teasing out the interplay between macro and micro levels, or the:

> socio-economic, structural and temporal processes that shape[s] the experiences and practices of care at various spatial sites and scales, from the personal and private through to public settings, and from local to regional and national levels, and beyond.
> *Ibid., 736*

Recent debate reflects how gendered landscapes of care are being transformed in an 'age of austerity' (Boyer et al. 2017). This scholarship examines the relationship between the changing socio-spatial dynamics of care, the shifting relations between waged work and care work and the gendered politics of care. Drawing on evidence from the US and Scandinavia that men (usually fathers) are increasingly engaging in social reproduction, with transformative potential in the regendering of care, Boyer et al. (2017) consider the impact of austerity in the UK on new gendered patterns of care. They conclude that in the aftermath of the 2008 and 2010/2011 double-dip global recessions, a combination of political and economic transformations is affording an increasing number of men with opportunities to assume greater responsibilities for care, albeit unevenly and influenced by class position. Their work provides the basis for further consideration of diversity in male caregiving in a number of contexts.

Spaces for caring masculinities

The history of geographical interest in masculinities and the critical turn towards considerations of spatial accounts of men *as men* (Jackson 1991) has been well documented (Hopkins and Noble 2009). It has carved an essential space for feminist considerations that are relational and considerate of the power relations both within and between groups of men, and between men and women (Calasanti 2003). Sparked by Jackson (1991), who challenged aspatial accounts of men's lives and drew attention to the socially and culturally constructed nature of masculinity, feminist geographical research has since examined the complex relationship between masculinities and place and the ways in which male identities vary both spatially and temporally (Berg and Longhurst 2003; van Hoven and Hörschelmann 2006). Drawing predominantly on Connell's (1995) framework of multiple masculinities, which has dominated the theoretical development of the field, geographers have demonstrated effectively the plurality and complexities of masculinities and their inextricable links with other social locations like race, age, ethnicity, sexuality and class, examining them empirically within an ever-wider variety of places and geographical locations (Hopkins and Noble 2009).

The breadth and extent of the subfield has expanded notably, covering a range of empirical topics related to key arenas of men's lives, from the intimate spaces of the body (Laurendeau 2014; Longhurst, 2005), the home (Pilkey 2014; Tarrant 2016a; Waitt and Gorman-Murray 2007, 2008), the family and domestic labour (Cox 2015; Tarrant 2013), community and sports (Evers 2009), employment (McDowell 2009), and men and migration (Walsh 2011). Many of these themes are consolidated in Gorman-Murray and Hopkins' (2015) extensive edited collection, *Masculinities and Place,* which is testament to the collective strength of the masculinities subfield and its feminist standpoint on the relationships between men, masculinities and place.

Many of the above arenas are also arenas of care. The intimate spaces of the body, home and family each relate to aspects of care and caregiving, even if care is not explored explicitly. This includes consideration of the spaces and institutions in which caring by men takes place; men's self-care and care of the body; caregiving and receiving by men; and care of places. Considerations of care and caregiving have also been prompted in part by parallel conceptual advances in the geographies of age, the lifecourse and intergenerationality (Hopkins and Pain 2007). These concepts have reinvigorated spatial accounts of men's roles in the family, including as fathers and grandfathers (Aitken 2009; Tarrant 2013, 2016a) and in the institutional contexts in which men give care and receive it, for example in old age (Milligan 2009; Milligan et al. 2013). The recognition that caring occurs across the lifecourse and is experienced from cradle to grave has prompted much of this research about older men. In ageing societies, older men are also increasingly engaging in informal economies of care (Wheelock and Jones 2002), challenging traditional expectations and norms about gender (Russell 2007).

While geographers have examined the complex relationship between masculinities, care and place, another interdisciplinary literature has made progress in conceptualizing 'caring masculinities' (Elliott 2016; Hanlon 2011) which geographical research might usefully advance. This largely theoretical conversation considers the analytical potential of applying a feminist ethics of care perspective to men and masculinities (Elliott 2016; Hanlon 2011), already with some useful application to studies of fatherhood (Doucet 2017; Held 2006; Philip 2014). This promising scholarship has undertaken important work in exploring how gendered binaries in relation to work and care might be decoupled by emphasizing that, like women, men also are involved in, and need, care. Elliott's (2016, 17) conceptualization of 'caring masculinities' explicates how the combination of an ethics of care perspective with masculinities scholarship might aid in reconfiguring 'masculine identities away from values of domination and aggression and toward values of interdependence and care'. Caring masculinities, therefore, require men to reject the practices of domination most associated with hegemonic masculinity, such as 'physical strength, aggression, virility, professional success, wealth, heterosexual prowess, and self-control over such emotions as hurt, grief, or shame' (Calasanti 2003, 307). According to Elliott (2016), the embrace of caring masculinities and the rejection of hegemonic masculinities are constitutive of men's commitment to gender equality, with prospects for social change.

There is great potential for these ideas to be taken forward productively by geographers, who are well placed to develop complex accounts of the social contexts in which caring masculinities might flourish and be supported. In what follows, I describe two empirical studies that are drawn on to consider the contexts in which men's caring responsibilities for others are being opened up, in this case later in the lifecourse.

The empirical studies

Data and findings from two research studies are presented in this chapter. The first, conducted between 2008 to 2011, explored contemporary grandfatherhood and grandfather identities and examined their everyday care practices, lived experiences and intergenerational relationships. The 31 men interviewed were aged 51 to 88, were relatively well resourced financially and had accrued some wealth from secure employment across the lifecourse. For these men, grandfathering reflected the 'leisure/pleasure' grandparenting typically associated with a middle-class demographic (May et al. 2012). This meant that their practices of caregiving for grandchildren were readily incorporated into their everyday leisure time post-employment, including some episodic and temporary childcare provided in line with the needs of their adult children.

The second study was a qualitative longitudinal (QL) research project called 'Men, Poverty and Lifetimes of Care' (MPLC),[1] which explored men's care responsibilities across the lifecourse in low-income families. It began with a qualitative secondary analysis (henceforth QSA) of two QL datasets stored in the Timescapes archive,[2] 'Following Young Fathers' and 'Intergenerational Exchange', and a subsequent additional empirical phase of data generation, including in-depth, qualitative, semi-structured interviews with 26 men aged between 14 and 76. These men were engaging in a diverse range of fathering practices in a number of familial generational positions, including as teenage and young fathers, uncles, grandfathers and great-grandfathers. The recruitment of men across this broad age range produced a dataset that supported an examination of men's care practices in a range of familial generations. Extending from the first study about grandfatherhood, MPLC explored men's changing patterns of care over the lifecourse and across households in contexts of financial constraint.

Substantive conclusions about men's care responsibilities using QSA have been reported elsewhere (Tarrant 2016b; Tarrant and Hughes 2018), and are not the main focus of this chapter. Instead, brief findings are presented from the biographical interviews conducted with eight grandfathers from MPLC. The emerging analyses indicated that these men were typically younger than those in more socio-economically resourced families and that they narrated a 'rescue and repair' style of grandparenting (Emmel and Hughes 2014), which differs from the 'leisure/pleasure' style observed in the previous study. 'Rescue and repair' grandparenting is characterized by the rescue of children from vulnerability and attempts by grandparents to repair intergenerational relationships across familial generations. Caregiving is more intensive and sometimes includes greater dependence on formal health and social care services and/or legal mediation.

Brief analyses from both studies are now presented, demonstrating how spatial and temporal analyses of men's caregiving and investments in intergenerational relations later in the lifecourse and in different socio-economic positions open up and shut down possibilities for new, gendered patterns of care, influencing contexts for the production of caring masculinities.

The carescapes of grandfatherhood

In the first study, the carescapes concept (Bowlby et al. 2011) helped to explicate the ways in which grandfathering is organized spatially and temporally (Tarrant 2013). Grandfatherhood was described as a social arena in which men variously draw on particular constructions of masculinity in response to their practices of caregiving and processes of ageing. Grandfathers do this in complex and diverse ways, referring to distinctly masculine spaces of care later in life as a response to the spatially embedded ageism that constructs their identities and through comparisons of grandfathering to previous lifecourse subjectivities, such as fathering. While grandfathering constitutes a multitude of performances reflecting ageing (and caring) masculinities, it was notable that some men adhered to the maintenance of a gendered division of labour in family care work, while others were performing alternative masculinities that offered the potential to transform and redefine gendered practices and spaces of care.

Most notable was that, for some men, perceptions of masculinity mattered. In describing their practices as grandfathers, many reinforced the ideals of hegemonic or traditional masculinity to make sense of their identities in later life. In alignment with existing research (Barker 2008; Brandth and Kvande 2001) concerning fathering, for example, grandfathers constructed a distinctly masculine concept of caring (Barker 2008) that did not replicate the care given by their wives. Many, for example, drew on spatial metaphors and gendered divisions of labour to distinguish gendered differences between their own practices and those of their wives:

> I think the children see … my wife, more as the provider of food and home comforts if that's? and I'm more the Butlins Redcoat[3] you know, like I always take them swimming or, riding their bike on the prom, well we sometimes do … I'm hopeless at dressing them and bathing them.
>
> *Philip, age 61, still married*

While a dominant narrative, some of the participants also indicated the potential for change. Some men had primary-care responsibilities for children, meaning that fathering and grandfathering opened up opportunities for engaging in care practices later in the lifecourse that they had not been able to engage in as younger fathers. Many described their involvement in practices of social reproduction, like changing nappies and giving feeds, although they often framed themselves as last in line to conduct these practices if wives, daughters and sons were present. In some instances, having children later in life with new partners opened up these opportunities further. For these men in particular, their caring practices were highly visible and sometimes considered unusual by others, particularly in public spaces:

> when [daughter, age 7] was born, I used to be the one off a lot [working from home] but, went to playgroups with her and so, you can imagine that was slightly difficult, not only because I was a man, that was difficult enough because there were very few men main carers. There might be one other man or two other men, but of course they were 20 years younger than me, so I was this older man! So that was you know, nine years ago I was, already then 50 whatever I was, 54 or something, in amongst all these young women with babies and I was odd, I got some funny looks, and people say 'Who's that old grandad?'
>
> *Gerald, age 63, remarried*

Being viewed with caution in public places was not uncommon. Here, David describes this caution when showing an interest in children:

> I like children you know. Unfortunately in this day and age you're not really allowed to like children in the same way, and it's spoiling ... it's a shame, you know, because like I see children in buggies and they look cute and you know, pull faces at them and you know [current partner] will say 'Stop that, you'll get arrested, 'What, for pulling faces?!'
>
> *David, age 51, remarried*

Gerald and David highlight the different ways in which their care responsibilities and interactions with children become subject to surveillance by others in public spaces. For these men, the school playground and the street are key arenas in which their ageing bodies and assumed heterosexuality are viewed with suspicion. While these men are in positions to regender care in the family and in the homespace, the gender and age-related norms and expectations about caregiving intersect and remain pervasive in public spaces.

Low-income grandfathering

In contrast to the first study, the MPLC study examined fathering and grandfathering in low-income family contexts. In line with previous studies (Emmel and Hughes 2014), this study found that, in poverty contexts, older generations are increasingly required to step in to provide care for children. Yet such contexts can also reduce men's abilities to invest in their grandchildren's futures effectively. The intensification of family hardships and responsibilities to provide care for children in the decade following the imposition of austerity is particularly prevalent in this study. Austerity is impacting on families deeply, as Paul, a 61-year-old grandfather with legal responsibility as a kinship carer for three grandchildren, states: 'It's just a hard life and the government's going to make it harder ... I mean financially, you know, to look after a child.' Despite evidence that men in low-income families are taking on unanticipated care responsibilities for children later in life, many are still judged and assessed on their capabilities to care on the basis on their identities as older men. While the more-resourced grandfathers experience this marginalization in public spaces, the participants in the MPLC study experience it in their homes as well. Here, Sam, age 51, discusses being assessed by social services to gain legal responsibility for his 4-year-old grandson. He had a long and difficult battle with social services to be considered for assessment, having raised concerns of possible abuse and neglect when his grandson was originally placed with his mother and maternal grandmother. Despite actively raising concerns and stating his desire to become the primary carer for his grandson, the process had been fraught with challenges to his identity and, in the following discussion, his home. In the following excerpt, he describes an encounter with a social worker during an assessment:

> 'Right, OK, something else that might go against you ...', she starts throwing all these things up. She said it was a 'very manly house'. I said that there was flowers in the vase. She said it was full of boys. And I said – 'Of course, I ain't got no daughters.' I asked if she wanted me to put make-up on and do me hair, and to tell me what she meant by that. She just said it didn't matter.

Sam's narrative indicates that gender biases and assumptions still permeate interventions in social work (e.g. Scourfield 2010), here in relation to who is recognized as capable of taking on the care of a child and in relation to the physical presentation and gendered character of Sam's home. He is the most-resourced person in his family and has continued to express a desire to be recognized as the prime carer for his grandson, yet his capability to provide a good home was judged on the material presentation of his home.

Despite experiencing a range of challenges in becoming legally recognized as a kinship carer, the decision to become primary caregiver can be beneficial. Here Pearce, age 57, explains the pragmatic decision that he and his wife made about caring for their two grandchildren:

> [wife] were on a lot more money than me, and I were having a stressful job. So I just said right, I will jack it in. I'll look after the kids. But it were a massive learning curve because I've never done it before. I mean, when my kids were growing up, I were working on nights for nine years and I went full time on days, but I never had to deal with small ones. At the time, my kids were 9 and 10, when I came off nights. So it were a massive learning curve for me and really, really hard. I can see what women sort of complain about when they've got kids round their feet all day, and it's nice when my wife comes home. We get these to bed and we can talk and I've got adult conversation. It's really, really strange.

Poignant here is the notable impact that becoming the primary carer for his grandchildren had had on Pearce's relationship with his wife. He explained that, in switching roles, they have come to understand each other better. As reported elsewhere (Giesbrecht et al. 2016), this involves learning and a revaluing of care by men, highlighting the possibilities of the regendering of care for supporting processes of gender equality.

Conclusion

This chapter considers how 'caring masculinities' might be taken forward in productive ways by feminist geographers through empirical considerations of the spaces and times in which men provide care. The findings presented highlight the range of contexts in which the production and practices of caring masculinities might be possible and encouraged to flourish, and also where they continue to be viewed with caution. In examining older men's carescapes, the gendered norms and expectations in relation to space and place are made visible, highlighting some, albeit modest, changes in men's care practices in public and intimate arenas such as the home, and also some evidence of continued caution in public spaces.

In mapping out the ways in which geographers have problematized care, this chapter confirms that there is still a need for a critical focus on care in what are our challenging contemporary socio-economic and political times and contexts (Lawson 2007). In low-income families in particular, austerity has produced the conditions in which men as grandfathers are increasingly required to take on primary-care responsibilities for children, yet this is done so in a neoliberal context that is antithetical to care in its emphasis on personal responsibility and individualization (Cox 2010; Jarvis 2007; McDowell 2004). The challenges of providing full-time care for children in unanticipated circumstances are further exacerbated by the retrenchment of the welfare state and government policy, which the men anticipate will make their caring responsibilities harder to manage. This is worrying, given that, as Pearce's narrative demonstrates, when men have the opportunity to engage in caring this can lead to more egalitarian relationships between men and women. At present, the gender equality project is highly individualized and is not being supported to flourish at the structural level (see also Tarrant 2018).

Taken together, these two studies indicate that there is great potential to take forward a geographical and feminist agenda that focuses on the contexts in which caregiving is regendered (Boyer et al. 2017). This might include attention to the intimate spaces and stages of the lifecourse in which caring masculinities are practised, as well as the sustainability of the broader structural processes at play that either hinder or support men's engagements in caregiving across the lifecourse. There is also value in extending the focus from Western societies to those in the Global South or in situations of classic patriarchy. Given the possibilities that attention to care practices among older men have to invert patriarchal power relations within households, there is scope to challenge the absence of studies on men and care in these contexts, including developing understandings of how both caring and masculinity are constructed, viewed and maintained across global contexts.

Notes

1 This study was funded by the Leverhulme Trust between October 2014 and June 2018.
2 www.timescapes.leeds.ac.uk.
3 Butlins is a chain of holiday camps in the UK. 'The Redcoats' are its frontline staff.

Key readings

Gorman-Murray, A., and P. Hopkins. 2015. *Masculinities and Place*. Surrey: Ashgate.
Lawson, V. 2007. "Geographies of Care and Responsibility." *Annals of the Association of American Geographers* 97 (1): 1–11.
Tarrant, A. 2016. "The Spatial and Gendered Politics of Displaying Family: Exploring Material Cultures in Grandfathers' Homes." *Gender, Place and Culture* 23 (7): 969–982.

References

Aitken, S. 2009. *The Intimate Spaces of Fatherhood*. Farnham: Ashgate.
Atkinson, S., V. Lawson, and J. Wiles. 2011. "Care of the Body: Spaces of Practice." *Social & Cultural Geography* 12 (6): 563–572. doi:10.1080/14649365.2011.601238.
Barker, J. 2008. "'Manic Mums' and 'Distant Dads'? Gendered Geographies of Care and the Journey to School." *Health & Place* 17 (2): 413–421. Available at: http://doi:10.1016/j.healthplace.2010.04.001.
Berg, L.D., and R. Longhurst. 2003. "Placing Masculinities and Geography." *Gender, Place and Culture* 10 (4): 351–360. https://doi.org/10.1080/0966369032000153322
Bowlby, S., L. McKie, S. Gregory, and I. MacPherson. 2011. *Interdependency and Care Across the Lifecourse*. London: Routledge.
Boyer, K., E. Dermott, A. James, and J. MacLeavy. 2017. "Regendering Care in the Aftermath of Recession?" *Dialogues in Human Geography* 7 (1): 56–73. https://doi.org/10.1177/2043820617691632.
Brandth, B., and E. Kvande. 2001. "Flexible Work and Flexible Fathers." *Work, Employment & Society* 15 (2): 251–267. Available at: https://doi.org/10.1177/09500170122118940.
Calasanti, T. 2003. "Feminist Gerontology and Old Men." *Journals of Gerontology: Series B* 59 (6): 305–314. https://doi.org/10.1093/geronb/59.6.S305.
Connell, R. 1995. *Masculinities*. Berkeley: University of California Press.
Cox, R. 2010. "Some Problems and Possibilities of Caring." *Ethics, Place & Environment* 13 (2): 113–130. https://doi.org/10.1080/13668791003778800.
Cox, R. 2015. "Working on Masculinity at Home." In: *Masculinities and Place*, edited by A. Gorman-Murray and P. Hopkins, 227–238. Farnham: Ashgate.
Doucet, A. 2017. "The Ethics of Care and the Radical Potential of Fathers 'Home Alone on Leave': Care as Practice, Relational Ontology, and Social Justice." In: *Comparative Perspectives on Work-Life Balance and Gender Equality*, edited by M. O'Brien and K. Wall, 11–28. Online, Springer. Available at: https://repositorio.ul.pt/bitstream/10451/25537/1/ICS_KWall_Comparative_LEI.pdf (accessed 6 February 2018).

Elliott, K. 2016. "Caring Masculinities: Theorizing an Emerging Concept." *Men & Masculinities* 19 (3): 240–259. https://doi.org/10.1177/1097184X15576203.
Emmel, N., and K. Hughes. 2014. "Vulnerability, Intergenerational Exchange, and the Conscience of Generations." In: *Understanding Families over Time: Research and Policy,* edited by J. Holland and R. Edwards, 161–175. Basingstoke: Palgrave.
Evers, C. 2009. "'The Point': Surfing, Geography and a Sensual Life of Men and Masculinity on the Gold Coast, Australia." *Social & Cultural Geography* 10 (8): 893–908. https://doi.org/10.1080/14649360903305783.
Giesbrecht, M., A. Williams, W. Duggleby, J. Ploeg, and J.M. Markel-Reid. 2016. "Exploring the Daily Geographies of Diverse Men Caregiving for Family Members with Multiple Chronic Conditions." *Gender, Place and Culture* 23 (11): 1586–1598. https://doi.org/10.1080/0966369X.2016.1219329.
Gorman-Murray, A. 2008. "Masculinity and the Home: A Critical Review and Conceptual Framework." *Australian Geographer* 39 (3): 367–379. https://doi.org/10.1080/00049180802270556.
Gorman-Murray, A., and Hopkins, P. 2015. *Masculinities and Place.* Farnham: Ashgate.
Hanlon, N. 2011. *Masculinities, Care and Equality,* Basingstoke: Palgrave Macmillan.
Held, V. 2006. *The Ethics of Care: Personal, Political, Global.* Oxford: Oxford University Press.
Hopkins, P., and G. Noble. 2009. "Masculinities in place: situated identities, Relations and Intersectionality." *Social & Cultural Geography* 10 (8): 811–819. https://doi.org/10.1080/14649360903305817.
Hopkins, P., and R. Pain. 2007. "Geographies of Age: Thinking Relationally". *Area* 39 (2): 287–294. https://doi.org/10.1111/j.1475-4762.2007.00750.x.
Jackson, P. 1991. "The Cultural Politics of Masculinity: Towards Social Geography." *Transactions of the Institute of British Geography* 16 (1): 199–213. doi:10.2307/622614.
Jarvis, H. 2007. "Home Truths about Care-less Competitiveness." *International Journal of Urban and Regional Research* 31: 207–214. https://doi.org/10.1111/j.1468-2427.2007.00715.x.
Laurendeau, J. 2014. "'Just Tape It Up for Me, OK?': Masculinities, Injury and Embodied Emotion." *Emotion, Space and Society* 12: 11–17. https://doi.org/10.1016/j.emospa.2013.03.010.
Lawson, V. 2007. "Geographies of Care and Responsibility." *Annals of the Association of American Geographers* 97 (1): 1–11. https://doi.org/10.1111/j.1467-8306.2007.00520.x.
Longhurst, R. 2005. "Man-breasts: Spaces of Sexual Difference, Fluidity and Abjection." In: *Spaces of Masculinities,* edited by B. van Hoven and K. Hörschelmann, 152–165. Abingdon: Routledge.
May, V., J. Mason, and L. Clarke. 2012. "Being There Yet Not Interfering: The Paradoxes of Grandparenting." In: *Contemporary Grandparenting: Changing Family Relationships in Global Contexts,* edited by A. Sara and V. Timonen, 139–158. Bristol: Policy Press.
McDowell, L. 2004. "Work, Workfare, Work/life Balance and an Ethic of Care." *Progress in Human Geography* 28 (2): 145–163. https://doi.org/10.1191/0309132504ph478oa.
McDowell, L. 2009. *Working Bodies: Interactive Service Employment and Workplace Identities.* Oxford: Wiley-Blackwell.
McEwan, C., and M. Goodman. 2010. "Place Geography and the Ethics of Care: Introductory Remarks on the Geographies of Ethics, Responsibility and Care." *Ethics, Place and Environment* 13 (2): 103–112. https://doi.org/10.1080/13668791003778602.
Milligan, C. 2009. *There's No Place Like Home: Place and Aare in an Ageing Society.* Farnham: Ashgate.
Milligan, C., and J. Wiles. 2010. "Landscapes of Care." *Progress in Human Geography* 34 (6): 736–754. https://doi.org/10.1177/0309132510364556.
Milligan, C., S. Payne, A. Bingley, and Z. Cockshott. 2013. "Place and Well-being: Shedding Light on Activity Interventions for Older Men." *Ageing & Society* 35 (1): 124–149. https://doi.org/10.1017/S0144686X13000494.
Philip, G. 2014. "Fathering after Separation or Divorce: Navigating Domestic, Public and Moral Spaces." *Families, Relationships and Societies* 3 (2): 219–233. https://doi.org/10.1332/204674314X14017856302453.
Pilkey, B. 2014. "Queering Heteronormativity at Home: Older Gay Londoners and the Negotiation of Domestic Materiality." *Gender, Place & Culture* 21 (9): 1142–1157. doi:10.1080/0966369X.2013.832659.
Power, A., and E. Hall. 2017. "Placing Care in Times of Austerity." *Social & Cultural Geography* 19 (3). https://doi.org/10.1080/14649365.2017.1327612.
Russell, R. 2007. "Men Doing 'Women's Work': Elderly Men Caregivers and the Gendered Construction of Care Work." *Journal of Men's Studies* 15 (1): 1–18. https://doi.org/10.3149/jms.1501.1.

Scourfield, J. 2010. "Gender and Child Welfare in Society: Introduction to Some Key Concepts." In: *Gender and Child Welfare in Society,* edited by B. Featherstone, C.-A. Hooper, J. Scourfield, and J. Taylor, 1–25. Oxford: Wiley-Blackwell.

Tarrant, A. 2013. "Grandfathering as Spatio-temporal Practice: Conceptualizing Performances of Ageing Masculinities In Contemporary Familial Carescapes." *Social & Cultural Geography* 14 (2): 192–210. https://doi.org/10.1080/14649365.2012.740501.

Tarrant, A. 2016a. "The Spatial and Gendered Politics of Displaying Family: Exploring material Cultures in Grandfathers' Homes." *Gender, Place and Culture* 23 (7): 969–982. https://doi.org/10.1080/0966369X.2015.1073703.

Tarrant, A. 2016b. "Getting out of the Swamp? Methodological Reflections on Using Qualitative Secondary Analysis to Develop Research Design." *International Journal of Social Research Methodology*, 20 (6): 599–611. https://doi.org/10.1080/13645579.2016.1257678.

Tarrant, A. 2018. "Care in an Age of Austerity: Men's Care Responsibilities in Low-income Families." *Ethics and Social Welfare* 12 (1): 34–48. https://doi.org/10.1080/17496535.2017.1413581.

Tarrant, A., and K. Hughes. 2018. "Qualitative Secondary Analysis: Building Longitudinal Samples to Understand Men's Generational Identities in Low Income Contexts." *Sociology.* https://doi.org/10.1177/0038038518772743.

van Hoven, B., and K. Hörschelmann. 2006. *Spaces of Masculinities.* London: Routledge.

Waitt, G., and A. Gorman-Murray. 2007. "Homemaking and Mature-age Gay Men 'Down Under': Paradox, Intimacy, Subjectivities, Spatialities, and Scale." *Gender, Place and Culture* 14 (5): 569–584. https://doi.org/10.1080/09663690701562305.

Walsh, K. 2011. "Migrant Masculinities and Domestic Space: British home-making Practices in Dubai." *Transactions of the Institute of British Geographers* 36: 516–529. https://doi.org/10.1111/j.1475-5661.2011.00442.x.

Wheelock, J., and K. Jones. 2002. "Grandparents are the Next Best Thing: Informal Child Care for Working Parents in Urban Britain." *Critical Social Policy* 31 (3): 441–463. https://doi.org/10.1017/S0047279402006657.

33

GIVING BIRTH TO GEOGRAPHIES OF YOUNG PEOPLE

The importance of feminist geography beyond feminist geography

Ann E. Bartos

Introduction

Figuratively and discursively, throughout time and space, women and children have been conflated in policy, popular media, academic scholarship, everyday violence and geopolitical warfare. Cynthia Enloe referred to this as the 'womenandchildren' phenomenon, in which women and children are relegated to the private sphere and, therefore, are void of political agency, resulting in their similar states of 'vulnerability' (1990). Such vulnerabilities can also be pitted against one another in times when material or political resources as scarce: 'womenandchildren' can lead to a 'woman versus children' paradigm (e.g. Rosen and Twamley 2018; Twamley, Rosen and Mayall 2016). Rosen and Twamley (2018) suggest that the 'woman–child' problematic is important to unpack, because the 'everyday entanglements of women and children ... could add to the other's understandings of the dynamic processes whereby inequalities are made, replicated and challenged' (Rosen and Twamley 2018, 18). The authors suggest that, to better illuminate and untangle these relationships, scholarship on gender and age should focus on 'looking in, looking out, looking back, widening the frame and breaking away' (ibid., 12). This chapter heeds their suggestion to demonstrate that within the discipline of geography, feminist geographic scholarship has played a significant role in the development of geographies of children, youth and families (GCYF). I 'look in, out and back' within these two geographic subfields to highlight some specific insights that feminist geography has offered to research on young people.

In the following section, I provide a brief overview of feminist geography's origins and the influence that it has had on the development of GCYF. Feminist geographers were pushing the boundaries of geographic knowledge production in the discipline in the 1980s to include the voices and experiences of those beyond the white, male, rational political actor; women and subaltern voices and perspectives were the keystones of feminist geography. It was soon after the introduction of gender and feminist issues into the discipline that attention began to be paid to some of the least-recognized knowledge producers at the time: children.

I trace these origins with the intention of highlighting points of connection between these two subdisciplines. Building on this brief historical overview, I then discuss three specific areas of research within GCYF that show evidence of feminist scholarship: the body, emotions, and care. The conclusion of this chapter 'breaks away and widens the frame' to consider other areas of inquiry informed by contemporary feminist geographic research and pertinent to GCYF.

Feminist foundations of geographic research on youth

The feminist geography flagship journal, *Gender, Place and Culture* (*GPC*), celebrated its twenty-fifth anniversary in 2018. To honour this success, the journal focused on publishing its own history and, in the process, the history of feminist geographies (e.g. Gorman-Murray and Hopkins 2017; Lynch 2018; Yeoh and Ruwanpura 2018). The journal's inaugural year published two issues, in 1994, under the editorship of Liz Bondi and Mona Domosh. In the first paragraph of the journal's first issue, Bondi and Domosh wrote:

> To the pioneers of the 1970s, a journal devoted to feminist geography must have seemed a distant dream. Although the women's liberation movement motivated many to take action in the streets, few dared to move that action into the halls of the academy. Yet feminist voices persisted and slowly gathered momentum. And from these modest yet courageous beginnings, feminist geography has now permeated most, if not all, corners of the discipline and has become a force that cannot be ignored in accounts of contemporary geographical knowledge.
>
> *1994, 3*

At the time when Bondi and Domosh were writing about this disciplinary shift, prominent feminist geographers were already serving on editorial boards for top journals and holding faculty positions in geography departments (Bondi and Domosh 2003), and feminist analysis was becoming key to framing debates across the discipline (e.g. Dowler and Sharp 2001; England 2003; Rose 1993; Staeheli 1996). While it may not have been 'normal' to discuss 'gender' or 'feminism' (what Domosh referred to as the 'f-word' [Domosh and Ruwanpura 2017]), feminist geographers called attention to the subtle and overt ways that the discipline favoured and reproduced masculinist discourses of knowledge production. In other words, the foundations of feminist geography focused on power relations.

Inspired by poststructural and postmodern theories, feminists were interested in the ways that knowledge and power were inextricably related (Gibson-Graham 1994). Traditional (masculinist) notions of power focused on the 'capacity to control or shape an event, person or process' (Staeheli, Kofman and Peake 2004, 7). Feminist geographers worked with an expanded understanding of power to include attention to how it is 'multifaceted, diffuse, and relational, even as it is understood that power is not equally distributed or always available in its different expressions' (ibid., 8). This more amplified definition of power makes space for exploring power *relations*. Power relations are manifested in various ways through multiple sites and scales, such as nation states, institutions, bodies and discourses. Concerned about the power relations inherent in geographic discourse, feminist geographers exposed that some people, predominantly white, heterosexual and cis-gendered men set the agenda in the academy while others worked hard to revise the agenda toward more inclusivity (e.g. Domosh 1991; Massey 1991; Monk and Hanson 1982). Such scholarship was necessary for researchers to convincingly argue for the value of including female and marginalized perspectives and politics into mainstream research and practice. Importantly, the more comprehensive and nuanced understandings of power relations were fundamental to the development of contemporary geographic research on children and youth (Aitken 2018; Holloway 2014; Tisdall and Punch 2012).

Prior to the poststructural turn, research on children in geography focused on their spatial cognition and map-reading abilities (e.g. Blaut and Stea 1971). Despite 'the rather obvious point … that maps and everything they contain are referents to systems of power-knowledge' (Aitken 2001, 49), early geographical research on children was not about power, per se. However, early studies on children's mapping abilities laid the foundation for more theoretically informed research challenging traditional developmental psychological assumptions about children and youth at the time. Insights on children's mapping abilities contradicted the assumed linear progression of childhood and adolescence in which significant 'milestones' determined 'progress' and '(ab)normalcy'; children demonstrated physical and intellectual capabilities through maps that rose beyond traditional developmental psychology metrics. In effect, these studies demonstrated that children and young people are important knowledge producers that, at the time, were essentially unbeknownst to traditional human geography. Over the next few decades, the 'new social studies of childhood' (NSCC) influenced and catapulted further research into more critical and comprehensive geographies of children, youth and families.

In 2003, *Children's Geographies* was launched. It marked a turning point in the discipline which 'signal(led) that children's geographies has finally "come of age"' (Matthews 2003a, 5). The journal 'sets out to unravel the complexities and ironies of childhood(s) and to challenge common (mis)conceptions conveniently perpetrated by the selective adult lens' (Matthews 2003b: 147). *Children's Geographies* and research across the discipline that focused on children and youth-based research was guided by the three basic (and interrelated) tenets of the NSCC: that children are *beings* in their own right; a recognition of the social construction of childhood(s); and that there is no universal 'child' or 'childhood'. This new paradigm effectively reconceptualized childhood and paved the way for a wide array of qualitative (and frequently ethnographic) empirical studies aimed at providing and privileging the voices of children and young people.

Since the journal's origins, geographic research on young people has expanded across the discipline (Robson, Horton and Kraftl 2013). A variety of geographers who engage with issues around children and youth publish in the top journals in the discipline (e.g. Ansell 2009; Hopkins et al. 2011; Kraftl 2015; Ruddick 2003) and have produced key manuscripts (e.g. Aitken 2001; Hopkins 2010; Katz 2004) and edited texts with notable publishers on child/youth specific research (e.g. Ergler, Kearns and Witten 2017; Holloway and Valentine 2000; Holt 2010; Jeffrey and Dyson 2008; Skelton and Valentine 1998). While youth issues may not be at the forefront of all pressing questions across the discipline, it is not uncommon to find important and provocative research on children and youth within diverse fields such as urban geography, political geography, health geography and migration studies.

Despite the decade separating the seminal issues in *GPC* and *Children's Geographies*, an overlap between these two subdisciplinary threads was apparent from their earliest publications, and this continues today. As discussed above, both feminist geography and GCYF view power as relational. This theme has been evident in various and overlapping ways over time. For example, GCYF and feminist geography share an interest in how space and scale are produced rather than accepted a priori. Individual bodies shape space and scale differently, and hence the everyday embodied experiences of place are important. Similarly, axes of difference are neither universal nor 'natural'; difference is dynamic and contested. Therefore, both subdisciplines argued that notions of difference influence and impact on political agency, which challenges traditional approaches to formal politics and opens avenues for more inclusive and transgressive research. In the early issues of both journals, the focus on space, scale and difference were evident through empirical investigations around, for example, race, sexuality, the body, technology and urban and rural dynamics (e.g. Aitken and Marchant 2003; Beazley 2003; Bell et al. 1994; Jackson 1994; Jones, Williams and Fleuriot 2003; Kobayashi and Peake 1994; Longhurst 1995; Morris-Roberts 2004; Pratt and Hanson 1994; Vanderbeck and Dunckley 2003).

The brief histories of feminist geography and GCYF outlined above attempt to provide a window into their complementary relationship over time, as opposed to one of competition (c.f. Rosen and Twamley 2018; Twamley, Rosen and Mayall 2016). Attention to the important feminist geographic research occurring prior to, and in tandem with, the origins of GCYF provides insights into the foundations and research directions in geography around children, youth and their families. In the next section, I bring attention to three areas of research in GCYF that emerge from the application of feminist methods and theories to contexts of children and youth: research on the body, emotions, and care.

Feminist and youth research on bodies, emotions and care

Feminist research on bodies and embodiment has been central to the development of feminist geographies (Longhurst 1997). Feminists have played an important role in questioning the Cartesian duality between the mind and the body, and have legitimately argued that the body is a site of power relations. Therefore, the geographic scale of the body lends insights into place and space that disembodied territories or locations fail to recognize. Bodies are not universal, nor are they unproblematic. Bodies are sexed, raced, gendered and categorized in various ways that stratify society in both empowering and dangerous, or harmful, ways. For example, the problematic woman–child dyad that opened this chapter is further troubled through the lens of embodiment. In that dyad, at least one body, if not both, is erased and enfolded into the other. A focus on embodiment brings attention to the different spaces, scales and sites that both these bodies occupy and their ensuing agentic potential.

GCYF highlights that young people occupy liminal spaces beyond the radar of traditional research on youth, such as the playground or the home. However, exploring the various spatialities of young people through embodied relationships with place provides novel perspectives on, for example, identity formation, mundane political relations and geographies of inclusion and exclusion. In fact, Horton and Kraftl (2006) encouraged children's geographers to provide 'fresh and rich' insights into the subfield through an explicit focus on the body and embodiment. Horschelmann and Colls (2010) provide a notable contribution with their edited collection to demonstrate that research on children and youth's bodies needs to focus on young people's imagination of what is possible in the context of wider social structures such as ethnicity, gender or class; on how youth embody and transgress deviance; and how young bodies are relationally situated to other bodies and, therefore, immersed in wider political relations.

In contrast to the erasure resulting from the women-child dyad, children's bodies can be used as a trope in popular media and discourse. The child-body trope is evident particularly in discourses of risk: children's and youth's bodies are categorized as either at risk or risky (e.g. Pain 2004). For example, the child is both victim and perpetrator in their grim future, and these are narrated through specific bodies such as the starving child, the obese toddler, the teenage mother and the juvenile detainee (c.f. Aitken 2001; Katz 2017). Aitken argued that 'what focuses the moral panics that surround the activities of children and youths are problematic social constructions of young people and the simultaneously disembodied and disembedded contexts of their lives' (2001, 25). In such discursive projections, disembodied young people contradictorily lack agency yet remain somewhat responsible for their dire predicament. GCYF has made a concerted effort over the years to discount such misconceptions about children and youth with research that seeks to draw attention to the multiplicity, complexity and diversity of children's embodied experiences of their life-worlds (Evans 2010; Herrera et al. 2009; Woodyer 2008).

Feminist geographies have also shaped an important trajectory in research on the relationship between place, space and emotions. Davidson, Bondi and Smith (2005) and others (e.g. Olson 2016; Pile 2010) developed a research agenda that took emotions seriously as a topic of inquiry. As the number of empirically rich case studies of children's lives grew, Horton, Kraftl and Tucker (2008) made a strong argument that children's geographers were well placed to contribute to the 'emotional turn'; studying the emotional lives of children and young people seemed a logical extension to the vast trove of research on young people's bodies and embodiment. Such investigations of young people's everyday embodied and emotional engagements with place and space have offered new theoretical insights in the subfield and closely align with feminist geographies. For example, Holt et al. (2013) drew on Judith Butler and Pierre Bourdieu to discuss how children's social relations are paramount to the development of their 'emotional capital'. Bartos (2013) integrated Sara Ahmed's work with traditional geographic research on 'sense of place' to better understand how children experience place through direct sensory engagements felt through their bodies. Blazek (2013) drew on psychoanalysis and the research of Anna Freud to unpack how children use a variety of 'defence mechanisms' to emotionally manage complicated entanglements between a child's lived experience and her imagined realities. These and other studies attempt not only to prioritize the emotional lives of children and young people but to bring attention to the ways that, regardless of the life stage, emotions may reflect wider social relations and confluences than disembodied and non-emotional approaches to research.

Thirdly, feminist geographic insights have played a pivotal role in contemporary research on the intersections of care and young people. The traditional and problematic portrayal of women and children as the quintessential care-giving relationship inspired robust interdisciplinary feminist scholarship, aimed at challenging the presumed apolitical, non-confrontational, and 'naturalness' of such a relationship (e.g. Ruddick 1989). Feminist care theorists such as, Tronto (1993, 2013), Held (2006) and Robinson (2011), argue that while caring practices are performed by gendered and sexed bodies, these activities occur within a capitalist patriarchal society that fundamentally devalues care work and care providers. Therefore, research has focused on illuminating the many actors, spaces and scales at which care is given and received to expose how caring activities are necessary for our personal and collective survival.

A key issue across much geographic research on care is a focus on global care chains and the myriad actors responsible for, and often subservient in, Western frameworks of care provisions and practices (Raghuram, Madge and Noxolo 2009). However, Elizabeth Olson (2019) argues that such global care chains need to be reconceptualized with attention not only to gender, race, class and ethnicity but also to age. Her research illuminated how, historically, children caregivers have played and important role in curbing infant mortality and will remain key actors in the modern global economy. Similarly, in Ruth Evans' research with young Sub-Saharan Africans and young African migrants in the UK, youth have been key caregivers not only for their own siblings but also for other family members, including HIV+ parents (Evans 2011, 2014; Evans and Thomas 2009). These studies and others (e.g. Horton and Pyer 2017; Robson 2004) illuminate the often hidden roles that young people play in caregiving relationships and how their experiences and practices of care are interrelated with wider economic geographies of social reproduction.

Beyond research on the practices of care, feminist research on care has expanded the ways in which we understand political relations. Feminist geographers have drawn on the insights of care theorists to argue that care challenges the dominant assumptions of the autonomous, self-made, masculine political subject (Brown 2003; Lawson 2007). Feminist research on the politics of care sheds light on how we are all interdependent and relational. All humans require care and all of us, at one time or another, will also provide care. Tronto (1993, 2013) has argued that this is fundamental to what it means to be human, and therefore should not continue to be ignored or denied.

Those studying children and youth have found the lens of care to be a particularly useful and important framework to bring attention to children's political subjectivities. As feminist care scholars have argued, caring is a political act, because it requires a fundamental reframing of how we understand power relations. In fact, while the context of care, caring practices and dominant caring actors are geographically specific, care, in general, is underappreciated and undervalued across the globe because of an incomprehensive understanding of how essential caring is to our individual and collective success (Raghuram 2016). When children are introduced into the conversation and understood as legitimate caring agents, as competent social actors capable of providing and also receiving care, their political potential is realized (e.g. Bartos 2012; Kallio and Bartos 2017; Olson 2019).

While feminist geographic insights on the body, emotions and care have been highly influential to GCYF research, it is also important to acknowledge that youth research on the body, emotions and care has the potential to give back to feminist geography. For example, research on the child-body and embodiment can offer nuance between the human and the nonhuman; Olson (2018) argues that when children's bodies are the centre of popular discourse, their bodies are also deemed *less than*. This *less than* has implications for questions around responsibility and participation. Concomitantly, while children's politics and political relations may be dismissed as unimportant or non-existent in traditional academic research or popular discourse, children and youth remain central to a wide variety of formal political decisions and policies, and are largely governed without their consent. Questioning where and how children's politics are silenced can potentially illuminate and problematize the practices and approaches to building an effective social movement around, for example, sexual violence. Yet, this would require a commitment to revisiting some of the basic assumptions about what and who is worthy of citizenship (e.g. Staeheli et al. 2013). Such feminist geographic scholarship has inspired children and youth researchers with frameworks for investigating how politics develops and takes shape in young people's lives, and the future implications of these politics (e.g. Kallio and Häkli 2016). Such insights can be shared *back* with feminist geographies, and can lead to new and fruitful inquiries made visible through the lens of youth.

The three themes raised in this section provide only a snapshot of some of the intersections between GCYF and feminist geographies. They are certainly not the only examples that I could have chosen to elaborate. Nonetheless, this brief account of the themes of embodiment, emotions and care demonstrates the important work that feminist geography offers to the discipline as a whole, not only to the subfield of GCYF. In the concluding section, I consider how 'breaking away and widening the frame' can help to identify some potential areas for expansion.

Conclusion

This chapter 'looked in, out and back' at the fields of feminist geography and GCYF to highlight their complementary relationship over time (c.f. Rosen and Twamley 2018). Both subdisciplines demonstrate the value and importance of exploring and exposing the voices and experiences of subaltern and other disadvantaged actors to academic discourse. This commitment builds on a recognition that power and knowledge are deeply interrelated and immersed in a wide array of (often) unequal relationships. Both feminist and GCYF geographers bring attention to such relationships to offer new insights into more traditionally (masculine) approaches to human geography, which tended to ignore the context that age, gender identity and other axes of difference offer to the construction of knowledge. The birth of *Gender, Place and Culture* and *Children's Geographies*, in 1994 and 2003 respectively, continue to grow and flourish, a testament to the importance of both subdisciplines to geography.

Looking in, out and back provided a lens to explore how the feminist topics of embodiment, emotions and care have informed some GCYF research. For example, both subfields have raised interesting questions around the politics of research, representation and methods; issues of sexuality; the intersections between the human and the nonhuman in relation to agency; and the significance of intersectionality as method, theory and practice. While I do not suggest that all children and youth geographers are feminists, and I acknowledge the apprehension of academic labels (e.g. Holloway 2014), it is short-sighted to imagine academic knowledge as arising in silos. Rather, a cross-pollination of ideas, people, discourses and arguments is necessary for academic knowledge to expand, develop and stretch boundaries. I have suggested that feminist geographic insights have been invaluable to the development of research in geography on children, youth and their families. Other scholars may wish to draw out the intersections between GCYF with other important geographic subfields such as urban geography, health geography, political geography and development geography.

Rosen and Twamley (2018) suggest that 'widening the frame and breaking away' are two further vantage points for scholars to interrogate the relationships between gender and age-based research. Scholarship that 'widens the frame' could include research that incorporates the people and places tangentially to the intended research participants, for instance, women and children. Research from this vantage point would include those who do *not* identify as 'woman', or people *beyond* a young person's immediate surroundings. Importantly, both feminist and youth geographers are already engaged with this vantage point. For example, from the early issues of *GPC*, men and masculinity were understood as legitimate actors and topics of investigation within feminist geographic research (e.g. Bell et al. 1994). Similarly, insights on children's agency as not only *beings* but also as *becomings* required attention to their wider social, cultural and economic relationships (e.g. Katz 2004).

Finally, 'breaking away' includes:

> the possibility of becoming otherwise in ways which dismantle not only the [woman-child] binary but subject positions and power relations altogether. [...] to jettison both femininity and childhood as well as their dualisms: masculinity and adulthood [...] to do away with historically sedimented categories and their differential privileges.
>
> *Rosen and Twamley 2018, 13–14*

In order to 'break away' effectively, more focus is needed on exploring the power imbalances that undergird and maintain oppressions, regardless of age or gender. For example, feminist geographies and GCYF have gained insights from critical race scholarship and the importance of an intersectionality and anti-essentialism approach: a recognition that there are no universal categories of identity and that our various experiences of oppression are not transferrable to other bodies. 'Breaking away' encourages scholarship that dismantles the power relations that enable patriarchy, racism and capitalism to persist and to structure society in disempowering ways for both women *and* children, especially as they embody other bases of oppression (Grillo 1995). Feminist geographies and GCYF research on intersectionality will be an important area of inquiry for all future critical research in these subfields (Hopkins 2018; Johnston 2018; Konstantoni and Emejulu 2017; Mollett and Faria 2018; Rodó-de-Zárate 2017). Focusing on intersectionality has the potential to revisit and reframe current debates within both subfields, and will continue to push the boundaries of academic knowledge production in ways that remain focused on the fundamental theme that guided the origins of both feminist and youth geographies: the relevance of the interconnections between people, place, power and knowledge.

Key readings

Holloway, S.L., and G. Valentine, eds. 2000. *Children's Geographies: Playing, Living, Learning*. London: Routledge.

Holt, L., ed. 2010. *Geographies of Children, Youth and Families: An International Perspective.* Oxford: Routledge.

"Intersectionality and Childhood Studies: A Critical Dialogue across Time, Space and Place." *Children's Geographies* 2017 Special issue 15 (1).

References

Aitken, S.C. 2001. *Geographies of Young People: The Morally Contested Spaces of Identity*, edited by K. Konstantoni, M. Kustatscher, and A. Emejulu. London: Routledge.

Aitken, S.C. 2018. "Children's Geographies: Tracing the Evolution and Involution of a Concept." *Geographical Review* 108: 3–23.

Aitken, S.C., and R.C. Marchant. 2003. "Memories and Miscreants: Teenage Tales of Terror." *Children's Geographies* 1: 151–164.

Ansell, N. 2009. "Producing Interventions For AIDS-affected Young People on Lesotho's Schools: Scalar Relations and Power Differentials." *Geoforum* 40: 675–685.

Bartos, A.E. 2012. "Children Caring for Their Worlds: The Politics of Care and Childhood." *Political Geography* 31: 157–166.

Bartos, A.E. 2013. "Children Sensing Place." *Emotion, Space and Society* 9: 89–98.

Beazley, H. 2003. "Voices from the Margins: Street Children's Subcultures in Indonesia." *Children's Geographies* 1: 181–200.

Bell, D. et al. 1994. "All Hyped Up and No Place to Go." *Gender, Place & Culture* 1: 31–47.

Blaut, J., and D. Stea. 1971. "Studies of Geographic Learning." *Annals of the Association of American Geographers* 61: 387–393.

Blazek, M. 2013. "Emotions as Practice: Anna Freud's Child Psychoanalysis and Thinking – Doing Children's Emotional Geographies." *Emotion, Space and Society* 9: 24–32.

Bondi, L., and M. Domosh. 1994. "Editorial." *Gender, Place & Culture* 1: 3–4.

Bondi, L., and M. Domosh. 2003. "Gender, Place and Culture: Ten Years On." *Gender, Place & Culture* 10: 3–4.

Brown, M. 2003. "Hospice and the Spatial Paradoxes of Terminal Care." *Environment and Planning D* 35: 833–851.

Davidson, J., L. Bondi, and M. Smith, eds. 2005. *Emotional Geographies*. Aldershot: Ashgate.

Domosh, M. 1991. "Beyond the Frontiers of Geographical Knowledge." *Transactions of the Institute of British Geographers* 16: 488–490.

Domosh, M., and K.N. Ruwanpura. 2017. "A Conversation between Mona Domosh and Kanchana N. Ruwanpura: Reflections on the Past, Present and Future on GPC's 25th Anniversary." *Gender, Place & Culture* 25: 4–12.

Dowler, L., and J.P. Sharp. 2001. "A Feminist Geopolitics?" *Space and Polity* 5: 165–176.

England, K. 2003. "Towards a Feminist Political Geography?" *Political Geography* 22: 611–616.

Enloe, C. 1990. "Womenandchildren: Making Feminist Sense of the Persian Gulf Crisis." *Village Voice*, 25 September.
Ergler, C., R. Kearns, and K. Witten, eds. 2017. *Geographies of Children's Health and Wellbeing in Urban Environments*. London: Routledge.
Evans, B. 2010. "Anticipating Fatness: Childhood, Affect and the Pre-Emptive 'War on Obesity'." *Transactions of the Institute of British Geographers* 35: 21–38.
Evans, R. 2011. "'We Are Managing Our Own Lives': Life Transitions and Care in Sibling-headed Households Affected by AIDS in Tanzania and Uganda." *Area* 43: 384–396.
Evans, R. 2014. "Children as Caregivers." In: *Handbook of Child Well-Being: Theories, Methods and Policies in a Global Perspective*, edited by A. Ben-Arieh, F. Casas and J.E. Korbin, 1893–1916. Dordrecht: Springer.
Evans, R., and F. Thomas. 2009. "Emotional Interactions and an Ethics of Care: Caring Relations in Families Affected by HIV and AIDS." *Emotion, Space and Society* 2: 111–119.
Gibson-Graham, J.K. 1994. "'Stuffed If I Know!': Reflections on Post-Modern Feminist Social Research." *Gender, Place & Culture* 1: 205–224.
Grillo, T. 1995. "Anti-Essentialism and Intersectionality: Tools to Dismantle the Master's House." *Berkeley Journal of Gender, Law & Justice* 10: 16–30.
Gorman-Murray, A., and P. Hopkins. 2017. "Andrew Gorman-Murray and Peter Hopkins in Conversation: Reflections on Masculinities and Sexualities Research on GPC's 25th Anniversary." *Gender, Place & Culture* 25: 317–324.
Held, V. 2006. *The Ethics of Care: Personal, Political and Global*. Oxford: Oxford University Press.
Herrera, E., G.A. Jones, and S.T. De Benitez. 2009. "Bodies on the Line: Identity Markers among Mexican Street Youth." *Children's Geographies* 7: 67–81.
Holloway, S.L. 2014. "Changing Children's Geographies." *Children's Geographies*, 12: 377–392.
Holloway, S.L., and G. Valentine, eds. 2000. *Children's Geographies: Playing, Living, Learning*. London: Routledge.
Holt, L., ed. 2010. *Geographies of Children, Youth and Families: An International Perspective*. Oxford: Routledge.
Holt, L., S. Bowlby, and J. Lea. 2013. "Emotions and the Habitus: Young People with Socio-Emotional Differences (Re)producing Social, Emotional and Cultural Capitali Family and Leisure Space-Times." *Emotion, Space and Society* 9: 33–41.
Hopkins, P. 2010. *Young People, Place and Identity*. New York: Routledge.
Hopkins, P. 2018. "Feminist Geographies and Intersectionality." *Gender, Place & Culture* 25: 585–590.
Hopkins, P. et al. 2011. "Mapping Intergenerationalities: The Formation of Youthful Religiosities." *Transactions of the Institute of British Geographers* 36: 314–327.
Horschelmann, K., and R. Colls, eds. 2010. *Contested Bodies of Childhood and Youth*. London: Palgrave.
Horton, J., and P. Kraftl. 2006. "What Else? Some More Ways of Thinking and Doing 'Children's Geographies'." *Children's Geographies* 4: 69–95.
Horton, J., P. Kraftl, and F. Tucker. 2008. "The Challenges of 'Children's Geographies': A Reaffirmation." *Children's Geographies* 6: 335–348.
Horton, J., and M. Pyer. 2017. "Introduction: Children, Young People and 'Care'." In *Children, Young People and Care,* edited by J. Horton and M. Pyer, 1–25. London: Routledge.
Jackson, P. 1994. "Black Male: Advertising and the Cultural Politics of Masculinity." *Gender, Place & Culture* 1: 49–59.
Jeffrey, C., and J. Dyson, eds. 2008. *Telling Young Lives: Portraits of Global Youth*. Philadelphia, PA: Temple University Press.
Johnston, L. 2018. "Intersectional Feminist and Queer Geographies: A View from 'Down-Under'." *Gender, Place & Culture* 25: 554–564.
Jones, O., M. Williams, and C. Fleuriot. 2003. "'A New Sense of Place?' Mobile 'Wearable' Information and Communications Technology Devices and the Geographies of Urban Childhood." *Children's Geographies* 1: 165–180.
Kallio, K.P., and A.E. Bartos. 2017. "Children's Caring Agencies." *Political Geography* 58: 148–150.
Kallio, K.P., and J. Häkli. 2016. "Geosocial Lives on Topological Polis: Mohamed Bouazizi as a Political Agent." *Geopolitics* 17 (1): 1–19.
Katz, C. 2004. *Growing up Global: Economic Restructuring and Children's Everyday Lives*. Minneapolis: University of Minnesota Press.
Katz, C. 2017. "Revisiting Minor Theory." *Environ Plan D* 35: 596–599.
Kobayashi, A., and L. Peake. 1994. "Unnatural Discourse. 'Race' and Gender in Geography." *Gender, Place & Culture* 1: 225–243.

Konstantoni, K., and A. Emejulu. 2017. "When Intersectionality Met Childhood Studies: The Dilemmas of a Travelling Concept." *Children's Geographies* 15: 6–22.
Kraftl, P. 2015. "Alter-Childhoods: Biopolitics and Childhoods in Alternative Education Spaces." *Annals of the Association of American Geographers* 105: 219–237.
Lawson, V. 2007. "Geographies of Care and Responsibility." *Annals of the Association of American Geographers* 97: 1–11.
Longhurst, R. 1995. "Viewpoint: The Body and Geography." *Gender, Place and Culture* 2: 97–106.
Longhurst, R. 1997. "(Dis)Embodied Geographies." *Progress in Human Geography* 21: 486–501.
Lynch, C. R. 2018. "25th Anniversary Retrospective: A Postcapitalist Politics." *Gender, Place & Culture* 25 (7): 1089–1091. doi: 10.1080/0966369X.2018.1462756.
Massey, D. 1991. "Flexible Sexism." *Environment and Planning D: Society and Space* 9: 31–57.
Matthews, H. 2003a. "Coming of Age for Children's Geographies." *Children's Geographies* 1: 3–5.
Matthews, H. 2003b. "Editorial." *Children's Geographies* 1: 147–149.
Mollett, S., and C. Faria. 2018. "The Spatialities of Intersectional Thinking: Fashioning Feminist Geographic Futures." *Gender, Place & Culture* 25: 565–577.
Monk, J., and S. Hanson, S. 1982. "On Not Excluding Half of the Human in Human Geography." *Professional Geographer* 34: 11–23.
Morris-Roberts, K. 2004. "Girls' Friendships, 'Distinctive Individuality' and Socio-Spatial Practices of (Dis) Identification." *Children's Geographies* 2: 237–255.
Olson, E. 2016. "Geography and Ethics II: Emotions and Morality." *Progress in Human Geography* 40: 830–838.
Olson, E. 2018. "Geography and Ethics III: Whither the Next Moral Turn." *Progress in Human Geography* 42 (6): 937–948.
Olson, E. 2019. "'The Largest Volunteer Life Saving Corps in the World': Centering Child Caregiving in Histories of U.S. Human Security through the Little Mothers' League." *Social & Cultural Geography* 20 (4): 445–464.
Pain, R. 2004. "Introduction: Children at Risk." *Children's Geographies* 2: 65–67.
Pile, S. 2010. "Emotions and Affect in Recent Human Geography." *Transactions of the Institute of British Geographers* 35: 5–20.
Pratt, G., and S. Hanson. 1994. "Geography and the Construction of Difference." *Gender, Place & Culture* 1: 5–29.
Raghuram, P. 2016. "Locating Care Ethics beyond the Global North." *ACME* 15: 511–533.
Raghuram, P., C. Madge, and P. Noxolo. 2009. "Rethinking Responsibility and Care for a Postcolonial World." *Geoforum* 40: 5–13.
Robinson, F. 2011. *The Ethics of Care: A Feminist Approach to Human Security*. Philadelphia, PA: Temple University Press.
Robson, E. 2004. "Hidden Child Workers: Young Carers in Zimbabwe". *Antipode* 36: 227–248.
Robson, E., J. Horton, and P. Kraftl. 2013. "Children's Geographies: Reflecting on Our First Ten Years." *Children's Geographies* 11: 1–6.
Rodó-de-Zárate, M. 2017. "Who Else Are They? Conceptualizing Intersectionality for Childhood and Youth Research." *Children's Geographies* 15: 23–35.
Rose, G. 1993. *Feminism and Geography: The Limits of Geographical Knowledge*. Minneapolis: University of Minnesota Press.
Rosen, R., and K. Twamley. 2018. "The Woman–Child Question: A Dialogue in the Borderlands." In: *Feminism and the Politics of Childhood: Friends or Foes?* edited by R. Rosen and K. Twamley, 1–20. London: UCL Press.
Ruddick, S. 1989. *Maternal Thinking: Towards a Politics of Peace*. Boston, MA: Beacon Press.
Ruddick, S. 2003. "The Politics of Aging: Globalization and the Restructuring of Youth and Childhood." *Antipode* 35: 334–362.
Skelton, T., and G. Valentine, eds. 1998. *Cool Places: Geographies of Youth Cultures*. London: Routledge.
Staeheli, L.A. 1996. "Publicity, Privacy and Women's Political Action." *Environment and Planning D: Society and Space* 14: 601–619.
Staeheli, L.A., K. Attoh, and D. Mitchell. 2013. "Contested Engagements: Youth and the Politics of Citizenship." *Space and Polity* 17: 88–105.
Staeheli, L.A., E. Kofman, and L.J. Peake, eds. 2004. *Mapping Women, Making Politics: Feminist Perspectives on Political Geography*. New York: Routledge.

Tisdall, E.K.M., and S. Punch. 2012. "Not So 'New'? Looking Critically as Childhood Studies." *Children's Geographies* 10: 249–264.
Tronto, J.C. 1993. *Moral Boundaries: A Political Argument for an Ethic of Care*. New York: Routledge.
Tronto, J C. 2013. *Caring Democracy: Markets, Equality, and Justice*. New York: NYU Press.
Twamley, K., R. Rosen, and B. Mayall. 2016. "The (Im)Possibilities of Dialogue across Feminism and Childhood Scholarship and Activism." *Children's Geographies* 15: 249–255.
Vanderbeck, R.M., and C.M. Dunckley. 2003. "Young People's Narratives of Rural-Urban Difference." *Children's Geographies* 1: 241–259.
Woodyer, T. 2008. "The Body as a Research Tool: Embodied Practice and Children's Geographies." *Children's Geographies* 6: 349–362.
Yeoh, B.S.A., and K.N. Ruwanpura. 2018. "A Conversation between Brenda S.A. Yeoh and Kanchana N. Ruwanpura: Reflections from the 'Margins'." *Gender, Place and Culture's* 25th Anniversary." *Gender, Place & Culture* 25, 159–165.

34
GENDERED GEOGRAPHIES OF DEVELOPMENT

Paula Meth

Introduction

The relationship between gender and development is controversial. In recent years, two examples of the gendered practices of development have caught the world's attention. The first is the scandal surrounding aid giant, Oxfam, regarding the sexual exploitation by its staff of vulnerable women and girls in Haiti and the organization's seeming failure to discipline staff appropriately. The second is the reports that victims of the Syrian war are being sexually exploited in order to access aid, and the women are refusing to collect the aid resources, including food, because they fear such exploitation and the associated stigma. Women without male protectors are seen as particularly vulnerable. These stories point to the dark side of the development industry and the politics of development more broadly, and women's ongoing vulnerability at multiple geographic scales in crisis contexts. They support Pearson's (2005) critiques of conflict management, more generally, as a subfield of academic and humanitarian development that fails to take gender into account in any meaningful way. Crucially, they reveal how significant is the ongoing interrogation of gender and development.

This chapter focuses on gendering the geographies of development, and it explores the multiple ways in which development, a concern with 'global poverty and inequality' (Potter 2014, 17) – as an intellectual project, a practical set of interventions and policy measures and as a political construct – is gendered; namely, how it works to shape social relationships between men and women. The geographies of development call also for an appreciation of the spatialities of development. This reveals how economic, political, social and cultural variations across national scales are key to the meaning of development, with the historical (and problematic) association of the need both *for* development with countries and conditions in and of the Global South and responses *to* development emerging from such contexts and from the beneficent Global North. These wider geographies have significant implications for gender relations. The term 'geographies of development' also refers to spatial specificities within cities, regions and households that work to shape gendered outcomes in various ways. Similarly, the geographies of development are shaped by gender, with crude use of the male/female binary being central to many assessment frameworks and mainstreaming agendas.

This chapter first explores definitions of gender in the context of development and charts the areas of development that are commonly concerned with gender, including the key policy response of mainstreaming. It then moves on to consider dominant historical and current approaches to the project of gender and development cognizant of the geographies of these intellectual and policy trends, with a particular focus on questions of difference, representation and power.

How gender shapes development concerns and vice versa

Gender is a concept used to describe the social differences between men and women and to analyse the notion of what it means to be female or male in a given social context. It is recognized as a political, intellectual and practical concept, and this multidimensional nature is especially evident when analysed in relation to development. In a paper on gender and mobility, Hanson (2010) contrasts two broad approaches; namely, 'how mobility shapes gender' and 'how gender shapes mobility', with the latter often revealed through quantitative research that takes the unreconstructed binary variables of male/female at face value and uses them to evidence substantive patterns and changes in mobility over time. Hanson critiques this body of work for failing to take context into account and for revealing very little about how mobility differences actually affect and shape women's lives – yet such work is all-powerful and dominates the policy discourse and academic work. This chapter is not claiming that work on how gender shapes development fails to consider the complex impacts on women's lives, but it does argue that, at the scale of the global development or financial institution (see ADB 2018 and WHO 2018, for example), gender is almost always used as a binary variable (male/female) through which to understand different aspects of development as a problem or, perhaps, an opportunity. This binary variable is quite often homogenizing and heteronormative (Cornwall 2017) and, most commonly, gender in these contexts refers to women and girls, although some evidence is emerging of a widening focus on vulnerable boys and men.

Quantitative accounts of how gender shapes development are evident in the key publication by the United Nations (UN), *The World's Women* (UN 2015), which employs the male/female binary as well as, at times, the developing/developed regions distinction. Focusing on these trends here is purposeful. Despite the many problems associated with such broad-brush essentializing interpretations, these have shock value and reach and, as such, they are useful to underscore the global message that gendered inequalities do persist and, hence, that gender continues to matter in the context of development, recognizing that gendered inequalities are critical in so-called developed contexts, too.

Three key trends are noted. The first is that economic disparities persist between men and women, with significant disparities in labour-force participation in Northern Africa and Western and Southern Asia. Globally, women suffer from a gendered pay gap, earning between 70 to 90 per cent of what men earn in full-time positions. Women experience higher levels of unemployment (UN 2015). A second trend pertains to violence, and the data show that around a third of all women around the world have experienced some form of physical and/or sexual violence at the hands of partners or of sexual violence by other men, with up to 30 per cent of women in half of all developing countries experiencing a full lifetime of such violence (UN 2015). The responses to violence against women reveal distressing trends, and many women are not seeking any sort of help following experiences of violence and even fewer are turning to the police for support (UN 2015).

Finally, the dire inequities in gendered terms, in relation to positions of power, political and corporate leadership and influence, are summarized, revealing a male domination in judiciaries and governments globally. Women occupy positions as chief executive officers (CEOs) in less than 4 per cent of the world's top 500 corporations (UN 2015). These damning statistics reveal mixed trends across developed/developing contexts, with some Global South countries scoring high in the representation of women in certain powerful sectors, underscoring the message that gender inequalities do not necessarily map onto the so-called 'development' indicators.

Alongside this global recognition of gendered development inequities has been the corresponding rise in gender mainstreaming (GM) as a policy response. Ideas around GM originated in the UN World Conference on Women in Nairobi in 1985, followed by a commitment by the European Union in 1995 to incorporate a gender dimension into policy-making. GM and the adoption of a gendered perspective were required across the UN from 2001 onwards (UN 2002), defined as follows:

> the process of assessing the implications for women and men of any planned action, including legislation, policies or programmes, in all areas and at all levels. It is a strategy for making women's as well as men's concerns and experiences an integral dimension of the design, implementation, monitoring and evaluation of policies and programmes in all political, economic and societal spheres so that women and men benefit equally and inequality is not perpetuated. The ultimate goal is to achieve gender equality.
>
> *UN 2002, v*

GM has been adopted across most international development organizations and large global NGOs (Pearson 2005), with GM policies commonly employed in a largely bureaucratic and technical manner through the use of checklists, gender impact assessments, training manuals, meetings, data collection and analytical tools (Parpart 2014). Implementing meaningful GM has, however, proved challenging. The papers in a special issue of the *Journal of International Development* (2014, 26 [3]) critically analyse the reasons. Van Eerdewijk and Davids (2014, 304), for example, note the 'excessive technocratization and depoliticisation' of GM and the tendency to burden those least able with the task of transformation (ibid., 308). Parpart (2014) argues that the proponents of GM failed to consider the multiple resistances that would emerge in response to the implementation of GM policy, which are often widespread yet unacknowledged, including by senior staff. Commonly, GM is considered to be challenging neither gender relations nor the structural inequality, with 'men and masculine privilege ... left off the hook, as are cultural, economic and political institutions' (2014, 390). Simply put, GM is seen as a continuation of the much-critiqued women in development (WID) approach discussed below.

This quick summary, which rests on a relatively uncontested understanding of what gender itself means, opens the door to a more nuanced analysis of how gender and development intersect and, in particular, how changing development trajectories and geographies work to construct gendered identities and lives.

Gender and development: key debates over time

As in the academic disciplines of geography, sociology, and so on, gender emerged as a key focus for analysis and critique in the academic field of development studies from the 1970s onwards, recognizing that the concept and subject of development, itself, has evolved since it first emerged in the 1940s (Potter 2014). Because of the discipline's close ties with the policy community and the practice of development, the concept of gender quickly made its way into interventions by development institutions, although its passage was fraught and contested, as the experiences of mainstreaming above illustrate (see Pearson 2005). Westernized accounts of debates pertaining to gender and development adopt a standard narrative summarizing the intellectual path using the all-too familiar WID, WAD and GAD terminology, referring to 'women in development', 'women and development' and 'gender and development' (see Young 2002 for a useful summary). These are briefly outlined here as they are fundamental, recognizing how very particularly located the perspectives are.

'Women in development' (WID) captures an approach to gender and development aligned more closely with the liberal feminist politics that emerged from the 1970s onwards. The liberal emphasis was around critiquing and resolving the absence of women from development programmes and on offering a far more complex interpretation and understanding of women's economic roles than hitherto, often with a view to enhancing the productivity of women's activities. This work was significant and sought to justify women's inclusion in terms of the likely impact on poverty and its alleviation. Critics pointed to the lack of emphasis on gender equality in this approach, and WID has been persistently critiqued for its unwillingness to 'rock the boat' or challenge norms in any meaningful way, particularly patriarchal norms and capitalist systems of production. Brown summarizes the problems of the application of WID approaches through often male-dominated policy formulations and institutions, which tend to 'be geared towards bringing women "into" the economic sphere, but without challenging the roles already ascribed to them in a division of labour established during the process of colonization' (2006, 62). As such, WID is often belittled as the 'add women and stir' approach (Moser 2016), with Moser previously noting that such interventions illustrated an 'efficiency approach' to gender and development (Chant and Sweetman 2012). Despite decades of critique, WID is not irrelevant; on the contrary, authors such as Cornwall (2017) note, with palpable frustration, just how prevalent a WID perspective is, encapsulating much gender and development policy currently. Brown (2006) concurs but, drawing on findings in Tanzania, she addresses critiques of WID head on and makes the important point that gendered interests shift over time and, for poor women, WID-oriented programmes can meet immediate practical needs and also result in unanticipated forms of empowerment. She uses her research to complicate 'our understandings of a distinct WID/GAD dichotomy, along with the practical/strategic distinction(s) that helps define it' (2006, 79).

As hinted at in critiques of WID, subsequent 'women and development' and 'gender and development' approaches emerged in the late 1970s, both as a response to WID and as a more sustained and forceful critique of how gender was understood in development. McIlwaine and Datta (2003) describe this critical shift in how gender and development is theorized as one from the 'feminization' of development to the 'engendering' of development, and they chart the opening out of debates, research and focus of work on gender and development. As with wider feminist debates around the 1970s, the rise of GAD in particular served to unleash a sustained critique of patriarchy alongside capitalism, placing power at the epicentre of analyses. Critical questions were raised over historic and current economic models, including the introduction of structural adjustment programmes across much of the Global South and the impact of global financial crises, such as that experienced in 2008 (see Kabeer 2015 for an effective summary of research on this). Gendered impacts of these sustained shocks are not linear or simple, and male loss of employment is often evident alongside rising pressure on women and a squeezing of their time and resources. The evidence points to women taking on low-waged informal work and 'low status and physically demanding jobs like cleaning, laundry and sex work' as they strive to feed children and support families (Kabeer 2015, 199).

The burden of women's double roles, in productive and reproductive terms, emerged as a central concern of a GAD focus, with Kabeer (2015) drawing on Palmer's earlier work to describe asymmetrical sexual divisions of labour in the household and the ensuing 'reproductive tax' (in

Kabeer 2015, 195), forcing women to spend less time on paid employment and increasing their dependence on men. Other work reveals how development interventions themselves can fuel further challenges for women: 'By not recognizing inequalities in the sexual division of labour, whereby women perform the bulk of labour associated with "the home," women's burden increases with their "integration" into the development process' (Brown 2006, 62). Feminists continue to note this burden emerging from development interventions themselves, as women and girls are increasingly heralded as a development resource. In their work on smart economics, Chant and Sweetman critique the politically problematic practice of forgoing women's interests for the sake of the efficiency and the outcomes of the development interventions:

> In smart economics, lack of an essentially political critique of what is wrong with the world at the level of analysis results in programming which focuses solely on the agency of individual women and girls to deliver development goals – changing the world with minimal or no support from other actors.
>
> *Chant and Sweetman 2012, 526*

Feminists note a rise in the feminization of responsibility (adding to women's burdens) as a result of macro-economic trends (Chant and Sweetman 2012), which ties into the concerns over the 'feminisation of poverty' also proposed by Chant. As is evident from the above, the critiques and questions asked by feminists working from a GAD perspective are unsettling. GAD approaches put forward an agenda and a strong series of critiques centred on transformation, empowerment and the wholesale change of structures shaping society (see Young 2002 for details). Chant and Sweetman (2012) emphasize relationality, shifting the emphasis from the problem of and for women to society more broadly: 'A gender and development approach recognizes gender inequality as a relational issue, and as a matter of structural inequality which needs addressing directly and not only by women, but by development institutions, governments and wider society' (2012, 518). This more critical approach is evident in Oxfam's *Gender and Development,* which publishes scholarly papers interrogating a wide range of development-related issues from a feminist perspective, including the Sustainable Development Goals, resource extraction, fragile states, and so on. Much of this work does, however, focus on the experiences of women rather than men.

Difference, representation, knowledge, power and masculinities

Feminists adopting a GAD approach also pushed concepts of gender and development in other key directions, including centring questions of difference. In the first instance, researchers highlighted the evident inequality of women in the 'third world', and proponents of GAD advanced understandings of geographically inscribed inequalities and differences between women in first- and third-world contexts. This opened up discussion about the differences between women at a global scale in a meaningful way although, as will be considered below, in a somewhat problematic manner. However, work progressed on exploring other axes of difference between women in societies that are shaped by ethnicity, religion, age, class, caste, sexuality and disability differences. This field of work is extensive and reflects wider 'cultural turns' in the social sciences that take difference and a more fluid understanding of identity as central to understanding how gender in development must be understood.

Debates over 'gender' and 'development', together and separately, have also benefited from and been challenged by the application of extensive critiques from post-colonial feminist scholars, often working within wider fields such as literary criticism, including Gayatri Spivak,

Lata Mani, Trinh Minh-Ha, Chandra Mohanty and bell hooks (Kothari 2001; and see McEwan 2001 for a detailed analysis). These scholars have questioned and critiqued the hegemony of Western feminism and, in particular, the arrogance and presumption of such feminists to speak on behalf of 'others' (McEwan 2001). Post-colonial feminists also critique the failure to theorize class and race adequately in feminist work, particularly that arising from colonial histories and neo-colonial presents, as well as the tendency to homogenize and essentialize women in the Global South (McEwan 2001) through Western feminist scholarship. A post-colonial feminist approach often begins with a critique of the relationships between 'gender, race and imperialism' (Kothari 2001, 46) and forces a re-examination of 'how other places and people are constructed and problematized' (ibid.). Their work has much to offer a critique of development as a concept, discipline and project, questioning the very logic of development as inherently problematic. Kothari (2001) argued that the varied works of post-colonial feminists, alongside others, have shaped debates over participation, ideas of expertise and questions of representation – particularly by White women of Black women, usually in homogenous terms. Kothari also questions the use of the singular category 'woman', as well as the reliance in Western policy and academia on Western standards and frameworks when studying 'others' (ibid., 49). Interestingly, in a subsequent reflection on her original 1986 essay 'Under Western Eyes', Mohanty notes that she remains 'committed to reengaging in the struggles to criticize openly the effects of discursive colonization on the lives and struggles of marginalized women' (2003, 509), pointing to the persistence of homogenizing and othering discourses over women in the 'third world'. However, given the shifting nature of global politics and economics, Mohanty's focus is not simply on the distinctions between Western versus third-world feminist practices but increasingly on an 'anticapitalist transnational feminist practice' and using this to build a critique of capitalism more broadly (ibid.). At the same time, there has been a significant set of feminist critiques of a post-colonial feminist agenda that describe the latter as 'elitist and removed from reality' and call for a focus on the material conditions of women rather than on discursive debates (see McEwan 2001, 103 for a fuller analysis).

Nonetheless, this wide body of post-colonial feminist critique has influenced a generation of authors writing on the intersection of gender and development, often through a geographical lens, questioning in more complicated ways its epistemological dominance by thinking and writing located in the Global North and West, yet also interrogating and critiquing the notions of women, gender, identity, place and representation, including in the Global South. Some of this work engages with the spatiality of a post-colonial feminist critique through a more global analysis of North/South relations and the politics of knowledge production. Other work uses the scale of the gendered and racialized body, the home, the academy or practices of travel and tourism to rethink how gender is bound up in complex ways within development work. Narayanaswamy (2016), for example, throws open the question of discursive exclusion by critiquing the silencing practices of Southern elite feminists in the context of India who commonly represent the views and interests of upper (class and caste)/middle-class urban Indian women. She questions the process of professionalization and expertise in relation to a transnational development discourse and, in doing so, destabilizes the very idea of a Southern-originated discourse as one that is inclusive and representative (2016, 2170). Syed and Ali (2011), in an effort to understand the mistrust of such projects by those in the Global South, critically examine the idea of the 'White woman's burden' underpinning much (gendered) Western development work. They interrogate Whiteness and trace histories of White women's engagement in the South as part of the colonizing missions (2011, 356) and locate that practice in current development work, including the rescue of oppressed Muslim woman: 'By virtue of her race, class and gender, the white feminist occupies a privileged place with a moral high ground over development and welfare, thus establishing her position in a public realm of power' (2011, 357). Bandyopadhyay and Patil (2017), in contributing to work on tourism geographies, adopt a post-colonial feminist approach and employ the notion of 'White saviour complex' to critique the role of White women in volunteer tourism who work in the Global South, using concepts such as 'contact zones' (2017, 648) to understand socio-spatial relations imbued with power differentials. They locate the colonial mission as largely masculine and trace the gendered changes in such civilizing missions through to development interventions and seemingly apolitical practices, including volunteer tourism. They question the ways in which such work by young White women continues colonial-era racialization and query how it ties to Western women's own subjectivities, particularly through the cultural space of the internet and electronic media (2017, 651). They highlight the significance of Christianity that underpins many volunteer organizations and note its continuity from colonialism and its missionary efforts. This work also questions the focus of those located in the Global North on problems in the South, asking why such women do not attend to the domestic inequalities that are evident in the US, for example.

This question about the geography of gender and development is also picked up by Kishwar (2014), who questions the ways in which the Global South is over-selected for analysis and study as a site of 'gender' problems, while Eurocentric scholarship tends to overlook the gendered inequalities in the West: 'The assumption is that they have solved all their problems and have provided benchmarks and a road map for countries in the South' (2014, 403). Kishwar goes on to critique the development industry's focus on 'gainful employment' as a desirable outcome for all women, including the discourses of feminists who treat women who choose to work as homemakers and mothers as unproductive. This argument is used to make the point that feminism should be about 'respecting women's choices, not imposing a pre-set, ideologically determined road map for all' (2014, 407). More recently, Cornwall (2017) has called for the 'decolonisation of gender and development', insisting that the frameworks and assumptions used during the colonial era about households, nuclear families and gender are still evident in the approaches to gender and development, and which work to limit knowledge and marginalize understandings of gender further. Cornwall calls for a '3Es' approach to counter this: emplacement; empowerment; and encroachment (see Cornwall 2017 for details).

Work on gender and development is also influenced by the idea of intersectionality, after Crenshaw (1989), although the reach of this concept is still limited:

> social relations and multiple, complex identities, which lie at the basis of persisting inequalities, are seldom taken into account in either development theories or development practice. Where they are, 'difference' is often compartmentalized and not integrated.
>
> *Bastia 2014, 237*

However, Bastia goes on to note that some key works in gender and development had an intersectional analytic yet did not refer to it by that name. Bastia turns to migration studies and their use of intersectionality to highlight their potential for development studies. The idea of 'postcolonial intersectionality' is put forward by Mollett (2017), following Mollett and Faria (2013), to facilitate an analysis of feminist political ecology and to examine human rights discourse in relation to Afro-descendant women in Latin America. Explaining the concept, Mollett and Faria argue that:

> postcolonial intersectionality acknowledges the way patriarchy and racialized processes are consistently bound in a postcolonial genealogy that embeds race and

> gender ideologies within nation building and international development processes. This concept reflects the way people are always marked by difference whether or not they fit nicely in colonial racial categorizations, as cultural difference is also racialized … Postcolonial intersectionality addresses Mohanty's warnings against the construction of a 'third world woman' and prioritizes a grounded and spatially informed understanding of patriarchy constituted in and through racial power.
>
> *Mollett and Faria 2013, 120*

Mollett and Faria's work seeks to 'mess with gender' by insisting on a focus on race, alongside gender, in relation to a spatial imagination. Geographical processes work alongside these racial and gendered practices to produce particular experiences of inclusion, privilege, exclusion and oppression. These include the global-scale yet uneven discourses of human rights and 'modernity', colonial racisms and patriarchies that structure relations across the Global North and South, as well as in their case study in Honduras, and the practices of everyday exclusion from society and space through the implementation of legal measures. These measures include changes to land titling that are both individuated and gendered. Mollett and Faria's work reveals that gender, race and geographical context are explicitly intertwined (Mollett and Faria 2013).

Alongside this focus on difference are the efforts by development feminists to emphasize *gender* rather than *women*, noting the relationality of the concept. As is evident in its title, GAD spelled moves to incorporate a focus on masculinity and development, recognizing the roles played by varied groups of men in shaping outcomes and also as key subjects of a development agenda. From around the 1990s onwards, particularly into the early 2000s, burgeoning publications on this topic (see edited collections by Cleaver 2002 and Cornwall, Edström and Greig 2011, for example) analysed gender roles and the construction of gendered identities, focusing on masculinities in relation to politics, work, violence, poverty and inequality. Some of this work drew on the foundational framings of masculinities as relational, as advanced by Connell (1995), distinguishing between hegemonic, complicit, subordinate and marginalized masculinities (Connell 1995, 76–80), as well as recognizing how masculinities are constructed in relation to femininities, themselves intersected by axes of difference such as race and class. Relational conceptualizations served to assist in the theorization of men in the Global South (for example, see Meth 2009), and beyond the discipline of development studies (see Ratele's 2016 work, for example).

At the same time, Connell's framework, particularly the concept of hegemonic masculinity, has come under scrutiny and critique for being too narrow (Gökarıksel and Secor 2017, 385). In response, Connell and Messerschmidt address various critiques with a particular focus on the 'geography of masculinities' (2005, 849), in which they centre the significance of global, regional and local scales as critical to the ways in which masculinity unfolds. Recognizing the interrelationships between these scales, they advise against assuming a power relation between the three that places the global scale as dominant and note the significance of place in shaping masculinities more generally, thereby avoiding a 'monadic' interpretation of place or masculinity uniqueness (2005, 850). Despite this explicit attention to geography, Gökarıksel and Secor (2017) argue that other, more recent, academic work continues to employ more complex approaches to questions of 'scale and place' than Connell and Messerschmidt achieve (385). Work on masculinities in the Global South placed histories of colonial domination, as well as subsequent regimes of marginalization – the Apartheid state, for example (Morrell 1998) – at the centre of analyses working to understand the significance of domination, occupation, exclusion, racism and violence for understanding masculinities. The lack of employment opportunities for men in such contexts is argued to be critical to masculine identity (Morrell 1998). Masculinity theorists working across the Global South also highlight the interplay of race, ethnicity and class (Morrell et al. 2012) as central to men's identities, noting how they are relationally constructed through domination by White men as well as White women. In my own work, I have explored how living in informal housing in South Africa shapes marginalized men's masculinities (Meth 2009) and, subsequently, how the receipt of state-subsidized housing, in turn, reshapes men's experiences and identities (Meth and Charlton 2016).

Conclusions

Gender and development, both as individual and relational concepts, are an ongoing practical, political and intellectual issue: gender and development is not done! On a practical front, there is some cause for celebration, depending on what indicators of 'improvement' are used. There has been some improvement in education, health and access to political life, however the evidence globally in relation to the private sphere (violence, in particular) and the labour market is certainly depressing. There has often been no significant improvement, alongside evidence that situations for many women are worsening. Development is frequently not linear, and progressive initiatives can have detrimental consequences alongside positive outcomes (see Meth 2015).

Importantly, gendered equality has been achieved neither in countries in the Global North and West nor in many of the rapidly transforming contexts of the Global South. This is despite decades of intervention at policy level in terms of education, and training and investment. Cornwall and Rivas (2015) argue that, despite the inclusion of gender in development policy and practice, it actually lacks any meaningful political clout: 'Gender equality and women's empowerment are, we contend, frames that have led feminist activists into a cul-de-sac and away from a broader-based alliance of social change activists' (ibid., 397). They call instead for a focus on accountability, non-discrimination and inclusion (ibid.) echoing the words of Mohanty in 2003.

Furthermore, as argued by many GAD theorists yet also post-colonial scholars, more attention needs to be directed towards the real aspects of transformation and structural change, particularly of global economic practices, which frequently undermine women's experiences. Analyses of gender and development also need to remain alert to the specificities of the world's fluid political climate and the increasing absence of security in certain parts of the world, parts of cities and also some households. Related to this are the ways in which the rule of law is articulated in gendered terms and how it is implemented, abused and protected.

Key readings

Kabeer, N. 2015. "Gender, Poverty, and Inequality: A Brief History of Feminist Contributions in the Field of International Development." *Gender & Development* 23 (2): 189–205. doi:10.1080/13552074.2015.1062300.

Moser, C. 2016. "Gender Transformation in a New Global Urban Agenda: Challenges for Habitat III and Beyond." *Environment and Urbanization* 29 (1): 1–16. doi:10.1177/0956247816662573.

Narayanaswamy, L. 2016. "Whose Feminism Counts? Gender(ed) Knowledge and Professionalisation in Development." *Third World Quarterly* 37 (12): 2156–2175. doi:10.1080/01436597.2016.1173511.

References

ADB. 2018. Asian Development Bank. https://blogs.adb.org/subjects/gender (accessed 9 March 2018).

Bandyopadhyay, R., and V. Patil. 2017. "'The White Woman's Burden' – The Racialized, Gendered Politics of Volunteer Tourism." *Tourism Geographies* 19 (4): 644–657. doi: 10.1080/14616688.2017.1298150.

Bastia, T. 2014. "Intersectionality, Migration and Development." *Progress in Development Studies* 14 (3): 237–248.

Brown, A.M. 2006. "WID and GAD in Dar es Salaam, Tanzania: Reappraising Gender Planning Approaches in Theory and Practice." *Journal of Women, Politics & Policy*, 28 (2): 57–83. doi: 10.1300/J501v28n02_03.

Chant, S., and C. Sweetman. 2012. Fixing Women or Fixing the World? 'Smart Economics', Efficiency Approaches, and Gender Equality in Development." *Gender & Development* 20 (3): 517–529. doi: 10.1080/13552074.2012.731812.

Cleaver, F., ed. 2002. *Masculinities Matter: Men, Gender and Development*. London: Zed Books.

Connell, R.W. 2005. *Masculinities*, second edition. Cambridge: Polity Press.

Connell, R.W., and J.W. Messerschmidt. 2005. "Hegemonic Masculinity: Rethinking the Concept." *Gender and Society* 19 (6): 829–859.

Cornwall, A. 2017. *Decolonising Gender and Development*. Sussex Development Lecture, 17 November 2017. Available at: www.ids.ac.uk/events/decolonising-gender-and-development (accessed 20 November 2017).

Cornwall, A., J. Edström, and A. Greig, eds. 2011. *Men and Development: Politicising Masculinities*. London: Zed Books.

Cornwall, A., and A.-M. Rivas. 2015. "From 'Gender Equality' and 'Women's Empowerment' to Global Justice: Reclaiming a Transformative Agenda for Gender and Development." *Third World Quarterly* 36 (2): 396–415. doi: 10.1080/01436597.2015.1013341.

Crenshaw, K. 1989. "Demarginalizing the Intersection of Race and Sex: A Black Feminist Critique of Antidiscrimination Doctrine, Feminist Theory and Antiracist Politics." *University of Chicago Legal Forum* 1 (8): 139–167.

Gökarıksel, B., and A.J. Secor. 2017. "Devout Muslim Masculinities: The Moral Geographies and Everyday Practices of Being Men in Turkey." *Gender, Place and Culture* 24 (3): 381–402.

Hanson, S. 2010. "Gender and Mobility: New Approaches for Informing Sustainability." *Gender, Place and Culture* 17 (1): 5–23.

Journal of International Development. 2014. *Journal of International Development*. Special Issue: Rethinking Gender Mainstreaming 26 (3): 303–408.

Kabeer, N. 2015. "Gender, Poverty, and Inequality: A Brief History of Feminist Contributions in the Field of International Development." *Gender & Development* 23 (2): 189–205. doi: 10.1080/13552074.2015.1062300.

Kishwar, M.P. 2014. "Feminism and Feminist Issues in the South: A Critique of the 'Development' Paradigm." In: *The Companion to Development Studies*, third edition, edited by V. Desai and R. Potter, 402–407. London: Routledge.

Kothari, U. 2001 "Feminist and Post-colonial Critiques of Development." In: *Theory and Practice: Critical Development Perspectives*, U. Kothari, and M. Minogue, 35–51. London: Palgrave Macmillan.

McEwan, C. 2001. "Postcolonialism, Feminism and Development: Intersections and Dilemmas." *Progress in Development Studies* 1 (2): 93–111.

McIlwaine, C., and K. Datta. 2003. "From Feminising to Engendering Development." *Gender, Place and Culture* 10 (4): 369–382.

Meth, P. 2009. "Marginalised Men's Emotions: Politics and Place." *Geoforum* 40: 853–863.

Meth, P. 2015. "The Gendered Contradictions in South Africa's State Housing: Accumulation Alongside Erosion of Assets through Housing." In: *Gender, Asset Accumulation and Just Cities*, edited by C. Moser, 100–116. London: Routledge.

Meth, P. and Charlton, C. 2017. "Men's Experiences of State-sponsored Housing in South Africa: Emerging Issues and Key Questions." *Housing Studies* 32 (4): 470–490. doi: 10.1080/02673037.2016.1219333.

Mohanty, C. 2003. "'Under Western Eyes' Revisited: Feminist Solidarity through Anticapitalist Struggles." *Signs* 28 (2): 499–535.

Mollett, S. 2017. "Irreconcilable Differences? A Postcolonial Intersectional Reading of Gender, Development and Human Rights in Latin America." *Gender, Place & Culture* 24(1): 1–17. doi: 10.1080/0966369X.2017.1277292.

Mollett, S., and C. Faria. 2013. "Messing with Gender in Feminist Political Ecology." *Geoforum* 45: 116–125.

Morrell, R. 1998. "Of Boys and Men: Masculinity and Gender in Southern African Studies." *Journal of Southern African Studies* 24: 605–630.

Morrell, R., R. Jewkes, and G. Lindegger. 2012. "Hegemonic Masculinity/Masculinities in South Africa: Culture, Power, and Gender Politics." *Men and Masculinities* 15 (1): 11–30.

Moser, C. 2016. "Gender Transformation in a New Global Urban Agenda: Challenges for Habitat III and Beyond." In: *Environment and Urbanization*, International Institute for Environment and Development, 1–16. doi: 10.1177/0956247816662573.
Narayanaswamy, L. 2016. "Whose Feminism Counts? Gender(ed) Knowledge and Professionalisation in Development." *Third World Quarterly* 37 (12): 2156–2175. doi: 10.1080/01436597.2016.1173511.
Parpart, J. 2014. "Exploring the Transformative Potential of Gender Mainstreaming in International Development Institutions." *Journal of International Development* 26: 382–395.
Pearson, R. 2005. "The rise and rise of gender and development." In: *A Radical History of Development Studies: Individuals, institutions and ideologies*, edited by Uma Kothari, 157–179. London: Zed Books.
Potter, R. 2014. "The Nature of Development Studies." In: *The Companion to Development Studies*, edited by V. Desai and R. Potter, third edition, 16–20. London: Routledge.
Ratele, K. 2016. *Liberating Masculinities*. Cape Town: HSRC Press.
Syed, J., and F. Ali. 2011. "The White Woman's Burden: From Colonial Civilisation to Third World Development." *Third World Quarterly* 32 (2): 349–365.
UN. 2002. *Gender Mainstreaming: An Overview*. New York: United Nations. Available at: www.un.org/womenwatch/osagi/pdf/e65237.pdf (accessed 9 March 2018).
UN. 2015 https://unstats.un.org/unsd/gender/worldswomen.html (accessed 12 December 2017).
Van Eerdewijk, A., and T. Davids. 2014. "Escaping the Mythical Beast: Gender Mainstreaming Reconceptualised." *Journal of International Development* 26: 303–316.
WHO. 2018. World Health Organisation. Available at: www.who.int/gender-equity-rights/understanding/gender-definition/en/ (accessed 9 March 2018).
Young, K. 2002. "WID, WAD and GAD." In: *The Companion to Development Studies*, edited by V. Desai and R. Potter, 321–324. London: Arnold.

35

FEMINIST VISCERAL POLITICS

From taste to territory

Allison Hayes-Conroy, Jessica Hayes-Conroy, Yoshiko Yamasaki and Ximena Quintero Saavedra

Introduction

Feminist visceral politics refers to the elaboration of political ways of thinking, acting and being that take seriously the visceral realm of life: the sensations, moods and states born of our corporeal engagement with the material world. The notion and impulse of visceral politics build on diverse legacies of interest in and engagement with bodily experience from within feminist geography and wider feminist scholarship (e.g. Bondi 2005; Longhurst 2001, 2005; Longhurst et al. 2008; McWhorter 1999; Moss and Dyck 2003; Probyn 2001; Thien 2005). The (first) authors of this chapter began thinking about feminist visceral politics in their research on food–body relationships (e.g. Hayes-Conroy and Hayes-Conroy 2008, 2010). Examining the visceral experience of food meant digging into the ways in which social concepts like identity and difference 'mattered' to how bodies reacted to, tasted or were otherwise drawn towards (or away from) different kinds of foods. And, it meant paying attention to the situatedness of food experience – to the ways in which structures of power enter into bodily judgments of food (as well as food spaces, ideas and events). In this chapter, we ask what it means to 'do' feminist visceral politics, both in and beyond research. The chapter evaluates the actions and outcomes of three distinct projects, all of which were influenced by theories and methods broadly related to feminist political theory, corporeal feminism and body studies.

This chapter is not focused on research findings per se; that is, we are not interested in portraying the data that these projects generated in ways that argue for a singular, coherent story of social scientific discovery (Davies and Dwyer 2007). Instead, we are interested in practice. The outcomes we aim to consider are flexible, transferrable models for action that build from feminist theory and feminist geography – they are the embodied, relational and, indeed, the visceral 'things', both intended and unintended, that (can) happen in the research process that allow us eventually to arrive at a conclusion of sorts: the deepening of relationships; the reframing of concepts; the creation of new ideas or paths forward; the witnessing of complexity and contradiction. We call these 'things' models, rather than findings, because we want to emphasize them as examples (though not blueprints) of how one might 'do' feminist visceral politics. These models for action come both from our own research events and practices and from the events and practices that we set out to study – and usually from a combination of both. As such, they are illustrations of what might emerge when we focus our energies on recognizing and harnessing the politics of the visceral body.

Before we turn to discuss the examples from our research, however, a note on complexity and contradiction: over a decade ago, Gail Davies and Claire Dwyer (2007) identified some important shifts in the impulses that drive (some) qualitative research in human geography. In particular, they noted – especially among researchers interested in questions of affect, phenomenology and materiality – a rejection of the idea that the purpose of research is to generate clarity and reduce uncertainty. 'In place of the pursuit of certainty in generating representations of the world,' Davies and Dwyer noted,

> there is recognition that the world is so textured as to exceed our capacity to understand it, and thus to accede that social science methodologies and forms of knowing will be characterized as much by openness, reflexivity and recursivity as by categorization, conclusion and closure.
>
> *2007, 258*

Certainly, in our own work, as well as in much of the scholarship that we have been inspired by, themes of hybridity, relationality and the rhizome feature prominently. In this work, the binary between the empirical and theoretical seems to fall away, replaced by an (open, partial) attempt to witness and explain complexity and contradiction within particular, embodied contexts.

But (why) is it so important to witness and explain complexity? What does a recognition of contradiction lend to the doing of feminist visceral politics? When we focus on research findings, 'complexity' is arguably neither surprising nor particularly helpful. Indeed, everything is complex (and many things are in contradiction). But when we focus on the praxis of complexity – on what to do with it – the outcomes of our work can be far more valuable. For example, the recognition that taste preferences are a complex amalgam of structural inequality, cultural knowledge and messy biosocial relationships can be immediately valuable to someone who is struggling to understand their own food–body relationship. And such a recognition can also be valuable, at a policy level, to nutrition practitioners or public health professionals who seek to intervene in particular food habits or behaviours. However, the practice of complexity takes both creativity and courage. Many in the academy are so well trained in the craft of critique that *doing* seems at times formidable, even unthinkable (lest we be critiqued ourselves). And although many of us *do*, via our research, we position our *doing* as methods and findings in order to justify and clarify our interactions with the world and hide the 'messy bits' that might make our research seem less conclusive (Jones and Evans 2011). Similarly, the practice of policy writing (at least for nutritional policy) seems wholly oriented to the task of simplification (such that the food messages that we receive are predictable and repetitive, and often also ineffective). What would it look like to 'do' complexity in the practice of research? What would it look like to embrace, even encourage, contradiction in the articulation of research outcomes and policy recommendations? Our research experiences have motivated us to ask these questions, if not to begin to seek some situated answers.

The three examples that we discuss in this chapter – a project on critical nutrition in Philadelphia, PA; a project on the food adequacy of displaced women in Medellin, Colombia; and a project of peacebuilding in Colombia that grew out of the second example – together provide a glimpse of feminist visceral politics as both vital and mundane. While we provide clear evidence of why doing feminist visceral politics is important – for example, it remains sensitive to diverse experiences of real bodies and it uncovers aspects of life often overlooked by other approaches – we also express why it challenges us to want to 'do' more. Especially as 'feminism' and 'feelings' are further vilified in corporate, conservative and mainstream discourses, the most achievable and effectual outcomes may be idiosyncratic and fleeting rather than generalizable and sustained.

Critical nutrition: co-producing a different approach to nourishment in Philadelphia, PA

The authors have a long-standing interest in nutrition, not just the science but the ways that nutrition knowledge and practice become pulled into the reproduction of social identity, the methods and motivations behind attempts to intervene in people's food habits and behaviours, and especially the biosocial mechanisms through which bodies come to be differently nourished. Critical nutrition, as a field of study, has emerged from the recognition that 'it is vital to defamiliarize nutrition, to undo its taken-for-grantedness in order to understand better its sociological and cultural underpinnings, as well as the effects that it has beyond improving or failing to improve dietary health' (Guthman 2014, 2). More specifically, much of our own impulse in pursuing critical nutrition work came from the recognition that 'hegemonic nutrition' – in short, the mainstream discourses and practices of nutrition, which promote standardization, decontextualization and expert knowledge (see Hayes-Conroy and Hayes-Conroy 2013) – is largely ineffective and sometimes even harmful to the advancement of bodily well-being. Thus, we began to wonder about how we might 'do nutrition differently' (ibid.), in ways that are open to the complexities and contradictions of everyday food-body relationships. In other words, what would it look like to practise nutrition through the lens of feminist visceral politics?

Over the course of a few years, Allison and Jessica Hayes-Conroy developed an ongoing, working relationship with several staff members of the Norris Square Community Alliance (NSCA) of Philadelphia, PA, including especially Yoshiko Yamasaki (a collaborator on this chapter). NSCA is a community development corporation that was founded over thirty years ago by a group of women who wanted to ensure a safe and healthy neighbourhood for their children. Residents of Norris Square are encouraged to be active at all levels of the organization. The scope of the NSCA is wide, including issues such as affordable housing, employment training, early childhood education, community organizing and more (NSCA 2015). Recently, through Yoshiko's guidance, the NSCA also began a project to enhance nutrition education for children and families. Allison and Jessica's connection to the NSCA emerged through our work in critical nutrition as both researchers and teachers. Through conversations with Yoshiko, we began to develop a series of dialogue-based activities (workshops) that would benefit both students of critical nutrition (at Temple University) and the families and staff of NSCA who are interested in nutrition education.

The overarching idea of the critical nutrition workshops (which were held at the main NSCA building over several days) was to experiment with the practice of critical nutrition and to encourage dialogue about the potential utility of doing nutrition differently – critically, contextually and viscerally. Participants included staff and parents, with students and researchers facilitating. The workshop format sought to counter the expectation of expert-led nutrition education by enrolling participants as experts of sorts – of their own desires and cravings, daily life habits, obstacles to change, and so on – and the researchers (us and our students) as the anti-experts (Heyman 2010). The dialogue first centred on participants' own experiences and concerns regarding bodily nourishment, then on the idea of critical nutrition itself and, finally, on the potential utility of employing critical, embodied perspectives in nutrition education.

The most significant outcome was arguably not the data itself (although we did record the conversations, transcribe and analyse them) but rather in the (ephemeral) energy that the workshops evoked in many of the participants. There was a tangible dynamism as we worked through various issues, ranging from doctors and fat stigma, to the draw of a slice of pizza after work, to the hidden assumptions behind the term 'food desert'. The dynamism came not from a sense that we were 'getting somewhere' (we weren't, really; there were no clear next steps) but rather from the doing itself – the fact that the dialogue enrolled participants (however briefly) in the production of critical knowledge about bodily nourishment, especially about feelings surrounding the nourishment of their own bodies. In this context, our own attempts to share critical nutrition ideas with the participants functioned as way 'to diffuse power and knowledge production, [so as to] not perpetuate a closed realm of privileged knowledge' (Heyman 2010, 315). This point is particularly important, given that much of the scholarship in geography about bodies, food and affect can feel impenetrable to a non-expert.

Regarding complexity and contradiction, we might note that the dialogue itself, as well as the broader relationship between us (scholars) and NSCA staff contained much of both. As critical theorists with an interest in 'practice', we frequently find ourselves needing to work alongside and through structures of knowledge that we are – well – critical of. Many funding streams for food-based interventions, for example, require outcomes assessments to test the effectiveness of nutrition education. Are students able to identify a vegetable correctly? Did staff initiate regular use of nutrition lessons? We attempted to translate our own critical nutrition work into such metrics with only minimal success. Similarly, obesity prevention has become a dominant frame and motivating force for many nutrition intervention initiatives, often including standardizing metrics such as the Body Mass Index (BMI). These ideas and impulses circulated our collective conversations in a variety of incongruous affective capacities – as disdain, hope, envy, anxiety, scepticism, and so on – that both disrupted and affirmed the metrics themselves. Although our own critical instincts drive us to interrupt these modes of assessing bodies and health, our scepticism comes not (only) because we have witnessed how alienating and oppressive these metrics can be but, more basically, because they tend to ignore and discount the ephemeral, affective work that can happen when we provide a space for critical praxis.

It is likely impossible to trace the micro-effects of the NSCA workshops that we convened, but this does not indicate that they were not meaningful. We know, for example, that the events provided broader meaning to Yoshiko's work. (She said so.) We know, too, that the events provided the space for intellectual curiosity and emotional sharing that some community members had been desiring. (Again, they said so.) Part of the work of 'doing complexity', then, involves providing a communal space for critical praxis, where complexity and contradiction are not just inevitable but encouraged. This might seem a mundane point but, if so, it is telling that so much nutrition policy aims at the opposite: simplification, universal messaging and cookie-cutter protocols. What would nutrition policy look like if it embraced – even facilitated – complexity?

Food adequacy: cooking up affective partnerships in Medellin, Colombia

In a small project on food adequacy in the city of Medellin, Colombia, Allison and another researcher partnered with leaders from a community-based organization, La Colonia de San Luis, which accompanies families from rural areas of Antioquia, Colombia, who were violently displaced to the city. Part of the initial idea of the work was to broaden the scope of 'adequacy' in food security policy and academic conversations (Hayes-Conroy and Sweet 2015). Food policy work seldom considers the complexities of embodied experience, and food adequacy is often assessed on the basis of calorie counts and macro/micronutrients. So, the work wanted to understand how the displaced women encountered food in the city as distinct from the rural areas from which most had been displaced, and specifically how they evaluated the adequacy of the food that they found there. Hayes-Conroy and Sweet talked with displaced women who were charged with procuring food for their households. They also planned, shopped, cooked and ate meals with the women. They returned a year later to create a 'body-mapping' event, which tried to understand the bodily impact of violent displacement to the city including, yet moving beyond, food insecurity. Body mapping (see Figure 35.1) offered a way to visualize, express and materially understand the ways in which displaced women judged and assessed their bodily experiences with displacement – documenting everything from blistering feet and paused menstruation through heartache and concern for children to emotions like relief and nostalgia.

The stories that the work uncovered contained plenty of complexity and contradiction. For example, the women generally viewed the food that they could access in the city as less fresh and less healthy and/or tasty as the food back in their rural homes, but certain brands were viewed favourably. Also, almost all the women had moderate to severe household food insecurity yet demonstrated a hesitancy about eating food from the 'urban industrial food system' or feeding it to children. Many food-insecure women communicated the importance of food in terms much broader than alleviating hunger or providing necessary nutrition – the importance of sharing meals, of having connection with or control over food production and exchanging ideas through cooking, eating and feeding all were significant aspects of food adequacy, for these women. And, finally, women had had both positive and negative emotions surrounding their experiences in the city since their displacement.

In light of such complexity, the writing that came out of this project pushed not only for an expansion and recentring of how we define and value (adequate) food, beyond that which is assessable through nutrition status alone, but also, and more broadly, for more sensitive and adaptable kinds of social policy. We asked how we document, understand and create more a flexible metrics for adequacy that speaks to women's diverse visceral experiences of food insecurity. How do we create social policy that is sensitive to different embodied experiences? Such musings were a small part of the visceral politics that were sought and (partially) accomplished through the project. To be clear, the 'doing' of feminist visceral politics in this case does not refer to the type of intellectual output, nor to the research events or to the participatory methods per se, but to the ways in which specific thoughts, actions and connections were invigorated by the work with particular people in a particular place. In other words, feminist 'doing' is always specific to context and is neither contained in certain methods (like group interviews or body-mapping), nor given in scholarly approaches.

So the 'doing' of feminist visceral politics in this case looked much like relationship-building: the researchers and participating women dreamed up delicious meals, went shopping and worried about our children together; we listened and spoke, we cooked and learned, we tasted and ate, we defined and redefined, we decided and changed our minds, we misunderstood each other, we showed joy, fear, anger and appreciation. At the end of the small research project, what did the 'doing' of such feminist visceral politics actually *do*? Certainly, it did much less than anyone had hoped with respect to challenging and resolving food insecurity. Sharing women's experiences and stories was valuable to all, but offered little change. Instead, the most important outcome of the work may be an outcome that is hard to quantify in academic terms: it inaugurated an *affective partnership* – a connection through which some participants shared the motivation to continue to work together. This partnership – centred through the

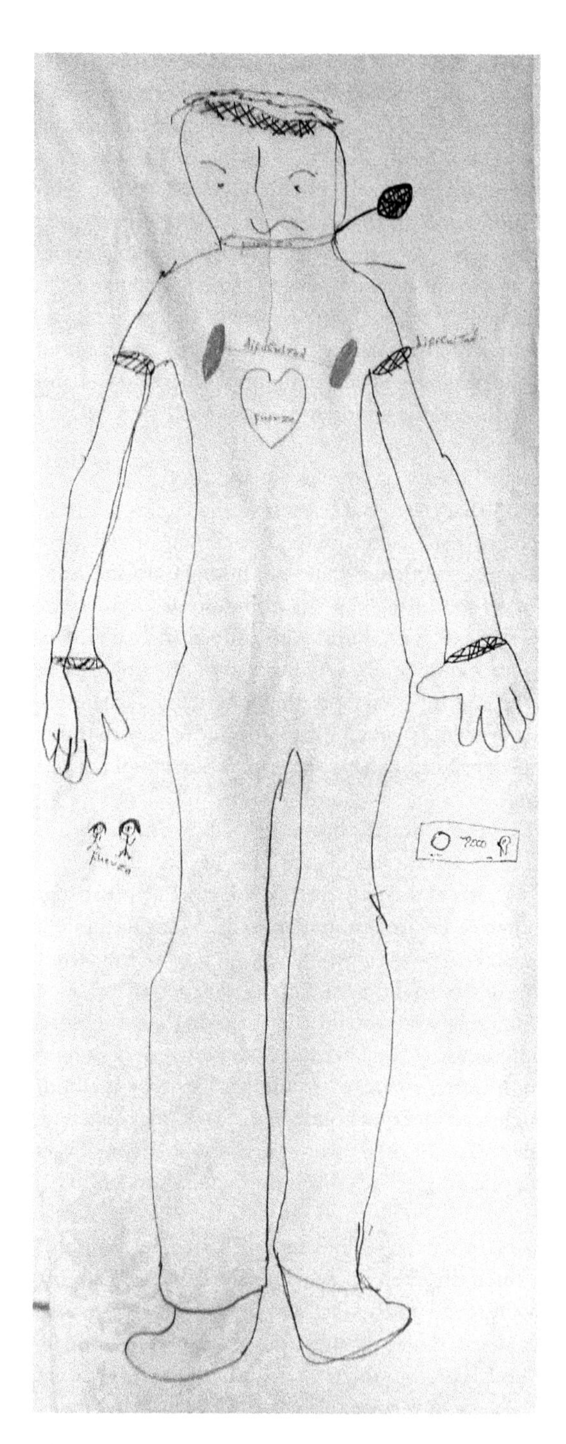

Figure 35.1 Body map.

organization La Colonia de San Luis – has shifted in unexpected ways over the last six years and has enabled new achievements and new goals (some of which are described below), as well as some false starts and dead ends. It has been unsurprisingly complex and contradictory, just like the stories that we set out to understand, yet has permitted the expansion of an affective network of people accompanying other people in their life work.

The next section takes off from this idea of an *affective partnership* and describes how prioritizing feminist visceral politics allowed us to nurture a vexingly unspecific vision for change that came alongside one particular set of affective partnerships.

Embodied social transformation across borders: the case of the Legión del Afecto

Growing from the partnership inaugurated through the research described above, Allison Hayes-Conroy began working with community leaders and youth (both women and men) from La Colonia de San Luis and, later, with others from a partner organization called Casa Mia (including Ximena Quintero Saavedra, a collaborator on this chapter). Allison's work with these groups has focused on a peacebuilding initiative known as La Legión del Afecto (hereafter the Legión), which La Colonia and Casa Mia helped to build.[1] The Legión could be described as a network of affective partnerships on a large scale. It is a social initiative with much history, piloted in the early 2000s as a collaboration between these community organizations and progressive academics who were interested in building an everyday kind of peace in violence-afflicted rural and urban territories. As it began to mobilize many young people across Colombia, the Legión grew to become both a social movement and a programme of the national government in over 35 different locations (see Hayes-Conroy and Saenz Montoya 2017; Hayes-Conroy, Saenz Montoya and Buitrago 2017). Much of this network was built by paying attention to the corporeal need for affection and coming up with diverse ways – for instance, dance, music, ritual, theatre and shared meals – to communicate it in order to rebuild urban and rural territories in and through peace. Today it exists, unfunded, as an uncertain but potent network built through thousands of individual efforts on non-violent, affective relationship-building across the country in some of its most precarious locations. Allison's work on the Legión has been largely ethnographic, accomplished over five years with much support from other collaborators close to the initiative, students and colleagues, including Jessica Hayes-Conroy and Ximena Quintero Saavedra.

To be clear, our work with the Legión has centred on the youth and leaders in the network, who share the experience of economic and social precarity, rather than with the top officials who came to manage the initiative as a government programme. The research events were largely determined by Legión praxis and involved many dialogues, journeying and visiting, and witnessing and acting in celebratory and/or commemorative events.

In what ways has our work on peacebuilding in and around the Legión carried out feminist visceral politics? Through the work, we have tried to take seriously the affective model that the Legión has developed, learning how to describe it accurately, emphasizing the Legión's legacy and history (the good and the bad), understanding its power structures and working with its youth to write about, talk about, reflect upon and otherwise support and improve the potential of its peacebuilding. The model itself is an example of visceral politics; it is a way towards a social transformation that emphasizes sensations, feelings and corporeal dispositions. *Lo efectivo es lo afectivo* ('the affective is effective') was the motto that the Legión gleaned from a radically non-violent leftist priest who himself practised peacebuilding through the 1980s and 1990s along similar lines. The Legión has developed *lenguajes alternativos* (alternative languages) that use the body (through things like dance, music, sport and more) to communicate feelings, ideas and visions for the future. While the Legión has largely organized around generating positive affects and feeling through its public assemblies, the realities of its organizing have been far from rosy. Not only have youth had to find ways to collectively cope with trauma and loss but their participation generated unexpected challenges and hard social, economic and emotional choices. Thus, although our work has sought to shore up the potential of the Legión's model, practising feminist visceral politics has also meant asking hard questions about power, racism, sexism and economic inequity in its past and present. It has meant learning to be silent and learning to listen to stories and reflections told through multiple means (not all verbal). It has meant being present with young leaders as they talk about the future; and it has meant *making places* where we can be present together to have those conversations. And it has meant giving up any chance that all of this will fit together neatly into one intellectual package containing a perfectly crafted set of publications about what it all means.

Our writings on the Legión inevitably contradict each other. The doing of feminist visceral politics with the Legión has meant using our academic and empirical expertise to tell stories in ways that are valuable to our partners and that continue to build strategic partnerships for our work towards a 'fuzzy' common goal (peace/non-violence/anti-precarity/social transformation). The action plans and goals of the actors in and around the Legión are themselves ripe with contradiction and complexity. While non-violent social transformation is a commonly held ideal, in practice the work has been full of political ambiguity, economic tension, passions and ideas, challenges and opportunities, each of which pulls actors apart as well as together (c.f. Butler 2015). How do you tell the story and convey the needs of the Legión as a powerful social initiative in such a context? In the academic world, what theoretical framework do you pick? In the policy world, what strategy do you push? Work in and with the Legión has been a continual negotiation of contradictory ideas and practices and complex needs. What do we do with that complexity? Externally and internally, a common expectation of social initiatives is that they have clearly defined goals and political or social intentions. Yet, one of the magical aspects of the Legión as an affective network has been the way that it draws in diverse, often once-conflicting actors (e.g. people on different sides of the armed conflict), who tend to have differing visions about the future of the initiative. The praxis of complexity in this case has thus meant working with youth and leaders to try to understand and build a future for the Legión that does not try to erase this complexity but to remain an effective and influential model for social transformation.

Conclusions

Our intention in this chapter has been to focus on the *doing* of feminist visceral politics rather than on the findings that emerge from such doing. These two things are not mutually exclusive, but researchers undoubtedly tend to privilege the latter. Our suggestion is simply that the outcomes of this doing – the ephemeral happenings, the affective traces, the partnerships built and the futures imagined – *matter*, and that indeed they may be as meaningful as the findings themselves. Perhaps feminist visceral politics, then, is a different way to understand and to build 'place' as it is understood in geography, with connotations of nuance and situated meaning-making; or, perhaps feminist visceral politics is another form of privileging 'small data' (Delyser and Sui 2013), as a counter (or counterpart) to big data. Through this chapter we have argued for a way of recognizing, for those of us who desire our research to do something (to effect change), that the *doing* of feminist work need not always be measurable, concrete and academically admissible for it to be important. We should not let the intellectual push for momentous and transformational work overtake the diverse possibilities of the messier *doings* that we so rarely consider as research outcomes yet which actually do the work of connecting discrete projects or publications to actual bodies and places. At the same time, in each of the three research examples that we discuss, feminist visceral politics expands the very importance of the body beyond its traditional bounds; from the tastes of community members in Philadelphia and Medellin to the rebuilding of territory through attentiveness to feeling, feminist visceral politics also demonstrates the power of the body to shape both research and community practice.

Note

1 The material described is based on work supported by the National Science Foundation (NSF) under Grant #1452541. Any opinions, findings, and conclusions or recommendations expressed in this material are those of the authors and do not necessarily reflect the views of the NSF.

Key readings

Hayes-Conroy, A., and J. Hayes-Conroy. 2010. "Visceral Difference: Variations in Feeling (Slow) Food." *Environment and Planning A* 42 (12): 2956–2971.

Hayes-Conroy, A., and A. Saenz Montoya. 2017. "Peace Building with the Body: Resonance and Reflexivity in Colombia's Legión del Afecto." *Space and Polity* 21 (2): 144–157.

Longhurst, R., E. Ho, and L. Johnston. 2008. "Using 'the Body' as an 'Instrument of Research': Kimch'i and Pavlova." *Area* 40 (2): 208–217.

References

Bondi, L. 2005. "Making Connections and Thinking Through Emotions: Between 'Geography and Psychotherapy'." *Transactions of the Institute of British Geographers* NS 30 (4): 433–448.

Butler, J. 2015. *Notes Toward a Performative Theory of Assembly*. Cambridge, MA: Harvard University Press.

Davies, G., and C. Dwyer. 2007. "Qualitative Methods: Are You Enchanted or Are You Alienated?" *Progress in Human Geography* 31 (2): 257–266.

DeLyser, D., and D. Sui. 2013. "Crossing the Qualitative–Quantitative Divide II: Inventive Approaches to Big Data, Mobile Methods, and Rhythmanalysis." *Progress in Human Geography* 37 (2): 293–305.

Guthman, J. 2014. "Introducing Critical Nutrition." *Gastronomica: Journal of Critical Food Studies* 14 (3): 1–4.

Hayes-Conroy, A., and J. Hayes-Conroy. 2008. "Taking Back Taste: Feminism, Food and Visceral Politics." *Gender, Place and Culture* 15 (5): 461–473.

Hayes-Conroy, A., and J. Hayes-Conroy. 2010. "Visceral Difference: Variations in Feeling (Slow) Food." *Environment and Planning A* 42 (12): 2956–2971.

Hayes-Conroy, A., and J. Hayes-Conroy. 2013. *Doing Nutrition Differently: Critical Approaches to Diet and Dietary Intervention*. Farnham: Ashgate.

Hayes-Conroy, A., and E.L. Sweet. 2015. "Whose Adequacy? (Re)imagining Food Security with Displaced Women in Medellín, Colombia." *Agriculture and Human Values* 32 (3): 373–384.

Hayes-Conroy, A., and A. Saenz Montoya. 2017. "Peace Building with the Body: Resonance and Reflexivity in Colombia's Legion del Afecto." *Space and Polity* 1–14.

Hayes-Conroy, A., A. Saenz Montoya, and C. Buitrago. 2017. "La Legión del Afecto, Colombia's Powerful Network for Peace." 11 August, Sustainable Security. Oxford Research Group. Available at: www.oxfordresearchgroup.org.uk/blog/la-legin-del-afecto-colombias-powerful-network-for-peace (accessed 6 February 2020).

Heyman, R. 2010. "People Can: The Geographer as Anti-Expert." 4th Annual James Blaut Memorial Lecture Las Vegas, Nevada, 24 March. *ACME: An International Journal for Critical Geographies* 9 (3): 301–326.

Jones, P.J., and J. Evans. 2011. "Creativity and Project Management: A Comic." *ACME: An International Journal for Critical Geographies* 10 (3): 585–632.

Longhurst, R. 2001. *Bodies: Exploring Fluid Boundaries*, vol. 11. London: Routledge.

Longhurst, R. 2005. "Situating Bodies." In: *A Companion to Feminist Geography*, edited by L. Nelson and J. Seager, 337–349. Oxford: Blackwell.

Longhurst, R., E. Ho, and L. Johnston. 2008. "Using 'the Body' as an 'Instrument of Research': Kimch'i and Pavlova." *Area* 40 (2): 208–217.

McWhorter, L. 1999. *Bodies and Pleasures: Foucault and the Politics of Sexual Normalization*. Bloomington: Indiana University Press.

Moss, P., and I. Dyck. 2003. *Women, Body, Illness: Space and Identity in the Everyday Lives of Women with Chronic Illness.* Lanham, MD: Rowman & Littlefield.

NSCA (Norris Square Community Alliance). 2015. "About Us – Sobre Nosotros." Available at: http://nscaonline.org/?page_id=36 (accessed 3 January 2017).

Probyn, E. 2001. *Carnal Appetites: Foodsexidentities*. New York: Verso.

Thien, D. 2005. "After or Beyond Feeling? A Consideration of Affect and Emotion in Geography." *Area* 37 (4): 450–456.

36

FEMINIST PERSPECTIVES ON NEOLIBERAL GLOBALIZATION, (POST)FEMINISMS AND (HOMO)NORMATIVITIES

Shirlena Huang and Qian Hui Tan

Introduction

Proponents of neoliberal globalization believe that market liberalization – typically premised on a free-market economy and characterized by deregulation, privatization and other forms of economic restructuring (Harvey 2005) – is beneficial to societies across the globe in gender-neutral ways. For women, an early promise of neoliberal economic reforms at the macro-scale included the creation of more employment opportunities, both locally and abroad. It was argued that this would lead to women's empowerment and gender equality, as access to the productive (work) sphere would free them of their oppression in the reproductive (domestic) realm. At the micro-scale, neoliberalism's emancipation of women from patriarchal oppression was to be experienced through the ability and right to make choices, for example through entrepreneurship and consumerism.

Lately, however, 'even the most devoted believers in the neoliberal paradigm will have had their convictions shaken', as markets have been found to be neither self-correcting nor efficient allocators of resources (Cornwall et al. 2008, 1–2; Newman 2013). Neoliberal policies have been ineffective, arguably because of how economic activities remain structured by a rigid oppositional dichotomy whereby the market is equated with the masculine and the household with the feminine (Cameron and Gibson-Graham 2003). A more radical proposition for the failure of neoliberalism to effect change is that (Western) feminism has become, either unknowingly, voluntarily or by force, complicit with neoliberalism, despite the effects of the latter going against feminism's general aims (Fraser 2013; Korolczuk 2016; McRobbie 2009; Thorpe et al. 2017).[1] Thus, rather than counteracting neoliberalism's negative effects on the feminist movement, neoliberal feminism is seen as contributing to the project of corporate globalization and reinforcing heteronormative notions of womanhood alongside their socially ascribed roles as altruistic caretakers (Lind, 2010) in society and the family.

Neoliberalism's contradictory relationship with feminism is partially attributed to its conceptual ambiguity. 'Neoliberalism' is often used loosely by proponents and critics as a catch-all term for a miscellany of ideas that underscore private enterprise as well as the 'rolling back' of the state (Stiglitz 2008). Discussions of neoliberalism often fail to specify which of its facets – for example, global macroeconomic doctrine; political ideology or a mode of governmentality cultivating certain cultural dispositions; or the selective appropriation of technological assemblages – is being referenced (Calkin 2015; Ferguson 2010; Gregor and Grzebalska 2016; Newman 2013). Yet, as Larner (2000, 6) argues, attempting to differentiate between the various inferences is not just an intellectual exercise; it also affects our understanding of the 'scope and content of possible political interventions'.

In the rest of this chapter, we consider some key issues in scholarship at the intersections of neoliberal globalization and feminism, with respect to three interventions: development and women as neoliberal subjects; the proliferation of (post-)feminist identities; and queer politics under neoliberalism. The first relates to a fundamental issue that has been at the heart of most feminist projects, while the latter two pertain to more recent concerns.

Global development and the ideal neoliberal subject

Arguably, women have often been positioned as integral to neoliberal strategies. Their place in neoliberal economies and the impact of developmental policies on women, however, have not been uncontested. In the Global North, feminists' critique of the male-breadwinner model has contributed to women's greater participation in the economy (Newman 2013). In the Global South, 'the third-world woman' featured prominently on the development agenda in the 1970s as part of liberal feminists' 'win-win' narrative, whereby integrating women into development through the 'feminization of policy' was argued to be beneficial not just for the women and their families but for their communities and nations (Calkin 2015; see also Roy 2010). In this view, the archetypal 'third-world woman' is commonly (re)presented as the 'solution' to poverty, because her 'gendered subjectivity [renders her] self-sacrificing' (Wilson 2011, 325). For instance, with respect to Latin America's foreign debt crisis in the 1980s, the establishment of numerous community-based organizations, such as communal kitchens and day-care centres, pivoted on women's voluntary contributions to poverty management (Lind 2010). While neoliberalism no longer casts women as passive victims, their new and unique 'entrepreneurial potential' derives from and is dependent upon their altruistic 'maternal nature' (Calkin 2015, 301). In other words, in reality, the costs of neoliberal restructuring (e.g. cuts in spending on social welfare) are absorbed by women's unpaid work, thereby entrenching long-standing gender inequalities, even if some women have been politically empowered in the process.

Accordingly, neoliberalism re(de)fines gender norms by constructing the 'good woman', especially of the Global South, as the ideal neoliberal subject whose dormant economic agency, once tapped, will reinvest any gains received from the market into her family and community (while her male counterpart spends on himself). Thus, women have been the targets of transnational NGO and corporate initiatives, such as Oxfam's 'Oxfam Unwrapped', Goldman Sach's '10,000 Women' and Nike Foundation's 'Girl Effect' campaigns, in international development efforts (Wilson 2011). The 'good woman' is also the selfless 'third-world' mother/daughter/sister, migrating in search of work as an entertainer, domestic worker or other form of low-level service work in response to (transnational) employment opportunities opened up by neoliberal globalization. Arat-Koc (2006) contends that migrant domestic workers especially are ideal subjects of the neoliberal state because, first, the economic, social and psychic costs of their work are largely transferred to a different location/country; second, they enable both their own and their middle-class female employers' social reproduction in the private sphere to remain invisible; and third, by enabling wealthier women to participate in the workforce, they deflect the need to change gender relations in society.

Critics have highlighted the negative outcomes of neoliberal economic policies on women's entry into the labour force. In countries of the Global North, such as the UK, state reforms to administrative and governance structures have resulted in, *inter alia*, low wages and poor labour standards in sectors associated with female labour, especially at the lower end of the job market (Cornwall et al. 2008). Further, while the large number of migrant women employed in developed countries challenge traditional notions of 'masculinist hypermobility' (Pratt and Yeoh 2003), they typically become 'disenfranchised diasporic citizens' (Hawkesworth 2006, 202; Moghadam 2005). Instead of alleviating the burden of domestic care on women, global restructuring has intensified the privatization of social reproduction by engendering an international transition towards the abandonment of the family wage (in lieu of the living wage) and declining welfare regimes (Bakker and Silvey 2008).

Criticism has also been levelled at the construction of women as self-regulating entrepreneurs and responsible decision-makers who, in the face of neoliberal reconfigurations of the global economy, always 'choose' to act in the best interests of their families; this is seen as simply entrenching a 'feminization of responsibility' (Sharp 2007; Wilson 2011). Relatedly, critical commentators have noted how terms originally associated with feminist activism – such as 'empowerment', 'agency' and even 'gender' – have been appropriated by the mainstream lexicon on development and cited 'in the context of strategies for survival rather than transformation, and … of the individual, rather than the collective' (Thorpe et al. 2017, 375). This empties them of their activist content while contributing to an unprecedented evisceration of feminist politics (Batliwala 2007; Prügl 2015; Wilson 2011, 318). While highlighting how feminist theorists have introduced new frameworks – such as market feminism (Kantola and Squires 2012), corporate feminism (Gill 2016; Prügl 2015), hegemonic feminism (Eisenstein 2009), transnational business feminism (Roberts 2012, 2015) and faux feminism[2] (McRobbie 2009) – to 'rehabilitate feminist discourses and goals', Calkin (2015, 302) cautions against simple claims of co-option; instead, she calls for the need to recognize 'the multiple and sometimes contradictory strands of feminism'.

Neoliberalizing (post-)feminism

A more salient theme that has surfaced since the 1990s, especially in feminist media and cultural circles, centres on the profound links between neoliberalism and post-feminism. Post-feminism has been an interesting topic of intellectual investigation, one that gestures towards a (con)fusion of feminist as well as anti-feminist idea(l)s by encompassing, yet also repudiating, the key objectives of second-wave feminism. Set against this backdrop of 'neoliberal feminism' (Braidotti 2005), the social status of women is now narrowly defined by their financial capabilities, which would, in turn, augment liberalist goals such as choice and independence.

Moran (2017, 123) contends that post-feminism can be subsumed under neoliberalism because both valourize 'privatization, deregulation and deinstitutionalization', which then encourages individualism, optimization and responsibilization. Additionally, these post-feminist proclivities have proliferated rapidly through sophisticated media technologies and virtual relationalities, thereby equipping geographically distant individuals with the interpretative resources to reconstruct their personhood. Relatedly, post-feminism has become synonymous with a 'weightless' consumer culture that does little to defy heteropatriarchal regimes or sexual–legal frameworks. For example, 'retail therapy', as a means of sedimenting one's self-worth, perpetuates a tyranny of slenderness, while the media obsession with spectacular performances of beauty has exacerbated the self-objectification and surveillance of women's bodies (Gill 2016). Women are lauded as 'aesthetic entrepreneurs', exerting 'aesthetic labour' while internalizing the male gaze in order to appear more youthful, hyperfeminine and heterosexy (Elias, Gill and Scharff 2017), especially in the face of a sex-positive porno-chic or raunch culture (Attwood 2014). Moreover, neoliberal sensibilities conjure up feminism in a way that projects it as an anachronistic spent force, as denoted in the prefix 'post-' (McRobbie 2004; Scharff 2016; also, see Gill et al. 2017 for a discussion on how the 'post-' in 'post-feminism' may be interpreted). Consequently, young women who see themselves as forerunners of social change are attending to feminism in ways that render it a 'cheer word' – 'unimpeachable, but also devoid of substance' (Gill 2016, 619, 623; 2017; McRobbie 2007, 2009). For instance, the freedom to consume has become conflated with the freedom of (sexual) expression, both of men's rights to procure (foreign) women's bodies and of women's rights to put them up for sale. Market values have subsequently attained a moralistic status, with strategic consumption acting as a panacea for various social problems. The 'good female' consumer is therefore abstracted from grids of power in such a reductionist conceptualization of agency (Ferguson 2010; Gill 2007).

Concomitantly, feminists have expounded on the contradictions of post-feminine identities in two ways. First, its premature pronouncement of gender equality entails a full accounting of feminism in order that it be disavowed. McRobbie (2007) calls this a post-feminist masquerade in which gender norms are reified and patriarchy is reaffirmed. Second, women are now ironically disempowered by the clichéd discourses of empowerment that they are being offered (Fahs 2011; McRobbie 2009; Tan 2014). Within a neoliberal imperative of choice, a woman's voluntaristic will is inflected through 'compulsory individuality' (Cronin 2000, 277).

Feminists have also questioned whether membership in the 'global sisterhood' of post-femininity is exclusive to White, conventionally good-looking women. They have cast doubts on how post-feminism may be relevant to women with intersectional identities who may be 'globally scattered' outside the metropolitan core (Butler 2013; Lazar 2006; Nast 2002; Tasker and Negra 2007). Nonetheless, post-feminism as a transnational culture has circulated widely, with 'post-colonial elites' in the Global South (Dosekun 2015, 966), adapting it vis-à-vis vernacular socio-cultural practices. Examples include well-educated women in India (Grewal 2005; Parameswaran 2004; Reddy 2006), Nigeria (Dosekun 2015) and Singapore (Lazar 2009).

Crucially, however, post-colonial feminist scholars have taken issue with a monolithic view of (post)feminine and (post-)feminist subjectivities that reinforces the dichotomy between the 'West and the rest', thereby implying that what transpires elsewhere (outside the West) is merely a watered-down derivative of a more legitimate version. Ultimately, this perspective 'traps non-Western [women] in a double-bind': that they should either adhere to a neoliberalist tyranny of 'emancipation' or adopt victim-posturing as a strategy (Giraldo 2016, 165). For example, McRobbie (2007, 733) presents the image of a hypothetical non-Western 'global girl' imitating 'her Western counterpart' and whose citizenship is shaped by her consumption habits. 'Global girl' as a governable subject may feature in the local edition or local equivalent of an international women's magazine – she does not necessarily hope to live in the West, yet desires the commodities linked to Western femininity and sexuality (McRobbie 2009). Invocations of 'global' brands and subscriptions to Anglo-American yardsticks of beauty in non-Western contexts may serve to reposition post-feminism squarely in the West yet also cite a Western aesthetic with a different spin (Sensoy and Marshall 2010; Switzer 2013; Wilson 2011). In a close reading of a magazine for English-reading Indian women, *Femina*, Reddy (2006, 75) notes the tension between a 'nationalist naturalization' of Indian beauty (plump lips, darker skin) and the 'cultural appropriation' of Western beauty as Indian (big, lined eyes, fairer skin, bleached blonde hair).

Regardless, it is still possible for local feminist histories displaying a different trajectory from White Western ones to be discounted in both the developed and the developing world, where settler colonialism is writ large. It is still unclear how post-feminism can exist in tandem with more traditional, grassroots-oriented forms of politics.

Queering/querying neoliberalism

The literature on queer politics and neoliberalism, the third theme of this chapter, has developed along two strands: how sexual politics 'filter[s] through a multi-scalar system' via a global–local nexus (Podmore 2013, 265); and investigations of LGBT-friendly cities beyond a homonormative logic. In recognizing the spread of a globally identifiable gay lifestyle, scholars have argued that Eurocentric practices are not easily replicated in other cultural contexts (Altman 2004a). The perception of equality and progressiveness in Western cities, vis-à-vis the liberalization of sexual laws, has been vehemently critiqued alongside the assumption that LGBT activism necessarily gets diffused from the (Western) core to the (Eastern) periphery. Whereas transnational LGBT coalitions may have (in)directly influenced LGBT movements all over the world, Asian academics have asserted that Southern cities are not mere imitations of their Western counterparts. Moreover, they have evinced that the process of queer globalization is not unidirectional or universal; neither is there a stark dichotomy between the wholesale acceptance and rejection of sexual democracies from the Anglo-American world. Instead of homogenization, hybrid regional identifications, such as 'Queer Asia', may emerge (Leung 2009). Concomitantly, they have cautioned against a reactionary rush towards the celebration of Indigenous sexual cultures, one that abhors anything non-local, or the adoption an occidentalist approach that perceives the non-West as being completely distinct from the West (Altman 2004b).

In the past two decades, queer theorists and critical geographers have identified the rise of new homonormativities and homonationalisms. Homonormative tendencies refer to an assimilated homosexuality that incorporates gays and lesbians into state-sanctioned institutions, such as monogamous marriage, thereby transforming them into exemplary citizens (Duggan 2003). Homonationalism takes its cue from a post-9/11 moment in the United States' history. It marshals the 'acceptance' of gays and lesbians on a larger scale, but this sexual inclusivity is consolidated by racial exclusivity and xenophobia (Puar 2013). Taken together, such proclivities and their overtones of 'queer liberalism' (Eng et al. 2005) perpetuate 'a demobilised gay constituency' (Duggan 2002, 179), whereby aspirations for sexual diversity have been supplanted by fervent claims to full citizenship (for example, equalizing the age of sexual consent and the reform of discriminatory laws). More pertinently, these claims have perpetuated a privatized gay culture that situates the burden of social welfare firmly within the domestic sphere. For Nast (2002), a penchant for privatization and consumerism has clearly shored up the patriarchal and racist lifestyle of gay White men. In the West, the integration of 'gaybourhoods' into plans for urban regeneration and global reimaging, primarily for the purpose of capital accumulation, has been well studied by human geographers (Gorman-Murray and Waitt 2009). Central to this accommodationist form of neoliberalism is that of 'pink washing', a normalization of gay/lesbian leisure and recreational landscapes resting on the differentiation between queer-friendly and queer-phobic establishments (Puar 2013).

Academics, however, have started to complicate the simplistic and American-centric conceptualizations of homonormativity, which are mainly informed by studies of politically liberal cities in the Global North. Critical geographers have attempted to disrupt the univocality of commercialism while being wary of totalizing representations of homonormativity as a seemingly unassailable force (Brown 2008, 2009; Lewis 2013). Inspired by Gibson-Graham's (1996, xi) call for an 'anti-capitalist politics of economic invention', Brown (2009) avoids rehashing trite discourses on 'pink washing' by emphasizing the socio-cultural roles played by pro-LGBT sites in terms of service provision, community building and political recognition. In so doing, he teases out the not-for-profit 'queer commons' (Brown 2009, 1504) or 'community economies' that are evident even during commercialized hallmark activities such as pride parades, where information and resources may be exchanged on the premise of goodwill or reciprocity. Additionally, he points out alternative sites of faith-based LGBT socialization, steeped in activist work as well as care-giving centres.

Likewise, Squires (2017) surmises that certain LGBT events may exhibit some neoliberal elements yet are not stereotypically or reducibly so, thereby calling for a more sustained academic engagement with localized manifestations of homonormativity. This may entail a closer examination of racially inflected 'parties with a politics' (Browne 2007), the spatialization of gender variations in consumption and other kinds of 'the queer unwanted' that remain understudied. For Squires (2017), the scholarship on homonormal spaces has positioned sexual minorities of colour at the two extreme ends of a spectrum, as either hapless victims or heroic freedom-fighters in a capitalist economy.

This parallels some of arguments of Oswin (2005, 2008), who opines that most critiques of the White, middle-class gay man perpetuate a binary between conventional versus radical sexual subjectivities. Rather than fixing the analytical spotlight on the distinctly or authentically counter-hegemonic, she attends to complicit queer subjectivities or spaces that have always been, to varying degrees, embroiled in reinforcing not just neoliberal ideologies but also other dominant forms of performing gender, ethnicities and nationalities. Moreover, Brown (2009) contends that if complicity highlights how seemingly 'transgressive' sites can shore up the hegemons, practices that appear to be 'assimilationist' may similarly harbour unexpected potential for expanding alternative scripts for queer living.

Outside the West, attempts have been made to appreciate the cultural methods of queer Asian activism on its own terms. Thus, unlike an American queer politics, which is marked by being out, loud and proud, the LGBT landscape in East Asia is characterized by a 'reticent poetics' that stresses harmonious familial/communal relations (Liu and Ding 2005). For example, scholars have noted how social movements in Singapore's censorious political climate necessitate calculated manoeuvres and prudent discretion. Thus, Pink Dot, the city-state's most high-profile LGBT event, does not opt for either belligerent protests or flamboyant pride parades but is infused with a pro-family rather than a pro-gay rhetoric to avoid a clampdown by the government. Far from 'disabl[ing] political analysis', as Duggan (2003, xx) propounds, the dissemination of a culturally resonant maxim of 'every LGBT person a family member' has helped to garner the support of straight allies in the city-state. From a Western lens, Pink Dot may appear to be a manifestation of homonormativity and homonationalism but through the lenses of Asia, it epitomizes tacticality and inventiveness on the part of a 'queer subaltern constituency' (Lazar 2017, 421).

Concluding thoughts

By gathering current debates on development, post-feminism and homonormativity, we have shown how feminist and queer culture has been 'colonized' by neoliberal discourses of gender/sexuality. By assuming that individuals possess the capacity to mitigate the inequalities that they confront on an everyday basis just by working harder, for example, neoliberal feminism and queer liberalism promote a false consciousness that obviates the motivation for real changes. In sum, these three themes highlight the need for a closer inspection of a purportedly celebratory rhetoric to avoid taking them at face value, and to reassess these ideologies of progress vis-à-vis actual improvements in people's lives. This may mean acknowledging the feminist/queer project is an unfinished one. As such, critical scholarship must resist the 'pasting' (Tasker and Negra 2007) and 'overing' (Ahmed 2012) of sexism and homophobia and, instead, seek out opportunities for animating a collective resistance that is antithetical to neoliberal individualism (Gill, Kelan and Scharff 2017; Motta 2013). Simultaneously, it will have to be attuned to the intersectionalities of gender with, *inter alia* race, class, nationality and sexuality that produce inequalities between and within groups of women and men.

Apart from staging a facade of public engagement in a transformative politics, these homogenizing perspectives on gender–sexual relations have dismissed the significance of geographical particularity in that neoliberal forces do not affect every woman or sexual dissident in the same way (Gregor and Grzebalska 2016; Roberts 2015). If we are to trouble 'empowerment, choice and agency' as superficial discourses that buttress the status quo, then we need to demonstrate clearly the uneven implementation of neoliberal policies as well as its complex outcomes (Cornwall et al. 2008; Moran 2017). Specifically, there is a dire need for more theoretically rigorous and empirically nuanced research on how 'neoliberalized feminisms [can] provide openings to challenge oppressive power relations' in the Global South (Prügl 2015, 627). Just as neoliberalism is a mobile ideology, researchers are aware that post-feminist and homonormative sensibilities interpellate women and sexual subjects beyond the West, yet this has not been adequately or systematically studied. Media scholars like Dosekun (2015) have argued that fine-grained studies of how post-feminism may unfold within localities can help to shed light on how 'new' femininities are reconstituted, especially in places where feminism clashes with the vernacular culture or is deemed as a form of Westoxification. Likewise, Lazar (2006, 2009, 2011) situates the genesis of post-feminism in Western popular culture but offers a local analysis of how it is hybridized or tweaked in Singapore. This line of thought provides a 'fertile ground for the post-feminist distancing of feminism to take root' (Lazar 2011, 39). Moving forward, we need more of such studies that theorize post-feminism as a transnational culture that is packaged and marketed multidirectionally across borders.

Finally, we will have to work harder at rethinking the gender(ing) of globalization. On a practical level, we will also have to propel a progressive politics that not only disrupts but dissolves 'capitalocentric conceptions' of the economy, ones that do not dissociate work from the feminine domain (Cameron and Gibson-Graham 2003, 146; Freeman 2001). One way to proceed is for further 'feminist theorizations that see the body, nation and global as indicative of the same processes rather than as different scales, however well connected' (Sharp 2007, 382). This may entail perceiving social reproduction/consumption to be as central to global flows as economic production, despite the latter operating on a much larger scale. Another involves a queer praxis plugged into broader calls for social moments, beyond tired debates on the pink economy or a myopic focus on only the gay/lesbian part of the LGBT spectrum. Overall, it is our contention that feminist perspectives of/on neoliberalism remain invaluable to social scientific research, without which we would have an impoverished view of how social justice may be achieved.

Notes

1 In her much-cited essay, Fraser (2013) outlines how feminism has legitimized neoliberalism and urges feminists to break off this 'dangerous liaison'. See also Newman (2013) and Korolczuk (2016) for summaries of how Fraser's view has been problematized.

2 'Lean-in feminism' (Sandberg 2013) may be regarded as an example of faux feminism. In advocating that any woman who is willing to work hard can get to the top of the corporate ladder, it ignores the various structural obstacles that women face in their respective societies.

Key readings

Duggan, L. 2011. "After Neoliberalism? From Crisis to Organizing for Queer Economic Justice." *Scholar & Feminist Online* 10 (1–2).
Gill, R. 2016. "Post-postfeminism?: New Feminist Visibilities in Postfeminist Times." *Feminist Media Studies* 16 (4): 610–630.
Prügl, E. 2015. "Neoliberalising Feminism." *New Political Economy* 20 (4): 614–631.

References

Ahmed, S. 2012. *On Being Included: Racism and Diversity in Institutional Life*. London: Duke University Press.
Altman, D. 2004a. "Sexuality and Globalization." *Sexuality Research and Social Policy* 1 (1): 63–68.
Altman, D. 2004b. "Queer Centers and Peripheries." *Cultural Studies Review* 10 (1): 119–128.
Arat-Koc, S. 2006. "Whose Social Reproduction? Transnational Motherhood and Challenges to Feminist Political Economy." In: *Social Reproduction: Feminist Challenges to Neo-liberalism*, edited by K. Bezanson and M. Luxton, 75–92. Montreal: McGill-Queens University Press.
Attwood, F. 2014. *Mainstreaming Sex: The Sexualization of Western Culture*. London: IB Tauris.
Bakker, I., and R. Silvey. 2008. *Beyond States and Markets: The Challenges of Social Reproduction*. London: Routledge.
Batliwala, S. 2007. "Taking the Power Out of Empowerment – An Experiential Account." *Development in Practice* 17 (4): 557–565.
Braidotti, R. 2005. "A Critical Cartography of Feminist Post-postmodernism." *Australian Feminist Studies* 20: 169–180.
Brown, G. 2008. "Urban (homo) Sexualities: Ordinary Cities and Ordinary Sexualities." *Geography Compass* 2 (4): 1215–1231.
Brown, G. 2009. "Thinking beyond Homonormativity: Performative Explorations of Diverse Gay Economies." *Environment and Planning A* 41 (6): 1496–1510.
Browne, K. 2007. "A Party with Politics? (Re)making LGBTQ Pride Spaces in Dublin and Brighton." *Social & Cultural Geography* 8: 63–87.
Butler, J. 2013. "For White Girls Only? Post-feminism and the Politics of Inclusion." *Feminist Formations* 25 (1): 35–58.
Calkin, S. 2015. "Feminism, Interrupted? Gender and Development in the Era of 'Smart Economics'." *Progress in Development Studies* 15 (4): 295–307.
Cameron, J., and J.K. Gibson-Graham. 2003. "Feminising the Economy: Metaphors, Strategies, Politics." *Gender, Place and Culture* 10 (2): 145–157.
Cornwall, A., J. Gideon, and K. Wilson. 2008. "Introduction: Reclaiming Feminism, Gender and Globalisation." *IDS Bulletin: Transforming Development Knowledge* 39 (6): 1–9.
Cronin, A. 2000. "Consumerism and Compulsory Individuality: Women, Will and Potential." In: *Transformations: Thinking Through Feminism*, edited by S. Ahmed, J. Kilby, C. Lury, M. McNeil, and B. Skeggs, 273–287. London: Routledge.
Dosekun, S. 2015. "For Western Girls Only? Post-Feminism as Transnational Culture." *Feminist Media Studies* 15 (6): 960–975.
Duggan L. 2002. "The New Homonormativity: The Sexual Politics of Neoliberalism." In: *Materialising Democracy: Towards a Revitalized Cultural Politics*, edited by R. Castronovo and D. Nelson, 175–194. Durham: Duke University Press.
Duggan, L. 2003. *The Twilight of Equality? Neoliberalism, Cultural Politics, and the Attack on Democracy*. Boston, MA: Beacon Press.
Eisenstein, H. 2009. *Feminism Seduced: How Global Elites Use Women's Labor and Ideas to Exploit the World*. Boulder, CO: Paradigm.
Elias, A., R. Gill, and C. Scharff. 2017. *Aesthetic Labour: Rethinking Beauty Politics in Neoliberalism*. London: Palgrave Macmillan.
Eng, D., J. Halberstam, and E. Munoz. 2005. "What's Queer about Queer Studies Now?" *Social Text* 23 (3/4): 1–17.
Fahs, B. 2011. *Performing Sex: The Making and Unmaking of Women's Erotic Lives*. New York: State University of New York Press.
Ferguson, J. 2010. "The Uses of Neoliberalism." *Antipode* 41 (1): 166–184.

Fraser, N. 2013. "How Feminism Became Capitalism's Handmaiden – And How to Reclaim It." *The Guardian,* 14 October. Available at: www.theguardian.com/commentisfree/2013/oct/14/feminism-capitalist-handmaiden-neoliberal (accessed 13 December 2017).

Freeman, C. 2001. "Is Local: Global as Feminine: Masculine? Rethinking the Gender of Globalization." *Signs* 26 (4): 1007–1037.

Gibson-Graham, J-K. 1996. *The End of Capitalism (As We Knew It): A Feminist Critique of Political Economy.* Oxford: Blackwell.

Gill, R. 2007. "Critical Respect: The Difficulties and Dilemmas of Agency and 'Choice' for Feminism: A Reply to Duits and Van Zoonen." *European Journal of Women's Studies* 14 (1): 69–80.

Gill, R. 2016. "Post-postfeminism? New Feminist Visibilities in Postfeminist Times." *Feminist Media Studies* 16(4): 610–630.

Gill, R. 2017. "The Affective, Cultural and Psychic Life of Postfeminism: A Postfeminist Sensibility 10 Years On." *European Journal of Cultural Studies* 20 (6): 606–626.

Gill, R., E. Kelan, and C. Scharff. 2017. "A Postfeminist Sensibility at Work." *Gender, Work & Organization* 24 (3): 226–244.

Giraldo, I. 2016. "Coloniality at Work: Decolonial Critique and the Postfeminist Regime." *Feminist Theory* 17 (2): 157–173.

Gorman-Murray, A., and G. Waitt. 2009. "Queer-friendly Neighbourhoods: Interrogating Social Cohesion across Sexual Difference in Two Australian Neighbourhoods." *Environment and Planning A* 41 (12): 2855–2873.

Gregor, A., and W. Grzebalska. 2016. "Thoughts on the Contested Relationship between Neoliberalism and Feminism." In: *Solidarity in Struggle: Feminist Perspectives on Neoliberalism in East-Central Europe,* edited by E. Kováts, 11–20. Budapest: Friedrich-Ebert-Stiftung.

Grewal, I. 2005. *Transnational America: Feminisms, Diasporas, Neoliberalisms.* Durham, NC: Duke University Press.

Harvey, D. 2005. *A Brief History of Neoliberalism.* Oxford: Oxford University Press.

Hawkesworth, M. 2006. "Feminists v. Feminization: Confronting the War Logics of the Bush Administration." *Asteriskos* 1 (2): 117–142.

Kantola, J., and J. Squires. 2012. "From State Feminism to Market Feminism?" *International Political Science Review* 33 (4): 382–400.

Korolczuk, E. 2016. "Neoliberalism and Feminist Organizing: From 'NGO-ization of Resistance' to Resistance against Neoliberalism." In: *Solidarity in Struggle: Feminist Perspectives on Neoliberalism in East-Central Europe, edited by* E. Kováts, 32–41. Budapest: Friedrich-Ebert-Stiftung.

Larner, W. 2000. "Neo-liberalism: Policy, Ideology, Governmentality." *Studies in Political Economy* 63 (1): 5–25.

Lazar, M. 2006. "'Discover the Power of Femininity!': Analyzing Global Power Femininity in Local Advertising." *Feminist Media Studies* 6 (4): 505–517.

Lazar, M. 2009. "Entitled to Consume: Postfeminist Femininity and a Culture of Post-Critique." *Discourse & Communication* 3 (4): 371–400.

Lazar, M. 2011. "The Right to be Beautiful: Post-Feminist Identity and Consumer Beauty Advertising." In: *New Femininities: Post-Feminism, Neoliberalism and* Subjectivity, R. Gill and C. Scharff, 37–51. Basingstoke: Palgrave Macmillan.

Lazar, M. 2017. "Homonationalist Discourse as a Politics of Pragmatic Resistance in Singapore's Pink Dot Movement: Towards a Southern Praxis." *Journal of Sociolinguistics* 21 (3): 420–441.

Lewis, N. 2013. "Ottawa's le/the Village: Creating a 'Gaybourhood' Amidst the Death of the Village." *Geoforum* 49: 233–242.

Leung, H.H.S. 2009. *Undercurrents: Queer Culture and Postcolonial Hong Kong.* Vancouver: University of British Columbia Press.

Lind, A. 2010. "Gender, Neoliberalism and Post-neoliberalism: Reassessing the Institutionalisation of Women's Struggle for Survival in Ecuador and Venezuela." In: *The International Handbook of Gender and Poverty: Concepts, Research, Policy,* edited by S. Chant, 649–654. Cheltenham: Edward Elgar.

Liu, J.P., and N.F. Ding. 2005. "Reticent Poetics, Queer Politics." *Inter-Asia Cultural Studies* 6 (1): 30–55.

McRobbie, A. 2004. "Post-feminism and Popular Culture." *Feminist Media Studies* 4 (3) 255–264.

McRobbie, A. 2007. "Top Girls? Young Women and the Post-feminist Sexual Contract." *Cultural Studies* 21 (4): 718–737.

McRobbie, A. 2009. *The Aftermath of Feminism: Gender, Culture and Social Change.* London: Sage.

Moghadam, V. 2005. *Globalizing Women: Transnational Feminist Networks.* Baltimore, MD: Johns Hopkins University Press.
Moran, C. 2017. "Re-positioning Female Heterosexuality within Postfeminist and Neoliberal Culture." *Sexualities* 20 (1–2): 121–139.
Motta, S.C. 2013. "'We Are the Ones We Have Been Waiting For': The Feminization of Resistance in Venezuela." *Latin American Perspectives* 40 (4): 35–54.
Nast, H. 2002. "Queer Patriarchies, Queer Racisms, International." *Antipode* 34 (5): 874–909.
Newman, J. 2013. "Spaces of Power: Feminism, Neoliberalism and Gendered Labour." *Social Politics: International Studies in Gender, State and Society* 20 (2): 200–221.
Oswin, N. 2005. "Towards Radical Geographies of Complicit Queer Futures." *ACME: An International E-Journal* 3 (2): 79–86.
Oswin, N. 2008. "Critical Geographies and the Uses of Sexuality: Deconstructing Queer Space." *Progress in Human Geography* 32 (1): 89–103.
Parameswaran, R. 2004. "Global Queens, National Celebrities: Tales of Feminine Triumph in Post-Liberalization India." *Critical Studies in Media Communication* 21 (4): 346–370.
Podmore, J. 2013. "Critical Commentary: Sexualities Landscapes beyond Homonormativity." *Geoforum* 49: 263–267.
Pratt, G., and B.S.A. Yeoh. 2003. "Transnational (Counter) Topographies." *Gender, Place and Culture* 10 (2): 159–166.
Prügl, E. 2015. "Neoliberalising Feminism." *New Political Economy* 20 (4): 614–631.
Puar, J. 2013. "Homonationalism as Assemblage: Viral Travels, Affective Sexualities." *Jindal Global Law Review* 4: 23–387.
Reddy, V. 2006. "The Nationalization of the Global Indian Woman." *South Asian Popular Culture* 4 (1): 61–85.
Roberts, A. 2012. "Financial Crisis, Financial Firms..., and Financial Feminism? The Rise of 'Transnational Business Feminism' and the Necessity of Marxist-Feminist IPE." *Socialist Studies/Études Socialistes* 8 (2): 85–108.
Roberts, A. 2015. "The Political Economy of 'Transnational Business Feminism': Problematizing the Corporate-led Gender Equality Agenda." *International Feminist Journal of Politics* 17 (2): 209–231.
Roy, A. 2010. *Poverty Capital: Microfinance and the Making of Development.* London: Routledge.
Sandberg, S. 2013. *Lean In: Women, Work, and the Will to Lead.* New York: Alfred A. Knopf.
Scharff, C. 2016. *Repudiating Feminism: Young Women in a Neoliberal World.* London: Routledge.
Sensoy, O., and E. Marshall. 2010. "Missionary Girl Power: Girl at a Time." *Gender and Education* 22 (3): 295–311.
Sharp, J. 2007. "Geography and Gender: Finding Feminist Political Geographies." *Progress in Human Geography* 31 (3): 381–387.
Squires, K. 2017. "Rethinking the Homonormative? Lesbian and Hispanic Pride Events and the Uneven Geographies of Commoditized Identities." *Social & Cultural Geography* 20 (3): 1–20.
Stiglitz, J. 2008. "The End of Neo-Liberalism." *Daily News*, Cairo, 7 July.
Switzer, H. 2013. "(Post)Feminist Development Fables: The Girl Effect and the Production of Sexual Subjects." *Feminist Theory* 14 (3): 345–360.
Tan, Q. H. 2014. "Postfeminist Possibilities: Unpacking the Paradoxical Performances of Heterosexualized Femininity in Club Spaces." *Social & Cultural Geography* 15 (1): 23–48.
Tasker, Y., and T. Negra. 2007. *Interrogating Post-Feminism: Gender and the Politics of Popular Culture.* Durham, NC: Duke University Press.
Thorpe, H., Toffoletti, K., and Bruce, T. 2017. "Sportswomen and Social Media: Bringing Third-Wave Feminism, Postfeminism, and Neoliberal Feminism into Conversation." *Journal of Sports and Social Issues* 41 (5): 359–383.
Wilson, K. 2011. "Race, Gender and Neoliberalism: Changing Visual Representations in Development." *Third World Quarterly* 32 (2): 315–331.

PART 4

Doing feminist geographies

37

EMBODIED TRANSLATIONS

Decolonizing methodologies of knowing and being

Beaudelaine Pierre, Naimah Petigny and Richa Nagar

Entangling our voices, feeling our grounds: three points of departure

One: h(a)unting

This writing begins in May 2017, as Donald Trump's administration hunts for evidence of crimes committed by Haitian immigrants in order to justify the non-renewal of Temporary Protection Status or TPS, an immigration status that has allowed more than 50,000 Haitians to stay in the US after the earthquake of 2010. How can the US government justify the ways in which it arbitrarily authorizes itself to decide the fate of lives within its geographical borders? Calling this removal violence does not make it such, in the eyes of the US government, nor is it an adequate intervention on my part. As someone very close to the 2010 earthquake and its aftermath, the question that haunts me is not only how to make sense of my own and my children's situation as TPS holders but, more importantly, how to ethically account for injustices that happen within and beyond the reach of my arms? It demands that I look at the Haitian TPS issue not only through the ways that people of both Haiti and the US theorize each other, but also the resonances and entwining of individual stories of oppression that are often starkly separated, such as that of Philando Castile, who was gunned down by a police officer in July 2016 in Saint Paul, Minnesota; or Trump's Muslim travel ban during the first months of his presidency; or the killing in May 2017 in Portland, Oregon, of two men on a train who tried to intervene when a man was yelling racial slurs at two women who appeared to be Muslim.

Two: reinvesting

trace your lines of intention
slow and steady, yet with fervor
push your fingers into the divots of your borderless body
feel your body push back

follow the winding stretches of bone, muscle, sinew
this wholeness that is actually a continuous scattering of material
identify where the cuts have been made
split them open – wide – once again

recall all those sites of injury
the broken backs, sullen cheeks, severed fingers of our great grandmothers
the scatterings of self across the break

coherence is a fallacy
the brokenness of being is what we must own

reinvest in the aliveness of your breath
the deep resonances of heel, ball, toe
heel, ball, toe
heel, ball, toe
on wet earth

know that in this treading – the hips, ankles, spine are grounded once again, anchored in the flesh

the overflows, the spills, the uncontained agents of our bodies
are hard at work, like they always have been
squirreling away, burrowing in deep, preparing for this break

in this break, lies our expansion
in this tension lies our liberation

Reinvest in this aliveness of your breath,
the weight of flesh on your bones,
the curve of your back,
the articulation of your hips.
Reinvest. Reinvest. Reinvest.

Naimah Petigny 2016, 5

Three: war-recording

As the world prepares to celebrate Eid-ul-fitr in June 2017, 15-year-old Hafiz Junaid Khan boards a train with his brothers to return home to his village after buying clothes in Delhi. An argument over a seat turns into slurs against Muslims for wearing skullcaps, for eating cow meat and for being 'anti-national'. The men pull Junaid's beard, fling the brothers' skullcaps and slap them. The teenagers tell the mob that cow meat was not even eaten in their village, but the men pull out their knives and one stabs Junaid until he dies. Junaid's injured brother Hashim recalls in shock: 'Instead of saving us, the crowd was egging the attackers on. They held us by our arms, while the men pierced our bodies with their knives.' A few days after this incident, India's Prime Minister Narendra Modi wraps his arms around Donald Trump in the White House, and India purchases drones worth $2 billion from the US. Twelve years ago, the same Modi was denied a visa to enter the US for his role in the 2002 pogrom that killed more than 1,000 Muslims when he was the Chief Minister of Gujarat. Even as these events unfold, I learn that the world's forcibly displaced people now number approximately 65.6 million, making such people the equivalent of the twenty-first largest country in the world, at the same time as the National Public Radio announces that the United States' longest war on Afghanistan is expected to continue for years. In Minnesota, my other home, the jury investigating the murder of Philando Castile acquits the police officer who gunned him down seven blocks from where I live, dashing the hopes of his family and thousands of protesters who were somehow confident that justice could not be denied this time, especially in the face of the damning evidence against the police officer.

***

Recording each war. Enunciating each displacement. Gesturing towards every haunting. Naming every lynching … So that we can gain the strength to reinvest, to fight, to overcome, to breathe, to dance … without identifying that which has already happened as belonging to the 'past'. As Michel-Roplh Trouillot (1995, 15) puts it, 'The past – or, more accurately, pastness – is a position. Thus, in no way can we identify the past as past'. We must push back on the temptation to forget – the sweet drawl of clean, confined pasts. Instead, we claim pasts that spill over into present futures and conjure the buried agents beneath. We demand 'discontinuous, contradictory, multifarious legacies', without requirement of resolution (Hong 2015, 3), so that our translations or retellings (see Merrill 2009) can try to do justice to landscapes like the ones we describe above. We cannot let these landscapes slip away, because, within them – rooted deep – are lessons about how to be in community, again and again. We begin here because – like others who have written, danced, rallied and performed before us – this is where decolonial praxis must begin: we must interrogate modes of knowledge that place past-present-future, mind-body-spirit, being-doing-knowing into neat compartments (Keating 2016; Anzaldúa 1987). We must interrogate and challenge the systems of power that excise specific streams of thought from the realms of knowledge that are pronounced to be valid or superior. We must insist on feeling, embodying and relearning the knowledges that have been erased or foreclosed due to ongoing projects of colonization, displacement and ethnic cleansing. Armed with such commitments, our praxis must insist on unearthing a set of maps that chart a different movement of bodies – of our multiple selves, of our ancestors and of multiple others – through times and spaces that both acknowledge and refuse borders (Hartman, 2007).

These are only some of the many possible starting points for co-authoring and stringing together a movement – an agitation of words, passions and commitments – that seeks to articulate what it might mean to decolonize methods of knowing and being. We embed our reflections in not only a world of wars, displacements and lynchings but also in a landscape of neo-colonial and neoliberal institutions of formal learning, activism, within and despite which we grow, struggle and build dreams and solidarities for justice. Neoliberalism is only interested in the selective protection of life; it offers up remembrance as a form of containment, and it disallows knowledges that hold multiple and overlapping 'modes of being, affects, memories, temporalities' in suspension (Hong 2015, 16). These institutionalized landscapes show us the limits of the knowledges imprisoned in them, even as they give us the reasons for imagining past them.

To decolonize the methods of knowing and being requires us to reform, even revolutionize, the relationships among multiply situated knowledges and knowers in incommensurable worlds and journeys. One way to begin embracing such labour is by recognizing the epistemic agency of those whose bodies and beings are relegated to the 'margins' or declared to be 'past' (Smith 1999). Making this choice means learning to learn from doing and dreaming, being and moving, remembering and relating in deeply embodied ways. It also implies intervening in the dominant academic politics of knowledge production by learning when and how to refuse citational practices that fragment the bodies (of knowledge) that constitute our consciousness and conscience and that reduce modes of creating knowledges to recognizable and nameable individual sources, while at the same time erasing the necessarily complex collective processes from which we all come to know and be known. The questions, then, are: how to co-imagine and co-create alternative citational practices where knowledge can also be understood as a political-spiritual-activist force that flows out of fragmentation, reduction and uprooting, a force that is ever partial, ever irreducible and ever embracing of the tensions and frustrations that emerge between and across incongruent and vastly unequal sites, epistemes and bodies (Alexander 2005; Collins 2000; Simpson 2014)? Grappling with the (im)possibility of accounting for all that goes on in knowledge making is a commitment and a process that cannot be formalized or mastered. By its very nature, this labour must be political, spiritual and in(di)visible (Keating 2016; Moraga and Anzaldúa 1983). It demands that we meditate from a place of knowing and unknowing, a place of inexplicability and love, that is forever open to embracing new co-travellers in the journey so that we can continue to yearn for justice: for bodies, histories, places and rhythms that often remain hidden, uncounted, unacknowledged or dismissed in our worlds (Nagar, in journeys with Sangtin Kisan Mazdoor Sangathan and Parakh Theatre, 2019). It is such a commitment that informs what we offer here.

We dwell in the entanglements of spaces, identities and languages that search for ethics, justice and solidarities in at least two ways. First, we centre as a site of knowledge and struggle bodies that have been multiply marked, violated and erased, including along the axes of race, religion, caste, gender, sexuality, place and citizenship. Second, we consider how feminists' searches for decolonizing methods have inspired dynamic engagements with translations or retellings that fully engage our embodied beings. We underscore the need for embodied translations that fight geographies that keep the so-called 'margins' partitioned and that reclaim the stories, places, paradigms and methodologies of knowing, being, protesting and (re)creating that have been repeatedly erased by institutionalized systems. In addressing these themes, we also collapse, blur and stitch the borders among Brown/Black/Indigenous/China@x/transnational/women of colour feminisms, without reifying the boundaries that have created these as disciplined fields or subfields. As well, disagreements and disruptions are essential ingredients of this full-bodied agitation so that, even as we invoke situated solidarities across multiple borders, we are aware of the ever-present need to attend to the faults and fractures that inevitably shape the collectivities we forge (Nagar 2006, 2014).

Below, we provide glimpses of instances where embodied engagement through dance, theatre and writing have animated our efforts to decolonize the dominant methods of knowing and being in academic, activist and artistic spaces. By bringing these instances into a conversation, we embrace a praxis of translation or retelling that can enliven flattened renderings of space into lived geographies. Our intimate rendering of home, historical memory and landscapes inserts certain bodies into the very spaces that have attempted to erase their existence; it insists that landscapes are never closed off from the energies – dead and alive, animate and inanimate – that circulate within them.

Decolonizing bodies

In September 2017, Naimah spent a week moving with Moroccan dancer and choreographer Bouchra Ouizguen and her contemporary dance company based in Marrakech, Morocco. The discussion in this section draws upon Naimah's writing about this experience. The 'I' here refers to Naimah, then, even as all three of us co-own the ways in which we sew, stitch and knot her individual reflections with our collective churnings.

Joined by performance artists across the Twin Cities, Naimah participated in workshops for Ouizqguen's new piece, *Corbeaux*. *Corbeaux*, or 'Crows' in English, is both a living sculpture and a rapturous performance. Marked by its uninhibited and serendipitous nature, *Corbeaux* enlivens difference amid universality, all the while contesting partitions that divide performers from audience, movement from knowledge and bodies from landscapes. Since its premiere in 2014, *Corbeaux* has toured the world with an intergenerational company made up of professional dancers from Marrakech and local women from each city in which the piece is performed. A singular gesture inspired the *Corbeaux* score: the sharp, backwards thrusting of the head, tilting toward the sky with a broad open chest, accompanied by a guttural outcry – deep and resounding. From Marrakech to the Cour Carrée at the Louvre to the Walker Art Center in Minneapolis, *Corbeaux* has offered an intimate engagement for dancers and audience-cum-witnesses alike. Although the piece does not draw upon 'traditional' Moroccan movement forms, it integrates Mediterranean styles of dress and Moroccan–Senegalese ritual gestures into varying city landscapes across the globe.

Centering a variety of forms of cultural production, Black performance theory attends to ways of knowing and movements of being that engender fluid, dissident underpinnings of Black social life (Nyong'o 2015; Williamson 2016). This Blackness insists on a mode of embodied engagement that conjures histories, hauntings and possibilities of embodied resistance across multiply violated geopolitical sites, and it hungers for justice. Performance, then, is marked by that which 'subverts cultural norms … blurring the lines between action, performance, and works of art' and, in the process, constitutes powerful practices and pedagogies of everyday life (McMillan 2015, 4).

Even though *Corbeaux* arises from a particular set of impulses that were birthed in a Moroccan context, it necessitates a co-constitutive enlivening of individual life and collective (after)life (Sharpe 2016, 3). In this co-constitutive enlivening, the dancers encounter new terrains of self, contoured by the affects, passions and complicities that structure who we are and how our bodies show up in the world (Cox 2015). We expand approaches to contemporary performance that allow individual dancers to harness the elasticity of choreography while still working from within the same movement repertoire. We enliven the bodies, breath and terrain in concert with the discursive – referents, utterances and other communicative practices – as a means of intervening in and re-imagining the world (Taylor 2003, 15). No two of us execute the movement in the same way, and our collective virtuosity curates a rich, multiplicitous performance. A week spent together in rehearsal and performance is a week full of creative and political lessons – it allows us to teach one another new ways of moving, new methods of breathing and new ways of being fully in our bodies.

Artistic exchanges like this one hinge upon remembrance, vulnerability and release of expectations that affix to bodies and mark them as incommensurable across disparate geopolitical locations. And yet, the challenge is to collaboratively string our movement *together* across moments of both agitation and collective embrace. This collective agitation and embrace must reject simplistic narratives of multicultural alliance and instead take up the much more difficult task of committing one's desire for freedom to movement – in all the senses of that term.

The methodology, then, is in the practice – in the opportunity to create something afresh while rearticulating and re-energizing new ways of collapsing the 'I's', eyes and the many provisional 'we's' that make us at (in) any given time, place and struggle. In decolonizing our bodies in this manner, corporal and aesthetic risk co-constitutes more than just artistic vision (McMillan 2015, 7). This risk embraces radical vulnerability as a collective mode of unlearning and relearning (Nagar, in journeys with Sangtin Kisan Mazdoor Sangathan and Parakh Theatre, 2019). It dares to leave things undone and to allow the body to be an engine of that undoing. If decolonized bodies are to exist across fragmented sites, splintered histories and embodied memories, then our methods must also resist concretizing or systematizing movement. It is precisely because we are entangled with those who live while also being inhabited by our dead (Simpson and Smith 2014) that we do not, we cannot, seek neat resolutions. We look to what survives abjection, exclusions and ontological negation. A decolonizing mode of studying and being demands that we surrender to movement and collective action through a mode of radical vulnerability that defines a non-individualistic ethic of engaging one another and being together – so that we can plot and map; so that we can recollect our souls and beings in order to reimagine the meanings of accountability and justice.

Decoloniality and politics of living

That the world's displaced people approximate 65.6 million; that the growing mobility of people carrying multiple citizenship erodes the boundaries of the nation state; that some refuse passports and shun symbols of freedom and mobility; and that more than a third of Haiti's population lives outside of its territory. These are only a few resonances of struggles, projects and worlds that demand that we turn away from reading these struggles as commensurable with the modern praxis of political democracy or as stifled by imposed states of backwardness in need of civilizing. Through our interbraiding of multiple sites of embodied translations, our readings deploy an understanding of the broader and deeper political charges these struggles carry – struggles that refuse to be read within frameworks that are contained and worded through the ongoing practices and processes of slavery and colonization.

As one of the many possible ways of extending the search for decolonizing methods, we ask: what might a just and ethical story of Haiti within our present local, national and transnational contexts look like? This question summons us to place ourselves in the faults between lived experiences and the stories made possible within them. The dominant ways of knowing sometimes make us feel as if stories and bodies exist outside of the languages that narrate them and bring them into being. An alternative way of knowing is by learning to feel how one's being is tied to others – both human and other than human – in ways that engage both the story and the body as conditions of possibilities. There are stories the body writes as an open-sided sphere of entangled cultural–natural phenomenon, as a site of articulation that enables a coiling up of everything – languages, signs, logics, histories, myths and thoughts – without precise intentions or genealogies: only the pretence to a bliss, a dream, a vision that tends towards knowing. Let's pay attention to the Haitian American poet, Valerie Deus, in her poem 'Haiti Unfinished':

I want to write you another note about
feeling like a jack-o'- lantern hollow
with the seeds and threads missing
with the soup and the guts gone
there's no independence day long enough
or revolution deep enough to save me
from writing a poem about watching novellas with your mother
while drinking tea
or picking hazelnuts in her backyard
Valerie Deus 2011, 77

The poem offers a space of contemplation and inquiry from which both the poet and the narration co-emerge. Such contemplation suggests that the awareness of one's experience within the collective and of the telling of that experience contribute to the movements and possibilities of history/ies. It troubles the question of how Haiti, or any location or place for that matter, should be thought. The poem crafts a poetics of retelling that exposes the intensities of living between forces and energies of all sorts: the moving back and forth across times, the crossing of geographic boundaries and of protagonists; the intimate and confessional tone in and through which the narrator makes an entrance within webs of beings. For Deus, her location within the project called Haiti is one that is constantly in the making – open, 'unfinished' and marked by playful shifts between interpellating and being interpellated. In offering a poetic narrative that exposes how one's body engages painfully, purposefully, as well as creatively within complex webs of relationships, Deus emphasizes the kind of labour most people undertake in the face of embedded multilayered violence. She retells the coming into one's own identity. Aimé Césaire terms this impulse 'poetic knowledge', the sole force capable of effecting a 'co-naissance', a knowledge from which emerge both the self and the narration altogether (Césaire and Pinkham 2012).

Mimerose Beaubrun in her book, *Nan Dòmi,* deploys a similar approach to the body that is creative, aesthetic and spiritual. She conceptualizes the body as a site of open-ended systems in interaction and in differentiation with the material discursive environment. Beaubrun begins her journey with the goal of learning about the Lakou project and its importance in Haitian political struggles. She ends up engaging with the Lakou through her own body and that of others as a kind of vital space and a place of multidimensional life. The author's journey, then, is an ongoing process of being and becoming at the level of the body as a site of knowledge. The body in *Nan Dòmi* is an ontological condition that depicts a means of being and a means of knowing (Beaubrun 2013). This intervention is reminiscent of Maria Lugones's work (2007), in which she draws from Quijano (2000) to propose a reading of how lived experiences negotiate the arrangements of colonial relations and, in so doing, make possible elaborate responses to oppression. Both Lugones and Beaubrun rethink the possibilities of the self and selves by emphasizing the logic of difference and multiplicity. Lugones, in particular, insists on a mode of theorizing that demands a body-to-body engagement and that attends to the ways in which colonial relations cut across everyday practices, ecology, economics, government relations and spirituality to evolve modes of being and knowing that stand in their own right as struggles that enact refusals. Such refusals make possible worlds, visions and movements with transformative and political consequences (Lugones 2003).

The body, in this sense, is inseparable from the complexities of the everyday through which power relations of all sorts are forged and articulated. A turn to the body propels us to ask what kinds of embodied knowledges emerge within the dynamic balance of diverse spiritual, economic, cultural energies and worlds within which a person and their personhood are rooted. Our search for decolonizing methods and ethics of retelling is a similarly unsettling inquiry accompanied by a basic demand – the demand to remove the focus from dominant epistemes and to direct it towards alternative epistemic forms of being through which new politics can be imagined to nourish the lives of all of us who have been variously colonized.

Continuing co-tellings

To decolonize methodologies is to insist on the necessarily entangled and inseparable nature of embodied pedagogy, research, artistry and movements that strive for connection and justice across communities, worlds and struggles. For those of us creating knowledge from a location of such power and privilege as a US research university, such methodologies must necessarily involve agitating against the ways in which the academy's rhetoric of interdisciplinarity often allows for a coming together of disciplined fragments, albeit without requiring a transformation of those fragments. The co-authored retellings or translations we offer here are a part of an anti-definitional agitational praxis through which unplanned freedoms and serendipitous movements for justice can be imagined and enacted (Nagar, in journeys with Sangtin Kisan Mazdoor Sangathan and Parakh Theatre, 2019). In embracing the idea of spiritual activism as an undefinable and non-reductive mode of co-travelling and co-making knowledges across worlds and struggles, we join many other feminist scholars and activists who simultaneously remain grounded in the structures, processes, languages and feelings that constitute those worlds and struggles. We embrace our responsibility as bearers and co-creators of knowledge in ways that appreciate knowledge as an unfixable living and throbbing force, without an easily identifiable beginning or end, yet always partially within the reach of our hearts and minds. This possibility of reaching and feeling knowledge inspires us to reclaim and reword, to remember and retrace, to redo points of reference and bring them into tension – so that we can dodge and dismantle the traps that choke the truths that have been violated. Knowledges emerge from different voyages that involve singing, naming and mourning; playing and laughing and dancing. These journeys ask that we linger with the possible meanings of not only that which is utterable but also of that which is silent. For us, being silent is a state in which one might silence all thought; it is also a refusal of predetermined frameworks, which enables us to more responsibly witness those modes of living, being, fighting and knowing that are consistently rendered invisible or invalid.

Whether our attention is turned towards the Haitian TPS in the era of Trump, or the ways in which Black bodies navigate deathly terrains with the softness of 'heel-ball-toe' or the threats leveled against Muslims for being 'anti-national' in Modi's India, a commitment to decolonize knowledges involves a shared thirst to know the multiple geographies, bodies and scars of these hauntings so that our movements may work through varied levels of intimacy and so that we may realign commitments and practices with all co-living bodies that summon us. We, then, continue to search for such practices of discontent that will push us to patiently and steadily unearth the modes of dwelling in and linking all those traumas and scars that the prevailing maps present before us in isolated forms. It is from here that we reopen ourselves to each other so that we may continue traversing known and unknown terrains in our collective search to re-imagine, undo and redo the ways in which we come to know.

Acknowledgements

All the co-authors contributed equally to the writing. Thanks go to the editors of this Handbook for giving us the opportunity to co-create this chapter and to David Faust for his helpful comments on an earlier draft.

A performed and expanded version of this chapter appeared in *Commoning Ethnography* 2 (1): 113–131 (2019).

Key readings

Smith, L.T. 1999. *Decolonising Methodologies: Research and Indigenous Peoples.* London: Dunedin.

Alexander, M.J. 2005. *Pedagogies of Crossing: Meditations on Feminism, Sexual Politics, Memory, and the Sacred.* Durham, NC: Duke University Press.

Keating, A.L. 2016. "Spiritual Activism, Visionary Pragmatism, and Threshold Theorizing." *Departures in Critical Qualitative Research* 5 (3): 101–107. doi: 10.1525/dcqr.2016.5.3.101.

References

Alexander, M.J. 2005. *Pedagogies of Crossing: Meditations on Feminism, Sexual Politics, Memory, and the Sacred.* Durham, NC: Duke University Press.

Anzaldúa, G.E. 1987. *Borderlands/La Frontera: The New Mestiza.* San Francisco: Spinsters/Aunt Lute.

Beaubrun, M. 2013. *Nan dòmi, an Initiate's Journey into Haitian Vodou.* San Francisco: City Lights.

Césaire, A., and J. Pinkham. 2012. *Discourse on Colonialism.* Marlborough: Adam Matthew Digital.

Collins, P.H. 2000. *Black Feminist Thought: Knowledge, Consciousness, and the Politics of Empowerment.* New York: Routledge.

Cox, A.M. 2015. *Shapeshifters: Black Girls and the Choreography of Citizenship*. Durham, NC: Duke University Press.
Deus, V. 2011. "Haiti Unfinished." In: *How to Write an Earthquake: comment écrire et quoi écrire mo pou 12 Janvye: Sixteen Haitian Writers Respond: seize écrivains haïtiens parlent: sèz ekriven ayisyen reponn*, edited by B. Pierre and N. Ďurovičová, 77–79. Iowa City: Autumn Hill Books.
Hartman, S.V. 2007. *Lose Your Mother: A Journey along the Atlantic Slave Route*. New York: Farrar, Straus and Giroux.
Hong, G.K. 2015. *Death beyond Disavowal: The Impossible Politics of Difference*. Minneapolis: University of Minnesota Press.
Keating, A.L. 2016. "Spiritual Activism, Visionary Pragmatism, and Threshold Theorizing." *Departures in Critical Qualitative Research* 5 (3): 101–107. doi: 10.1525/dcqr.2016.5.3.101.
Keeling, K. 2007. *The Witch's Flight: The Cinematic, the Black Femme, and the Image of Common Sense*. Durham, NC: Duke University Press.
Lugones, M. 2003. *Pilgrimages/Peregrinajes: Theorizing Coalition against Multiple Oppressions*. Landham, MA: Rowman & Littlefield.
Lugones, M. 2007. "Heterosexualism and the Colonial/Modern Gender System." *Hypatia* 22 (1): 186–209. doi: 10.1111/j.1527-2001.2007.tb01156.x.
McMillan, U. 2015. *Embodied Avatars: Genealogies of Black Feminist Art and Performance*. New York: New York University Press.
Merrill, C.A. 2009. *Riddles of Belonging: India in Translation and Other Tales of Possession*. New York: Fordham University Press.
Moraga, C., and G.E. Anzaldúa. 1983. *This Bridge Called My Back: Writings by Radical Women of Color*. New York: Kitchen Table Women of Color.
Nagar, R. 2006. "Introduction." In: Sangtin Writers and Richa Nagar, *Playing with Fire: Feminist Thought and Activism Through Seven Lives in India*. Minneapolis: University of Minnesota Press.
Nagar, R. 2014. *Muddying the Waters: Co-authoring Feminisms Across Scholarship and Activism*. Champaign: University of Illinois Press.
Nagar, R. in journeys with Sangtin Kisan Mazdoor Sangathan and Parakh Theatre, 2019. *Hungry Translation: Relearning the World Through Radical Vulnerability*. Champaign: University of Illinois Press.
Nyong'o, T. 2015. "Performance and Technology." *What is Performance Studies*. Durham, NC: Duke University Press. Available at: http://scalar.usc.edu/nehvectors/wips/performance-and-technology
Petigny, N. 2016. *Fugitive Flesh: The Queer Embodiment of Black Liberation*. Conference paper presented at National Women's Studies Association Conference, Montreal, November.
Petigny, N. 2017. "Rapturous Groundings: Bouchra Ouizguen's *Corbeaux* and the Performance of Release." Fourth Wall, *Walker Art Center Magazine*. Available at: https://walkerart.org/magazine/rapturous-groundings-bouchra-ouizguens-corbeaux-and-the-performance-of-release#_edn1.
Quijano, A. 2000. "Coloniality of Power, Eurocentrism, and Latin America," translated by Michael Ennis. *Nepantla: Views from South* 1 (3): 533–580.
Sharpe, C.E. 2016. *In the Wake: On Blackness and Being*. Durham, NC: Duke University Press.
Simpson, A. 2014. *Mohawk Interruptus: Political Life across the Borders of Settler States*. Durham, NC: Duke University Press.
Simpson, A., and A. Smith. 2014. *Theorizing Native Studies*. Durham, NC: Duke University Press.
Smith, L.T. 1999. *Decolonising Methodologies: Research and Indigenous Peoples*. London: Dunedin.
Taylor, D. 2003. *The Archive and the Repertoire: Performing Cultural Memory in the Americas*. Durham, NC: Duke University Press.
Trouillot, M.-R. 1995. *Silencing the Past: Power and the Production of History*. Boston, MA: Beacon Press.
Williamson, T.L. 2016. *Scandalize My Name: Black Feminist Practice and the Making of Black Social Life*. New York: Fordham University Press.

38

'STILL WE RISE'

Critical participatory action research for justice

Caitlin Cahill, David Alberto Quijada Cerecer, Leticia Alvarez Gutiérrez, Yvette Sonia González Coronado, José Hernández Zamudio, Jarred Martinez and Alonso R. Reyna Rivarola

(ONE VOICE) We have abandoned the shadow to proudly speak our minds to challenge 'commonsense', which claims that we are apathetic and careless.
(ANOTHER VOICE) It is obvious we will not melt into the pot, but rather will savour our flavours in a bowl of *pico de gallo*.
(ONE VOICE) We are, most importantly, the people who make up Utah.
(ONE VOICE) We share the same land, breathe the same air, live on the same soil. Why are we not treated the same?
(ALL) We have come – our time is now! Today! Not yesterday!
(ONE VOICE) We are rejecting the crumbs that we have been given and demand that we be given a piece of the pie – a piece with which we will nourish our communities and counteract the hunger that we have been plagued with.
(ALL) We are hungry! Hungry for change!
(ONE VOICE) Meaningful change!

from We the People, *a spoken-word piece performed by Mestizo Arts & Activism Collective*

Taking up space in the rotunda of the Utah State Capitol, we performed the spoken word piece *We the People* (excerpt above), demanding our rights in loud, sing-song voices, claiming home, a sense of belonging and community. Referring to the first line of the US Constitution, we rearticulated what this means through the lens of immigrant experiences. In our participatory performance, we joined those organizing for migrant justice across the US (and the world), whether on the frontline calling for an end to attacks on immigrants, protesting at detention centres and senators' offices or working behind the scenes to care for and support each other, engaging in a quiet politics of transformation and resistance (Askin 2015; Solorzano and Delgado Bernal 2001). Bearing in mind the wisdom of the anti-apartheid movement, 'Nothing about us, without us, is for us', we engage with critical participatory action research (PAR) to address structural injustice through collective knowledge production and activism.

The political economic context of globalization is experienced by many as a state of ongoing crisis. These are harrowing times indeed. Undocumented immigrants are under threat, including community members and the extended family with whom we research, create, and work alongside. This crisis is not new, but it has deepened exponentially over

Figure 38.1 Homeland Security © 2010 Ruby Chacón.

the past few years. Stark images of children at the US–Mexico border sleeping in cages and separated from their families, the growing military presence at the border, the amplification of aggressive xenophobic rhetoric and the US government shutdown over funding the border wall stake out the contours of the ongoing crisis. Within this context, Appadurai (2006, 168) argues that researchers' 'right to research' is the right to 'systematically increase the stock of knowledge which they consider most vital to their survival as human beings'. With a commitment not only to study injustice but to do something about it, we are inspired by the participation and action of immigrant young people and their families, who work together in order to open up an expansive conversation about what kind of world we are fighting for. With the knowledge that 'the master's tools will never dismantle the master's house' (Lorde 2003, 2), we are committed to doing research differently, in a way that honours our whole selves, our culture and our community. Drawing upon our work with the Mestizo Arts & Activism Collective (https://maacollective.org), an intergenerational social justice think-tank based in Salt Lake City, Utah, we discuss the relationships between critical PAR and activism drawing on specific projects.

In what follows, we offer a brief overview of critical participatory action research (PAR), exploring the principles of this epistemological approach and the commitments involved. Next, we discuss the productive tensions between theory and practice in our work focused on the concerns of immigrant communities.

Critical participatory action research

Offering a meaningful framework for researchers committed to social justice and change, critical PAR is an epistemological approach to collaborative knowledge production that draws upon feminist, critical race and Indigenous theories, community development and legacies of grassroots organizing, antiracist and social justice movements (Delgado Bernal 2001; Collins 2000; Crenshaw 2005; Crenshaw et al. 1995; Delgado 1983; Du Bois, 1898; Hale 2008; hooks 1999; Horton 1990; Kretzmann and McKnight 1996; Smith 1999; Torre et al. 2012; Tuck 2009). Tracing interdisciplinary lineages from around the world (in particular, Latin and South America), PAR is informed by wide-ranging thought from across the social sciences, including liberation theology, critical pedagogy, psychology, sociology and geography (Bell 2001; Bunge et al. 2011; Bunge and Bordessa 1975; Cammarota and Fine 2010; Fals Borda 1979; Fine 2017; Freire 1997; Hale 2008; Hart 1997; Lewin 1948; Pain and Kindon 2007; Torre et al. 2015, 2017; Torre and Ayala 2009; Zeller-Berkman 2014). Feminist geographers have played a critical role in theorizing PAR as an alternative ontology, drawing on situated knowledge and taking seriously what it means to do *social* science (mrs c. kinpaisby-hill 2011; Askins and Pain 2011; Breitbart 2003; Cahill 2007; Cameron and Gibson 2005; Donovan 2014; Elwood 2006; Gilmore 2008; Katz 1994; Kesby 2005; Kindon et al. 2007; Nagar 2014; Pratt 2003; Pain 2004; Pain et al. 2011; Pratt 2010; Ritterbusch 2019). Recognizing the connectedness of knowing, doing and being, critical participatory praxis may potentially destabilize binaries of theory and practice to offer 'an alternative ontology of theorizing' that moves between social constructions and embodied and emotional experiences (mrs c. kinpaisby-hill 2011). What this looks like in practice is a process that starts with the investigation of personal experiences, moves towards social theorizing and structural analysis, and ultimately, action, as will be described in more detail.

Putting the emphasis upon process, PAR recognizes the power of knowledge produced in collaboration and action, whereby community members can 'make meaningful contributions to their own well-being and not serve as objects of investigation' (Breitbart 2003, 162). PAR shifts accountability to how research might be 'of use' to communities (Fine and Barreras 2004) outside the 'ivory tower' and 'beyond the journal article' (Cahill and Torre 2007). Critical scholars suggest that we need to be wary of broad applications of the term 'participation', as it masks tokenism and the illusion of consultation that may, in fact, advance dominant interests (Arnstein 1969; Cooke and Kothari 2001; Kesby 2005). This critique is especially important as the rhetoric of participation dovetails with the neoliberal agenda's emphasis upon local control and personal responsibility, offloading obligations onto communities who 'participate' and take on the work of the state (Cahill 2007; Harris 2004; Herbert 2005; Purcell 2006; Wilson 2004). This concern is exacerbated when participation is presented as a set of techniques rather than as a political and epistemological commitment to working *with*, not for or on, communities. A *critical* PAR approach centres an analysis of power, 'taking into account that all people are embedded within complex social, cultural, and political systems historically defined by structural inequalities and privilege' (Public Science Project, n.d.), as we discuss further below. For us, this involves countering the logic of global neoliberalism that shifts the accountability for economic inequalities to undocumented immigrants. In our analysis, we join critical scholars who plot the intersections of neoliberalism, dispossession, racism and xenophobia in analyse of racial capitalism (De Genova 2013; Gilmore 2002; Loyd 2012; Melamed 2015; Pulido and Lloyd 2010; Robinson 2000). This is the contradictory and slippery ground upon which we locate critical participatory theory and practice; not as an ideal but as a fraught, urgent engagement of ideology, power, politics and context that is in conversation with activist movements (Cahill 2007).

Mestizo Arts & Activism Collective

Over the last decade, the Mestizo Arts & Activism Collective has developed under the leadership of young people who are committed both to each other and to sustaining a creative alternative space of community-based participatory action research and activism. We engage in liberatory, *critical* PAR. By putting the emphasis upon the *critical*, we signal our commitment to documenting and challenging 'the grossly uneven structural distributions of opportunities, troubling ideological categories projected onto communities, demonstrating how science has been recruited to legitimate dominant policies and practices' (Torre et al. 2017, 2012, 171).

The Mestizo Arts & Activism Collective centres the critical insights and understanding that young people bring to understanding their everyday lives within an intergenerational, multiracial community context. Each year, approximately 10 to 25 young people aged 14 to 20 participate, representing ethnically diverse backgrounds (Latinx, Chicanx, Mixed race, African–American, Asian and White). While our work has focused on many issues, ranging from community development to media representations to sustainability issues, among others, what has remained consistent over the years is the ongoing concern about immigration rights, specifically focused on undocumented students' educational rights and resisting what we call the 'school-to-sweatshop pipeline' (Cahill et al. 2016, 2019; Quijada Cerecer et al. 2019). Our conceptualization is informed by the framing of the 'school-to-prison pipeline', which sheds light on how punitive policies target students of colour to push them out of school and into the criminal legal system (Bahena et al. 2012; Fasching-Varner et al. 2016; Morris 2016). The 'school-to-sweatshop pipeline' tracks how immigrant students are steered into a shadow state of exploitative labour practices (including a denial of rights and protection), from mowing lawns to taking care of other people's children, unregulated construction work and seasonal farm work (Alvarez Gutiérrez 2016; Pratt 2012; Pulido 2007; Reyna Rivarola 2013; Varsanyi 2008). Reflecting a dialectic of dreams and dispossession, the 'school-to-sweatshop pipeline' conjoins with the activist DREAMER movement of undocumented students at this precarious political moment when DACA (Deferred Action for Childhood Arrivals) and other rights are under threat as immigrant communities are targeted (Alvarez Gutiérrez 2015; Cahill et al. 2016; Delgado-Bernal and Alemán Jr 2017; Diaz-Strong et al. 2014; Gonzales 2011; Patel 2012; Reyna Rivarola 2017).

As scholars who have been working together for the past decade, we note how our Collective's questions, concerns and consciousness have transformed in response to the changing political context and public debate at local and national level and as aggressive attacks upon immigrants have increased. Committed to the guiding principles of critical PAR, we shift what we define as 'problems' off the backs of individuals and onto the structures, systems and policies (Torre et al. 2012). Paying attention to how we frame the unit of analysis, we document how 'the global and the intimate intertwine' (Pratt and Rosner 2012) in young immigrant students' everyday lives at school. For example, in our theorizing of the school-to-sweatshop pipeline, we trace how the production of 'illegality' (De Genova 2002) functions dialectically to produce a disposable (and deportable) reserve of labour while obfuscating the role of global neoliberal restructuring policies (e.g. NAFTA) (Cahill et al. 2016). At the same time, our inquiry considers the deeply personal experiences of the political (Pain and Staeheli 2014) – what it feels like 'living in this skin' (Gonzalez Coronado 2009) – as global neoliberalism informs young people's subjectivities and aspirations, as expressed by undocumented student Rafael in his reflection upon 'learning to be illegal' (Gonzales 2011):

> I am smart. I am hard-working ... But the moment that you shut me out, you cut my arms and legs off, and I cannot move. And there is nothing left, sometimes, but a feeling of desperation. Because then you are reminded that you don't belong here.

In the context of state-sanctioned dispossession and violence (another way we might theorize the impact of the 'production of illegality'), we co-create a social and shared context for witnessing each other's private experiences of discrimination (Cahill et al. 2019; Torre et al. 2017). And crucially, collectively we take action to transform oppressive conditions. Informed by an 'ethics of care' in its most profound sense (Bartolomé 2008; Cahill et al. 2007; DeNicolo et al. 2017; Ellis 2007; Gilligan 1982; Valenzuela 2000), critical PAR reflects a deep respect for relationships, humanity and collective well-being (Cahill et al. 2007; Tuck and Guishard 2013).

As activists and scholars engaged with critical PAR, we are in conversation with social movements and elders in our communities, who share insights through testimonials and counter-stories of survival (Delgado-Bernal et al. 2012; Solorzano and Yosso 2001; Yosso 2013). Centring an intersectional racial justice perspective, we engage in what Solorzano and Delgado-Bernal (2001) identify as 'transformational resistance', embracing diverse strategies of engaging in structural critique, including critical PAR. What this means is that our forms of resistance include engaging collectively in inquiry and reflection, working behind the scenes and caring for each other in ways that are often under-acknowledged in activism. While at the same time, organizing and demanding our rights, standing on the frontlines of protest. Critical PAR is not, in this sense, separate from the other ways in which we are active and engaged in our community.

Critical PAR emphasizes how we are mutually implicated in each other's lives, engaging our diverse standpoints and differential relationships to structural conditions (Torre et al. 2017, 467). With a shared commitment to migrant justice, we work across our differences within an intentionally diverse and democratic space of inquiry: what María Elena Torre (2010; Torre et al. 2017) identifies as a 'participatory contact zone' (cf. Askins and Pain 2011). This encompasses 'excavating disjunctures rather than smoothing them over in the interest of consensus', as they provide insight into critical insights about the larger social and political dynamics at play in the research (see Public Science Project n.d.; Torre et al. 2017). Attending to our intersectional relationships and positionality involves valuing and foregrounding the wisdom and concerns of those without documentation while strategically working with the (unearned) privileges afforded to those born on this side of the border.

We write as a 'strategic we', as co-founders, facilitators, current and former advisors and youth researchers of the Mestizo Arts & Activism Collective, highlighting our political stance and solidarity while embracing dissent and negotiation in our collective process. Challenging the neoliberal emphasis upon the individual, we write, research and create art together, signalling that all knowledge is produced collectively, whether this is acknowledged or not (mrs c. kinpaisby-hill 2011; Cahill et al. 2019). Our starting point is taking seriously the participatoriness of our

Figure 38.2 *Caution* © Mestizo Arts & Activism Collective.

collective and the relationships that we have with each other. Honouring our intersectionality (Collins 2000; Crenshaw 1995; Hopkins 2018) and creating a space of radical acceptance for our whole selves feels especially important as scholars and activists at this political moment, time and space. We pay attention to how the site-specificity of growing up and living in Utah informs our subjectivities, our relationships and our research, acknowledging the impact of Utah's long history of racism, xenophobia, homophobia and conservatism associated with the Mormon Church and the history of white-settler colonialism. With this in mind, we attend both theoretically and practically to how this impacts both our research and the members of the Mestizo Arts & Activism Collective, asking questions within the framework of ethics (Cahill et al. 2007; Fine et al. 2000; Manzo and Brightbill 2007; Tuck and Guishard 2013): Who is made vulnerable by our research? Who has the 'authority' to represent a community's point of view? Is there a 'we' within the community being represented? In what voice or language should we speak? How might the research provoke action? Which publics do we privilege or prioritize in our research? And, in what ways do we honour the integrity of our commitments to justice in the presentations of our work? (Cahill and Torre 2007).

Caution/*Cuidado*: we have power

> One of the interesting things is that this image has always been interpreted as, and even called, *Caution: We have power.* In reality, the conversation within the group never headed in that direction. It was more of a 'We are peaceful and we care about each other, our communities, our families and we have a responsibility to each

other': *Estamos junt@[x]s pase lo que pase* (we are together whatever happens). Which, now that I come to think about it, could be interpreted as 'Caution, we have power'.

Alonso R. Reyna Rivarola, Youth Researcher (2008–2011)/Co-Director (2012–2015), Mestizo Arts & Activism Collective

What happens when underrepresented perspectives (for example, those of women, people of colour, immigrants, Indigenous peoples) enter the academy and participate in the production of 'official' knowledges? Not only might they transform themselves; they may also transform the academy. This is the grounds upon which new knowledge takes root, pushing scholarship in new directions, asking new questions, challenging old assumptions and looking beyond the privileged perspectives of the ivory tower (Cahill 2007). As Delgado-Bernal and Villalpando (2002) argue, there is an 'apartheid of knowledge' in the academy, as the scholarship, epistemologies and cultural resources of communities of colour have been consistently devalued and marginalized in the context of institutional racism. Committed to foregrounding and centring underrepresented voices in the academy, critical PAR opens up a space committed to the production of knowledge. What Robin Kelley (2018) calls, 'love, study, struggle', in all of its complexity as a 'site of contestation, a place of refuge, and a space for collective work', while attending to the contradictions of institutions as spaces of the neoliberal order (cf. Harney and Moten 2013).

Challenging what Foucault (1980) identified as the 'subjectifying social sciences', critical PAR has profound implications for rethinking the politics of representation and contesting epistemological violence (Kelley 1997; Teo 2010). And, as we know all too well, epistemological violence is not just an academic matter but intertwines with ontological violence (Butler 2004, 1999; Kelley 1997). As we have written about in more detail elsewhere (Cahill et al. 2019), the artwork *Caution* (Figure 38.2) was created in response to xenophobic attacks upon our Collective on online news comment boards. It offers an example of how our participatory research and cultural praxis open up a space for us to 'theorize from the flesh' (Anzaldúa 1987/1999; Anzaldúa and Moraga 1983), literally positioning our bodies to rework cultural narratives that misrepresent Latinx immigrant communities as criminal and dangerous. Instead, we turn our backs to the camera, engaging in 'refusal' as a 'generative, analytic practice' (Tuck and Yang 2014, 817). This is not just a matter of optics but a project to transform social relations on our own terms. Foregrounding our relationships with each other, we hold hands, raising our arms up in power and in community (for more discussion, see Cahill et al. 2019, 577; Quijada Cerecer et al. 2011). In our debriefing of *Caution*, we draw upon our research focused on resisting the school-to-sweatshop pipeline, while calling attention to how young immigrant bodies have become sites for constructing national borders and at the same time justifying the exploitation of undocumented communities (Flores 2003). Drawing connections between representations, political economic conditions and our own subjectivities and agency to make change, in our critical PAR praxis we follow Freire, who states: 'people develop their power to perceive critically the way they exist in the world with which, and in which, they come to see the world not as a static reality, but as a reality in process, in transformation' (Cahill et al. 2019; Freire 1974, xx).

One of the most significant insights of our engagement together in critical PAR over many years is that shared inquiry and action critically creates moments in which we see ourselves in one another's struggles. *Caution* was created by us and for us. Privileging our own communities, we rearticulate the dominant narrative for ourselves. We did not create this image for outsiders to think differently about who we are; instead, we focused upon our own collective agency (Quijada Cerecer et al. 2019). Similarly, reflecting upon the experience of performing *We the*

People (excerpt at start of chapter) at the State Capitol, surrounded by our community members who came to support us, Yvette González Coronado explains:

> For me it felt like the power was in owning our stories, and in telling them from our perspectives. And I think that's the political act itself. Implicit in the process of owning our experiences, is the process of self-inquiry. That was the transformative piece.

Through the process of self-inquiry and 'owning our stories', we transform our understanding of ourselves and relationships with each other. Through critical collective praxis, we 'stitch together different ways of knowing' (Fox 2015, 6) and consider our distinct relationships to the structural injustices of anti-immigrant racism and exploitation.

By opening up the potential for what Butler (2004) describes as 'the constitutive sociality of the self', critical PAR offers a process for thinking about building political community and solidarities with each other. As opposed to thinking of social and political change as happening out there and investing in others the power to transform our world, our emphasis instead is upon recognizing the power that we possess to create change in our own lives. Robin D. G. Kelley explains (2014, 93): 'change requires a set of simple breaks in structure – structural power breaks … Perhaps most importantly of all, those breaks are also conceptual because you cannot design a different future unless you can think through the current one.' In this sense, Kelley argues, '*We are* the breaks'. Through critical PAR, we rearticulate research, bearing in mind Audre Lorde's provocation that 'the master's tools will never dismantle the master's house' (Lorde 2003, 2). As we have written elsewhere (Quijada Cerecer et al. 2019), we engage in critical PAR not just to dismantle but to rebuild the house. While considering the publics and purposes of our research, we consider how we build, with whom and why. In this way, we 'work the minor, to decompose the major from within' (Katz 1996, 2017), attending to how we collectively respond to the ongoing crisis of neoliberal racial capitalism and how, in this process, we come to understand our own sense of agency as a collective. As Cindi Katz explains (2017, 598), engaging minor theory is doing theory differently, working from the inside out: 'of fugitive moves and emergent practices interstitial with "major" productions of knowledge.' Understanding critical PAR within this frame, we undo and rework to transform how we know and act, building our capacity to address urgent concerns in our community.

We conclude with a statement from the youth organizers of the public action 'Still We Rise' in Salt Lake City, Utah, inspired by Maya Angelou's (1978) poem, *Still I Rise*:

> We rise for the student who works against all odds to stand for an equal education.
> We rise for the family that fought to live an American dream and despite the realities, continued to contribute to a society that did not accept them …
> We rise because we are the sons and daughters of those who believed in you and despite the betrayals, despite the torments, this is our home.
> We rise for those who live in silence, scared to bring their struggle to light …
> *Excerpt from statement 'Still We Rise', written by Salt Lake City, Utah, youth organizers*[1]

As one of the youth activists explained: 'We are still here. We are not going away. Our community has always had to fight for equal rights, and here we are still. We still rise.' As scholars and activists, we look forward with dreams and visions for a more just collective future, while looking back and remembering that we are part of a long history of struggle over the kind of world we are living in; for us, critical PAR is one way that we produce knowledge in this struggle.

Acknowledgments

We are grateful to Peter Hopkins and all the editors for their generous work in editing this much-needed Handbook. We dedicate this chapter to undocumented immigrants everywhere who are engaged in the struggle for recognition, justice and freedom. We honour Matt Bradley (1970–2012), a passionate educator, committed activist, critical scholar and co-founder of the Mestizo Arts & Activism Collective.

Note

1 The 'Still We Rise' public action was organized by youth leaders from the FACE movement and Brown Berets in Salt Lake City, Utah, with the *Movimiento Estudiantil Chicano de Aztlan* (MEChA), Mestizo Arts & Activism Collective (MAA), Mestizo Institute of Culture and Art (MICA), Family School Partnership (FSP) and the Utah Coalition of La Raza (UCLR).

Key readings

Cahill, C., L. Alvarez Gutiérrez, and D.A. Quijada Cerecer. 2016. "A Dialectic of Dreams and Dispossession: The School-to-Sweatshop Pipeline." *Cultural Geographies* 23: 121–137.

Cahill, C., D.A. Quijada Cerecer, J. Hernández Zamudio, A. Reyna Rivarola, and L. Alvarez Gutiérrez. 2019. "'Caution, We Have Power': Resisting the School-to-Sweatshop Pipeline through Participatory Artistic Praxes and Critical Care." *Gender & Education* 31: 576–589.

Kindon, S., R. Pain, and M. Kesby. 2007. *Participatory Action Research Approaches and Methods: Connecting People, Participation and Place.* New York: Routledge.

Torre, M.E., B. Stoudt, E. Manoff, and M. Fine. 2017. "Critical Participatory Action Research on State Violence: Bearing Wit(h)ness Across Fault Lines of Power, Privilege, and Dispossession." In: *The SAGE Handbook of Qualitative Research*, edited by N.K. Denzin and Y.S. Lincoln, 492–523. Los Angeles: SAGE.

Torre, M.E., M. Fine, B.G. Stoudt, and M. Fox. 2012. "Critical Participatory Action Research as Public Science." In: APA Handbook of Research Methods in Psychology, edited by H. Cooper, P.M. Camic, D.L. Long, A.T. Panter, D. Rindskopf, and K.J. Sher, vol. 2: 171–184. Washington, DC: American Psychological Association.

Quijada Cerecer, D.A., C. Cahill, Y.S. Gonzalez Coronado, and J. Martinez. 2019. "'We the People': Epistemological Moves Through Cultural Praxis." *Cultural Studies – Critical Methodologies* 19: 214–221.

References

Alvarez Gutiérrez, L.A. 2015. "Poder en las voces y acciones comunitarias: Immigrant Young People and their Families' Transformative Engagement with High School." *Association of Mexican American Educators Journal* 9 (2): 31–44.

Alvarez Gutiérrez, L.A. 2016. "'!Ya basta con la ciudadanía restrictiva!': Undocumented Latina/o Young People and Their Families' Participatory Citizenship." *International Journal of Multicultural Education* 18: 107–125.

Angelou, M. 1978. *And Still I Rise.* New York: Random House.

Anzaldúa, G. 1987/1999. *Borderlands La Frontera: The New Mestiza*, 2nd edition. San Francisco, CA: Aunt Lute Books.

Anzaldúa, G., and C. Moraga. 1983. "This Bridge Called My Back." *Writings by Radical Women of Color*, 2nd edition. New Jersey: Kitchen Table: Women of Color Press.

Appadurai, A. 2006. "The Right to Research." *Globalisation, Societies and Education* 4: 167–177.

Arnstein, S.R. 1969. "A Ladder of Citizen Participation." *Journal of the American Institute of Planners* 35 (4): 216–224.

Askins, K. 2015. "Being Together: Everyday Geographies and the Quiet Politics of Belonging." *ACME: International E-journal for Critical Geographies* 14: 461–469.

Askins, K., and R. Pain. 2011. "Contact Zones: Participation, Materiality, and the Messiness of Interaction." *Environment and Planning D: Society and Space* 29: 803–821. Available at: https://doi.org/10.1068/d11109

Bahena, S., N. Cooc, R. Currie-Rubin, P. Kuttner, and M. Ng. 2012. *Disrupting the School-to-Prison Pipeline.* Cambridge, MA: Harvard Education Press.
Bartolomé, L.I. 2008. "Authentic *cariño* and Respect in Minority Education: The Political and Ideological Dimensions of Love." *International Journal of Critical Pedagogy* 1 (1): 1–17.
Bell, E.E. 2001. "Infusing Race in the US Discourse on Action Research." In: *Handbook of Action Research: Participative Inquiry and Practice,* edited by P. Reason and H. Bradbury, 48–58. London: Sage.
Breitbart, M. 2003. "Participatory Research." In: *Key Methods in Geography*, edited by N. Clifford and G. Valentine, 161–178. London: Sage, London.
Bunge, W., and R. Bordessa. 1975. *The Canadian Alternative: Survival, Expeditions and Urban Change.* Toronto: York University.
Bunge, W., N. Heynen, and T. Barnes. 2011. *Fitzgerald Geography of a Revolution.* Athens: University of Georgia Press.
Butler, J. 1999. *Gender Trouble: Feminism and the Subversion of Identity*, 2nd edition. New York: Routledge.
Butler, J. 2004. *Undoing Gender.* New York: Routledge.
Cahill, C. 2007. "The Personal is Political: Developing New Subjectivities Through Participatory Action Research." *Gender, Place and Culture* 14: 267–292.
Cahill, C., L. Alvarez Gutiérrez, and D.A. Quijada Cerecer. 2016. "A Dialectic of Dreams and Dispossession: The School-to-Sweatshop Pipeline." *Cultural Geographies* 23: 121–137.
Cahill, C., D.A. Quijada Cerecer, A. Reyna Rivarola, J. Hernández Zamudio, and L. Alvarez Gutiérrez. 2019. "'Caution, We Have Power': Resisting the School-to-Sweatshop Pipeline through Participatory Artistic Praxes and Critical Care." *Gender & Education* 31: 576–589.
Cahill, C., F. Sultana, and R. Pain. 2007. "Participatory Ethics: Politics, Practices, Institutions." *ACME: An International E-journal for Critical Geographies* 6: 304–18.
Cahill, C., Torre, M., 2007. "Beyond the Journal Article: Representations, Audience, and the Presentation of Participatory Action Research." In: *Connecting People, Participation and Place: Participatory Action Research Approaches and Method,* edited by S. Kindon, R. Pain, and M. Kesby, 196–206. London: Routledge.
Cameron, J., and K. Gibson. 2005. "Participatory Action Research in a Poststructuralist Vein." *Geoforum* 36: 315–331.
Cammarota, J., and M. Fine. 2010. *Revolutionizing Education: Youth Participatory Action Research in Motion.* New York: Routledge.
Collins, P.H. 2000. *Black Feminist Thought: Knowledge, Consciousness, and the Politics of Empowerment.* New York: Routledge.
Cooke, B., and U. Kothari. 2001. *Participation: The New Tyranny?* London: Zed Books.
Crenshaw, K. 1995. "The Intersection of Race and Gender." In: *Critical Race Theory: The Key Writings That Formulated the Movement,* edited by K. Crenshaw, N. Gotanda, G. Pellow, and K. Thomas, New York: New Press.
Crenshaw, K. 2005. "Mapping the Margins: Intersectionality, Identity Politics, and Violence against Women of Color." In: *Violence Against Women: Classic Papers*, edited by R.K. Bergen, J.L. Edleson, and C.M. Renzetti, 282–313. Auckland: Pearson Education New Zealand.
Crenshaw, K., N. Gotanda, and G. Peller, G. 1995. *Critical Race Theory: The Key Writings That Formed the Movement.* New York: New Press.
De Genova, N. 2002. "Migrant 'Illegality' and Deportability in Everyday Life." *Annual Review of Anthropology* 31: 419–447.
De Genova, N. 2013. "Spectacles of Migrant 'Illegality': The Scene of Exclusion, the Obscene of Inclusion." *Ethnic and Racial Studies* 36: 1180–1198.
Delgado, R. 1983. "Imperial Scholar: Reflections on a Review of Civil Rights Literature." *University of Philadelphia Law Review* 132: 561.
Delgado Bernal, D. 2001. "Learning and Living Pedagogies of the Home: The *mestiza* Consciousness of Chicana Students." *International Journal of Qualitative Studies in Education* 14: 623–639. Available at: https://doi.org/10.1080/09518390110059838.
Delgado Bernal, D., and E. Alemán Jr. 2017. *Transforming Educational Pathways for Chicana/o Students: A Critical Race Feminista Praxis.* New York: Teachers College Press.
Delgado Bernal, D., R. Burciaga, and J. Flores Carmona. 2012. "Chicana/Latina testimonios: Mapping the Methodological, Pedagogical, and Political." *Equity & Excellence in Education* 45: 363–372.
Delgado Bernal, D., and O. Villalpando. 2002. "An Apartheid of Knowledge in academia: The Struggle over the 'Legitimate' Knowledge of Faculty of Color." *Equity & Excellence in Education* 35: 169–180.

DeNicolo, C.P., M.Yu, C.B. Crowley, and S.L. Gabel. 2017. "Reimagining Critical Care and Problematizing Sense of School Belonging as a Response to Inequality for Immigrants and Children of Immigrants." *Review of Research in Education* 41: 500–530.
Diaz-Strong, D., C. Gomez, M. Luna-Darte, and E.R. Meiners. 2014. "Out for Immigration Justice: Thinking Through Social and Political Change." In: *Youth Resistance Research and Theories of Change*, edited by E. Tuck and K.W. Yang, 218–229. New York: Routledge.
Donovan, G.T. 2014. "Opening Proprietary Ecologies: Participatory Action Design Research with Young People." In: *Methodological Challenges When Exploring Digital Learning Spaces in Education*, edited by G.B. Gudmundsdottir and K.B. Vasbø. Rotterdam: Sense Publishing.
Du Bois, W.B. 1898. "The Study of the Negro Problems." *Annals of the American Academy of Political and Social Science* 11 (1): 1–23.
Ellis, C. 2007. "Telling Secrets, Revealing Lives: Relational Ethics in Research with Intimate Others." *Qualitative Inquiry* 13: 3–29.
Elwood, S. 2006. "Negotiating Knowledge Production: The Everyday Inclusions, Exclusions, and Contradictions of Participatory GIS Research." *Professional Geographer* 58: 197–208.
Fals Borda, O. 1979. "Investigating the Reality in Order to Transform It: The Columbian Experience." *Dialectical Anthropology* 4: 33–55.
Fasching-Varner, K.J., L.L. Martin, R.W. Mitchell, K. Bennett-Haron, and A. Daneshzadeh. 2016. *Understanding, Dismantling, and Disrupting the Prison-to-School Pipeline.* Lanham, MD: Lexington Books.
Fine, M. 2017. *Just Research in Contentious Times: Widening the Methodological Imagination.* New York: Teachers College Press.
Fine, M., and R. Barreras. 2004. "To Be of Use." *Analyses of Social Issues and Public Policy* 1: 175–182.
Fine, M., L. Weis, S. Weseen, and L. Wong. 2000. "'For Whom': Qualitative Research, Representations, and Social Responsibilities." In: *Handbook of Qualitative Research,* edited by N.K. Denzin and Y.S. Lincoln. 2, 107–131. Thousand Oaks, CA: Sage.
Flores, L.A. 2003. "Constructing Rhetorical Borders: Peons, Illegal Aliens, and Competing Narratives of Immigration." *Critical Studies in Media Communication* 20: 362–387.
Foucault, M. 1980. *Power/knowledge: Selected Interviews and Other Writings, 1972–1977.* New York: Pantheon.
Fox, M. 2015. "Embodied Methodologies, Participation, and the Art of Research." *Social and Personality Psychology Compass* 9 (7): 321–332.
Freire, P. 1974. *Education: The Practice of Freedom.* London: Writers & Readers Publishing Cooperative.
Freire, P. 1997. *Pedagogy of the Oppressed.* Harmondsworth: Penguin.
Gilligan, C. 1982. *In a Different Voice: Psychological Theory and Women's Development.* Cambridge, MA: Harvard University Press.
Gilmore, R.W. 2002. "Fatal Couplings of Power and Difference: Notes on Racism and Geography." *Professional Geographer* 54: 15–24.
Gilmore, R.W. 2008. "Forgotten Places and the Seeds of Grassroots Planning." In: *Engaging Contradictions: Theory, Politics, and Methods of Activist Scholarship,* edited by C. Hale, 31–61. Berkeley: University of California Press.
Gonzales, R.G. 2011. "Learning to Be Illegal: Undocumented Youth and Shifting Legal Contexts in the Transition to Adulthood." *American Sociological Review* 76: 602–619.
Gonzalez Coronado, Y.S. 2009. *Living in this Skin in the Classroom.* Salt Lake City: University of Utah.
Hale, C.R. 2008. *Engaging Contradictions: Theory, Politics, and Methods of Activist Scholarship.* Berkeley: University of California Press.
Harney, S., and F. Moten. 2013. "The Undercommons: Fugitive Planning and Black Study." *Research Collection.* Lee Kong Chian School of Business. Available at: https://ink.library.smu.edu.sg/lkcsb_research/5025.
Harris, A. 2004. *Future Girl: Young Women in the Twenty-first Century.* London: Taylor and Francis.
Hart, R. 1997. *Children's Participation: The Theory and Practice of Involving Young Citizens in Community Development and Environmental Care.* New York: UNICEF.
Hart, R. 2008. "Stepping Back from 'The Ladder': Reflections on a Model of Participatory Work with Children." In: *Participation and Learning,* 19–31. Netherlands: Springer.
Herbert, S. 2005. "The Trapdoor of Community." *Annals of the Association American Geographers* 95: 850–865.
hooks, b. 1999. "Choosing the Margin as a Space of Radical Openness." *Yearning: Race, Gender, and Cultural Politics.* Cambridge, MA: South End Press.
Hopkins, P. 2018. "Feminist Geographies and Intersectionality." *Gender, Place & Culture* 25: 585–590.
Horton, M. 1990. *The Long Haul: An Autobiography.* New York: Doubleday.
Katz, C. 1994. "Playing the Field: Questions of Fieldwork in Geography." *Professional Geographer* 46: 67–72.

Katz, C. 1996. "Towards Minor Theory." *Environment and Planning D: Society and Space* 14: 487–499. Available at: https://doi.org/10.1068/d140487.
Katz, C. 2017. "Revisiting Minor Theory." *Environment and Planning D: Society and Space* 35 (4): 596–599.
Kelley, R.D. 1997. *Yo' Mama's Disfunktional!: Fighting the Culture Wars in Urban America.* Boston: Beacon Press.
Kelley, R.D. 2014. "Resistance as Revelatory." In: *Youth Resistance Research and Theories of Change*, edited by E. Tuck and K.W. Yang, 82–96. New York: Routledge.
Kelley, R.D. 2018. "Black Study, Black Struggle." *Ufahamu: Journal of African Studies* 40 (2).
Kesby, M. 2005. "Re-theorising Empowerment-Through-Participation as a Performance in Space: Beyond Tyranny to Transformation." *Signs: Journal of Feminist Theory* 30 (4): 2037–2065.
Kindon, S., R. Pain, and M. Kesby. 2007. *Participatory Action Research Approaches and Methods: Connecting People, Participation and Place.* New York: Routledge.
Kretzmann, J., and J. McKnight. 1996. "Asset-Based Community Development." *National Civic Review* 85 (4): 23–29.
Lewin, K. 1948. "Action Research and Minority Problems." *Resolving Social Conflicts* 143–152. New York: Harper.
Lorde, A. 2003. "The Master's Tools Will Never Dismantle the Master's House." *Feminist Postcolonial Theory: A Reader* 25: 27.
Loyd, J. 2012. "Borders, Prisons, and Abolitionist Visions." In: *Beyond Walls and Cages: Prisons, Borders and Global Crisis*, edited by J.M. Loyd, M. Mitchelson, and A. Burridge, 1–15. Athens: University of Georgia Press.
Manzo, L., and N. Brightbill. 2007. "Towards a Participatory Ethics." In: *Connecting People, Participation and Place: Participatory Action Research Approaches and Methods* 33–40. London: Routledge.
Melamed, J. 2015. "Racial Capitalism." *Critical Ethnic Studies* 1: 76–85.
Morris, M. 2016. *Pushout: The Criminalization of Black Girls in Schools.* New York: New Press.
mrs. c-kinpaisby-hill. 2011. "Participatory Praxis and Social Justice: Towards More Fully Social Geographies." In: *A Companion to Social Geography,* edited by V. Casino, M. Thomas, P. Cloke, and R. Panelli, 214–234. Oxford: Blackwell.
Nagar, R. 2014. *Muddying the Waters: Coauthoring Feminisms Across Scholarship and Activism.* Chicago: University of Illinois Press.
Pain, R. 2004. "Social Geography: Participatory Research." *Progress in Human Geography* 28: 652–663.
Pain, R., M. Kesby, and K. Askins. 2011. "Geographies of Impact: Power, Participation and Potential." *Area* 43: 183–188.
Pain, R., and S. Kindon. 2007. "Participatory Geographies." *Environment and Planning A* 39: 2807–2812.
Pain, R., and L. Staeheli. 2014. "Introduction: Intimacy-geopolitics and Violence." *Area* 46: 344–347.
Patel, L.L. 2012. *Youth Held at the Border: Immigration, Education, and the Politics of Inclusion.* New York: Teachers College Press.
Pratt, G. 2010. "Collaboration as Feminist Strategy." *Gender, Place & Culture* 17: 43–48.
Pratt, G. 2012. *Families Apart: Migrant Mothers and the Conflicts of Labor and Love.* Minneapolis: University of Minnesota Press.
Pratt, G., and E. Kirby. 2003. "Performing Nursing: The BC Nurses Union Theatre Project." *ACME: An International E-Journal for Critical Human Geographies* 2: 14–32.
Pratt, G., and V. Rosner. 2012. "Introduction: The Global and the Intimate." In: *The Global and the Intimate: Feminism in Our Time*, edited by G. Pratt and V. Rosner, 1–27. New York: Columbia University Press.
Public Science Project. n.d. Online. Available at: www.publicscienceproject.org.
Pulido, L. 2007. "A Day without Immigrants: The Racial and Class Politics of Immigrant Exclusion." *Antipode* 39: 1–7.
Pulido, L., and D. Lloyd. 2010. "From La Frontera to Gaza: Chicano-Palestinian Connections." *American Quarterly* 62: 791–794.
Purcell, M. 2006. "Urban Democracy and the Local Trap." *Urban Studies* 43: 1921–1941.
Quijada Cerecer, D.A., C. Cahill, and M. Bradley. 2011. "Resist This! Embodying the Contradictory Positions and Collective Possibilities of Transformative Resistance." *International Journal of Qualitative Studies in Education* 24: 587–593. Available at: https://doi.org/10.1080/09518398.2011.600269.
Quijada Cerecer, D.A., C. Cahill, Y.S. Gonzalez Coronado, and J. Martinez. 2019. "'We the People': Epistemological Moves Through Cultural Praxis." *Cultural Studies – Critical Methodologies* 19: 214–221.
Reyna Rivarola, A.R. 2013. *Aguantamos porque no nos queda de otra (We Hold Up Because We Have No Other Option): Everyday Resistance in the Lives of Latina Immigrant US Hospitality Industry Workers.* Salt Lake City: University of Utah.

Reyna Rivarola, A.R. 2017. "'Undocumented' Ways of Navigating Complex Sociopolitical Realities in Higher Education: A Critical Race Counterstory." *Journal of Critical Scholarship on Higher Education and Student Affairs* 3: 101–125. Available at: http://ecommons.luc.edu/jcshesa/vol3/iss1/6.

Ritterbusch, A.E. 2019. "Empathy at Knifepoint: The Dangers of Research and Lite Pedagogies for Social Justice Movements." *Antipode*. doi:10.1111/anti.12530.

Robinson, C.J. 2000. *Black Marxism: The Making of the Black Radical Tradition*. Chapel Hill: University of North Carolina Press.

Smith, L.T. 1999. *Decolonizing Methodologies: Research and Indigenous Peoples*. New York: Zed Books.

Solorzano, D.G., and D. Delgado Bernal. 2001. "Examining Transformational Resistance Through a Critical Race and LatCrit Theory Framework: Chicana and Chicano Students in an Urban Context." *Urban Education* 36: 308–342.

Solorzano, D.G., and T.J. Yosso. 2001. "Critical Race and LaCrit Theory and Method: Counter-story Telling." *International Journal of Qualitative Studies in Education* 14: 471–495.

Teo, T. 2010. "What is Epistemological Violence in the Empirical Social Sciences?" *Social and Personality Psychology Compass* 4: 295–303.

Torre, M.E. 2010. "Participatory Action Research in the Contact Zone." In: *Revolutionizing Education*, edited by J. Cammarota and M. Fine, 31–52. New York: Routledge.

Torre, M.E., and J. Ayala., 2009. "Envisioning Participatory Action Research Entremundos." *Feminism & Psychology* 19: 387–393. Available at: https://doi.org/10.1177/0959353509105630.

Torre, M.E., C. Cahill, and M. Fox. 2015. "Participatory Action Research in Social Research." *International Encyclopedia of the Social & Behavioral Sciences*, J.D. Wright, 540–544. Amsterdam: Elsevier.

Torre, M.E., M. Fine, B.G. Stoudt, and M. Fox. 2012. "Critical Participatory Action Research as Public Science." In: *APA Handbook of Research Methods in Psychology*, edited by H. Cooper, P.M. Camic, D.L. Long, A.T. Panter, D. Rindskopf, and K.J. Sher, vol. 2: 171–184. Washington, DC: American Psychological Association.

Torre, M.E., B. Stoudt, E. Manoff, and M. Fine. 2017. "Critical Participatory Action Research on State Violence: Bearing Wit(h)ness Across Fault Lines of Power, Privilege, and Dispossession." In: *The SAGE Handbook of Qualitative Research*, edited by N.K. Denzin and Y.S. Lincoln, 492–523. Los Angeles: SAGE.

Tuck, E. 2009. "Re-visioning Action: Participatory Action Research and Indigenous Theories of Change." *Urban Review* 41: 47–65. Available at: https://doi.org/10.1007/s11256-008-0094-x

Tuck, E., and M. Guishard. 2013. "Scientifically Based Research and Settler Coloniality: An Ethical Framework of Decolonial Participatory Action Research." In: *Challenging Status Quo Retrenchment: New Directions in Critical Qualitative Research*, edited by T. Kress, C. Malott, and B. Porfilio. Charlotte, NC: Information Age.

Tuck, E., and K.W. Yang. 2014. "Unbecoming Claims: Pedagogies of Refusal in Qualitative Research." *Qualitative Inquiry* 20: 811–818.

Valenzuela, A. 2010. *Subtractive Schooling: US–Mexican Youth and the Politics of Caring*. New York: Suny Press.

Varsanyi, M.W. 2008. "Immigration Policing Through the Backdoor: City Ordinances, the 'Right to the City' and the Exclusion of Undocumented Day Laborers." *Urban Geography* 29: 29–52.

Wilson, D. 2004. "Towards a Contingent Urban Neoliberalism." *Urban Geography* 25: 771–783.

Yosso, T.J. 2013. *Critical Race Counterstories Along the Chicana/Chicano Educational Pipeline*. New York: Routledge.

Zeller-Berkman, S. 2014. "Lineages: A Past, Present and Future of Participatory Action Research." In: *The Oxford Handbook of Qualitative Research*, edited by P. Leavy, 518–532. New York: Oxford University Press.

39

SPACES AND SCALES OF FEMINIST ACTIVISM

Claire Hancock, Roxane Bettinger and Sofia Manseri

I would say that activism is, at one point, to ask the question of what world we live in. And to start trying to change it, at any scale.

Sofia Manseri 27 April 2016[1]

Introduction

It is by no means new for women to be publicly claiming space and making their voices heard, as the recent commemoration of the suffragette movement in the UK made clear; the 1789 women's march on Versailles also comes to mind (Zimmermann 2017). Women's activism has taken centre-stage in the West since the election of Trump as President of the United States, in ways that are both unprecedented in scale and largely announced by other forms of activism that have been gaining momentum for decades (Currans 2017; Gökarıksel and Smith 2017). In the US context, as Nancy Whittier writes, 'women of color and queer people have been leading some of the most vibrant protests of the past few years, such as Black Lives Matter, the Standing Rock Pipeline Protests, and the Dreamers movement' (Whittier 2018). In other parts of the world, women have become the face of fights against oppression: the figure of Ahed Tamimi, the Palestinian girl who became an internet sensation for the footage that showed her slapping Israeli soldiers, is a case in point. Social media have given more women and girls, throughout the world, a voice likely to be heard worldwide and the ability to stage their own appearances and frame their own messages (Paveau 2017). Girls and women are made visible in such struggles, and they have moved beyond being mere abstract symbols (such as the bare-breasted Liberty, leading the way to freedom) to become actual political subjects, with names, voices and fully-fledged identities. Kurdish fighters in Rojava, north-east Syria, have been anointed 'the most feminist' revolution by world media. In many cases, however, it is still the images of voiceless females that capture the West's imagination and go viral, especially when they revive well-known tropes and play on obsessions, as do the pictures of Iranian women removing their veils in a form of protest against the regime. How much women's visibility amounts to 'feminism' remains contested – as it should, in a movement that is as diverse as the countries in which it emerges.

Feminist activism is a pursuit that is both situated and grounded in local contexts and is increasingly globally networked. How do we make sense of the forms, scales and spaces of action and knowledge-sharing that are essential to feminist activism? This chapter attempts to map the spaces and places of feminist activism, speaking from the specific microcosm of French, particularly Parisian, feminist activism. It important to state the place we speak from, per se, but we also want to argue that many discussions of activism from Anglophone perspectives to some extent miss the many spatial barriers that operate to keep feminist activism locally rooted and necessarily tactical in responding to local concerns and challenges (which may include imperialism from Anglophone countries).

This chapter also considers the increasing circulation of feminist knowledge online, and the relations between cyberactivism and other forms of activism. Drawing on material by activists from outside academia, within academia or lurking on the boundary (as is the case of an increasing number of researchers) makes it possible to question the 'inside/outside' division of feminism and the risks and gains from identifying as feminist in different environments and to document individual and collective strategies. This chapter deliberately references not only scholarly work on activism but work by activists, journalists and bloggers, all of whom, we argue, contribute to the advancement of both feminist knowledge and the entire cause. Despite the importance of intellectual pursuit and the thriving virtual mobilizations that take place online, we want to show how space still matters and that a geography of feminist activism is still needed.

Feminist tactics in academia spaces and knowledge production

The production and sharing of knowledge (as part of 'awareness raising') has long been a crucial part of feminist geography mobilizations. This raises the question of the specific role played by academics in the production and exchange of knowledge about women's rights and feminist theories. Arguably, feminism is one of the strands of thought for which the overlap between academic knowledge production and activism is the most crucial and the most fraught, and it is also one of the sources of a major epistemological overhaul of social sciences generally. An important literature has developed over the years on teaching, researching and generally acting as feminists in academia, which questions universities as sites that reproduce a sexist and racist division of labour (Ahmed 2012), while emphasizing the potentialities for imparting 'stealth feminism' (Laliberté et al. 2017) and 'dismantling hegemonic human geography knowledges' (Johnston 2017, 650).

Much received wisdom and many canonical texts are steeped in patriarchy, making teaching against the grain particularly exhausting (if not outright dangerous, in some instances). Some take advantage of the fact that patriarchal structures are acknowledged as such in 'other' cultures and, therefore, work with this prejudice to illuminate patriarchy abroad before bringing the focus closer to home (where the assumption is that gender equality is a given, in the classical geography of sexism that we are brought up to believe in). Other instructors use similar prejudice about 'the past' being inherently more sexist to emphasize historical continuities and stories not only of progress but also of stagnation and backlash (Laliberté et al. 2017). The point is to take the familiar and show it in a different light and to question the normalization of everyday behaviours, and many pedagogical tricks are being shared over the internet and between colleagues. One example of this is the suggestion to begin a class on everyday mobilities with a question to both male and female students about what precautions they take to avoid rape and sexual assault when going out. The responses are likely to open people's eyes about the very unequal degree of freedom that women and men experience in public space.

Feminist geography pedagogy does not occur in classrooms only, and much of it takes place over the internet and social networks. Though masculinist and sexist ideology also has massive platforms there, those willing to educate themselves are likely to find ample resources in many forms. Our specific challenge as academics is to shape this knowledge into formats that will be accepted as valid in scholarly settings, even as many of our colleagues are teaching differently and might try to undermine what we teach. This, however, should not come at the cost of the vital critical component of feminist activism. Scholars of colour such as Sirma Bilge have expressed concern about the 'whitening of intersectionality' (Bilge 2015) and the ways in which it has become the object of abstract scholarly discussions that side-line women of colour and marginalize racialized knowledge producers, which is a critique to be taken seriously.

The separation between academic and non-academic knowledge production and activism is blurred, to some extent, by the relative accessibility and low status of universities and research institutions in the French context. They are not where French elites are reproduced, and the knowledge produced there is often disregarded or dismissed by policymakers or people in power. But, simultaneously, there has been an overbearing tradition of 'axiologic neutrality' enforced in French social research, based on a partial and distorted reading of Max Weber to discredit the then-powerful Marxists, which a new translation is helping to lay to rest. This means that it has been easy to disqualify feminist research and challenge its status as valuable scholarly contribution on the basis that it was 'biased'. Though it may seem counter-intuitive to people outside France, who perceive the country as a purveyor of critical and feminist theory, there is a striking illiteracy among French decision-makers about gender, in general, and a corresponding side-lining of academic research.

In 2017, an academic conference on intersectionality in research in the field of education came under threat because, in the context of French elections to the presidency and the Assembly, political groups claimed that it was spreading a dangerous ideology (in particular toward high-school teachers, who were supposed to attend as part of their training).[2] This controversy, like many others, began with attacks through social networks. The head of research at the university of Paris-Est Créteil claimed that the organizers had been gratuitously provocative by including the image of three 'Rosie the Riveter'-inspired figures on the conference flyer (see Figure 39.1), and asked for it to be removed. This instance of censorship goes to show that feminism, in particular of the intersectional persuasion, is all but normalized in our academic environment, and remains controversial.

Cyberspace and social networks: a new public space for activism?

There is consensus among scholars that new means of communication, such as social networks and the internet, have worked wonders to counter the silencing of oppressed minorities and to allow for efficient networking of resistance movements (Tufekci 2016). Arguably, a worldwide campaign such as #MeToo would have been unthinkable before the global spread of such networks and the development of specific skills to gain attention and exert pressure, thanks to international public opinion. They do, however, also create an unprecedented means by which to target, survey and harass those very movements and minorities, as the example above shows, and there are difficulties inherent in translating viral internet campaigns into successful mobilizations (Tufekci 2016). These developments have brought about new and more complex ways of thinking about space(s) and how activism is likely to unfold spatially: cyberspace reflects issues that play out in all sorts of other spaces, in terms of access, authority, who is widely heard and who is harassed into withdrawal or silence. While many long-overdue conversations have been started by the #MeToo movement, the backlash is strong, organized and with unprecedented access to traditional channels of communication: The Dutch activist Flavia Dzodan has discussed how the public and the private, formerly thought of as discrete

Figure 39.1 Flyer for May 2017 conference at Université Paris-Est as initially designed (the final version had only the image of the children).

entities, are becoming blurred as women in the public space of cities are designated as 'fair game' and alt-right representatives make themselves at home on the internet (Dzodan 2016). There is an ongoing discussion on the efficiency of 'cyberfeminism' in respect of traditional forms of feminist mobilizations (Blandin 2017). The case of the Parisian collective *Le Seum* offers some perspectives on the part played by social media in feminist organizing.

The *Le Seum* collective takes its name from the slang word *seum*, from an Arabic word meaning venom (*avoir le seum* is a phrase commonly used to express anger and disgust). It was created in 2016 during follow-up discussions after the *Marche de la Dignité* of 31 October 2015,[3] and gathers people who experience one or several forms of systemic oppression based on gender, sexual orientation, race, class, religion,[4] and so on. This membership responds to the need, in the French context, to oppose the strategic deployment of concerns that claim to be feminist against racialized minorities (Farris 2017) and to couch feminist engagements in terms that resonate with working-class populations and counter the othering of people of Arab descent (hence the slang). Both Sofia and Roxane are part of the subgroup *Le seum des meufs*, 'women's anger', which has been actively taking part in demonstrations, actions and elaborating feminist knowledge in the shape of collective texts, podcasts, and so on, for the past two years. The group initially met on Twitter, sharing anger and disgust as France rushed headlong into a nightmare of Islamophobic repression after the November 2015 attacks in Paris and Saint-Denis. *Le Seum* therefore addresses specific French challenges (how to talk about racial discrimination when the word 'race' is considered taboo, for instance) and local issues (in particular, the divide between Paris and its suburbs, and the difficulty of organizing across this divide), and discusses Anglophone references such as Anne Fausto-Sterling or Valerie Solanas (along with French classics such as Colette Guillaumin). Both Sofia Manseri and Roxane Bettinger are sceptical of the efficiency of international networking or hashtags such as #MeToo, and emphasize the importance of acting locally, debating and convincing locally and exerting pressure on decision-makers, rather than relying on a surge in awareness of issues (which always takes place once-removed, because of the linguistic barrier and the fact that such movements do not feel immediately relevant to their struggles).

The group communicates by private subgroups of social networks and, though it holds meetings once a month, Sofia Manseri emphasizes that, for its members, 'compulsory physical presence is what kills movements, and allows some people who can always be present to seize power'. Roxane Bettinger praises online private communication as offering a 'space for expression, exchange, support and activist production' and considers the internet to be 'crucial': in this space she first came in contact with feminist struggles, developed knowledge and made a large number of friends.

This does not entail a withdrawal from more classical forms of organizing and mobilizing. Both Sofia Manseri and Roxane Bettinger try to forward feminist ideas in their professional environments and make them part of their personal engagements (Manseri as a local elected official, Bettinger as a teacher). Besides demonstrations, members of *Le Seum* are engaged in tactical tagging of pavements to advocate for legalization of reproductive assistance (ART) for same-sex couples, a creative claiming of urban space inspired by similar tactics by *La Manif pour Tous*, a strongly reactionary movement that mobilized against same-sex marriage in 2012 and 2013. Using stencils, members of *Le Seum* very cheaply and effectively cover up the anti-abortion tags that proliferate in the wealthier parts of Paris (Figure 39.2). Roxane underlines the profoundly gratifying result of marking space and putting a message out in physical space (as well as contesting the perceived enemy's dominance over space on its own territory).

Figure 39.2 Messages stencilled over anti-abortion slogans on Paris pavements: 'Procreation without fear or father', 'Access to abortion' and 'Reproductive technologies for all'.
Photograph: R. Bettinger.

How space still matters

Knowledge production has been tremendously accelerated and has benefited from a large number of contributions from many sources. However, it seems important to emphasize the distortion that comes from many 'global' conversations, which take place mostly in English. In France, for instance, this is off-putting for many, who feel that they cannot participate fully in the conversation since it is dominated by Anglo hegemony and answers to political contexts too alien to mean much in the French national context. The failure on the part of US-based critics to appreciate the extent to which making sense locally might be more important than fostering a worldwide discussion is one of the aspects that Currans underlines in her chapter about the Toronto slutwalks (Currans 2017). The issue is all the more important in countries in which English is not the most commonly used language, because the issues translate differently and the slogans need to take on other forms to function effectively. Cross-cutting questions such as reproductive rights, access to abortion or protection from sexual violence gain from international solidarity, but they have to be reworked in response to the dissimilar local priorities, legal frameworks or political and media discourses. Thus the construction of transnational coalitions is fraught, as is the building of local solidarities. Language constitutes a barrier to fully globalized conversations. Debates or issues that are thrashed out in the English-speaking parts of the internet enter our academic and activist environments only partly, and remain highly contested (with an ongoing feud, in France, between those who have adopted the perspectives of queerness and those who find it depoliticizing and insist on materialistic approaches).

As an example, much was made of the fact that the Women's March was replicated 'worldwide', when in fact the size of the demonstrations in those countries outside the US to a large extent correlated to the proportion of English-speakers and/or size of the US expat community there. In many countries, the demonstration served as an opportunity not only to showcase local issues related to women's rights but also to protest US imperialism. In Paris, it was considered irrelevant by most of the women's movements from deprived neighbourhoods, as only mainstream and White-identified movements had been called on to take part in the *Trocadéro* gathering and march. Its very location in the wealthy Western part of Paris, familiar to US expats but alien to most working-class women, spoke volumes about the way in which it was planned and received locally.

There is a sense in which the fragmented social geography of the French capital is translated into fragmented mobilizations that seem impossible to reconcile. The overarching yet unacknowledged colonial matrix of the entanglements of sexism and racism reinforces this fragmentation. Racial divides loom large, and with the added twist of official colour-blindness. Thus, in the summer of 2017, when the Afrofeminist collective Mwasi organized a festival in which some sessions were exclusively for women experiencing racial discrimination, it was dragged through the mud as a 'racial discriminator' and communitarian (with Paris's nominally left-wing [Socialist] mayor, Anne Hidalgo, weighing in to declare the event unacceptable and trying to have it prohibited).[5]

There is also a demonstrated difficulty for women or gender minorities to gain a significant part in movements in Western cities that rely on a lengthy occupation of public space and/or high-risk protest tactics in the face of police violence, with similar problems arising in the US Occupy movement and the Parisian *Nuit Debout* (Hancock 2017b; Hurwitz and Taylor 2018) While it is particularly challenging to have bodies socially constructed as vulnerable or illegitimate out in space, there is also difficulty in gaining admittance for minority voices in some supposedly 'umbrella' feminist movements. Some prominent figures in the Parisian administration who claim to be feminists have publicly reviled intersectionality as an abomination and criticized the import into French debates of concepts that they claim are alien or servile imitations of Anglophone thought. Thus, the very term 'genre', the French translation of gender, is considered unacceptable by many Parisian decision-makers, who stick to the essentializing phrase 'equality of women and men' and refuse to consider contestations of gender binarism to be relevant to feminism.

Those who trouble these understanding of women, such as openly Muslim women, racialized, queer or transgender people and sex workers, remain *persona non grata* in the institutional demonstrations that take place on 8 March. While officially sanctioned demonstrations such as the Women's March take place peacefully and with police protection (especially when they comprise a majority of White faces: see Zimmermann 2017), in France at the moment the less-consensual and more-radical feminist demonstrations are exposed to police brutality and intimidation. For instance, at the Parisian '8 mars pour touTEs' alternative march in 2016, the police arrested some demonstrators who were wearing BDS (Boycott, Divestment, Sanctions) T-shirts, with the clear objective of intimidating all participants. What counts as acceptable feminist intervention and what is to be violently repressed is the object of unclear ideological decisions: the FEMEN, who bare their breasts to protest about various figures and movements,[6] seem mostly to escape unscathed, as does the collective *La Barbe,* whose members wears fake beards to protest against men-only or particularly homogeneous male events of various sorts.[7] Conversely, when the demonstrators include other political agendas (such as anti-racism, gay and lesbian and trans rights), they are more likely to be considered dangerous radicals and be disciplined for it. This is in conformity with the official government line on feminism, which stresses (White, middle-class) women's rights to public space, even as it stigmatizes 'less respectable' women.

This echoes the transformation of many urban spaces by gentrification, which has co-opted 'women's safety' under the pretext of enforcing greater surveillance and eradication of 'undesirables' (Listerborn 2016). A new urban space being produced and marketed as 'safe', ostensibly in response to feminist claims of a 'right to the city', may be spatially reproducing other exclusions. For instance, the city of Paris's drive to fight sexual harassment targets specific areas and populations on the frontline of gentrification, where the issue is perceived to be most acute, while downplaying the harassment taking place behind closed doors in places of work or study.

Becoming mainstream, becoming universal?

There is a threat inherent in feminism becoming 'mainstream' and co-opted by (diluted in?) social justice movements – as it is likely to become subordinated to other struggles and objectives, even as it is being paid lip service and de-politicized by institutions through 'gender mainstreaming' policies. Nancy Fraser has expressed fears that feminism has become 'capitalism's handmaiden' (Fraser 2009), and there are also legitimate concerns that, in Europe particularly, it plays into the hands of nationalism and racism (Farris 2017). Beyond these most unsavoury tendencies, there is a general risk for feminism to be co-opted and de-politicized by institutions and governments and, increasingly, to take forms that are at odds with the everyday lives and struggles of those women for whom it means the most.

The varying degree of responsiveness by local governments to feminist advocacy accounts for the strikingly different ways in which the European Charter for the Equality of Women and Men in Local Governments is being implemented. This transformation sometimes takes a reassuring form, for instance in the case of Barcelona, where Ada Colau, a former activist for housing rights, is defending a need to 'feminize politics', which has become a slogan for municipalist movements. This is to oppose the virilist face-off between Spain's central government and Catalan independence advocates, for instance, but also to welcome migrants, in contrast to nationalistic and xenophobic tendencies across Europe. The municipality is currently implementing an ambitious and well-funded plan for 'gender justice' and fighting the 'feminization of poverty', and its organization includes a service explicitly devoted to 'feminism and LGBT'. The specificity of the situation in Barcelona arguably owes much to the fact that the mayor herself and many on her team come from activism and have brought their concerns and ideals effectively into city management.

Thinking more generally about what seems new in current activism, one fascinating and relatively recent development is the way in which movements initiated and led by women (not all of whom would necessarily define themselves as feminist) have increasingly become inclusive 'umbrella' movements for many struggles against injustice, not restricted to women's rights or issues coded as feminine. In addition to the ones listed by Whittier, above, the Women's March federated a large number of protesters with a variety of concerns, among them the environment, social welfare, racial justice, immigration and LGBTQ issues (Fisher, Dow and Ray 2017; see also the collection of 'rapid responses', edited by Moss and Maddrell 2017). Another local example is the Parisian *Marche de la Dignité*, in 2015, a women-led movement protesting against police violence, racial discrimination and the stigmatizing of working-class neighbourhoods (Hancock 2017a).

This suggests that women's activism, or feminist activism, is no longer seen as 'specific' or niche but has actually gained a place in universality (Husquin 2017). This might point to a resolution of what was termed the 'Wollstonecraft dilemma' by Carole Pateman (1988), or the intrinsic paradox of feminist claims, according to Joan Wallach Scott (1997); that is, the need to embrace the specificity of women's condition in order to erase it or make it irrelevant to citizenship rights. At the time of writing in early 2018, it seems that the cause of equal rights for women has become a central part of every struggle for equality, rights and social justice. But maybe the sense of novelty is also part and parcel of what the feminist movement is about: as Jo Reger (2018) points out, in the movement's history both its death and rebirth have been reported regularly.

Notes

1 From her interview, available at: http://quartiersxxi.org/on-a-besoin-avant-tout.
2 See: https://bibliobs.nouvelobs.com/idees/20170518.OBS9602/comment-un-colloque-sur-l-intersectionnalite-a-failli-etre-censure.html for a further discussion of these events (accessed 16 February 2018).
3 For a discussion of the *Marche de la Dignité,* see Hancock 2017a.
4 See: https://leseumcollectif.wordpress.com/2016/02/10/le-seum-collectif/ (accessed 6 February 2018).
5 See: https://mwasicollectif.com/2017/06/04/festival-nyansapo-avoir-laudace-detre-une-organisation-noire-politique-et-autonome/ (accessed 16 February 2018).
6 See: https://en.wikipedia.org/wiki/Femen (accessed 16 February 2018).
7 See: http://labarbelabarbe.org/ (accessed 16 February 2018).

Key readings

Currans, E. 2017. *Marching Dykes, Liberated Sluts and Concerned Mothers. Women Transforming Public Space.* Urbana: University of Illinois Press.
Johnston, L. 2017. "Gender and Sexuality II: Activism." *Progress in Human Geography* 41 (5): 648–656.
Scott, J.W. 1997. *Only Paradoxes to Offer: French Feminists and the Rights of Man.* Cambridge, MA: Harvard University Press.

References

Ahmed, S. 2012. *On Being Included: Racism and Diversity in Institutional Life.* Durham, NC: Duke University Press.
Bilge, S. 2015. "Le blanchiment de l'intersectionnalité." *Recherches féministes* 28 (2): 9–32. doi 10.7202/1034173ar.

Blandin, C. 2017. "Le web: de nouvelles pratiques militantes dans l'histoire du féminisme?" *Réseaux*, special issue *Féminisme*, online 201. doi: 10.3917/res.201.0009.
Currans, E. 2017. *Marching Dykes, Liberated Sluts and Concerned Mothers. Women Transforming Public Space*. Urbana: University of Illinois Press.
Dzodan, F. 2016. "Private Internet and Public Streets." Available at: https://medium.com/this-political-woman/private-internet-and-public-streets-38c1e3f80e0d (accessed 12 February 2018).
Farris, S. 2017. *In the Name of Women's Rights. The Rise of Femonationalism*. Durham, NC: Duke University Press.
Fisher, D.R., D.M. Dow, and R. Ray. 2017. "Intersectionality Takes It to the Streets: Mobilizing Across Diverse Interests for the Women's March." *Science Advances* 3 (9). doi: 10.1126/sciadv.aao1390.
Fraser, N. 2009. "Feminism, Capitalism and the Cunning of History." *New Left Review* 56, March–April: 97–117.
Gökarıksel, B., and S. Smith. 2017. "Intersectional Feminism Beyond US Flag Hijab and Pussy Hats in Trump's America." *Gender, Place and Culture* 24 (5): 628–644.
Hancock, C. 2017a. "Feminism from the Margin: Challenging the Paris/*banlieues* Divide." *Antipode* 49 (3): 636–656. doi: 10.1111/anti.12303.
Hancock, C. 2017b. "Why Occupy *République*? Redefining French Citizenship from a Parisian Square." *Documents d'Anàlisi Geogràfica* 63 (2): 427–445. Available at: http://dx.doi.org/10.5565/rev/dag.443.
Husquin, H. 2017. "Quand le féminisme catalyse les luttes. De la *Women's March* à l'Internationale féministe." *Contretemps*, 27 September. Available at: www.contretemps.eu/womens-march-internationale-feministe/ (accessed 13 February 2018).
Hurwitz, H.M, and T. Vertua. 2018. "Women Occupying Wall Street: Gender Conflict and Feminist Mobilization." In: *100 Years of the Nineteenth Amendment: An Appraisal of Women's Political Activism*, edited by L.A. Banasak and H.J. McCammon, 334–355. New York: Oxford University Press.
Johnston, L. 2017. "Gender and Sexuality II: Activism." *Progress in Human Geography* 41 (5): 648–656.
Laliberté, N., A. Bain, G. Lankenau, M. Bolduc, A. Mansson McGinty, and K. Sziarto. 2017. "The Controversy Capital of Stealth Feminism in Higher Education." *ACME: An International Journal for Critical Geographies* 16 (1): 34–58.
Listerborn, C. 2016. "Feminist Struggle of Urban Safety and the Politics of Space." *European Journal of Women's Studies* 23 (3): 251–264.
Moss, P., and A. Maddrell. 2017. "Emergent and Divergent Spaces in the Women's March: The Challenges of Intersectionality and Inclusion." *Gender, Place and Culture* 24 (5): 613–620.
Pateman, C. 1988. "The Patriarchal Welfare State." In: *Democracy and the Welfare State,* edited by A. Gutmann, 231–260. Princeton, NJ: Princeton University Press.
Paveau, M.-A. 2017. "Les poings d'Ahed Tamimi ou de la vidéo comme unité de sens." *La pensée du discours*, 23 December. Available at: https://penseedudiscours.hypothses.org/15470 (accessed 15 February 2018).
Reger, J. 2018. "Is the Women's Movement New Again?" *Gender and Society*, 1 February. Available at: https://gendersociety.wordpress.com/2018/02/01/is-the-womens-movement-new-again/ (accessed 16 February 2018).
Scott, J.W. 1997. *Only Paradoxes to Offer: French Feminists and the Rights of Man*. Cambridge, MA: Harvard University Press.
Tufekci, Z. 2016. *Twitter and Teargas: The Power and Fragility of Networked Protest*. New Haven, CT: Yale University Press.
Whittier, N. 2018. "Activism against Sexual Violence is Central to a New Women's Movement: Resistance to Trump, Campus Sexual Assault and #MeToo." *Gender and Society*, 7 February. Available at: https://gendersociety.wordpress.com/2018/02/07/activism-against-sexual-violence-is-central-to-a-new-womens-movement-resistance-to-trump-campus-sexual-assault-and-metoo/ (accessed 16 February 2018).
Zimmermann, J. 2017. "The Myth of the Well-behaved Women's March." *New Republic*, 24 January. Available at: https://newrepublic.com/article/140065/myth-well-behaved-womens-march (accessed 16 February 2018).

40

AN ARTFUL FEMINIST GEOPOLITICS OF CLIMATE CHANGE

Sallie A. Marston, Harriet Hawkins and Elizabeth Straughan

Introduction: a feminist geopolitics of climate change?

For feminist geographers, the body is a foundational site through which to counter abstractions, to challenge and, ideally, to replace such distanced renderings with more materially empowering formulations (Longhurst and Johnston 2014). Climate change has become one of the most preoccupying generalizations of our times, thus critical geopolitics has sought to challenge one of its most dominant facets – the techno-scientific narrative (O'Lear 2016; O'Lear and Dalby 2015) – in order both to critique and replace the 'dangerous myths of climate science' underpinning much of the governance and governmental practice of climate security. In place of such 'geometrics' such as carbon targets and tipping points that 'reduce climate to measurable, quantifiable observations about environmental systems' (Liverman 2009; O'Lear 2016, 5), critical geopolitics has sought to reframe 'humanity's place and role in the biosphere' (Dalby 2007, 113). In this chapter, we imagine a feminist geopolitics of climate change and the contribution that it might make to (and the productive critique it might make of) this wider geopolitical project.

In what follows, we draw together three rich sets of feminist resources, from geographers and others, to challenge the 'delusions of hyper separation, transcendence and dominance' that tend to underpin the prevailing techno-scientific accounts of climate change and that might be seen to deny global environmental crises while also hindering effective and engaged responses (Gibson-Graham 2011, 28; see also Plumwood 2002). We draw inspiration from feminist geographers and activists Gibson-Graham (2011), feminist new materialist Stacy Alaimo (2010, 2016) and philosopher Luce Irigaray (1992, 1993), including work from feminist geography that bears her influence. In doing so, we imagine a feminist geopolitics of climate change that opens up abstractions through two intersecting dimensions: first, an engagement with embodied experiences; and second, through taking the 'geo' in geopolitics seriously (Adey 2015; Elden 2012; Squire 2016). In doing so, we offer a feminist geographical take on concerns to re-couple the scientific debate and study of climate change with 'the social and political contexts of its material production and cognitive understanding'. Through the study of artworks, we are concerned in particular with approaches that foreground how climate change must be understood through proximate relations attuned to embodied and, specifically, located and felt manifestations (Brace and Geoghegan 2010; Hawkins et al. 2015; Yusoff and Gabrys 2011).

Our feminist geography colleagues offer us a range of resources through which to think further about these proximate relations. We are inspired by Gibson-Graham's feminist project of belonging in the Anthropocene and its imperative for connection with the nonhuman. These are connections, they argue, that are felt or otherwise sensed, rather than simply seen, and built through attunement to the lively animate matters of the world. Attending to animate earthly matters as political agents reframes the 'geo' in geopolitics, but also requires a rethinking of politics beyond intra-human relations and an acknowledgment of the critical importance of a human politics within and alongside any politics of the nonhuman. Such debates have been proceeding apace within geopolitics and beyond (see, for example, Clark 2013; Dalby 2013; DeLoughrey, Didur and Carrigan 2015; Elden 2012; Squire 2016, 547; Swyngedouw 2013). We are surprised, however, that geopolitics has yet to turn to feminist materialisms as a site from which to rethink the 'geo' and to theorize the intra-human and human–nonhuman relations that such a rethinking involves. Here, then, we turn to both Alaimo (2010, 2016) and Irigaray (1992, 1993), and their understanding of vibrant, volatile matters, bodily vulnerabilities and the becoming of body-worlds to evolve the basis for the kinds of embodied, animated accountings of climate change we seek.

While the triad of feminist thinkers offers us the theoretical means to unsettle and replace abstract understandings of climate change, we develop these ideas through an examination of the art–science collaborations of Cape Farewell, a UK-based organization working internationally with artists and scientists to produce a 'cultural response to climate change' (Cape Farewell n.d.). From the wide range of Cape Farewell's work, we focus here on its interdisciplinary Arctic voyages, of which it has conducted eight since 2003, involving over a hundred artists and scientists. The voyagers spend between several weeks and several months on board a 1910 two-masted schooner – *The Noorderlicht* – conceptualizing and developing work that engages with climate change and its impacts in the polar regions. That we would choose arts practice as a site through which to glimpse felt and otherwise-sensed attunements to the environment and the body will come as no surprise to anyone following the evolution of aesthetic theory and art within geography (Colls 2012; Hawkins 2010; Hawkins and Straughan 2014a). Geographers have increasingly sought to understand the role that experience, encounter and embodiment have played in twentieth-century art, recognizing the valuable resource that aesthetics offers to some of our most pressing contemporary concerns (Hawkins 2013). Cape Farewell's icy imaginaries contribute directly to the diverse Western imaginaries of the Arctic as a site for 'new nature'. While many of these visions are shaped by encounters with atmospheric, optical and acoustic landscapes (Bloom 1993; Bravo 2000, 2009; Brunn and Medby 2014; Craciun 2016; Dittmer et al. 2011; Dodds 2008), they are also interwoven with debates concerning the north-polar region as a site for knowledge, whether of imperial and contemporary climate science or of Indigenous inhabitants (Anderson 2009; Bravo and Triscott 2011; Cameron 2015; Gearhead et al. 2017; Kusugak 2002).[1]

Our chapter proceeds through two analytic dimensions. First, we explore the experience of artists and scientists in the Arctic as an embodied challenge to abstractions of climate change and as cultivating attunement to animate 'geos'. Second, we investigate the embodied experiences of *High Arctic* to appreciate the vulnerable, transcorporeal bodies this installation instils in galleries distant from the Arctic experiences. Concluding, we probe the limitations of our particular feminist materialist approach to climate change, which incorporates a geopolitics that not only accounts for the intra-human as well as nonhuman but, perhaps most importantly, seeks a more expansive appreciation of the 'geo'.

Animate 'geos' – A felt politics of climate change

What happens when you come to 78 degrees North to document your experience of a vulnerable environment in a changing climate – which at this temperature seems unchanging, immutable, eternal. Your experience tells you one thing but 15 years absorbing the contents of peer-reviewed journals and computer models suggests something very different. In a sense both are right, the reality of standing on the ice confronts you with the power of the Arctic.

Charlie Kronick, Director of Greenpeace, Cape Farewell Participant 2011
(Cape Farewell n.d.)

Walking on those glaciers was the most magical moment for me. When I was standing on them, one of the scientists said: 'In 50 years' time, these won't be here.' It is this beauty, scale and fragility and sense of loss which we are trying to embody in this exhibition ... We would really like the visitor to slow down, listen, watch and think about the consequences of human behavior and how it will affect the Arctic. However we don't want to bombard people with facts, statistics and preach how it should be done.

Matthew Clark, artist, Cape Farewell Participant 2011
(Effects of Global Warming 2011)

The Cape Farewell voyagers, like seafarers before them, kept ships logs in the form or b/logs (Cape Farewell nd.). Reading, looking and listening to the poems, interviews, cartoons, sound works, impromptu songs, sketches, reflections on ethnographic visits to local Inuit services and settlements, photographs, scientific expositions and personal diary entries that constitute these blogs, we encounter a range of Arctic experiences. These are dissonant Arctics, with minor scenes of struggle to reconcile the conflicted, embodied experiences of frozen, seemingly static expanses that are simultaneously dynamic, rapidly changing environments characterized by cyclical freezing and thawing over a constantly churning ocean. Such dissonance is set against those Arctic abstractions – graphs, statistics, models – that assert the vulnerability of these environments and act as a harbinger of climate change writ large.

Western icy imaginaries have long been premised on comparisons of environmental stasis and the dynamisms at work in these frozen places. Indeed, since the nineteenth century, ice has shifted in the Western cultural imaginary from being dangerous 'evil matter' to a 'vehicle for and revelation of vital energy by both scientists and artists alike' (Wilson 2003). As Wilson observes of the Romantics: 'If the cosmic poles and massive glaciers reveal the life coursing within miniscule man, the tiny ice crystal opens into forces pervading the solar system' (2003, 5). In short, to Westerners, ice has been appreciated not only as a physical force but also a creative one that extends humans' capacities for knowledge as:

frozen forms pattern and reveal invisible, imponderable, holistic, causal, vital powers, ranging from electromagnetic waves that can be measured to psychological energies, vague yet discernible, to cosmological principles beyond fact and image.

Wilson 2003, 23

For Cape Farewell artist Matthew Clark, creator of *High Arctic*, the response to such corporeal and cognitive dissonances was a cultivation of particular bodily comportments; a 'slowing down, a listening, a watching and a thinking'. In these desires, he shares an affinity with the bodily dispositions that Gibson-Graham (2011) seek for engendering the worldly connections at the heart of their environmental politics. Their project of 'connecting' calls for 'subsuming ourselves within others', often by way of incorporating our own materiality with nonhuman others and attuning ourselves to forces and dynamics that do not originate in human action. They draw on Latour to detail the emergence of these connections as a learning to be affected by non-human others. For them, as for Clark and other Cape Farewell travellers, stillness and slowness become a key form of intra-action with the surroundings. While this stance risks a romantic relationship to the landscape, what Gibson-Graham foreground in such affective environmental encounters are the possibilities for 'becoming other': these are 'process[es] of constitution that produces a new body world' (2011, 322). These connections are built through a feminist materiality and an ontology that resists sorting into human and nonhuman categories. As such, they challenge any sense of distance, of vertical hierarchy, with a relational horizontality between 'humans, biota and abiota' (Bennett 2010, 112). As such, we might recognize the embodied encounters with environments that these Western artists experienced in the Arctic as being brought about by constitutive entanglements with the nonhuman, and their post-human condition. For Gibson-Graham, such an awareness opens up positive possibilities for reducing capitalist practices of production and consumption that recklessly harm or endanger those non-human elements (2011, 5).

Regarding the Cape Farewell voyagers, we are interested in how their experiences, in particular their art-making practices, demonstrate this becoming affected by their surroundings. Guided by discussions of matter and feminist new materialisms (e.g. Coole and Frost 2010; Grosz 1994), we note the importance of not simply admitting the relevance and force of the 'geo' (the complex physical world of Earth, the land, air and water and the organisms that live in and on them). We also argue the need for keen attention to detail, and for comprehension of the localized expressions of 'geos', materials and properties, including the nature of the entangled relations between human and nonhuman bodies (Adey 2015, 59; McCormack 2015; Steinberg and Peters 2015).

We turn now to the Cape Farewell artists' accounts of making work in the Arctic through their exploration of nonhuman forces. While the use of ice, snow and variegated solar power was a common theme, we discuss here Tracey Rowledge's Arctic drawings (Figure 40.1), which are intricate framings of geopower, such that their compositional concerns and their material manifestation in the final work demand but also cultivate refinements and understandings of the localized variations in elemental forces.[2] As Rowledge explains:

> I'm making drawings with coloured felt-tips on paper. These works respond directly to the sea, working with the impact in terms of movement the waves have on the ship and then using the Arctic seawater to impregnate the drawings, causing the images to bleed and fade. I have also rigged up an automatic drawing system to make drawings from the motion of the boat. This system has created interesting drawings, that for me explore movement, time, place and permanence.
>
> *Cape Farewell n.d.*

Titled by date, time and geographical coordinates, these are drawings of and by the force of motion. They document periods of calm, registered in minuscule marks on vast, white paper landscapes, and record rough seas when sharp zigzags and jagged lines sketch the movements of a ship in the chaotic grasp of weather, water and being buffeted by ice. Such documentations render visual a 'geo' of flows, connections, liquidities and becomings, enabling us to imagine a world in perpetual motion (Steinberg and Peters 2015). Such work is an example from an animated archive of the 'geo', enabling an account of climate change that requires not only registering the 'distinctively non-human materialities of the earth' but also attending to their agencies, 'releas[ing] the materialities of the earth from being seen as a stockpile of inert stuff or

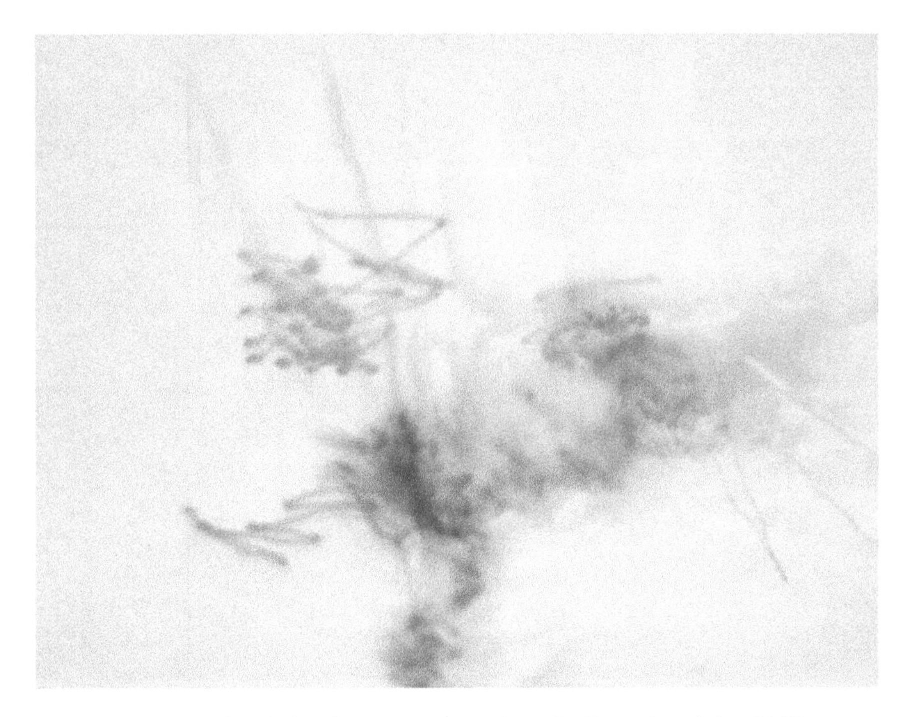

Figure 40.1 *No. 4, CF 2008 Disko Bay Expedition: 1 October,* Tracey Rowledge (2008).

a destructive threat', or as a sublime landscape backdrop through human attention to these non human materialities (Bennett 2010, xi).

We turn, in the next section, to the Cape Farewell-inspired installation, *High Arctic,* to elaborate on this relationship between human and nonhuman materialities. Inspired by feminist ideas of transcorporeality, we discuss how artworks might cultivate dispositions that instil bodies, of artists and audiences, as vulnerable, porous entities, and thus open to the kinds of environmental ethics that Gibson-Graham seek.

Vulnerable bodies

> Now all we have left to do is get together our own personal kit. Plenty of warm and waterproof clothes, wellingtons, lots of dry socks … And very important: sunscreen and sunglasses. The sun is very bright in all that snow, and the ozone layer is too thin to protect us properly from harmful ultraviolet radiation.
>
> *Sarah and Val, 2003*

This recounting of bodily threat and the action taken by the Cape Farewell voyagers to mitigate it echoes older Arctic narratives, where the crucial imperative for survival was for the 'frontiers of one's body to be rigorously established and maintained' (Wylie 2002, 259). Below, we recast these bodily vulnerabilities in our discussions of the body-worlds formed through the experiences of Cape Farewell art. We seek an understanding of how art might attune us both to our constitutive entanglement with the nonhuman and to a heightened awareness of the ethics of our post-human condition. Describing audience experiences of *High Arctic,* we recognize the

Figure 40.2 *High Arctic,* Matthew Clark and United Visual Artists, view of installation at the National Maritime Museum, London 2011.

entwining of human and nonhuman bodies in a becoming body world: what we might call, after Alaimo (2010), transcorporealities, where to be human is to be open – both voluntarily and involuntarily – to the world, rather than bounded from it. To understand how the installation creates these transcorporeal Arctics, we find inspiration in the feminist philosopher Luce Irigaray's (1992, 1993) writings on touch and hearing and the way her writings give meaning to experiences of light and sound in *High Arctic*:

> The edges of the gallery-cum-landscape are hard to make out, the collections of white glacial columns seem to rise up out of the black sea and fade gradually towards the horizon. Exploring the darkened environment visitors are guided through the 65 separate islands of some 3000 white columns – some at ankle level, some towering above them – that constitute the archipelago created in the space. The tones of the Shipping Forecast theme tune, slightly flattened, sound through the space, the almost monotonal voice of the newscaster begins the familiar roll call of places, Dogger, Fisher, German Bight ... With UV torches in hand, visitors sweep their blue spotlights around the installation, activating the white light animations ahead of them as they go. Navigating around the white columns, each named for a glacier in Svalbard that will disappear due to global warming, they step into pools of swirling snow, pin-pricks of white that collect in flurries around the base of the white columns before moving off again, swirling around the visitors' feet before dissolving into the black masses. A cacophony of shifts, pops, cracks, and creaks echoes through the space. Sweeping torches across contour-line maps of light, blue spotlights alight on a large black pool between towering white masses, as torches move over the glacial columns a mass of geometric

> light-based icebergs calve off into the blackness. Driven by invisible currents directed by the sweep of the torches, the pool becomes a seething mass of geometric shapes, coming together, splitting, moving off across the floor as great white masses, only to disappear into the distance. Directing the flows and intensities of these geometric bergs with sweeps of torches, animating the waxing and waning of snow flurries across the space, the audiences interact with abstract landscapes of this installation.[3]

The immersive installation of *High Arctic* creates a dynamic encounter for its interactive audiences. Its materiality offers a complex recreation of the tensions between static and dynamic and the agentive and vulnerable found in the Cape Farewell b/logs. The white columns, each named after a decaying glacier in Greenland, stand in for the Arctic's white expense, over, alongside and around which ephemeral icebergs of light and sound are scattered, composing and decomposing, calving, detaching, drifting and melting amid snowstorms of light.

We argue that the transcorporeal Arctics developed by the installation depend on these experiences of sound and light. These are worldly connections, formed less through an attention to *what* we see and hear and more through an appreciation of *how* we sense. As such, bodies, human or otherwise, are rendered as vulnerable porous entities, thoroughly interleaved in the decaying icy forms affected by those same forces.

Alaimo's (2010, 2012, 2016) idea of 'transcorporeality' provides a particularly materially and forcefully inclined understanding of the 'attunements' to lively matter we seek. She assembles feminist materialist theory, art analysis, literary theory and archival work to explore a 'sense of precarious, corporeal openness to the material world' that she argues fosters an environmental politics (2012, 23). She argues for environmental relations as ones of bodily vulnerability, performed as 'transcorporeal condition[s] in which the material interchanges between human bodies, geographical places and vast networks of power provoke ethical and political actions' (2012, 23). Alaimo (2012) employs Kirsten Justensen's artwork, *Ice Pedestal* (2000), to expand upon art's challenging of bounded, inviolate human subjects, interpreting Justensen's placing of her naked body in contact with ice as: renouncing the 'boundaries of the human … allowing us to imagine corporeality not as a ground of static substance but as a place of possible connections, interconnections and ethical becomings' (Alaimo 2012). The artist's vulnerability cultivates a 'sense of precarious, corporeal openness to the material world' that is founded in an environmental politics (2009, 23). Appreciating the volatile matter of the human body and its connections with other-than-human volatile matters is not a new perspective for feminist geographers. Indeed, attention to bodies as fleshy matter, as porous spatialities open and vulnerable to the wider world, is a central tenet of feminist geographical theorizing over the last thirty years (Colls 2012; Grosz 1994; Irigaray 1992; Longhurst and Johnston 2014). Here, we argue that experiences of the light and sound created by *High Arctic* produce in the gallery a series of transcorporeal Arctics that bring us to attend to our own vulnerable bodies and their implication and entanglement with the 'geo'.

Light is the key to audience experiences of the work. Most obvious are the lighted animations and the UV (ultraviolet) torches that the audience members brandish around the space, with their false-blue spotlight (UV light is not visible to the human eye). These torches trigger the exposure of Arctic destruction, whether the calving of light icebergs, the revelation of the glaciers' names or the changing patterns and movement of snowstorms. Such audience-directed agency references human-induced global warming, the result of increased amounts of UV light penetrating the polar regions. It is not just Arctic landscapes that are rendered vulnerable to UV light, however, but also human bodies. The potency of UV rays at the poles is exacerbated by the reflective capacities of snow, requiring humans to protect their skin and eyes from the extreme photic effects of the summer season. Recognizing the phenomenality of light and the vulnerability of the bodies that it penetrates is to move us beyond human subjectivities constituted through seeing. As Barker explains, 'the emission and reception of light [is] usually considered to be a phenomenon of vision, [whereas it is] actually a matter of light falling on and reflecting off or being absorbed by an object before it ever becomes visible' (2009, 30). The light-object relation is especially pronounced with UV light, which is invisible to the naked human eye yet permeates the skin, with the potential to cause serious damage.

The touch of light, for Irigaray, creates corporeal space-times that demand acknowledgement of bodily limits and horizons, thus 'disrupt[ing] the distinction between self and other' (Shildrick 2001, 392), at the same time replacing bodies as surfaces reflective of light with corporeal imaginaries that foreground porosity and vulnerability. In such an appreciation, Irigaray overturns a logics of light premised on truth and reason, and a world rendered knowable through distance, a separation of seer and seen and of the sensible from the intelligible (Irigaray 1992; Vassaleu 1998). Instead, her reframed logics of light implicate touch and vision in creating corporeal intimacies and proximities (Hawkins 2015). The materiality and phenomenality of UV light deployed in *High Arctic* demonstrate this effectively. Rather than rendering bodies visible to the human eye, UV light penetrates bodies, challenging the accepted view of their enclosed, stable nature and reforming them as permeable, susceptible bodies, thoroughly mingled with the world. As Irigaray writes, such 'touchings' of light preclude the possibility of ever 'closing … off [of the world] or closing off of the self (Irigaray 1993, 141). Indeed, feminist geographers have developed incisive accounts of how touch might evolve corporeal space-times that demand our encounter with bodily limits and horizons and 'disrupt the distinction between self and other' (Shildrick 2001, 392; see also Johnston 2012; Straughan 2012).

It is not just the interactive light components of *High Arctic* that render audiences as vulnerable humans, however, but also the audio experience. If sound objects are understood to 'engender the communication of urgent climate effects in a more accessible and publicly compelling way' (Kanngieser 2015), hearing creates the conditions for becoming-connected world bodies. The materiality and physicality of sound and the effect of listening are premised on the transcorporeal. Sound art is typically produced as an immersive experience, creating an audience that is 'engulfed, enveloped, absorbed, enmeshed, [such that] the subject loses itself … *touched* physically and emotionally' (Hawkins and Straughan 2014b, 132). Sound artist and theorist Brendon La Belle elaborates this point: sound 'deliver[s] a dynamic phenomenal weave in and around the individual body, creating points of contact … sound moves in and through us to immerse the body in energetic motion that is equally socially and politically charged' (2010, 5).

LaBelle's formulation reflects Irigaray's theorization of co-mingled body-worlds, this time through the bodily passage of sound. The Irigarayan body is a system of spaces and passages, of depths and membranes, through which sound as a set of vibrations passes. The result is a pervious spatiality of bodies, fully immersed in the materiality of the world of which they are an inseparable part. Penetrated by soundwaves, 'the open horizon of [the] body … [is a] living, moving border' (Irigaray 1993, 51). For Kanngieser, such properties of sound render it transversal, cutting across matter and beings, 'render[ing] apparent that the world is not *for* humans. The world is rather *with* humans – a relation that is not without antagonism' (Kanngieser 2015, 83, italics added). Attending to the materialities and the phenomenality of sound and hearing exposes those antagonisms as ontological vulnerabilities and epistemological perplexities (ibid.).

Conclusion: towards a feminist geopolitics of climate change

Feminist geography's orientation to the body has become almost a cliché. But in the rush to get comfortable or to dismiss what may seem 'done', we risk overlooking the complexities and intricacies of what it means to reterritorialize abstraction across the terrains of that most intimate 'space'. In this chapter, we have explored one iteration of what a feminist geopolitics of climate change might look like – or better, perhaps, feel like – and we have done so by using examples drawn from the icy archives of Cape Farewell's Arctic voyages, exploring both the making of art and the audience's experiences of it. We have sought to go beyond simply 'fleshing' out geopolitics, by way of taking seriously intersections of earthly elements and forces – animated geos – and the volatile matters of human bodies. We have drawn on three sets of feminist resources to further specify embodied engagements with climate change, ones that seek attunements of humans to the nonhuman and recognize the importance of embodiment to the ontological entanglements of a vital world.

What does it mean, though, to foreground the creation of these interpenetrated body-worlds in the face of climate change abstractions? One implication is that they might divorce the effects of climate change from other forms of politics or marginalize the cultural understanding of the elements proffered by the Indigenous inhabitants of the region, or ignore a range of intra-human politics, social relations and earthly entanglements. The soundings and visionings of sea, ice and air that the Cape Farewell artists offer are very much nonhuman ones. The more human-focused encounters present in the b\logs rarely find form in art. As well, in a centuries-long aesthetic and ideological manipulation of images, sounds (such as the snowmobiles of Arctic Indigenous people) deemed to 'rupture the tranquility' of the traveller's Arctic are largely absent from the Arctic encounters framed for audiences. We acknowledge the profound risks that these artworks negotiate when they create icy imaginaries that proffer a *terra nullius* (DeLoughrey 2015) and suggest that it remains an open challenge for Cape Farewell artists and others to reconcile both intra- and nonhuman relations. In their recognition of the role of the nonhuman in understanding climate change, DeLoughrey and others have, in the Pacific context, reflected on 'the practice and philosophical implications of thinking politically beyond intra-human concerns', while at the same time being 'critically attentive to the intra-human power relations and different cultural understandings and practices of history and the environment' (DeLoughrey, Didur and Carrigan 2015, 14). What we have offered here is but one iteration of a feminist geopolitics of climate change, built on taking seriously feminist materialist concerns with entwined body-worlds, wherein bodies emerge as open, porous and vulnerable. The result is two-fold: a disruption of distinctions between self and other that undermine ethical connections between humans and nonhuman worlds; and the proposition of the means through which such ethical connections might be made.

Notes

This research is part of a larger project titled "Art/Science: Collaborations, Bodies, and Environments" (http://artscience.arizona.edu/), co-funded under a joint agreement by the US National Science Foundation (Grant No. 86908) and the UK Arts and Humanities Research Council (Grant No. AH/I500022/1).

1 See for example 'Arctic Geopolitics and Autonomy', developed by geographer Michael Bravo and Nicola Triscott, details of which can be seen at: www.artscatalyst.org/arctic-geopolitics-autonomy-2010 (accessed 6 August 2016).

2 See also Boetzkes (2010) and Hawkins (2014) for how art has been understood in this way.
3 Account developed through authorial experience of the exhibition at the Greenwich Maritime Museum (July 2011–January 2012) and accounts in reviews by Dixon (2011) and Letchet (2011).

Key readings

Alaimo, S. 2010. *Bodily Natures: Science, Environment, and the Material Self.* Bloomington: Indiana University Press.
Gibson-Graham, J.K. 2011. "A Feminist Project of Belonging for the Anthropocene." *Gender, Place, and Culture* 18 (1): 1–21.
Longhurst, R., and L. Johnston. 2014. 'Bodies, Gender, Place and Culture: 21 Years On." *Gender, Place and Culture* 21 (3): 267–278.

References

Adey, P. 2015. "Air's Affinities: Geopolitics, Chemical Affect and the Force of the Elemental." *Dialogues in Human Geography* 5: 54–75.
Alaimo, S. 2009. "Insurgent Vulnerability and the Carbon Footprint of Gender." *Women Gender and Research*, Special issue on Gendering Climate Change 3–4: 22–34.
Alaimo, S. 2010. *Bodily Natures: Science, Environment, and the Material Self.* Bloomington: Indiana University Press.
Alaimo, S. 2012. "Insurgent Vulnerability and the Carbon Footprint of Gender". *Kvinder, Køn & Forskning* 34: 22–35.
Alaimo, S. 2016. *Exposed: Environmental Politics and Pleasures in Posthuman Times.* Minneapolis: University of Minnesota Press.
Anderson, A. 2009. *After the Ice: Life, Death and Politics in the New Arctic*, London: Virgin.
Barker, J. 2009. *The Tactile Eye: Touch and the Cinematic Experience.* London: University of California Press.
Brace, C., and H. Geoghegan. 2010. "Human Geographies of Climate Change: Landscape, Temporality, and Lay Knowledges." *Progress in Human Geography* 35 (3): 284–302
Bennett, J. 2010. *Vibrant Matter: A Political Ecology of Things.* Durham, NC: Duke University Press.
Bloom, L. 1993. *Gender on Ice: American Ideologies of Polar Expeditions.* Minneapolis: University of Minnesota Press.
Boetzkes, A. 2010. *The Ethics of Earth Art.* Minneapolis: University of Minnesota Press.
Bravo, M.T. 2000. "Cultural Geographies in Practice – The Rhetoric of Scientific Practice in Nunavut." *Ecumene* 7: 468–474.
Bravo, M.T. 2009. "Sea Ice Mapping: Ontology, Mechanics, and Human Rights at the Ice Floe Edge." In: *High Places: Cultural Geographies of Mountains and Ice*, edited by D.E. Cosgrove and V.D. Dora, 161–176. London: I B Tauris.
Bravo, M.T., and N. Triscott, eds. 2011. *Arctic Geopolitics and Autonomy.* London: Hatje Cantz.
Brunn, J.M., and I.A. Medby, 2014. "Theorising the Thaw: Geopolitics in a Changing Arctic." *Geography Compass* 8: 915–929.
Cameron, E. 2015. *Far Off Metal River: Inuit Lands, Settler Stories, and the Making of the Contemporary Arctic.* Vancouver: University of British Columbia Press.
Cape Farewell. nd. Ships log. Cape Farewell online archive. Available at: www.capefarewell.com/the-expeditions.html (accessed 6 August 2016).
Clark, N. 2013. "Geopolitics at the Threshold.' *Political Geography* 37: 48–50.
Colls, R. 2012. "'Bodies Touching Bodies': Jenny Saville's Over-life Sized Paintings and the 'Morpho-logics' of Fat, Female Bodies." *Gender, Place and Culture* 19: 175–192.
Coole, D., and S. Frost. 2010. *New Materialisms: Ontology, Agency and Politics.* Durham, NC: Duke University Press.
Craciun, A. 2016. *Writing Arctic Disaster, Authorship and Exploration.* Cambridge: Cambridge: University Press.
Dalby, S. 2007. "Anthropocene Geopolitics: Globalisation, Empire, Environment and Critique." *Geography Compass* 1: 103–118.
Dalby, S. 2013. "The Geopolitics of Climate Change." *Political Geography* 37: 38–47.
DeLoughrey, E. 2015. "Ordinary Futures: Interspecies Worldings in the Anthropocene." In: *Global Ecologies & the Environmental Humanities: Postcolonial Approaches,* edited by E. DeLoughrey, J. Didur, and A. Carrigan, 352–372. London: Routledge.

DeLoughrey, E., J. Didur, and A. Carrigan, eds. 2015. *Global Ecologies and the Environmental Humanities: Postcolonial Approaches.* London: Routledge.

Dittmer, J., S. Moisio, A. Ingram, and K. Dodds. 2011. "Have You Heard the One About the Disappearing Ice? Recasting Arctic Geopolitics." *Political Geography* 30: 202–214.

Dodds, K. 2008. "Icy Geopolitics." *Environment and Planning D: Society and Space* 26: 1–6.

"Effects of Global Warming." 2011. *Huffington Post,* 15 July 2011. Available at: www.huffingtonpost.com/2011/07/15/global-warming-arctic-london-maritime-museum_n_899985.html (accessed 1 March 2020).

Elden, S. 2012. "Secure the Volume: Vertical Geopolitics and the Depth of Power." *Political Geography* 34: 35–51.

Gearhead, S.F., L.K. Holm, H. Huntington, J. Mello-Leavitt, and A.R. Mahoney. 2017. *The Meaning of Ice: People and Sea Ice in Three Arctic Communities.* Hanover, New Hampshire: International Polar Institute Press.

Gibson-Graham, J.K. 2011. "A Feminist Project of Belonging for the Anthropocene." *Gender, Place & Culture* 18 (1): 1–21.

Gibson-Graham, J.K., and G. Roelvink. 2011. "An economic ethics for the Anthropocene." *Antipode* 41: 320–346.

Grosz, E. 1994. *Volatile Bodies: Towards a Corporeal Feminism.* Bloomington: Indiana University Press

Hawkins, H. 2010. "The Argument of the Eye? Cultural Geographies of Installation Art." *Cultural Geographies* 17, 3: 321–340.

Hawkins, H. 2013. *For Creative Geographies: Geography, Visual Art and the Making of Worlds.* London: Routledge.

Hawkins, H. 2015. "It's All Light: Pipillotti Rist's Feminist Language and Logics of Light." *Senses and Society* 10 (2): 158–178.

Hawkins, H., and E. Straughan. 2014a. *Geographical Aesthetics: Imagining Space, Staging Encounters.* London: Routledge.

Hawkins, H., and E. Straughan. 2014b. "Nanoart, Dynamic Matter and the Sight/Sound of Touch." *Geoforum* 51 (1): 130–139.

Hawkins, H., S. Marston, M. Ingram, and E. Straughan. 2015. "The Arts of Socio-Ecological Transformation." *Annals of the Association of American Geographers* 105 (2): 331–341.

Irigaray, L. 1992. *Elemental Passions.* London: Althone.

Irigaray, L. 1993. *An Ethics of Sexual Difference.* London: Althone.

Johnston, L. 2012. "Sites of Excess: The Spatial Politics of Touch for Drag Queens in Aotearoa New Zealand." *Emotion, Space and Society* 5 (1): 1–9.

Kanngieser, A. 2015. "Geopolitics and the Anthropocene: Five Propositions for Sound." *GeoHumanities* 1 (1): 80–85.

Kusugak, J. 2002. "Foreword: Where a Storm is a Symphony and Land and Ice Are One." In: *The Earth is Faster Now: Indigenous Observations of Arctic Environmental Change,* edited by I. Krupnik and D. Jolly, v–vii. Fairbanks, AK: Arctic Research Consortium of the United States.

LaBelle, B. 2010. *Acoustic Territories: Sound Culture and Everyday Life.* London: Continuum Books.

Letchet, J. 2011. "Review: High Arctic at the National Maritime Museum, Greenwich." Available at: www.jletchet.com/2011/08/review-high-arctic-at-national-maritime.html (accessed 27 July 2013).

Liverman, D. 2009. "The Geopolitics of Climate Change: Avoiding Determinism, Fostering Sustainable Development." *Climatic Change* 96, (1–2): 7–11.

Longhurst, R., and L. Johnston. 2014. 'Bodies, *Gender, Place and Culture*: 21 Years On." *Gender, Place and Culture* 21 (3): 267–278.

McCormack, D. 2015. "Envelopment, Exposure, and the Allure of Becoming Elemental." *Dialogues in Human Geography* 5 (1): 85–89.

O'Lear, S. 2016. "Climate Change and Slow Violence: A View From Political Geography and STS on Mobilising Technoscientific Ontologies of Climate Change." *Political Geography* 52: 4–13.

O'Lear, S., and Dalby, S. 2015. *Reframing Climate Change: Constructing Ecological Geopolitics.* London: Taylor & Francis.

Plumwood, V. 2002. *Environmental Culture: The Ecological Crisis of Reason.* London: Routledge

Sarah and Val, 2003. "All Tested and Packed!". Cape Farewell, Expedition Blog, 26 May, 8:25. Available at: https://photos.google.com/share/AF1QipPSfLYvCScn0q9oERx0H2Jbo7hs5iyTtSGIKUYkjTMf49VCyvfjwBVjhblU0pnr8A?key=Z19DNEI2Rm5DYVhRZjIwMWlLWkZtWjd4al93SGJn (accessed 1 March 2020).

Shildrick, M. 2001. "Some Speculations on Matters of Touch." *Journal of Medicine and Philosophy* 26: 387–404.
Squire, R. 2016. "Rock, Water, Air, and Fire: Foregrounding the Elements in the Gibraltar Spain Dispute." *Environment and Planning D: Society and Space* 34 (3): 545–564.
Steinberg, P., and K. Peters. 2015. "Wet Ontologies, Fluid Spaces: Giving Depth to Volume Through Oceanic Thinking." *Environment and Planning D: Society and Space* 33: 247–264.
Straughan, E.R. 2012. "Touched by Water: The Body in Scuba Diving." *Emotion, Space and Society* 5 (1): 19–26.
Swyngedouw, E. 2013. "The Non-political Politics of Climate Change." *Acme* 12 (1): 1–8.
Vassaleu, C. 1998. *Textures of Light Vision and Touch in Irigaray, Levinas and Merleau-Ponty.* London: Routledge.
Wilson, E. 2003. *The Spiritual History of Ice: Romanticism, Science and the Imagination.* London: Springer.
Wylie, J.W. 2002. "Becoming-icy: Scott and Amundsens's Polar Voyages." *Cultural Geographies* 9(3): 249–265.
Yusoff, K., and J. Gabrys. 2011. "Climate Change and the Imagination." *Wiley Interdisciplinary Reviews: Climate Change* 2 (4): 516–534.

41
FEMINIST GEOGRAPHY IN THE ANTHROPOCENE
Sciences, bodies, futures

Kai Bosworth

Introduction

According to many Earth scientists, the Earth has entered a new geologic epoch, age or era, defined by the global impact of humans on the planet. The suggested name for this new time period is the Anthropocene: 'the Age of Man'. Not all those who study the Earth's layered strata, such as physical geographers, geologists and stratigraphers, agree that the Anthropocene deserves to be recognized. Scientific arguments continue over the year in which the so-called 'golden spike' indicates that the beginning of the Anthropocene should be placed and what geologic 'signal' indicating this origin will be speculatively read in rock by a future scientist. The Anthropocene designation has gripped popular social imaginations as well, for the entry of humanity into the geologic record seems to indicate the end of any pure area of nature separate from social relations, a moment that Jaime Lorimer calls 'the public death of the modern understanding of Nature' (2012, 606).

The Anthropocene thesis that humanity and its actions are a determinate geological force has raised immediate questions for feminist geographers about *which* human social formations are responsible for this situation and how different actors might know or act upon that responsibility. This position deviates from two others: first, for some techno-optimists, the optimistic possibility of a 'good Anthropocene' is in sight, in which science and technology are returned to their supposedly apolitical role in arbitrating the management of natural resources and social systems. Second, for dystopian thinkers, the 'Anthropocene apocalypse' signals the end of human hubris and a return to what they see as a primordial war over scarce resources. Feminists have been central to deconstructing these highly differentiated social and political diagnoses and imaginaries and the manner in which these narratives rely on gendered and racist tropes, figures and absences. Recognizing the inability of the universal 'man' and singular planetary future to practise a just future, feminist geographers, scientists, social scientists and humanists ask: What futures and what pasts are called into being by the Anthropocene? Who is authorized to know and speak for this crisis, and with what political effects? What careful collaborations and experiments might be necessary to live with each other on a damaged planet?

In some ways, the Anthropocene thesis rekindles important insights that feminist geographers have long been debating. Feminist geographers have long studied the manner in which 'Woman stands as metaphor for Nature' (Massey 1994, 11). They attend to the manner in which an abstract 'mankind' or 'humanity' that would stand in for presumed universal experiences is capable of masking uneven gendered and racialized ecologies. The Anthropocene thesis is capable of extending many of these historic inequalities, solidifying their hold on imaginations of the future and the social roles available to different humans and nonhumans. So, too, are global climate change and social relationships with various environments structured by patriarchy and exceeded by practices of resistance led by women, trans, femme and other subaltern and marginalized people around the world.

While feminists have paid critical attention to the limitations of the 'good Anthropocene', the 'Anthropocene apocalypse' and to universal understandings of man, they have used this opportunity to create dialogue among the scientific, political and aesthetic responses to the Anthropocene. Careful attention to the practices, inquiries and institutions at stake in relation to the Anthropocene reveals openings in which feminist geographers are developing modes of thought capable of travelling to and through other genres of feminist inquiry and political practice. In part because the Anthropocene demands interdisciplinarity, the genres in which the creativity of feminist responses to the Anthropocene could be explored are vast, including political ecology and conservation (Collard et al. 2015), oceanography (Lehman 2016), urbanism (Derickson 2017), aesthetics (Yusoff 2015) and political economy (Gibson-Graham and Roelvink 2010).

Interdisciplinary feminist scholarship on the Anthropocene has been helpfully summarized in a number of reviews (Colebrook and Weinstein 2015; Ebron and Tsing 2017; Grusin 2017), to which this chapter adds a specifically geographical contribution. First, I present key arguments in the debate over the role of scientific practices in naming and presenting evidence for the Anthropocene. Then, I turn to approaches from feminist geophilosophy and new materialism, which encourage the examination of a differentiated corporeality as it meets geologic forces. Finally, I examine how Anthropocene conversations could benefit from critiques made by Black, Indigenous and subaltern feminisms, which each offer complementary differences in their provocation of and to the Anthropocene. It is with this hope that, as J.K. Gibson-Graham argues, one might 'find innovative and creative ways of inviting ourselves and earth others into a different developmental relationship, one that denies domination and explores mutuality and interdependence' (2011, 15).

Feminist approaches to Anthropocene science

The Anthropocene thesis has garnered such interest in part because it compellingly wraps together key questions about socio-natural relations and histories with a signal of the gravity of contemporary ecological catastrophes, including global climate change, species extinction and ocean acidification. Developed by Paul Crutzen and Eugene Stoermer in the early 2000s, 'the Anthropocene' is meant to indicate the passage of the planet into a new geologic era, period or epoch in which 'the effects of humans on the global environment have escalated' (Crutzen 2002, 23). While it is something of a spur-of-the-moment designation, geologists and stratigraphers have since debated the term in considerable detail through speculation on which geologic signals – such as the appearance of plastics, a change in carbon dioxide concentration or increased radioactivity – will be able to be read in the future stratigraphic record (Zalasiewicz et al. 2017). Key metaphors from this debate, such as the 'golden spike' and 'humans as a geologic force' (Rickards 2015b), have travelled well beyond esoteric scientific journals to become the subjects of widespread academic deliberation.

As Heather Davis and Zoe Todd note, the scientific debates surrounding the Anthropocene can appear 'a universalizing project [which] serves to re-invisibilize the power of Eurocentric narratives, again re-placing them as the neutral and global perspective' (2017, 763). This is most evident in the seemingly undifferentiated *anthropos* – man – at the heart of the term of Anthropocene. While, undoubtedly, planetary ecological systems have been modified by humans, the Anthropocene designation seems to suggest that it is humanity at large rather than specific social formations which can be named. Donna Haraway and others have proposed a series of possible alternatives, including the Capitalocene, Chthulucene and Androcene, which would indicate either the root cause of the crisis in capitalism and patriarchy or, in the case of the Chthulucene, the underlying weirdness of the world (Altvater et al. 2016; Haraway 2016). The feminist geologist Jill Schneiderman has argued that geologists should 'remain open to alternative nomenclature' (2015, 197). On the other hand, the supposed unity of the *anthropos* underlying the Anthropocene thesis could be complicated if the name is understood not as an indication of power, force and homogeneity but, instead, as the evacuation of those qualities from the figure of the human. As Stephanie Wakefield argues, 'the Anthropocene names liberal humanity, but does so only in the moment of its historical collapse' (2014). In her interpretation, the Anthropocene is a name that stages the end and impossibility of 'man' as the defining feature of our time. Instead, it opens a space for creative exploration of possible worlds beyond the techno-managerial status quo.

The debate over nomenclature is not only about the name itself but also the role of science in universalizing Eurocentric methods of determining ecological crisis and political response. An approach informed by feminist science studies might use the moment to resituate sciences as productive not of universal knowledge or evidence but partial and situated modes of knowing (Harding 1986; Haraway 1988). Feminist science studies scholars have long examined questions of nature and culture, difference and knowledge, futurity and history, seeking to break apart the narrative whereby a mythically neutral and homogeneous 'science' provides objectivity. In examining Anthropocene science from a partial perspective, Lauren Rickards queries whether scientists could be understood as 'an eclectic group of highly diverse thinkers squeezing their planetary rewriting in among marking and admin, not strikingly dissimilar to geography', rather than 'a group of men in short-sleeved shirts with clipboards gathered intently around a pulsing replica of the Earth' (Rickards 2015a, 340). If stratigraphy, geology and physical geography are partial, situated practices, as Rickards argues, perhaps they could be better engaged directly by critical feminist scholars. This is not to neglect that some of the sciences at stake in the Anthropocene (including geography) tend to be overwhelmingly demographically and perhaps irrevocably epistemologically patriarchal and European (Raworth 2014). Instead, I suggest that a resituating of scientific practices could help to build interdisciplinary feminist collaborations that are prepared to act responsibly in messy worlds.

Critical physical geography is one approach towards building the 'critical but interested, sceptical but open' practice that Rickards calls for (2015a, 340). Critical physical geography is defined by an approach that combines critical analysis of power – as in human geography – with the specific and situated investigations of physical geography. As Lave et al. argue, 'Put bluntly, to understand the Anthropocene we must attend to the co-production of socio-biophysical systems' (Lave et al. 2014, 6). Rebecca Lave and Brian Lutz offer one model of critical physical geography through their collaborative investigation into hydraulic fracturing for natural gas (Lave and Lutz 2014). In a careful, side-by-side review of both scientific and social scientific literatures and methods of studying its effects, Lave and Lutz find that the rise in fracking in the US has spatially redistributed environmental injustices to areas that have not previously experienced energy production. Consequently, the environmental impacts of fracking can be overemphasized in comparison to the evidence in physical geography, while the social effects of the fracking economy tend to be downplayed in solely physical geographic analyses. As the authors argue, 'Reviewing the physical and social literature on hydraulic fracturing together enables us to reconsider current research approaches to hydraulic fracturing, demonstrating the utility of a critical physical geography approach' (750).

The Civic Laboratory for Environmental Action Research (CLEAR) is another project that draws on the epistemological approaches in the sciences with a critical and situated inquiry. 'A feminist marine science and technology lab', CLEAR is grounded in grassroots- and action-oriented scientific research, meaning that the research questions and goals are designed with the impacted communities. Consequently, its research outcomes can better promote social and environmental transformation. One area in which the lab has been especially active is in monitoring plastics in marine fish populations. While maintaining that 'pollution is colonialism', CLEAR recognizes that scientific practice can contribute to fighting such a situation by publicizing its findings in policy White Papers, by sharing results with communities that rely on fish as a major food source and by making its results accessible to and readable by a broad audience. The lab's anti-colonial, collective and feminist orientation also affects how it practises science, recognizing the power imbalances in activities such as the organization of a lab, the publication of a multi-authored journal article or holding a disciplinary conference (Civic Laboratory 2017).

The methods of critical physical geography and feminist science labs create results that can circulate in social spheres different from those for traditional scientific or humanistic research. This can foment new connections among researchers and social, political and activist practices. Nonetheless, such practices are certainly marginalized in the broader politics of the Anthropocene. There is no comparison between the feminist 'ecology of practices' that Donna Haraway describes as a 'mundane articulating of assemblages through situated work and play in the muddle of messy living and dying' (Haraway 2016, 42) and the supposedly sleek, hubristic technofixes, such as eco-cities or geoengineering projects. Yet the latter are increasingly dominant, for the thesis of the 'good Anthropocene' that is promoted by techno-optimist entrepreneurs has a grip on the imagination, circulating in institutions of environmental governance. Rather than recognizing that the Anthropocene might signal something fundamentally unjust, with entrenched capitalism and the histories and present of global colonialism, the thesis promises 'the active, assertive, and aggressive participation of private sector entrepreneurs, markets, civil society, and the state' (Asafu-Adjaye et al. 2015, 30) coming together to accelerate technological innovation and modernize the planet's energy and food systems. Given the newness of the Anthropocene designation and its prominence in the public sphere, it is a matter of considerable debate whether feminists should attempt to reinvent and reclaim the meaning of the Anthropocene or abandon the term in favour of a more specific name.

Feminist approaches to corporeality and the inhuman

As Kathryn Yusoff (2013, 2015) and Elizabeth Povinelli (2016) have argued, the naming of the Anthropocene is but one historical event that signals the broader role of the inhuman and the non-living in constituting the conditions of possibility for what we recognize as life. Departing from the flat ontologies of actor-network theory and the living/dead distinction that governs biopolitics, their line of inquiry, along with many other corporeal or feminist materialisms, instead holds that the inhuman radically exceeds thought and knowledge. Relationships with more-than-human landscapes such as caves, mines and even landfills offer sites of intimacy or pleasure. Contrary to the idea of the Anthropocene as a homogenizing genre, this line of thought seeks to open new methods for 'learning to be affected' (Gibson-Graham and Roelvink

2010) by Earthly others, a feminist practice that can build the new relationships of responsibility required to weather socioecological crises.

Through differentiated bodies, individuals and collectives come to sense geologic forces. Much like the Earth itself, bodies are permeable in and through their generous and constitutive relationship with others (Bosworth 2017). In mines, caves and other underground spaces, this relationship is especially notable, as the interior spaces of the Earth generatively reverberate from encounters with human bodies. In their studies of caving practices, both Sarah Cant (2003) and María A. Pérez (2016) find that cavers create a kind of vulnerable and sensuous intimacy with and through the Earth, one that can have a demasculinizing edge on what might otherwise be a practice of adventure or competition. Drawing on a concept from Hélène Cixous, Yusoff (2016) finds a form of 'blood knowledge' among the coal miners who were on strike in 1980s UK as they built relationships of solidarity with LGBTQ activists. By this, she indicates the manner in which an 'intimate sensibility' of the wounds of coal mining becomes 'a point of access to studying the geologic matrix as it pertains to both power and possibility' (2016, 207). This common form of sensibility can also be seen among the women labourers of the 1800s, who donned asexual attire to retain their jobs after being banned from coalmines (McClintock 1995, 114–118, 408n126). These situations could be understood conceptually, in Elizabeth Grosz's words as 'a temporary detour of the forces of the earth through the forces of a body' (Yusoff et al. 2012), provided it is also recognized that such avenues of accessibility are part and parcel of differential power relations.

Through practices of Earthly investigation, the planet re-emerges not as the Whole Earth, a mother figure to be defended, but a porous, permeable or cracked relation of difference through which bodies compose their commons and commonalities (Bosworth forthcoming; Yusoff 2018). Engaging in such material and corporeal relationships with Earth forces – ethnographically, scientifically or artistically – requires learning to be affected in new ways. As environmental justice struggles have long taught us, however, the toxic effect of exposure to the circulation of materials through landscapes creates different capacities and capabilities among different bodies and populations divided among historical categories of race, class, and gender, among others. Toxins in drinking water and food sources can affect differently men and women, LGBTQ people, workers, Black, Indigenous and other marginalized populations, and humans and nonhumans. Environmental justice struggles further precipitate the situated narratives of survivance and persistence in the face of the debilitating effects of pollution (Houston 2013).

Rather than impelling us to lionize a normative healthy subject, Alexis Shotwell (2016) and Heather Davis (2015) query whether inhabiting the language of toxicity and impurity might provide feminist grounds for living together on a damaged planet. Instead of reacting with abhorrence to new, mutated creatures and ecologies emerging from the detrital landscapes produced by capitalist exploitation, what responsibilities would be created if acted with humility, care and responsibility? In conversation with Donna Haraway and Isabelle Stengers, María Puig de la Bellacasa (2017) asks how such relationships of care with nonhumans could be built *without innocence*. It is necessary to recognize that various populations are obligated and implicated in differential ways in the survival and persistence of nonhuman worlds. By attending to materiality, corporeality and the inhuman as specific conditions of existence, we could better engage with an Anthropocene-in-the-making, rather than one of finality.

Such narratives of careful consideration are lost in hegemonic Westphalian geopolitics, which takes the Anthropocene as a crisis featuring disrupted or insecure state borders, threatened national populations and unpredictable economies. Deborah Dixon (2016) argues that these geopolitical practices are overly reliant on anthropocentric configurations of states and peoples, thus seeing both problem and solution in masculine narratives of war and competition. At the border of art and science, she suggests a practice of feminist geopolitics that would seek an 'aesthetics in the apprehension of, and living with, the Anthropocene' (2016, 146). Developing a worldly aesthetic framework that re-composes multiple sites of resistance in science, art and politics is not an easy task, and it requires a re-composition that takes leave of Eurocentric aesthetic modes, as much as of scientific or political ones. Drawing on Guadeloupian writer Daniel Maximin's concept of geopoetics, Last (2015, 58) defines such practices as:

> a poetics that takes geographical features and geophysical forces seriously as an element of geopolitics, while seeking to constructively reinscribe them as a means to counter imperialist aspirations and hegemonic worldviews. In short, they represent a materialist, decolonial process of rewriting geopolitics.

As Last argues elsewhere, geopoetics would reconsider 'not only human-planet relations but the multiple asymmetry of such relations' (Last 2017, 149). Geopoetics is thus not just a deconstruction and critique of existing Eurocentric geopolitical relationships but also a rewriting or re-graphing of geopolitics towards decolonial ends. In this manner, the discipline and practice of geography implicate itself and its history, thus taking responsibility for its own toxic progeny, beyond critique and deconstruction. Such a practice, as I outline below, requires a proactive openness to going beyond the hegemonic Eurocentric frameworks of geographic thought.

Race and coloniality in the Anthropocene

It is surprising, then, that much literature – feminist or otherwise – fails to recognize the sustained manner in which the Anthropocene has racial difference at its core. Recognition of the consequences of the planetary spread of racial difference is too frequently a footnote to the discussion of the Anthropocene, rather than its central thesis. This has especially important implications for the figuring of planetary or global thought if, as Denise Ferreira da Silva (2007) argues, the global is itself a site of the construction of racial difference. While similar processes structure geographic inquiry more broadly (Esson et al. 2017; Pulido 2002), the absence of sustained attention to racism and coloniality structures the types of questions, knowledges, narratives and participants who are seen to be welcome in academic debate around the Anthropocene. One consequence is that the broader geopolitics of the Anthropocene, the pedagogies and epistemologies that it (re)authorizes and the apocalyptic or redemptive stories that it elicits remain largely disconnected from struggles for racial justice and decolonization.

Rarely is decolonial politics or the practice of solidarity seen to be a central plank of scientific, social scientific or humanistic Anthropocene inquiry. Vastly different peoples and their thoughts, practices, art and modes of life are frequently used as mere objects, tools or data to inject diversity into intellectual discussion, without having a stake or voice in the outcomes or dialogue taking place therein. As the Métis anthropologist Zoe Todd argues, this practice recreates discussions of the Anthropocene as a 'white public space' (2015). This sense is reproduced frequently in many academic conferences on the Anthropocene, in which questions of epistemology, research and representation are seen to be trivial in the face of either a trendy concept or the catastrophic event that it signifies (for another view on some of these conferences, see Swanson et al. 2015). In my experience, at the fringes of many Anthropocene conferences and conversations are 'feminist killjoys' (Ahmed 2010), prepared to counter the mythic tale of the dissolving boundaries between man, nature and technology with one of man's use of such boundaries to recode not-humans and nonhumans for death. The feminist killjoy draws attention to the manner in which the 'racist, gendered and classed dimensions' of the Anthropocene and its various crises are 'produced in the academic realm, through academic practices' (Kanngieser and Last 2016). It is precisely by ignoring these violences – or refiguring them as romantic – that the dual discourse of the 'good Anthropocene'/'Anthropocene crisis' can be sustained.

Drawing on the work of Cedric Robinson, Françoise Vergès asks of the 'racial Capitalocene' how we might write 'a history of environment that takes into account the history of racial capitalism' (Vergès 2017, 74). In this analysis, what was previously named the Anthropocene signifies not the sudden realization of humans' impact on the environment but, instead, precisely the long and slow European and Euro-American exploitation of cheap nature and disposable, racialized bodies worldwide. Such a framing resonates with the analysis of Potawatomi philosopher Kyle Powys Whyte (2016), who contrasts the sudden feeling of planetary crisis among non-Indigenous peoples to the centuries-long dystopia in which the Potawatomi and other Native nations have been living and surviving. The Anthropocene could then be reframed as 'the continuation of practices of dispossession and genocide … that have been at work for the last five hundred years' (Davis and Todd 2017, 761). We might understand such practices as fomenting an 'interim politics of resourcefulness' (Derickson and MacKinnon 2015), or something akin to what Anishinaabe scholar Gerald Vizenor has called 'survivance' (2008). This aesthetics and politics is reliant on the storytelling and other narratives of environmental justice to reframe the grand narratives of violence and victimhood of the Anthropocene (Houston 2013). Such practices hold a promise for recombinatory futures, centralizing the epistemologies and politics of marginalized peoples without trivializing or universalizing their grounded normativities.

By engaging questions of global coloniality and racial capitalism, new avenues of collaboration between physical sciences and humanistic frameworks can also be re-opened. The physical geographers Simon L. Lewis and Mark A. Maslin have argued that it is 'colonialism, global trade and coal' that best signify the beginning of the Anthropocene. For them, as for Vergès and other feminist geographers, 'this highlights social concerns, particularly the unequal power relationships between different groups of people, economic growth, the impacts of globalized trade, and our current reliance on fossil fuels' (Lewis and Maslin 2015, 177). However, outside, Black feminists such as Sylvia Wynter (Wynter and McKittrick 2015) and Christina Sharpe (2016) draw on the physical sciences to contest and refigure what it means to be human. Drawing on insights from investigations of human origins, climate and environmental science and oceanography, science opens up both the dominant narratives of man and the erosion of that very figure. Wynter asks whether, in the face of such narratives, we could take up the *species-oriented* project of reinventing communal life, an excavation of other genres, of being human as a political praxis. Such a project could be undertaken, in part, with reference to 'the species-threatening nature … of the relentlessly increasing fossil fuel-driven climate instability's ongoing catastrophe' (Wynter and McKittrick 2015, 43). Doing so requires recognition of the historic consolidations of 'man' through slavery and coloniality as central to that very catastrophe.

Such conversations provide a more productive set of parameters for investigation, while taking into account the centrality of race in the production of 'the fears of the Anthropocene' (Gergan et al. 2020). Ongoing mobilization of these arguments within or against discussions of the Anthropocene demonstrates that the concept is still an open site of struggle, not merely another concretion of the contemporary patriarchy. What stories might we tell of the Anthropocene if it named 'the death of an anti-human geopolitical project that dominates "other" humans, societies and potentialities of culture, philosophy and creativity, largely in the global "south"' (Tolia-Kelly 2016, 790)?

Conclusion

Debate, dialogue and critique concerning the Anthropocene and its implications among feminist geographers is certainly only just beginning. Contemporary environmental politics provides further opportunity for situated research and political contestation of resurgent narratives of nationalism and ecology, overpopulation and the subjugation of women's bodies, and settler coloniality and the violence of resource extraction. Research and writing that advance not only conversations but also the widespread acceptance of social and political movements for climate justice are desperately needed. If the Anthropocene is to have meaning for feminist geographers beyond an object of critique, it will be through reworking the relations among science, materiality and political resistance to coloniality and racial capitalism. What is at stake is the development of feminist practices that create and sustain space for learning to be affected, for becoming-human-otherwise and for new considerations of what geo-social flourishing might be possible.

Key readings

Davis, H., and Z. Todd. 2017. "On the Importance of a Date, or, Decolonizing the Anthropocene." *ACME: An International Journal for Critical Geographies* 16 (4): 761–780.
Gibson-Graham, J.K. 2011. "A Feminist Project of Belonging for the Anthropocene." *Gender, Place and Culture* 18 (1): 1–21.
Yusoff, K. 2015. "Geologic Subjects: Nonhuman Origins, Geomorphic Aesthetics and the Art of Becoming Inhuman." *cultural geographies* 22 (3): 383–407.

References

Ahmed, S. 2010. "Killing Joy: Feminism and the History of Happiness." *Signs: Journal of Women in Culture and Society* 35 (3): 571–594.
Altvater, E., E. Crist, D. Haraway, D. Hartley, C. Parenti, and J. McBrien. 2016. *Anthropocene or Capitalocene?: Nature, History, and the Crisis of Capitalism*. Oakland, CA: PM Press.
Asafu-Adjaye, J., L. Blomquist, S. Brand, B.W. Brook, R. DeFries, E. Ellis, C. Foreman, D. Keith, M. Lewis, and M. Lynas. 2015. "An Ecomodernist Manifesto." Available at: www.ecomodernism.org/manifesto-english/ (accessed 1 December 2017).
Bosworth, K. 2017. "Thinking Permeable Matter through Feminist Geophilosophy: Environmental Knowledge Controversy and the Materiality of Hydrogeologic Processes." *Environment and Planning D: Society and Space* 35 (1): 21–37.
Bosworth, K. 2020. "The Crack in the Earth: Environmentalism After Speleology." In: *Into the Void,* edited by A. Secor and P. Kingsbury. Omaha: University of Nebraska Press.
Cant, S.G. 2003. "The Tug of Danger with the Magnetism of Mystery: Descents into 'the Comprehensive, Poetic-Sensuous Appeal of Caves'." *Tourist Studies* 3 (1): 67–81.
Civic Laboratory. 2017. Online. Available at: https://civiclaboratory.nl/ (accessed 1 December 2017).
Colebrook, C., and J. Weinstein. 2015. "Introduction: Anthropocene Feminisms: Rethinking the Unthinkable." *philoSOPHIA* 5 (2): 167–178.
Collard, R.-C., J. Dempsey, and J. Sundberg. 2015. "A Manifesto for Abundant Futures." *Annals of the Association of American Geographers* 105 (2): 322–330.
Crutzen, P.J. 2002. "Geology of Mankind." *Nature* 415 (6867): 23.
Davis, H. 2015. "Toxic Progeny: The Plastisphere and Other Queer Futures." *philoSOPHIA* 5 (2) | : 231–250.
Davis, H., and Z. Todd. 2017. "On the Importance of a Date, or, Decolonizing the Anthropocene." *ACME: An International Journal for Critical Geographies* 16 (4): 761–780.
Derickson, K.D. 2017. "Urban Geography III: Anthropocene Urbanism." *Progress in Human Geography* 41 (2): 230–244.
Derickson, K.D., and D. MacKinnon. 2015. "Toward an Interim Politics of Resourcefulness for the Anthropocene." *Annals of the Association of American Geographers* 105 (2): 304–312.
Dixon, D.P. 2016. *Feminist Geopolitics: Material States*. Burlington: Routledge.
Ebron, P., and A. Tsing. 2017. "Feminism and the Anthropocene: Assessing the Field through Recent Books." *Feminist Studies* 43 (3): 658–683.

Esson, J., P. Noxolo,, R. Baxter, P. Daley, and M. Byron. 2017. "The 2017 RGS-IBG Chair's Theme: Decolonising Geographical Knowledges, or Reproducing Coloniality?" *Area* 49 (3): 384–388.
Gergan, M., S. Smith, and P. Vasudevan. 2020. "Earth beyond Repair: Race and Apocalypse in Collective Imagination." *Environment and Planning D: Society & Space* 38 (1): 91-110.
Gibson-Graham, J.K. 2011. "A Feminist Project of Belonging for the Anthropocene." *Gender, Place and Culture* 18 (1): 1–21.
Gibson-Graham, J.K., and G. Roelvink. 2010. "An Economic Ethics for the Anthropocene." *Antipode* 41 (1): 320–346.
Grusin, R., ed. 2017. *Anthropocene Feminism*. Minneapolis: University of Minnesota Press.
Haraway, D. 1988. "Situated Knowledges: The Science Question in Feminism and the Privilege of Partial Perspective." *Feminist Studies* 14 (3): 575–599.
Haraway, D.J. 2016. *Staying With the Trouble: Making Kin in the Chthulucene*. Durham, NC: Duke University Press.
Harding, S.G. 1986. *The Science Question in Feminism*. Ithaca, NY: Cornell University Press.
Houston, D. 2013. "Crisis is Where We Live: Environmental Justice for the Anthropocene." *Globalizations* 10 (3): 439–450.
Kanngieser, A. and A. Last. 2016. "Five Propositions: Critiques for the Anthropocene." *GeoCritique*. Available at: www.geocritique.org/five-propositions-critiques-anthropocene/ (accessed 27 November 2017).
Last, A. 2015. "Fruit of the Cyclone: Undoing Geopolitics through Geopoetics." *Geoforum* 64: 56–64.
Last, A. 2017. "We Are the World? Anthropocene Cultural Production between Geopoetics and Geopolitics." *Theory, Culture & Society* 34 (2–3): 147–168.
Lave, R., and B. Lutz. 2014. "Hydraulic Fracturing: A Critical Physical Geography Review." *Geography Compass* 8 (10): 739–754.
Lave, R., M.W. Wilson, E.S. Barron, C. Biermann, M.A. Carey, C.S. Duvall, L. Johnson, K.M. Lane, N. McClintock, and D. Munroe. 2014. "Intervention: Critical Physical Geography." *Canadian Geographer/ Le Géographe canadien* 58 (1): 1–10.
Lehman, J. 2016. "A Sea of Potential: The Politics of Global Ocean Observations." *Political Geography* 55: 113–123.
Lewis, S.L., and M.A. Maslin. 2015. "Defining the Anthropocene." *Nature* 519 (7542): 171–180.
Lorimer, J. 2012. "Multinatural Geographies for the Anthropocene." *Progress in Human Geography* 36 (5): 593–612.
Massey, D. 1994. *Space, Place, and Gender*. Minneapolis: University of Minnesota Press.
McClintock, A. 1995. *Imperial Leather: Race, Gender and Sexuality in the Colonial Contest*. New York: Routledge.
Pérez, M.A. 2016. "Yearnings for Guácharo Cave: Affect, Absence, and Science in Venezuelan Speleology." *cultural geographies* 23 (4): 693–714.
Povinelli, E.A. 2016. *Geontologies: A Requiem to Late Liberalism*. Durham, NC: Duke University Press.
Puig de la Bellacasa, M. 2017. *Matters of Care: Speculative Ethics in More than Human Worlds*. Minneapolis: University of Minnesota Press.
Pulido, L. 2002. "Reflections on a White Discipline." *Professional Geographer* 54 (1): 42–49.
Raworth, K. 2014. "Must the Anthropocene be a Manthropocene?" *The Guardian*, 20 October.
Rickards, L. 2015a. "Critiquing, Mining and Engaging Anthropocene Science." *Dialogues in Human Geography* 5 (3): 337–342.
Rickards, L.A. 2015b. "Metaphor and the Anthropocene: Presenting Humans as a Geological Force." *Geographical Research* 53 (3): 280–287.
Schneiderman, J.S. 2015. "Naming the Anthropocene." *philoSOPHIA* 5 (2): 179–201.
Sharpe, C. 2016. *In the Wake: On Blackness and Being*. Durham, NC: Duke University Press.
Shotwell, A. 2016. *Against Purity: Living Ethically in Compromised Times*. Minneapolis: University of Minnesota Press.
da Silva, D.F. 2007. *Toward a Global Idea of Race*. Minneapolis: University of Minnesota Press.
Swanson, H.A., N. Bubandt, and A. Tsing, A. 2015. "Less than One But More Than Many: Anthropocene as Science Fiction and Scholarship-in-the-Making." *Environment and Society* 6 (1): 149–165.
Todd, Z. 2015. "Indigenizing the Anthropocene." In: *Art in the Anthropocene: Encounters Among Aesthetics, Politics, Environments and Epistemologies*, edited by H. Davis and E. Turpin, 241–254. London: Open Humanities Press.
Tolia-Kelly, D.P. 2016. "Anthropocenic Culturecide: An Epitaph." *Social & Cultural Geography* 17 (6): 786–792.

Vergès, F. 2017. "Racial Capitalocene." In: *Futures of Black Radicalism,* edited by G.T. Johnson and A. Lubin, 72–82. London: Verso.

Vizenor, G. 2008. *Survivance: Narratives of Native Presence.* Omaha: University of Nebraska Press.

Wakefield, S. 2014. "Man in the Anthropocene (as Portrayed by the Film *Gravity*)." *May Revue* online. Available at: www.mayrevue.com/en/lhomme-de-lanthropocene-tel-que-depeint-dans-le-film-gravity/ (accessed 27 November 2017).

Whyte, K.P. 2016. "'Our Ancestors' Dystopia Now: Indigenous Conservation and the Anthropocene." In: *Routledge Companion to the Environmental Humanities*, edited by U. Heise, J. Christensen, and M. Niemann, 206–215. New York: Routledge.

Wynter, S., and K. McKittrick. 2015. "Unparalleled Catastrophe for Our Species? Or, to Give Humanness a Different Future: Conversations." *Sylvia Wynter: On Being Human as Praxis*, 9–89. Durham, NC: Duke University Press. https://doi.org/10.1215/9780822375852-002.

Yusoff, K. 2013. "Geologic Life: Prehistory, Climate, Futures in the Anthropocene." *Environment and Planning D: Society and Space* 31 (5): 779–795.

Yusoff, K. 2015. "Geologic Subjects: Nonhuman Origins, Geomorphic Aesthetics and the Art of Becoming Inhuman." *cultural geographies* 22 (3): 383–407.

Yusoff, K. 2016. "Anthropogenesis: Origins and Endings in the Anthropocene." *Theory, Culture & Society* 33 (2): 3–28.

Yusoff, K. 2018. "Politics of the Anthropocene: Formation of the Commons as a Geologic Process." *Antipode* 50 (1): 255–276.

Yusoff, K., E. Grosz, N. Clark, A. Saldanha, and C. Nash. 2012. "Geopower: A Panel on Elizabeth Grosz's 'Chaos, Territory, Art: Deleuze and the Framing of the Earth'." *Environment and Planning D: Society and Space* 30 (6): 971–988.

Zalasiewicz, J., C.N. Waters, A.P. Wolfe, A.D. Barnosky, A. Cearreta, M. Edgeworth, E.C. Ellis, I.J. Fairchild, F.M. Gradstein, and J. Grinevald. 2017. "Making the Case for a Formal Anthropocene Epoch: An Analysis of Ongoing Critiques." *Newsletters on Stratigraphy* 50 (2): 205–226.

42

QGIS IN FEMINIST GEOGRAPHIC RESEARCH

Its merits and limits

Nazgol Bagheri

Introduction

This chapter explores the more recent and innovative applications of Qualitative Geographic Information Systems (QGIS) and aims to illustrate their merits and limits. The term refers to the array of methodological efforts to incorporate into GIS more qualitative data than have traditionally been included. The goal is not only to appreciate the methodological opportunities that QGIS offers but also to examine its conceptual and technical boundaries. This chapter is written from a feminist, critical geographer's perspective in which mapping is seen as a subjective, geopolitical tool reinforcing global and local hierarchies. Therefore, while this chapter confirms that 'GIS can open new avenues of knowledge to feminist geographers by aiding in the analysis of the spatial and social contexts of women's lives' (Bagheri 2014a, 1287), it still emphasizes the significance of doing ethnographic methods. I draw upon the methodological portion of my ongoing research on the production of public space in Tehran, a process that is highly gendered and classed. Photographs, sketches and narratives were linked to GIS data gathered from the Iranian Census Organization, National GIS Database and Tehran municipality. This case study contextualizes QGIS applications in a perhaps unfamiliar cultural and political setting, which facilitates an examination of its limitations. The intensive literature review and findings from the case study illustrate not only the opportunities that QGIS offers to researchers but also the boundaries of the technology at its current level.

Defining QGIS

Both feminist geography and Geographic Information Systems (GIS) have been among the fastest-growing theoretical and practical research themes in geography in the last few decades (McLafferty 2002). Since the late 1990s, scholars have made significant efforts to connect the two in creative combinations such as qualitative GIS, feminist GIS, grounded visualization and geo-ethnography. Such hybrid epistemologies can enhance the subjective dimensions of GIS, in which the researcher has a richer understanding of everyday social life through observation and participation. In what follows, I provide a history of QGIS – rooted in feminist critical geographers' works in the late 1990s – and review the attempts to link qualitative and quantitative epistemologies that are often treated as distinctive, even incompatible perspectives. In other words, QGIS combines the feminist emphasis on participatory and qualitative methods with the spatial analysis and visualization tools of GIS.

GIS is a computer-based information application that uses both geographically referenced and non-spatial data to support spatial visualization and analysis, and is widely used in various disciplines such as geography, geology, environmental sciences, social studies and criminology. GIS researchers and critics have defined the rapidly progressing field of GIS according to their applications, subdisciplines and interests (Pickles 2006). For many reasons, GIS is far easier to imagine or apply than to define. Cope and Elwood (2009, 2) define GI systems as 'digital technologies for sorting, managing, analyzing, and representing geographic information'. GIS is also understood as a collection of practices for producing and negotiating geographic knowledge through the representation and analysis of spatial data. For Star and Estes (1990, 2–3) GIS is 'an information system that is designed to work with data referenced by spatial or geographic coordinates'. In other words, a GIS is both 'a database system with specific capabilities for spatially referenced data, as well [as] a set of operations for working with data. In a sense, a GIS may be thought of as a higher-order map'. William Huxhold (1991, 27, in *Introduction to Urban Geographic Information Systems,* suggests that 'the purpose of a traditional GIS is first and foremost spatial analysis. Therefore, capabilities may have limited data capture and cartographic output. Capabilities of analyses typically support decision making for specific projects and/or limited geographic areas.' Central to the concept of GIS is its capability of overlaying spatial and non-spatial data, making what geographers call a cartographic sandwich.

GIS has often been associated with positivist epistemology, instrumental rationality and quantitative and data-led methods, and it has been criticized for its limitations in terms of its representation of space, movement and subjectivity, particularly by critical human geographers. Among these geographers, GIS is seen, applied and examined as a geospatial tool, not a remedy. Many geographers (see Bell and Reed 2004; Cope and Elwood 2009; Elwood 2006a, 2006b; Knigge and Cope 2009; Kwan 2002; Matthews, Detwiler and Burton 2005; McLafferty 2002; Pavlovskaya 2002; Sheppard 2001; Wakabayashi and Nishimura 2010) have challenged the dualism of qualitative and quantitative geography and have successfully opened doors to human geographers to explore the benefits of GIS in their technology-societal studies. Sheppard (2001, 532) highlights that 'the distinction drawn between quantitative and qualitative geography has developed since the late 1970s, after humanistic and radical human geographers began to develop critiques of the philosophical and methodological foundations of the geography of the quantitative revolution'. Sheppard suggests that this dualism is both reductionist and paradoxical. Several critical feminist geographers agree with Sheppard and consider QGIS to be a way to bring quantities and qualitative geography back together.

An increasing number of researchers integrate GIS-based spatial visualization and analysis into qualitative methods, such as ethnography. Photographs, sketches, mental maps, audio and video records and narratives gathered by qualitative methods can be linked to spatially referenced data in GIS software to offer new ways of understanding. Such endeavours in connecting qualitative and quantitative data in GIS application emerged in response to the mid-1990s critiques of GIS, which presented it merely as a quantitative method rooted in positivist epistemology (Cope and Elwood 2009, 1). Reacting to those critiques, mixed-methods approaches, such as qualitative GIS (Bell and Reed 2004; Cope and Elwood 2009; Kwan 2002; McLafferty 2002; Pavlovskaya 2002), grounded-visualization (Knigge and Cope 2006), feminist-visualization (Kwan 2002) and geo-ethnography (Matthews, Detwiler and Burton 2005) have emerged and transformed the 'inherent' quantitative perspective of GIS as a research tool. These approaches, particularly those suggested and used by feminist and critical geographers, integrate multiple forms of data (both GIS-based spatial data and non-spatial data such as photographs, sketches, narratives, field notes and interviews) in order to create a bigger picture and greater insight into what is happening in the study area and why.

Since the late 1990s, many scholars have made significant efforts to connect qualitative research and GIS (see Cope and Elwood 2009 for a comprehensive history of QGIS). This combination has largely enriched feminist geography research methods by introducing GIS geo-visualization and spatial analysis techniques, and it has reduced the role of positivism in GIS. Although feminist geographers have typically preferred qualitative methods, such as ethnographies that capture 'situated knowledge' and account for 'lived experiences' (Haraway 1991), quantitative methods such as GIS can open new avenues of knowledge to feminist geographers by aiding in the analysis of the spatial and social contexts of women's (and other minorities') lives.

QGIS can be placed among post-positivist, poststructuralist and non-representational approaches in contemporary cultural geography that strive to reconcile 'social-cultural and spatial-analytical geographies' (Kwan 2004, 756). One major outcome of incorporating qualitative data with GIS, in the words of Cope and Elwood (2009, 4), is promoting the 'use [of] GIS in research that emerges from multiple or hybrid epistemologies, and theorizing previously unrecognized forms of social knowledge that may be present in GIS applications'. QGIS that includes ethnographic data can serve as a bridge, linking what often have often become separate perspectives based on distinctive quantitative and qualitative epistemologies. In this approach, research becomes truly an open-ended, emergent process, rather than a set of predictable, pre-determined steps (Lofland and Lofland 1994). Finding spatial patterns on maps can result in new and more questions; this invites the researcher back to the field to continue the data-gathering phase. Perhaps seeing maps as living and not as a final product brings about a more interactive and cyclical research process and more nuanced findings.

Although many critical and feminist geographers have already challenged the dualism separating quantitative and qualitative geography, innovative mixed-methods approaches such as QGIS are still in their early stages, perhaps in what I call the second generations of QGIS. Unlike the first generation of QGIS, this generation focuses on recognizing QGIS' limits while celebrating its merits. The majority of feminist geographers have emphasized the opportunities that QGIS, as an innovative method, introduces to feminist geography; however, little attention has been given to its limitations, particularly in non-Western contexts, where socio-cultural subjectivity appears to be more difficult to access. This chapter represents an attempt to go beyond the first generation of QGIS literature and to critically explore its limitations. To do so, I review several studies by geographers in an intensive examination of works that have applied QGIS, then turn to my own work in Tehran.

Applying QGIS

The application of QGIS varies greatly in the subject matter, as well as the extent to which GIS visualization and spatial analysis tools is employed. However, the pioneers (mothers) of QGIS have been feminist geographers who bravely pushed the boundaries of GIS technologies and integrated qualitative data into their GIS analysis. QGIS applications include diverse examples, ranging from economic geography, environmental sustainability and political geography to phenomenological research of space and sense of place. Researchers have also applied GIS at various levels: some use GIS as a framework to organize and visualize spatial data; others go beyond that by applying more advanced GIS spatial analysis tools to examine patterns among quantitative and qualitative variables.

For instance, McLafferty (2002) tells a story of how women concerned about breast cancer used GIS to explore the relationship between environmental hazards and the geographical distribution of disease in suburban Long Island, New York. She concludes that GIS is 'more than a computer-based technology' and can serve not only as 'a vehicle for women's empowerment' (McLafferty 2002, 268) but to empower other underrepresented populations. In another study asserting the value of grounded theory in urban cultural research, Knigge and Cope (2006, 2009) apply what they call 'grounded visualization' as an integrated method to enable them to explore spatially referenced data in close relationship to ethnographic and multi-media data for the study of community gardens and the meanings of vacancy in Buffalo, New York.

A common characteristic of these innovative applications has been the focus on mixed-methods approaches. Cieri (2003) combines three-dimensional GIS and qualitative data in order to study lesbian public spaces in Philadelphia. What makes her successful in such a critical endeavour, however, is her application of a mixed-methods approach similar to Knigge and Cope's grounded visualization. In an example from natural resources management, Nightingale (2003) claims to go beyond the limitations that partial and situated knowledges impose on geography by using QGIS in tracing forest change in Nepal. Kwan (2002), through her study of gender and individual access to urban employment opportunity in Columbus, Ohio, introduces 'feminist visualization'. Kwan and Ding (2008) and Kwan (2008) provide excellent examples of emotional geographic studies in which the 'geo-narrative' method, integrating GIS's spatial analysis and narratives, is used to examine the effects of anti-Muslim hate violence on Muslim women's lives in Columbus, Ohio, in the post-9/11 period. In another study, Watts (2010) reconciles on-the-ground qualitative data with point-distribution GIS data to investigate participants' socio-spatial behaviours in Los Angeles's public protests. Jones and Evans (2012) encourage mobility scholars to take advantage of QGIS application to examine better the role of space in data creation.

QGIS applications often emphasize the important role played in the research process by the researched subjects, their stories and their lives. Hence, many geographers have included ethnography in their data gathering. Coupling GIS techniques with ethnographic methods is referred to as 'geo-ethnography' by Matthews, Detwiler and Burton (2005) in their ongoing study of low-income families and their children and welfare reform in Boston, Chicago and San Antonio. Boschmann and Cubbon (2014), in their examinations of job accessibility and feelings of safety, respectively, among members of the working poor in Columbus, Ohio, and LGBT in St Louis, Missouri, digitized and transformed qualitative sketch maps using GIS. They conclude that, while such a process is time consuming, it can result in the following QGIS objectives: 'collecting unique spatial data of individual experiences, visualizing socio-spatial processes, breaking down particular barriers of positionality in research, and developing new uses of GIS' (Boschmann and Cubbon 2014, 3).

Based on its diverse applications in geography research, QGIS can be considered an umbrella term for the mixed-methods approaches described above. 'The qualitative' can be associated with the qualitative nature of either the data gathered (e.g. interviews or sketches) or the analysis methods used in the research (e.g. content analysis, triangulation and grounded theory) (Jung and Elwood 2010, 67). These approaches, though different, share an emphasis on integrating multiple forms of data to create a bigger picture and greater insight into what is happening (and why) in the study area. These approaches can also be empowering, action-oriented research methods, through which the stories of subaltern and/or marginalized groups including women, children, the disabled, LGBT and racial and religious minorities can be heard and included in the research process. QGIS offers both methodological and ethical advancements. Methodologically, QGIS can create more comprehensive maps and visualize various types of data. Ethically, it can offer more inclusive and democratic research process in which previously marginalized or ignored voices may be represented. Often, these methods are used in community-based and participatory-planning projects in which a deeper understanding of context and the groups that are being represented is essential to ensure the goals of justice, equit, and social sustainability. QGIS can enhance geography from a merely inductive and objective science to a more subjective one, in which the researcher has a richer understanding of individuals' experiences in everyday life through ethnography. In this approach, research becomes truly an open-ended emergent process and 'partial and situated' (Haraway 1991), rather than the set of predictable, pre-determined steps often advocated by GIS scientists.

In the following section, I develop these arguments by giving examples of using QGIS in a research project conducted in Tehran that examined women's presence in and preferences for certain spaces, and of the symbolic meanings that they attach to what urban planners might call modern and traditional public spaces. These examples are not intended to present a full account of women's conceptualization of modern and traditional public spaces in Tehran, but rather to illustrate how QGIS as a mixed-methods approach can be applied.

Contextualizing QGIS in Tehran: merits and limits

To emphasize the differences between social construction of public spaces in non-Western and Western contexts and to understand Muslim women's socio-spatial behaviours in public spaces, I studied women's presence and preferences in Tehran (see Bagheri 2014a, 2014b). One of the main reasons I chose to focus on female users was the prevailing societal norms and my own positionality: I am an Iranian–American urban, feminist geographer with a background in architecture and urban design. For most of my life I have lived in Tehran, Iran. While undertaking my research from 2011–2015, I came to understand my insider–outsider positionality (as an Iranian privileged by living in the US and with freedom of speech, among other freedoms) as a researcher. Due to my diverse educational background and research interests, I have borrowed methods from both architecture and urban design, including social behaviour mapping, sketching, photography, urban documents and plans, geography, such as GIS analysis, and urban anthropology, including participant observation and intensive interviews, to illustrate what is happening on the ground and why, giving particular attention to women's lived experiences (instead of the oppressed representations of Muslim women in Western mass media). As a female researcher, I found it more feasible to approach and talk to women than to men. In addition, since I grew up in Tehran and was therefore familiar with the local symbolic politics, this affected how I presented myself as a researcher, approached women in public spaces and interpreted their narratives. For instance, I chose to wear a typical academic/student *hijab* (*Maghnae*) instead of a more relaxed, everyday *hijab* (*Roosari*) during my field visits.

The goal of my research was to illustrate how modern and traditional public spaces in Tehran are used, even though the women's narrative did not recognize the often-used binary categorization of modern and traditional public spaces by urban professionals. Modern and privatized public spaces in North American contexts have been criticized for their highly commercialized atmosphere, controlled behaviour and design and their segregation of certain racial and class groups (see Arjmand 2016; Banerjee and Loukaitou-Sideris 1992; Crawford 1992; Low 2000; Low and Smith 2006; Madanipour 2003; Soja 1989; Sorkin 1992; Zukin 2003). Critics suggest that the privatized and modern, often standard, designs of public spaces have degraded them to more exclusive places and negatively influenced the social life taking place in them. I selected two public spaces in Tehran with contrasting locations, functions and architectural designs to examine these criticisms in a non-Western context. While there was no clear boundary between traditional and modern public spaces according to the women with whom I talked, the selected public spaces' contrasting locations in Tehran (one in the very northern and one in the south-central part of the city) and their socio-political history, as well as their architectural designs and functions, allowed me to represent and compare a more modern and a more traditional public space.

I chose a mixed-methods approach with a rich ethnographic quality and took advantage of socio-spatial data visualization and analysis in GIS. First, I collected photographs to illustrate visual details, such as women's style of *hijab* and make-up, and created spatial behaviour maps to track women's numbers, activities and locations, as well as characteristics that are readily observable, such as their approximate age, sex and whether they were alone or in a group. I drew architectural sketches to capture a sense of place. In addition, I conducted in-depth semi-structured interviews with 83 women who were using two seemingly contrasting, or what urban planners might call modern and traditional, public spaces in Tehran (discussed above). The women talked about their feelings, experiences and preferences in those public spaces, and I also examined the various ethno-cultural and symbolic meanings that the women attached to the spaces.

In contrast to the dominant critics of modern public spaces in the literature, I found that Iranian women enjoyed a unique sense of freedom and equity with men in Tehran's modern and semi-private public spaces. Women explained that they felt more comfortable doing certain activities, such as holding hands with their boyfriends, smoking, talking louder or laughing, in modern public spaces than in the more traditional public spaces. Women's narratives and details of their *hijab* styles, gathered during participant observation, also suggested that women had more freedom in their dress in the more modern public space; many interviewees emphasized that they would wear more make-up and fashionable, colourful clothes in such places. However, women did not prefer either of the public spaces; they enjoy using both, for different reasons. Women's narrative and experiences suggested that they enjoyed a sense of freedom in the more modern public space while appreciating the sense of nostalgia and connection to their Iranian–Islamic identity in the more traditional one. The study also highlighted that women's feelings about public space accurately reflected their personal life experiences, such as their socio-economic status, education, employment and home location, rather more than their gender. I suggest caution in using 'gender' as an essential category or independent social construct in feminist studies (for a detailed description data interpretations and findings, see Bagheri 2013), yet this kind of strategic essentialism allows me to highlight the QGIS application's merits and limits that I encountered in my on-the-ground work in Tehran.

While QGIS offers a valuable approach to many studies that deal with the concept of space and socio-cultural production of space, it is neither easy nor complete. I used diverse methods of data gathering and they were all, more and less, based on the socio-spatial experiences of women's everyday lives. I followed Lofland and Lofland's (1994) suggestion and began fieldwork visits with short-term covert participant observation. This gradual, 'bottom-up' process let the research interests and questions grow from the field itself, rather than from theory. The benefits of this approach include data management, discovery of spatiality of social processes and an emphasis on the importance of ethnographic data in cultural geographic research. Limitations consist of potential data exclusion and data scale incompatibility. For the remainder of this chapter, I consider these benefits and limitations in turn, while describing in more detail the techniques employed for my analysis.

I used ESRI's ArcGIS 10 to integrate the qualitative and quantitative data that I had gathered in the field. I first digitized the spatial behaviour maps and transferred the population counts to

GIS maps. Using Overlay, arguably the most important GIS analysis function, I examined the relationship between the different demographic and socio-economic data layers gathered from the Iranian Census Organization, National GIS Database, Tehran Municipality and population counts by gender on the maps. I used the Kernel Density Estimation tool in GIS to illustrate female and male density and spatial distribution on maps. The maps showed interesting spatial patterns of female and male users in selected public spaces; however, they did not explain why such patterns existed. For instance, both the geo-visualization and statistical spatial analysis undertaken using the GIS software suggested that women in the more modern public space were less segregated from men than those in the more traditional space. There were no regulations regarding male and female user segregation in any of those public spaces. Why, then, did GIS maps suggest such spatial segregation? The GIS results also suggested that the density of female users was relatively similar in both public spaces, particularly on weekends, when the places witnessed their highest percentage of female users. How, then, did these places' differences in design, location and, more importantly, attached symbolic meaning influence women's presence and preferences? My supplementary women's narratives and field notes were essential in answering this question. Through talking with women, I examined their ways and values and it became clear that 'culture ... is at the heart of everything' (Mitchell 2004, 155). Women attached different symbolic and ethno-cultural meanings to the selected public spaces, and these influenced their presence and behaviours in places.

Meanwhile, I created a geodatabase to record the location of each interview and to store the interviewees' characteristics (such as age, education and home location). Using data management tools such as Query and Attribute Table in ArcGIS allowed me to be more organized in the gathering phase. For instance, I digitized the location of each interview to a point feature and used the hyperlink tool to connect the point to qualitative data that I had gathered in the field (e.g. key points of interviews and field notes). I encountered various challenges in mapping my narratives. For example, women named different places, either similar or contrasting to the one they were in, to explain their feelings and preferences further. This raised several questions: Was mapping narratives to the specific interview location enough? Should I include the other numerous places named in the interviews on GIS maps without actually being there or studying them? Would that be a useful or even readable map? What kinds of data gathered during field observation and interviewing could be transformed into GIS? Basically, I was questioning to what extent the rich and complex qualitative data, such as interview transcriptions, were transferrable into quantitative GIS codes and symbols.

Examining possible answers to these types of questions, Jung and Elwood (2010, 68) recognize what I call an inevitable disadvantage of QGIS – potential data exclusion – and argue that 'a host of other potentially significant details are lost' in the transformation process.

A similar challenge appears when we consider the meaning of visual and spatial representation, one of the most exciting prospects of QGIS reported in earlier celebratory works on bridging the qualitative and quantitative domains. To this end, hyperlinking was enough to store the data yet not to represent them visually. To represent ethnographic data, I had to summarize and select from each interview and often 'exclude' data that I, as a researcher, thought was less important. It was almost impossible to include, even partially, the narrative from every interview. GIS is not an independent tool; rather, it becomes part of the storytelling itself. In other words, in the process of 'embedding' the individuals' lived experiences, the GIS user's agency (his or her personal characteristics and research goals) and the specific spatial analysis tools used affect the reality being represented and, consequently, the research results. While GIS can incorporate diverse forms of data, this case study illustrated how the two worlds of complex reality and GIS representation cannot be seamlessly overlapped. Here, I question Krygier and Wood's claim (2005, 51) that 'you can map just about any data you can collect from the environment'. I suggest that you can map almost all types of data, but not necessarily in an inclusive, efficient and/or meaningful way. There remain some realms of human experiences and stories that are not 'mappable', in a meaningful manner, at least not with current GIS technology. For example, women showed specific facial expressions and other body languages when they talked about their feelings in public spaces. How women felt while we were talking, why they attached certain symbolic meanings to those public spaces or how such meanings changed their presence and behaviours are just few examples of qualitative data that cannot be mapped in GIS. This highlights the importance of ethnographic data in cultural geographic research.

The second limitation I experienced with QGIS was data scale incompatibility. It is argued that QGIS allows researchers to link existing datasets (e.g. from censuses) to one's own specific dataset (Bell and Reed 2004; Jung and Elwood 2010). However, this linkage can be meaningful or meaningless, depending on the difference in the dataset scales. In my case study, the data gathered during participant observation and interviews were space- and time-sensitive and were based on individuals' opinions, while the background GIS data (e.g. census) were based on official urban planning, social and/or demographic data that were gathered and normalized at the scale of the *mahaleh* (neighbourhood), *mantagheh* (district) or city of Tehran. I found it difficult and sometimes impossible to analytically link qualitative data to quantitative data layers in GIS, considering their different scales. For example, while there could be a relationship between the interviewee's *hijab* style and her age, education and median household income, there was no meaningful relationship between her *hijab* and the similar data aggregated at the level of the district in which I interviewed her, or even at the level of the district in which she lived. Multiple GIS representations of different data can lead to incorrect or incomplete conclusions: in this case, to an incorrect correlation between women's *hijab* and the quantitative attributes of the district in which the interview was conducted (particularly for non-local scholars who may not be familiar with the symbolic politics of *hijab* in Iran). This issue of mis-connecting qualitative data to contextual/location layer is not limited to QGIS users; GIS users have long faced this dilemma. However, because it is more difficult to 'locate' emotions, feelings, preferences, stories and in general qualitative data to one specific location, this problem has become more apparent in QGIS applications. This data scale incompatibility can cause a major challenge in the data analysis phase, because GIS application can both facilitate and hinder efforts to represent complex social and spatial realities. While Knigge and Cope (2009) found it useful in their case study, linking GIS and ethnographic data gathered at different scales was not always helpful or even meaningful in my case study. Therefore, I suggest caution in overlapping the often-quantitative census, urban planning and land-use data that are gathered by census organizations, city and district governments or other state officials and the qualitative data collected by a researcher for scholarly purposes. Depending on one's research questions, this capability of QGIS can become an advantage or disadvantage.

I am by no means suggesting that QGIS is not useful in understanding complex social and spatial phenomena, patterns or processes. On the contrary, I acknowledge the advantages that QGIS offers researchers, particularly in public participatory-planning projects, by giving voice to under-represented groups. Mitchell and Elwood's collaborative mapping works (see Elwood and Mitchell 2015; Mitchell and Elwood 2012) have beautifully illustrated how new socio-spatial technologies including GIS have transformed our understanding and memories of space and time. As they suggest: 'Going forward, the presence of these persistent ambiguities and disparities underscores the necessity of moving beyond the utopian/dystopian opposition that has often framed discourse about new technologies.' (Elwood and Mitchell 2015, 152). Another inspiring example is Yoshida's (2016) work on integrating life stories and narrative of minorities in Japan, in which she advocates that, with geography, it has become possible to point out issues that are found only 'here' and cannot easily be generalized by questioning 'where', including micro-scale spaces that cannot be mapped or visualized (Yoshida 2016, 10).

In my case study, the geo-visualization capability of QGIS clarified that the specific geographical locations of selected public spaces (one in northern and one in south-central Tehran) played important roles in the symbolic meanings and ethno-cultural values that the women attached to them. In addition, users' spatial patterns on GIS maps also pointed out new aspects of public spaces and led me to modify my interview questions for the next round of field visits. For example, in the first round of my interviews I did not question women on how they felt about their movements in a place with regard to the male users, but after I observed spatial segregation in the more traditional public space on GIS maps, I added new, relevant questions to my interviews. Indeed, QGIS enhanced and, in some cases, expedited my data interpretation and analysis, allowing me to see more, yet not necessarily 'all'.

Conclusion

Qualitative GIS aims to incorporate into GIS more qualitative data than have traditionally been included. It pushes feminist geographers to go beyond traditional, qualitative methodologies and introduce more multi-methodological, radical (re)presentations of the world. However, little attention has been given to the limitations of its application, particularly in non-Western contexts, where socio-cultural subjectivity appears to be harder to access and more complex to understand. If QGIS combines GIS with ethnography as promised in its core, it inherits the politics and problems of doing ethnography, including the partiality, subjectivity and fluidity of multiple truths. Ethnography and its discursive practices have conventionally looked at 'others' – defined as primitive, tribal, native, subaltern or non-Western. But, as Clifford and Marcus (1986, 23) remind us, 'now, ethnography encounters others in relation to itself, while seeing itself as other'. Similar to any ethnographic work, QGIS applications may suffer from losing the perspective of the local, combining it with a more rigid, quantitative GIS that may risk misinterpretation and misrepresentation of the other cultures. A contribution of this chapter thus lies in its efforts to emphasize the ethnography component of QGIS, its irreplaceable role and inevitable limits and the technical and conceptual problems of connecting them to GIS.

My goal has been to go beyond the first generation of QGIS literature and to critically explore QGIS' limitations. While briefly reviewing its merits, I have focused on GIS' limitations through examples from research on Iranian women's presence and preferences in Tehran's public spaces. Besides its data management capabilities, QGIS application enabled me to record and present various types of data by engaging the different senses of my audience, among them non-local scholars who would otherwise have no access to the group under investigation, as well as Iranian women themselves. However, I encountered two major limitations during the course of my fieldwork: data exclusion and data scale incompatibility. I argue that there remain inevitable disadvantages to using QGIS, or any form of GIS application. First is the potential data exclusion in transforming rich, complex ethnographic data to GIS summaries, codes and symbols. Second, scale incompatibility among multiple layers of data in GIS can be misleading, particularly regarding the overlap of quantitative data aggregated from official urban divisions (e.g. district or neighbourhood) with ethnographic data collected at the individual level. This misconnection of data across different scales may result in skewed presentations of reality and/or misinterpreting cultural rituals, gender-based practices or behaviours, which become extracted from their geo-political context. This may be more common among non-local scholars who are less familiar with the larger contexts of systematic power inequality in the field. Hence,

I encourage cultural geographers to consider the limitations of these methods, particularly those of potential data exclusion and data scale incompatibility.

I illustrate that GIS applications allow geographers, especially those who often deal with the symbolic and ethno-cultural meanings people attach to places, to see more yet not necessarily 'all'. While I agree with Jackson (2008, 346) that 'digital technology [including GIS] can … reduce critical reflection on the enhanced slippage occurring between map, experience, and context', I believe that feminist/critical geographers' recent endeavours to think out of the box and act innovatively to make GIS more accessible to diverse participants are plausible. Wood (1992) contends that GIS is often used to promote distinctive interests or convince others; however, he emphasizes that the interest can be anyone's, including researchers' and researched groups'. In this regard, critical GIS applications including QGIS can give voice to those under-represented groups, which are often the subjects of studies conducted by feminist/critical geographers. Without any doubt, QGIS, compared to traditional applications of GIS, promotes more and crucial engagement with the people who are being studied than using solely GIS or solely ethnography. While QGIS can enable certain visualization and analysis, I suggest caution in assuming that it (re)presents a complete picture of any social life.

Finally, as a feminist urban geographer striving to represent a more realistic picture of people and places in underrepresented contexts such as Muslim-majority Iran, I believe that productive conceptual and methodological exchanges between cultural and feminist geographers will benefit both subdisciplines. Even though the application of QGIS to understanding the mutual construction of gendered identities and public spaces in Tehran encountered some limitations of methodology, I hope this chapter presents a step toward subdisciplinary communications in geography and enhances discussion of the efficacy of QGIS often used in the gender and queer geographies.

Key readings

Bagheri, N. 2014. "What Qualitative GIS Maps Tell and Don't Tell: Insights from Mapping Women in Tehran's Public Spaces." *Journal of Cultural Geography* 31 (2): 166–178. doi:10.1080/08873631.2014.906848.

Cope, M., and S. Elwood, eds. 2009. *Qualitative GIS: A Mixed Methods Approach*. Los Angeles: Sage.

Kwan, M.P. 2002. "Feminist Visualization: Re-envisioning GIS as a Method in Feminist Geographic Research." *Annals of the Association of American Geographers* 92 (4): 645–661. doi:10.1111/1467-8306.00309.

References

Arjmand, R. 2016. *Public Urban Space, Gender and Segregation: Women-only Urban Parks in Iran*. London: Routledge.

Bagheri, N. 2013. "Modernizing the Public Space: Gender Identities, Multiple Modernities, and Space Politics in Tehran." PhD dissertation, University of Missouri.

Bagheri, N. 2014a. "Mapping Women in Tehran's Public Spaces: A Geovisualization Perspective." *Gender, Place & Culture: A Journal of Feminist Geography* 21 (10): 1285–1301. doi:10.1080/0966369X.2013.817972.

Bagheri, N. 2014b. "What Qualitative GIS Maps Tell and Don't Tell: Insights from Mapping Women in Tehran's Public Spaces." *Journal of Cultural Geography* 31 (2): 166–178. doi:10.1080/08873631.2014.906848.

Banerjee, T., and A. Loukaitou-Sideris. 1992. *Private Production of Downtown Public Open Space: Experiences of Los Angeles and San Francisco*. Los Angeles: University of Southern California.

Bell, S., and M. Reed. 2004. "Adapting to the Machine: Integrating GIS to Qualitative Research." *Cartographica* 39 (1): 55–66. doi:10.3138/Y413-1G62-6H6G-0L3Q.

Boschmann, E., and E. Cubbon. 2014. "Sketch Maps and Qualitative GIS: Using Cartographies of Individual Spatial Narratives in Geographic Research." *Professional Geographer* 66 (2): 236–248. doi:10.1080/00330124.2013.781490.

Cieri, M. 2003. "Between Being and Looking: Queer Tourism Promotion and Lesbian Social Space in Greater Philadelphia." *ACME: An International E-Journal for Critical Geographies* 2 (2): 147–166.

Clifford, J.S., and G.E. Marcus. 1986. *Writing Culture: The Poetics and Politics of Ethnography*. Berkeley: University of California Press.

Cope, M., and S. Elwood, eds. 2009. *Qualitative GIS: A Mixed Methods Approach*. Los Angeles, CA: Sage.

Crawford, M. 1992. "The World in a Shopping Mall." In: *Variations on a Theme Park: The New American City and the End of Public Space*, edited by M. Sorkin, 3–30. New York: Hills & Wang.

Elwood, S. 2006a. "Critical Issues in Participatory GIS: Deconstructions, Reconstructions, and New Research Directions." *Transactions in GIS* 10 (5): 693–708. doi:10.1111/j.1467-9671.2006.01023.x

Elwood, S. 2006b. "Negotiating Knowledge Production: The Everyday Inclusions, Exclusions, and Contradictions of Participatory GIS Research." *Professional Geographer* 58 (2): 197–208. doi:10.1111/j.1467-9272.2006.00526.x

Elwood, S., and K. Mitchell. 2015. "Technology, Memory, and Collective Knowing." *cultural geographies* 22 (1): 147–154. doi:10.1177/1474474014556062

Haraway, D. 1991. *Simians, Cyborgs, and Women: The Reinvention of Nature*. New York: Routledge.

Huxhold, W.E. 1991. *Introduction to Urban Geographic Information Systems*. New York: Oxford University Press.

Jackson, S. 2008. "The City from Thirty Thousand Feet: Embodiment, Creativity, and the Use of Geographic Information Systems as Urban Planning Tools." *Technology and Culture* 49 (2): 325–346. doi:10.1353/tech.0.0039.

Jones, P., and J. Evans. 2012. "The Spatial Transcript: Analyzing Mobilities through Qualitative GIS." *Area* 44 (1): 92–99. doi:10.1111/j.1475-4762.2011.01058.x.

Jung, J.K., and S. Elwood. 2010. "Extending the Qualitative Capabilities of GIS: Computer-aided Qualitative GIS." *Transactions in GIS* 14 (1): 63–87. doi:10.1111/j.1467-9671.2009.01182.x.

Knigge, L., and M. Cope. 2006. "Grounded Visualization: Integrating the Analysis of Qualitative and Quantitative Data through Grounded Theory and Visualization." *Environment and Planning A* 38 (11): 2021–2037. doi:10.1068/a37327.

Knigge, L., and M. Cope. 2009. "Grounded Visualization and Scale: A Recursive Analysis of Community Spaces". In: *Qualitative GIS*, edited by M. Cope and S. Elwood, 95–104. Los Angeles, CA: Sage.

Krygier, J., and D. Wood. 2005. *Making Maps: A Visual Guide to Map Design for GIS*. New York: Guilford Press.

Kwan, M.P. 2002. "Feminist Visualization: Re-envisioning GIS as a Method in Feminist Geographic Research." *Annals of the Association of American Geographers* 92 (4): 645–661. doi:10.1111/1467-8306.00309.

Kwan, M.P. 2004. "Beyond Difference: From Canonical Geography to Hybrid Geographies." *Annals of the Association of American Geographers* 94 (4): 756763. doi:10.1111/j.1467-8306.2004.00432.x.

Kwan, M.P. 2008. "From Oral Histories to Visual Narratives: Re-presenting the Post-September 11 Experiences of the Muslim Women in the United States." *Social and Cultural Geography* 9 (6): 653–669. doi:10.1080/14649360802292462.

Kwan, M.P., and G. Ding. 2008. "Geo-narrative: Extending Geographic Information Systems for Narrative Analysis in Qualitative and Mixed-method Research." *Professional Geographer* 60 (4): 443–465. doi:10.1080/00330120802211752.

Lofland, J., and L. Lofland. 1994. *Analyzing Social Settings: A Guide to Qualitative Observation and Analysis*. Belmont, CA: Wadsworth.

Low, S.M. 2000. *On the Plaza: The Politics of Public Space and Culture*. Austin: University of Texas Press.

Low, S.M., and N. Smith. 2006. *The Politics of Public Space*. New York: Routledge.

Madanipour, A. 2003. *Public and Private Spaces of the City*. London: Routledge.

Matthews, S., J.E. Detwiler, and L.M. Burton. 2005. "Geo-ethnography: Coupling Geographic Information Analysis techniques with Ethnographic Methods in Urban Research." *Cartographica* 40 (4): 75–90. doi:10.3138/2288-1450-W061-R664.

McLafferty, S.L. 2002. "Mapping Women's Worlds: Knowledge, Power and the Bounds of GIS." *Gender, Place and Culture* 9 (3): 263–269. doi:10.1080/0966369022000003879.

Mitchell, D. 2004. "Cultural Geography at the Millennium: A Call to (Intellectual) Arms." *Journal of Cultural Geography* 22 (1): 155–157. doi:10.1080/08873630409478254.

Mitchell, K., and A. Elwood. 2012. "Mapping Children's Politics: The Promise of Articulation and the Limits of Nonrepresentational Theory." *Environment and Planning D: Society and Space* 30 (5): 788–804. doi:10.1068/d9011.

Nightingale, A. 2003. "A Feminist in the Forest: Situated Knowledges and Mixing Methods in Natural Resource Management." *ACME: An International E-Journal for Critical Geographies* 2 (1): 77–90.

Pavlovskaya, M. 2002. "Mapping Urban Change and Changing GIS: Other Views of Economic Restructuring." *Gender, Place & Culture: A Journal of Feminist Geography* 9 (3): 281–289. doi:10.1080/0966369022000003897

Pickles, J. 2006. "Ground Truth 1995–2005." *Transactions in GIS* 10 (5): 763–772. doi:10.1111/j.1467-9671.2006.01027.x.

Sheppard, E. 2001. "Quantitative Geography: Representations, Practices, and Possibilities." *Environment and Planning D: Society and Space* 19 (5): 535–554. doi:10.1068/d307.

Soja, E. 1989. *Postmodern Geographies: The Reassertion of Space in Critical Social Theory*. New York: Verso Books.

Sorkin, M. 1992. *Variations on a Theme Park: The New American City and the End of Public Space*. New York: Hills & Wang.

Star, J., and J. Estes. 1990. *Geographic Information Systems: An Introduction*. Englewood Cliffs, NJ: Prentice Hall.

Wakabayashi, Y., and Y. Nishimura. 2010. "Problems Concerning 'GIS and Society': Critical GIS as Another Geographical Information Science." *Geography Criticism Series A* 83 (1): 60–79. doi:10.4157/grj.83.60.

Watts, P. 2010. "Mapping Narratives: The 1992 Los Angeles Riots as a Case Study for Narrative-based Geo-visualization." *Journal of Cultural Geography* 27 (2): 203–227. doi:10.1080/08873631.2010.494401.

Wood, D. 1992. *The Power of Maps*. New York: Guilford Press.

Yoshida, Y. 2016. "Geography of Gender and Qualitative Methods in Japan: Focusing on Studies That Have Analyzed Life Histories." *Geographical Review of Japan Series B* 89(1): 4–13. doi:10.4157/geogrevjapanb.89.4.

Zukin, S. 2003. *Point of Purchase: How Shopping Changed American Culture*. New York: Routledge.

43

DOING GENDER IN THE DIGITAL

Feminist geographic methods changing research?

Jessica McLean, Sophia Maalsen and Nicole McNamara

Introduction

Feminist methods offer much to geographic research on and with the digital. In particular, over the last decade feminist approaches to the digital have developed in productive and exciting ways. In doing so, they have illustrated the possibilities for engaging with digital spaces, spaces whose sometimes messy materiality has challenged traditional geographic approaches. Here we describe ways in which feminist methods have enabled opportunities to investigate the digital and their role in geography's digital turn.

First, we ask whether feminist geographic methods have contributed to a digital turn. Second, we see how feminist approaches' recognition of knowledge as always situated and partial – requiring a reflexive research practice – has contributed to nuanced understandings of the digital. Third, we demonstrate how the critical perspective that feminism has contributed to studies of labour – the value of emotional and affective labour – brings important insights to both the advantages and exploitation of people's labour in digital spaces. We suggest that the value of feminist geographic methods for understanding the digital is a project that should continue and, also, inspire researchers to persist in creating innovative methodologies for understanding and working with/in digital spaces.

For the purposes of this chapter on feminist methods, we define the digital in a broad sense: it encompasses diverse technologies supported by the internet, including social media, new spatial media, digital research methods and tools like video conferencing, as well as software and devices that facilitate digital spaces. Drawing on Ash et al. (2017), we agree that the digital is more than just computer technologies as it has certain aesthetic, ontological and discursive aspects that are produced by and circulate through and with the networks and tools that are simultaneously enabling digitality.

Feminist geographic methods: contributing to the digital turn?

The question of whether it is possible to identify a geographic method as feminist remains challenging, especially given that feminisms continue to change and shift. As Moss (2003, 2) notes, despite feminist analysis being a project of early radical geography, questions around a feminist methodology did not appear in the literature until the 1990s, and debates on feminist methodology continue. In a thoughtful review for *Progress in Human Geography*, Sharp (2005) stated that certain geographic methods are more or less suited to feminist approaches and that there was uncertainty and debate over even whether a method *could* be framed as feminist. Despite this, Sharp (2005, 306) claimed: 'it is the feminist turn towards GIS [Geographic Information Sciences] that has perhaps presented the most significant change in methodological approach to gender issues in the last few years.' The concern that Sharp was hinting at in the mid-2000s was in respect of the additive approach to incorporating feminist methods into ways of doing geographic research, rather than an integrative approach that recognizes the messiness, complexity and embodied qualities of research that draws on feminist epistemologies. Over a decade later, the debate has shifted, with research emerging on a range of digital contexts, including feminist geographies of new spatial media (e.g. Leszczynski and Elwood 2015), feminist geographies of social media (de Jong 2015; McLean and Maalsen 2013; McLean, Maalsen and Grech 2016) and feminist geographies of digital work (Richardson 2018). By focusing on digital spaces, this body of research is redefining 'the field' and is expanding the scope of feminist geographic methods.

How does gender matter in the digital? Leszczynski and Elwood (2015, 12) frame gender as relevant to new spatial media spaces through: 'i) new practices of data creation and curation; ii) affordances of new technologies; and iii) new digital spatial mediations of everyday life.' We expand these observations of how gender matters in the digital beyond new spatial media to other digital spaces, such as social media and online activism. New modes of data creation and curation that foreground gender relations in and of the digital include collaborative approaches with feminist organizations, as demonstrated by McLean, Maalsen and Grech (2016). 'Destroy the Joint' (DTJ), an Australian online feminist group that counters sexism and misogyny, worked with McLean, Maalsen and Grech (2016) to design and implement a survey and participatory mapping activity to obtain a snapshot of its supporters' interests and goals. This research showed that the digital participants were evenly spread across Australia, had diverse interests in feminist action and pursued intersectional issues. Further, research participants valued engaging in online activism. There was no sense of this form of feminist intervention as existing in isolation to other more traditional modes of activism – the offline and online actions were constructed as intertwined. In another digital geographic context, collaborative mapping that disrupts heteronormativity was the subject of important research by Ferreira and Salvador (2015), who found that digital spaces have the potential to allow lesbian self-expression in the co-production of gender, sex and space.

The extent to which material and discursive changes are emerging from digital interventions is of concern to participants and organizers alike. What is interesting is that the work that goes into making these spaces function – the behind-the-scenes moderation and direction of micro-campaigns – is driven by emotion and affect (Gleeson 2016). From interviews with the people leading DTJ, Gleeson (2016, 82) found that:

> a number identified that they did it for the 'love' of the job (Irene), that they were 'passionate' about the cause (Pam), and were 'really offended' (Hannah) or 'pissed off' by media representations of women (Gina).

The methods, therefore, that allow for exploration of how emotion, affect and gender intertwine to produce the digital are broad. They include participatory mapping, surveys, content analysis of online spaces, discourse analysis and interviews. In another interesting contribution to the growing body of research that focuses feminist concerns on digital analyses, Drüeke and Zobl (2016) describe their mixed methods approach to researching the hashtag *#aufschrei* (German for 'outcry'), which brought together stories of the everyday sexism and harassment that German women experience. Drüeke and Zobl (2016) use quantitative methods to gather the tweets following *#aufschrei* and then qualitative content analysis to examine the blogs analysing the moment.

The digital turn in geographic research involves researchers taking account of the broadening and deepening of digital engagement in everyday lives, as well as challenging the structural dynamics that may reproduce inequalities and intensify corporatization and commodification processes. For instance, uneven geographies of representation and participation in the digital are commonplace (Graham, De Sabbata and Zook 2015), as the Global North dominates digital information technologies, both in terms of making and of being the subject of most digital data. Further, in Google searches, the Global South is more frequently represented *by* the North than self-represented: this form of digital hegemony skews the digital data and could potentially misrepresent the Global South (Ballatore et al. 2017). Problematic gender relations can be similarly (re)produced in the digital, as Jarrett (2016, 2018) has established and as we will discuss later in this chapter.

Queering digital spaces forms an important part of geographers' engagement with the digital. For instance, Gieseking (2017a) argues that queer digital spaces can be better conceptualized as messy spaces that are filling, or producing, gaps in face-to-face meeting spaces for queer people. In other work, Gieseking (2017b) offers a critical perspective on GIS technological structures and argues that we need 'good enough' software that is open and democratic, challenging corporate control of the digital on multiple levels. Other research on digital spaces demonstrates that heteronormative gender binaries are sometimes exaggerated due to demands for self-categorization, which lead to hyper-masculine and hyper-feminine expressions of identity (Maalsen and McLean 2015). The particular forms of gender and sexuality expression that are made possible in online spaces are contingent on the structures and agency of those producing and engaging with them. Methodologies need to respond to and to understand these contingencies and hence adapt and bend to the context under examination. These situated approaches call on critical reflexivity, something that feminist geographers have been advocating in a range of research areas.

Recognition of partial knowledges, reflexive and situated engagement with the digital

Feminist geographers working with/in the digital and elsewhere have long recognized that knowledges are partial (after Haraway 1988, 1991), and they have employed novel combinations of methods to remain sensitive to this understanding of knowledge production (see, for instance, Nightingale 2003, 2016). While geographers commonly recognize that the way in which methods are employed is imperative, feminist geographers have taken this understanding further by encouraging a deeper engagement with the way in which methods are selected and employed. Indeed, by first acknowledging that knowledges are partial and situated – that is, context- and theory-dependent (Nightingale 2003) – feminist geographers' methodological toolkit has grown to encompass well-used geographic methods in novel combinations. These novel offerings produce reflexive and situated engagements with research subjects.

For instance, Nightingale (2003, 2016) suggests employing mixed-methods approaches to feminist geographic research in order to draw attention to what is missing from the data collection process. For Nightingale (2003, 80), mixing methods allows for what she terms the 'silences and incompatibilities … [to] become evident when data sets produced by diverse methodologies are brought together'. It is the bringing together of partial knowledges, produced through different methods inspired by different epistemological traditions, that generates more nuanced and, ultimately, situated engagements for feminist geographers. Although Nightingale is not referring to research undertaken in or on the digital, her notion of bringing methods into proximity has the potential to be translated into the digital and, as the aforementioned feminist mixed-methods digital research demonstrates, has already been drawn upon.

Engaging with the digital can present practical and theoretical challenges for researchers, as digital technologies are in constant change and the digital is most often viewed as a site from which to take data (Morrow, Hawkins and Kern 2015, 526), rather than to engage with/ through. Rather than viewing the internet and digital technologies as sites from which to gather information, feminist geographers have found innovative ways to engage with the digital more meaningfully. For example, feminist geographers have explored the digital through theorizing posthuman agency within digitally mediated spaces of the city (Rose 2017) or through unpacking the design, production and application (and hence affect) of digitally produced images as part of urban design processes (Degan, Melhuish and Rose 2017). In these and other instances, feminist geographies can offer a distinctive epistemological understanding of the potentials that the digital offers. Other feminist geographers have encountered the digital by participating in the production of code/ing (Maaslen and Perng 2018), in online feminist activism (McLean and Maalsen 2013) and through conducting research about specific digital technologies (see, for example, Adams-Hutcheson and Longhurst 2017; Kwan 2002; Longhurst 2013, 2016, 2017).

Researching the digital can involve face-to-face interviews as well as observations of our interaction with digital technologies. For example, Longhurst's (2013, 2016) work on mothering and digital technologies, most notably Skype, illuminates how the social relationships between people are enabled and enacted through the digital. Longhurst (2013) conducted semi-structured interviews with, and observed, women who mothered via Skype and utilized digital media to support their mothering practices. By observing and engaging with the women as they interact with digital technologies and asking them about their use of digital technologies, feminist geographers are able to reveal interesting insights into the ways in which social relationships are maintained through the digital. For Longhurst (2013), it is the combination of video and audio, as well as a Skype user's sense of comfort with the technology, that allows for more meaningful engagements to take place across time and space. Recently, Longhurst (2017) has extended her research methods to include content analysis of the digital spaces themselves, including online spaces, to examine digital mourning practices.

Feminist geographers are pursuing methods that directly engage with digital spaces, for example moving the interview itself into the digital. Adams-Hutcheson and Longhurst (2017) suggest that interviews can be carried out using digital technologies such as Skype with audio and video capabilities. Adams-Hutcheson and Longhurst (2017, 148) describe interview methods carried out via digital technologies as 'stretch[ing] [fieldwork] in new directions across time and space'. A feminist geographic perspective here is concerned with the affective and emotional dimensions of the interview process rather than the medium used to conduct the interview. An acknowledgement of context, rather than scrutiny of the format in which an interview is conducted, is important here. The flexibility of being able to conduct interviews by online video and/or audio also works with the multiple demands of diverse research participants, who may have intense and layered care and work roles.

Feminist geographers frequently place emphasis on the positionality of the researcher, especially with regard to how one's positionality and politics can impact on fieldwork (Parker 2017). With respect to the digital, one's positionality or very identity is caught up in the politics of digital spaces, where it is easy to obscure or even change one's identity. Indeed, Parker (2017), drawing on Ahmed's notion of the feminist researcher as a killjoy (see Ahmed 2010), suggests that feminist geographers should play with how we choose to reveal (or not reveal) our feminist agenda when engaging with research participants. For Parker (2017), these situated engagements present both challenges and opportunities for feminist geographers with regard to positioning themselves as *feminist* researchers.

Following Haraway (1988), feminist geographers working with and on the digital appreciate that knowledges or perspectives are intrinsically partial and often acknowledge the limitations of their research. Rather than being bogged down in the realities of doing research, a feminist geographic perspective of partial knowledges offers ways of better revealing and navigating these realities through careful method selection and analysis. Indeed, methods are central to knowledge production, and multiple methods are often used to reveal 'partial perspectives, each demonstrating the strengths and limitations of the other', rather than a 'totalising account' (Sharp 2005, 305) of a research topic. When carried out simultaneously, different methods can reveal different, yet not irreconcilable, insights into the same research problem/topic.

Critical perspectives on emotion and affect in digital labour

Feminists within geography and beyond problematize 'labour' and provoke researchers to consider emotion and affect as work. Increasingly, this applies to digital labour. Central to these arguments is questioning sites of production and reframing what is considered to be labour. While seeing value in Marx's theories on labour and production, feminist scholarship critiques the absence of 'feminized' labour and domestic spaces as productive. Feminist thought challenges the naturalizing and subsequent devaluing of 'women's work', including predominantly unpaid care giving and domestic labour, and argues instead that these activities are critical mechanisms of capitalism. These feminist critiques inform debates on social reproduction and emotional labour within the digital. As both Jarrett (2014) and Staples (2007) eloquently observe, women's work and feminist thought are 'spectral presences' that haunt our engagement with capitalism and the affective and emotional nature of digital labour.

A useful place to start on digital labour is Gibson-Graham's (1996) feminist critique of capitalism and, while not directly situated in the digital, it provides a solid base for this analysis. The authors use postmodern Marxist and poststructuralist feminism to critique traditional Marxist frameworks. In particular, they see capitalism not as the homogenous, dominant and totalizing system that it is often portrayed but as one that is heterogenous. Heterogeneity within subjectivities, participations and activities eschews the structures of traditional Marxist analysis, for example the working class and big 'C' Capitalism, to show that subjectivities of production and sites of labour are varied. This opens up novel lines of enquiry to see what constitutes labour, where this takes place and what acts of resistance can emerge – the performance of alternative economies that Gibson-Graham identify in their work from the mid-2000s (2006). We suggest that feminist geographers could bring this critique to the exchanges of emotion and affect in diverse sites of exchange in the digital. Further, we suggest that the question posed by Gibson-Graham's (2011) exploratory piece, 'A Feminist Politics for the Anthropocene', on the possibly generative aspects of digital tools, is crucial to how theorizing gender and doing feminism in the digital plays out: 'Might we belong differently now that the vibrant materiality of the Internet and open source software allow for new interconnections in a potentially democratized world?' Vibrant materiality is an important component of feminist work on digital spaces, whether through considering new feminist geospatial imaging, through ways of being feminist online or through collaborating as feminist praxis. An example of these new interconnections comes from interesting empirical research by Strengers and Nicholls (2018). This has found that the new interconnections that 'smart homes' facilitate, including for work and leisure, seem to be gendered in that more men are taking up these tasks than women.

Like other subdisciplines, feminist geography is changing and shifting, and it necessarily draws upon other disciplines to inform its approach. In the context of the digital, media and cultural studies offer much to those geographers interested in feminist approaches. Geographers can both learn from this and can contribute much to the feminist agendas of other disciplines in return. For example, scholars such as Fortunati (2011), Jarrett (2016, 2018), Duffy (2015), Duffy and Hund (2015) and others have offered much to the debate on immaterial labour. Citing the influence of the Autonomist Marxists on framing the language of exploitation and work in digital industries, digital labour is seen as something that producers within the industry do, for example coders, developers and designers, but also something that users of digital media themselves participate in (Jarrett 2018). This argument is based in the idea that online activity – social media interactions, search history, purchasing behaviour – produces data that are converted to value for digital companies, Facebook and Google being key examples (Jarrett 2018). Affective or emotional energy is expended in these online activities. The value that this creates – as data for companies and a host of third parties – frames this activity as labour or immaterial labour (Jarrett 2018). Duffy illustrates how digital culture industries require 'aspirational labour' (2015). The labour of entrepreneurial social media platforms, such as blogs, is often framed in terms of love and romance for your work – 'getting paid to do what you love' – yet, as Duffy argues, the rewards are experienced unevenly and obscure class, gender and race relations, which continue to produce inequality (Duffy 2015; Duffy and Hund 2015).

We caution, however, that the immaterial labour associated with the digital is not new. Importantly, and using Magdalene Laundries as an example, Jarrett (2018) shows that immaterial labour's role in capitalism is not specific to the digital age, an important qualification for those interested in labour in the digital. That there is an increase in work that occurs online, however, leads Richardson (2018) to describe such workplaces as emergent and ambivalent – they can produce both affirmative and negative outcomes for participants and are inherently neither good nor bad. To understand how they emerge, argues Richardson, requires understanding through intimacy based in anti-essentialist (challenging the idea that the family unit and household are not sites of 'productive' labour, and arguing that the often-feminized labour in these spaces is indeed productive) and anti-normative (challenging social norms around work) politics. 'Making like a feminist' and applying intimacy as an analytical tool enables researchers to understand better the way that technology can extend and intensify the workplace and how technology mediates 'post-work spaces', combinations of 'bodies and machines at work' and the 'feeling' of work. These can all have benefits and detractions, such as flexible working hours and spaces, increased precarity and a blurring of lines between work and domestic life.

Other geographers look at the labour involved in producing the digital, focusing on cultures that support female engagement with or collective support through the digital. For example, Maalsen and Perng (2016, 2018) bring to light the crafting of community and programming skills through the all-female coding collective, PyLadies. Similarly, McLean and Maalsen (2013, 2016) show how the digital can mobilize collective resistance and activism. And it would be remiss not to highlight what feminist geography brings to understanding the work that we do in the field – the labour of our fieldwork and the relationships that this entails.

Collaborating with both other researchers and the groups within which the research is based is one way of doing the work of feminist geography and challenging the authority of the sole author (Monk et al. 2003; Sharp 2003, 2005). Doing so is not without additional emotional labour, however. Collaborations require affective input into negotiating outcomes and expectations within the academy and among co-researchers. McLean, Maalsen and Grech (2016) draw upon their research within digital spaces to illustrate the challenges and benefits of studies that 'give back' to those being researched. Research within these more-than-real spaces required letting go of expectations on research engagement, as McLean, Maalsen and Grech (2016) encountered varying intensities of engagement and paradoxes inherent to digital spaces, nonetheless found digital spaces and platforms to be rewarding places in which to work.

Conclusion: developing innovative ways of encountering the digital

Digital spaces provide new challenges and opportunities for geographers, and are also productive sites for research. This is illustrated in the development of feminist debates on the digital over the last decade. The attention that feminist approaches pay to situated and partial knowledges and the importance that they/we place on reflexive and collaborative research are well suited to investigating the digital. Here we have outlined some existing approaches that feminism brings to digital geographies: a focus on affect and emotion; embodied grounded accounts; destabilizing binaries of the offline/online, local/global; and highlighting the uneven geographies of digital representation and participation. These are necessary investigations, as the digital has reframed the field and the complexities of gendered, emotional and affective relationships within it. The approaches canvassed here are not prescriptive. Instead, we hope to inspire geographers and others to continue to experiment, to mix methods and to work with the digital in innovative ways.

Key readings

Gibson-Graham, J.K. 2011. "A Feminist Project of Belonging for the Anthropocene." *Gender, Place and Culture* 18 (1): 1–21.

McLean, J., S. Maalsen, and A. Grech. 2016. "Learning about Feminism in Digital and Paradoxical Spaces: Online Methodologies and Participatory Mapping." *Australian Geographer* 47 (2): 157–177.

Nightingale, A. 2016. "Adaptive Scholarship and Situated Knowledges? Hybrid Methodologies and Plural Epistemologies in Climate Change Adaptation Research." *Area* 48: 41–47.

References

Adams-Hutcheson, G., and R. Longhurst. 2017. "'At Least in Person There Would Have Been a Cup of Tea': Interviewing via Skype." *Area* 49 (2): 148–155.

Ahmed, S. 2010. "Killing Joy: Feminism and the History of Happiness." *Signs*, 35 (3): 571–594.

Ash, J., R. Kitchin, and A. Leszczynski. 2017. "Digital Turn, Digital Geographies?" *Progress in Human Geography*. doi:10.1177/0309132516664800.

Ballatore, A., M. Graham, and S. Sen. 2017. "Digital Hegemonies: The Localness of Search Engine Results." *Annals of the American Association of Geographers* 107 (5): 1194–1215. doi:10.1080/24694452.2017.1308240.

Degan, M., C. Melhuish, and G. Rose. 2017. "Producing Place Atmospheres Digitally: Architecture, Digital Visualisation Practices and the Experience Economy." *Journal of Consumer Culture* 17 (1): 3–24.

de Jong, A. 2015. "Using Facebook as a Space for Storytelling in Geographical Research." *Geographical Research* 53 (2): 211–223.

Duffy, B. 2015. "The romance of work: Gender and Aspirational Labour in the Digital Culture Industries." *International Journal of Cultural Studies* 19 (4): 441–457.

Duffy, B., and E. Hund. 2015. "'Having it All' on Social Media: Entrepreneurial Femininity and Self-Branding Among Fashion Bloggers." *International Journal of Cultural Studies*. doi: 10.1177/2056305115604337.

Drüeke, R., and E. Zobl. 2016. "Online Feminist Protest Against Sexism: The German-Language hashtag# aufschrei." *Feminist Media Studies* 16 (1): 35–54.

Ferreira, E., and R. Salvador. 2015. "Lesbian Collaborative Web Mapping: Disrupting Heteronormativity in Portugal." *Gender, Place & Culture* 22 (7): 954–970.

Fortunati, L. 2011. "ICTs and Immaterial Labor from a Feminist Perspective." *Journal of Communication Inquiry* 35 (4): 426–432.
Gibson-Graham, J.K. 1996. *The End of Capitalism (As We Knew It): A Feminist Critique of Political Economy*. Oxford: Blackwell.
Gibson-Graham, J.K. 2006. *A Postcapitalist Politics*. Minneapolis: University of Minnesota Press.
Gibson-Graham, J.K. 2011. "A Feminist Project of Belonging for the Anthropocene." *Gender, Place and Culture* 18 (1): 1–21.
Gieseking, J.J. 2017a. "Messing with the Attractiveness Algorithm: A Response to Queering Code/space." *Gender, Place & Culture* 24 (11): 1659–1665.
Gieseking, J.J. 2017b. "Operating Anew: Queering GIS with Good Enough Software." *Canadian Geographer / Le Géographe canadien* 62 (1): 55–66. doi:10.1111/cag.12397.
Gleeson, J. 2016. "'(Not) Working 9–5': The Consequences of Contemporary Australian-based Online Feminist Campaigns as Digital Labour." *Media International Australia incorporating Culture and Policy* 161 (1): 77–85.
Graham, M., S. De Sabbata, and M.A. Zook. 2015. "Towards a Study of Information Geographies: (Im) mutable Augmentations and a Mapping of the Geographies of Information." *Geo: Geography and Environment* 2 (1): 88–105. doi: 10.1002/geo2.8.
Haraway, D. 1988. "Situated knowledges: The Science Question in Feminism and the Privilege of Partial Perspective." *Feminist Studies* 14 (3): 575–599.
Haraway, D. 1991. *Simians, Cyborgs, and Women: The Reinvention of Nature*. New York: Routledge.
Jarrett, K. 2014. "The Relevance of 'Women's Work': Social Reproduction and Immaterial Labor in Digital Media." *Television & New Media* 15 (1): 14–29.
Jarrett, K. 2016. *Feminism, Labour and Digital Media: The Digital Housewife*. New York: Routledge.
Jarrett, K. 2018. "Laundering Women's History: A Feminist Critique of the Social Factory." *First Monday* 23 (3–5).
Kwan, M.P. 2002. "Feminist Visualization: Re-envisioning GIS as a Method in Feminist Geographic Research." *Annals of the Association of American Geographers* 92 (4): 645–661.
Leszczynski, A., and S. Elwood. 2015. "Feminist Geographies of New Spatial Media/Les géographies féministes des nouveaux médias spatiaux." *Canadian Geographer* 59 (1): 12–28. doi:10.1111/cag.12093.
Longhurst, R. 2013. "Using Skype to Mother: Bodies, Emotions, Visuality, and Screens." *Environment and Planning D: Society and Space* 31: 664–679.
Longhurst, R. 2016. "Mothering, Digital Media and Emotional Geographies in Hamilton, Aotearoa New Zealand." *Social & Cultural Geography* 17 (1): 120–139.
Longhurst, R. 2017. "Mourning: Skype, Screens, Bodies and Touch." Paper presented at the American Association of Geographers" AGM, Boston, April. Available at: http://app.core-apps.com/aagam2017/event/43ead77870c1a1a3b56ef2e500e8a03a? (accessed 14 March 2018).
Maalsen, S., and J. McLean. 2015. "Digging up Unearthed Down-Under: A Hybrid Geography of a Musical Space That Essentialises Gender and Place." *Gender, Place Culture* 23 (3): 1–17. doi:10.1080/0966369X.2015.1013443.
Maalsen, S., and S. Perng. 2016. "Encountering the City at Hackathons." In: Code *and the City*, edited by R. Kitchin and S.Y. Perng, 190–199. London: Routledge.
Maalsen, S., and S. Perng. 2018. "Crafting Code: Gender, Coding and Spatial Hybridity in the Events of Dublin Pyladies." In: *The Craft Economy*, edited by S. Luckman and N. Thomas, 223–232. London: Bloomsbury.
McLean, J., and S. Maalsen. 2013. "Destroying the Joint and Dying of Shame? A Geography of Revitalised Feminism in Social Media and Beyond." *Geographical Research* 51 (3): 243–256.
McLean, J., S. Maalsen, and A. Grech. 2016. "Learning about Feminism in Digital and Paradoxical Spaces: Online Methodologies and Participatory Mapping." *Australian Geographer* 47 (2): 157–177.
Monk, J., P. Manning, and C. Denman. 2003. "Working Together: Feminist Perspectives on Collaborative Research and Action." *ACME: An International E-Journal for Critical Geographies* 2 (1): 91–106.
Morrow, O., R. Hawkins, and L. Kern. 2015. "Feminist Research in Online Spaces." *Gender, Place & Culture* 22 (4): 526–543.
Moss, P. 2003. *Feminist Geography in Practice: Research and Methods*. Malden, MA: Blackwell.
Nightingale, A. 2003. "A Feminist in the Forest: Situated Knowledges and Mixing Methods in Natural Resource Management." *ACME: An International E-Journal for Critical Geographies* 2 (1): 77–90.
Nightingale, A. 2016. "Adaptive Scholarship and Situated Knowledges? Hybrid Methodologies and Plural Epistemologies in Climate Change Adaptation Research." *Area* 48: 41–47.

Parker, B. 2017. "The Feminist Geography as Killjoy: Excavating Gendered Urban Power Relations." *Professional Geographer* 69 (2): 321–328.
Richardson, L. 2018. "Feminist Geographies of Digital Work." *Progress in Human Geography.* Online. doi:10.1177/0309132516677177.
Rose, G. 2017. "Posthuman Agency in the Digitally Mediated City: Exteriorization, Individuation, Reinvention." *Annals of the American Association of Geographers* 107 (4): 779–793.
Sharp, J. 2003. "Feminist and Postcolonial Engagements." In: *A Companion to Political Geography,* edited by J. Agnew, K. Mitchell, and G. Toal, 59–74. Oxford: Blackwell.
Sharp, J. 2005. "Geography and Gender: Feminist Methodologies in Collaboration and in the Field." *Progress in Human Geography* 29: 304–309.
Staples, D. 2007. "Women's Work and the Ambivalent Gift of Entropy." In: *Affective Turn: Theorizing the Social,* edited by P.T. Clough and D.D. Halley, 175–194. Durham, NC: Duke University Press.
Strengers, Y., and L. Nicholls. 2018. "Aesthetic Pleasures and Gendered Tech-work in the 21st-century Smart Home." *Media International Australia* 166 (1): 70–80. doi: 10.1177/1329878X17737661.

44

DRONE QUEEN OF THE HOMELAND

The gendered geopolitics of television drama in the age of media convergence

Julie Cupples and Kevin Glynn

Introduction

Of the many feminist disruptions and contributions to the discipline of geography and its cognate areas of scholarly inquiry, some of the most significant have come from feminism's insights into the gendered dimensions of contemporary geopolitics. Geopolitics, in both its formal political and academic forms, has traditionally embodied male and masculinist concerns that emerge from and focus on 'statesmanship', conflict, war and the control of borders and territories. Feminist geopolitical scholarship has, however, drawn our attention to the ways in which foreign and domestic policy, security and international relations are discursively and materially gendered, including in their embeddedness in and responsiveness to various gendered dynamics of everyday life. Feminist scholars have emphasized the inseparability of top-down geopolitical decision-making (by government leaders, military commanders and supranational bodies, such as the UN and NATO) from the micropolitics and spaces of everyday life, where people of all genders form relationships, engage in paid and unpaid work, reproduce their households and immediate social relations, engage in practices of story- and truth-telling, encounter and resist sexual violence, consume products and services and participate in various kinds of political activism.[1]

Existing feminist scholarship in critical geopolitics works across a range of texts, arenas and sites, including the realm of media and popular culture, which has coalesced into a subfield known as popular geopolitics. Post-9/11 television offers one productive route to insights into the entanglements of geopolitics and gender. Numerous scholars of media and cultural studies and American studies have analysed post-9/11 television in a range of ways (see, for example, Dixon 2004; Martin and Petro 2006; Spigel 2004; Takacs 2012; Tryon 2016), but geographers, even those associated with popular and critical geopolitics, have been slower to engage with television (Glynn and Cupples 2015) – though some have taken up our encouragement to do so (see, for example, Saunders 2017, 2018). Nevertheless, television's neglect by many geographers is surprising, as the medium continues to be widely consumed and influential; is as much part of the process of digitalization and media convergence as other media forms commonly studied by geographers; and is undergoing cultural and political transformations that are of geographical significance. Furthermore, television's engagement with contemporary geopolitical issues spans a range of generic territories, from news and current affairs to political dramas, sitcoms, cartoons, talk shows, satirical 'fake news' programmes, dramedies, and so on. Television remains as much a key site for popular engagement with the geopolitics of the post-9/11 world as it was during and immediately after the Cold War (cf. Curtin 1995; Doherty 2003; Kackman 2005; Schwoch 2009). However, in the current conjuncture, 'post-network' television (Lotz 2007; Spigel and Olsson 2004) has ushered in new formations of interactive popular geopolitics that engage with questions of gender in interesting ways. The remainder of this chapter will discuss these developments and conclude with an analysis of Showtime's *Homeland* (broadcast 2011–present), a series that is a fixture in the new televisual geopolitics that explores gendered dimensions of the 'War on Terror'.

Televisual transformations

Post-9/11 interactions between entertainment television and geopolitics have occurred in parallel with, and are facilitated by, a range of contemporary transformations within the practices of television production, distribution and reception. Dramatic changes in the contemporary television and wider media environment are closely associated with, but not reducible to, technological innovations and processes of media convergence (cf. Jenkins 2008). Hence, there has been an expansion in various forms of user-driven interactivity and transmedia practices, facilitated in part by social media and the spread of devices such as tablets and smartphones. New modes of media consumption, modification and recirculation have enabled audiences to appropriate and rework content produced by media companies and to 'speak back' to these massive global megaconglomerates. In television's network and 'multi-channel transition' eras, spanning from roughly the 1950s to the start of the new millennium (Lotz 2007, 8), many viewers watched television shows at the time of broadcast and discussed them in person with family, friends and colleagues. These audience reception practices are still widespread, but viewing is now increasingly unmoored from broadcasting schedules and accompanied or even replaced by online conversations. The 'channel surfing' of television's early age of abundance has given way to the binge-watching associated with DVD box sets, smart televisions and digital streaming, for instance. Today, many viewers are likely to have watched a series that their family, friends and colleagues have not (yet) seen, but are able to listen to podcasts about it and engage in discussions over it with distant friends and family members, or with people whom they have never met, on Facebook, Twitter, Tumblr, series-related wikis, internet forums, digital newspaper comment sections and online dedicated fan forums such as Digital Spy, TV Time, Overclockers and the once-brilliant Television without Pity. Furthermore, the mass audiences of the old media environment have given way to a post-Fordist logic of extreme fragmentation of audiences and revenue streams that has, in turn, led to an expansion of nichification, cultification and greater openness toward experimentation within the television industries (see, for example, Gwenllian-Jones and Pearson 2004; Mittell 2006; Reeves et al. 1996). While audiences of commercially successful series are now much smaller than they once were, they are also likely to be more loyal, to stick with their series over multiple seasons and to engage with programming on multiple media platforms. To promote audience loyalty and even cultish devotion, producers have intensified the complexity of narrative forms and serialization practices (see, for example, Clarke 2013; Dunleavy 2018; Mittell 2015a).

In the multiplatform television era, then, narrative forms and reception practices that scholars have long associated with cult television and media fandom have expanded and become increasingly mainstream. While television has long trafficked in hybrid narratives that combine and experiment with various forms and permutations of serialization, in the age of media convergence it has departed farther than ever from both linearity and the episodic structures that once defined its most recognizable conventions. Like the cult television that forms one node in the prehistory of television's new modes of complex seriality, the latter often prominently foreground moral ambiguity, transgression, supernaturalism, characters who defy normalization, and multiple, simultaneously unfolding narrative arcs over numerous series and entire seasons (Dunleavy 2018). This complex seriality also often plays with alternative epistemologies that constitute the otherworldly televisual equivalent of magic realism, which Glynn (2003) has characterized as 'popular subjunctivity'. Television's complex seriality has developed alongside new practices of 'forensic fandom', whereby 'dedicated fans embrace' narrative complexities 'and seek to decode a programme's mysteries, analyze its story arc and make predictions' (Mittell 2015b). The series *Lost* (2004–2010), for instance, which attracted a large, multinational audience that stretched across many parts the world, is often held to exemplify the complex seriality of the new multimedia environment. *Lost* engaged amply, albeit often in displaced and allegorical ways, with myriad aspects of post-9/11 geopolitics and gave rise to a sprawling online repository of diegetic knowledge and fan speculation known as Lostpedia. In the words of one blogging devotee, Lostpedia:

> became an incredible resource for fans to keep track of all the twists and storylines within the show as they obsessively documented everything from the number of crash survivors still on the Island, to more bizarre stuff like how many times a season Hurley [a favorite character of fans] said 'dude'.
>
> *Baker1000 2015*

Through their narrative and fan engagement practices, *Lost* and its viewers, who vigorously and extensively discussed all aspects of the programme online, participated in constituting and reshaping the discursive and imaginary terrains of post-9/11 geopolitics.

One important feminist critique of the discourses and practices associated with the age of media convergence concerns the gendered distinctions that formed around what has often been called television's 'third golden age' of the past 15 or 20 years. During this period, the cutting-edge technologies and narrative forms associated with digitalization and media convergence have been frequently touted as 'cinematic', which has thus become a key term in the contemporary discourses that reproduce cultural hierarchies, and therefore inequalities, by differentiating between television's texts and audiences who are worthy and those who are less so (cf. Newman and Levine 2012). As Newman and Levine (2012, 10) write: 'that television has been classified as feminine, and therefore as a less worthy, significant, and serious medium, has been a fact of its history.' Television has, perhaps more than any other medium, been consistently identified with cultural decline, 'massness' and feminization. Prominent discourses of television's 'third golden age' thus reinscribe gendered distinctions and hierarchies within domestic spaces of media consumption by mobilizing the emergence of new televisual technologies and narrative forms that ostensibly transcend (mere) television-ness to achieve cinematic heights and dimensions.

There are parallel hierarchicizing discourses that differentiate the 'serious' and worthy from the insignificant and trivial in the realm of geopolitical analysis. Traditional geopolitics has historically emphasized statecraft, the realm of the traditional public sphere and policymaking, and the terrains of war and conflict – in short, the core concerns of 'statesmanship' and of the 'hard news', in masculinist journalistic parlance. As regards televisual taste and 'quality programming,' the often-derisively used terms 'soap opera' and melodrama neatly capture the confluence of feminization and trivialization widely invoked to derogate programming that falls outside the category of 'serious' (masculine-oriented) fare of one sort or another. In this way, *gender* and *genre* form intertwined and co-constitutive systems of categorization and distinction in relation to television production, circulation and consumption. Series such as *Homeland* are of interest in part for their deconstruction of such categories and distinctions. As Diane Negra and Jorie Lagerway (2015, 131) argue, *Homeland* is 'a consummate example of melodramatic political discourse [...] that does not seek to and never can quell the anxieties of the present moment with which it so forcefully engages.'

Post-9/11 television and gender

Since the 9/11 attacks, television has engaged amply with the gendered consequences of the 'War on Terror'. Fictional television's activities in these regards have been particularly important in light of the 'absence, erasure and invisibility' (Stubblefield 2014, 4) characteristic of the approach by conventional news and current affairs programming in the face of the Bush administration's extreme attempts at image and information management (Glynn 2009; Mirzoeff 2005; Takacs 2010). During the Bush years, for instance, journalists were banned from accessing, photographing or filming the arrival of coffins of fallen soldiers at any US military base (Vann 2003). Paradoxically, Thomas Stubblefield (2014, 4) argues, this absence, erasure and invisibility was a product of the logic of spectacularization that engulfed the 9/11 event itself and engulfed, by implication, many events that followed in its wake, including the spectacle of the Iraq War:

> Following the logic of implosion rather than explosion, the World Trade Center withheld its contents from view as it fell; its stories 'pancaked' on top of one another rather than turning themselves inside out. With the vast majority of the dead dying behind the curtain wall of the towers' facades, 'the most photographed disaster in history' failed to yield a single noteworthy image of carnage.

Nevertheless, although sometimes succumbing, television drama found ways to circumvent spectacularization and invisibility and to interrogate dominant post-9/11 geopolitical logics and images. It often did this by exploring how geopolitical dynamics reshape intimate relationships and position men, women, boys, girls, brothers and sisters differently in relation to significant events. A 2006 episode of long-running hospital drama *ER*, for example, plumbs the emotional depths surrounding the loss of a major character, Dr Michael Gallant, when a roadside bomb explodes near him while he is offering medical assistance in the war zone near Mosul (Season 12, Episode 22 'The Gallant Hero and the Tragic Victor'). Gallant's young widowed wife, Dr Neela Rasgotra, then finds herself in the same position as many other post-9/11 US military families, who are left emptily to mourn fallen loved ones with only a funereal military spectacle offered to them in lieu of adequate explanation, compensation, justification or political accountability (Season 12, Episode 23 'Twenty-One Guns'). Consequently, Neela joins the anti-war movement (Season 13, Episode 23, 'The Honeymoon is Over'). Similarly, *Six Feet Under*, a series about a family-run funeral parlour in Los Angeles that is, in many respects, taboo-breaking (see Akass and McCabe 2005), began to connect with gendered dynamics of the 'War on Terror'. In a 2005 episode (Season 5, Episode 11, 'Static'), the daughter/younger sister of the family, Claire Fisher, who is high on drugs at the time, yells at the grieving sister of a deceased soldier who was euthanized in the hospital after losing all of his limbs in Iraq. The soldier's sister is driving an SUV sporting a 'Support Our Troops' bumper sticker, which provokes Claire into an enraged tirade that covers a range of political angles on the war in Iraq: 'Why don't you try driving something that doesn't consume quite so much gas for starters, if you're so fucking concerned?', 'Dozens of fucking Iraqis are dying every day. The whole world hates us for going in there in the first place!', 'They bring the wounded soldiers back at night, so the press can't even film it and nobody sees!' Viewer commentary in response to a clip of this scene on YouTube reveals the modes of civic engagement with the geopolitical that Claire's impassioned speech provoked:[2]

> Let's be honest, she made a good point. The Iraqi War was nothing but a bunch of bullshit, causing mass suffering and great hostility among the young generation of adults who have no way of stopping it. What have we done in Iraq besides given the middle east more motivation to oppose us? Oh and spending $3 trillion of tax.

> This episode affected me deeply at the time as a cousin of mine was serving in Iraq, and the fact that I was pissed about Bush being re-elected the year before.

The following comment, which was posted on the Internet Movie Database (IMDb) in response to the series' finale, captures the show's resonance with the gendered and familial realities of the Iraq War, including for those serving in Iraq:

> I've watched plenty of show finales in my lifetime. Many had great endings and some had decent ones. I recently finished watching every episode of all 5 seasons of this amazing show while deployed in Iraq. From start to finish, I was completely captivated by the characters as they went about their every day lives. After the credits commenced to roll in Season 5's Series Finale 'Everyone's Waiting.' I just sat there in my chair completely flabbergasted at 'Alan Ball's' incredible and yet honest ending. Of course, that is after I wiped the tears from my eyes and took an enormously deep and saddened 'SIGH.' I replayed that ending over five other times and still felt the same chills as though it was the first time. The Series finale has inspired me to regain my artistic composure, which I had temporarily lost for the last year and a half. It motivated me to call my wife and kids every day to tell them I missed and loved them so much. It very well has changed the way I look at life itself. I know it sounds kinda weird but it did. 'Everyone's waiting' is the best ending I've ever seen in any medium of entertainment.[3]

While *Six Feet Under* and *ER* were series that predated 9/11 but began to engage with its politics, *Brothers and Sisters* was launched in the aftermath of the 'War on Terror'. It explores the lives of the large and affluent Walker family of Pasadena, California, and emphasizes that, although post-9/11 military interventions may well be happening 'over there', the impacts of the 'War on Terror' nevertheless permeate family life at home in the US. In the Walker family, everyday dramas and arguments over such things as business and professional life, marital affairs and terminal illness are shot through with and shaped by the legacies and impacts of the 'War on Terror'. These include son Justin Walker's deployment to Iraq and subsequent battle with post-traumatic stress disorder; high-flying daughter Kitty Walker's professional and political entanglements with and passionate support for the Republican Party; and matriarch Nora Walker's staunch opposition to conservative politics in general and the Bush/Cheney administration in particular. These are a few of the television dramas that delve melodramatically into the complex modes of gendered and familial sense-making around sacrifice, security, nationhood and terrorist threats, and have invited substantial online (and presumably offline) audience engagement. While several have strong female protagonists and deal with some of impacts of the 'War on Terror' on family life, *Homeland* features a female CIA agent whose work involves putting the 'War on Terror' into practice.

Homeland

Homeland (broadcast on Showtime, 2011–), which at the time of this writing has completed seven seasons, is a series focused on US counter-terrorist strategy in the aftermath of 9/11 and, in particular, the work of the CIA. Its protagonist, Carrie Mathieson, begins the series as a CIA agent who struggles, unbeknownst to the agency, with bipolar disorder. Seasons 1 to 3 focused on her complicated entanglements with a former US marine turned Al-Qaeda sympathizer, Sergeant Nicholas Brody. Prior to his rescue and return to the US in 2011, Brody had been held hostage by Al-Qaeda for eight years, during which time he converted to Islam. He also lived for a time with Al-Qaeda commander Abu Nazir and taught English to Nazir's son, Issa, who was later killed by a US drone strike, an event that impacted profoundly on Brody. Carrie correctly suspects that Brody has turned against the US and plans a terrorist attack; she subsequently manages to recruit him to work for the CIA. Brody and Carrie then embark on a sexual relationship, as a consequence of which Carrie becomes pregnant and gives birth to Brody's child, Frannie, shortly after he is publicly executed in Iran for treason at the end of Season 3.

As Dunleavy (2018) notes, contemporary television narratives often feature morally complex and transgressive figures who invite awkward and uncomfortable forms of identification. Carrie does her best intelligence work when she is 'off her meds', at which times she experiences heightened states of both mental acuity and instability, and is capable of both making brilliant insights into cases that her colleagues and superiors have misunderstood and breaking open difficult and extraordinarily complex investigations. She is often sexually manipulative and sometimes sleeps with CIA assets and informants. She drinks a fair amount (we often see her pouring herself a very large glass of white wine to wash down her medication), and she routinely defies normative notions of 'good mothering' by, for instance, repeatedly abandoning Frannie to the care of Carrie's sister, Maggie. Indeed, in Season 4, Carrie takes a seductively challenging and dangerous CIA station chief posting in Afghanistan that precludes Frannie from joining her, even though a safer (but duller) chiefship in Istanbul was available to her, which would have allowed the two to live together. There is a chilling scene in Season 4 when Carrie almost drowns Frannie in the bath. At other times, Carrie's intelligence work actively puts Frannie in serious danger. In Season 6 (Episode 5, 'Casus Belli'), for example, Frannie is effectively taken hostage in her own home, which is surrounded by police snipers, after Carrie leaves her in the care of former CIA colleague and post-traumatic stress disorder sufferer Quinn, in an incident the *Baltimore Sun* characterizes as asking 'the unhinged professional assassin guy [to] watch your child' (Renner 2017). As Robyn Longhurst (2008, 117–29) argues, discourses of 'bad mothering' generate moral panics that target those seen to be 'lacking' in relation to norms of femininity and familiality. Carrie lacks a husband, lacks neurotypicality and lacks a normative 'maternal instinct' capable of regulating her threateningly excessive professional devotion and aspirations. She both lacks and exceeds maternal normativity, and thus poses a threat to the stable familial order that in popular geopolitical narratives and representations often stands metonymically for the nation.

Homeland's representational politics are highly complex, and many viewers object to its depictions of gender, race and religion, and some have carefully detailed the factual and geographical errors that they have identified in the series.[4] Carrie is a professional woman who works in the aggressive, dangerous and male-dominated world of anti-terrorism and national security. The fate of the nation, indeed even of Western democracy, often hangs on her ability to perform beyond competently. She sometimes puts her work ahead of her relationship with her child, and some suggest that *Homeland*'s depiction of Carrie's 'hysterical' mental instabilities dovetail too neatly with patriarchal discourses of women's madness (see, for example, Strauss 2014, who nevertheless recognizes, as we argue below, that such a reading is not easily sustained).

Kathleen McInnis (2012), who has worked in national security, believes that what she views as Carrie's blatant lack of professionalism ('making passes at the boss' and 'taking classified materials home and displaying them on her wall'), does a disservice to the women who work in this field. While Carrie resists domestication, she does so in the name of a racist and imperialist war. Thus, Carrie is an unstable figure who challenges both overtly feminist and anti-feminist readings; that she makes available a range of contradictory potential meanings is symptomatic of ongoing gender trouble that has been active for decades yet is roiled in particular ways in the post-9/11 geopolitical climate. As Bhattacharyya (2013, 378) argues, *Homeland*'s narrative suggests that:

> femininity and feminism – both female bodies and a discourse of women's rights – can become central elements of the project of securitization. What appears to be a development is the use of familiar tropes of women's unreason to serve as an alternative justification of irrational actions by the state. The femininity that serves the security state here is the unruly femininity of the hysteric.

As regards representations of race and the 'War on Terror', criticism of *Homeland* is widespread. The *Washington Post* claimed that *Homeland* was 'the most bigoted show on television' and accused the series of churning 'out Islamophobic stereotypes as if its writers were getting paid by the cliché' (Durkay 2014). *The Guardian* has asserted that *Homeland* offers nuance, as in its depiction of how Brody was 'turned', in part because of the murder of an innocent child by a US drone, yet also espouses 'a dangerous set of lies about terrorism, American omnipotence and the very nature of international politics' (Cohen 2013). Pakistani officials objected loudly to *Homeland*'s depiction of a devastating US drone attack on a wedding in Pakistan in an episode from Season 4 that we discuss in the next section. One diplomat complained to the *New York Post* that 'Islamabad is a quiet, picturesque city with beautiful mountains and lush greenery [...]. In *Homeland*, it's portrayed as a grimy hellhole and war zone where shootouts and bombs go off with dead bodies scattered around. Nothing is further from the truth' (Schram 2014). Such criticisms have been echoed in some of the academic media studies literature. James Castonguay (2015, 139), for example, writes that *Homeland*:

> successfully exploits post-9/11 insecurities, psychological trauma, and narrative complexity to produce 'quality' television propaganda for the Obama administration's 'overseas contingency operations' and its unprecedented domestic surveillance on the home front under the umbrella of an $80 billion US security state.

In Season 5, *Homeland*'s producers hired several street artists to embellish with Arabic graffiti a set showing a Syrian refugee camp. Unbeknownst to the producers, the graffiti artists inscribed '*Homeland* is racist' and other comments critical of the series on the walls of the set. After the scenes set in the camp were broadcast, one of the artists, Heba Amin, stated:

> We think the show perpetuates dangerous stereotypes by diminishing an entire region into a farce through the gross misrepresentations that feed into a narrative of political propaganda. It is clear they don't know the region they are attempting to represent. And yet, we suffer the consequences of such shallow and misguided representation.
>
> *Phipps 2015*

This act of creative subversion highlights that the modes of sensemaking associated with popular geopolitical texts are both dispersed across and contested at a variety of different sites and levels, including those of production and the industry, of the text and its multiple and often contradictory discourses, and of audiences and their multifarious reception practices. Indeed, commercially successful television dramas of the media convergence era tend to promote controversy and the formation of coalition audiences by staging 'confrontations between competing perspectives, discourses, and ideologies' (Glynn and Cupples 2015, 279) and eschew 'binary frameworks of understanding' (ibid., 275; see also Fiske 1987; Gledhill 1988). While it is thus possible to argue that *Homeland*'s storylines 'reinforce the need for increased homeland security and the use of force in counterterrorism operations' and open up alternative possibilities only to then shut them down (Castonguay 2015, 141), we are suspicious of forms of textual determinism that seek to read audience meaning-making practices from the media texts themselves. Indeed, one study of the *Homeland* audience's reception practices found that fans frequently read the text against the grain to produce a range of oppositional positions regarding CIA operations (Pears 2016).

In our view, *Homeland* continually evades black and white political binarisms. As Richard McHugh (2016, 163) writes, the story arc that deals with Brody's radicalization at the hands of Al-Qaeda reveals that the character 'was already radicalized as a US marine through the same government system that radicalizes the CIA agents as patriots'. The show makes clear that this patriotism involves the murder with drones of innocent children, which constitutes an 'act of violence that pushed Brody further into his alternate-radical self' (ibid., 164). In addition to its highly critical treatment of US drone warfare in Pakistan, *Homeland* also engages critically with Zionist attitudes toward Israel's West Bank occupation. It is difficult to argue that *Homeland* valourizes the CIA, whose agents the series routinely depicts as flawed, corrupt, duplicitous, predatory, racist and villainous.

The entanglement of and interplay between motherhood, mental health and counterterrorism function as key narrative drivers in *Homeland*. Some of the existing feminist literature on *Homeland* captures these textual complexities and underscores why the gendered dynamics of this series are of theoretical importance. Alex Bevan's (2015) discussion of *Homeland* emphasizes how gendered embodiment functions 'as a nexus point for geopolitical discourses' (151), as the pathological, 'reproductive and sexual currencies of Carrie's body are burdened' with the symbolic task of territorializing and representing the 'elusive terms of twenty-first-century warfare and geopolitical power' (148). The management of Carrie's bipolar disorder, for instance, volatilizes the distinction between being surveilled and doing surveilling in ways that raise but do not settle questions about the degree of surveillance needed in the post-9/11 world and the consequences of its intensification. *Homeland* reveals not only how women's bodies are central rather than incidental to the project of securitization, but also how the 'unruly hysteric is an unexpected complement to the securitized state' (Bhattacharyya 2013, 378). Audiences might thus come 'to inhabit the logic of securitization' (ibid., 382) through identification with Carrie, yet are just as likely to accept the text's implicit invitation to reflect critically on the security state and its gendered logics in ways that leave them more unsettled with regard to US foreign policy directions and actions. The text of *Homeland* permits a range of political positions with respect to both women's rights and the 'War on Terror', and so remains polysemous and multi-discursive in its constant engagements with the multifaceted relationships between gender and geopolitics. In the final section, we develop an analysis of one highly gendered geopolitical instrument, the drone, and its treatment in the show.

The drone queen

Unmanned Aerial Vehicles (UAVs), also known as drones, first entered the 'War on Terror' as a consequence of two US government surveillance and assassination programmes launched in 2002 to target Al-Qaeda sympathizers in Pakistan and Yemen – two countries *not* at war with the US. The use of drone attacks as an implement of US foreign policy was escalated dramatically during the Obama presidency and was mostly shrouded in secrecy until 2012, when the US government admitted the existence of the programme but refused to say how many terrorists and civilians had been killed by UAV strikes, which the government defended by invoking Orientalist, masculinist and medicalized discourses. Obama propounded the view that Pakistan's Federally Administered Tribal Areas (FATA), near the border with Afghanistan, constituted the world's most threatening and dangerous locale and must be brought to order; he thus established a space of exception amenable to missile penetration from above (Gregory 2017, 31) through the mobilization of 'new visibilities' that 'produce a special kind of intimacy that consistently privileges the view of the hunter-killer' (Gregory 2011, 193). As one drone operator put it, 'sometimes I feel like a God hurling thunderbolts from afar' (Gregory 2011, 192). US officials have also commended drones for the 'surgical precision' that enables them 'to eliminate the cancerous tumor called an al-Qaida terrorist while limiting damage to the tissue around it' (John Brennan, Obama's counterterrorism advisor, quoted in Crawford 2012). Nevertheless, US drone strikes have, of course, killed hundreds of civilians, including hundreds of children (Gregory 2017, 29), and provoked substantial protests and political opposition in areas subject to attack.

The CIA's use of drones for both surveillance and targeted killing has been a recurring feature of *Homeland*'s narratives. The first episode of Season 4 ('The Drone Queen') deals with faulty 'intel' that leads Carrie to order a late-night missile strike on a farmhouse in Pakistan, thus killing 40 innocent civilians attending a wedding there. Just before she turns in for the night, Carrie's colleagues surprise her with a birthday cake inscribed with 'The drone queen'. The scene of Carrie getting into bed for a restful night's sleep cuts to images of community members frantically searching through piles of rubble for the bodies of the dead and injured at what is left of the farmhouse. The next day, Carrie scrutinizes live images relayed by drones trawling the scene of the previous night's strike, and is clearly able to see row upon row of corpses, along with a young medical student, Aayan Ibrahim, who is looking through the bodies for his mother and sister, both of whom were killed in the attack. In a chilling moment that literalizes the reversal of surveillance practices that occurs when the watched becomes watcher, Aayan stares bitterly up at the encircling drone and into its camera as if to lock eyes on Carrie and the entire US military industrial apparatus with an accusatory look that clearly, momentarily, unsettles Carrie. The previous night, Aayan had been using his mobile phone to film children dancing at the wedding party when the US missiles struck. After Washington issues vehement denials of claims that their missiles hit a wedding party and anti-drone protests ensue outside the US Embassy in Islamabad, Aayan's roommate uses a proxy server to upload Aayan's phone video of the attack on the celebrants to YouTube without Aayan's consent. The video quickly goes viral and generates a major political conflict for Carrie and the CIA. In the storylines that ensue, Carrie seduces and becomes sexually involved with Aayan in her efforts to use him to gain access to his uncle, the US-trained terrorist, Haissam Haqqani, who was the intended target of the missile strike on the farmhouse yet survived the attack. Carrie's deception and betrayal of Aayan leads ultimately to his murder at the hands of Haqqani, which takes place in full view of an overhead drone as Carrie and her colleagues look on.

Just as drones respatialize war, so does *Homeland* respatialize drones and the techniques and consequences of their use in the media. Drones destabilize the boundaries between battlezone and non-battlezone and render ambiguous the difference between combatants and non-combatants. Furthermore, they participate in the manipulation of 'the visibility and concealment of socially sanctioned forms of killing' (Asaro 2017, 286). Peter Asaro writes that:

> even while the work of drone operators has become increasingly important to the military, and to national and international politics, the actual work of drone operators has remained largely hidden from public view and increasingly protected from the prying eyes of journalists and social scientists. And even within the military, drone warriors are subject to powerful social pressures not to reveal or discuss their work or its psychological or emotional stresses.
>
> *Ibid.*

Consequently, surveillance and killing by drone are rendered resistant to sustained public debate and made 'difficult to perceive, bear witness to, or even conceive of' (Bevan 2015, 148). But while neither drones nor anti-drone protests in Pakistan have received much media coverage in the West (Parks 2017, 23), they have been subject to sustained fictional exploration in *Homeland.* In 'Drone Queen', for example, Carrie is directly confronted and challenged by the distressed and irate US soldier responsible for delivering the missile strikes on the wedding party and others, who tells Carrie that he is 'sick to the stomach' at these killings and calls those who ordered the attacks 'Fucking monsters, all of you'. The respatialization of drones in television drama thus helps audiences

> imagine and speculate about covert US drone attacks in Pakistan through multiple positions and modalities – the air or ground, perpetrator or victim … as part of the process of grappling with the killing of thousands of people, including civilians and children, that US officials have refused to account for.
>
> *Parks 2017, 23*

Lisa Parks (2017, 15–6) discusses what she calls 'drone media,' a category that includes grassroots and activist-made 'photographs, video, maps, data visualizations, and infographics' that circulate on internet sites such as YouTube and 'convey grounded dimensions of drone attacks' and 'challenge the widely-held assumptions that US military drones enable a remote and precise form of warfare that minimizes casualties and collateral damage'. Drone media participate in exposing 'how deeply and profoundly this "surgical" method of warfare has affected lifeworlds on the ground' and thereby 'model the kinds of knowledge practices that are needed when democratic states fail'. Drone media can thus be considered as a form of counter-media that generates an alternative scopic regime, wherein the very apparatus of state-sanctioned visualization, surveillance and killing becomes the object of a critical and interrogatory gaze. While Parks does not include television drama in the category of drone media, our view is that a series such as *Homeland* participates in and extends a similar project of counter-visualization that constructs spaces amenable to a critical and interrogatory gaze, but also, through television's emergent narrative modalities, to forms of identification and affective engagement with remote sufferers and suffering. Lilie Chouliaraki (2013) argues for the importance of the 'mundane acts of mediation' that, through everyday storytelling on television and elsewhere, offers a kind of 'moral education' in the form of 'a series of subtle proposals of how we should feel and act towards distant suffering' (57). Such acts of mundane mediation are necessary, she argues, to the production of a 'humanitarian imaginary' capable of stimulating forms of 'sympathetic identification that may lead to action' (44). Such identifications and solidarities are routinely provoked through television's new modes of narrative complexity and participatory reception practices.

As scholars have noted, television in the age of media convergence is marked by the development of new strategies of diegetic elaboration that facilitate intensified modes of audience engagement and the creation of 'worlds that viewers gradually feel they inhabit along with the characters' (Cupples and Glynn 2013; Sconce 2004, 95). In the case of

Homeland, we experience the devastation and suffering that missile strikes wreak on the ground through the eyes and voices of characters whom we have come to know and maybe care about, such as Aayan Ibrahim or the son of Abu Nazir. Moreover, as the camera's gaze shifts between the often abstract, top-down, surveillant perspective of the CIA command centre and the quotidian routines and relationships of the people living within the drone's field of vision – whom viewers have come to know as parents, friends, sons, daughters and lovers – we are invited to draw connections between the deterritorialized discourses of counterterrorism and the strategic exertion of control, on the one hand, and the grounded and embodied experiences of those unjustly targeted, on the other. We might read such interplays as the televisual and popular cultural equivalent of the kinds of connections that have long been at the heart of good feminist geography scholarship. As regards ongoing US military interventions in South and Central Asia, *Homeland* can thus be read as a kind of response, in its own way, to the fact that 'the space in which these continuing operations have been brought into public view remains strikingly limited' and that 'the space of the [drone] target has been radically underexposed' in Western media (Gregory 2011, 204).

Conclusion: popular cultural citizenship

Homeland helps us to illustrate some of the ways in which popular culture in general, television in particular, functions as a space of ideological mobilization, discursive struggle and gendered political deliberation in the age of media convergence. While expert and specialist knowledges are mobilized in the construction of official policy documents and proceedings in ways that preclude widespread participation and minimize opportunities for the involvement of 'ordinary people', popular culture circulates discourses of securitization and surveillance, war and conflict, gender and geopolitics, race and religion in ways that invite popular affective engagement and contestation. Joke Hermes (2005, 3–4) uses the term 'cultural citizenship' to explore the 'democratic potential of popular culture' as a terrain where, 'regardless of the commercial and governmental interests and investments that co-shape its forms and contents', space is nevertheless continually made for 'implicit and explicit social criticism' from a variety of locations and perspectives. Hermes' reading of popular culture as a space of affective engagement, discursive contestation and political negotiation follows a well-established scholarly tradition associated with cultural studies. Feminist cultural studies, like feminist geopolitics, has long sought to trace the complex and often contradictory connections between the micropolitics of everyday lives, struggles and pleasures, and the macropolitical realm where the structuring forces of political and economic institutions operate most powerfully and effectively. By problematizing the geopolitics of gender, surveillance, securitization and remote-controlled, state-sanctioned killing, television in the post-9/11 age of media convergence constitutes a key site of both popular cultural citizenship and critical scholarly investigation and analysis.

Notes

1 For introductions to feminist geopolitics, see Dixon 2016; Dowler and Sharp 2001; Glynn and Cupples 2015; Hyndman 2001; Sharp 2000, 2007.
2 Available at: www.youtube.com/watch?v=CewNhrRhOtM (accessed 27 August 2018).
3 Available at: www.imdb.com/title/tt0701989/reviews (accessed 27 August 2018).
4 See, for example, the mistakes identified by viewers in Season 4, Episode 1: www.imdb.com/title/tt3284012/goofs (accessed 27 August 2018).

Key readings

Glynn, K., and J. Cupples. 2015. "Negotiating and Queering US Hegemony in TV Drama: Popular Geopolitics and Cultural Studies." *Gender, Place and Culture* 22 (2): 271–287.
Parks, L. 2017. "Drone Media: Grounded Dimensions of the US Drone War in Pakistan." In: *Place, Space and Mediated Communication*, edited by C. Marvin and S. Hong, 13–28. London: Routledge.
Saunders, R.A., and V. Strukov, eds. 2018. *Popular Geopolitics: Plotting an Evolving Interdiscipline*. London: Routledge.

References

Akass, K., and J. McCabe. 2005. *Reading Six Feet Under: TV to Die For*. London: IB Tauris.
Asaro, P. 2017. "The Labor of Surveillance and Bureaucratized Killing: New Subjectivities of Military Drone Operators." In: *Life in the Age of Drone Warfare*, edited by Lisa Parks and Caren Kaplan, 282–314. Durham, NC: Duke University Press.
Baker100. 2015. "Lostpedia is 10 Years Old" (blog). *Lostpedia*, 22 September. Available at: http://lostpedia.wikia.com/wiki/User_blog:Baker1000/Lostpedia_is_10_years_old (accessed 20 February 2018).
Bevan, A. 2015. "The National Body, Women, and Mental Health in *Homeland*." *Cinema Journal* 54 (4): 145–151.
Bhattacharyya, G. 2013. "Narrative Pleasure in *Homeland*: The Competing Femininities of Rogue Agents and Terror Wives." In: *The Routledge Companion to Media and Gender*, edited by C. Carter, L. Steiner, and L. McLaughlin, 374–383. London: Routledge.
Castonguay, J. 2015. "Fictions of Terror: Complexity, Complicity and Insecurity in *Homeland*." *Cinema Journal* 54 (4): 139–145.
Chouliaraki, L. 2013. *The Ironic Spectator: Solidarity in the Age of Post-Humanitarianism*. Cambridge, UK: Polity Press. Kindle version.
Clarke, M.J. 2013. *Transmedia Television: New Trends in Network Serial Production*. New York: Bloomsbury.
Cohen, M. 2013. *Homeland* Isn't Just Bad TV, It Peddles the Worst Lies about US Foreign Policy. *The Guardian*, 16 December. Available at: www.theguardian.com/commentisfree/2013/dec/16/homeland-worst-lies-us-power-foreign-policy (accessed 6 March 2018).
Crawford, N.C. 2012. "'Targeted' Drones Strikes and Magical Thinking. *Huffpost*, 23 September. Available at: www.huffingtonpost.com/neta-crawford/drones-civilian-casualties_b_1907597.html (accessed 11 March 2018).
Cupples, J., and K. Glynn. 2013. "Postdevelopment Television? Cultural Citizenship and the Mediation of Africa in Contemporary TV Drama." *Annals of the Association of American Geographers* 103 (4): 1003–1021.
Curtin, M. 1995. *Redeeming the Wasteland: Television Documentary and Cold War Politics*. New Brunswick, NJ: Rutgers University Press.
Dixon, D. 2016. *Feminist Geopolitics: Material States*. London: Routledge.
Dixon, W.W., ed. 2004. *Film and Television after 9/11*. Carbondale: Southern Illinois University Press.
Doherty, T. 2003. *Cold War, Cool Medium: Television, McCarthyism, and American Culture*. New York: Columbia University Press.
Dowler, L., and J. Sharp. 2001. "A Feminist geopolitics?" *Space and Polity* 5 (3): 165–176.
Dunleavy, T. 2018. *Complex Serial Drama and Multiplatform Television*. New York: Routledge.
Durkay, L. 2014. "'Homeland' is the Most Bigoted Show on Television." *Washington Post*, 2 October. Available at: www.washingtonpost.com/posteverything/wp/2014/10/02/homeland-is-the-most-bigoted-show-on-television/?utm_term=.f862a7882e21 (accessed 6 March 2018).
Fiske, J. 1987. *Television Culture*. London: Routledge.
Gledhill, C. 1988. "Pleasurable Negotiations." In: *Female Spectators: Looking at Film and Television*, edited by E. Diedre Pribram, 64–89. London: Verso.
Glynn, K. 2003. "Challenging Disenchantment: The Discreet Charm of Occult TV." *Comparative American Studies: An International Journal* 1 (4): 421–447.
Glynn, K. 2009. "The 2004 Election Did Not Take Place: Bush, Spectacle and the Media Nonevent." *Television & New Media* 10 (2): 216–245.
Glynn, K., and J. Cupples. 2015. "Negotiating and Queering US Hegemony in TV Drama: Popular Geopolitics and Cultural Studies." *Gender, Place and Culture* 22 (2): 271–287.
Gregory, D. 2011. "From a View to a Kill: Drones and Late Modern War." *Theory, Culture and Society* 28 (7–8): 188–215.

Gregory, D. 2017. "Drones and Death in the Borderlands." In: *Life in the Age of Drone Warfare*, edited by L. Parks and C. Kaplan, 25–58. Durham, NC: Duke University Press.
Gwenllian-Jones, S., and R.E. Pearson, eds. 2004. *Cult Television*. Minneapolis: University of Minnesota Press.
Hermes, J. 2005. *Re-reading Popular Culture*. Malden, MA: Blackwell.
Hyndman, J. 2001. "Towards a Feminist Geopolitics." *The Canadian Geographer* 45 (2): 210–222.
Jenkins, H. 2008. *Convergence Culture: Where Old and New Media Collide*. New York: New York University Press.
Kackman, M. 2005. *Citizen Spy: Television, Espionage, and Cold War Culture*. Minneapolis: University of Minnesota Press.
Longhurst, R. 2008. *Maternities: Gender, Bodies and Space*. New York: Routledge.
Lotz, A.D. 2007. *The Television Will Be Revolutionized*. New York: New York University Press.
Martin, A., and P. Petro, eds. 2006. *Rethinking Global Security: Media, Popular Culture and the 'War on Terror'*. New Brunswick, NJ: Rutgers University Press.
McHugh, R. 2016. "Anarchism and Informal Informal Pedagogy: 'Gangs', Difference, Deference." In: *The Radicalization of Pedagogy: Anarchism, Geography, and the Spirit of Revolt*, edited by S. Springer, M. Lopes de Souza and R.J. White, 147–170. London: Rowman and Littlefield.
McInnis, K.J. 2012. "How 'Homeland' Undercuts Real Women in Government." *The Atlantic*, 16 November. Available at: www.theatlantic.com/sexes/archive/2012/11/how-homeland-undercuts-real-women-in-government/265242/ (accessed 6 March 2018).
Mirzoeff, N. 2005. *Watching Babylon: The War in Iraq and Global Visual Culture*. New York: Routledge.
Mittell, J. 2015a. *Complex TV: The Poetics of Contemporary Television Storytelling*, New York: New York University Press.
Mittell, J. 2015b. "Why has TV Storytelling Become So Complex?" *The Conversation*, 27 March. Available at: https://theconversation.com/why-has-tv-storytelling-become-so-complex-37442 (accessed 5 March 2017).
Mittell, J. 2006. "Narrative Complexity in Contemporary American Television." *The Velvet Light Trap* 58 (1): 29–40.
Negra, D., and J. Lagerwey. 2015. "Analyzing *Homeland*: Introduction." *Cinema Journal* 54 (4): 126–131.
Newman, M., and E. Levine. 2012. *Legitimating Television: Media Convergence and Cultural Status*. New York: Routledge.
Parks, L. 2017. "Drone Media: Grounded Dimensions of the US Drone War in Pakistan." In: *Place, Space and Mediated Communication*, edited by C. Marvin and S. Hong, 13–28. London: Routledge.
Pears, L. 2016. "Ask the Audience: Television, Security and *Homeland*." *Critical Studies on Terrorism* 9 (1): 76–96.
Phipps, C. 2015. "'Homeland is Racist': Artists Sneak Subversive Graffiti on to TV Show." *The Guardian*, 15 October. Available at: www.theguardian.com/tv-and-radio/2015/oct/15/homeland-is-racist-artists-subversive-graffiti-tv-show (accessed 6 March 2018).
Reeves, J.L., M.C. Rodgers, and M. Epstein. 1996. "Rewriting Popularity: The Cult Files." In: *Deny All Knowledge: Reading the* X-Files, edited by D. Lavery, A. Hague and M. Cartwright, 22–35. London: Faber and Faber.
Renner, E. 2017. "'Homeland' Recap: Babysitting Gone Wrong." *Baltimore Sun*, 21 February. Available at: www.baltimoresun.com/entertainment/tv/tv-lust/bal-homeland-recap-season-6–episode-5–2017 0221–story.html (accessed 7 March 2018).
Saunders, R.A. 2017. "Small Screen IR: A Tentative Typology of Geopolitical Television." *Geopolitics*. doi: 10.1080/14650045.2017.1389719.
Saunders, R.A. 2018. "Crossing the Boundary: 'Real-world' Geopolitical Responses to the Popular." In: *Popular Geopolitics: Plotting an Evolving Interdisciplinary*, edited by R.A. Saunders and V. Stropkov, 105–126. London: Routledge.
Schram, J. 2014. "Pakistani Officials Furious over 'Homeland'." *New York Post*, 27 December. Available at: https://nypost.com/2014/12/27/pakistani-officials-furious-over-countrys-portrayal-in-homeland/ (accessed 6 March 2018).
Schwoch, J. 2009. *Global TV: New Media and the Cold War, 1946–69*. Urbana: University of Illinois Press.
Sconce, J. 2004. "What If?: Charting Television's New Textual Boundaries." In: *Television after TV: Essays on a Medium in Transition*, edited by L. Spigel and J. Olsson, 93–112. Durham, NC: Duke University Press.
Sharp, J. 2000: "Remasculinizing Geo(-)politics? Comments on Gearóid Ó Tuathail's Critical Geopolitics." *Political Geography* 19: 361–364.

Sharp, J. 2007. "Geography and Gender: Finding Feminist Political Geographies." *Progress in Human Geography* 31 (3): 381–387.
Spigel, L. 2004. "Entertainment Wars: Television Culture after 9/11." *American Quarterly* 56 (2): 235–270.
Spigel, L., and J. Olsson, eds. 2004. *Television After TV: Essays on a Medium in Transition.* Durham, NC: Duke University Press.
Strauss, E. 2014. "Call Me a Bitch, Call Me a Slut, Just Don't Call Me Crazy." *Elle* 3 October. Available at: www.elle.com/life-love/a14808/just-dont-call-me-crazy/ (accessed 7 March 2018).
Stubblefield, T. 2014. *9/11 and the Visual Culture of Disaster.* Bloomington: Indiana University Press.
Takacs, S. 2010. "The Contemporary Politics of the Western Form: Bush, *Saving Jessica Lynch* and *Deadwood*." In: *Reframing 9/11: Film, Popular Culture and the 'War on Terror'*, edited by J. Birkenstein, A. Froula, and K. Randell, 153–165. New York: Continuum.
Takacs, S. 2012. *Terrorism TV: Popular Entertainment in Post-9/11 America.* Lawrence: University of Kansas Press.
Tryon, C. 2016. *Political TV.* New York: Routledge.
Vann, B. 2003. "White House Bans Coverage of Coffins Returning from Iraq." *World Socialist Web Site*, 23 October. Available at: www.wsws.org/en/articles/2003/10/bush-o23.html (accessed 26 June 2018).

45
HISTORICAL RESEARCH
Gender, politics and ethics

Laura Crawford and Sarah Mills

Introduction

Doing historical research can be illuminating and inspiring, as archives host unique and powerful opportunities for scholarship, activism, teaching and research. This type of work can also be challenging and complex, and present a number of methodological and ethical dilemmas. Historical geographers have written in detail about research practice, with rich accounts of how to source material and undertake archival fieldwork (e.g. Baker 1997; Gagen et al. 2007; Lorimer 2009) and, to a lesser extent, oral histories (Riley and Harvey, 2007). Miles Ogborn (2003) and, later, Ruth Craggs (2016) have provided excellent overviews of archival research, and there is a growing recognition of the increasingly creative engagement of geographers with archives (see also Mills 2013). A range of collections – national, regional, local, institutional, personal and private – offer researchers insights into past lives, places and events across different time periods and international contexts. Crucially, these different types of archives are all gendered in their construction, content and consumption. This chapter engages with these dynamics, focusing on the relationship between gender, politics and ethics in the context of feminist historical research, specifically on archival encounters.

Archives are institutions that collect and preserve written, visual and audio material that can be read as texts with discursive meaning, both in situ during fieldwork and on digital and online platforms. Across all types of institutional or personal archives, source material can be diverse in form, content and style, varying from comprehensive coverage to bit-part chronology. Archival work is fraught with challenges due to the inevitable distance from the 'live' events or lives themselves, the partial truths and memories that the archives house and the political creation and curation of what and who 'counts' as history. The dis/order of archival collections also reveals issues of representation and power connected to wider debates on knowledge, truth and evidence (see Lorimer and Philo 2009; Till 2001). In response to many of these challenges, and to the nature and politics of historical geography more broadly, feminist historical geography emerged as a research agenda and intellectual project in the late 1980s. It continues to grapple with the gendered construction of knowledge and to unearth hidden histories of the lived experience of women across historical periods. As Boyer notes, it:

> can entail both seeking out sources which shine light on the social construction of gender and other kinds of power relations in historical contexts, as well as seeking to understand the politics that structure what we can know of past worlds.
>
> *Boyer 2004, 170*

This chapter introduces this research area. It has an explicit focus on the methodological approaches to archival fieldwork in the spirit of Part 4 of this Handbook: 'Doing feminist geographies'. Specifically, we focus on methodological approaches by both engaging the literature and reflecting on our own attempts to develop an ethical practice in relation to archival work. The remainder of the chapter is structured in three core sections: first, we introduce feminist historical geography and explain how it has developed distinctive approaches to the gendered politics of the archive; second, we focus on the often-neglected ethical considerations of historical research, supporting the continued development of archival approaches that intersect with, or are sympathetic to, feminist theory; finally, we consider the three areas and current developments that should challenge and interest those working in this field.

Feminist historical geographies: gendered genealogies

In 1988, Rose and Ogborn drew attention to historical geography's neglect of 'the empirical achievements of feminism in increasing our understanding of the past' (405). Their call for greater attentiveness in historical geography to women's lives and gendered environments also highlighted that existing work of feminist influence was 'not being written by self-styled historical geographers' (Rose and Ogborn 1988, 405). Feminist historical geography has since grown, not merely as an act of seeking out the voices of past women but through engaging with new material and researching women's experiences in diverse spaces. Feminist geographers have conducted important historical work on race, colonialism and urban landscapes (e.g. Anderson 1991; Bressey 2002), women's labour (e.g. McDowell 2013), community and citizenship (e.g. Cope 1998), children and motherhood (e.g. Gagen 2000; Olson 2019), as well as the gendered history of geography itself (e.g. Maddrell 2009; Monk 2004). Many scholars have reviewed developments in feminist historical geography since Rose and Ogborn's landmark publication (e.g. Domosh and Morin 2003; Moore 2018; Morin and Berg 1999; Rothenberg et al. 2016) and these commentaries allude to the problems with its 'place' in wider disciplinary infrastructures. Indeed, Morin and Berg (1999) explain how, in some contexts, the type of work cited above can also be classed as political, urban or postcolonial geography. Domosh and Morin (2003) also described an 'undercover feminism' in historical geography, yet today this is more clearly visible. For example, a recent special issue of *Historical Geography* (2016), explicitly on feminist historical geographies, includes articles on women's participation in the Royal Geographical Society expeditions (Evans 2016), the relationship between gender and historical ecological reconstructions (Greer and Bols 2016), anarchist geographers and feminism (Ferretti 2016) and women making native space in Quebec (Desbiens and Lévesque 2016). And yet, the guest editors comment how there is still limited representation of feminist work in historical geography arenas, identifying instead a wider 'alive' nexus of feminism-history-geography (Rothenberg et al. 2016, 28). More broadly, the studies captured in this special issue echo Domosh's early argument that feminist historical geography is 'a way of interpreting landscape that does not establish authorship as the basis of meaning, but rather focuses on the social and cultural context in which that landscape is created' (1997, 232; see also Domosh 1991). Furthermore, this approach to archival collections demonstrates attentiveness to the gendered politics of the archive as a material space.

Gender plays into both the composition of the archives and the researchers' ability to extract meaning from them, as the content of collections is indicative of broader societal structures and pressures. As well as the institutional and financial burdens that influence what historical sources are kept by official archives and the value judgements made on whether certain material is worth preserving for posterity, wider social dynamics have influenced the historical record. McDowell recognizes the work of feminist scholars in drawing attention to the gendered division of urban space with, 'a so-called private arena associated with women and a public world of men' (1999, 73). At risk of oversimplifying a contentious and heavily critiqued set of debates, in crude terms this notion of separate spheres emerged at a particular historical moment, most notably in relation to the Industrial Revolution, and the home came to occupy a greater significance for women, often as a site of unpaid domestic labour (McDowell 1999). Her research importantly highlights that, while over a third of all women in Britain did engage in paid work between 1850 and 1950, they were largely excluded from more senior, higher-paid roles. This is just one example of the need to understand the place of women through space and time and to understand how this influences the historical record and knowledge production.

In a different historical and geographical context, Morin and Berg note the 'ideological and embodied gendered differences' in anti-colonial writing during the Maori Wars in New Zealand in the 1860s, recognizing that, 'while the men published legal documentation, pastoral letters to Anglican communities, and journals … the women primarily wrote (and published) letters, among themselves and to relatives and associates back in England' (2001, 208). This example demonstrates that women had 'limited access to the most "authoritative" venues of writing and publishing' (Morin and Berg 2001, 208). This gendered landscape of knowledge production has also shaped the writing of history and, while legal documentation by its very nature implies a level of authority, access to these spheres and professions was not equal across time and space; as such, women's contributions and access to what Domosh calls 'the formal record' have been sidelined. Domosh reflects on the subsequent challenges associated with finding women's voices and suggests that one of the consequences has been 'a broadening of the definition of reliable and accurate source materials, in order to allow those without access to the formal record to speak to us now' (1997, 228). It is this approach that has characterized the methodological drive of much feminist historical geography work to date.

This distinctive methodological approach that underpins feminist historical geography work has emerged in response to the gendered conditions and character of archives outlined above, challenging a wider politics of exclusion. This approach involves not only simply including more women in geographies of the past but also utilizing feminist theory more widely within the use of historical research methods. As such, feminist geographers have used archival research to document women's experiences, conducted oral histories and led historical Geographical Information System projects. Through an attentiveness to voices that have seldom been recorded in the dominant discourse, geographers have been able to challenge the existing narratives and create a more complex and nuanced picture of past lives. Boyer's practical guidance for conducting feminist historical geography in the archive is an insightful reflection on methodology. She advocates that researchers attend to the importance of power, suggesting that analysis of the influence of power can be determined by looking for certain types of source, those 'which document transgressions' (2004, 170). This is a useful guide to account for gaps in official and institutional records, and it is worth considering that the evidence of these transgressions is often deemed valuable and kept in archives on the basis that it documents something extraordinary, remarkable or radical. Yet the mundane and more 'ordinary' experiences can be difficult to find, and researchers run the risk of telling the history of marginalized communities only through the eyes of an extraordinary few. For example, in Mills' (2011) work on gender and youth organizations in early twentieth-century Britain, official archival material is dominated by accounts of a small group of 'rebel' girl scouts who demanded to join the Boy Scouts and were later institutionalized as Girl Guides. These extraordinary few are cemented in time via official archival records, yet little is known about the more ordinary or mundane acts of transgression or compliance within these uniformed organizations. Careful fieldwork across multiple archival sites, including personal collections and the triangulation of data, reveals (as much as possible) the everyday lived experience of women, girls and marginalized groups in these spaces. These dilemmas surrounding power and 'voices' in the archive require researchers to ask more questions of data, such as considering the potential motivations behind the creation of a source or choice to donate personal texts and objects to an archive. Boyer provides advice on how to come to terms with this challenge of source material that documents transgressions, concluding that 'this does not preclude the use of such texts as historical documents, but it does require recognizing how questions of textual authority shape the kind of story one constructs' (2004, 171; see also Gagen 2001). In a recent article, Moore (2018) goes further by advocating emancipatory and participatory archival work that is attentive to, and useful for, understanding the feminist histories of the present.

In the context of these debates on power, representation and archival methods, Bressey (2003) has importantly highlighted the challenges of scarce written source material, namely in recovering the historical geographies of Black people in Britain. Her archival research into these 'forgotten histories' (see also Bressey 2002) therefore drew heavily on photographic material from a number of collections. She describes how this visual methodology created an ethical 'paradox' for her, in relation to the 'persistence of race as a means to classify, identify and divide human beings' (2003, 221) in an anti-racist scholarly project. Bressey's work on Black women in late-Victorian London not only provokes a series of methodological and ethical reflections, a topic discussed in the next section, but demonstrates the importance of locating these historical geographies within a wider politics of exclusion, marginalization and intersectionality (Brown 2011; Hopkins 2017; Valentine 2007). Although Morin and Berg highlight that 'Feminist historical geographies are subverting the erasure of women and many Others effaced from geographies of the past' (1999, 326), more work can and should be done with an attentiveness to inequality that interrogates socio-spatially produced power structures, including patriarchal systems, yet also racism, disablism, homophobia and ageism. The drive for much of this work can be found in many public histories and critical scholarly works that seek to preserve and legitimize subaltern historical knowledge, responding to some of the absences from the historical record. Scott (2017) advocates the use of oral histories in disrupting dominant narratives within work on Black geographies to reveal stories of place and community, and McKittrick's (2006, 2011) research on Black geographies is a powerful example of trying to write the historical spaces that have literally been rubbed out because of violence.

As McGeachan notes in her latest *Progress in Human Geography* report on historical geography, archival research can reach into the 'darkest of terrains', including gendered violence (i.e. Fuentes 2016, on enslaved women in eighteenth-century Barbados), 'uncovering geographies that can be almost too painful to bear and too harrowing to witness' (McGeachan 2017, 10). This dimension to historical research reminds us of the urgent press to consider ethics, even when confronting data about people who are no longer alive. Indeed, the ethics of archival research is one of the central ways in which historical geography has, and can further be, inspired by feminist theory and praxis.

Archival fieldwork: ethical considerations

It is fair to say that, as a subdiscipline, historical geography does not have an extensive literature on ethical fieldwork practice, especially compared to social and cultural geography or the wider social sciences. An important exception is Moore's (2010) intervention. She writes powerfully about the ethical considerations in her archival fieldwork on abortion in Lancashire in the late-nineteenth and early-twentieth century (2010; see also Cameron 2001 and Legg 2016 on ethical debates). Moore details her fraught work with sensitive, private and personal material, and the subsequent legal and ethical implications. Moore calls for attentiveness to 'the conflict of interest that can exist between researcher and participant, even when the participant is dead' (2010, 268), concluding that all geographers, including those engaged in historical projects, should carefully consider the effects of their research. Moore's reflections have inspired our own approaches to ethical dilemmas and archival encounters.

In our respective research projects on residential homes for disabled people (Crawford) and youth organizations (Mills), we have both grappled with ethical dilemmas of consent, confidentiality, anonymity and privacy. Indeed, as most archival work involves tracing the lives of individuals either through detailed biography or as actors within a wider narrative, researchers have a responsibility to handle these (usually past) lives with care and to consider how they are (re)presented. In navigating these struggles, we have sought an approach that employs reflexivity, liaising closely with archivists (and, indeed, published work by archivists; see Schwarz 1992) and, when appropriate, consulting named individuals or family members, if possible. This logistics is often shaped by the nature of the historical material in question, for example public or private collections and the conditions of use imposed by owners and guardians. At times, concerns over privacy and sensitive data have led us to use pseudonyms for historical research subjects, where relevant. For example, in a recent project involving children's accounts of everyday home and school life from 1960s and 1980s London, pseudonyms for individuals and schools were used to protect the rights of these historical subjects, who may still be living at those addresses or connected to local areas as adults (Mills 2017). In other cases, anonymity is a more complex and political dilemma, with researchers having to rely on their own moral compass to navigate the most appropriate approach to handling data. In Crawford's ongoing research on residential homes, some individuals in the archival material are identifiable through their political activism, either by association or through their networks, meaning that anonymity can be incredibly difficult to achieve, not least as some were public figures and their stories are in the public domain. Furthermore, there is a wider political motivation in naming the research subjects frozen in time in the archival material. Wright and Saucier (2012) discuss how protecting anonymity and using pseudonyms can run counter to a desire to empower and give voice to marginalized individuals and groups, framing these ethical conundrums in the context of researching mental health, disability and medicine. Citing Iacovetta and Mitchinson (1998), they suggest that 'our legal obligations as researchers to protect the privacy of individuals in the past can lead us to write the marginal into history by writing their names and faces out of it' (Wright and Saucier 2012, 76). Anonymity, in this context, is seemingly at odds with the notion of agency, and it calls into question whether academic conventions are continuing to marginalize and silence voices that have previously been hidden.

In critically reflecting on our own archival practice, we have also tried to respond to Bailey, Brace and Harvey's (2009, 255) call in their discussion of collaborative archival work 'for geographers to maintain a reflexive approach to their historically grounded identities'. Indeed, we advocate an approach to doing historical research that explicitly considers positionality (see also Mills 2012). Although usually associated with social research and encounters with living subjects, positionality is also an important consideration in historical work. The selection of research topics and reading of archival material are unquestionably shaped by positionality, including but not limited to gender, race, religion and class.

Whether it is for collaborative or individual research, archival fieldwork is not neutral or somehow 'excused' from ethical considerations by virtue of its 'distance' from historical subjects. Indeed, many research projects use archival methods that involve subjects who are still alive or who are, when in combination with oral history data. As such, historical research does host ethical dilemmas, yet these are not always recognized within institutional and funding guidelines and frameworks (Tesar 2015). Ethical dilemmas occur in any research project, therefore the topic of doing ethical historical research should be a regular discussion among academic communities and as part of relevant forums and conferences. There is also a need for training postgraduate and early-career researchers in how to strike a balance between sharing material in the public interest as part of research projects against the rights of historical subjects to protection and privacy, and in what contexts this is appropriate. These issues are growing in importance, especially in the UK, where the boundaries between historical and contemporary research practices are blurring. First, historical researchers are increasingly creating their own 'archival' collections through their own fieldwork photographs and digital notes, perhaps collating and storing data from multiple sites and creating their own personal records. Second, geographers across the discipline who are engaged in contemporary research (e.g. interviews, focus groups or ethnography) are becoming guardians and custodians of 'new archives' of database material (e.g. interview transcripts, participatory artwork) that is increasingly cleaned and deposited to funding websites or held in institutional collections for future preservation, research use or public engagement (Mills 2017). Overall, these ethical and data management challenges are part of a wider politics of research practice and link to much deeper questions about the role, purpose and practice of research.

Changing practices, ongoing politics

This section discusses three areas where changing (and hidden) practices are beginning to push scholars to consider research practice anew. These selected examples signal important changes that open up a space for new and different kinds of feminist practice, as well as presenting new ethical challenges. In selecting these three particular areas, we highlight how each case aligns with and furthers feminist practice. They are important considerations for researchers engaged in archival fieldwork and represent new opportunities to expand feminist engagements by further considering who 'makes' history.

First, the practices of collecting and archiving, especially in the context of hidden labour, are vital considerations yet are often marginalized. Questions of who does the work of archiving (before a researcher even accesses any material) are important, especially as unanswered concerns for feminists. The labour of archivists is often hidden, especially the emotional labour of preserving, rescuing and collecting material that is often gendered in relation to the voluntary action of community and charity groups in civil society. A recent post on Twitter from Guy Walters – author, historian and journalist – claimed to have 'discovered' a letter from British philosopher Bertrand Russell, stating 'Of all the letters I've unearthed in archives, this is by far the best',[1] and his post received over a million impressions via different social media platforms. It did, however, provoke a backlash of responses, primarily from professional archivists and librarians, after McMaster Archives & Research Collections, Ontario, Canada (which holds the original letter) replied, 'What's vulgar is refusing to acknowledge the people who selected, bought, catalogued, and preserved the letter so you could read it – us ... we don't treat [material] as our exclusive discovery … they are our common heritage'.[2] In many ways, social media has given a voice to the often-hidden work of archivists, challenging the problematic notion that academics *discover* hidden gems in their collections rather than *accessing* the material through professionals' support. A further response on Twitter, from archivist Sian Collins, included images of historical papers in severe disrepair and a pile of hand-removed rusty paperclips, with the comment: 'This is what some collections look like pre-archivist, before they are "discovered" by others.' (Collins 2 October 2017).[3] This exchange is indicative of a wider debate on the construction and (re)production of knowledge and the past. Indeed, the practices of collection, preservation, recording and display are political. The processes of 'archivalization' (Ashmore et al. 2012; Ketelaar 2001), of how archives come into being and the labour involved in their creation and maintenance, are important concerns for feminist theorists, as outlined earlier in relation to the wider politics of the archive and gendered construction of knowledge.

Second, 'new' archival collections and spaces are critical to the placement of otherwise narrowly dictated histories. For example, within our own research areas on religious youth work and people with learning disabilities are emerging archives and collections that document the lives and experiences of minority groups, driven and shaped by communities themselves. These important new collections create opportunities for scholarship and educational activities, yet also have power through a politics of representation that is politically vital to contemporary society. A clear example is the recent opening of East London Mosque's new archives in November 2017. As the UK's first-ever Muslim purpose-built strongroom, the archive contains around 250,000 documents relating to one of the country's oldest mosques and the wider local Muslim community. At the opening, Mayor of London Sadiq Khan stated that 'familiarity with our history frees us from false narratives' and Dr Jamil Sherif, Chair of the Archives Project's Steering Committee, explained that the archives 'allow the Muslim experience to be woven into the tapestry of British social history … without that narrative, Muslims are always going to be considered as the "Other"' (cited in Hussain 2017). This example, although isolated, demonstrates the important role of historical material and feminist theorizing for a range of communities, both within and beyond academia.

Online platforms can also facilitate the creation of new archival collections, despite the challenges of the digital age and pressures for archives to digitize catalogues and material. These online platforms can lead to self-advocacy and representation by diverse communities through the creation and curation of non-traditional archival collections.

One such example is the Big 30 Project by the London-based creative arts company and charity, Heart n Soul: 'the first of its kind where people with learning disabilities tell their own stories in their own words, creating a unique archive of learning disability culture' (Heart n Soul, 2017). The archive is a celebration of artistic expression, whereby people with learning disabilities share their experiences through photography, stories and music. As Ono from the Big 30 Archive website states, 'People with a learning disability are slowly breaking the glass ceiling and showing what they can do' (The Big 30 n.d). This example shows that archives are multi-scalar, with local and community-based collections often documenting more intimate and personal experiences than do the national or government-funded collections. Furthermore, this example illustrates how the process of selecting and creating material to contribute to an archive can be a political act, designed to challenge established societal norms. It is not just our role as academic researchers to visit, read and critically engage with archival collections as part of our research and pedagogy, but to listen, support, advocate and collaboratively engage (where appropriate) with archivists, civil society groups and local communities in their plans for new collections or creative responses to existing material. Within the UK, this type of public engagement work will be even more vital in austere times, as archives and libraries are under extreme pressure from funding cuts and face growing precarity, particularly within the charity and voluntary sector (Brewis 2012).

Finally, archives have been important resources for activism and preserving the legacy of attempts to combat injustice. As Rose and Ogborn outline, 'there have always been struggles against such oppression and so it is also vital that historical geography maps the creation by women of these spaces of resistance' (1988, 408). Feminist activisms (see Hancock, Bettinger and Manseri, Chapter 39 in this volume) are an important part of social history, and the relationship between activism and archives includes the vital task of archiving the present contemporary moment and political engagements. For example, the collective efforts by archives and museums across the globe to preserve artwork, signs and banners from the Women's March in January 2017 that followed the inauguration of Donald Trump as President of the United States. While Moore (2018) has highlighted the significance of this event in terms of feminist histories of the present, here we briefly describe the (changing) archival practices at play during and after such a protest. Following these global protests, calls for photos, videos and signs were posted by the Smithsonian Museum in Washington DC, right through to smaller libraries and collections across the world. Newberry Library in Chicago was one of the archives that began 'crowdsourcing' material, mostly via social media requests, and it gave donors the option to submit short narratives alongside their posters and creative submissions (Levitt 2017). There were also more haphazard encounters with (future) historic material from the Women's March: for example, the 'accidental archive' in Boston after local residents rescued assembled banners from an impending rubbish collection (Deruy 2017). These examples demonstrate the changing nature of archival practice and shifts in methods of sourcing, depositing and collecting material, as well as the wider role of citizens (including researchers) in constructing and shaping future collections.

Conclusion

Historical geography has undoubtedly been enriched by the influence of feminist principles, theories and approaches and, likewise, feminist geography has expanded its scope and wider contributions through historical work and archival encounters. Indeed, as Boyer claims, 'historical geography can bring temporal depth and a different kind of texture to feminist geography; feminist geography can and should inform the study of past spaces and past worlds' (2004, 173). This chapter has provided an overview of 'doing historical research' and outlined important debates on the practice and politics of archival fieldwork. Although the chapter is most relevant to those already engaging with, or considering embarking on, historical research projects, its discussion about power and the ethics of research should also have a wider reach. This chapter has reflected on practices of activism, curation and advocacy. As such, this discussion only speaks to established or budding archival researchers, but to a broader audience concerned with practising human geography and the important relationship between gender, politics and ethics.

Notes

1 https://twitter.com/guywalters/status/913299775885365248.
2 https://twitter.com/MacResColls/status/913842397800534016.
3 https://twitter.com/SianECollins/status/914765553868443648.

Key readings

Boyer, K. 2004. "Feminist Geography in the Archive: Practice and Method." *Gender and Geography Reconsidered*. New York: Rensselaer Polytechnic Institute.

Domosh, M., and K.M. Morin. 2003. "Travels with Feminist Historical Geography." *Gender, Place & Culture* 10 (3): 257–264.
Rose, G., and M. Ogborn. 1988. "Feminism and Historical Geography." *Journal of Historical Geography* 14 (4): 405–409.

References

Anderson, K. 1991. *Vancouver's Chinatown: Racial Discourse in Canada, 1875–1980*. Montreal & Buffalo: McGill-Queen's University Press.
Ashmore, P., R. Craggs, and H. Neate. 2012. "Working-with: Talking and Sorting in Personal Archives." *Journal of Historical Geography* 38 (1): 81–89.
Bailey, A., C. Brace, and D.C. Harvey. 2009. "Three Geographers in an Archive: Positions, Predilections and Passing Comment on Transient Lives." *Transactions of the Institute of British Geographers* 34 (2): 254–269.
Baker, A.H.R. 1997. "The Dead Don't Answer Questionnaires: Researching and Writing Historical Geography." *Journal of Geography in Higher Education* 21 (2): 231–243.
Boyer, K. 2004. "Feminist Geography in the Archive: Practice and Method." *Gender and Geography Reconsidered* 169–174. Available at: www.gfgrg.org/resources/downloads/ (accessed 7 December 2017).
Bressey, C. 2002. "Forgotten Histories: Three Stories of Black Girls from Barnardo's Victorian Archive." *Women's History Review* 11 (3): 351–374.
Bressey, C. 2003. "Looking for Blackness: Considerations of a Researcher's Paradox." *Ethics, Place & Environment* 6 (3): 215–226.
Brewis, G. 2012. "Launching the Campaign for Voluntary Sector Archives" Available at: www.vahs.org.uk/2012/10/charity-archives-brewis/ (accessed 28 November 2017).
Brown, M. 2011. "Gender and Sexuality I: Intersectional Anxieties." *Progress in Human Geography* 36 (4): 541–550.
Cameron, L. 2001. "Oral History in the Freud Archives: Incidents, Ethics and Relations." *Historical Geography* 29: 38–44.
Cope, M. 1998. "'She Hath Done What She Could': Community, Citizenship, and Place among Women in Late 19th Century Colorado." *Historical Geography* 26: 45–64.
Craggs, R. 2016. "Historical and Archival Research." In: *Key Methods in Geography*, edited by N. Clifford, T. Cope, T. Gillespie, and S. French, 111–128. London: SAGE.
Deruy, E. 2017. "What Happens to Those Handmade Posters from the Women's Marches." Available at: www.theatlantic.com/education/archive/2017/01/what-happens-to-those-handmade-posters-from-the-womens-marches/514373/ (accessed 28 November 2017).
Desbiens, C., and C. Lévesque. 2016. "From Forced Relocation to Secure Belonging: Women Making Native Space in Quebec's Urban Areas." *Historical Geography* 44: 89–101.
Domosh, M. 1991. "Toward a Feminist Historiography of Geography." *Transactions of the Institute of British Geographers* 16 (1): 95–140.
Domosh, M. 1997. "With South Boots and a Stout Heart': Historical Methodology and Feminist Geography." In: *Thresholds in Feminist Geography: Difference, Methodology, Representation*, edited by J.P. Joans III, H.H. Nast and S.M. Roberts, 225–240. Lanham, MD: Rowman and Littlefield.
Domosh, M., and K.M. Morin. 2003. "Travels with Feminist Historical Geography." *Gender, Place & Culture* 10 (3): 257–264.
Evans, S.L. 2016. "Mapping *Terra Incognita:* Women's Participation in Royal Geographical Society-supported Expeditions 1913–1939." *Historical Geography* 44: 30–44.
Ferretti, F. 2016. "Anarchist Geographers and Feminism in Late-19th Century France: The Contributions of Elisée and Elie Reclus." *Historical Geography* 44: 68–88.
Fuentes, M.J. 2016. *Dispossessed Lives: Enslaved Women, Violence, and the Archive*. Philadelphia: University of Pennsylvania Press.
Gagen, E.A. 2000. "Playing the Part: Performing Gender in America's Playgrounds." In: *Children's Geographies: Playing, Living, Learning*, edited by S.L. Holloway and G. Valentine, 213–229. London: Routledge.
Gagen, E.A. 2001. "Too Good to Be True: Representing Children's Agency in the Archives of the Playground Movement." *Historical Geography* 29: 53–64.
Gagen, E.A., H. Lorimer, and A. Vasudevan, eds. 2007. *Practising the Archive: Reflections on Methods and Practice in Historical Geography*. Historical Geography Research Series 40. London: Royal Geographical Society.

Greer, K., and Bols, S. 2016. "'She of the Loghouse Nest': Gendering Historical Ecological Reconstructions in Northern Ontario." *Historical Geography* 44: 45–67.

Heart n Soul. 2017. "The BIG 30." Available at: www.heartnsoul.co.uk/category/artists/details/the_big_30 (accessed 8 December 2017).

Hopkins, P. 2017. "Social Geography I: Intersectionality." *Progress in Human Geography.* Online First. doi: 10.1177/0309132517743677.

Hussain, T. 2017. "Britain's First Mosque Archives Launched by Mayor of London." Available at: www.mysalaam.com/en/story/britains-first-mosque-archives-launched-by-mayor-of-london/SALAAM26112017014632 (accessed 29 November 2017).

Iacovetta, F., and W. Mitchinson. 1998. "Introduction: Social History and Case Files Research." In: *On the Case: Explorations in Social History*, edited by F. Iacovetta and W. Mitchinson, 3–22. Toronto: University of Toronto Press.

Ketelaar, E. 2001. "Tacit Narratives: The Meaning of Archives." *Archival Science* 1: 131–144.

Legg, S. 2016. "Empirical and Analytical Subaltern Space? Ashrams, Brothels and Trafficking in Colonial Delhi." *Cultural Studies* 30 (5): 793–815.

Levitt, A. 2017. "A Look inside the Newberry Library Protest Archive." Available at: www.chicagoreader.com/Bleader/archives/2017/03/15/a-look-inside-the-newberry-library-protest-archive (accessed 28 November 2017).

Lorimer, H. 2009. "Caught in the Nick of Time: Archives and Fieldwork." In: *The SAGE Handbook of Qualitative Research in Human Geography*, edited by D. DeLyser, S. Aitken, M.A. Crang, S. Herbert, and L. McDowell, 248–273. London: SAGE.

Lorimer, H., and C. Philo. 2009. "Disorderly Archives and Orderly Accounts: Reflections on the Occasion of Glasgow's Geographical Centenary." *Scottish Geographical Journal.* 125 (3–4): 227–255.

Maddrell, A. 2009. *Complex Locations: Women's Geographical Work in the UK 1850–1970.* Oxford: Wiley-Blackwell.

McDowell, L. 1999. *Gender, Identity and Place: Understanding Feminist Geographies.* Oxford: Blackwell.

McDowell, L. 2013. *Working Lives: Gender, Migration and Employment in Britain, 1945–2007* Oxford: Wiley-Blackwell.

McGeachan, C. 2017. "Historical Geography III: Hope Persists." *Progress in Human Geography.* Online First. doi: 10.1177/0309132517740481.

McKittrick, K. 2006. *Demonic Grounds: Black Women and the Cartographies of Struggle.* Minneapolis: University of Minnesota Press.

McKittrick, K. 2011. "On Plantations, Prisons, and a Black Sense of Place." *Social & Cultural Geography* 12 (8): 947–963.

Mills, S. 2011. "Scouting for Girls? Gender and the Scout Movement in Britain." *Gender, Place and Culture* 18 (4): 537–556.

Mills, S. 2012. "Young Ghosts: Ethical and Methodological Issues of Archival Research in Children's Geographies." *Children's Geographies* 10 (3): 357–363.

Mills, S. 2013. "Cultural-Historical Geographies of the Archive: Fragments, Objects and Ghosts." *Geography Compass* 7 (10): 701–713.

Mills, S. 2017. "Voice: Sonic Geographies of Childhood." *Children's Geographies* 15 (6): 664–677.

Monk, J. 2004. "Women, Gender and the Histories of American Geography." *Annals of the Association of American Geographers* 94 (1): 1–22.

Moore, F.P.L. 2010. "Tales from the Archive: Methodological and Ethical Issues in Historical Geography Research." *Area* 42 (3): 262–270.

Moore, F.P.L. 2018. "Historical Geography, Feminist Research and the Gender Politics of the Present." *Geography Compass* 12 (9). https://doi.org/10.1111/gec3.123988of8.

Morin, K.M., and L.D. Berg. 1999. "Emplacing Current Trends in Feminist Historical Geography." *Gender, Place and Culture* 6 (4): 311–330.

Morin, M.K., and L.D. Berg. 2001. "Gendering Resistance: British Colonial Narratives of Wartime New Zealand." *Journal of Historical Geography* 27 (2): 196–222.

Ogborn, M. 2003. "Finding Historical Data." In: *Key Methods in Geography*, edited by H.J. Clifford and G. Valentine, 101–115. London: SAGE.

Olson, E. 2019. "The Largest Volunteer Lifesaving Corps in the World': Centering Child Caregiving in Histories of U.S. Human Security through the Little Mothers' League." *Social & Cultural Geography* 20 (4): 445–464. https://doi.org/10.1080/14649365.2017.1362585.

Riley, M., and D. Harvey. 2007. "Talking Geography: On Oral History and the Practice of Geography." *Social & Cultural Geography* 8 (3): 345–351.

Rose, G., and M. Ogborn. 1988. "Feminism and Historical Geography." *Journal of Historical Geography* 14 (4): 405–409.

Rothenberg, T., M. Domosh, and K. Morin. 2016. "Introduction to the Special Issue: Feminist Historical Geographies." *Historical Geography* 44: 27–29.

Schwarz, J. 1992. "The Archivist's Balancing Act: Helping Researchers While Protecting Individual Privacy." *Journal of American History* 79 (1): 179–189.

Scott, D. 2017. "Oral History and Emplacement in 'Nowhere At All': The Role of Personal and Family Narratives in Rural Black Community-building." *Social & Cultural Geography.* Online First. https://doi.org/10.1080/14649365.2017.1413205.

Tesar, M. 2015. "Ethics and Truth in Archival Research." *History of Education* 44 (1): 101–114.

The Big 30. n.d. "Changes in Society." Available at: www.thebig30.com/changes/ (accessed 30 November 2017).

Till, K.E. 2001. "Fragments, Ruins, Artifacts, Torsos." *Historical Geography* 29: 70–73.

Valentine, G. 2007. "Theorising and Researching Intersectionality: A Challenging for Feminist Geography." *Professional Geographer* 59 (1): 10–21.

Wright, D., and R. Saucier. 2012. "Madness in the Archives: Anonymity, Ethics, and Mental Health History Research." *Journal of the Canadian Historical Association* 23 (2): 65–90.

46

TEACHING FEMINIST GEOGRAPHY

Practices and perspectives

Joos Droogleever Fortuijn

Introduction

Forty years ago, feminist geography teaching began on the first elective courses on 'women's studies in geography or 'feminist geography' at a few universities in Aotaroa New Zealand, Denmark, India, Netherlands, the UK and the US (Peake 1989). These courses were created in response to the demands of critical feminist geography students. They were taught by a few feminist geography lecturers, who were able to organize them despite the resistance of the mainstream geographers in their departments.

In the past forty years, feminist geography teaching has undergone significant changes. Nowadays feminist geography, although still absent from some areas of the world, is taught in many countries across the globe, partly as a separate elective course and partly integrated into core geography teaching. Feminist geography teaching is much more diversified than in the initial period. This chapter focuses on the diverse practices and perspectives on feminist geography teaching and argues that *what* we teach and *how* we teach are highly dependent on *where* we teach feminist geography (see overviews of feminist geography across the world in *Journal of Geography in Higher Education* (Peake 1989), *Espace, Populations, Sociétés* (Creton 2002), *Belgeo* (Garcia-Ramon and Monk 2007), *Australian Geographer* (Johnston and Longhurst 2008) and *International Research in Geographical and Environmental Education* (Monk 2011)).

What we teach: the changing content of feminist geography courses

In the initial period, 'making women visible' and a focus on inequalities between women and men were the main aspects of feminist geography teaching. Feminist geography teaching started as a critique on mainstream geography 'excluding half of the human in human geography' (Monk and Hanson 1982). In a report on the development of feminist geography in German-speaking countries, Buehler and Baechli (2007) distinguish four stages in the field of gender and geography: 1. Women's studies, aiming to include women's lives in the discipline of geography; 2. Gender relations studies with a focus on the analysis of the (re)production of gender relations; 3. Men's and masculinity studies; and 4. Gender studies, focusing on the construction of gender identities (see also Little 2007).

Early handbooks on feminist geography focused on feminism, gender and differences between men and women as its core concepts. Core themes at that time were the gender division of paid and unpaid work, women and the city, and women and development (see for example WGSG 1984). Handbooks and courses referred to women in general yet in practice focused on the lives of White, middle-class, heterosexual mothers with children in cities, and on the 'others': women in developing countries. A strong engagement with feminist movements and community activism and a commitment to social change characterized feminist geography teaching. In the initial period, feminist geography courses were taught by female staff for female students. Later, with the introduction of themes about men, masculinity and sexuality, male staff and students became engaged in feminist geography teaching as well.

Postmodernism and the 'cultural turn' in geography transformed feminist geography teaching, while at the same time feminist geography contributed to this 'cultural turn'. Nowadays, the construction of gender identities, political structures and power relations and the deconstruction of gender and other binaries usually constitute the content of feminist geography courses. Cultural diversity and intersectionality form the core concepts. Feminist geography is influenced by Kimberlé Crenshaw's conceptualization of intersectionality, which she defines as 'the various ways in which race and gender intersect in shaping structural, political and representational aspects of violence against women of color' (Crenshaw 1993, 1244). Crenshaw referred to two dimensions, gender and race, but the list of social dimensions has extended and includes now class, sexual orientation, age, ability, religion, geographic and linguistic origin. In line with the focus on cultural diversity and intersectionality, gender and geography teaching is, in many cases, integrated into broader courses on cultural diversity or development studies.

Definitions of intersectionality use geometric and geographical metaphors: crossroads (Hovorka 2015); axes (O'Neill Gutierrez and Hopkins 2015); lines (Metcalfe and Woodhams 2012); and mapping (Crenshaw 1993). These metaphors suggest that someone occupies one position on the line of gender and one on the line of race, and that a multiple identity is where the lines cross. This is a problematic way of presenting multiple identities and experiences of inequality and marginalization. As Mieke Verloo (2006, 211) argues, it 'assumes an unquestioned similarity of inequalities, to fail to address the structural level and to fuel political competition between inequalities', therefore runs the risk of marginalizing gender.

Feminist methodology forms a core element of courses in feminist geography. Feminist geography has developed a broad array of qualitative research methods: in-depth interviewing; ethnographies; action research; and a discursive analysis of spoken, written and visual material. Linda Peake (2015), however, criticizes both feminist geographers' dismissal of quantitative methods and the association of quantitative techniques with masculine research practices. She argues that quantitative methods and GIS can be very valuable to feminist geography, as long as:

> it can yield knowledge that is reliable, effective, and non-oppressive to women or to other socially marginalized or disempowered people; that it honours feminists' commitments to taking seriously the experiences of these people on their own terms; that it addresses differences while retaining a capacity to draw general, even law-like conclusions, but conclusions that derive from specific historical and geographical contexts and not a generalized notion of women; and that it produces knowledge that can lead to transformative action in the direction of more democratic social relations.
>
> *Peake 2015, 262*

In teaching about gender and intersectionality, we might therefore focus less on the dimensions and (often binary) categories and more on the power relations involved. Intersectionality matters because of the implications of inequality, marginalization, exclusion, discrimination, prejudice, power and hegemony. In teaching, I emphasize, it is important not only to focus on patterns and mechanisms of hegemony and subordination (who is marginalized, subordinated, how and why) but also to highlight the agency, the way that subordinated people or groups are using their power and counteracting hegemony and their activism to create spaces of resistance in which the dominant culture is contested and reversed. It is important to give students, as the future generations of geographers, a perspective on transforming society and space.

How we teach: feminist pedagogy in geography teaching

A common starting point in a gender course is an assignment in which students are invited to reflect on the concept of gender from their own personal experience. They are asked to reflect on their own geographical and social background and to give examples of gender bias, gender blindness or gender prejudice and discrimination in their personal life as a starting point for more theoretical discussions.

Feminist pedagogy is usually characterized as non-hierarchical, using personal experiences and intersectional knowledges of students in a diverse classroom, with learning and teaching as instruments of empowerment and social change (Brown et al. 2014; Webber 2006). Feminist classrooms can be seen as 'heterotopias' (Kannen 2014), as safe places in which students feel free to speak about issues that they do not feel comfortable to speak about in other classes, together with: readings from different perspectives (Spencer 2015); student-led classes and field-trips (Van Hoven et al. 2010); assessment based on essay writing and reflection journals, instead of written exams (Lee 2012); participatory action research as a teaching tool (Grant and Zwier 2011); engagement with diverse communities, community organizations and social movements (Burke et al. 2017) 'about, with and from' (Tallon 2011); and unorthodox modes of expression, such as video-making, rap and blog-writing.

Feminist geography teachers adhere to these characteristics of feminist pedagogy yet at the same time share some ambivalences with respect to these characteristics. The main ambivalences refer to issues of hierarchy, safe classrooms, diverse classrooms, the role of fieldwork and separation versus integration strategies.

Non-hierarchical feminist pedagogy implies that students are learners and teachers, and that teachers are also learning. Feminist classrooms are indeed usually more inclusive, interactive and collaborative, and students usually have more impact on shaping course outcomes than in 'mainstream' teaching. Kath Browne is nevertheless critical about incorporating private, personal experiences of students into feminist geography classrooms: 'The assumption that including the personal necessarily contests hegemonic power relations is problematic' (Browne 2005, 352). Feminist geography courses as part of a university curriculum are inevitably hierarchical (Webber 2006). The teacher is responsible for the course, is assessing it and has to decide whether or not a journal, blog or video is good, or whether or not students are successful in connecting personal experiences and theoretical concepts. Feminist geography teachers are embedded in an educational system in which boards of examiners, university managers and external assessment committees are meticulously controlling the quality of the education, assessment and procedures.

A second ambivalence regards the issue of 'safe' classrooms. Students in feminist classrooms are supposed to be free to speak about personal experiences and need a safe environment to do so. Several authors, however, refer to the feelings of discomfort, in particular for those students who are becoming aware of their privileged position for the first time of their life. Audrey Kobayashi reports on her experiences as a teacher of classrooms on racism: 'I feel sometimes as though I carry a bomb into class, and if I am unsuccessful in establishing the right degree of comfort (or discomfort) it will explode with irreversible results' (Kobayashi 1999, 180). Stephanie Simon (2009) reports similar experiences after showing *Paris is Burning,* a movie on drag balls in New York in the 1980s. She chose this movie to start a discussion on compulsory heterosexual urban spaces, but it created strong feelings of discomfort and silence in the classroom. Later, she was able to have a meaningful conversation, starting from the feelings of discomfort. Can classrooms be safe and uncomfortable at the same time? Or do we need (a certain degree of) discomfort to challenge existing stereotypes and hegemonies, and should we give up our claim of 'safe spaces'?

A third ambivalence concerns the issue of homogeneous versus heterogeneous classrooms. For geography in particular, as a discipline with a focus on diversity, to create a fruitful platform for discussing issues of gender and intersectionality a diverse classroom is necessary. As Glen Elder argues (1999, 88): 'queer theorists argue that geography is useful for thinking about identity because identity is more fluid than fixed … queerness captures a delightful sense of unbounded chaos and uncertainty and it helps me to think about identity.'

Other authors are more critical about the issue of diverse classrooms. William New and Michael Merry (2014) refer to Allport, who already in 1954 had formulated five conditions to avoid stereotypes: equal status between persons of different backgrounds; common goals; intergroup cooperation instead of competition; mutual recognition of authority; and personal interactions between persons of different backgrounds. New and Merry report on a pedagogical experiment by Alison Jones in Aotearoa New Zealand, which had separate subsections for Pākehā (White students, usually of European descent) and Māori (Indigenous Aotearoa New Zealand) students. The White students did not like it, because they missed the 'minority' experiences, while the Māori students felt relieved: in 'the segregated setting they felt much more comfortable expressing themselves, without the pressure of becoming someone else's "other"' (New and Merry 2014, 215). Victoria Kannen recounts the 'lone Other' in the class. As the only male student in her classroom complains: 'That's why I am always asked to talk from the "guy" perspective – but why should I know it?' Asking him to speak 'on behalf of' is problematic, because 'it requires these students to use one of their identities as if it is separate from the rest of how they understand themselves and politicize it in a way that may not be comfortable or appropriate for them' (Kannen 2014, 63).

In a session of the IGU Commission on Gender and Geography in 2013 in Kyoto, Ellen Hansen presented a paper on 'Gender, Ethnicity, and White Privilege'. She shared her difficulties in her efforts to deal with teaching about White privilege in the context of an overwhelmingly White student and staff population at her university in a small town in Kansas, using the 'privilege walk' as a teaching instrument. This was one of the many situations in which I, coming from a non-Anglophone country, felt myself to be an ignorant outsider. I had never heard of a 'privilege walk'; it looked like a very exotic teaching tool and I couldn't imagine how to use it in my own teaching. Ellen Hansen explained that one of the male students became angry and accused her of assigning him a role that he did not deserve, leaving out issues of socioeconomic class. Her presentation at the IGU conference resulted in a lively debate in which many people exchanged experiences and suggestions about how to deal with such a situation. My ambivalent feelings are connected with the necessity, on the one hand, to use categories (often binary) to explain and discuss intersectionality and, on the other hand, to avoid such categorization and the use of binaries. The work of Maria Rodó-Zárate on 'relief maps' as a methodological tool to analyse and understand intersectionality is innovative, in this respect (Rodó-de-Zárate 2014). This methodology does not necessarily start with preconceived dimensions or categories, but with everyday experiences or feelings, using questions such as 'Where do you usually go, which places do you like most/least, why don't you like this place', and only gradually introducing dimensions – 'Would you feel the same if you were a boy/White' – and so on. Students in diverse classrooms can compose, analyse and discuss individual as well as collective 'relief maps' (Baylina Ferré and Rodó-de-Zárate 2017).

A fourth ambivalence is the role of fieldwork in feminist geography teaching. Fieldwork is usually seen as fundamental to teaching geography, to 'deepen experiences with place, broaden student learning, facilitate methodological training and enable the transfer of soft skills and tacit knowledge on what it means to be a professional geographer' (Glass 2015, 485). Feminist geographers, however, criticize fieldwork practices in geography because of exclusionary processes, the tendency to exoticize and the priority of 'sight' above 'cite' (field observations above reading theories) (Monk 2000). Geographical fieldwork, in particular international fieldwork under demanding conditions, leads to the exclusion of specific groups of students such as female, gay, older or disabled students (Hall et al. 2004). According to Nairn (2005), fieldwork has the tendency to exoticize and to reinforce thinking in binary categories instead of contesting binaries. One way to counteract the 'exoticizing' outsider position of students during fieldwork is through a service-learning project of Ann Oberhauser (2008), in which students become insiders as volunteers in community organizations as a way to critically engage in and reflect on geographical concepts. Similar experiences are reported on the role of apprenticeships as a way of insider learning by Hanna Carlsson (2017), who used an apprenticeship to learn about the embodied construction of gender in the masculine spaces of boxing gyms.

A final ambivalence concerns the separation versus integration strategy already discussed in 1982 by Monk and Hanson: the ambivalence between teaching separate courses for groups of students who are really interested, and introducing gender issues into the usual courses for all geography students. They argue that integration is necessary. 'Only in this way, we believe, can geography realize the promise of the profound social change that would be wrought by eliminating sexism' (Monk and Hanson 1982, 11). Louise Johnson is more sceptical about integration: 'As a result, putting gender into geography could well just add "women's concerns" into an unaltered discipline and deflect the feminist focus on women's oppression and patriarchal power' (Johnson 1990, 16).

Feminist geographers usually have a dual strategy, and do both. They experiment with new forms of teaching and 'radical' literature in separate courses and use more mainstream pedagogy and 'acceptable' reading material, packed up in general courses on cultural, social, urban or development geography (Droogleever Fortuijn 2008). What is seen as experimental or mainstream and radical or acceptable, however, depends on the specific context in the faculty, university or country (Monk 2011). This brings me to my third point: *where* we teach feminist geography. How important is the learning environment? What are the differences between countries, universities within countries, faculties within universities and between monodisciplinary and interdisciplinary programmes?

Where we teach: the importance of the learning environment

Judith Timár and Éva Fekete report on the development of feminist geography in East-Central Europe. In the context of a traditional, male-dominated geography department, they use a strategy of 'adding' gender. They quote the Romanian experience of Sorina Voiculescu and Margareta Lelea, using mapping and counting as convincing and acceptable tools: 'At first the students were very sceptical as to how gender can be part of geography. In order to build on familiar terrain, the course started with maps of indicators relevant to gender by nation-state' (Timár and Fekete 2010, 779). The way that gender and intersectionality are taught and the way that dimensions of intersectionality are seen as important depend on the learning environment. Judit Timár (2007) challenges the Anglo-American hegemony in feminist geography that prioritizes theory above empirical work and applied research, and symbolism above a focus on material world and spatial differences of gender inequalities. Saraswati Raju (2002, 173) criticizes postmodern Anglo-American feminist geography as 'an academic luxury that we from the "Third World" cannot afford'.

As feminist geography courses in non-Anglophone countries make use of the abundant feminist geography literature in the English language, they are influenced by the way of thinking in Anglo-American feminist geography. At the same time, they struggle with the dominance of Anglo-American perspectives and the English language. Several authors refer to the linguistic challenge, in that the binary concepts of sex and gender do not exist in many other languages (Garcia-Ramon, Simonsen and Vaiou 2006; Huang et al. 2017; Louargant 2002), and a discussion with students in their own language about these concepts results necessarily in a different debate from that in a classroom with American or British students. Course content mirrors the specific focus of feminist geography in their own country, for example on: issues of work and everyday life in Nordic countries (Chardonnel and Sanders 2002); informal care relations and the family in South-European countries (Garcia Ramon, Simonsen and Vaiou 2006); poverty and livelihoods and gender and environment in Ghana (Awumbila 2007); social movements in Latin-American countries (Veleda da Silva and Lan 2007); rural–urban migration and transnationalism in East Asia (Chiang and Liu 2011; Yeoh and Ramdas 2014); the intersections between sexuality, post-coloniality, indigeneity and race in Australasia (Johnston and Longhurst 2008); or public and private spaces and migration and the diaspora in the Middle East (Fenster and Hamdan-Saliba 2013).

In 2009, I analysed the various ways that feminist geographers design assignments in which students are required to make observations on gendered behaviour and communication in public urban spaces (Droogleever Fortuijn 2009). The analysis resulted in two interpretations to understand the differences in teaching styles: one in connection with differences in national teaching cultures; and one in connection with disciplinary differences in teaching. The first interpretation identified both an Anglo-American informal, open, relational style with a focus on the research subject and theory, and a continental European more formal and protocolized style with a focus on the research object and methodology. The second interpretation focuses on differences in disciplinary embedding: the difference between geography embedded in a social sciences or humanities environment and geography in a planning, architecture or natural sciences environment.

Traditional academic cultures and male hegemony refer not only to the management and academic staff but to students, as well. Heidi Nast (1999) discussed the paradox of critical pedagogy and student-centred learning styles in a Catholic university with a traditional, male-dominated population of students who express their dissatisfaction in their student evaluations. In line with an integration strategy, Heidi Nast had included issues of gender, race, class and sexuality in her 'Social Movements of the Twentieth Century' course and found that the students complained: 'Frankly, I was offended by much of the readings we had to do. I feel that I am attending a Catholic university and am appalled that the teachings of the Church are completely denounced in this course' (Nast 1999, 105). A strategy of separation, however, is a vulnerable one in a neoliberal academic context in which the number of enrolled students represents the most important criterion in the decision on whether or not to keep an elective course on the programme. Robyn Longhurst (2011) and Maria Dolors Garcia-Ramon (2011) report on the tough competition of technical (GIS), planning, management, economics and business studies electives, at the expense of courses on gender. Robyn Longhurst and Lynda Johnston emphasize the openness of the newly founded universities to feminist geography teaching, compared to the older, traditional universities, and the importance of a social sciences and humanities faculty (instead of a natural sciences faculty) to embedding it to open up opportunities to teach sexualities, space and place in Australasia (Longhurst and Johnston 2005; Johnston and Longhurst 2008).

Several authors (Chiang and Liu 2011; Drucker 2016; Fenster 2011; Timár and Fekete 2010) demonstrate that interdisciplinary programmes are much more open to gender issues than disciplinary geography programmes. Interdisciplinary programmes are more open to innovative teaching styles and have a more diverse teaching staff and student population. Robyn Longhurst (2011) notes that geography departments with a high percentage of female staff are more open to teaching feminist geography courses and to including gender themes and perspectives in core courses.

Conclusion

Due to the long history of feminist geography in Anglophone countries and the availability of an extensive body of feminist geography literature in the English language, Anglo-American feminist geography has had a strong impact globally. In many places in the world, feminist geography developed from women's studies into postmodern feminist geography, with a focus on cultural diversity, intersectionality and identity.

Feminist geography teaching globally is characterized by a focus on gender relations and gender identities, a practice of using qualitative methodologies, non-hierarchical pedagogies and a strong commitment to social change. *What* we teach and *how*, however, are highly dependent on context. Teaching practices and perspectives differ between countries, linguistic communities, universities, faculties and (monodisciplinary or interdisciplinary) programmes. The course content in feminist geography teaching reflects the major social and political themes in a specific country. The context of the university (new or traditional, religious or secular), the faculty (geography embedded in a social sciences or a natural sciences faculty) and department (high or low proportion of female staff) have an impact on course content, methodology and pedagogy in feminist geography teaching. International contacts, meetings and exchanges of practices enable feminist geography teachers to learn about teaching practices and perspectives globally and to adjust them to fit the specific context in which they work.

Key readings

Baylina Ferré, M., and M. Rodó-de-Zárate. 2017. "New Visual Methods for Teaching Intersectionality from a Spatial Perspective in a Geography and Gender Course." *Journal of Geography in Higher Education* 40 (4): 608–620.

Droogleever Fortuijn, J. 2009. "Gender-sensitive Observations in Public Spaces as a Teaching Tool." *Geographica Helvetica* 64 (1): 37–44.

Monk, J., ed. 2011. "Perspectives on Teaching Geography and Gender in a Postsocialist, Neoliberal-Dominated World." *International Research in Geographical and Environmental Education* Special issue 20 (3).

References

Awumbila, M. 2007. "Challenging Contexts: Gender Studies and Geography in Anglophone African Countries." *Belgeo* 3: 261–274.

Baylina Ferré, M., and M. Rodó-de-Zárate. 2017. "New Visual Methods for Teaching Intersectionality from a Spatial Perspective in a Geography and Gender Course." *Journal of Geography in Higher Education* 40 (4): 608–620.

Brown, S., R.-C. Collard, and D. Hogeveen. 2014. "Pedagogical Declarations: Feminist Engagements with the Teaching Statement." *Journal of Geography in Higher Education* 38 (1): 148–154.
Browne, K. 2005. "Placing the Personal in Pedagogy: Engaged Pedagogy in 'Feminist' Geographical Teaching." *Journal of Geography in Higher Education* 29 (3): 339–354.
Buehler, E. and Baechli, K. 2007. "From Migration der Frau aus Berggebieten to Gender and Sustainable Development: Dynamics in the Field of Gender and Geography in Switzerland and in the German-speaking Context." *Belgeo* 2007 (3): 275–300.
Burke, S., A. Carr, H. Casson, K. Coddington, R. Colls, A. Jollans, S. Jordan, K. Smith, N. Taylor, and H. Urquhart. 2017. "Generative Spaces: Intimacy, Activism and Teaching Feminist Geographies." *Gender, Place & Culture* 24 (5): 661–673.
Carlsson, H. 2017. "Researching Boxing Bodies in Scotland: Using Apprenticeships to Study the Embodied Construction of Gender in Hyper Masculine Space." *Gender, Place & Culture* 24 (7): 939–953.
Chardonnel, S., and L. Sanders. 2002. "La question du genre en Suède: Enjeu de société et objet de recherche géographique." *Espace, Populations, Sociétés* 3 (3): 265–281.
Chiang, L.-H.N, and Y. Liu. 2011. "Feminist Geography in Taiwan and Hong Kong." Gender, *Place & Culture* 18 (4): 557–569.
Crenshaw, K. 1993. "Mapping the Margins: Intersectionality, Identity Politics, and Violence against Women of Color." *Stanford Law Review* 43: 1241–1299.
Creton, D. 2002. "Editorial. Questions de genre." *Espace, Populations, Sociétés* 3 (3): 253–255.
Droogleever Fortuijn, J. 2008. "Balancing the Margin and the Mainstream." In: *Feminisms in Geography; Rethinking Space, Place, and Knowledges*, edited by P. Moss and K.F. Al-Hindi, 75–83. Plymouth: Rowman and Littlefield.
Droogleever Fortuijn, J. 2009. "Gender-sensitive Observations in Public Spaces as a Teaching Tool." *Geographica Helvetica* 64 (1): 37–44.
Drucker, D.J. 2016. "Bringing Gender and Spatial Theory to Life at a German Technical University." *Gender, Place & Culture* 23 (11): 1560–1571.
Elder, G.S. 1999. "'Queering' Boundaries in the Geography Classroom." *Journal of Geography in Higher Education* 23 (1): 86–93.
Fenster, T. 2011. "Teaching Gender in Israel: Experiences at the Tel Aviv University." *International Research in Geographical and Environmental Research* 20 (3): 195–197.
Fenster, T., and H. Hamdan-Saliba. 2013. "Gender and Feminist Geographies in the Middle East." *Gender, Place & Culture* 20 (4): 528–546.
Garcia-Ramon, M.D. 2011. "Teaching Gender and Geography in Spain." *International Research in Geographical and Environmental Research* 20 (3): 185–188.
Garcia-Ramon, M.D., and J. Monk. 2007. "Gender and Geography: World Views and Practices." *Belgeo* 2007 (3): 247–260.
Garcia-Ramon, M.D, K. Simonsen, and D. Vaiou. 2006. "Guest Editorial: Does Anglophone Hegemony Permeate *Gender, Place and Culture*?" *Gender, Place & Culture* 13 (1): 1–5.
Glass, M.R. 2015. "'Introduction.' International Geography Field Courses: Practices and Challenges." *Journal of Geography in Higher Education* 39 (4): 485–490.
Grant, C.A. and E. Zwier. 2011. "Intersectionality and Student Outcomes: Sharpening the Struggle against Racism, Sexism, Classism, Ableism, Heterosexism, Nationalism, and Linguistic, Religious and Geographical Discrimination in Teaching and Learning." *Multicultural Perspectives* 13 (4): 181–188.
Hall, T., M. Healy, and M. Harrison. 2004. "Fieldwork and Disabled Students: Discourses of Exclusion and Inclusion." *Journal of Geography in Higher Education* 28 (3): 255–280.
Hovorka, A.J. 2015. "The Gender, Place and Culture Jan Monk Distinguished Annual Lecture. Feminism and Animals: Exploring Interspecies Relations through Intersectionality, Performativity and Standpoint." *Gender, Place & Culture* 22 (1): 1–19.
Huang, S., J. Monk, J. Droogleever Fortuijn, M.D. Garcia-Ramon, and J. Henshal Momsen. 2017. "A Continuing Agenda for Gender: The Role of the IGU Commission on Gender and Geography." *Gender, Place & Culture* 24 (7): 919–938.
Johnson, L. 1990. "New Courses for a Gendered Geography: Teaching Feminist Geography at the University of Waikato." *Australian Geographical Studies* 28 (1): 16–28.
Johnston, L., and R. Longhurst. 2008. "Queer (ing) Geographies 'Down Under': Some Notes on Sexuality and Space in Australasia." *Australian Geographer* 39 (3): 247–257.
Kannen, V. 2014. "These Are Not 'Regular Places': Women and Gender Studies Classrooms as Heterotopias." *Gender, Place & Culture* 21 (1): 52–67.

Kobayashi, A. 1999. "'Race' and Racism in the Classroom: Some Thoughts on Unexpected Moments." *Journal of Geography* 98 (4): 179–182.
Lee, M.R. 2012. "Teaching Gender and Intersectionality: A Dilemma and Social Justice Approach." *Psychology of Women Quarterly* 36 (1): 110–115.
Little, J. 2007. "Gender and Geography: Developments in the United Kingdom 1980–2006." *Belgeo* 3: 335–348.
Longhurst, R. 2011. "Teaching Gender Geography in Aotearoa New Zealand." *International Research in Geographical and Environmental Research* 20 (3): 179–183.
Longhurst, R., and L. Johnston. 2005. "Changing Bodies, Spaces, Places and Politics: Feminist Geography at the University of Waikato." *New Zealand Geographer* 61: 94–101.
Louargant, S. 2002. "De la géographie féministe à la 'gender geography': Une lecture francophone d'un concept Anglophone." *Espace, Populations, Sociétés* 3: 397–410.
Metcalfe, B.D., and C. Woodhams. 2012. "Introduction. New Directions in Gender, Diversity and Organization Theorizing: Re-imagining Feminist Post-colonialism, Transnationalism and Geographies of Power." *International Journal of Management Reviews* 14: 123–140.
Monk, J. 2000. "Looking Out, Looking In: The 'Other' in the *Journal of Geography in Higher Education*." *Journal of Geography in Higher Education* 24 (2): 163–178.
Monk, J. 2011. "Politics and Priorities: Placing Gender in Geographic Education." *International Research in Geographical and Environmental Education* 20 (3): 169–174.
Monk, J., and S. Hanson. 1982. "On Not Excluding Half of the Human in Human Geography." *Professional Geographer* 34 (1): 11–23.
Nairn, K. 2005. "The Problems of Utilizing 'Direct Experience' in Geography Education." *Journal of Geography in Higher Education* 29 (2): 293–310.
Nast, H.J. 1999. "'Sex', 'Race' and Multiculturalism: Critical Consumption and the Politics of Course Evaluations." *Journal of Geography in Higher Education* 23 (1): 102–115.
New, W.S., and M.S. Merry. 2014. "Is Diversity Necessary for Educational Justice?" *Educational Theory* 64 (3): 205–225.
Oberhauser, A. 2008. "Feminist Pedagogy: Diversity and Praxis in a University Context." In: *Feminisms in Geography; Rethinking Space, Place, and Knowledges*, edited by P. Moss and K.F. Al-Hindi, 215–220. Plymouth: Rowman and Littlefield.
O'Neill Gutierrez, C., and P. Hopkins. 2015. "Introduction. Young People, Gender and Intersectionality." *Gender, Place & Culture* 22 (3): 383–389.
Peake, L., ed. 1989. "The Challenge of Feminist Geography." *Journal of Geography in Higher Education* 13 (1): 85–121.
Peake, L. 2015. "The Suzanne Mackenzie Memorial Lecture: Rethinking the Politics of Feminist Knowledge Production in Anglo-American Geography." *Canadian Geographer* 59 (3): 257–266.
Raju, S. 2002. "We Are Different, But Can We Talk?" *Gender, Place & Culture* 9 (2): 173–177.
Rodó-de-Zárate, M. 2014. "Developing Geographies of Intersectionality with Relief Maps: Reflections from Youth Research in Manresa, Catalonia." *Gender, Place & Culture* 21 (8): 925–944.
Simon, S. 2009. "'If you Raised a Boy in a Pink Room…?' Thoughts on Teaching Geography and Gender." *Journal of Geography* 108 (1): 14–20.
Spencer, L.G. 2015. "Engaging Undergraduates in Feminist Classrooms: An Exploration of Professors' Practices." *Equity & Excellence in Education* 48 (2): 195–211.
Tallon, R. 2011. "Creating 'Little Sultans' in the Social Sciences: Learning about the Other through Benevolent Eyes." *International Research in Geographical and Environmental Education* 20 (4): 281–286.
Timár, J. 2007. "Differences and Inequalities: The 'Double Marginality' of East Central European Feminist Geography." *Documents d'Anàlisi Geogràfica* 49: 73–98.
Timár, J., and E. Fekete. 2010. "Fighting for Recognition: Feminist Geography in East-Central Europe." *Gender, Place & Culture* 17 (6): 775–790.
Van Hoven, B., W. Been, J. Droogleever Fortuijn, and V. Mamadouh. 2010. "Teaching Feminist Geographies in the Netherlands: Learning from Student-led Fieldtrips." *Documents d'Anàlisi Geogràfica* 56: 305–321.
Veleda da Silva, S., and D. Lan. 2007. "Geography and Gender Studies: The Situation in Brazil and Argentina." *Belgeo* 3: 371–382.
Verloo, M. 2006. "Multiple Inequalities, Intersectionality and the European Union." *European Journal of Women's Studies* 13 (3): 211–228.

Webber, M. 2006. "Transgressive Pedagogies? Exploring the Difficult Realities of Enacting Feminist Pedagogies in Undergraduate Classrooms in a Canadian University." *Studies in Higher Education* 31 (4): 453–467.
WGSG. 1984. *Geography and Gender. An Introduction to Feminist Geography*. Women and Geography Study Group of the Institute of British Geographers. London: Hutchinson.
Yeoh, B., and K. Ramdas. 2014. "Gender, Migration, Mobility and Transnationalism." *Gender, Place & Culture* 21 (10): 1197–1213.

47
AUTOGEOGRAPHY
Placing research in the first-person singular

Sophie Tamas

Here's the deal.

When I did my PhD, my supervisor had small babies. My two girls were already in elementary school. 'Writing,' she said, 'is like cooking a meal one grain of rice at a time.' I nodded sympathetically, playing with her infant on her living-room floor while she gave me feedback on chapter drafts, but I did not understand. I didn't know what each grain cost, or the vast difference between zero and one.

My girls are 17 and 20 now, and my son is four months old. I have not slept for more than two consecutive hours for so long that time has gone shapeless. One of the grandmas has taken the baby for a walk; I have perhaps an hour in which to do everything, but I am sitting on the swinging bench on the peeling front porch of my small-town Ontario century cottage, on a rare cool day in August, with a warm hand of sunshine on my back. Instead of weeding or washing or sorting or resting or reading my students' theses (like my supervisor did), I am writing, slowly and laboriously. The writer in me is submerged and dissolving in the drowning-pool of mothering. This line is one grain of rice. This chapter a desiccant, or maybe a snorkel, or a tracheotomy? I am too tired to keep my metaphors straight.

Reading academic prose makes me feel informed but inadequate. Writing it used to make me feel smart and armoured-up with citations, but now it feels impossible. The abstract for this chapter is accidentally ironic. I was just trying to be coherent, but ended up doing the god-trick (Harraway 2003) of talking from outside and above myself, as if I know what I'm doing: performing the useful service of passing on knowledge about something significant that I know about that might make you better somehow. But I don't know you. You don't need improvement. I am the one who needs this. Somehow, magically, stupidly, talking to my imaginary reader is making me feel a little less trapped or lonely. I will have a bit more room in my heart to lift my drooly boy out of his pushchair gratefully and graciously, because I have taken this time, put this before every other more urgent thing. This chapter, and maybe all of my writing, is about managing my abandonment issues. I am softening and smoothing the frontier between inside and out. Do you need to know that? Would you prefer an impersonal, abstract discussion of why a feminist geographer's writing could or should be personal and concrete? The do's and the don'ts? You'll read them in, even if I try to leave them out. There are only a few ways to shuffle the cards of the available discourses, and only a few games that we know how to play.

The best writing advice I ever got was 'Show, don't tell', so I won't just give you a list of the reasons and risks around bringing personal narrative into feminist geography. I'm being performative (I don't like that word) because, for you, being shown is often more useful than being told, and because, for me, it's all that I've got. I don't have a clever, armoured person on hand to explain everything. It's just me, on the porch, with the baby on his way back, the sun behind a cloud and the gears of social reproduction turning in my skin (Henry 2017).

(That's not true. There is also a dog beside me on the bench, jumping down now and then to ward off passing squirrels. He is roughly the size and colour of a fuzzy loaf of bread with dainty legs. His concerns about personal narrative are: where does it place his body, in relation to my lap? How does it smell? Does it carry my hands toward him? Significant issues that I tend to overlook.)

Spades: why write research in first person singular?

A spade is a useful tool for digging and planting, much like personal narrative.

I thought of the term *autogeography* while not-sleeping one night in mid-August. A search online the next day found it attached to one album, two poetry collections and a chapter by Barrie Jean Borich on creative nonfiction. She says:

> I coined this term *autogeography* to define the creative nonfiction project concerned with the ways we might map our bodies and places as interdependent historical strata … An autogeography is self-portrait in the form of a panoramic map of memory, history, lyric intuition, awareness of sensory space, research, and any other object or relic we pick up along the way that offers further evidence of what does or did or will happen here.
>
> *2013, 99*

I like this definition. It fits my feminist and geographic contention that bodies, things and places are not inert backdrop matter, and meaning does not emerge directly from the discursive ether. It suits my suspicion that, no matter how neatly it's presented, knowledge mostly arrives shambolically and slantwise. These ideas are not new, and not mine – ideas never are. I could give them an unslanted pedigree (new materialist this, poststructural that), but I don't really know where or when they slid from tacit to stated, and now they are everywhere in my field, like intellectual dandelions (see, for instance, Anderson and Harrison 2010).

In this chapter, I was asked to write something about autobiography. Most of my other publications are called autoethnography. Do we really need another word for writing about yourself? Probably not, but (of course) I think autogeography says something different and useful. Biographies and ethnographies – auto or not – often tend to stand-in and speak for whoever was written about. This tendency is amplified when we've written about ourselves – someone we presumably know intimately and well (more on that later). The primary verb that autogeography invokes is mapping. In mapping, the gap between representation and reality is intuitively obvious. You're not transposing one abstract thing (a self or culture) into another abstract thing (language). A good map might be useful, beautiful, original, fascinating, and so on, but it's not a tracing; we don't expect it to fully encapsulate complex realities (Deleuze and Guattari 2004). Whether they're actively wayfinding or passively browsing, the map reader is liable to know they're *doing* something, co-constructing meaning rather than simply ingesting it.

Borich offers autogeography as a type of creative non-fiction. Is it also a type of scholarship? My personal answer is 'Yes, but'. Creative non-fiction, including memoir and autobiography, can (and perhaps should) be written as an end in itself. It often aims to inform, influence, entertain, titillate, etc., but it has not necessarily failed in any fundamental way if the answer to the 'so what?' test is purely aesthetic or ambiguous. I believe academic work can (and perhaps should) be as carefully crafted, accessible and literary as good creative non-fiction, but I don't see it as an end it itself. In order to feel satisfied, I need to be able to plausibly impute some other purpose to it.

This purpose is not to reveal, guide or fix, although I often think that's what I'm doing, and I might indirectly or incidentally do so to some degree. When I write about myself, I am trying to trace how something constellates in experiences that are mine, but not only mine. This is a quixotic mandate; it takes just a whiff of psychoanalytic or poststructural theory to disrupt coherent, autonomous, self-aware subjectivity, let alone authoritative, singular meaning (see Anderson and Harrison 2010, Britzman 2002, Popke 2003 or Tamas 2013). I didn't invent myself. There are no clear lines between the personal and the social; the most detached third-person puppet-show of conventional academic prose is shaped by individual subjects, and the most intimate first-person performance is based on social scripts. But even in the absence of binary poles, there are still degrees of difference and writerly decisions to be made. Ostensibly dispassionate, detached prose is a good tool for some jobs, but not the best tool for every job. I've pinned my use of personal narrative on my abandonment issues, but here's a few other reasons why I believe it deserves more space in the academic repertoire.

Location matters. Humility is hard. Showing where you speak from (aka your standpoint) can minimize the potential trespass of speaking for others. It doesn't automatically do so, particularly if your personal coordinates are wedged into a breezy preface meant to allow readers to calculate and offset your biases so that the dream of objective knowledge can be sustained. We generally want authors to offer certainty, allowing us to either trust them or be certain they're wrong. Reminding you that I am just some idiot, now hiding in a bedroom while a sitter amuses the baby downstairs, holding back the need to pee so I can eke out a few more words, shakes up the snow-globe of authority. I have to resist my own inclination to confine this data to a parenthetical aside; see how easily the text veers into disembodied instruction? I'm not bringing my body back in (it was here all along), but I am deciding that it matters. This is partially about ethics and representation, but it's also about boredom. If I locate myself well, with precise, evocative images, I might also draw you into your body, which increases the odds that you'll feel something – curiosity, surprise, connection, amusement, irritation – any feeling could help sustain your attention.

I have 23 more minutes of freedom, but the baby is getting fussy downstairs and I'm losing momentum. I don't like the way I'm making pronouncements then calling attention to their artifice, having my cake and eating it. Noticing a habit is not the same as changing it.

Now it's a few days later and the sitter is here again, but I don't have enough faith in this to prioritize it over packing for the weekend away (volunteering with my partner's reno company to fix up a non-profit summer camp and going to his sister's surprise birthday party; trying grimly to hold on to the shreds of things that were possible and pleasurable in the life before baby). I should go have a nap, but I won't. I don't have much faith in sleep, either.

Hope matters. I can't write without it. Even the most crushingly bleak personal story carries an implicit hope in the fact that the narrator lived to tell the tale. Carolyn Ellis (2009), the Grande Dame of autoethnography, warns that our stories should leave room for hope. I agree, but this easily veers into saccharine lies of omission and gendered, raced, religious codes of quiet suffering (the meek will inherit oppression). Everything happens for a reason. Every cloud has a silver lining. What doesn't kill you makes you stronger. These tropes are supposed to help us choke down unbearable suffering, but they can also leave us chewing on things that are wrong as if it's our personal duty to render social problems digestible. Experience that cannot be hammered into a narrative arc that features clearly defined heroes and foes and ends on an upswing (or, failing that, in pitiful ruin) is rendered unspeakable. Trauma appears as an awful and fascinating departure from the imaginary security of normal life, its disruptive potential contained by facile explanations that estrange those others or aspects of ourselves in most acute need of witnessing.

The hope that my writing and I require is a darker thing with complex flavours, which lives beside pain without obscuring it. This isn't about a goth crush on being angsty; it's about respecting the gap between language and experience. It is, according to Theodor Adorno (1982), barbaric to represent the unthinkable as meaningful scenes in which humanity flourishes. The furore of writing by Holocaust survivors reveals both the inadequacy of language and its necessity. Its value lies primarily in what it does rather than what it denotes. The poet Celan explains that language can pass 'through its own answerlessness' and yield 'no words for what was happening', but still provide those who are 'stricken by and seeking reality' with 'a being underway, an attempt to gain direction', a movement 'toward something standing open, occupiable, perhaps a "thou" that can be addressed' (cited in Felman 1995, 34).

For me, ethical, sustainable hope requires resisting the urge to clip unmanageable things (including myself) into stable, familiar shapes. This is difficult, weak magic, with no miraculous healing powers, so I don't judge folks who stick by what Lather and Smithies (1997) call the 'comfort text'. I have no right to undermine the practices that make your life bearable if I don't have anything better to replace them with. I do, however, judge the cultivation of discomfort texts, in which the ability to find fault stands in for intelligence. The hyper-critical, compulsory anxiety of the academy doesn't make me happy. According to Eve Sedgewick (2002), it's a paranoid defence, clenched around our fears. According to Brian Massumi (2002), it's poor strategy:

> If you don't enjoy concepts and writing, and feel that when you write you are adding something to the world, if only the enjoyment itself, and that by adding that ounce of positive experience to the world you are affirming it, celebrating its potential, tending its growth, in however small a way, however really abstractly – well, just hang it up. It is not that critique is wrong. As usual, it is not a question of right and wrong (nothing important ever is). It is a question of dosage. It is simply that when you are busy critiquing you are less busy augmenting. You are that much less fostering. There are times when debunking is necessary. But if applied in a blanket manner, adopted as a general operating principle, it is counterproductive. Foster or debunk. It's a strategic question. Like all strategic questions, it is basically a question of timing and proportion. Nothing to do with morals or moralizing. Just pragmatic.
>
> *13*

This bruised hope feels like it lets me, the text and the reader breathe. (Every time I read that line, I breathe). It follows the Trickster. Its footprints are found in juxtapositions, evocative details, artistic beauty, vulnerable slips, bits of context or funny asides that gently recognize the ridiculous, especially when I'm moralizing, taking myself too seriously or being contradictory – by, for instance, writing defensively about defensive writing (Exhibit A: over-kill Holocaust references when I'm supposedly too spent to armour-up with citations).

Voice matters. I am scared, now, that this text is getting too long and loose, rattling with stops and starts. I want to rush, to pack it orderly and tight as commuters on a Japanese train. It takes patience to let it shamble along like we're taking a walk with a toddler. (Which makes me think of Mountz et al. 2015, on slow scholarship.)

I did a few radio stories for the Canadian Broadcasting Corporation back when my girls were small. They told me to write as if I was speaking into the ear of one person. People often listen to the radio alone; it's an intimate mass medium. People often read alone, too. Keeping this one-to-one voice on the page shapes the relationship between me and the reader. It's just us, in here, with voice managing our proximity and positions in the imaginary geography of the text. You can tell if I am being distant or letting you in, if I am making you comfortable. The voice I use shows you my spatial beliefs about knowledge: is it achieved by standing back and looking (the standard academic model) or by leaning in, listening and feeling? The feminist philosopher Lorraine Code (1995) calls this intimacy a method of knowing responsibly and well. Voice can intimidate and exclude, telling readers they are too ignorant to read this. It can welcome or manipulate, enrage or bore. It can re-centre the hegemonic (White, unimpaired, straight, neurotypical, middle-class, masculinist, disembodied) body, if I assume that's who I'm speaking to.

You may think there is only one kind of voice that is acceptable for important, adult speech. You may write in that voice, even if you don't like it. I still do (did you read my abstract?). Sometimes the normative academic voice is necessary, but it can also be a habitual hiding place with political consequences. When I stop speaking that way, nothing bad happens. I feel relief, a release of the pressure to inhabit a barely recognizable version of myself. Speaking from my heart doesn't mean I stop using my head; it's a false binary.

I want to write in the voice that occurs on long drives, often at night, when you are sitting nearby but facing outwards, scanning terrain that we are crossing together but seeing differently. I love the kind of hearing it allows, and the feel of those conversations.

Power matters. In this text I am behind the wheel, and you are a passenger. You've climbed in and closed the door based on trust that I'll take you where the abstract promised, not straight into a wall or backwards around traffic circles. The trespasses inherent in research that treats others like lab rats may be obscured but are not erased by my intimate tone. They might be exacerbated; if I don't think I am exercising power, I am unlikely to be mindful of its effects.

If I write well, I can probably make you feel. This power must be conserved and used carefully. I am writing things into existence, making them appear more or less real and significant by rendering them relatable. I don't control the meaning you take from the text – what you see along the way – but I choose the route and speed, and your trip is framed by my intentions.

Friendly feminist research methods often aim to flatten power hierarchies but remain ethically treacherous; as Stacey (1996) points out, we become the lover who feels more free to leave. When my fieldwork happens at home, I have to live with its relational consequences indefinitely. This makes it more difficult to play the curious colonial tourist, but also exposes us to more complicated forms of harm (see Tuhiwai Smith 1999; Simmonds 2011). Because my experiences are never only mine, my personal narrative inevitably touches others, in ways they might not like. Deciding if this is the kind of pain that means *stop* or the kind that means *keep going* is always a power move. (I'm not pulling over now, you can hold it). Having more skin in the game can make me more careful and attentive, but it can also make me more cowardly. I now spend more time on the page fussing about power implications than I do telling stories. Wondering whether my over-shares are fuelled by masochistic repetition compulsion is likely both ethically necessary and a defensive retreat from more precarious, difficult journeys – but I appreciate the way personal narrative puts these power dynamics in high relief.

Scale matters. We tend to imagine that the macro is big, distant, general or abstract, and important, while the micro is small, up close, specific or concrete, and relatively unimportant.

Patterns and trends matter; outliers are 'noise' we filter out to render data useful. In one of his many clever publications, John Law (2004) turns this on its head, arguing that the macro is actually only ever found within the micro. As an example, he digs into the minutiae of a bit of British aerospace history and finds the entire Cold War packed into scraps of hallway chit-chat.

This makes me think of black holes, where vast things are compressed into vanishing points, and the wormholes that form inside them, where nothingness takes you elsewhere, and Mobius strips, where this surface becomes that, and Leibniz's monadology, in which every particle contains the whole (see Law 2004). It also makes me think of personal narrative and its aspirations. The idea that important things are big and far can make us feel, and therefore be, disempowered. If I will never have read enough or know enough or be lofty enough to see a proper macro overview, I might feel like I have nothing worth sharing, no gift to give. My smallness feels weak, vulnerable and shameful, like something I could and should overcome or conceal prior to publication. Everyone else seems so big.

I can make myself feel bigger by calling in a posse of citations to bully you into seeing my point, but I'd rather pull them out of my pocket like a child playing 'show and tell' with linty bits of treasure. When I speak from a place of smallness, more interesting things happen. Sometimes, its humble particularity unfolds and opens out into a spacious wonder.

Clubs: what are the risks of using personal narrative?

A club is a thing used to hit or exclude.

Back to the Holocaust. Primo Levi wrote about his experiences in the camps and reflected on that writing. He said, 'There is only one risk, of writing badly'. For him, that meant writing that is useless (cited in Probyn 2010, 87, in her lovely discussion of writing and shame).

What makes writing useless? I don't know, but here's a list of possibilities. (Ah, lists: you make endless things seem finite and feasible.)

- Lacking generosity toward the reader and failing to make good use of their time. This usually manifests as a reluctance to cut text, because I like it whether or not the reader needs it. If I am writing for myself, it's a journal, not an outward-facing publication. It is self-centred to approach the page as an opportunity to show off my cleverness, vocabulary, theoretical erudition, likeability, travails or other exceptional qualities. It is similarly inconsiderate to expect readers to stumble through bad grammar, awkward phrasing and choppy structure. Learn how to write well: read how-to books, find authors with admirable style, write lots of drafts and find honest friends to read them.
- Losing readers' interest and trust because my narrator is too unreliable or too authoritative. If meaning is too open-ended and diffuse or too closed and static, I've overstepped and lost my balance on the knife-edge of (minor) crisis that makes education possible. As Britzman and Pitt (2006) advise, my job is 'provoking, not representing, knowledge' (394). This requires trust. Readers might need handrails on steep bits and bridges over gaps, but they don't need to be carried, and don't want me to tell them exactly where to look. Beware of too much expository prose, even in handbook chapters ☺.
- Preaching. Aiming to help seems nice, but it positions us as external and superior to circumstances, not as participants. As my current therapist puts it, 'We're all just bozos on the bus'. Trying to instruct and change others, as I am doing now, places us above our readers and places readers above whomever we have written about. This can

temporarily relieve anxiety, but it is manipulative, coercive, disrespectful, unproductive and ultimately boring (see Mamet 1998). Compelling, believable stories are inhabited by complex, imperfect characters, not angels and demons.

- Carelessness. Readers will learn more from the ethic that I enact than from the ethic that I describe. I aim for compassionate curiosity. This starts with self-care. You cannot know how you (or others) will feel about the text, and those feelings will change over time; but the more volatile it is, the longer I sit with it before publication. One book has been resting for three years while I let things settle and feel out the tangle of impulses that produced it. Fear is not a reliable guide, but don't write into the dark unless you have strong relational ropes to follow back out to here and now. Don't pimp your own suffering; you deserve your own kindness. Our worst moments don't need to define us. Don't conflate yourself with your narrator. I love the part of me that writes, but she's not all of me, and her mania for observation and analysis is an inadequate way of being in the world. Don't displace the people you're ostensibly writing about. Don't worry too much about how you are read. 'People are going to judge', my former therapist once said. 'This tells you something about them. It doesn't say much about you.' Don't over-estimate or abdicate your responsibility for what you render imaginable; I can't tell the whole truth, but I try to write with integrity. Don't count on the false comfort of binaries; this is neither a good example nor a cautionary tale. Don't rely on other people's lists of do's and don'ts; there is no recipe for innocence.

Hearts

A heart feels and circulates the things that sustain us.

I am standing at the counter again, rocking the fitfully sleeping baby in a backpack. He might have thrush now, or reflux, or maybe he's just grumpy with the way I've written him as a burden and obstacle to things that matter more. There isn't enough of me to go around, my needs won't be met, everything isn't okay: these fears have travelled with me through all sorts of circumstances, so I am learning to take them more lightly. If the sky is always falling, it's not likely to touch down today.

They are not the only feelings, nor the most useful or precious legacy, but they often surface on the page. They're readily speakable. Part of my job is to notice their peculiarity, to denaturalize my own nature and expand the fullness of my account, because it is liable to be read as if it stands in for and reproduces reality. The monarch butterflies on the tall white marigolds I grew from seed, and the chipmunk (named Kevin) skittering over to the water bowl outside my kitchen window: do I make space for their significance?

The baby, who is awake now and restless: do I give you the warm scent of his wheat-blonde hair? We lose too much of what matters when we let academic habit confine our recognition of meaning.

Diamonds

Made from carbon plus undisturbed time, as is graphite (used in pencils) and organic life.

Some precious things form only under pressure. You squeeze them out (babies, publications). It hurts. They bear so little resemblance to what went into them that where they came from seems mysterious. Who knows where they will go, what they will do? Their value is barely articulable. Why do we make them? We just do. You lose sleep over them, obsess over finding their best possible shape: the one that will capture the most light.

Key readings

Code, L. 1995. "How Do We Know? Questions of method in Feminist Practice." In: *Changing Methods: Feminists Transforming Practice*, edited by S. Burt and L. Code, 13–44. Peterborough, ON: Broadview.
Law, J. 2004. *After Method: Mess in Social Science Research*. New York: Routledge.
Richardson, L. 2003. "Writing: A method of inquiry." In: *Turning Points in Qualitative Research: Tying Knots in a Handkerchief*, edited by N.K. Denzin and Y.S. Lincoln, 379–396. Walnut Creek, CA: AltaMira.

References

Adorno, T. 1982. "Commitment." In: *The Essential Frankfurt School Reader*, edited by A. Aratot and E. Gebhardt, 300–318. New York: Continuum.
Anderson, B., and P. Harrison. 2010. "The Promise of Non-representational Theories." In: *Taking Place: Non-representational Theories and Geography*, edited by B. Anderson and P. Harrison. Farnham: Ashgate.
Borich, B.J. 2013. "Autogeographies. In: *Bending Genre*, edited by M. Singer and N. Walker, 97–102. London: Bloomsbury.
Britzman, D. 2002. "Theory Kindergarten." In: *Regarding Sedgewick: Essays on Queer Culture and Critical Theory*, edited by S. Barber and D.L. Cook, 121–142. New York: Routledge.
Britzman, D., and A.J. Pitt. 2006. "Speculation on Qualities of Difficult Knowledge in Teaching and Learning: An Experiment in Psychoanalytic Research." In: *Doing Educational Research*, edited by K. Tobin and J.L. Kincheloe, 379–402. Boston, MA: Sense Publishers.
Code, L. 1995. "How Do We Know? Questions of Method in Feminist Practice." In: *Changing Methods: Feminists Transforming Practice*, edited by S. Burt and L. Code, 13–44. Peterborough, ON: Broadview.
Deleuze, G., and F. Guattari. 2004. *A Thousand Plateaus*, translated by B. Massumi. London: Continuum.
Ellis, C. 2009. *Revision: Autoethnographic Reflections of Life and Work*. Walnut Creek, CA: Left Coast Press.
Felman, S. 1995. "Education and Crisis, or the Vicissitudes of Teaching." In: *Trauma: Explorations in Memory*, edited by C. Caruth, 13–60. Baltimore, MD: Johns Hopkins University Press.
Harraway, D. 2003. "Situated Knowledges: The Science Question in Feminism and the Privilege of Partial Perspective." In: *Turning Points in Qualitative Research: Tying Knots in a Handkerchief*, edited by N.K. Denzin and Y.S. Lincoln, 21–46. Walnut Creek, CA: AltaMira.
Henry, C. 2017. "The Abstraction of Care: What Work Counts?" *Antipode*. doi: 10.1111/anti.12354.
Lather, P., and C. Smithies . 1997. *Troubling the Angels: Women living with HIV/AIDS*. Boulder, CO: Westview Press.
Law, J. 2004. "And If the Global Were Small and Noncoherent?' Method, Complexity and the Baroque." *Environment and Planning D: Society and Space* 22: 13–26.
Mamet, D. 1998. *Three Uses of the Knife: On the Nature and Purpose of Drama*. New York: Vintage.
Massumi, B. 2002. *Parables for the Virtual: Movement, Affect, Sensation*. Durham, NC: Duke.
Mountz, A., et al. 2015. "For Slow Scholarship: Feminist Politics of Resistance through Collective Action in the Neoliberal University." *ACME: An International E-Journal for Critical Geographies* 14 (4): 1235–1259.
Popke, J.E. 2003. "Poststructural Ethics: Subjectivity, Responsibility, and the Space of Community." *Progress in Human Geography* 27 (3): 298–316.
Probyn, E. 2010. "Writing Shame." In: *The Affect Theory Reader*, edited by M. Gregg and G. Seigworth, 71–92. Durham, NC: Duke University Press.
Sedgewick, E.K. 2002. *Touching Feeling*. Durham, NC: Duke University Press.
Simmonds, N. 2011. "Maha wahine: Decolonizing Politics." *Women's Studies Journal* 25 (2): 11–25.
Stacey, J. 1996. "Can There Be a Feminist Ethnography?" In: *Feminism and Social Change: Bridging Theory and Practice*, edited by H. Gottfried, 206–224. Urbana, IL: University of Chicago Press.
Tamas, S. 2013. "Who's There? A Week Subject." In: *The Handbook of Autoethnography*, edited by T. Adams and C. Ellis, 186–201. Walnut Creek, CA: Left Coast Press.
Tuhiwai Smith, L. 1999. *Decolonizing Methodologies*. London: Zed Books.

48

NARRATING NEW SPACES

Theories and practices of storytelling in feminist geographies

Sarah de Leeuw and Vanessa Sloan Morgan

Introduction: the stories we will tell you

Storytelling. The word simultaneously suggests something innocent and something just a little tainted. After all, the *Oxford English Dictionary* (*OED*) offers definitions of the root word, 'story', including 'a piece of gossip, a rumour; a false statement, a lie' (*OED* https://en.oxforddictionaries.com/definition/story). 'Story', the *OED* goes on to note, is a word used in 'resigned acknowledgement' (as in, 'It's the story of my life') or to indicate that 'one does not want to expand on it for now' (as in, 'But that's another story') (*OED* https://en.oxforddictionaries.com/definition/story). Somewhat differently, but still underscoring how 'stories' and 'storytelling' are slightly tainted as terms, feminist theorists have long observed that stories, as fables or fairy tales, as everyday fictions in films and books or in popular and mass media, have discursively produced sociocultural conditions deeply unfavourable to women (Eagleton 2011). From positioning women as the catalyst of original sin through to producing tropes about evil stepmothers, *femme fatales*, ditzy beauties and jilted lovers (Cohan and Shires 1988; Eagleton 2011), stories are practices and projects that we tell ourselves. As such, and according to many feminist scholars, including geographers (see for instance Domosh 1991; Price 2010), stories and storytelling can narrate patriarchy by legitimating the specific, gendered, sexually oriented, expected, prescribed and enforced ways in which people behave toward others in various spaces across times. Stories have also been understood as cultural architectures of inequities and injustices. When gender intersects in stories with racialization, ethnicity, sexuality, sexual orientation, health and ability or class, many of the tropes narrated in story-spaces become the lived and forcibly marginalized realties of embodied subjects in place. These extend beyond women and men to include colonial constructs of Euro-White supremacy, insistences that only heterosexual women are normal or good, legitimations of racial power and authority, and enduring pathologies about people (especially women) with mental and physical illnesses (see, for instance, McClintock 2013; Said 2012). Indeed, according to Indigenous author and writer Thomas King (2003), 'the truth about stories' is that, as humans, 'that's all we are' (4): stories are wonderous, they are dangerous and they form our very existence, both good and bad.

Dian Million (2014, 32–33) shares that 'stories, unlike data, contain the affective legacy of our experiences. They are a felt knowledge that accumulates and becomes a force that empowers stories that are otherwise separate to become a focus, a potential for movement'.

Speaking from her Tanana Athabascan perspective, Million (2014) emphasizes not only the power of story but also the way in which stories are often decontextualized when subsumed in academic discourse. Stories, and how and where and by whom and to whom they are told (that is, storytelling), are thus powerful forces with long and complicated histories that vary from place to place and from community to community. Increased use of the power of stories and storytelling by corporations, for instance, has led to 'capturing' stories as decontextualized scripts (de Leeuw et al. 2017), further complicating the power, history, intent and placed-ness of stories. Indeed, the specificity of place in storytelling and the decision about who has the right to 'tell someone's story' have long been politicized by women, speaking especially from the perspectives of Indigenous peoples and women of colour (for example Keeshig-Tobias 1997).

Speaking to the power of Black women's stories as active resistance, Katherine McKittrick (2006) reminds us that geography, 'with its overlapping physical, metaphorical, theoretical, and experiential contours … overlap[s] with subjectivities, imaginations, and stories' (xxii). Writing about how geographers and geography as a discipline have narrated spaces, McKittrick observes further that:

> Geography's and geographers' well-known history in the Americas, of white masculine European mappings, explorations, conquests, is interlaced with a different sense of place, those populations and their attendant geographies that are concealed by what might be called rational spatial colonization and domination: the profitable erasure and objectification of subaltern subjectivities, stories, and lands.
>
> *McKittrick 2006, x*

Storytelling can be an act of telling, retelling or re-storying to make visible yet, equally, to invisibilize peoples, places and histories. What, then, is the history of geography's interest in stories and storytelling? Where and by whom have stories in geography circulated? What are the implications of stories and storytelling for contemporary geographers, especially feminist and critical geographers? How might critical feminist geographers (re)use stories and storytelling as tools of scholarship in new ways? How might we re-narrate stories and storytelling methods and methodologies to allow ourselves to think differently about the spaces, places, genders, racializations, movements and mobilities, politics and other themes that matter to many geographers?

The work of this chapter addresses these questions. We do this both by reflecting on select (and in no way comprehensive) works by geographers and by examining stories and storytelling, as considered by systemically marginalized voices and subjectivities. We begin with a brief survey of story and storytelling in the discipline of geography, answering in part some very fundamental questions about what constitutes a story and/or the practices of storytelling. We then explore theories about stories and storytelling, offering an overview of how geographers have theorized the two, how the two have travelled within and across the discipline and why stories have been considered important or, conversely, how they have been critiqued. Next, we turn to the applied ways that the stories and storytelling have been, and are currently being, deployed in geography. Indigenous geographies, for instance, have long included storytelling as a culturally and politically rooted means to narrate embodied experiences and knowledges through space and over time (Hunt 2013a); queer geographies include LGBTQ2S+ peoples' stories to highlight intersectional critiques of neocolonial discourses and spatialities in queer tourism (Puar 2002); health geographers use storytelling methods to work with research participants (de Leeuw et al. 2017); and emotional geographers are telling new stories in order to consider affective and experiential ways of being in the world (Munt 2012). Finally, we turn to some concluding thoughts, including where stories and storytelling might innovatively and creatively go in the future, especially for feminist geographers working on critical geographies. We pay some concerted attention to geography's relatively recent and, we feel, rapidly ascending turn toward geohumanities and creative practices. Considering these two trending disciplinary trajectories, we think, provides a rich preface to and fodder for careful reflection on the importance of thinking through where, how and why feminist (and anti-racist and critical) lenses can and might be applied to any emerging modalities of thought and practice about stories and storytelling in our discipline. With this in mind, then, throughout the chapter we are cognizant of King's (2003) observation that stories and storytelling are always, and simultaneously, both wonderous and dangerous.

Once upon a time: histories of stories and storytelling in geography

As a discipline focused on understanding space and place as fundamental powers that structure social, cultural, emotional and physical worlds, geography is historically anchored in stories and storytelling. As Emilie Cameron (2012, 573) points out, however, geographers have not always identified storytelling as a powerful expression of experiences and knowledges:

> There was a time, not too long ago, when few geographers were interested in telling stories. Stories, it seemed, were at best a quaintly humanistic preoccupation and at worst understood as the building blocks of oppression and inequality.

From its inception, though, geographers have observed that the word 'geography' is translated literally from the Greek as 'earthwriting'; put another way, geography as a discipline is founded on telling stories about the world (Clark and Martin 2013; Springer 2017). Despite geographers' hesitancy at times to admit to their practice of 'storytelling', Cameron (2012, 573) asserts that 'geographic engagement with theories of discourse, power, and knowledge led geographers to understand stories as fundamentally implicated in the production of cultural, economic, political, and social power'.

That geography is a practice of telling the world's stories – of earthwriting – is also and especially true if stories and storytelling are broadly understood as ways of making sense of and organizing the world through narrative practices (see, for instance, de Leeuw et al. 2017) and of drawing theory from embodiment (Hunt 2013a; Million 2014). Stories, in other words, are meant to convey knowledge to a recipient through an ordered chronicling, a rendering and retelling (through a variety of means, oral, written and visual) of events, people or places (real or imagined). The practices of telling stories (graphing) about the earth (geo) and the ways in which those stories have been narrated and structured have actually formed and dictated the discipline's very existence (de Leeuw 2017; Springer 2017). Some of geography's earliest pioneers (mostly White men) fervently followed the work of (almost always White male) poets, believing poetry to tell some of the most powerful stories about human–land interactions (Marston and de Leeuw 2013). Perhaps because, in its early incarnations, geography paid attention to sources from both poetry *and* quantitative scientific inquiry and because the story of the discipline is dependent on who is telling it, geography is understood either to have been left out of histories of science or seen as a science unto itself (see, for instance, Livingstone and Rogers 1996). Geography might be a social science or it might be a humanities discipline, depending on the stories being told about it: whatever the discipline's designation, however, it is still a discipline storied mostly by White men (Kobayashi 2014). Its interdisciplinary nature means that stories have been understood differently by different geographers, with social science geographers broadly understanding their stories as concerned with thematic patterning across time and place, while humanities-informed geographers have perhaps been more preoccupied with the uniqueness and creativity of storied materials (Price 2010). Indeed, and again depending on who is telling the story or what story is given validity, geography might have reason to be taught at universities such as Oxford and Cambridge or be cut completely from the curriculum at Harvard. The discipline itself, in other words, is storied: it remains, however, that the stories being told and circulated in the discipline are gendered and racialized. And those stories have structured the ways and places that the discipline is configured, understood and practised.

Geographers, those practitioners within and of the discipline, have for a long time worked with and told stories. Telling and working with stories were the domain of early humanistic geographers striving to theorize and explore sensorial relationships between place and people. Many of these humanist and phenomenological geographers, most whom would not have identified as feminist geographers, turned to stories as source material about deep, affective, thick and emotive relationships with landscape (Hawkins and Straughan 2016; see also Meinig 1983), some going so far as to say that the discipline of geography would be lifeless and dead without poetry, narrative, story and storytelling (Meinig 1983). Alongside and often in conversation with anthropologists and ethnographers in the mid- to late-twentieth century, geographers working against the primacy of quantitative inquiry in the discipline turned to stories and storytellers as phenomenological source materials about how people felt, experienced, built or lived with and in landscapes. Sometimes this turn to stories took its impetus from a search for beauty and deep feeling as an antidote to worries about modern industrialization and the dehumanization of landscapes (Bunkse 2004; Watson 1983). Travel stories were early tropes of geographers, including narratives published in colonial journals or in magazines such as *National Geographic* (Crang 1998). Indeed, storytelling in geography itself has played a role in the discipline's masculinizing and colonizing tendencies. Tools long thought to be core to geography, such as the map, the survey and the grid, as Nick Blomley (2003) points out, have been shared in tandem with stories about race, Eurocentric visions of progress and land (Harley 1989). In other words, while geographers were writing stories about the world they were authoring *their* stories from their positions, often as White, able-bodied, cis-gendered, heterosexual men (Choi 2018; Kobayashi 2014; Kobayashi and Peake 1994).[1] What is a picture or a map, if not a thousand (storied) words?

Stories were also categorized as teachings and enfolded into geographical pedagogies. A number of early human geographers named themselves as storytellers, sharing personal narratives and contributing to creative worldings (Hawkins and Straughan 2016; Watson 1983). Sometimes, in the hands of physical geographers, stories are understood and circulated as scientific and objective evidence or as quantitatively informed exploration – as such, they form the ways in which environment and physical geographies were and are conceptualized (Yusoff and Gabrys 2011). Stories were the mainstay of cultural geography, in some ways, as it established itself as a subdiscipline of human geography; widespread or vernacular cultural artifacts, often in the form of popular texts and creative expressions, were its primary sources, replete with representations and symbols through which to read and understand people, places and productions of power. Contemporary cultural geographers continue to draw on tropes and tools of literary theorists and turn to stories in the form of fiction, poetry and films or, more recently, YouTube videos and blogs as insights into how and why humans behave in the places (or landscapes and regions) where they live (Cosgrove and Domosh 1993; Crang 2013). Social and political geographers turn to stories, often drawn from qualitative sources, as evidence about trends and themes circulating through the subdisciplines, including through feminist geography (Hyndman 2004). Stories and storytelling, in other words, have long been taken up by geographers and have been deployed by geographers for many purposes.

Despite the variability of the discipline across time and place, feminist geographers have observed common themes in the use of stories and storytelling in geography. Many stories have been told by men or are focused on tales of masculinity and men's relationship with space and place (see Domosh 1991). Women's stories, or even just the voices and experiences of women as stories that mattered to geography and for geographers, have either been long overlooked or omitted altogether from the building of the discipline (Domosh 1991). Stories about women of colour (for example, Gilmore 2007), Indigenous women (Goeman 2013; Simmonds 2016), queerly sexually oriented and/or gender non-conforming and trans people (Doan 2010) have also been storied. Women's voices and geographers who are writing from these situated positions also use storytelling as one way to rewrite the stories that have historically been told about 'us' (whoever 'us' may be), without us and by 'others'. Those engaged in such rewriting and critical, reflexive engagement in storytelling have developed unique methods and methodologies that are now central to feminist inquiry in geography.

That's a good story! Stories and storytelling as methods, methodologies and means of situating ourselves in geography

Broadly speaking, a *methodology* is a theoretical or philosophical framework that informs a practice. Storytelling is often conceptualized in this way. Stories and storytelling are also often considered to be *methods* in geography, as things or actions or practices that are the outcomes of an activity or an inquiry and as a way of relating to one another to share situated knowledges, perspectives and experiences. Feminist geographers, for instance, use storytelling as a way to work in alliance, to come together and to co-produce knowledge. Richa Nagar (2013), for instance, discusses how 'co-authoring' stories is itself a form of feminist alliance. Nagar (2013, 1) explains how, in doing so, feminists have worked in relation with people in places to 'mobilize experience and memory work in ways that connect questions of feminist subject/subjectivity with those of representation in organization, leadership, and movement politics'. Emphasizing that 'alliance work demands a constant rethinking of the relationship between authorship of words and authorship of struggles' (Nagar 2013, 9), Nagar outlines how co-authoring and storytelling can mean working with one another for action: '[c]o-authoring stories is a chief tool by which feminists working in alliances across borders mobilize experience to write against relations of power that produce social violence, and to imagine and enact their own visions and ethics of social change' (2013, 1). Storytelling as co-authoring demonstrates not only how feminist geographers can methodologically approach a topic but also how the co-production of knowledge as storytelling is key to every aspect of embodiment involved in feminist research praxis. Nagar shares five 'truths' that arose from her feminist alliances, emerging as stories themselves through storytelling (see Nagar 2013, 8–13). Themselves generated through reflexive engagement with a story that came about over a decade of alliance building, these truths (to paraphrase) include: 1) the need for continual check-ins as struggles and stories evolve; 2) the potential for stories and co-authoring to share and generate connections across differences in light of power relations; 3) the obligation to pay critical attention to how stories are used in different contexts and in light of different listeners; 4) the urgency to negotiate between the individual and collective in sharing and co-creating stories; and 5) the need for every co-author to consider their implication, embeddedness and complicity in the very structures of power at the crux of feminist alliances and inquiry.

Geographers Ellen Kohl and Priscilla McCutcheon explain how the 'everyday talk' of what they explore as kitchen-table reflexivity can encourage self-reflexivity both in our research endeavours and in ourselves as researchers, imbued as we are in the power geometries (to borrow Doreen Massey's 1993 term) that support interlocking structures of power. Indeed,

Black feminists, especially, have for decades used 'everyday talk' to share experiences and positionalities. Scholars have argued that the key to Black feminist theory is storytelling and narrative itself; as Jewel Amoah (1997, 84) states: 'Storytelling is not merely a means of entertainment. It is also an educational tool, and for many, it is a way of life. For others, it is the only way to comprehend, analyze, and deal with life'.

Geographers have used storytelling among themselves to share their everyday experiences and how, as researchers, these experiences shape us and are perceived and prescribed in social contexts. For instance, Kohl and McCutcheon (2015, 748) introduce their stories as White and Black feminist researchers working on issues of social and environmental justice respectively, explaining how they employ everyday talk 'as a methodological tool qualitative researchers can use to interrogate their positionalities through formal and informal conversations'. In so doing, they broach tough and, at times, awkward conversations with one another on positionality and power, while emphasizing the different places and histories from which their individual, situated stories emerge.

Geographers of many stripes and in many subdisciplines and specialized streams of geography, from Indigenous geography to queer geography, from health geography to emotional geography, from cultural geography and postcolonial and feminist geography to research that is focused on non-representational geography and more-than-human geography, are returning in the opening decades of the twenty-first century to stories and storytelling. The purposes of using storytelling as method varies, however. Emilie Cameron (2012) characterizes how storytelling has been used in geography as small stories, telling stories and storying (for) change. Small stories, Cameron (2012, 576) proposes, were of particular interest to cultural and historical geographers who were intent on exploring 'the small, the local, the specific, the particular, the intimate, and the mundane', thus shedding light on interest in place. Telling stories is an invitation to 'take up the role of storyteller and call for greater fertilization between geography and the literary and creative arts' (Cameron 2012, 583). Wielding stories in this way has varied purposes, although geographers interested in affect have been particularly keen to take up storytelling in this vein. Storying for change has been a method and methodology of particular interest to feminist geographers. Whether storying for change through imagined futures or with a critical eye to the tendency for small stories to invisibilize the structures from which storylines are influenced (or, more specifically, the interlocking systems of power that have created ongoing forms of injustice and inequities), storying for change draws on what Cameron (2012, 580) sees as geographers' 'relatively longstanding interest in the capacity for stories to create social, political, and intellectual change'.

Storytelling has a varied story across geography's subdisciplines. Indigenous geographers and Indigenous geographies have a long-standing history, with storytelling acting more as a methodology than a practice. Kwagiulth scholar and geographer Sarah Hunt (2013b, 27) highlights how, in 'looking to Indigenous epistemologies for ways to get beyond the ontological limits of what is legible as western scholarship, a number of Indigenous scholars have pointed to stories, art, and metaphor as important transmitters of Indigenous knowledge'. 'Stories and storytelling', Hunt (ibid.) continues, 'are widely acknowledged as culturally nuanced ways of knowing, produced within networks of relational meaning-making'. Citing other Indigenous scholars, such as Linda Tuhiwai-Smith (1999), Margaret Kovach (2009) and Shawn Wilson (2008), Hunt explains how concepts in geography, such as 'ontology' (see also Todd 2016), are insufficient to convey Indigenous peoples' often embodied knowledges adequately and in respectful ways. Storytelling has also been employed in community-based participatory approaches to research, especially between and across Indigenous geographies, conducted by non-Indigenous researchers (Christensen 2012a). Indeed, Indigenous peoples' stories of experiences with homelessness

(Christensen 2012b, 2016), colonial violence (de Leeuw 2016; Hunt 2013b), Indigenous and non-Indigenous collaborations and solidarities (Johnson and Larsen 2013; Larsen and Johnson 2016) and Indigenous map-making (Goeman 2013; Louis 2007; Louis, Johnson and Pramono 2012) are but a few of the ways that storytelling has been used in Indigenous geographies.

Much like storytelling is used in Indigenous geographies to point to systemically entrenched colonial violence, anti-racist geographers of colour have used storytelling to highlight personal narratives that provide in-depth knowledges and experiences of how race has been perceived in the discipline (Gilmore 2002; Hudson, McKittrick and Caribbean Philosophical Association 2014; Mahtani 2014). Andrea Choi's (2018) research on the life histories of early geographers of colour, such as Harold Rose, Audrey Kobayashi and Bobby Wilson, links how geographers' long-standing roots in social movements have impacted on their scholarship, thus shaping anti-racist and activist geographies today. Recalling Amoah's (1997, 84) powerful statement that 'storytelling is not merely a means of entertainment … it is a way of life' and that, for some, 'it is the only way to comprehend, analyze, and deal with life', geographers have engaged the stories of people of colour to provide in-depth storying about the nuanced ways that dispossession, racialization and racial capitalism impact peoples differently and how these structures of power are also resisted. For instance, Ruthie Gilmore's (2007) influential work on the stories of people of colour in California's prison industrial complex sheds light on how resistance can take place even in an incarcerated space. Katherine McKittrick's reading of Black women's stories as resistance and place-making (2006, 2016) and Black sense of place (2011) highlights the overlooked ways of reading concepts that have long been central to the discipline. Beverley Mullings' (2011) engagement with the experiences of the Jamaican diaspora sheds light on labour-market experiences and social networks. Anti-racist scholars (such as Pulido 2015, 2017) have also challenged geography to rethink concepts such as environmental justice and racism through the experiences and stories of those living in place.

Queer theorists and geographers challenging heteronormativity and gender-normativity have also confronted geography's engagement with sexuality and gender by using personal narratives and storytelling. Commenting on the canon that is often considered 'geography', Judith Jack Halberstam (2005) observes that 'queer work on sexuality and space, like queer work on sexuality and time, has had to respond to canonical work on "postmodern geography" by Edward Soja, Fredric Jameson, David Harvey, and others that has actively excluded sexuality as a category …' (5–6). Halberstam continues, stating that 'this foundational exclusion, which assigned sexuality to body/local/personal and took class/global/political as its proper frame of reference, has made it difficult to introduce questions of sexuality and space into the more general conversations about globalization and transnational capitalism' (6). That Halberstam draws from the dominant 'canon' of geography – for example, predominantly White men – is telling in and of itself, demonstrating that the external views of geography as a discipline (and, perhaps, the internal views of many geographers) may very well invisibilize geographies and geographers who stem from queer, feminist and critical race perspectives and experiences. Queer geographies are continuously emerging, drawing on queer theory's reading of story to share an in-depth, often critically inspired reading of spatialities (e.g., Oswin 2008). Queer stories have thus provided insight into the ways in which queer people both experience and navigate space and also contribute to and challenge taken-for-granted concepts that are core to the discipline – what Halberstam frames as bringing questions of sexuality into conversations key to geography, such as globalization and transnationalism. Jack Giesking (2016, 264), for instance, relates 'the everyday stories of lesbians and queer women experiences in public spaces in New York City' and, in so doing, he pushes geographers to consider how 'territory and borders are spatial forms that we cannot assume to be self-evident'. J.P. Catungal and Eugene McCann (2010) analyze discourse to critically view how the stories and experiences of gay men in parks in Vancouver, Canada, are not only regulated but re-storied in the dominant media. May Farrales (2018), also focusing on Vancouver and in settler colonial contexts, presents stories of Filipinos/as to show how sexuality and gender paradigms manifest and are performed through pageantry. In so doing, Farrales links contemporary debates on diaspora, globalization, settler colonialism and sexuality in nuanced and storied ways.

Geography, then, has been rich ground for new stories and storytelling, at the same time as being problematic terrain for many voices and experiences. Stories and storytelling thus remain complicated and rightly contested, functioning in various discursive and material ways to form and produce ontological and epistemological realities for different subjects. With this understanding, radical and critical geographers are telling new stories and theorizing storytelling in new ways in order to story creatively the discipline of geography and other epistemological spaces.

A storied turn and an ending: the rise of geohumanities and of creative geographies, and some final words

A remarkable thing about stories and storytelling is how they shift with telling: how they change and transform in their (re)telling by different people, in different spaces, in their being listened to and thought about by different audiences in and from different places. Stories and storytelling are thus, in many ways, alive and living, evolving, adapting and shape-shifting forces that move across the various times and spaces of geography as a discipline. While, as explored throughout this chapter, stories and storytelling have a long history in geography, they have recently (for less than a decade) made something of a comeback, twinned with a more general rise of humanities-focused work in geography. Evoked in part by the 'creative re-turn' and our turn toward 'geohumanities', geographers are increasingly lauding the importance of deep, thick, affective, interdisciplinary and imaginative orientations to critical questions in our discipline and beyond (Cresswell et al. 2015; Dear 2015). Stories and storytelling, taking forms such as poems, essays, visual narratives, graphic novels, artistic curations, embodied interventions and choreographed performances (to name just a few), are no longer occupying fringe spaces in geography (Donovan and Ustundag 2017; McLean 2018). While geography has long considered biographic stories as integral to knowing and understanding place and space, biography, too, is being re-embraced by geographers in efforts to understand more fully the personal, tactile and subjective natures of being in geography (Daniels and Nash 2004). Some of the discipline's largest professional associations (including the International Geographic Union, the Association of American Geographers, the Canadian Association of Geographers and the Royal Geographical Society) are increasingly present at international conferences and, in the form of specialty study groups, dedicated to creative practices, including stories and storytelling. These trends might be understood as culminating, disciplinarily, in the Association of American Geographers launch in 2015 of a new flagship journal, *GeoHumanities,* which features an 'in practice' section. This is often populated by creative storytelling and geographic stories told in creative ways, including as poems and photo-essays. *GeoHumanities* is not alone in its publication of stories and creative storytelling works: the journal *cultural geographies* has long published poetic works, biographically infused considerations of subjectivity and place, creative writings and storied engagements with cultural expressions, such as landscape painting or film (see, for instance, DeLyser and Hawkins 2014), while the newer journal *Literary Geographies* is almost exclusively devoted to geographic analysis of stories and storytelling in literature and literary studies.

The recent ascendance, the force and the potential perils of stories and storytelling mean that they are rich ground for feminist and other critical geographers: stories and storytelling, like so many practices and epistemological framings in geography, deserve interrogation by feminist scholars, by queer, anti-racist, anti-colonial, Indigenous, racialized scholars and by activists of all ages, abilities and career stages. Telling new stories, embracing difference and radicality in storytelling – this will transform both the discipline of geography and the ways in which different geographies of the world are lived and conceptualized. Feminist geographers have long been at the helm of disciplinary transformation. We have long occupied and argued for transformative spaces, doing so in great part by storying places and peoples in new and radical ways. In the opening decades of the twenty-first century, we are continuing what we have long been doing. Still. There are so many stories yet to be told; so many as-yet unimagined ways of storytelling. New generations of geographer activists, artists, scholars and citizens of the world will story geography – and spaces beyond geography – with new voices and visions, voices and visions that will hopefully continue to 'earthwrite' transformative potential and social change from their unique, inspiring, embodied and storied perspectives. After all, as we began, stories really are all that we really are.

Note

1 See Deloria Jr. 1997 for work on storytelling, myth, Whiteness and violence. For more on racism in geography, see Gilmore 2002; Mahtani 2014.

Key readings

Hunt, S. 2013. "Ontologies of Indigeneity: The Politics of Embodying a Concept." *cultural geographies* 21 (1): 27–32.

Keeshig, L. 2015. "Stop Stealing Native Stories." In: *Introduction to Indigenous Literary Criticism in Canada*, edited by H. Macfarlane and A. Garnet Ruffo, 33–36. Cambridge: Broadview Press.

King, T. 2003. *The Truth about Stories: A Native Narrative*. Toronto: House of Anansi.

Moss, P., and C. Donovan, eds. 2017. *Writing Intimacy into Feminist Geography*. London: Taylor & Francis.

References

Amoah, J. 1997. "Narrative: The Road to Black Feminist Theory." *Berkeley Women\s Law Journal* 12: 84–102. Available at: https://doi.org/10.15779/z38fp3b.

Blomley, N. 2003. "Law, Property, and the Geography of Violence: The Frontier, the Survey, and the Grid." *Annals of the Association of American Geographers* 93 (1): 121–141.

Bunkśe, E.V. 2004. *Geography and the Art of Life*. Baltimore, MD: Johns Hopkins University Press.

Cameron, E. 2012. "New Geographies of Story and Storytelling." *Progress in Human Geography* 36 (5): 573–592. Available at: https://doi.org/10.1177/0309132511435000.

Catungal, J.P., and E.J. McCann. 2010. "Governing Sexuality and Park Space: Acts of Regulation in Vancouver, BC." *Social and Cultural Geography* 11 (1): 75–94. Available at: https://doi.org/10.1080/14649360903414569.

Choi, A. 2018. *Geography, Geographers, and the Geographies of Antiracism*. Queen's University at Kingston, Kingston, ON. Available at: https://qspace.library.queensu.ca/bitstream/handle/1974/23959/Choi_Andrea_J_201802_PhD.pdf?sequence=2&isAllowed=y (accessed 23 June 2018).

Christensen, J. 2012a. "Telling Stories: Exploring Research Storytelling as a Meaningful Approach to Knowledge Mobilization with Indigenous Research Collaborators and Diverse Audiences in Community-based Participatory Research." *Canadian Geographer / Le Géographe Canadien* 56 (2): 231–242. Available at: https://doi.org/10.1111/j.1541-0064.2012.00417.x.

Christensen, J. 2012b. "'They Want a Different Life': Rural Northern Settlement Dynamics and Pathways to Homelessness in Yellowknife and Inuvik, Northwest Territories." *Canadian Geographer / Le Géographe Canadien* 56 (4): 419–438. Available at: https://doi.org/10.1111/j.1541-0064.2012.00439.x.

Christensen, J. 2016. "Indigenous Housing and Health in the Canadian North: Revisiting Cultural Safety." *Health and Place* 40: 83–90. Available at: https://doi.org/10.1016/j.healthplace.2016.05.003.

Clark, J., and C. Martin. 2013. *Anarchy, Geography, Modernity: Selected Writings of Elisée Reclus*. Oakland, CA: PM Press.

Cohan, S., and L. Shires. 1988. *Telling Stories*. London: Routledge.

Cosgrove, D., and M. Domosh. 1993. "Author and Authority: Writing the New Cultural Geography." *Place/Culture/Representation* 25: 31.

Crang, M. 1998. *Cultural Geography*. London: Psychology Press.

Crang, M. 2013. *Cultural Geography*, 3rd edition. London: Routledge.

Cresswell, T., D.P. Dixon, P.K. Bol, and J.N. Entrikin. 2015. "Editorial." *GeoHumanities* 1 (1): 1–19.

Daniels, S., and C. Nash. 2004. "Lifepaths: Geography and Biography." *Journal of historical Geography* 30 (3): 449–458.

Dear, M. 2015. "Practicing Geohumanities." *GeoHumanities* 1 (1): 20–35.

de Leeuw, S. 2016. "Tender Grounds: Intimate Visceral Violence and British Columbia's Colonial Geographies." *Political Geography* 52: 14–23. Available at: https://doi.org/10.1016/j.polgeo.2015.11.010.

de Leeuw, S. 2017. "Writing as Righting: Truth and Reconciliation, Poetics, and New Geo-graphing in Colonial Canada." *Canadian Geographer/Le Géographe canadien* 61 (3): 306–318.

de Leeuw, S., M.W. Parkes, V. Sloan Morgan, J. Christensen, N. Lindsay, K. Mitchell-Foster, and J. Russell Jozkow. 2017. "Going Unscripted: A Call to Critically Engage Storytelling Methods and Methodologies in Geography and the Medical-health Sciences." *Canadian Geographer / Le Géographe Canadien*. Available at: https://doi.org/10.1111/cag.12337.

Deloria Jr., V. 1997. *Red Earth, White Lies: Native Americans and the Myth of Scientific Fact*. Colorado: Fulcrum.

DeLyser, D., and H. Hawkins. 2014. "Introduction: Writing Creatively – Process, Practice, and Product." *cultural geographies* 21 (1): 131–134.

Doan, P. 2010. "The Tyranny of Gendered Spaces – Reflections from Beyond the Gender Dichotomy." *Gender, Place & Culture* 17 (5): 635–654.

Domosh, M. 1991. "Toward a Feminist Historiography of Geography." *Transactions of the Institute of British Geographers* 16 (1): 95–104.

Donovan, C., and E. Ustundag. 2017. "Graphic Narratives, Trauma and Social Justice." *Studies in Social Justice* 11 (2): 221–237.

Eagleton, M., ed. 2011. *Feminist Literary Theory: A Reader*. London: John Wiley and Sons.

Farrales, M. 2018. "Repurposing Beauty Pageants: The Colonial Geographies of Filipina Pageants in Canada." *Environment and Planning D: Society and Space*. October. Available at: https://doi.org/10.1177/0263775818796502.

Gieseking, J. 2016. "Crossing Over into Neighbourhoods of the Body: Urban Territories, Borders and Lesbian-queer Bodies in New York City." *Area* 48 (3): 262–270. Available at: https://doi.org/10.1111/area.12147.

Gilmore, R. 2002. "Fatal Couplings of Power and Difference: Notes on Racism and Geography." *Area* 54 (1): 15–24.

Gilmore, R. 2007. *Golden Gulag: Prisons, Surplus, Crisis, and Opposition in Globalizing California*. Berkley: University of California Press.

Goeman, M. 2013. *Mark My Words: Native Women Mapping Our Nations*. Minneapolis: University of Minnesota Press.

Halberstam, J. 2005. *In a Queer Time and Place: Transgender Bodies, Subcultural Lives*. New York: New York University Press.

Harley, J. 1989. "Deconstructing the Map." *Cartographica* 26 (2): 1–20.

Hawkins, H., and E.R. Straughan. Eds. 2016. *Geographical Aesthetics: Imagining Space, Staging Encounters*. New York: Routledge.

Hudson, P.J., K. McKittrick, and Caribbean Philosophical Association. 2014. "The Geographies of Blackness and Anti-Blackness: An Interview with Katherine McKittrick." *CLR James Journal* 20 (1): 233–240. Available at: https://doi.org/10.5840/clrjames201492215.

Hunt, S. 2013a. "Witnessing the Colonialscape: Lighting the Intimate Fires of Indigenous Legal Pluralism." PhD dissertation, Simon Fraser University, Vancouver, BC. Available at: http://summit.sfu.ca/item/14145#310.

Hunt, S. 2013b. "Ontologies of Indigeneity: The Politics of Embodying a Concept." *cultural geographies* 21 (1): 27–32. Available at: https://doi.org/10.1177/1474474013500226.

Hyndman, J. 2004. "Mind the Gap: Bridging Feminist and Political Geography Through Geopolitics." *Political Geography* 23 (3): 307–322.
Johnson, J.T., and S.C. Larsen, eds. 2013. *A Deeper Sense of Place: Stories and Journeys of Indigenous-academic Collaboration*. Corvallis: Oregon State University Press.
Keeshig-Tobias, L. 1997. "Stop Stealing Native Stories." In: *Borrowed Power: Essays on Cultural Appropriation*, edited by B. Ziff and P. Rao, 71–73. New Brunswick, NJ: Rutgers University Press.
King, T. 2003. *The Truth about Stories: A Native Narrative*. Minneapolis: University of Minnesota Press.
Kobayashi, A. 2014. "The Dialectic of Race and the Discipline of Geography." *Annals of the Association of American Geographers* 104 (6): 1101–1115. https://doi.org/10.1080/00045608.2014.958388.
Kobayashi, A., and L. Peake. 1994. "Unnatural Discourse. 'Race' and Gender in Geography." *Gender, Place and Culture* 1 (2): 225–243.
Kohl, E., and P. McCutcheon. 2015. "Kitchen Table Reflexivity: Negotiating Positionality Through Everyday Talk." *Gender, Place and Culture* 22 (6): 747–763. https://doi.org/10.1080/0966369X.2014.958063.
Kovach, M. 2009. *Indigenous Methodologies: Characteristics, Conversations and Contexts*. Toronto, ON: University of Toronto Press.
Larsen, S.C., and J.T. Johnson. 2016. "The Agency of Place: Toward a More-Than-Human Geographical Self." *GeoHumanities* 2 (1): 149–166. https://doi.org/10.1080/2373566X.2016.1157003.
Livingstone, D.N., and A. Rogers. 1996. *Human Geography: An Essential Anthology*. London: Blackwell.
Louis, R.P. 2007. "Can You Hear Us Now? Voices from the Margin: Using Indigenous Methodologies in Geographic Research." *Geographical Research* 45 (2): 130–139. https://doi.org/10.1111/j.1745-5871.2007.00443.x.
Louis, R.P., J.T. Johnson, and A.H. Pramono. 2012. "Introduction: Indigenous Cartographies and Counter-mapping." *Cartographica: International Journal for Geographic Information and Geovisualization* 47 (2): 77–79. https://doi.org/10.3138/carto.47.2.77.
Mahtani, M. 2014. "Toxic Geographies: Absences in Critical Race Thought and Practice in Social and Cultural Geography." *Social and Cultural Geography* 15 (4): 359–367. Available at: https://doi.org/10.1080/14649365.2014.888297.
Marston, S.A., and S. De Leeuw. 2013. "Creativity and Geography: Toward a politicized Intervention." *Geographical Review* 103 (2): iii–xxvi.
Massey, D. 1993. "Power-Geometry and a Progressive Sense of Place." In: *Mapping the Futures: Local Cultures, Global Change*, edited by J. Bird, B. Curtis, T. Putnam, G. Robertson, and L. Tickner, 59–69. London: Routledge.
McClintock, A. 2013. *Imperial Leather: Race, Gender, and Sexuality in the Colonial Contest*. New York: Routledge.
McKittrick, K. 2006. *Demonic Grounds: Black Women and the Catographies of Struggle*. Minneapolis: University of Minnesota Press.
McKittrick, K. 2011. "On Plantations, Prisons, and a Black Sense of Place." *Social and Cultural Geography* 12 (8): 947–963. Available at: https://doi.org/10.1080/14649365.2011.624280.
McKittrick, K. 2016. "Rebellion/Invention/Groove." *Small Axe* 20 (149): 79–91. Available at: https://doi.org/10.1215/07990537-3481558.
McLean, H. 2018. "Regulating and Resisting Queer Creativity: Community-engaged Arts Practice in the Neoliberal City." *Urban Studies* 55 (16): 3563–3578. doi:0042098018755066.
Meinig, D. 1983. "Geography as an Art." *Transactions of the Institute of British Geographers* 8: 314–328.
Million, D. 2014. "There is a River in Me: Theory from Life." In: *Theorizing Native Studies*, edited by A. Simpson and A. Smith, 31–42. North Carolina: Duke University Press. Available at: https://doi.org/10.1215/9780822376613-002.
Mullings, B. 2011. "Diaspora Strategies, Skilled Migrants and Human Capital Enhancement in Jamaica." *Global Networks* 11 (1): 24–42. Available at: https://doi.org/10.1111/j.1471-0374.2010.00305.x.
Munt, S.R. 2012. "Journeys of Resilience: The Emotional Geographies of Refugee Women." *Gender, Place & Culture* 19 (5): 555–577.
Nagar, R. 2013. "Storytelling and Co-authorship in Feminist Alliance Work: Reflections From a Journey." *Gender, Place and Culture* 20 (1): 1–18. Available at: https://doi.org/10.1080/0966369X.2012.731383.
Oswin, N. 2008. "Critical Geographies and the Uses of Sexuality: Deconstructing Queer Space." *Progress in Human Geography* 32 (1): 89–103.
Price, P.L. 2010. "Cultural Geography and the Stories We Tell Ourselves." *cultural geographies* 17 (2): 203–210.
Puar, J. 2002. "A Transnational Feminist Critique of Queer Tourism." *Antipode* 34 (5): 935–46.

Pulido, L. 2015. "Geographies of Race and Ethnicity 1: White Supremacy vs White Privilege in Environmental Racism Research." *Progress in Human Geography* 39 (6): 809–817. Available at: https://doi.org/10.1177/0309132514563008.

Pulido, L. 2017. "Geographies of Race and Ethnicity II: Environmental Racism, Racial Capitalism and State-sanctioned Violence." *Progress in Human Geography* 41 (4): 524–533. Available at: https://doi.org/10.1177/0309132516646495.

Said, E.W. 2012. *Culture and Imperialism*. New York: Vintage.

Simmonds, N. 2016. "Transformative Maternities: Indigenous Stories as Resistance and Reclamation in Aoteaora New Zealand." In: *Everyday Knowledge, Education and Sustainable Futures*, edited by M. Robertson and P.K.E. Tsang, 71–88. Singapore: Springer.

Springer, S. 2017. "Earth Writing." *GeoHumanities* 3 (1): 1–19.

Smith, L.T. 1999. *Decolonizing Methodologies: Research and Indigenous Peoples*. Dunedin, New Zealand: Zed Books.

Todd, Z. 2016. "An Indigenous Feminist's Take on the Ontological Turn: 'Ontology' is Just Another Word for Colonialism." *Journal of Historical Sociology* 29 (1): 4–22. Available at: https://doi.org/10.1111/johs.12124.

Watson, J.W. 1983. "The Soul of Geography." *Transactions of the Institute of British Geographers* 8: 385–399.

Wilson, S. 2008. *Research is Ceremony: Indigenous Research Methods*. Halifax, NS: Fernwood.

Yusoff, K., and J. Gabrys. 2011. "Climate Change and the Imagination." *Wiley Interdisciplinary Reviews: Climate Change* 2 (4): 516–534.

INDEX

Note: Page locators in *italic* refer to figures.

www.ingramcontent.com/pod-product-compliance
Ingram Content Group UK Ltd.
Pitfield, Milton Keynes, MK11 3LW, UK
UKHW050846300425
458020UK00008B/78

* 9 7 8 1 0 3 2 5 7 0 0 2 0 *